THE
INTERNATIONAL SERIES
OF
MONOGRAPHS ON PHYSICS

International Series of Monographs on Physics

Quantum Chromodynamics

High Energy Experiments and Theory

GÜNTHER DISSERTORI

Institute for Particle Physics, ETH Zürich, Switzerland

IAN G. KNOWLES

Formerly University of Edinburgh, Scotland, U.K.

MICHAEL SCHMELLING

Max-Planck-Institute for Nuclear Physics, Heidelberg, Germany

CLARENDON PRESS • OXFORD

2003

OXFORD

UNIVERSITY PRESS

~~Great Clarendon Street, Oxford OX2 6DP~~

Oxford University Press is a department of the University of Oxford.
It furthers the University's objective of excellence in research, scholarship,
and education by publishing worldwide in

Oxford New York

Auckland Bangkok Buenos Aires Cape Town Chennai
Dar es Salaam Delhi Hong Kong Istanbul Karachi Kolkata
Kuala Lumpur Madrid Melbourne Mexico City Mumbai Nairobi
São Paulo Shanghai Taipei Tokyo Toronto

Oxford is a registered trade mark of Oxford University Press
in the UK and in certain other countries

Published in the United States
by Oxford University Press Inc., New York

© Oxford University Press, 2003

A catalogue record for this title is available from the British Library

Library of Congress Cataloging in Publication Data
(Data available)

ISBN 0 19 850572 8

10 9 8 7 6 5 4 3 2 1

Typeset by the authors using LaTeX

Printed in Great Britain
on acid-free paper by
T. J. International Ltd,
Padstow, Cornwall

PREFACE

Quantum Chromodynamics (QCD) was formulated as a non-abelian field theory in 1973 and in the meantime has evolved to the generally accepted theory describing strong interactions. This book is intended to give a comprehensive overview over QCD studies in e^+e^- annihilation, lepton–nucleon scattering and hadron–hadron scattering. The book comes at a time when the analysis of the LEP data is being finalized and more than ten years of most fruitful research deserve to be summarized. One of the goals of the book is to bridge the gap between theory and experiment, combining into a single volume theoretical description and the experimental measurements.

Experimental results are discussed with emphasis on their relevance with respect to strong interactions physics at high energies. Although not necessary for understanding the physics, it might occasionally be helpful to have some background knowledge with respect to the analysis of experimental data, such as the least-squares method or the interpretation of experimental errors and confidence level intervals. For an introduction to this subject, which is beyond the scope of this book, we would like to refer the reader to any of the many excellent textbooks on the subject.

Dealing with particle physics we will express energy in units of eV or multiples of it, such as, for example, MeV or GeV. In addition the so-called *natural units* will be used, that is, a system of units where $\hbar = c = 1$. Thus energy, momentum and mass are measured in units of energy, lengths in units of energy^{-1}. Only in cases where we want to be explicit about momentum or mass we will use GeV/c or GeV/c^2, respectively. If desired, conversion to MKSA-units is achieved by multiplying with appropriate powers of $\hbar$ and c. Here it is helpful to remember that $\hbar c \approx 0.197$ GeV fm, $c \approx 3 \times 10^8$ m/s and $e \approx 1.602$ As. Electrical charges are measured in units of the positron charge e and may carry an index specifying the identity of a particle, for example, e_q for the charge of a quark. An overview of the mathematical notation used in this book is given on page x.

The presentation of the subject should be understandable for graduate students, and to some extent even undergraduates, but also contains material useful for a research physicist. The basics of perturbative QCD are presented in quite some detail. Anybody who has heard an introduction to field theory should be able to follow also the derivations and learn how the building blocks of the QCD Lagrangian can be probed in high energy physics experiments. The book focuses on basics. Nevertheless, some space in the presentation of the theory of QCD is also dedicated to advanced topics aiming to convey at least some impression of the general picture. For an in-depth discussion of, for example, renormalization or the path-integral formalism the reader is referred to specialized textbooks.

In order to achieve a deeper understanding, some of the concepts discussed in the main text are illustrated by means of problems. The solutions are given in the appendix. Depending on the background of the reader the apparent difficulty of the problems may vary considerably. As a rough guideline we introduced a star-rating system for grading the complexity. A single star signals moderate difficulty, two stars are attached to what we consider really hard problems.

After a short historical introduction, summarizing the steps which led to the formulation of QCD, the book focuses on the tests of QCD. First the theory and phenomenology of perturbative QCD for different kinds of reactions are explained at Born level, then higher order corrections are discussed. Considerable weight is also given to non-perturbative effects such as the hadronization process. The currently available models are discussed in some detail — both concerning the different phenomenological ideas and their implementation in Monte Carlo models.

The experimental section starts with a short overview about accelerator and detector techniques and a detailed description of the general strategies of QCD analyses. Then structure function measurements are discussed, which were vital in establishing QCD as a field theory and still are a thriving field of research. The focus then shifts to measurements done mainly in e^+e^- annihilation, which test the detailed structure of stong interactions as they can be probed by perturbative QCD. This covers measurements of the strong coupling constant as well as tests of the structure of QCD, which finally led to the unambiguous proof for the existence of the gluon self-interaction.

Going from hard scattering to semi-soft processes, the book describes how interference effects in higher order amplitudes affect the properties of hadronic final states. Finally studies of the hadronization process, the conversion from coloured partons to colour neutral hadrons, are presented, as it can be probed by means of particle multiplicities or particle–particle correlations.

Clearly, within the limited space available it is only possible to cover a subset of all topics connected to QCD. The emphasis is put on hard-scattering processes. Other subjects such as quark mass determinations, polarization phenomena, photon–photon scattering, QCD at thresholds, non-relativistic QCD, nuclear collisions, lattice gauge theories, chiral perturbation theory, quark–gluon plasma, or exclusive processes, if at all, are only briefly mentioned. Again, for an in-depth discussion the reader is referred to external references specializing on the respective subject.

Even after almost 30 years of research there are still many open questions and plenty of opportunity for significant contributions both in theory and experiment. Having worked through the book, we hope that the reader will have gained an overview of how QCD developed in the twentieth century and where we stand with respect to a quantitative understanding after the turn of the millenium. Many of the results collected for this book are likely to be superseded by improved measurements in the future, but the basic facts of QCD and the methods to extract its defining parameters will remain valid.

In collecting the material for this book we tried to do as complete as possible a survey of the relevant publications. Still, some important paper will have been missed, and we would like to apologize to all authors whose work is not yet fully appreciated in this book. Updates and amendments to the book, covering new developments as well as improvements to the exercises or error corrections can be found on the book's World Wide Web site. The link can be located via the catalogue of the Oxford University Press home page at:

http://www.oup.co.uk/

November 2002   G.D., I.G.K., M.S.

ACKNOWLEDGEMENTS

Our thanks are due above all to Sönke Adlung of Oxford University Press for encouraging us to write this book, as well as to Anja Tschörtner and Marsha Filion of Oxford University Press for their support during its making and completion.

G.D. would like to thank the following people for their comments on the manuscript: Cristiano Borean, Silvia Bravo, Barbara Clerbaux, Maria Hörndl, Frederic Teubert, Andrea Valassi and Valeria Tano.

I.G.K. thanks Sinéad Farrington and the numerous other colleagues who have helped inform this work.

M.S. would like to thank many colleagues and friends who directly or indirectly contributed to the results presented in this book, in particular Glen Cowan, Thomas Lohse, Ramon Miquel, Cristobal Padilla and Ron Settles, with very special thanks to Anke-Susanne Müller for continued support and valuable input during the entire writing phase.

For permissions to reproduce various figures and diagrams we are indebted to the authors cited in the text and figure captions and to the following experimental collaborations and publishers:

- ALEPH Collaboration, for Figures 5.7, 6.1 and 10.5.
- American Institute of Physics, publishers of *AIP Conf. Proc. 531 Particles and Fields: Seventh Mexican Workshop 1999*, ©*2000 American Institute of Physics* , for Figure 7.1.
- American Physical Society, publishers of *Phys. Rev. Lett. 78, 79*, ©*1997 by the American Physical Society, Phys. Rev. Lett. 80, 81 and Phys. Rev. D58*, ©*1998 by the American Physical Society*, for Figures 7.13, 7.14, 7.19(a), 7.19(b), 7.20, 7.21, 7.25(b), 10.9(a) and 10.9(b).
- Annual Reviews, publishers of *Annual Review of Nuclear and Particle Science, Volume 49* ©*1999 by Annual Reviews www.annualreviews.org*, for Figures 6.9, 6.10, 6.11 and 6.12.
- Elsevier Science Ltd., publishers of *Nucl. Phys. B (Proc. Suppl.) 65* ©*1998, Nucl. Phys. B470* ©*1996 and B545* ©*1999, Phys. Lett. B456, B499* ©*1999 and B487* ©*2000, Phys. Rept. 294* ©*1998 and Nucl. Instr. and Meth. A360* ©*1995*, for Figures 6.2, 6.8, 6.13, 7.9, 7.12(b), 8.6, 9.1, 9.3, 9.4, 10.1, 10.4, 10.7, 12.3, 13.1, 13.2 and 13.3.
- H1 Collaboration, for Figures 7.2 and 7.5.
- Institute of Physics Publishing Ltd., publishers of *J. Phys. G: Nucl. Part. Phys. 23*, for Figures 3.29, 3.33, 3.34, 3.35 and 3.36.
- Prof. J. Friedman for the MIT-SLAC Collaboration, for Figures 2.2 and 2.3.

- Shaker Verlag, publishers of *ISBN 3-8265-7436-2*, for Figures 5.1, 5.2, 5.3, 5.4, 5.5 and 5.6.
- Springer Verlag, publishers of *Z. Phys. C57* ©*1993, C63* ©*1994, C73, C75 and C76* ©*1997 and Eur. Phys. J. C1* ©*1998, C7, C11* ©*1999, C12, C13, C14 and C17* ©*2000 and C19* ©*2001*, for Figures 6.3, 6.5, 7.3, 7.4, 7.6, 7.7, 7.10, 7.11, 7.12(a), 7.22, 7.23, 7.24, 7.25(a), 10.8, 11.6, 11.7, 12.1, 12.2, 12.4, 12.5, 12.6 and F.5.
- The Royal Swedish Academy of Sciences, publishers of *Physica Scripta, Volume 51*, for Figures 2.9, 9.2, 11.1, 11.2 and 11.3(left).
- World Scientific Publishing, publishers of the *Proceedings of the 9th International Workshop DIS 2001, Bologna, Italy*, for Figure 7.8.

NOTATION

$\mu, \nu, \rho \ldots$	Lorentz four-vector indices $\{0,1,2,3\}$
$i, j, k, \ldots$	Euclidean three-vector indices $\{1,2,3\}$,
	or colour indices of the fundamental representation
$a, b, c, \ldots$	colour indices in the adjoint representation
v	Lorentz four-vector
$\boldsymbol{v}$	Euclidean three-vector
$v = (v^0, \boldsymbol{v})$ or $v^\mu = (v^0, \boldsymbol{v})$	explicit components of a four-vector
$p = (E, \boldsymbol{p})$	four-momentum
$\eta_{\mu\nu} = \eta^{\mu\nu} = \mathrm{diag}(1, -1, -1, -1)$	metric tensor of special relativity
$p_\mu = \eta_{\mu\nu} p^\nu$	associated covector to p^μ
$\boldsymbol{p} \cdot \boldsymbol{q} = p_1 q_1 + p_2 q_2 + p_3 q_3$	scalar product of two three-vectors
$p \cdot q = p^\mu q^\nu \eta_{\mu\nu} = p^0 q^0 - \boldsymbol{p} \cdot \boldsymbol{q}$	scalar product of two four-vectors
$[\,A, B\,] = AB - BA$	commutator of two operators or matrices
$\{A, B\} = AB + BA$	anti-commutator of two operators or matrices
$\mathbf{1}$	unit matrix
λ^a	Gell-Mann matrices, $a = \{1, \ldots, 8\}$
$T^a = \lambda^a / 2$	generators of SU(3), $a = \{1, \ldots, 8\}$
f^{abc} with $[T^a, T^b] = \mathrm{i}\, f^{abc} T^c$	structure constants of SU(3)
γ^μ with $\{\gamma^\mu, \gamma^\nu\} = \eta^{\mu\nu}$	Dirac matrices satisfying the Clifford algebra
$\not{p} = \gamma^\mu p_\mu = \gamma_\mu p^\mu$	'slash' notation
$\delta_{ij} = \begin{cases} 1 & i = j \\ 0 & \text{otherwise} \end{cases}$	Kronecker-delta
$\epsilon_{ijk\ldots}$ with $\epsilon_{(0)123\ldots} = +1$	fully antisymmetric Levi-Civita tensor
$\delta(x)$ with $\int \mathrm{d}x\, f(x)\delta(x) = f(0)$	Dirac delta-function
$\Theta(x) = \begin{cases} 0 & x < 0 \\ 1 & \text{otherwise} \end{cases}$	Heaviside step-function
$M^\dagger$	Hermitian conjugate matrix, $M^\dagger_{ij} = M^\star_{ji}$
$\ln(x)$	natural logarithm, base e
∂_μ	shorthand for $\partial / \partial x^\mu$
$\gamma_E = 0.577\,215\,664\,901\ldots$	Euler–Mascheroni constant
$\zeta(x)$	Riemann zeta-function $\sum_{n=1}^{\infty} n^{-x}$
$\Gamma(x)$	Euler gamma-function
$f \otimes g(x)$	convolution $\int \mathrm{d}y / y\, f(y) g(x/y)$
$\tilde{f}(n)$	Mellin transformation of f

Summation over repeated indices is implied by default.

CONTENTS

1

INTRODUCTION

Particle physics is the search for the fundamental constituents of matter and the understanding of their interactions. The concept of basic building blocks for the entire physical universe dates back to the Greek philosopher Demokritos, 460–370 B.C., who postulated that everything is built from indivisible entities, the so-called 'atoms', from the Greek word for indivisible or void, and that all conventional properties such as colour, taste, hardness etc. are a consequence of interactions between different arrangements of atoms.

With the advancement of science different types of interactions between matter were identified, such as gravity or the electromagnetic forces, and continued study led to consistent theories providing a mathematical description of many phenomena. In addition it was realized that all known substances could be synthesized from a finite number of chemical elements, which in themselves appeared immutable. The fact that chemical reactions always happened for fixed mass ratios was the first experimental hint that the material world indeed was made of discrete entities, which again were called atoms. The final proof that the chemical elements indeed are made of atoms then came around the beginning of the twentieth century together with the development of statistical mechanics and quantum physics.

After the discovery of radioactivity it was, however, quickly realized that the atoms which appeared immutable in chemical reactions are not the fundamental quantities of the Greek philosophers. Instead, they appeared to be complex objects where light electrons orbit around a heavy compact nucleus, which was later found to be composed of protons and neutrons. A closer look at the atomic nucleus soon revealed that in addition to the long known and well established electromagnetic interactions new short range forces were also present: a strong force which binds the nucleons together, and a weak force which mediates the radioactive beta-decay.

Today we know that the strong force observed between nucleons is only a van-der-Waals type residual force of a more fundamental interaction between their constituents. The field theory of strong interactions is known as *Quantum Chromodynamics* (QCD). Like the other known forces, it belongs to the class of so-called *gauge theories*. In a gauge theory, the fields are described by representations of an abstract symmetry group, and the interaction between the fields, mediated by the *gauge bosons*, is induced by the requirement that the Lagrangian is invariant with respect to arbitary local transformations of the fields. How this works technically is shown in the theory section of the book.

Table 1.1 *Qualitative picture of the fundamental forces with the relative strengths evaluated for $Q = 1$ GeV*

Interaction	Approx. potential	Parameter values	Relative strength
strong	$\dfrac{12\pi/23}{Q^2 \ln(Q^2/\Lambda^2)}$	$\Lambda \approx 0.2$ GeV	1
electromagnetic	$\dfrac{\alpha_{\mathrm{em}}}{Q^2}$	$\alpha_{\mathrm{em}} \approx 1/137$	1.4×10^{-2}
weak	$\dfrac{\alpha_{\mathrm{em}}}{Q^2 - M_{\mathrm{W}}^2}$	$M_{\mathrm{W}} \approx 80$ GeV$/c^2$	2.2×10^{-6}
gravity	$\dfrac{G_{\mathrm{N}} m_1 m_2}{Q^2}$	$G_{\mathrm{N}} \approx \dfrac{6.7 \times 10^{-39}}{\mathrm{GeV}^2}$	1.2×10^{-38}

In the language of gauge theories electromagnetic interactions are character-ized by an U(1) gauge symmetry, weak interactions between left-handed fermions by an SU(2) symmetry and strong interactions by an SU(3) symmetry. All these symmetries are internal in the sense that they do not act on space–time coordi-nates. A gauge symmetry which is based on invariance with respect to arbitary local coordinate transformations finally leads to the decsription of gravity in the framework of general relativity.

A quantitative comparison of the known interactions is not entirely trivial. Taking for instance two test particles and asking for the relative strength of the different forces between them, the answer depends both on the particle types and the distance of the two. The picture becomes most transparent when analyzing the situation in momentum space and looking at the Fourier transform of the potential which mediates the interaction. The absolute value of the transformed potential can then be taken as a measure for the strength of the interaction. In momentum space the distance is replaced by the momentum transfer Q between the charges, which for this comparison is set to $Q = 1$ GeV. The charges are unit charges of the respective interaction. For the strong, the electromagnetic and the weak force these are the electric charge, weak isospin and colour charge, respectively. In case of gravity, the relevant charge is the mass of a particle, which is a free parameter of the theory. In the comparison the proton mass is used. Table 1.1 gives a qualitative overview for the known forces.

At the scale of $Q = 1$ GeV of the comparison one observes a marked hierarchy when going from strong interactions to gravity. One also sees some characteristics which neatly summarize the current picture of the Standard Model of particle physics. Electromagnetic and weak interactions are unified in the sense that at asymptotic energies both become similar, with a coupling strength evolving pro-portional to α_{em}. In the low energy limit the weak interactions are suppressed by the M_{W}^2 term, which is due to the large mass of the gauge bosons of the in-teraction. In QCD there is no such low energy suppression. One rather observes a growth of the interaction strength with a divergence in the $\ln(Q^2/\Lambda^2)$-term, which implies that around and below a certain cut-off energy Λ this simple expression no longer describes the physics of strong interactions. On the other hand the coupling strength decreases with increasing Q^2 and eventually approaches the same value as that of the unified electroweak interaction. The exact point of this grand unification depends on the details of the theory. A hint that it might exist can already be inferred from the rough sketch given by Table 1.1.

The fundamental fields known today are *leptons* and *quarks*, which are both spin-1/2 fermions, and spin-1 gauge boson fields such as the *gluon* (g), the *photon* (γ) and the $W^\pm$ and the Z bosons, which mediate strong, electromagnetic and weak interactions, respectively. Isolated free quarks have never been observed experimentally. Bound states of three quarks form the so-called *baryons*, such as the proton or the neutron, combinations of a quark and an antiquark yield a *meson*, such as, for example, the pion or the kaon. Mesons and baryons are collectively referred to as *hadrons*, heavy particles which are subject to strong interactions. In comparison to quarks, leptons, with the electron and its neutrino being the most prominent representatives, are rather light particles which do exist as free fields and are oblivious to strong interactions.

Quarks carry colour charge, electric charge and weak isospin and thus couple to gluons, photons and $W^\pm$ and Z bosons. All leptons carry weak isospin and thus are subject to the weak interaction, but only the charged leptons have electric charge and thus also interact electromagnetically. In the context of the Standard Model, all massive particles acquire their mass by coupling to the scalar Higgs field H. Finally, all energy couples to the spin-2 graviton field (G), which in a quantum theory of gravity is responsible for the gravitational interaction. Table 1.2 summarizes basic properties of the fields of the Standard Model. The masses quoted in the table are taken from the 'Review of Particle Properties' (PDG, 2000). The considerable ranges given for the quark masses reflect the difficulties in dealing with masses of strongly interacting particles which are not observable as free fields. They have to be understood as mass parameters of the theory rather than mass contributions to the bound states corresponding to observable hadrons.

Table 1.2 shows that we currently have to deal with 18 fundamental fields, and the natural question arises whether there might be a more fundamental level at which the picture becomes much simpler. Unification of the different interactions including gravity addresses this question, for example, in the context of the so-called *string theories*. In such theories also the apparent symmetry between leptons and quarks and the repetition of weak-isospin doublets in both sectors may find a natural explanation. Promising alternative models to extend the minimal Standard Model exist, based on very good theoretical arguments. However, all experimental findings are perfectly described by the Standard Model so far.

The main focus of this book is QCD in hard interactions. Structurally, QCD is a very straightforward theory, being a Yang–Mills gauge theory based on an

Table 1.2 *Fields of the mimimal Standard Model. The first group contains the quarks, the second the leptons and the last one the bosonic fields. All charges are given in units of the positron charge.*

Field	Mass/MeV/c^2	Spin	Charge/e	$(I, I_3)_{\text{weak}}$	Colour states
d	$3 - 9$	1/2	$-1/3$	$(1/2, -1/2)$	3
u	$1 - 5$	1/2	$+2/3$	$(1/2, +1/2)$	3
s	$75 - 170$	1/2	$-1/3$	$(1/2, -1/2)$	3
c	$1150 - 1350$	1/2	$+2/3$	$(1/2, +1/2)$	3
b	$4000 - 4400$	1/2	$-1/3$	$(1/2, -1/2)$	3
t	174300	1/2	$+2/3$	$(1/2, +1/2)$	3
e^-	0.511	1/2	-1	$(1/2, -1/2)$	0
ν_e	< 0.000003	1/2	0	$(1/2, +1/2)$	0
μ^-	105.66	1/2	-1	$(1/2, -1/2)$	0
ν_μ	< 0.19	1/2	0	$(1/2, +1/2)$	0
τ^-	1777	1/2	-1	$(1/2, -1/2)$	0
ν_τ	< 18.2	1/2	0	$(1/2, +1/2)$	0
γ	0	1	0	$(0, 0)$	0
$W^\pm$	80419	1	± 1	$(1, \pm 1)$	0
Z	91188	1	0	$(1, 0)$	0
g	0	1	0	$(0, 0)$	8
H	> 114000	0	0	$(1/2, -1/2)$	0
G	0	2	0	$(0, 0)$	0

unbroken SU(3) symmetry. Historically, however, it was a long way from the realization that the binding energy of the atomic nucleus is due to a new type of interaction until QCD could be formulated, since the basic fields of the theory were never observed as free particles. As sketched in Table 1.1, the reason may be related to the divergent behaviour of the QCD potential at large distances, which causes perturbation theory to break down at low momentum transfers. Thus, despite QCD being based on a rather straightforward ansatz, the fact that the interaction is strong renders it quite difficult with respect to theoretical calculations.

In the following chapters we will thus first give an overview on how strong interaction physics evolved towards the formulation of QCD. Then we will discuss in detail the Lagrangian of the theory, the basic phenomenology of strong interactions in different environments, and how QCD allows to give a unified quantitative description of a wide variety of phenomena. Given the theoretical basis, we will briefly explain the basic concepts of high energy physics experiments, covering accelerator and detector technologies as well as data analysis, and then turn to actual measurements of structure functions, the strong coupling constant, tests of the structure of QCD and studies of the hadronization process. Since it is practically impossible to review all relevant results in one book, we focus on QCD studies performed at high energy colliders.

2

THE DEVELOPMENT OF QCD

2.1 Experimental evidence

The history of high energy physics in the second half of the twentieth century was driven by a sequence of increasingly more powerful particle accelerators, which allowed matter to be probed at ever smaller distances. In this chapter we will briefly recapitulate how the experimental evidence accumulated by experiments at those machines led to the theory of *Quantum Chromodynamics* (QCD). In the end, this chapter will bring us to the discussion of the QCD Lagrangian and its physics implications.

Schematically, the road from nuclear physics to QCD can be separated into three phases. The first phase can be characterized as the era of hadron spectroscopy, which culminated in the formulation of the quark model. Then came a first series of deep inelastic scattering experiments which established the physical reality of quarks in the context of the *quark parton model* (QPM). Finally followed a set of improved measurements which probed the interactions between the quarks and allowed the first quantitative tests of QCD.

2.1.1 *The quark model*

The historical foundations of QCD date back to the early days of nuclear physics, when the binding energy of the nucleus was realized to be due to a new kind of interaction between protons and neutrons. Scattering experiments soon showed that the interaction is not only very strong, but also that it acts only over very short distances. It was Yukawa's insight that such a short range force could be understood by assuming that the interaction is mediated by a heavy boson, a so-called *meson*. Scattering experiments also revealed a certain symmetry between protons and neutrons which was encapsulated in the isospin formalism introduced by Heisenberg and others. The meson theory of Yukawa did very well account for the phenomenology of those days by introducing the π-mesons as force carriers for strong interactions. The picture was nicely confirmed when those pions were actually discovered as free particles in cosmic ray studies and in accelerator experiments in the 1940s. The mass was determined to be 140 MeV/c^2 and the lifetime around 2.6×10^{-8} s. What came as a surprise was the discovery of many other new particles which all could be produced in interactions between nuclear matter. Interestingly, some of those particles were found to decay within time spans as short as 10^{-23} s, while others had lifetimes many orders of magnitude larger. Since short lifetimes are related to a strong interaction mediating the decay, it followed that the short-lived particles decay via the strong force and the others through weak interactions. The strange behaviour of those long-lived particles, which were produced in strong interactions and then decayed weakly, was explained formally by Pais and Gell-Mann who introduced a new quantum number, *strangeness*, which is conserved in strong interactions and can be violated by the weak force.

The originally rather simple picture of the world of elementary particles had changed completely by the mid 1960s, when so many different species were known that people were talking of a *particle zoo*. Fortunately, at that point, sufficient information had been gathered for some structure to emerge. While different schemes were being tried to quantify this, the most successful one was the ansatz by Gell-Mann and Ne'eman, who showed that the known hadrons could be classified in group-theoretical terms as multiplets of the special unitary Lie group SU(3), using isospin I and hypercharge $Y = B + S$, the sum of baryon number and strangeness, as the relevant quantum numbers. Thus, some kind of periodic table for elementary particles could be constructed. That this was more than just an abstract mathematical game became clear when vacant positions in the multiplets were filled by new particles. The most spectacular event was certainly the discovery of the Ω^-, a baryon with strangeness $S = -3$, at Brookhaven in 1964.

An introduction to the theory of Lie groups is given in Appendix A.1. Using group theory for the classification of the known hadrons was rather successful in explaining the observed regularities in the particle zoo, but there was the disturbing fact that all known particles were sitting in higher dimensional representations of SU(3), while the fundamental one remained empty. In order to resolve this dilemma and against all experimental evidence Gell-Mann and Zweig finally made the step to postulate that a set of three particles corresponding to the fundamental representation of SU(3) should also exist. These new particles were called *quarks* by Gell-Mann, where, as he described in his book *The Quark and the Jaguar* (Gell-Mann, 1994), the sound was first and the spelling was adopted later from the line "Three quarks for Muster Mark" in James Joyce's book *Finnegan's Wake*. Taken seriously, the quarks would constitute the basic building blocks of all hadronic matter. Unfortunately, according to SU(3) they would have to carry electric charges $\pm 1/3$ and $\pm 2/3$ of the electron charge, something that had never been observed. Thus elementary particle physics at the time was faced with a situation, where the world of hadrons was beginning to be understood in terms of some hypothetical particles with absolutely no experimental evidence in support of their existence. The properties, that is, quantum numbers, of those quarks, commonly called 'u' (up), 'd' (down) and 's' (strange) and their antiparticles are given in Table 2.1. Note that the antiquarks of the various types or *flavours*, as the types are also referred to, have the signs of their additive quantum numbers reversed and, being spin-1/2 particles, opposite parities.

Baryons are constructed by combining three quarks; mesons are obtained as a combination of a quark and an antiquark. With proper assumptions concerning the spin and orbital angular momentum, the actual hadrons and their excited

Table 2.1 *Quantum numbers of the light quarks. The electric charge e_q of the quarks is given in units of the positron charge.*

Quark	Spin	Parity	e_q	I	I_3	S	B
u	1/2	+1	+2/3	1/2	+1/2	0	+1/3
d	1/2	+1	−1/3	1/2	−1/2	0	+1/3
s	1/2	+1	−1/3	0	0	−1	+1/3
ū	1/2	−1	−2/3	1/2	−1/2	0	−1/3
d̄	1/2	−1	+1/3	1/2	+1/2	0	−1/3
s̄	1/2	−1	+1/3	0	0	+1	−1/3

states can be constructed, like, for example, the particles in the spin-3/2 baryon decuplet. Pertaining to the ground state, the orbital angular momentum vanishes, the parity is positive and the wavefunctions are fully symmetric in both flavour and spin.

Baryon decuplet				**Quark content**			
Δ^-	Δ^0	Δ^+	Δ^{++}	ddd	ddu	duu	uuu
	Σ^-	Σ^0	Σ^+	dds	dus	uus	
		Ξ^-	Ξ^0	dss	uss		
		Ω^-		sss			

Note that going horizontally from left to right the d-quarks are substituted by u-quarks, one at a time, going down parallel to the left edge d-quarks are exchanged for s-quarks and parallel to the right edge u-quarks are replaced by s-quarks. In each direction one encounters an SU(2) sub-symmetry, horizontally the familiar isospin or I-spin symmetry and in the other directions the so-called U-spin and V-spin symmetries. Using the quark model it is very easy to construct the known particles in a systematic way, which helped to keep the concept alive until firm experimental evidence for the existence of quarks was found. It is also worth noting that not all possible representations of the flavour-SU(3) are realized in nature. Only those multiplets are allowed where the difference between the number of quarks and antiquarks is a multiple of three, which ensures that all observed hadrons have integer electric charge.

2.1.2 The quark parton model

As higher energy accelerators became available the resolution at which matter could be probed also increased. When the momentum per particle passed

1 GeV/c, according to de Broglie's relation

$$\lambda = \frac{h}{p} \, , \tag{2.1}$$

structures smaller than 1 fm, the size of a proton, could be resolved for the first time. The point was reached where one actually could see the charge distribution inside the nucleus of hydrogen atoms and address the question whether there are pointlike constituents which serve as scattering centres. In other words, it became possible to do a Rutherford experiment for the nucleon rather than the atom. The kinematics of such a deep inelastic lepton–nucleon scattering experiment is sketched in Fig. 2.1. A high energy electron with initial energy E and four-momentum l, via exchange of a virtual space-like photon, is scattered off a nucleon with mass M and four-momentum p, which is at rest in the lab system. The scattering angle of the electron is θ. The final state is characterized by the four-momenta l' of the scattered electron and p' of the hadronic system with an invariant mass W.

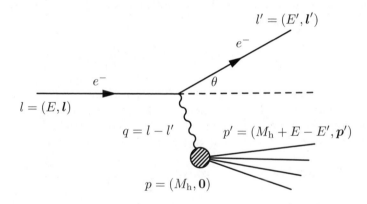

FIG. 2.1. Kinematics of deep inelastic electron–nucleon scattering

Starting from this comparatively simple diagram we will now show explicitly what kind of phenomenology one expects if the proton is a bound state of pointlike charged objects. As a first step it is convenient to introduce two new quantities, the energy transfer ν from the electron to the hadronic system in the hadron's rest frame

$$\nu = E - E' = \frac{q \cdot p}{M_{\rm h}} \tag{2.2}$$

and the squared and the squaredmomentum transfer Q^2 carried by the virtual photon,

$$Q^2 = -(l - l')^2 = -(p' - p)^2 = 2p \cdot p' - p^2 - p'^2 \, . \tag{2.3}$$

With the quantities defined in Fig. 2.1 one finds

$$Q^2 = 2M_{\rm h}(E + M_{\rm h} - E') - M_{\rm h}^2 - W^2 = 2M_{\rm h}\nu + M_{\rm h}^2 - W^2 \leq 2M_{\rm h}\nu, \quad (2.4)$$

where the equality for the last term is obtained in the limiting case $M_{\rm h}^2 = W^2$, that is, for the case of elastic scattering. The deviation from elastic scattering thus can be described by the Bjorken-variable x_B

$$x_B = \frac{Q^2}{2M_{\rm h}\nu} \quad \text{with} \quad 0 \leq x_B \leq 1. \quad (2.5)$$

To proceed one needs to know the cross section for elastic scattering of an electron with a spin-1/2 fermion of mass $M_{\rm h}$ and charge $e_{\rm q}$. See Ex. (2-1) for the elementary but rather lengthy calculation. One obtains

$$\frac{\mathrm{d}\sigma}{\mathrm{d}Q^2} = \frac{4\pi\alpha_{\rm em}^2 e_{\rm q}^2}{Q^4}\frac{E'}{E}\left\{\cos^2\frac{\theta}{2} + \frac{Q^2}{2M_{\rm h}^2}\sin^2\frac{\theta}{2}\right\}. \quad (2.6)$$

From this the double differential cross section with respect to Q^2 and ν can be derived. Starting with the trivial case of elastic scattering, where the relation $Q^2 = 2M_{\rm h}\nu$ must be satisfied, one gets

$$\frac{\mathrm{d}^2\sigma}{\mathrm{d}Q^2 d\nu} = \frac{4\pi\alpha_{\rm em}^2 e_{\rm q}^2}{Q^4}\frac{E'}{E}\left\{\cos^2\frac{\theta}{2} + \frac{Q^2}{2M_{\rm h}^2}\sin^2\frac{\theta}{2}\right\}\delta\left(\nu - \frac{Q^2}{2M_{\rm h}}\right). \quad (2.7)$$

The δ-function guarantees that integration over ν yields eqn (2.6) with only one value of ν contributing. In case of non-pointlike particles the double differential cross section has the same structure as eqn (2.7) and, introducing two so-called *structure functions* $W_1(Q^2, \nu)$ and $W_2(Q^2, \nu)$, can be written as

$$\frac{\mathrm{d}^2\sigma}{\mathrm{d}Q^2 d\nu} = \frac{4\pi\alpha_{\rm em}^2}{Q^4}\frac{E'}{E}\left\{W_2(Q^2, \nu)\cos^2\frac{\theta}{2} + 2W_1(Q^2, \nu)\sin^2\frac{\theta}{2}\right\}. \quad (2.8)$$

2.1.2.1   *Elastic scattering*   From the derivation of eqn (2.8) the structure functions for elastic scattering of pointlike particles with charge $e_{\rm q}$ are read off immediately as

$$W_1^{\rm el}(Q^2, \nu) = e_{\rm q}^2\frac{Q^2}{4M_{\rm h}^2}\delta\left(\nu - \frac{Q^2}{2M_{\rm h}}\right) \quad \text{and} \quad W_2^{\rm el}(Q^2, \nu) = e_{\rm q}^2\delta\left(\nu - \frac{Q^2}{2M_{\rm h}}\right). \quad (2.9)$$

2.1.2.2   *The parton model*   Much more interesting is the case of the parton model, where inelastic electron–nucleon interactions are understood in terms of incoherent elastic scattering processes between the electron and pointlike constituents of the nucleon. In other words, one assumes that a single interaction does not happen with the nucleon as a whole, but with exactly one of its constituents. Physically this picture makes sense when the energy of the projectile is sufficiently large to resolve the inner structure of the target. To describe this situation one has to partition the total four-momentum of the nucleon between its constituents. Each constituent i thus carries the fraction x_i with a probability density $f_i(x_i)$, the so-called *parton density function* p.d.f., meaning that the probability for x_i to fall into the infinitesimal range $[x, x+dx]$ is given by $f_i(x)dx$. From these assumptions the structure functions W_1 and W_2 can be calculated as superpositions of the elastic structure functions eqn (2.9) with weights $f_i(x)$. Summing over all constituents, taking into account that the masses M_i of the constituents are given by $M_i = x_i M_h$ and integrating out the δ-functions then yields

$$W_1(Q^2, \nu) = \sum_i \int_0^1 dx_i f_i(x_i) e_i^2 \frac{Q^2}{4x_i^2 M_h^2} \delta\left(\nu - \frac{Q^2}{2M_h x_i}\right) = \sum_i e_i^2 f_i(x_B) \frac{1}{2M_h}$$

(2.10)

and

$$W_2(Q^2, \nu) = \sum_i \int_0^1 dx_i f_i(x_i) e_i^2 \delta\left(\nu - \frac{Q^2}{2M_h x_i}\right) = \sum_i e_i^2 f_i(x_B) \frac{x_B}{\nu}, \quad (2.11)$$

with x_B as defined in eqn (2.5). It follows that in the parton model the variable x_B can be identified with the four-momentum fraction x carried by the struck parton. Note the subtle point that x_B as defined above is an experimental observable, while x is a parameter of the theoretical description of the nucleon. That the two are the same is a highly non-trivial finding. In the rest of this chapter and later on, we will usually use x and write x_B only in cases where we want to emphasize an experimental measurement.

Finally, absorbing the total target mass M_h and the energy transfer ν into a redefinition of the structure functions, one sees that deep inelastic scattering processes between a charged unpolarized lepton and an unpolarized nucleon can be described in terms of two functions

$$F_1(x) \equiv M_h W_1 = \frac{1}{2} \sum_i e_i^2 f_i(x) \tag{2.12}$$

and

$$F_2(x) \equiv \nu W_2 = \sum_i e_i^2 x f_i(x) \tag{2.13}$$

which only depend on the sharing of the target nucleon's four-momentum between its constituents. The structure function F_1 measures the parton density as function of x while F_2 describes the momentum density, both weighted with the coupling strength to the photon probe.

The fact that the observed cross sections depend only on a single dimensionless variable x is also referred to as *scaling* behaviour. As shown above, it is a direct consequence of having pointlike dimensionless scattering centres. Extended objects would introduce a new energy scale into the problem.

2.1.2.3 *Experimental findings* The experimental observation of scaling was the first clear evidence for a partonic sub-structure in the nucleon, giving support to the concept of quarks as the building blocks of hadronic matter. Some early results (MIT-SLAC Collab., 1970) are shown in Fig. 2.2 and Fig. 2.3. That the structure functions for deep inelastic electron–nucleon scattering are mainly a function of x_B and essentially independent of Q^2 is illustrated by Fig. 2.2. An alternative way of showing scaling is to plot F_2 at, for example, $x_B = 0.25$, as a function of Q^2. This is done in Fig. 2.3, where one sees that F_2 is indeed independent of Q^2.

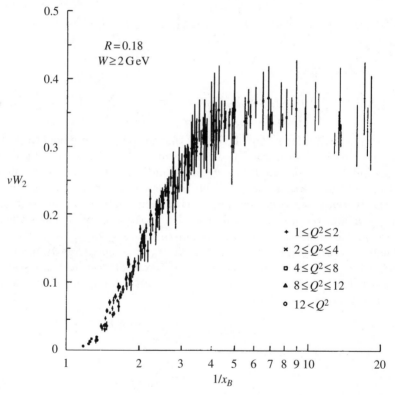

FIG. 2.2. Scaling behaviour of $\nu W_2(\omega) = F_2(\omega)$, $\omega = 1/x_B$, for various Q^2 ranges. Figure from MIT-SLAC Collab.(1970).

Once it was possible to look into the nucleon, one could also try to determine the properties of those partons. While generic scaling behaviour is a universal feature of any parton model, the details of course depend on the properties of

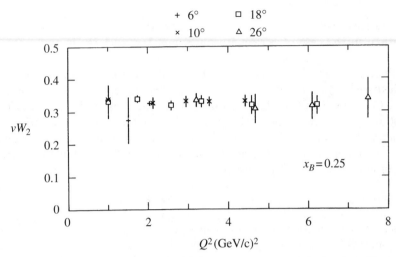

FIG. 2.3. Value of $\nu W_2(Q^2) = F_2(Q^2)$ for $x_B = 1/\omega = 0.25$. Figure from MIT-SLAC Collab.(1970).

the particles involved. The above derivation for electron–nucleon scattering, for example, explicitly assumed that the partons are spin-1/2 particles. In this case a definite relation has to hold between F_1 and F_2, the so-called *Callan–Gross relation*

$$2xF_1(x) = F_2(x) . \tag{2.14}$$

The fact that the lepton–nucleon data are in very good agreement with this prediction shows that the struck partons indeed are spin-1/2 fermions.

Additional possibilities arise when different probes are used in deep inelastic scattering processes. Using, for example, neutrinos instead of electrons, the interaction is mediated by $W^\pm$ or Z bosons rather than photons. Now the coupling strength is given by the third component of the weak isospin, which unlike the electric charge gives the same coupling strength to all quarks. Comparing electron–nucleon and neutrino–nucleon cross sections thus allows to probe the electric charge of the partons. Introducing quark densities for up and down quarks and their antiparticles in the proton,

$$u = u_{\rm p}(x) \quad d = d_{\rm p}(x) \quad \bar{u} = \bar{u}_{\rm p}(x) \quad \bar{d} = \bar{d}_{\rm p}(x) \tag{2.15}$$

and assuming the quark-charges as predicted by the quark model, the structure functions F_2 for electron–proton and electron–neutron scattering are given by

$$F_2^{\rm ep}(x) = x\left\{\frac{4}{9}(u + \bar{u}) + \frac{1}{9}(d + \bar{d})\right\} \tag{2.16}$$

$$F_2^{\rm en}(x) = x\left\{\frac{4}{9}(d + \bar{d}) + \frac{1}{9}(u + \bar{u})\right\} . \tag{2.17}$$

The transition from eqn (2.16) to eqn (2.17) is done by a simple isospin transformation: in the case of perfect isospin symmetry u-quarks in the proton are equivalent to d-quarks in the neutron and *vice versa*. It is also worth noting, that the ansatz for the structure functions takes antiquarks into account, the so-called *sea-quarks*, which are expected to contribute because of vacuum fluctuations. Contributions from heavier quarks are neglected at this stage, although they, too, are present in the nucleon. Doing a scattering experiment with a target material having equal numbers of protons and neutrons, like, for example, ^{40}Ca, the effective nucleon structure function seen is the arithmetic average of the proton and the neutron contributions

$$F_2^{\mathrm{eN}}(x) = \frac{5}{18} x \left\{ u + \bar{u} + d + \bar{d} \right\}.$$
(2.18)

In the structure function describing neutrino–nucleon scattering, where, for example, an incident muon neutrino interacts via a charged W boson and is transformed into a μ^-, because of charge conservation only negatively charged quarks contribute. One finds

$$F_2^{\nu \mathrm{p}}(x) = 2x \left\{ \bar{u} + d \right\}$$
(2.19)
$$F_2^{\nu \mathrm{n}}(x) = 2x \left\{ \bar{d} + u \right\},$$
(2.20)

where the overall factor of 2 follows from the theory of weak interactions. Averaging the neutrino–nucleon structure functions, one sees that there is a fixed ratio to the electron–nucleon structure function which is determined by the electric charges of the partons,

$$F_2^{\nu \mathrm{N}}(x) = \frac{5}{18} F_2^{\mathrm{eN}}(x).$$
(2.21)

Confirmed by experimental measurements, this relation gave further support to the assumption that the partons found inside the nucleon indeed are the quarks inferred from hadron spectroscopy.

The physics potential of neutrino–nucleon scattering experiments is, however, not yet exhausted. In high energy weak interactions only left-handed fermions and right-handed antifermions participate. Since the incident neutrinos are also left-handed, parity is maximally violated. From the angular distributions of the scattered leptons one can thus disentangle the contributions from quarks and antiquarks, which allows the extraction of an additional structure function $F_3(x)$,

$$F_3(x) = (u + d) - (\bar{u} + \bar{d}),$$
(2.22)

which is the difference of the quark and the antiquark densities. Integrating over F_3 thus allows us to count the net number of quarks in the nucleon. An early result for this Gross–Llewellyn-Smith sum rule was

$$\int_0^1 \mathrm{d}x F_3(x) = 2.5 \pm 0.5,$$
(2.23)

consistent with the expectation from the quark model.

To summarize, up to this point the observation of scaling in deep inelastic scattering processes which allowed us to probe the interior structure of the nucleon has shown that it contains pointlike constituents, the so-called partons. Further studies then revealed that those partons carry the quantum numbers predicted by the quark model, so that the two merged into the QPM.

2.1.3 Colour

Despite its successes the QPM still left open questions, which indicated that it was not yet the complete story. In the following pages we will go through the most important ones and show how the evidence accumulated that finally led to the field theory of Quantum Chromodynamics. We will start with the findings that point towards a new internal quantum number, which later received the label *colour*.

2.1.3.1 The spin-statistics problem

In the quark model the particles of the baryon decuplet have an s-wave spatial wavefunction and are fully symmetric in spin and flavour. This is most easily seen for the particles occupying the corner positions in the $J_z = +3/2$ state:

$$
\begin{aligned}
|\Delta^{++}; +3/2\rangle &= |u \uparrow\rangle \, |u \uparrow\rangle \, |u \uparrow\rangle \\
|\Delta^{-}\ ; +3/2\rangle &= |d \uparrow\rangle \, |d \uparrow\rangle \, |d \uparrow\rangle \\
|\Omega^{-}\ ; +3/2\rangle &= |s \uparrow\rangle \, |s \uparrow\rangle \, |s \uparrow\rangle \, .
\end{aligned}
\tag{2.24}
$$

If this were the whole story, the complete wavefunction would be totally symmetric for identical fermions, which is a blatant violation of the Pauli-principle. A possible way out is to assume that the quarks carry an additional degree of freedom, *colour*, which can take on three distinct values. Then the Pauli-principle can be restored by assuming that the wavefunction is completely antisymmetric in this new degree of freedom, which usually is labelled 'red', 'green' or 'blue'. Denoting the colour state of a quark by an index to the flavour symbol, the wavefunction of, for example, the Δ^{++} can be written as

$$
|\Delta^{++}; +3/2\rangle = \frac{1}{\sqrt{6}} \sum_{ijk=1}^{3} \varepsilon_{ijk} |u_i \uparrow\rangle |u_j \uparrow\rangle |u_k \uparrow\rangle ,
\tag{2.25}
$$

where ε_{ijk} is the completely antisymmetric tensor with $\varepsilon_{123} = +1$. A baryon thus is described by a totally antisymmetric superposition of all arrangements of the three basic colours between the constituent quarks.

The name 'colour' is taken from the everyday experience that all ordinary colours can be composed from three basic colours. For ordinary colours, a superposition of equal amounts of the basic colours red, green and blue yields white, and something similar also holds for the quark colours. If one assumes that those colours exhibit an SU(3) symmetry, then the colour part of the baryon wavefunction can be shown to transform as an SU(3) singlet, that is, a baryon does not have any net colour. It is 'white'. This is a very important observation, since it implies that the new quantum number colour is effectively hidden inside the baryons and becomes visible only when it is probed at a momentum transfer which allows to resolve the individual partons. Postulating that net colour is always confined inside hadrons immediately gives a heuristic explanation why free quarks are never observed. A plausible dynamical explanation for colour confinement appeared later in the context of QCD. At the current level of understanding, the confinement hypothesis was, however, consistent in the sense that also the mesons could be understood as colour singlets. Made from a quark and an antiquark, its colour-wavefunction is a superposition of colour–anticolour states of the form

$$|\pi^+\rangle = \frac{1}{\sqrt{6}} \sum_{ij=1}^{3} \delta_{ij} \left[\, |u_i \uparrow\rangle |\bar{d}_j \downarrow\rangle + |u_i \downarrow\rangle |\bar{d}_j \uparrow\rangle \, \right]. \qquad (2.26)$$

So far we have just postulated that quarks come in three colours. Below we will now continue to discuss some of the evidence which supports this assumption.

2.1.3.2   *The Adler–Bell–Jackiw anomaly*   Theories such as the Standard Model of electroweak interactions distinguish between left- and right-chirality fields. Technically this means that the couplings depend on the Dirac-matrix γ_5. The treatment of γ_5 in loop diagrams is very delicate and, as illustrated schematically below, can lead to unexpected results. Consider the reaction of $Z \to \gamma\gamma$ which in leading order proceeds via the triangle diagram given in Fig. 2.4. Note that in order to contribute to a physical process at least one of the final state photons would have to be off mass-shell, since otherwise the reaction would be forbidden by angular momentum conservation. Here, however, we can ignore these details since we only want to show how such loop diagrams can give rise to problems.

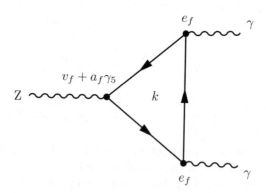

FIG. 2.4. Triangle diagram giving rise to the Adler–Bell–Jackiw anomaly. Note that all charged fermions f contribute in the loop.

The coupling of the Z to the fermion-loop is proportional to $v_f + a_f\gamma_5$, the sum of the vector and the axial vector coupling; the coupling to the photons is proportional to the fermion charge e_f. According to Furry's theorem the vector coupling of the Z does not contribute to the closed fermion loop, that is, only the axial contribution proportional to γ_5, summed over all fermions, remains. The result is a term that would violate the gauge symmetry of the Standard Model since it is only associated with the γ_5-coupling. Thus, for the Standard Model to be consistent, the sum over all fermions must cancel. Up-type and down-type fermions have $a_f^{\mathrm{up}} = +1/2$ and $a_f^{\mathrm{down}} = -1/2$, respectively. Taking only particles from the first generation one obtains

$$\sum_f e_f^2 a_f = \frac{1}{2}\left[-1 + N_c\left(\frac{4}{9} - \frac{1}{9}\right)\right], \qquad (2.27)$$

where N_c is the number of colours for the quarks. The condition that the sum cancels immediately translates into $N_c = 3$, that is, the requirement that the Standard Model of electroweak interactions is consistent with fractionally charged quarks implies that there has to be an additional internal degree of freedom for the quarks which can take on three different values. It is also interesting to note, that a cancellation of the Adler–Bell–Jackiw anomaly generation by generation implies that the existence of the bottom and the top quark could be inferred already from the discovery of the tau lepton.

2.1.3.3 *The π^0 decay rate* A triangle diagram of the type discussed above also describes the leading order decay amplitude of the neutral pion (Fig. 2.5). Here, however, the diagram does not violate a gauge symmetry and gives rise to physical effects. With the main contributions coming from u- and d-quarks in the loop, the width of the π^0 is given by

$$\Gamma(\pi^0 \to \gamma\gamma) = N_c^2\left(e_u^2 - e_d^2\right)^2 \frac{\alpha_{\mathrm{em}}^2 m_\pi^3}{64\pi^3} \frac{1}{f_\pi^2}, \qquad (2.28)$$

where e_u and e_d are the electric charges of the u- and d-quark, respectively, expressed in units of the positron charge, and f_π the pion decay constant. As pointed out by Abbas (2000), these charges have to be known from some external source in order to infer the number of colours N_c. Taking them either from the static quark model or measurements performed in deep inelastic scattering, one finds

$$\Gamma(\pi^0 \to \gamma\gamma) = 7.63\,\mathrm{eV}\left(\frac{N_c}{3}\right)^2. \qquad (2.29)$$

The experimental result $\Gamma(\pi^0 \to \gamma\gamma) = 7.84\pm0.56$ eV again gives strong evidence for the existence of a three-valued internal degree of freedom for the quarks.

2.1.3.4 *Electron–positron annihilation into hadrons* Studying electron–positron annihilation into hadronic final states, we are confronted with a situation

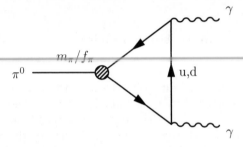

FIG. 2.5. Diagram of the dominant contribution to the π^0 decay

where a very simple initial state of two light pointlike fermions is transformed into a complex multi-particle system of mostly pions, some kaons and a few baryons and antibaryons. The description of such a process by means of perturbation theory at first glance appears to be rather hopeless. The picture, however, becomes much simpler when ignoring the details of the multi-hadron final state. Assuming that Z production is negligible, the dominant contribution to hadron production must start with the creation of a quark–antiquark pair from a virtual photon, which later on evolves into the complex system observed in the detector.

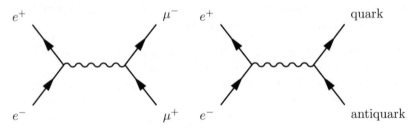

FIG. 2.6. Leading order contributions to μ-pair and multi-hadron production

Details of this process will be discussed at greater length later on. From Fig. 2.6 it is evident that the initial phase of multi-hadron production is very similar to the creation of muon pairs in e^+e^- annihilation. The cross sections can be compared directly. Taking the ratio, everything except the coupling strengths cancels. The ratio R_γ thus directly measures the sum of the squares of the quark charges, that is, given the contributing flavours, it allows us to determine the number of quark colours. In the energy region above the Upsilon resonances, where five quark flavours contribute, we therefore expect

$$R_\gamma = \frac{\sigma(e^+e^- \to \text{hadrons})}{\sigma(e^+e^- \to \mu^+\mu^-)} = N_c \sum_q e_q^2 = N_c \frac{11}{9} \; . \tag{2.30}$$

The actual measurement yields $N_c \approx 3.2$. Again there is experimental evidence for three states of the quarks, although it seems that a value slightly above $N_c = 3$

is preferred. Later this excess will find a natural explanation in the context of higher order QCD corrections. Here it is a first indication that colour may be more than just a quantum number which is needed to get some book keeping right.

2.1.4   *Other puzzles*

Having dealt with the evidence for colour as a new quantum number carried by the quarks, we now turn to those findings which led to the realization that quarks are in fact strongly interacting fields.

2.1.4.1   *The momentum sum rule*   As described earlier, deep inelastic electron–nucleon scattering experiments allowed us to measure the momentum weighted probability density function, $F_2^{\mathrm{eN}}(x)$, of quarks and antiquarks in the nucleon. Integration over x then yields the fraction of the nucleon momentum carried by the charged partons. The experimental finding was

$$\frac{18}{5} \int_0^1 \mathrm{d}x \, F_2^{\mathrm{eN}}(x) = \int_0^1 \mathrm{d}x \left[u(x) + d(x) + \bar{u}(x) + \bar{d}(x) \right] \approx 0.5 , \qquad (2.31)$$

indicating that the charged partons which are probed by the scattering process carry only about one half of the total momentum. Apparently, there exist other components in the nucleon in addition to the quarks, which do not carry electric charge and thus are invisible when using an electromagnetic probe. They also are invisible in neutrino–nucleon scattering, that is, they do not carry weak charges either. This means that those components are either subject to an altogether new type of force, or, staying within the catalogue of known interactions, must be specific to the strong interaction.

2.1.4.2   *Scaling violations*   Scaling in deep inelastic scattering was derived from the assumption that inside the nucleon there are non-interacting pointlike scattering centers. Although phenomenologically very successful, it is obvious that this simple picture can only hold approximately. Since the partons were found to be charged particles at least electromagnetic interactions between the constituents of the nucleon have to be taken into account.

With increasing Q^2 the spatial and temporal resolution of the probe will also increase and become able to resolve vacuum fluctuations. This means that a quark which at lower Q^2 is just seen as a pointlike particle will be resolved into more partons at higher momentum transfers. A pictorial representation of this scenario is given in Fig. 2.7. As a consequence, the total four-momentum of the nucleon is distributed over more constituents, which implies a softening of the structure function. With increasing momentum transfer the average fractional momentum $\langle x \rangle$ per parton will decrease as sketched in Fig. 2.8. The amount of change in the structure functions will be proportional to the strength α of the interaction between the partons, that is, we might expect a qualitative behaviour like

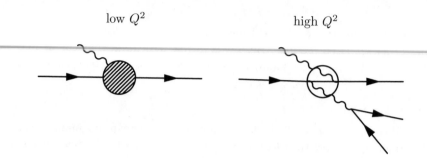

FIG. 2.7. Resolution of vacuum fluctuations at large Q^2

$$\frac{\mathrm{d}F}{F} \sim \alpha \frac{\mathrm{d}Q^2}{Q^2} \, . \tag{2.32}$$

Due to electromagnetic interactions we thus would expect scaling violations

$$\frac{\mathrm{d}\ln F}{\mathrm{d}\ln Q^2} \sim \alpha_{\mathrm{em}} \approx \frac{1}{137} \, . \tag{2.33}$$

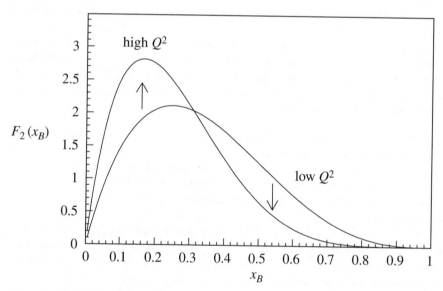

FIG. 2.8. Qualitative behaviour of the Q^2-evolution of structure functions

Scaling violations were indeed found experimentally. Taking for example measurements of F_2 from deep inelastic electron– or muon–nucleon scattering, one has $F_2(x = 0.5, Q^2 = 10 \text{ GeV}^2) \approx 0.1$ and $F_2(x = 0.5, Q^2 = 100 \text{ GeV}^2) \approx 0.07$, which gives

$$\left| \frac{\mathrm{d} \ln F}{\mathrm{d} \ln Q^2} \right| \approx 0.16. \tag{2.34}$$

Apparently, there is a strong force acting between the quarks, much stronger than electromagnetic interactions, which has to be explained theoretically. As an aside it may be worth noting, that the almost perfect scaling observed in Fig. 2.3 results from the fact that $x_B = 0.25$ appears to be just the fixed point with respect to the softening of F_2: for values $x_B > 0.25$ the probability density is shrinking with increasing Q^2, for values $x_B < 0.25$ it is growing.

2.2 The QCD Lagrangian

In a nutshell the evidence presented in the previous sections can be summarized as follows:

- hadrons are composed of fractionally charged quarks
- quarks are spin-1/2 fermions
- they come in three distinct colours
- there is evidence that colour exhibits an SU(3) symmetry
- quarks are subject to a strong interaction
- besides quarks there are additional partons in the nucleon
- those partons feel neither the electromagnetic nor the weak force

Note that the colour SU(3) is distinct from and must not be confused with the flavour SU(3) discussed in Section 2.1.1. The puzzles posed by these findings can be solved in a very elegant way by assuming that colour is a charge-like quantum number, conceptually similar to the electric charge or the weak isospin. Originally proposed only as an index to get the book keeping right, it is then understood as the source of a colour field. If the interaction mediated by this field is strong enough, then the large scaling violations observed in the structure functions of the nucleon find a natural explanation. The colour field apparently glues the quarks together to form the observed hadrons, motivating the name *gluons* for the quanta of the colour field. If those gluons couple only to colour charge, then they are invisible in all deep inelastic scattering experiments using lepton probes and also the missing momentum found from the integral over $F_2(x)$ can be accounted for.

Evidently there are good arguments in favour of a field theory of strong interactions based on the colour charge of the quarks. The requirement that the theory be renormalizable suggests a Yang–Mills gauge theory (Yang and Mills, 1954). Assuming an unbroken gauge symmetry the general form of the Lagrangian is

$$\mathcal{L}_{\mathrm{QCD}} = -\frac{1}{4} F_{\mu\nu}^a F^{a\mu\nu} + \sum_q \bar{q}_i (\mathrm{i}\gamma^\mu D_\mu - m_q)_{ij} q_j \tag{2.35}$$

where sums over repeated indices are implied. The field strength tensor $F_{\mu\nu}^a$ and the covariant derivative D_μ are given by the following expressions

$$F_{\mu\nu}^a = \partial_\mu A_\nu^a - \partial_\nu A_\mu^a - g f^{abc} A_\mu^b A_\nu^c$$

$$(D_\mu)_{ij} = \delta_{ij}\partial_\mu + i\,g_s T^a_{ij} A^a_\mu \tag{2.36}$$

$$(m_q)_{ij} = m_q \delta_{ij} \ ,$$

where A^a_μ are the gluon fields, g_s the gauge coupling, f^{abc} the structure constants and T^a_{ij} the generators of the Lie group which defines the gauge symmetry. Details of the group-theoretical foundations are explained in Appendix A.1. Note that the mass of a quark is independent of its colour. The free parameters of the theory are the mass terms and the coupling constant g_s. Given these parameters, the only input needed to define the Lagrangian of the theory is the gauge symmetry for the fundamental charges. In case of Quantum Electrodynamics (QED) this is a U(1)-symmetry, for weak interactions it is the group SU(2)$_{\rm L}\otimes$U(1)$_{\rm Y}$. From what has been said before, it is clear that the most natural assumption for the underlying symmetry in a Yang–Mills theory of strong interactions would be SU(3), with the quark states transforming under the fundamental representation of the group. For SU(3) there are eight generators $T^a = \lambda_a/2$, with λ_a the Gell-Mann matrices

$$\lambda_1 = \begin{pmatrix} 0 & 1 & 0 \\ 1 & 0 & 0 \\ 0 & 0 & 0 \end{pmatrix} \quad \lambda_2 = \begin{pmatrix} 0 & -i & 0 \\ i & 0 & 0 \\ 0 & 0 & 0 \end{pmatrix} \quad \lambda_3 = \begin{pmatrix} 1 & 0 & 0 \\ 0 & -1 & 0 \\ 0 & 0 & 0 \end{pmatrix}$$

$$\lambda_4 = \begin{pmatrix} 0 & 0 & 1 \\ 0 & 0 & 0 \\ 1 & 0 & 0 \end{pmatrix} \quad \lambda_5 = \begin{pmatrix} 0 & 0 & -i \\ 0 & 0 & 0 \\ i & 0 & 0 \end{pmatrix} \tag{2.37}$$

$$\lambda_6 = \begin{pmatrix} 0 & 0 & 0 \\ 0 & 0 & 1 \\ 0 & 1 & 0 \end{pmatrix} \quad \lambda_7 = \begin{pmatrix} 0 & 0 & 0 \\ 0 & 0 & -i \\ 0 & i & 0 \end{pmatrix} \quad \lambda_8 = \begin{pmatrix} 1 & 0 & 0 \\ 0 & 1 & 0 \\ 0 & 0 & -2 \end{pmatrix} \frac{1}{\sqrt{3}} \ .$$

The structure constants of the group, $f^{abc} = f_{abc}$, defined through the commutation relations

$$[T^a, T^b] = i f^{abc} T^c \ , \tag{2.38}$$

are totally antisymmetric in their indices. For SU(3) the non-vanishing values are

$$f_{123} = 1$$
$$f_{458} = f_{678} = \sqrt{3}/2 \tag{2.39}$$
$$f_{147} = f_{165} = f_{246} = f_{345} = f_{376} = f_{257} = 1/2 \ ,$$

with permutations of the indices being understood. A more inituitive represensentation of the QCD Lagrangian is given in Fig. 2.9. The gluonic part derived from the field strength tensor consists of a free field term and two interaction terms where gluons couple to gluons. This coupling between gauge bosons is characteristic of a gauge theory based on a non-abelian group where the gauge bosons carry the charge of the interaction, colour in case of QCD, and thus are able to couple directly to themselves. The fermionic part of the Lagrangian is a sum over all quark flavours, again featuring a free field term and a term for the quark–gluon coupling. The triple-gluon and the quark–gluon coupling are proportional to the gauge coupling g_s, the four-gluon coupling is proportional to g_s^2. In addition the amplitudes associated with the individual couplings depend on the detailed structure of the underlying symmetry group. Quark colours are indexed by $i, j = 1, 2, 3$, gluon colours by $a, b, c, d, e = 1, \ldots, 8$. The three-gluon coupling between gluons of colour states a,b and c is proportional to the structure constant f^{abc}, and the coupling between two quarks of colours i and j to a gluon of type a is proportional to the matrix element T_{ij}^a.

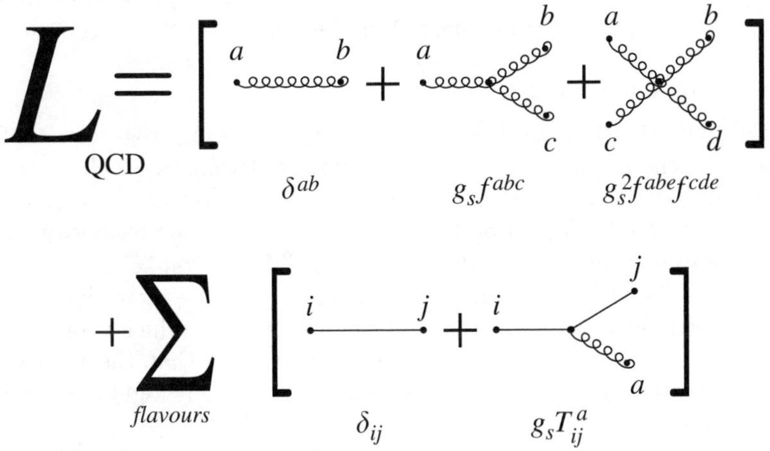

FIG. 2.9. Pictorial respresentation of the QCD Lagrangian. Figure from Schmelling(1995 a).

The physics content of the QCD Lagrangian is further discussed in the following chapter and in the problems Ex. (2-2) and Ex. (2-3) given below. It is shown explicitly, that there is a full symmetry in all colours with respect to physics, which is maybe not entirely obvious from the representation of the Gell-Mann matrices or the numerical values of the structure constants. One finds that the probability for gluon emission is the same for all quark colours, that the probability for gluon splitting into quark pairs is the same for all gluon states as is the probability of a gluon splitting into secondary gluons. Denoting the relative strengths of the splitting probabilities with C_F, C_A and T_F for gluon radiation off a quark, gluon splitting into two gluons and gluon splitting into two quarks, respectively, QCD predicts

$$C_F = \frac{4}{3}, \quad C_A = 3 \quad \text{and} \quad T_F = \frac{1}{2}. \tag{2.40}$$

Since those numbers are proportional to the normalization of the generators of
the group, only ratios have a physical meaning. A convenient choice is

$$\frac{C_A}{C_F} = \frac{9}{4} \quad \text{and} \quad \frac{T_F}{C_F} = \frac{3}{8} \,. \tag{2.41}$$

The first of the ratios can be interpreted as the probability of gluon emission off
a gluon relative to that of gluon emission off a quark, that is, as the ratio of the
colour charges of gluons and quarks. For an abelian gauge theory such as QED,
where the gauge bosons do not carry any charge, this ratio would be zero. QCD
predicts it to be $C_A/C_F = 2.25$, which means that the gluon has a colour charge
more than twice as large as that of a quark. As will be discussed later, the two
ratios introduced in eqn (2.41) are characteristic of the gauge group chosen in
the Lagrangian. Measuring them thus provides a way to probe experimentally
the gauge structure of the fundamental interactions.

Exercises for Chapter 2

2–1 Calculate the Born level prediction for the scattering cross section of
an high energy electron and another spin-1/2 fermion with charge e_f
and mass M at rest in the laboratory system. Neglect the mass of the
electron. (Hint: Skip to Section 3.3 to learn about the technicalities
of the evaluation of Feynman diagrams.) $(\bigstar\bigstar)$

2–2 Express the Gell-Mann matrices associated with the eight gluons as
linear combinations of operators of the type $|C'\rangle\langle C|$, which transform
a quark of colour state $|C\rangle$ into $|C'\rangle$, and show that the relative
probabilities for a red, green and blue quark to emit a gluon are all
the same.

2–3 From the structure constants of SU(3) calculate explicitly the relative
splitting probabilities for all gluon states into secondary gluons and
secondary quark–antiquark pairs.

3

THE THEORY OF QCD

3.1 QCD as an SU(3) gauge theory

Quantum Chromodynamics (QCD) is the gauge theory of coloured quarks and gluons (Fritzsch $et\ al.$, 1973; Gross and Wilczek, 1973; Weinberg, 1973a). It is an example of a non-abelian Yang–Mills theory. Its action is defined in terms of a Lagrangian density which for a single flavour of non-interacting quark is given by

$$S = \mathrm{i} \int \mathrm{d}^4 x \, \mathcal{L}(x) \quad \text{with} \quad \mathcal{L}(x) = \bar{q}_j(x)(\mathrm{i}\,\partial\!\!\!/ - m)q_j(x) \ . \tag{3.1}$$

The index j on the Dirac four-spinors runs over the N_c quark colours. In practice we have $N_c = 3$ but it is useful to leave it free and maintain generality. This expression is clearly invariant under a linear transformation, $q_j \mapsto U_{jk}q_k$ with $U^\dagger U = 1$. That is, when U is an element of the fundamental representation of the unitary group $U \in U(N_c)$. The unitary group is given by a direct product of groups $U(N_c)=U(1)\otimes SU(N_c)$, each of which can be treated separately. Here, we will concentrate on the $SU(N_c)$ group. As shown in Appendix A, the group elements can be written as a function of $N_c^2 - 1$ real parameters θ_a

$$U(\theta) = \exp(\mathrm{i}\,\theta_a T^a) \tag{3.2}$$

where the index a runs over all generators T^a of the group. When the parameter vector θ is position independent we refer to $U(\theta)$ as a global symmetry transformation. Similarly, if $\theta(x)$ is position dependent then we refer to $U(x) = U(\theta(x))$ as a local symmetry transformation. Due to the derivative term in eqn (3.1) it is not invariant under local gauge transformations. To make the Lagrangian density invariant we begin by introducing $N_c^2 - 1$ real valued, gauge fields A_a^μ and replace ∂^μ by the covariant derivative D^μ,

$$D^\mu = \partial^\mu + \mathrm{i}\,g_s A^\mu \quad \text{with} \quad A^\mu = A_a^\mu T^a \ . \tag{3.3}$$

The parameter g_s is called the gauge coupling. Local gauge invariance requires the transformation property

$$D^\mu(A') = U(x)D^\mu(A)U(x)^{-1}$$

$$\text{equivalent to} \quad D^\mu(A')q'(x) = U(x)D^\mu(A)q(x) \tag{3.4}$$

which is realized if the A^μ field transforms as:

$$A^\mu \mapsto U(x)A^\mu U(x)^{-1} + \frac{\mathrm{i}}{g_s}\big[\partial^\mu U(x)\big]U(x)^{-1} \,. \tag{3.5}$$

Interestingly, the presence of the second, inhomogeneous term means that non-vanishing gauge field configurations can be generated from the vacuum, $A^\mu = 0$.

The above considerations lead to the locally gauge invariant Lagrangian density for the quark fields.

$$\begin{aligned}\mathcal{L}_{\text{quark}} &= \bar{q}_j(x)[\mathrm{i}\not{D} - m]_{jk}q_k(x)\\ &= \bar{q}_j(x)\big[(\mathrm{i}\not{\partial} - m)\delta_{jk} - g_s\not{A}_a T^a_{jk}\big]\,q_k(x)\end{aligned} \tag{3.6}$$

Unfortunately, due to the lack of derivative terms, $\partial^\nu A^\mu_a$, in eqn (3.6) the gauge fields can be regarded only as auxiliary fields associated with external sources. In order to make them dynamical we must find a new, gauge and Lorentz invariant term to add to eqn (3.6) which contains derivatives of A^μ_a. In order to find a combination of $\partial^\nu A^\mu_a$, and possibly A^a_μ, terms that has a simple behaviour under a gauge transformation we investigate the non-commutativity of successive covariant derivatives:

$$[D_\mu, D_\nu] \equiv \mathrm{i}\,g_s F_{\mu\nu} \implies F_{\mu\nu} = \partial_\mu A_\nu - \partial_\nu A_\mu + \mathrm{i}\,g_s[A_\mu, A_\nu]$$
$$\text{or, taking components} \quad F^a_{\mu\nu} = \partial_\mu A^a_\nu - \partial_\nu A^a_\mu - g_s f^{abc} A^b_\mu A^c_\nu \,. \tag{3.7}$$

This defines the gauge field strength (Lorentz) tensor, $F_{\mu\nu}$. The action of a gauge transformation on $F_{\mu\nu}$ is easily derived by applying eqn (3.4) to its definition as a commutator, see also Ex. (3-1), giving

$$F_{\mu\nu} \mapsto U(x)F_{\mu\nu}U(x)^{-1} \,. \tag{3.8}$$

Whilst $F_{\mu\nu}$ transforms non-trivially, as a tensor under $\mathrm{SU}(N_c)$, it is now easy to construct a suitable Lorentz and gauge invariant term to add to the Lagrangian density,

$$\mathcal{L}_{\text{gauge}} = -\frac{1}{2}\mathrm{Tr}\{F_{\mu\nu}F^{\mu\nu}\} = -\frac{T_F}{2}F^a_{\mu\nu}F^{a\mu\nu} \,. \tag{3.9}$$

The pre-factor, $\frac{1}{2}$, is purely conventional. The terms eqns (3.6) and (3.9) together define the classical Lagrangian density for QCD.

Now eqn (3.9) is not the only invariant term which could be added to the Lagrangian density. Possible extra terms include

$$\bar{q}(x)\left(D_\mu D^\mu\right)^2 q(x)\,, \quad [\bar{q}(x)F_{\mu\nu}D^\mu D^\nu q(x)]^3\,, \quad (F_{\mu\nu}F^{\mu\nu})^4 \quad \text{etc.} \tag{3.10}$$

If added to the Lagrangian density then all these examples would require co-efficients carrying negative mass dimension, in order to ensure that the action remains a dimensionless number. However, the requirement of renormalizability, discussed in Section 3.4, forbids all such terms. Thus gauge invariance and renormalizability prove to be highly restrictive with respect to the construction of a Lagrangian density. In particular, gauge invariance also implies that gluons must be massless, since a mass term for the gauge fields, $m_A^2 A_\mu^a A^{a\mu}$, would not be gauge invariant due to the inhomogeneous term in eqn (3.5).

In addition to eqn (3.9) there is one other invariant term involving the gauge fields of mass dimension four or less which could be added to the standard Lagrangian density. In terms of the dual field strength tensor,

$$\tilde{F}_{\mu\nu}^a = \frac{1}{2}\epsilon_{\mu\nu}{}^{\sigma\tau} F_{\sigma\tau}^a \quad \text{normalized such that} \quad \tilde{\tilde{F}} = F , \qquad (3.11)$$

the so-called θ-term is given by

$$\begin{aligned}
\mathcal{L}_\theta &= \theta \frac{g_s^2 T_F}{16\pi^2} F_{\mu\nu}^a \tilde{F}^{a\mu\nu} \\
&= \theta \frac{g_s^2 T_F}{16\pi^2} \frac{\partial}{\partial x^\mu} \left[2\epsilon^{\mu\nu\sigma\tau} \left(A_\nu^a \partial_\sigma A_\tau^a + \frac{2}{3} g_s f_{abc} A_\nu^a A_\sigma^b A_\tau^c \right) \right] . \qquad (3.12)
\end{aligned}$$

The parameter θ appearing above has nothing to do with the parameters θ_a in eqn (3.2). As the second form makes clear, $\mathcal{L}_\theta$ can be expressed as the total divergence of a gauge dependent current. As such it contributes only a surface term to the action which naïvely may be neglected. Unfortunately, life is not so simple and the surface integral is related to a topological invariant, called the Pontryagin index.[1] The non-trivial topological structure of the vacuum in QCD is such that in practice the θ-term does give a non-perturbative contribution. This represents a serious problem since the θ-term violates both the discrete symmetries parity (P) and time reversal (T), which are known to be respected by QCD to high accuracy, along with charge conjugation (C) invariance (Cheng, 1988). Since T-violation is equivalent to CP-violation, one would expect a contribution to the CP-violating electric dipole moment of the neutron

$$\text{e.d.m.} = \bar{\theta} \times 10^{-(15-16)} \, e\cdot\text{cm} , \qquad (3.13)$$

where $\bar{\theta}$ is the sum of θ and the electroweak, CP-violating phase in the quark mass matrix. Given the measured value, e.d.m. $\leq 10^{-25} \, e\cdot\text{cm}$ (PDG, 2000), this requires $\bar{\theta} \leq 10^{-10}$; far smaller than the CP-violation observed in weak interactions. This is the 'strong CP problem' for which several putative solutions are available in the literature, most prominent of which is the axion. However, we adopt a pragmatic approach and simply set $\theta = 0$; in any case the θ-term does not give rise to any perturbative physics.

The classical QCD Lagrangian density $\mathcal{L}_{\text{class}} = \mathcal{L}_{\text{quark}} + \mathcal{L}_{\text{gauge}}$, described so far, is constructed to be invariant under local gauge transformations. However, this requirement leads to difficulties in formulating the quantum theory. The crux of the problem is the large degeneracy between sets of gluon field configurations which are all equivalent under gauge transformations. The treatment of this

---

[1] Loosely speaking, the number of twists in the mapping of the 3-sphere, at infinity, into the SU(3) gauge space.

problem requires the apparatus of gauge fixing and ghost fields. Here we provide
a heuristic discussion of the solution using the Feynman path integral method;
more complete details can be found in any modern quantum field theory text
book, such as the one by Peskin and Schroeder (1995).

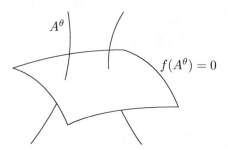

FIG. 3.1. A schematic diagram of gauge field space showing the 'fibres' of gauge
equivalent field configurations, A^θ, and the surface defined by the gauge fixing
condition, $f(A^\theta) = 0$

In the Feynman path integral approach, quantities of interest are evaluated
as averages over all configurations of the quark and gluon fields weighted by the
exponential of the action for the fields. This is similar to the use of the partition
function in statistical mechanics. It is the naïve functional integral over all the
gluon fields, including the gauge equivalent copies, which causes a divergence.
This divergence is completely unrelated to those discussed later in the context
of renormalization. The basic resolution, due to Faddeev and Popov (1967), is to
split the functional integral into an integral over unique elements representing
the sets of gauge equivalent field configurations and a common integral over
the space of gauge transformations. The latter integral represents a constant
(infinite) factor which can be safely dropped. The gauge degeneracy is broken
by imposing a gauge fixing condition of the form $f(A^\theta) = 0$. Here A^θ is the
transform of the gauge field A under the action of $U(\theta)$, eqn (3.5), and $f(A)$
is a function such that for a given A a solution exists for only one value of the
gauge parameters θ. In a non-abelian theory this may be true only if we exclude
topologically non-trivial gauge field configurations, which in any case give only
very small contributions to the action and do not affect perturbation theory
(Gribov, 1978). The situation is illustrated in Fig. 3.1. By inserting the identity
in a suitable form, c.f. $1 = \int dx |df/dx| \delta(f(x))$, and not showing source terms,
the fundamental partition function can be symbolically written as

$$Z = \int \mathcal{D}\bar{\psi}\mathcal{D}\psi\mathcal{D}A \exp\left(\frac{i}{\hbar}\int d^4x \mathcal{L}_{\text{class}}(\bar{\psi},\psi,A)\right) \times \int \mathcal{D}\theta \det\left|\frac{\delta f(A^\theta)}{\delta\theta}\right| \delta\left(f(A^\theta)\right)$$

$$\rightarrow \int \mathcal{D}\bar{\psi}\mathcal{D}\psi\mathcal{D}A\mathcal{D}\eta^\dagger\mathcal{D}\eta \exp\left(\frac{i}{\hbar}\int d^4x \left\{\mathcal{L}_{\text{class}} - \frac{1}{2\xi}f(A)^2 + \eta^\dagger\frac{\delta f(A^\theta)}{\delta\theta}\eta\right\}\right).$$

$$(3.14)$$

Source terms are not shown. In the second line the divergent θ integral has been discarded as the remaining terms are actually θ-independent. Formally, this means that we have redefined the integration measure. The δ-function has been implemented in the action as the quadratic term. The parameter ξ is arbitrary, contributing only to the overall normalization, and as such it cannot enter into any physical quantity, like S-matrix elements, though it may appear in intermediate expressions. As is made clear below, particular choices, such as $\xi = 1$, are often preferred due to the relative simplicity of the resulting gluon propagator. The determinant of the Jacobian matrix is incorporated into the action as an integral over the octet of ghost fields, η^a. These are unphysical, complex valued, Lorentz scalars which obey Fermi–Dirac statistics, that is, they are represented by Grassmann variables, and transform under the adjoint representation of the gauge group. Ghost fields only appear internally in loop diagrams, their physical rôle is discussed in a less abstract fashion in Section 3.3.3.1. The result of these manipulations is the addition of gauge fixing and ghost terms to the Lagrangian density.

The gauge divergence in the path integral which is associated with the gauge degeneracy of the gluon fields manifests itself perturbatively in the lack of a gluon propagator. The addition of a gauge fixing term allows this propagator to be defined. To see how this works consider the popular choice of covariant gauge: $\partial_\mu A^{a\mu} = 0$. As indicated above, this requires two new terms to be added to the classical Lagrangian.

$$\mathcal{L}_{\text{fix+ghost}} = -\frac{1}{2\xi}(\partial_\mu A^{a\mu})(\partial_\nu A^{a\nu}) + \partial_\mu \eta^{a\dagger}\left(\partial^\mu \delta^{ab} + g_s f_{abc} A^{c\mu}\right)\eta^b \qquad (3.15)$$

Observe that the bracketed term in the ghost Lagrangian is the appropriate generalization of the covariant derivative for the adjoint representation: $T^a(A)_{bc} = -\mathrm{i} f_{abc}$. It provides a kinetic term for the ghost fields and in this covariant gauge a ghost–gluon coupling. Propagators are derived from the quadratic, free particle, terms in the action; for the A^a_μ field these are

$$2S_{\text{gauge}}$$
$$= -\mathrm{i}\int \mathrm{d}^4x \left\{(\partial_\mu A^a_\nu - \partial_\nu A^a_\mu)(\partial^\mu A^{a\nu} - \partial^\nu A^{a\mu}) + \frac{1}{\xi}\partial_\mu A^{a\mu}\partial_\nu A^{a\nu} + \mathcal{O}(A^3)\right\}$$
$$= +\mathrm{i}\int \mathrm{d}^4x\, A^a_\mu(x)\left[\eta^{\mu\nu}\partial^2 - \left(1 - \frac{1}{\xi}\right)\partial^\mu\partial^\nu\right]\delta_{ab}A^b_\nu(x) \qquad (3.16)$$
$$= -\mathrm{i}\int \mathrm{d}^4p\, \tilde{A}^a_\mu(p)\left[\eta^{\mu\nu}p^2 - \left(1 - \frac{1}{\xi}\right)p^\mu p^\nu\right]\delta_{ab}\tilde{A}^b_\nu(p)\,.$$

In the second line, integration by parts is used whilst in the third a Fourier transform, $\partial_\mu \mapsto -\mathrm{i} p_\mu$, is used to go to momentum space. The gluon propagator, $\prod(p)^{\mu\nu ab}$, is given by the inverse of the bracketed term.

$$\eta^{\mu\nu}\delta^{ab} = \prod(p)^\mu{}_{\sigma}{}^a{}_c \cdot i \left[\eta^{\sigma\nu}p^2 - \left(1 - \frac{1}{\xi}\right) p^\sigma p^\nu \right] \delta^{cb}$$

$$\Longrightarrow \quad \prod(p)^{\mu\nu ab} = \frac{i}{p^2 + i\epsilon} \left[-\eta^{\mu\nu} + (1 - \xi)\frac{p^\mu p^\nu}{p^2} \right] \delta^{ab} \tag{3.17}$$

It is easy to see that this inverse would not exist in the absence of the gauge fixing term, that is, in the limit $\xi \to \infty$. Since then the momentum-vector p^μ would be an eigenvector of the inverse propagator with eigenvalue zero, this results in a matrix with at least one vanishing eigenvalue which cannot be inverted. The $i\epsilon$ term enforces causality. It can be traced to adding a term $+i\epsilon A^a_\mu A^{a\mu}$ to the action to ensure that the action integral is convergent.

Another popular choice is the axial or physical gauge defined by $n \cdot A^a = 0$ where n is a fixed Lorentz four-vector. Sometimes, the additional restriction $n^2 = 1$ or $n^2 = 0$ is applied. The required gauge fixing term is

$$\mathcal{L}_{\text{fix}} = -\frac{1}{2\xi}(n \cdot A^a)(n \cdot A^a) . \tag{3.18}$$

Since in this axial gauge the corresponding ghost term only contains the kinetic piece and does not couple ghosts to any other fields, the ghosts may be trivially integrated out and need not be considered further. The corresponding, momentum space, gluon propagator is given by

$$\prod(p)^{\mu\nu ab} = \frac{i}{p^2 + i\epsilon} \left[-\eta^{\mu\nu} + \frac{n^\mu p^\nu + p^\mu n^\nu}{n \cdot p} - (n^2 + \xi p^2)\frac{p^\mu p^\nu}{(n \cdot p)^2} \right] \delta^{ab} . \tag{3.19}$$

Now, in any gauge the gluon propagator can be decomposed into a weighted sum of direct products of polarization vectors $\epsilon(p)^\mu$ for the off mass-shell gluon:

$$\prod(p)^{\mu\nu} = \frac{i}{p^2 + i\epsilon} \sum_{T,L,S} C_i \cdot \epsilon(p)^\mu_i \epsilon^\star(p)^\nu_i , \tag{3.20}$$

where, in general, the sum includes contributions from two transverse (T), one longitudinal (L) and one scalar component (S). Significantly, for an axial gauge in the on mass-shell limit, $p^2 = 0$, only the physical, transverse polarizations propagate $(C_L = C_S = 0)$. This proves to be very useful in situations where physical arguments are to be used. The price to be paid is the relative complexity of the propagator, in particular the presence of the spurious singularities in $(n \cdot p)^{-1}$ which require a careful treatment in terms of principal values (Leibbrandt, 1987). In practice, it proves popular to use the Feynman gauge for higher order calculations and the axial gauges to gain physical insight.

Finally, we give the complete QCD Lagrangian density, in the covariant gauge:

$$\mathcal{L}_{\text{QCD}} = \bar{q}_j(x)[i\,\partial\!\!\!/ - m]q_j(x)$$
$$- \frac{1}{2}\left[(\partial_\mu A^a_\nu - \partial_\nu A^a_\mu)(\partial^\mu A^{a\nu} - \partial^\nu A^{a\mu}) + \frac{1}{\xi}(\partial^\mu A^a_\mu)(\partial^\nu A^a_\nu) \right]$$

$$+(\partial^\mu \eta^{a\dagger})(\partial_\mu \eta^a) - g_s T^a_{jk} \bar{q}_j(x) A^a q_k(x) + g_s f_{abc}(\partial_\mu \eta^{a\dagger}) \eta^b A^{c\mu}$$

$$+g_s f_{abc}(\partial_\mu A^a_\nu) A^{b\mu} A^{c\nu} - \frac{g_s^2}{4} f_{abc} f_{ade} A^{b\mu} A^{c\nu} A^d_\mu A^e_\nu \qquad (3.21)$$

This has been separated into the quadratic parts and the remaining 'perturbations' which are proportional to either g_s or g_s^2. The first three terms give rise to the quark, gluon and ghost propagators, the fourth and fifth the quark–gluon and ghost–gluon vertices, whilst the sixth and seventh terms give rise to triple- and quartic-gluon vertices. The corresponding Feynman rules are detailed in Appendix B, together with those for the axial gauge.

3.2 The QCD description of basic reactions

In the following section, we attempt to give an overview of three basic aspects of hadron production in collider experiments. These are: a summary of the actual properties of the events that are seen in lepton–lepton, lepton–hadron and hadron–hadron collisions; an outline of these events' formal description using QCD; and an insight into the physical pictures which guide people's thinking.

In general, the use of QCD to describe a reaction means the use of perturbative QCD (pQCD). This restriction is purely practical and merely reflects our present inability to calculate more than a few non-perturbative properties within QCD. The applicability of perturbation theory relies on the strong coupling being small. A very important property of QCD is that the size of the strong coupling varies with the size of the characteristic momentum transfer in a process. The coupling runs in such a way that it is small for large momentum transfers, $Q \gg \Lambda_{\mathrm{QCD}}$, and large for small momentum transfers. To leading order one has

$$\alpha_s(Q^2) \equiv \frac{g_s^2(Q^2)}{4\pi} = \frac{1}{\beta_0 \ln(Q^2/\Lambda_{\mathrm{QCD}}^2)} . \qquad (3.22)$$

Here Λ_{QCD} is an energy scale at which non-perturbative effects become important. Experimentally, it is found to be $\mathcal{O}(200)$ MeV, that is, the mass scale of hadronic physics as given by the pion mass or equivalently the inverse of a typical hadron size R_0. The coefficient $\beta_0 > 0$ in eqn (3.22) is defined in Section 3.4.5. The appearance of the scale Λ_{QCD} and the running of the coupling is a subtle aspect of renormalizable theories, such as QCD, which we shall discuss later. As a consequence, the major part of this book is dedicated, necessarily, to discussing hard processes that involve a large momentum transfer. This may arise naturally, as for example in the production of a heavy particle, or may be engineered, by, for example, only considering jets with large transverse energies. Of course, we do need to discuss non-perturbative aspects of QCD, in particular hadronization. Here, when detailed descriptions are needed, we must mainly rely on models rather than theoretically secure QCD predictions. Fortunately, the effects of hadronization on pQCD predictions appear to be modest.

Restricting our attention to large momentum transfer processes, $Q^2 \gg \Lambda_{\mathrm{QCD}}$ with $\Lambda_{\mathrm{QCD}} \sim R_0^{-1}$, implies, by virtue of the uncertainty principle, that we see nature on a small, sub-nuclear scale. At these scales hadrons appear to be composed of the (anti)quarks and gluons which appear in the QCD Lagrangian. Furthermore, they are only weakly self-interacting thanks to the running strong coupling. This allows the individual, target hadrons to be characterized by parton density functions (p.d.f.s) describing the distributions of partons as a function of the fraction of their parent hadron's momentum that they carry. In the parton model, cross sections for hard processes are calculated in terms of the tree-level scattering or annihilation of individual (anti)quarks and gluons convoluted with the appropriate p.d.f.s. What gives this statement its power is the fact that the p.d.f.s are independent of the hard subprocess. In essence what we have is that the cross section can be factorized into a process dependent, short-distance, hard subprocess, involving partons, and a process independent, long-distance part, the p.d.f.s, describing the hadrons involved.

Now, many of the hard subprocesses of interest are electroweak in nature so that QCD really only enters via the higher order corrections. Two important features of this QCD improved parton model are the dependence of the p.d.f.s on the hard scale of the interaction and the appearance of multiparton final states. It is the QCD improved parton model that provides the framework for most of what follows.

As we have just said, calculations in pQCD are carried out in terms of the quark and gluon degrees of freedom appearing in the QCD Lagrangian rather than the colourless hadrons observed in experiments. The confinement transition from the almost free partons to the bound state hadrons is still not well understood but must be addressed before making comparisons with experiment. Fortunately, given the necessary restriction to hard processes, it is believed that non-perturbative effects, which involve small momentum transfers, $Q \leq \Lambda_{QCD}$, do not spoil parton level predictions. This can be seen in two complementary ways. First, the disparity in momentum transfers argues that the perturbative features of an event can not be modified significantly by hadronization without introducing a new, perturbative scale. Second, the uncertainty principle can be used to relate the four-momentum of a virtual particle, Q^μ, to the space–time distance it travels, Q^μ/Q^2; see Ex. (3-2). Thus, perturbative physics takes place on short-distance scales, whilst non-perturbative effects are long range in nature and can only have limited effect on the widely separated hard partons.

Two basic approaches are available to calculate hadronic event properties within pQCD. One approach is fixed-order perturbation theory, the other one is based on a summation of leading logarithms.

To describe a given type of event using fixed order perturbation theory, its dominant features are identified, typically collimated sprays of hadrons known as jets, and these are associated with well separated primary partons. In the absence of flavour tagging these may be either quarks or gluons. In this way the event is matched to a scattering amplitude containing the primary partons as external states. This amplitude is described by a sequence of ever more complex Feynman diagrams which may be grouped into sets according to how many gauge couplings, $g_s = \sqrt{4\pi\alpha_s}$, they contain. The simplest set of (tree) diagrams contribute to the cross section, which is proportional to the amplitude squared, at $\mathcal{O}(\alpha_s^n)$ where the power n is characteristic of the process. In gluon–gluon scattering, for example, one has $n = 2$, whilst for three-jet production in e^+e^- annihilation it is $n = 1$. This is the *leading order* (LO) approximation. The next simplest set of (one-loop) diagrams contribute at $\mathcal{O}(\alpha_s^{n+1})$; this is the *next-to-leading order* (NLO) approximation, etc. Given a sufficiently small coupling, this perturbation series should converge to the correct answer as more terms are added. In practice, the series is expected to be only asymptotically convergent so that beyond a certain order the numerical evaluation of the series begins to diverge from the true answer.

A complication arises in this approach because tree-level diagrams diverge whenever external partons become soft or collinear and related divergences arise in virtual (loop) diagrams. This is in addition to the ultraviolet divergences treated by renormalization. Fortunately, in sufficiently inclusive measurements, such as the total hadronic cross section, it is guaranteed that the two sets of divergences cancel. Unfortunately, in more exclusive quantities, which involve restricted regions of the external partons' available phase space, the cancellation is less complete and large logarithmic terms remain, generically of the form $L = \ln(Q^2/Q_0^2)$. Since $\alpha_s(Q^2)L$ is of order unity for $Q^2 \gg Q_0^2$, see eqn (3.22), this can spoil the convergence of finite order perturbation theory.

In the second approach, the original perturbation series is rearranged in terms of powers of $\alpha_s L$.

$$d\sigma = \sum_n a_n(\alpha_s L)^n + \alpha_s(Q^2) \sum_n b_n(\alpha_s L)^n + \cdots \qquad (3.23)$$

The first, infinite set of terms represent the *leading logarithm approximation* (LLA), then comes the genuinely α_s-suppressed next-to-LLA (NLLA) and so on. Since the enhanced regions of phase space involve near collinear or soft gluon emission, it is favourable for the primary partons to dress themselves with a shower of near collinear or soft partons. These are the parton precursors of hadronic jets. An important feature of such multiparton matrix elements is that in the enhanced regions of phase space they factorize into products of relatively simple expressions allowing significant simplifications in the treatment of leading logarithms. In some cases, it is actually possible to sum analytically the LLA- and NLLA-series to all orders in α_s.

The emerging picture of an event follows a sequence of decreasing scales. A genuinely hard subprocess produces a number of primary partons which then undergo semi-hard gluon radiation resulting in showers of soft partons which ultimately hadronize. The main features of an event are determined during its perturbative stages, thereby allowing tests of (p)QCD. In the following subsections we describe the basic phenomenology of the three main types of particle collision and how QCD applies to them.

3.2.1   *Electron–positron annihilation*

Electron–positron annihilation to hadrons provides the simplest colliding beam processes that can be described using pQCD. The simplicity follows from both the well-defined energies of the initial state particles and the fact that the leptons interact via a weakly coupled, colour singlet, virtual photon. This allows a clean separation of the initial and final state particles. The combined momentum of the incoming leptons provides a large scale justifying the use of pQCD. In the parton model the basic interaction is an electroweak process, $e^+e^- \rightarrow \gamma^*/Z \rightarrow q\bar{q}$; this has essentially the same cross section as the well established process $e^+e^- \rightarrow \mu^+\mu^-$. It is usually adequate to consider single photon exchange due to the small value of the electromagnetic coupling $\alpha_{em} = e^2/(4\pi) \approx 1/137$. The structure of the hadronic final state depends only on the centre-of-momentum (C.o.M.) energy, $\sqrt{s}$, of the collision and if polarized the polarizations of the incoming leptons. The C.o.M. system is often also referred to as 'centre-of-mass' system, since in the system where the momenta balance, the centre-of-mass of the interacting particles is at rest. Dealing with relativistic particles, however, the name 'centre-of-momentum' is more to the point.

At low C.o.M. energies, $0 \leq \sqrt{s} \leq 5\,\text{GeV}$, the most interesting quantity is the total hadronic cross section. This shows a lot of structure characterized by 'steps' at quark thresholds together with strong resonances, associated with $q\bar{q}$ bound states that possess the same quantum numbers as the exchanged photon. In essence the off mass-shell photon behaves as a $J^{PC} = 1^{--}$ vector meson: ρ, ω, ϕ, J/ψ, $\Upsilon(1S)$, etc. The hadronic final state is characterised by low multiplicities and only modest structure. It can be described adequately by a mix of isotropic phase space and resonance decays.

As the C.o.M. energy increases, the final state hadrons show a tendency to align along an axis and a back-to-back two-jet structure begins to appear. This is followed, at around $\sqrt{s} = 30\,\text{GeV}$, by the emergence of three-jet features in a fraction, $\mathcal{O}(10\%)$, of the events. By identifying these jets with primary partons it is possible to test the nature of QCD's basic constituents and their couplings. For example, three-jet events are believed to be a manifestation of vector gluon emission in the process $e^+e^- \rightarrow q\bar{q}g$. An example of a three-jet event is shown in Fig. 6.1. The rate of this three-jet production gives a measure of the strong coupling, α_s, whilst the angular distribution of the jets reflects the spin-1 nature of the gluon. At even higher energies, small fractions of well separated four, five and more jet events appear, allowing tests of the triple and quartic gluon couplings. Note that these jets are required to be well separated to avoid the collinear and soft enhancements that would invalidate fixed order perturbation theory, thereby complicating any comparisons to theory. A more precise definition of a jet is given in Section 6.2.

On dimensional grounds the total cross section must take the form

$$\sigma(s) = \frac{1}{s}f\left(\frac{m_i^2}{s}\right)\,, \tag{3.24}$$

where the $\{m_i\}$ represent the relevant masses, such as quark or hadron masses. The function $f(x_i)$ tends to a non-zero constant as $x_i \to 0$. Since the quark and hadron masses are mostly small, their effect becomes negligible as s increases and the cross section falls as s^{-1}, as prescribed by the photon propagator. This remains true until around $\sqrt{s} = 40\,\text{GeV}$ when deviations begin to be seen: this is the tail of the Z resonance which becomes dominant at $\sqrt{s} = 91\,\text{GeV}$. Apart from the large enhancement in the total cross section, the main effect of Z exchange is to modify the flavour mix of produced quarks and to introduce asymmetries into the polar angle distributions of the primary quarks, compared to pure photon exchange. Above $\sqrt{s} = 91\,\text{GeV}$, photon and Z exchange remain of comparable importance, but the total hadronic cross section continues to fall and becomes of less relative importance as other production channels, such as $e^+e^- \to W^+W^-$, open up.

An example of a less inclusive quantity in e^+e^- annihilation is the cross section for the production of a specific type of hadron in the final state. Suppose this hadron, h, has momentum p^μ, then the differential cross section can be written in the form of a convolution

$$d\sigma^{e^+e^- \to hX}(p, s) = \sum_a \int_0^1 \frac{dz}{z} d\hat{\sigma}^{e^+e^- \to aX}\left(\frac{p}{z}, s\right) D_a^h(z) . \qquad (3.25)$$

The first term, $d\hat{\sigma}$, is the hard cross section for the production of a parton a such that it carries momentum p^μ/z. The second term is a fragmentation function, $D_a^h(z)dz$, which gives the probability that the parton a produces the hadron h carrying a fraction z of the primary parton's momentum. This fragmentation function is the final state analogue of the previously mentioned p.d.f.s, to be discussed more fully in Section 3.2.2. The product of these two terms is summed over all the possible contributing partons and integrated over the momentum fractions. The factorization is between a perturbatively calculable, short-distance cross section and a non-perturbative fragmentation function. It is important to realize that $\hat{\sigma}$ does not depend on the identity of the hadron h, which would be a long-distance effect, but only on the parton a and the colliding beams. Conversely, D_a^h does not depend on the short-distance, hard subprocesses; in this sense it is universal and can be applied to any subprocess that produces the outgoing parton a.

At the lowest order the relevant hard subprocess is $e^+e^- \to q\bar{q}$, so that in eqn (3.25) the sum is over quarks with $2m_q < \sqrt{s}$. This gives the parton model prediction for which, as indicated, the fragmentation function depends only on the momentum fraction z. The inclusion of QCD corrections complicates matters, though the basic factorized form remains the same. In particular, renormalization requires the introduction of an arbitrary renormalization scale, μ_R, whilst the factorization procedure introduces a second, arbitrary factorization scale, μ_F. This acts as a cut-off on the virtuality of intermediate particles, equivalent to a cut-off on the (inverse) distance it travels. The exact origin of the scales μ_R and

μ_F will become clear in Sections 3.4 and 3.6. The QCD improved parton model prediction is

$$\mathrm{d}\sigma^{e^+e^-\to hX}(p,s) = \sum_a \int_0^1 \frac{\mathrm{d}z}{z} \mathrm{d}\hat{C}^{e^+e^-\to aX}\left(\frac{p}{z},s;\mu_R,\mu_F\right) D_a^h(z;\mu_R,\mu_F), \quad (3.26)$$

where the parton sum now includes contributions from gluons. Here the coefficient function $\hat{C}$ is derived from the partonic cross section for the subprocess, $e^+e^- \to aX$. Thanks to its short-distance nature $\hat{C}$ is calculable using pQCD and is devoid of any divergences. On the other hand, D_a^h is not calculable with today's technology and therefore must be determined from experiment. That said, the dependence on the scale μ_F, which is introduced in order to separate short- and long-distance effects, is calculable. Recall that both μ_R and μ_F are arbitrary as are aspects of the renormalization and factorization schemes. However, the physical cross section, on the left-hand side of eqn (3.26), is independent of the particular scales and schemes used, provided that the same choices are used consistently in both $\hat{C}$ and D_a^h. Whilst not necessary, it is common practice to only consider the case that $\mu_R = \mu_F(= \sqrt{s})$.

A simplification in the above description of electron–positron annihilation is the assumption that the colliding leptons are mono-energetic. This is not true. In a process known as bremsstrahlung they decelerate into the collision by emitting photons which reduces the effective C.o.M. energy. This *initial state photon radiation* (ISR) may be treated using structure functions (perhaps more properly called electron density functions) (Kleiss *et al.*, 1989). The idea is that the incident electron is really surrounded by a cloud of photons and further e^+e^- pairs. What the structure function, $f_{e/e}(x,\mu^2)$, gives is the probability density for finding an electron in this cloud of particles carrying a fraction x of the parent electron's momentum when it is probed at a scale μ. The electron–positron collision is then between these constituents. Summing up all the contributions gives

$$\mathrm{d}\sigma_{\mathrm{ISR}}(s) = \int_0^1 \mathrm{d}x_1 \int_0^1 \mathrm{d}x_2\, f_{e/e}(x_1,s) f_{e/e}(x_2,s)\, \mathrm{d}\hat{\sigma}(\hat{s}=x_1 x_2 s)\,. \quad (3.27)$$

On the assumption of massless electrons, the C.o.M. energy squared in the hard subprocess is given by $\hat{s} \equiv (x_1 p_{e^-} + x_2 p_{e^+})^2 = x_1 x_2 s$. It is possible to calculate this structure function in QED perturbation theory. An approximate form is

$$f_{e/e}(x,\mu^2) = \beta(1-x)^{\beta-1} \quad \text{with} \quad \beta(\mu^2) = \frac{2\alpha_{\mathrm{em}}}{\pi}\left[\ln\left(\frac{\mu^2}{m_e^2}\right)-1\right]. \quad (3.28)$$

In practice, one has $0 < \beta \ll 1$ so that $f_{e/e}(x,\mu^2)$ acquires an integrable singularity for $x \to 1$, which favours soft photon emission. Whilst the singularity can be treated analytically, its treatment in a numerical implementation takes some care. (Computers don't handle singularities very well ...)

The effect of initial state photon emission on a total cross section is strongly influenced by the C.o.M. energy dependence of the hard subprocess cross section. There is a trade-off between the two terms in eqn (3.27). If s is tuned to lie on a resonance then any ISR will reduce the effective C.o.M. energy, $\hat{s} = x_1 x_2 s$, and $d\hat{\sigma}(\hat{s})$ will be significantly reduced compared to $d\hat{\sigma}(s)$. In this case, the effect of ISR is modest, only distorting the resonance's line shape. However, if s is tuned to lie above a resonance then any ISR which reduces s to $\hat{s} \approx m_R^2$ is favoured by the increase in cross section. For an illustration see Fig. 6.3. Such 'radiative returns' can have a major impact on the line shape and lead to individual events being thrown to the left or right according to the energy imbalances in the post bremsstrahlung leptons actually entering the hard subprocess.

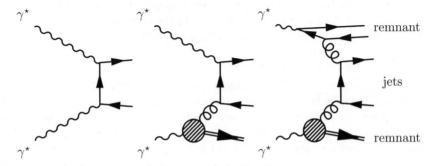

FIG. 3.2. Left, direct $\gamma\gamma$ interaction with QED photon–quark couplings. Centre, singly resolved $\gamma\gamma$ interaction with the lower photon behaving as a collection of partons characterized by its own p.d.f. Right, doubly resolved $\gamma\gamma$ interaction; in the upper photon the partons can be traced to a pointlike component originating from a perturbative $\gamma^\star \to q\bar{q}$ vertex, whilst in the lower photon the 'remaining' hadron-like partons have a non-perturbative origin.

Not all the photons emitted as bremsstrahlung escape without interacting. A rich variety of processes resulting in resonance pairs, jet events etc. can occur due to photon–photon interactions (Aurenche $et\ al.$, 1996). Indeed, the total cross section for $e^+e^- \to e^+e^- +$ hadrons grows as $\ln^2(s/m_e^2)$. A source of this complexity is that photons, as sketched in Fig. 3.2, are not as simple as might be naïvely thought. The reader is warned that the notations used to describe photons have become rather confused in the literature. Here we follow Chyla (2001). We are familiar with the idea of a direct photon which has pointlike couplings and can lead to hadron production via the hard subprocess $\gamma^\star\gamma^\star \to q\bar{q}$. In addition, there are resolved photons that behave as dense clouds of (anti)quarks and gluons. Such a photon behaves like a hadron and is characterized by parton density functions. A pointlike component of these partons can be traced back to an initial QED vertex $\gamma^\star \to q\bar{q}$ and the subsequent radiation of gluons which in turn may split into gluon-pairs or further $q\bar{q}$-pairs (Witten, 1977).

A hadron-like component accounts for the remaining partons which have their origins in non-perturbative physics, perhaps associated with the (negative virtuality) photon fluctuating into a vector meson. This opens up the possibility of effective lepton–hadron, known as singly resolved, and hadron–hadron, known as doubly resolved, collisions. These are discussed in detail below. All these types of $\gamma\gamma$-events are characterized by low C.o.M. energies and sizeable longitudinal momentum imbalances. Often these events constitute a hadronic background to the events of real interest in an experiment.

We also mention one further complication. At very high energies it is necessary to use beams with very small transverse sizes in order to increase the luminosity and compensate for the falling cross section. This gives very high charge density particle bunches whose intense electromagnetic fields can induce radiation in one another as they approach each other, the so-called beamstrahlung. The details of this depend on the specifics of the beam profile, but can be treated in a similar vein to bremsstrahlung (Palmer, 1990).

3.2.2   Lepton–hadron scattering

Lepton–hadron scattering is a traditional method of probing the structure of hadrons. Since hadrons are now known to be composite particles with partonic, (anti)quark and gluon, constituents, such collisions are more complex to describe than lepton–lepton collisions. In essence, we view the observed scattering, $\ell\,\mathrm{h} \to \ell' X$, as a manifestation of the hard subprocess $\ell\mathrm{q} \to \ell'\mathrm{q}'$. The advantage of lepton probes is that they undergo experimentally and theoretically clean, pointlike interactions which are describable in terms of the exchange of a single, virtual, gauge boson. Multiple boson exchange, whilst possible, is suppressed by additional factors of the electroweak couplings, α_{em}^2 or G_{F}^2. A basic classification of the events is based on the nature of the boson exchanged by the initial lepton and quark. In neutral current events, characterized by $\ell = \ell'$, a photon or Z is exchanged. Whilst in charged current events, characterized by $\ell = \mathrm{e}$, μ, τ and $\ell' = \nu_{\mathrm{e}}$, ν_μ, ν_τ or vice-versa, a $\mathrm{W}^\pm$ is exchanged. For charged leptons the exchanged particle is predominantly a photon. Weak boson, Z or $\mathrm{W}^\pm$, exchange is observed, however, at low $Q^2 \ll M_{\mathrm{W}}^2$ their contributions are many orders of magnitude lower, $\mathcal{O}(G_{\mathrm{F}}^2/\alpha_{\mathrm{em}}^2)$. If a neutrino beam is used, only weak interactions can occur, making them a useful probe to disentangle the contributions from quarks and antiquarks.

Figure 2.1 illustrates the basic process underlying lepton–hadron scattering as viewed from the target hadron's rest frame. This frame coincides with the laboratory frame for fixed target experiments. To date, only the HERA machine at DESY provides (asymmetric) colliding beams of electrons/positrons and protons. Once the square of the C.o.M. energy, $s = (\ell + p)^2$, is fixed, the most important quantity for describing the scattering is the momentum transfer $q^\mu = \ell^\mu - \ell'^\mu$ and the Lorentz invariant $Q^2 \equiv -q^2 > 0$. The significance of Q^2 is that it characterizes the wavelength, or resolving power, of the probe. Also of interest is ν, the energy transferred to the target hadron. For a charged lepton neutral current event these can be determined experimentally by measuring the energy and angular deflection θ of the scattered lepton. Neglecting the relatively small lepton masses one finds

$$
\begin{aligned}
Q^2 &\equiv -(\ell - \ell')^2 & \nu &\equiv (\ell - \ell') \cdot p/M_{\mathrm h} \\
&= +4E_\ell E_{\ell'} \sin^2(\theta/2) & &= (E_\ell - E_{\ell'})|_{\mathrm{rest}} .
\end{aligned}
\tag{3.29}
$$

In addition, experiments may also measure details of the hadronic final state, X. This can provide complementary, or indeed when $\ell' = \nu$ the only determination of Q^2 and ν. The invariant mass, W, of the hadronic final state is given by

$$
\begin{aligned}
W^2 &= (p + q)^2 \\
&= M_{\mathrm h}^2 + 2M_{\mathrm h}\nu - Q^2 .
\end{aligned}
\tag{3.30}
$$

$$
\text{If} \quad W^2 = M_{\mathrm h}^2 \quad \Longrightarrow \quad Q^2 = 2M_{\mathrm h}\nu .
$$

In the special case of elastic scattering, when the target hadron remains intact in the final state, eqn (3.30) implies that Q^2 and ν are not independent.

The formal description of lepton–hadron scattering is facilitated greatly by the 'factorized' nature of the interaction. Details of the calculations can be found in the following sections. In terms of the leptonic and hadronic currents the matrix element is given by

$$
\mathcal{M}(\ell \mathrm{h} \to \ell' X) = \langle \ell' | J_\mu | \ell \rangle g_{\ell V} \frac{-\eta^{\mu\nu}}{q^2 - M_V^2} g_{\mathrm h V} \langle X | J_\nu | \mathrm{h} \rangle
\tag{3.31}
$$

where the electroweak couplings have been factored out: $g_{fV}^2 = \kappa_V^2 (v_{fV}^2 + a_{fV}^2)$, see Appendix B. In (high-Q^2) neutral current events the matrix element should include both γ and Z exchange contributions. Equation (3.31) suggests writing the inclusive lepton–hadron scattering cross section in terms of two tensors $L_{\mu\nu}$ and $H^{\mu\nu}$ as

$$
\mathrm{d}\sigma^{\ell \mathrm h} = \frac{1}{4\ell \cdot p} \frac{(g_{\ell V} g_{\mathrm h V})^2}{(Q^2 + M_V^2)^2} L_{\mu\nu} H^{\mu\nu} (4\pi) \frac{\mathrm{d}^3 \ell'}{2E_{\ell'} (2\pi)^3} \quad \text{with}
$$

$$
L_{\mu\nu} = \frac{1}{2} \langle \ell | J_\mu^\dagger | \ell' \rangle \langle \ell' | J_\nu | \ell \rangle \quad \text{and}
$$

$$
H^{\mu\nu} = \frac{1}{2 \cdot 4\pi} \sum_X \langle \mathrm{h} | J^{\dagger\mu} | X \rangle \langle X | J^\nu | \mathrm{h} \rangle (2\pi)^4 \delta^{(4)}(p_X - k - p) .
\tag{3.32}
$$

The hadronic tensor is summed over all the allowed final states and by convention includes a factor $(4\pi)^{-1}$ and an overall four-momentum conserving δ-function. Also, the definition of both tensors includes an average over spins on the assumption that the incoming particles are unpolarized. As mentioned earlier, the simplicity of leptons means that the tensor $L_{\mu\nu}$ is calculated readily to be

$$
L_{\mu\nu} = 2 \left[\ell_\mu \ell'_\nu + \ell'_\mu \ell_\nu - (Q^2/2)\eta_{\mu\nu} + \mathrm{i} C_{\ell V} \epsilon_{\mu\nu}{}^{\sigma\tau} \ell_\sigma \ell'_\tau \right] + 2D_{\ell V} m_\ell^2 \eta_{\mu\nu} .
\tag{3.33}
$$

The last two terms are associated with parity violation. The coefficient $C_{\ell V} = 2a_{\ell V} v_{\ell V} / (v_{\ell V}^2 + a_{\ell V}^2)$ depends on the type of vector boson emitted by the incoming lepton. For a photon it is $C_{\ell\gamma} = 0$, for a W boson one has $C_{\ell W^+} = +1$ (ν_ℓ or ℓ^+ beam) and $C_{\ell W^-} = -1$ ($\bar{\nu}_\ell$ or ℓ^- beam), respectively. The last term, with coefficient $D_{\ell V} = (v_{\ell V}^2 - a_{\ell V}^2)/(v_{\ell V}^2 + a_{\ell V}^2)$, is suppressed by a relative factor m_ℓ^2/Q^2 and is almost always neglected. It is not straightforward to calculate the hadronic tensor. However, since it must be constructed from the only available four-vectors, p^μ and q^μ, and the two isotropic tensors, $\eta^{\mu\nu}$ and $\epsilon^{\mu\nu\sigma\tau}$, it is possible to write down its general form as

$$H^{\mu\nu} = -F_1\eta^{\mu\nu} + \left[F_2 p^\mu p^\nu + i F_3 \epsilon^{\mu\nu}{}_{\sigma\tau} p^\sigma q^\tau + (F_4 + i F_5)p^\mu q^\nu \right.$$
$$\left. +(F_4 - i F_5)q^\mu p^\nu + F_6 q^\mu q^\nu\right](p \cdot q)^{-1} . \qquad (3.34)$$

Here, the hadron specific structure functions, F_i, are dimensionless (thanks to the factor of $(p \cdot q)^{-1}$), Lorentz scalars. It is also common to see eqn (3.34) defined in terms of the equivalent structure functions $W_1 = F_1$ and $W_{2-6} = (M_h^2/p \cdot q)F_{2-6}$. We do not do this as it only adds unnecessarily to the notational burden. If the spin of the colliding particles is specified, then extra terms, containing S_ℓ^μ and S_h^μ, would be possible in eqn (3.34). Terms involving further four-momenta would arise also if measurements are made on the hadronic final state.

Quantum mechanics and symmetries impose important constraints on the F_i (Treiman et al., 1972). As defined in eqn (3.34) they are all real. The time reversal invariance of QCD implies that $F_5 = 0$. As we shall learn in Section 3.3.1, electromagnetic gauge invariance implies the following current conservation constraints

$$q_\mu \cdot H^{\mu\nu} = 0 \quad \text{and} \quad H^{\mu\nu} \cdot q_\nu = 0 . \qquad (3.35)$$

A similar, approximate constraint applies to the weak currents. By imposing eqn (3.35), see Ex. (3-3), the form of the hadronic tensor is restricted further to

$$H^{\mu\nu} = F_1\left(-\eta^{\mu\nu} + \frac{q^\mu q^\nu}{q^2}\right) + \frac{F_2}{p \cdot q}\left(p^\mu - \frac{p \cdot q}{q^2}q^\mu\right)\left(p^\nu - \frac{p \cdot q}{q^2}q^\nu\right) + i\frac{F_3}{p \cdot q}\epsilon^{\mu\nu\sigma\tau}p_\sigma q_\tau.$$
$$(3.36)$$

If we do not impose eqn (3.35) then we must add residual F_4' and F_6' structure functions, à la eqn (3.34), to eqn (3.36) but these would be suppressed as $(m_\ell m_q/Q^2)^2$, c.f. eqn (3.33), and will not subsequently trouble us (Jaffe and Llewellyn-Smith, 1973). Combining eqn (3.36) with eqn (3.33), which satisfies also the equivalent of eqn (3.35), gives

$$L_{\mu\nu} \cdot H^{\mu\nu} = F_1 2Q^2 + \frac{F_2}{p \cdot q}\left[4(p \cdot \ell)(p \cdot \ell') - M_h^2 Q^2\right] - C_{\ell V}\frac{F_3}{p \cdot q}p \cdot (\ell + \ell')Q^2 , \quad (3.37)$$

where we used $\epsilon_{\mu\nu\sigma\tau}\epsilon^{\mu\nu}{}_{\sigma'\tau'} = -2[\eta_{\sigma\sigma'}\eta_{\tau\tau'} - \eta_{\sigma\tau'}\eta_{\sigma'\tau}]$. Applying this result in eqn (3.32) gives the general expression for unpolarized, inclusive lepton–hadron scattering; see also Ex. (3-4):

$$\frac{d^2\sigma^{\ell h}}{dE' d\cos\theta} = 8\pi\frac{\alpha_{\ell V}\alpha_{hV}}{(Q^2 + M_V^2)^2}\frac{E'^2}{M_h}$$

$$\times \left\{ \left[2F_1 - C_{\ell V} F_3 \frac{E + E'}{E - E'} \right] \sin^2(\theta/2) + F_2 \frac{M_{\rm h}}{E - E'} \cos^2(\theta/2) \right\}$$

$$\frac{{\rm d}^2 \sigma^{\ell h}}{{\rm d}x {\rm d}Q^2} = \frac{4\pi}{x} \frac{\alpha_{\ell V} \alpha_{\rm h V}}{(Q^2 + M_V^2)^2}$$

$$\times \left\{ xy^2 F_1 + \left(1 - y - \frac{(xy M_{\rm h})^2}{Q^2} \right) F_2 - C_{\ell V} x \left(y - \frac{y^2}{2} \right) F_3 \right\}$$

$$= \frac{y}{Q^2} \times \frac{{\rm d}^2 \sigma^{\ell h}}{{\rm d}x {\rm d}y} . \tag{3.38}$$

Here, we have defined $\alpha_{fV} = g_{fV}^2/(4\pi)$. The first form uses the energy and angle of the scattered lepton in the target rest frame. The second and third form use the Lorentz invariant variables Q^2, defined in eqn (3.29), and x and y defined by

$$\left. \begin{array}{l} x = \dfrac{Q^2}{2 M_{\rm h} \nu} \\[2mm] y = \dfrac{q \cdot p}{\ell \cdot p} = \dfrac{E - E'}{E} \end{array} \right|_{\rm rest} \right\} \implies s = \frac{Q^2}{xy} + M_{\rm h}^2 . \tag{3.39}$$

In this framework all information on the possible scatterings resides in the structure functions F_{1-3}. These, in turn, may be only functions of dimensionless ratios of Q^2, $p \cdot q$ and 'M^2', where 'M' represents any mass (or inverse length) characteristic of the hadron.

At low C.o.M. energy and low Q^2, $\lesssim 0.01\,({\rm GeV})^2$, elastic, electromagnetic scattering is dominant (Taylor, 1975). Since $Q^2 = 2 M_{\rm h} \nu$ for an elastic scattering the structure functions $F_{1,2}$ have to be functions of $Q^2/{}'M^2{}'$ or be constant; $F_3 = 0$ for purely electromagnetic processes. Furthermore, the long wavelength of the exchanged photon means that the target hadron is seen as a coherent whole, so that 'M' must be a macroscopic property of the hadron. In this low-virtuality limit the form of the hadronic current is actually known to be

$$J^\mu = \frac{1}{2 M_{\rm h}} \bar{u}(p') \left[(p + p')^\mu \mathcal{F}_1(Q^2, {}'M{}') + {\rm i}\,(\mu_{\rm h} - 1) q_\nu \sigma^{\nu\mu} \mathcal{F}_2(Q^2, {}'M{}') \right] u(p) . \tag{3.40}$$

Here $\mu_{\rm h}$ is the magnetic moment of the hadron measured in units of the nuclear magneton, $e\hbar/(2 M_{\rm p})$, and $\mathcal{F}_1$ and $\mathcal{F}_2$ correspond directly to the electric and magnetic form factors of the hadron. For the proton and the neutron $\mu_{\rm p} = +2.793$ and $\mu_{\rm n} = -2.913$ respectively. These form factors can be related to the Fourier transform of the hadron's electric charge distribution. Using eqn (3.40) in eqn (3.32) gives the Rosenbluth formula, which takes the form of eqn (3.38) with F_1 and F_2 given in terms of $\mathcal{F}_1$ and $\mathcal{F}_2$. Empirically, the two form factors are both described well by the dipole formula which corresponds to a spherically symmetric, exponentially falling charge distribution. One has

$$\rho(\mathbf{r}) = \frac{{\rm e}^{-|\mathbf{r}|/\sigma}}{8\pi\sigma^3} \quad \Longleftrightarrow \quad \mathcal{F}_i(\mathbf{q}) = \frac{1}{(1 + \sigma^2 |\mathbf{q}|^2)^2} . \tag{3.41}$$

The parameter σ is related to the hadron's mean charge radius squared according to $\langle r_{\text{ch}}^2 \rangle = 12\sigma^2$. That the structure functions vanish for $Q^2 \to \infty$ reflects the lack of high frequency Fourier components in a smooth charge distribution.

At slightly higher C.o.M. energies, quasi-elastic scatterings become important. Here, the target hadron is excited and breaks up into a low multiplicity system of hadrons, for example, $\gamma^\star \text{p} \to \Delta^+ \to \text{n}\pi^+$. Again this process can be described by eqn (3.38) with form factors similar to those in eqn (3.41).

At larger C.o.M. energies, high Q^2 processes become kinematically possible, allowing the internal structure of the target hadron to be probed. In this regime, the fast falling (quasi-)elastic cross sections vanish and the majority of collisions become inelastic — the target hadron being broken up. This is *deep inelastic scattering* (DIS). Since the invariant mass of the hadronic final state, W, is not determined, Q^2 and $\nu = p \cdot q / M_{\text{h}}$ are independent variables. Again eqn (3.38) applies, but the form of the structure functions F_{1-3} undergo a qualitative change. Rather than vanish as $Q^2 \to \infty$ they remain finite and become practically a function of the single variable $x = Q^2/(2M_{\text{h}}\nu) \in (0,1)$ (MIT-SLAC Collab., 1972).

$$F_{1,2,3}(q^2, p \cdot q, \text{`}M\text{'}) \xrightarrow{\;Q^2 \to \infty\;} F_{1,2,3}(x) \neq 0 \qquad (3.42)$$

This Bjorken scaling (Bjorken, 1969) demonstrates that the exchanged vector boson now scatters off pointlike objects that have no mass scale 'M' associated with them. Furthermore, the effective constraint $Q^2 = x \times 2M_{\text{h}}\nu = 2xp \cdot q$ is reminiscent of elastic scattering, c.f. eqn (3.30). It is interpreted as being due to the lepton scattering elastically off a charged, constituent (anti)quark which carries a fraction x of its parent hadron's momentum.

In the parton model (Bjorken and Paschos, 1969; Feynman, 1972) the hadron is viewed as a collection of independent, that is, essentially non-interacting or free, (anti)quarks and gluons each carrying a fraction of the parent hadron's longitudinal momentum; any transverse momentum is taken to be small by comparison. The hadron is now described by giving the probability density distributions for the momentum fractions of its parton constituents

$$f(x)\text{d}x = \mathcal{P}\big(x' \in [x, x + \text{d}x]\big) \quad f = q, \ \bar{q} \text{ or } g . \qquad (3.43)$$

The $f(x)$ are known as *parton density functions* (p.d.f.) or also, somewhat confusingly, as 'structure functions'. Here, and in the following, we shall reserve the name structure function for physically observable quantities. These functions are similar to the fragmentation functions, which we met in Section 3.2.1, but in a reverse sense. The hadron cross section is then formed as a sum of pointlike (anti)quark cross sections weighted by their p.d.f.s, in direct analogy to eqn (3.25). Again, in this factorized form the long-distance p.d.f.s are universal, that is, independent of the particular hard subprocess. As the exchanged vector bosons only couple to (anti)quarks, the presence of gluons in the hadron is felt only indirectly in DIS experiments.

Depending on the nature of the exchanged particle the structure functions F_{1-3} measure different combinations of the p.d.f.s.

$$2xF_1 = F_2$$

$$F_2^\gamma = x \sum_{D,U} \left[\frac{1}{9}(D + \overline{D}) + \frac{4}{9}(U + \overline{U}) \right] \qquad xF_3^\gamma = 0$$

$$F_2^{W^+} = 2x \sum_{D,U} [D + \overline{U}] \qquad\qquad xF_3^{W^+} = 2x \sum_{D,U} [D - \overline{U}]$$

$$F_2^{W^-} = 2x \sum_{D,U} [U + \overline{D}] \qquad\qquad xF_3^{W^-} = 2x \sum_{D,U} [U - \overline{D}] \quad (3.44)$$

Here, D represents any down-type quark (d, s, b) and U represents any up-type quark (u, c, t). Whilst these formal sums include the heavy quarks (c, b, t) their practical contribution is negligible if the probing boson is unable to resolve them. In the transverse plane, which is unaffected by Lorentz boosts along the beam axis, the size of the heavy quark is given by $\sim 1/M_Q$, whilst the exchanged boson sees scales $\geq 1/Q$. Therefore, if $M_Q > Q$, the quark can be dropped from the summation. The coefficients in eqn (3.44) reflect the normalized electric and weak charges of the (anti)quarks.

The constituent quark model (Close, 1979) together with the conservation of flavour imposes a number of constraints on the p.d.f.s. For example, for a proton we have

$$\int_0^1 dx \left[u(x) - \bar{u}(x) \right] = 2 \qquad \int_0^1 dx \left[s(x) - \bar{s}(x) \right] = 0$$

$$\int_0^1 dx \left[d(x) - \bar{d}(x) \right] = 1 \qquad\qquad etc. \qquad (3.45)$$

These equations state that the proton contains two units of up-ness, one unit of down-ness and no net strangeness. There is no such constraint on the gluons, $\int_0^1 dx\, g(x)$, as the number of bosons is not conserved. It is usual to see the quark p.d.f.s separated into two components (Kuti and Weisskopf, 1971; Landshoff and Polkinghorne, 1971): the valence quarks which carry all of the proton's quantum numbers and the sea quarks and antiquarks which make up the remainder and carry no net charges. For example,

$$u(x) = u_v(x) + u_s(x) \qquad\qquad \int_0^1 dx\, u_v(x) = 2$$

$$\bar{u}(x) = \bar{u}_s(x) \qquad\qquad \int_0^1 dx \left[u_s(x) - \bar{u}_s(x) \right] = 0 \, . \qquad (3.46)$$

The sea quarks are commonly assumed to be produced in $g \to q\bar{q}$ splittings. This suggests the idea that the sea quarks are symmetric in the sense that $u_s = \bar{u}_s = d_s = \bar{d}_s = s_s = \bar{s}_s = \cdots$. Whilst this makes many formulae simpler, it is known empirically not to be exactly true, though a theoretical understanding of how this comes about remains elusive.

The parton model interpretation of deep inelastic lepton–hadron scattering
is only approximate and QCD corrections should be taken into account. In the
QCD improved parton model the DIS cross section again can be written in a
factorized form, but one which can now be proved formally to hold (Collins and
Soper, 1987), with the structure functions given by

$$F_i^{(Vh)}(x, Q^2) = \sum_{f=q,\bar{q},g} \int_x^1 \frac{dz}{z} f_h\left(\frac{x}{z}, \mu_F, \mu_R\right) \hat{F}_i^{(Vf)}\left(Q^2, z, \mu_F, \mu_R\right) . \quad (3.47)$$

Here $\hat{F}_i^{(Vf)}$ is a projection of the cross section for the partonic scattering $Vf \to f'$
appropriate to the ith structure function. Again, it has been necessary to introduce a factorization scale, μ_F, and scheme, plus a renormalization scale, μ_R,
and scheme. The (projected) parton cross sections, $\hat{F}_i^{(Vf)}$, contain only short-
distance physics and are calculable in perturbation theory. They do not depend
on the hadron h. By contrast, the p.d.f.s, f_h, know nothing of the hard subprocess and depend on the incoming hadron; they are not calculable using only
perturbation theory. The proof of eqn (3.47) also justifies the assumption of incoherent scattering and provides a formal definition of the p.d.f.s. This definition
shows that in a frame in which the target hadron has infinite momentum the
p.d.f.s reduce to the matrix elements, $\langle h|\hat{N}_f(x)|h\rangle$, where $\hat{N}_f(x)$ is the number
density operator for partons of type f with given momentum fraction.

Formally, eqn (3.47) only represents the first term in an operator product
expansion for $F_i^{(Vh)}(x, Q^2)$ (Altarelli, 1982). This means that it is only exact for $Q^2 \to \infty$. The expansion is organized in terms of the operators' twist
(= mass dimension − spin). Thus, at finite values of Q^2 there are higher twist
corrections which are suppressed as

$$\frac{[\ln(Q^2/Q_0^2)]^{m<n}}{Q^n}, \quad (3.48)$$

where $n = 4$ for DIS. In general, these non-perturbative corrections are neglected,
though there are situations where their effects should be taken into account.

In eqn (3.47) the factorization and renormalization scales are arbitrary. In
practice, it is common to set all scales equal, $\mu^2 \equiv Q^2 = \mu_R^2 = \mu_F^2$. This simplifies
the coefficient function, giving, for example, $\hat{F}_i^{(Vf)}(\mu^2, z; \mu, \mu) \propto \delta(1 - z)$ in the
so-called DIS factorization scheme, which allows combinations of the $f_h(x, \mu^2 =
Q^2)$ to be determined directly in an experiment. Whilst the f_h involve long-
distance physics, the scale μ may still be sufficiently large that $\alpha_s(\mu^2)$ is small
enough to allow the dependence on the scale to be calculable at least down
to some low scale $\mu_0 \gtrsim \Lambda_{QCD}$. This results in the p.d.f.s developing small, but
measurable, logarithmic dependences on μ^2. Such scaling violations are described
well by the coupled, integro-differential DGLAP equations (Altarelli and Parisi,

1977; Gribov and Lipatov, 1972; Dokshitzer, 1977).[2] At leading order these are given by

$$\mu^2 \frac{\partial q}{\partial \mu^2}(x,\mu^2) = \int_x^1 \frac{dz}{z} \frac{\alpha_s}{2\pi} \left[P_{qq}^{(0)}(z) q\left(\frac{x}{z},\mu^2\right) + P_{qg}^{(0)}(z) g\left(\frac{x}{z},\mu^2\right) \right]$$

$$\mu^2 \frac{\partial \bar{q}}{\partial \mu^2}(x,\mu^2) = \int_x^1 \frac{dz}{z} \frac{\alpha_s}{2\pi} \left[P_{qq}^{(0)}(z) \bar{q}\left(\frac{x}{z},\mu^2\right) + P_{qg}^{(0)}(z) g\left(\frac{x}{z},\mu^2\right) \right] \quad (3.49)$$

$$\mu^2 \frac{\partial g}{\partial \mu^2}(x,\mu^2) = \int_x^1 \frac{dz}{z} \frac{\alpha_s}{2\pi} \left[P_{gg}^{(0)}(z) g\left(\frac{x}{z},\mu^2\right) + \sum_{f=q,\bar{q}} P_{gq}^{(0)}(z) f\left(\frac{x}{z},\mu^2\right) \right] .$$

The kernel functions, $P_{ab}(z, \alpha_s(\mu^2))$, are known as Altarelli–Parisi splitting functions and are associated with the branchings $b \to aX$. They can be expanded as a power series in α_s. The leading order expressions are

$$P_{qq}^{(0)}(z) = C_F \left(\frac{1+z^2}{1-z} \right)_+$$

$$P_{qg}^{(0)}(z) = T_F[z^2 + (1-z)^2]$$

$$P_{gg}^{(0)}(z) = 2C_A \left(\frac{z}{(1-z)_+} + \frac{(1-z)}{z} + z(1-z) \right) + \frac{(11C_A - 4n_f T_F)}{6}\delta(1-z)$$

$$P_{gq}^{(0)}(z) = C_F \frac{[1 + (1-z)^2]}{z} . \quad (3.50)$$

Away from $z = 1$ these are ordinary functions, but at $z = 1$ the diagonal splitting functions, $P_{aa}^{(0)}$, must be regarded as distribution functions. Details are elaborated in Section 3.6.3, where also the meaning of the plus-prescription is explained. Since the virtualities involved in this initial state evolution are negative these are the space-like splitting functions. Equations very much like eqn (3.49) control the $\mu_F(= Q)$ behaviour of the fragmentation functions (Owens, 1978). The structure and interpretation of these sets of equations are essentially the same and to $\mathcal{O}(\alpha_s)$ so are the splitting functions. However, beyond this leading order the space-like and time-like splitting functions differ. The full NLO splitting functions for time-like evolution can be found in Appendix E.

The equations in eqn (3.49) have an appealing physical interpretation. We picture the (anti)quarks which make up the hadron as surrounded by clouds of virtual particles, constantly being emitted and absorbed. These virtual particles may in turn emit and absorb further virtual particles. Thus, as the Q^2 of the probing vector boson increases, the content of the hadron appears to change as it is seen on smaller distance scales. It is this evolution which is described by eqn (3.49). The terms $(\alpha_s/2\pi)P_{ab}^{(0)}(z)dz$ are interpreted as the probability

---

[2] These equations have quite a history and the name reflects the main contributors to their elucidation: Dokshitzer, Gribov, Lipatov, Altarelli and Parisi. In the past, the name was often shortened to Altarelli–Parisi equations.

densities that in the branching $b \rightarrow aX$ parton a will carry a fraction in the range $[z, z + dz]$ of its parent, b's, momentum, and any other products, X, a fraction $1 - z$. Strictly, the branching probability densities are given by the distribution functions $\delta(1 - z)\delta_{ab} + (\alpha_s/2\pi)P_{ab}^{(0)}(z)$ which are regular functions away from $z = 1$. Thus, for example, the probability that a high virtuality gluon, carrying momentum fraction x, came from a low virtuality gluon, with a larger momentum fraction y, is given by

$$\int_0^1 dz \int_0^1 dy \frac{\alpha_s}{2\pi} P_{gg}^{(0)}(z)g(y)\delta(x - yz) = \int_x^1 \frac{dz}{z} \frac{\alpha_s}{2\pi} P_{gg}^{(0)}(z)g\left(\frac{x}{z}\right) . \quad (3.51)$$

All the terms in eqn (3.49) have a similar interpretation. Figure 3.3 shows schematically this interpretation.

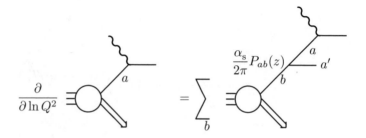

FIG. 3.3. A schematic interpretation of the space-like DGLAP equations whereby the scale dependence follows from the presence of partons within other higher momentum partons

Given the above interpretation of eqn (3.49) it is straightforward to anticipate how the p.d.f.s will change with Q^2. At low Q^2, one might expect that there are few partons in a hadron and that subsequently their p.d.f.s are skewed to high momentum fractions. This picture is not too far from saying, for example, that a proton consists only of two u-quarks and one d-quark, each with momentum fractions smeared around the value $x = 1/3$. As the Q^2 increases, the typical parton momentum fractions decrease as momentum is shared via parton branchings. Thus, we anticipate a growth in the small-x component of the p.d.f.s. Furthermore, we expect many of these small-x partons to be gluons, which have a high probability to undergo g $\rightarrow$ gg branchings, and sea quarks such as ū and s̄ which arise in g $\rightarrow$ qq̄ branchings. As the p.d.f.s shift towards small x and the sea grows we must respect the sum rules for the quark flavours, eqn (3.45), and the conservation of momentum,

$$1 = \int_0^1 dx\, x \left[g(x) + \sum_{f=q,\bar{q}} f(x) \right] . \quad (3.52)$$

It is the application of this sum rule which provides compelling evidence for the existence of gluons, carrying over 50% of the momentum, in a proton (Llewellyn-Smith, 1972).

Equation (3.49) allows us to calculate how the p.d.f.s change between the scales μ_0 and μ. However, since the $f(x, \mu_0)$ ($f = q, \bar{q}, g$) only involve a non-perturbative scale, we can not use pQCD to calculate them. In principle, non-perturbative techniques, such as lattice calculations, may allow them to be calculated. However, at the present time only a few moments of their distributions have been obtained and we must rely on experiment (Capitani *et al.*, 2002). Their determination relies on an interplay between the use of eqn (3.49) and the measurement of structure functions giving various combinations of p.d.f.s at different Q^2 scales. In essence, one tries to 'guess' a set of $f(x, \mu_0)$, evolve them to higher scales using eqn (3.49), and then optimize the fit to the measured combinations at the higher scales. Details of the procedure and results are discussed in Section 7.5. The overall consistency of this procedure gives evidence for the validity of the evolution equations and thereby pQCD. Many sets of p.d.f.s are available, for example the package PDFLIB (Plothow-Besch, 1993) contains a a compendium.

As mentioned above, the formal proof of the parton model can be achieved using the apparatus of field theory. However, we can gain insight into its motivation by considering the space–time structure of the collision. The hadron is pictured as a collection of partons sitting within clouds of further partons that are being emitted and absorbed constantly by one another . The virtualities involved must be low, $k^2 \lesssim M_{\mathrm{h}}^2$, if the hadron is to remain intact. Indeed, high momentum transfers are suppressed as $[\alpha_{\mathrm{s}}(Q^2)M_{\mathrm{h}}^2/k^2]^n$, where $n = 2$ for mesons and $n = 3$ for baryons. This, in turn, implies that the partons have lifetimes $\sim 1/M_{\mathrm{h}}$, whilst the incoming exchanged boson interacts for a mere $1/Q$. Thus, to the incoming boson the partons appear almost frozen having been formed well in advance of the near instantaneous collision. The struck parton has essentially no time to communicate with the other partons and therefore behaves as if it were free. This also implies that the hard scattering knows nothing of the target hadron beyond the probability that it contains the struck parton.

Returning to the struck parton, it is impulsively kicked out of the hadron and leaves behind its cloud of partons. These remaining partons have been 'shaken free' and as they have nothing to be re-absorbed by, they continue to fly forwards on near collinear trajectories. This initial state radiation continues to shower and hadronize, resulting in a target region jet. The struck parton behaves much like a quark produced in an e^+e^- collision and fragments to produce a current region jet. Between the colour charge on the scattered quark and the anticolour left behind on the hadron remnant is a colour field which converts into low energy hadrons lying between the two jets. Actually, since these intermediate hadrons are produced in a statistical Poisson-like process, it is possible that no hadrons form between the two jets, although the probability for such a gap is expected to be exponentially suppressed as the distance between the jets increases. A typical neutral current DIS event is shown in Fig. 7.2 and a typical charged current event is shown in Fig. 7.5.

Jet-like structures start to become apparent in DIS for $Q^2 \gtrsim (4\,\text{GeV})^2$. As the Q^2 (C.o.M. energy) increases multi-jet structures appear, just as in e^+e^- collisions. The LO hard subprocess is $Vq \to q'$, which results in a far forward, target region, beam remnant and a more central, current region, jet. At $\mathcal{O}(\alpha_\mathrm{s})$ the NLO subprocesses are the QCD Compton process, $Vq \to gq'$, for scattering off a(n anti)quark and boson–gluon fusion, $Vg \to q\bar{q}'$, for scattering off a gluon. Both of these processes can give rise to two central jets in addition to the forward jet. The type of vector boson exchanged is strongly dependent on the event's Q^2 and type of lepton involved. For charged leptons at low to intermediate $Q^2 \lesssim (40\,\text{GeV})^2$, the neutral current cross section, mediated by a photon, is very much larger than the charged current cross section, mediated by a $W^\pm$. Measurements are shown in Fig. 7.8. This difference essentially reflects the propagators of the exchanged bosons which lead to different Q^2 behaviour: photons give a $1/Q^4$ fall-off whilst W bosons give a nearly constant cross section for $Q^2 \ll M_\mathrm{W}^2$.

$$\frac{\mathrm{d}\sigma_\mathrm{NC}}{\mathrm{d}Q^2} \propto \alpha_\mathrm{em}^2 \frac{1}{Q^4} \quad \text{and} \quad \frac{\mathrm{d}\sigma_\mathrm{CC}}{\mathrm{d}Q^2} \propto \frac{\alpha_\mathrm{em}^2}{\sin^2\theta_\mathrm{w}} \frac{1}{(Q^2 + M_\mathrm{W}^2)^2} \approx \frac{2G_\mathrm{F}^2}{\pi^2} . \tag{3.53}$$

As the Q^2 increases further both cross sections begin to fall faster. This is because kinematics require higher Q^2 events to have higher x values and the p.d.f.s, $f_\mathrm{h}(x \sim 1, Q^2)$, vanish as $Q^2 \to \infty$. Also their difference diminishes until they become of equal magnitude for $Q^2 \gtrsim (80\,\text{GeV})^2$. An example of electroweak unification in action! Above $Q^2 = (40\,\text{GeV})^2$ Z exchange starts to visibly contribute to neutral current events. This is manifested by the appearance of the parity violating F_3 structure function through $\gamma - Z$ interference effects, which start to reduce $\sigma_\mathrm{NC}(\ell^+\mathrm{h})$ compared to $\sigma_\mathrm{NC}(\ell^-\mathrm{h})$. In charged current events $\sigma_\mathrm{CC}(\ell^+\mathrm{h})$ is always less than $\sigma_\mathrm{CC}(\ell^-\mathrm{h})$, and vice-versa for antihadrons, with the difference becoming more pronounced as Q^2 increases. This reflects the fact that W^- and W^+ couple to different constituents in the target hadron. Using eqn (3.38) and eqn (3.44) we have:

$$\frac{\mathrm{d}^2\sigma_\mathrm{CC}}{\mathrm{d}x\,\mathrm{d}Q^2}(e^+\mathrm{h}) \propto \left[\bar{u} + \bar{c} + (1-y)^2(d+s)\right]$$

$$\text{and} \quad \frac{\mathrm{d}^2\sigma_\mathrm{CC}}{\mathrm{d}x\,\mathrm{d}Q^2}(e^-\mathrm{h}) \propto \left[u + c + (1-y)^2(\bar{d}+\bar{s})\right] . \tag{3.54}$$

For a proton, we expect qualitatively $u(x) = 2d(x) > \bar{q}(x)$, which gives the hierarchy in the cross sections. For neutrino beams only Z exchange can contribute to the neutral current cross section, which is consequently not too dissimilar to the charged current cross section.

Before continuing our discussion we digress slightly in order to introduce a natural variable for describing an outgoing particle. Rapidity, y, and pseudorapidity, η, are defined with respect to an axis, typically the beam or a jet axis assumed to be pointing along the z-direction, by

$$y = \frac{1}{2} \ln \left(\frac{E + p_z}{E - p_z} \right) \xrightarrow{m \to 0} \eta = \ln[\cot(\theta/2)] , \tag{3.55}$$

where θ is the polar angle of the particle and m its mass; see also Ex. (6-2). Rapidity is small for central production, $\theta \sim \pi/2$, and large for far forward/backward production, $\theta \to 0$ or π. In a collision with C.o.M. energy $\sqrt{s}$ the allowed rapidity is restricted kinematically to the range $[-\ln(\sqrt{s}/m), +\ln(\sqrt{s}/m)]$. The usefulness of rapidity stems from its appropriateness in describing the Lorentz invariant phase space of the final state particle:

$$\frac{\mathrm{d}^3 \boldsymbol{p}}{2E} = \mathrm{d}p_\perp^2 \, \mathrm{d}\phi \, \mathrm{d}y . \tag{3.56}$$

The advantage of this form is the simplicity of the way in which each term transforms under a boost along the beam axis. In particular for a boost of velocity $\beta = v/c$,

$$y \longrightarrow y + \frac{1}{2} \ln \left(\frac{1 - \beta}{1 + \beta} \right) \tag{3.57}$$

so that $\mathrm{d}y$ is invariant, as are $p_\perp^2$ and ϕ. Also, as we shall learn, soft particle production typically has a flat distribution in rapidity.

In most DIS events the target hadron is 'blown apart', resulting in a trail of soft hadronic activity lying between the colour connected remnant, target region, jet and one or more current region jets. However, at HERA in a large fraction of those inelastic events with small x, and therefore large values of W, the total mass of the outgoing hadronic system, the distribution of hadronic activity is markedly different (Hebecker, 2000). The inverse relation $W^2 = M_h^2 + (1-x)Q^2/x \approx Q^2/x$ is easily derived from eqn (3.30). Whilst central 'jet' activity occurs, it is isolated from the target hadron which is only slightly deflected and appears not to break up. A rapidity gap, typically a region of size $\Delta y \gtrsim 3$ in which no hadrons are found, lies between the 'jet' and the scattered hadron. What is seen in practice is no forward activity in the main detector and, in the absence of specialized, far-forward detectors, a target hadron which can be inferred to have disappeared down the beam pipe. Compared to a regular DIS event, $\gamma^* h \to X$, this subset of events behave as $\gamma^* h \to XY$ where Y is the scattered target hadron or possibly a low mass excitation of it. Empirically, both the square of the four-momentum transferred to the forward hadron, $t = (p - p')^2 < 0$, and the mass of the observed hadronic system, $M_X^2 = (q + p - p')^2$, that is excluding the scattered hadron, are characteristically small, with a functional behaviour like

$$\frac{\mathrm{d}\sigma}{\mathrm{d}t} \sim \mathrm{e}^{bt} \quad \text{and} \quad \frac{\mathrm{d}\sigma}{\mathrm{d}M_X} \sim \frac{1}{M_X^2} . \tag{3.58}$$

The dependence on $W^2 = (q + p)^2$ and $Q^2 = -(\ell - \ell')^2$ is modest. In particular, the cross section for this type of event stays constant, or even grows, as $s = (\ell + p)^2$ or W^2 increase, in marked contrast to the rapidly falling cross sections of DIS events. These are the so-called diffractive DIS events, characterized by their rapidity gaps and almost constant cross sections. The situation is illustrated in Fig. 3.4.

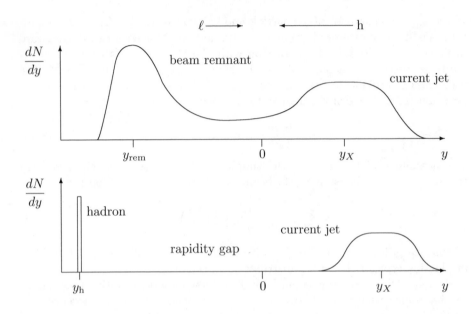

FIG. 3.4. Schematic diagrams of the rapidity distribution for the number of hadrons produced in: top, a regular deep inelastic scattering; bottom, a diffractive deep inelastic scattering

The requirement that the incoming hadron remains intact limits the momentum transfer to $|t| < -t_{\max} \lesssim (1/R_0)^2$, where R_0 is the hadron's size, which is related to the parameter b in eqn (3.58) via $b = R_0^2/6$. A second, lower bound on t is provided by the kinematics of the process, $-t_{\min} \approx M_{\mathrm{h}}^2 (M_X^2 + Q^2)^2/W^4$; see Ex. (3-10). Since we require $-t_{\min} < -t_{\max}$, the coherence requirement bounds M_X, in practice $M_X^2 < 0.2\, W^2$. This in turn implies a large separation in rapidity, $\ln(W^2/M_{\mathrm{h}} M_X)$, between the scattered hadron and the produced hadronic system. The picture which suggests itself is of the target hadron shedding a near collinear 'object' which is then struck by the virtual photon leading to central jet activity. This object carries a modest fraction, $x_{\mathbb{P}}$, of its parent's momentum and no quantum numbers. In particular, it is colour neutral. This ensures a clean separation, in rapidity, of the deflected hadron and the central activity. This object is often identified with the Pomeron, the exchange of which is

believed to dominate the hadron–hadron scattering cross section according to
Regge phenomenology: this is discussed in more detail in the next section.

The formal description of these rapidity gap events follows similar lines to
that of regular DIS events. The main difference is that the hadronic final state
contains a hadron of known quantum numbers and momentum p'. In addition
to the usual variables, Q^2, x and y, we introduce the quantities

$$
\begin{aligned}
t &= (p_{\mathrm{h}} - p')^2 \\
x_{\mathbb{P}} &= \frac{q \cdot (p - p')}{q \cdot p} &\approx \frac{M_X^2 + Q^2}{W^2 + Q^2} \\
\beta &= \frac{Q^2}{2q \cdot (p - p')} = \frac{x}{x_{\mathbb{P}}} &\approx \frac{Q^2}{Q^2 + M_X^2} \; .
\end{aligned}
\tag{3.59}
$$

The approximations hold for low values of $t < 0$ and high values of W^2, such as
are characteristic of high energy diffraction. A moment's reflection will convince
you that, in analogy to the usual phenomenology of DIS, $x_{\mathbb{P}}$ should be identified
with the momentum fraction of the object, $p_{\mathbb{P}}^\mu \equiv p^\mu - p'^\mu \approx x_{\mathbb{P}} p^\mu$, and β with the
fraction of the object's momentum carried by the struck constituent. Thus x, the
constituents momentum fraction with respect to the incoming hadron, is given
by $x = x_{\mathbb{P}} \beta$. Both $x_{\mathbb{P}}$ and β lie in the range $[0, 1]$. The four-fold differential,
diffractive DIS cross section is given by

$$
\frac{\mathrm{d}^4 \sigma_{\mathrm{D}}}{\mathrm{d}x_{\mathbb{P}} \, \mathrm{d}t \, \mathrm{d}x \, \mathrm{d}Q^2} = \frac{2\pi \alpha_{\mathrm{em}}^2}{xQ^4} \left[1 + (1-y)^2 \right] F_2^{\mathrm{D}(4)}(x_{\mathbb{P}}, t, x, Q^2) \; .
\tag{3.60}
$$

Here, for simplicity, we have neglected the small contribution from the longitu-
dinal, diffractive structure function, $F_L^{\mathrm{D}(4)} \equiv F_2^{\mathrm{D}(4)} - 2xF_1^{\mathrm{D}(4)}$. Integrating over
t, which is often not observed, gives a three-fold differential distribution, now in-
volving $F_2^{\mathrm{D}(3)}$ etc. As with ordinary DIS a factorization theorem has been proved
(Collins, 1998),

$$
\frac{\mathrm{d}^2 F_2^{\mathrm{D}(4)}}{\mathrm{d}x_{\mathbb{P}} \, \mathrm{d}t}(x_{\mathbb{P}}, t, x, Q^2) = \sum_{f=q,\bar{q},g} \int_{x_{\mathbb{P}}}^1 \frac{\mathrm{d}z}{z} \frac{\mathrm{d}^2 f^{\mathrm{D}}}{\mathrm{d}x_{\mathbb{P}} \, \mathrm{d}t} \left(x_{\mathbb{P}}, t, \frac{x}{z}, \mu_F \right) \hat{F}_2^{(\ell f)}(Q^2, z, \mu_F) \; .
\tag{3.61}
$$

Here μ_F is the factorization scale (we have suppressed the renormalization scale
μ_R) and $\hat{F}_2^{(\ell f)}(z)$ is the usual DIS structure function describing a photon scattering off a parton f carrying a fraction z of its parent hadron's momentum.
The remaining terms are the new diffractive parton density functions (Berera
and Soper, 1994), also known as (extended) fracture functions (Trentadue and
Veneziano, 1994). The diffractive p.d.f.s satisfy the usual DGLAP evolution equations.

Attempts have been made to go beyond eqn (3.61) using Regge factorization.
This assumes that the Pomeron is a real object whose coupling to the parent

hadron is described by a function of $x_{\mathbb{P}}$ and t and whose parton content is then described by functions of β and Q^2. This unproven assumption gives

$$\frac{\mathrm{d}^2 F_2^{\mathrm{D}(4)}}{\mathrm{d}x_{\mathbb{P}}\,\mathrm{d}t}(x_{\mathbb{P}}, t, x, Q^2) = f_{\mathbb{P}/\mathrm{h}}(x_{\mathbb{P}}, t) F_2^{\mathrm{D}(2)}\left(\beta = \frac{x}{x_{\mathbb{P}}}, Q^2\right), \qquad (3.62)$$

where $f_{\mathbb{P}/\mathrm{h}}$ is often referred to as the Pomeron flux factor. Measurements to date suggest that the Pomeron has a high gluon content and that there is a significant probability that a gluon carries nearly all of its momentum. Such a picture has also been promoted in the context of hadron–hadron collisions (Ingelman and Schlein, 1985). Unfortunately, it appears that the same Regge factorized structure functions as measured in DIS will not be applicable, without at least some modification, in the description of hadron–hadron collisions (Collins *et al.*, 1993).

Historically, diffractive events have long been known in hadron–hadron collisions where a well developed phenomenology has arisen. Indeed, this was used to predict that sizeable diffractive cross sections would occur at HERA (Donnachie and Landshoff, 1987). However, the discovery of such events at HERA (ZEUS Collab., 1993; H1 Collab., 1994) still came as surprise to many people and it has led to a resurgence of interest in the nature of diffraction.

Deep inelastic scattering events, whether diffractive in nature or not, are characterized by large values of $Q^2 \gtrsim (3\,\mathrm{GeV})^2$. There also exist events in which an incoming charged lepton emits via bremsstrahlung a quasi-real, $Q^2 \approx 0$, photon which interacts with the incoming hadron: the so-called photo-production events. As mentioned earlier, such photons appear to have a rich structure and variety of behaviours. They may behave as a hadron, giving effectively a hadron–hadron scattering. This in turn could be elastic, here meaning $\gamma^\star \mathrm{h} \to V\mathrm{h}$ with V a vector meson, diffractive, soft inelastic or hard inelastic. All these categories are elaborated below. The hard inelastic events are viewed as due to the scattering of (anti)quark or gluon constituents within both the hadron and photon. Thus we require p.d.f.s to describe even the photon. Of course, it is also possible that the photon remains intact and interacts directly with a quark or an antiquark.

3.2.3 *Hadron–hadron scattering*

Hadron–hadron collisions exhibit a rich variety of reactions. These can loosely be divided into two classes. The first class involves soft interactions which have only small momentum transfers so that they are sensitive to long-distance effects and see a hadron as a coherent whole. These have typically large, $\mathcal{O}(10\,\mathrm{mb})$, cross sections which change slowly (logarithmically) with the C.o.M. energy. Examples include the total, elastic and single/double diffractive cross sections discussed in more detail below. The second class involves hard interactions, defined by the presence of a large momentum transfer so that they probe the internal structure of a hadron. These have typically small to tiny cross sections and more pronounced C.o.M. energy dependencies. Examples include high transverse energy jet, heavy quark and high mass lepton pair production. The non-perturbative nature of the physics involved in the first class of reactions means that a more phenomenological approach is taken when describing them. Since pQCD can be applied directly to the second class of reactions these shall be our main concern.

The above classification of events is a little misleading. For the so-called soft events we intend that the characteristic momentum transfers are small in comparison to the C.o.M. energy $\sqrt{s}$. This leaves open the possibility that at high C.o.M. energies sufficiently large momentum transfers may occur to open up the possibility of applying pQCD. For example, this is under active study for hard diffractive events.

Whilst it is hard to apply QCD to the bulk of hadron–hadron reactions, it is nevertheless helpful to appreciate their basic properties. The general behaviour of the total cross sections is as follows. Initially, the cross section falls from $\mathcal{O}(100\,\text{mb})$ at very low C.o.M. energies to a broad minimum around $\sqrt{s} \sim 20\,\text{GeV}$ before rising slowly. Below $\sqrt{s} \lesssim 3\,\text{GeV}$ resonance structure is apparent. Above the resonance region simple quark counting rules give an indication of the relative cross sections. The rules posit that a total hadron–hadron cross section is proportional to the number of (anti)quarks in the projectile, as determined by the constituent quark model, times the number of (anti)quarks in the target. For $s \to \infty$ one expects, for example, $\sigma_{\text{tot}}(\pi^- p) \approx \sigma_{\text{tot}}(K^- p) \approx 2/3 \times [\sigma_{\text{tot}}(pp) \approx \sigma_{\text{tot}}(p\bar{p})]$. The asymptotic equality of $\sigma_{\text{tot}}(pp)$ and $\sigma_{\text{tot}}(p\bar{p})$ is also required by the Pomeranchuk theorem. Figure 3.5 shows these total cross sections as a function of the C.o.M. energy. The total cross section can be parameterized as

$$\sigma_{\text{tot}}(s) = \left[a_0 + a_2 \ln^2\left(\frac{s}{s_0}\right)\right][1 + F(s)] \qquad (3.63)$$

where $F(s)$, which vanishes as $s \to \infty$, describes the low energy behaviour. This parameterization automatically satisfies the requirement of unitarity as captured in the Froissart bound, $\sigma_{\text{tot}}(s) < (\pi/m_\pi^2) \ln^2(s/s_0)$ for some unknown s_0 (Froissart, 1961; Martin, 1963). We shall largely be concerned with high energy pp and $p\bar{p}$ collisions as this is where the search for new particles has focused the attention of experimentalists.

In elastic scatterings the hadrons remain intact without excitation of any internal degrees of freedom. They comprise a sizeable component of the total cross section, $\sigma_{\text{el}}(s) \approx 1/6 \times \sigma_{\text{tot}}(s)$. Elastic scatterings are specified by the space-like momentum transfer $t = (p_{\text{in}} - p_{\text{out}})^2 < 0$, which given s is equivalent to the C.o.M. scattering angle, $\theta^\star$, via $t = -4p^{\star 2}\sin^2(\theta^\star/2) \approx -s\sin^2(\theta^\star/2)$. A number of t-ranges can be identified according to whether electromagnetic or strong forces dominate: the Coulomb region, $|t| < 0.001\,\text{GeV}^2$; the interference region $0.001 < |t| < 0.01\,\text{GeV}^2$; and the diffraction region $0.01 < |t| < 0.15\,\text{GeV}^2$. Only the first region is well understood. The cross section is described by the t-channel exchange of a photon whose coupling to hadrons is described by two form factors, eqn (3.41). A typical behaviour for the differential cross section shows a strong peak below $|t| = 0.01\,\text{GeV}^2$, then a steady fall until reaching a sharp minimum as $|t| \sim 1.4\,\text{GeV}^2$ which is followed by a broad peak at $|t| \sim 2\,\text{GeV}^2$. Above

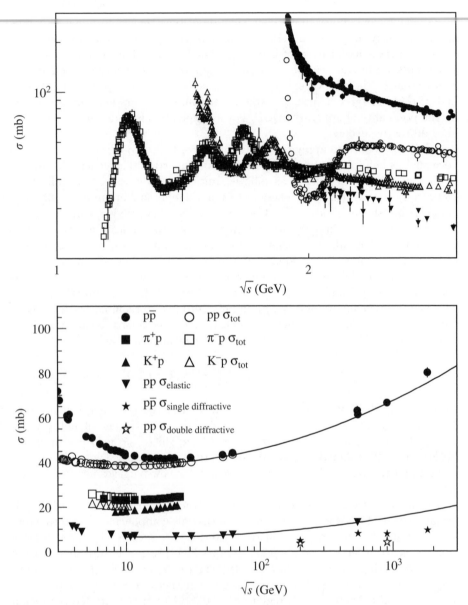

FIG. 3.5. Measured cross sections in hadron–hadron collisions as a function of the C.o.M. energy. Data are taken from the *Review of Particle Properties* (PDG, 2000) and from the Durham reactions database
http://durpdg.dur.ac.uk/hepdata/reac.html.

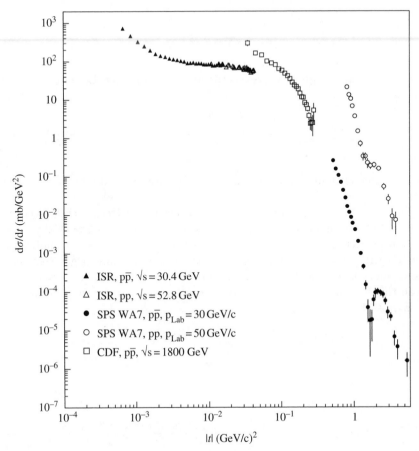

FIG. 3.6. The measured differential cross section in pp̄ and pp collisions as a function of the momentum transfer t for various C.o.M. energies. Data are taken from the Durham reactions database http://durpdg.dur.ac.uk/hepdata/reac.html.

$|t| = 2\,\text{GeV}^2$, the hard diffractive region, the differential cross section falls as t^{-n}. The so-called dimensional counting rules (Brodsky and Farrar, 1975) suggest values $n = 6, 8, 10$ for meson–meson, meson–baryon and baryon–baryon scattering, respectively, though more explicit calculations based on gluon exchange between the constituent quarks modify these simple exponents (Landshoff, 1974). Figure 3.6 shows the t-dependence of pp and pp̄ scattering. Approximate forms for the differential cross section in the diffractive region are given by

$$\frac{d\sigma_{\text{el}}}{dt} \propto \begin{cases} e^{At+Bt^2} & |t| < 0.4\ \text{GeV}^2 \\ t^{-n} & |t| > 3.0\ \text{GeV}^2 \ . \end{cases} \tag{3.64}$$

The 'dip–bump' structure can be described by interference between two exponentials. The t-distribution can be related, via a Fourier–Bessel transformation, to the impact parameter space distribution of the scattering centres in the hadron, $\exp(b_{\mathrm{eff}}t) \leftrightarrow \exp(-b_{\mathrm{eff}}b^2)$, where b is the impact parameter. Defining an effective slope by

$$\frac{\mathrm{d}\sigma_{\mathrm{el}}}{\mathrm{d}t}(s,t) = \frac{\mathrm{d}\sigma_{\mathrm{el}}}{\mathrm{d}t}(s;t_0)e^{b_{\mathrm{eff}}(s;t_0)t} \implies b_{\mathrm{eff}}(s;t_0) = \frac{\mathrm{d}}{\mathrm{d}t}\left(\ln\left(\frac{\mathrm{d}\sigma_{\mathrm{el}}}{\mathrm{d}t}\right)\right)\Bigg|_{t=t_0},$$

$$(3.65)$$

the measurements show that b_{eff} increases for large s. Thus the hadron shrinks at higher energies.

The very forward peaked nature of the elastic scattering cross section indicates that low momentum transfers are dominant. This essentially straight through behaviour means that specialized low angle detectors, usually in conjunction with low luminosity, are required to measure this large cross section. Interestingly the optical theorem provides a highly non-trivial connection between this forward ($t = 0$) differential cross section and the total cross section; see Ex. (3-11).

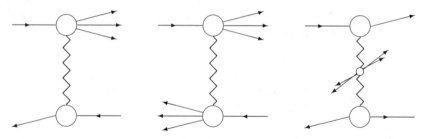

FIG. 3.7. Single diffractive dissociation, double diffractive dissociation and central diffraction. Experimentally these events are characterized by: a forward jet separated by a rapidity gap from an intact, scattered, incoming hadron; or two forward jets separated by a central rapidity gap; or central activity separated by two rapidity gaps from the scattered, incoming hadrons.

The next important class of reactions involve diffractive dissociation processes in which there is some break-up of the scattering hadrons. A possible way to view these events is as the t-channel exchange of a colour singlet object called a Pomeron, see Fig. 3.7. Unlike the case of elastic scattering, in a single/double diffractive dissociation event one/both of the hadrons is left in an excited state which then breaks up into a low multiplicity system of hadrons (jet), for example, $p \to \Delta^+ \to n\pi^+$. Typically the mass of the excited hadronic system is distributed as $\mathrm{d}\sigma \sim \mathrm{d}M_X/M_X$, whilst the t-dependence of the cross section falls away exponentially with a coefficient which decreases as M_X increases. Experimentally the key signature of these events is the lack of any particle production in between the scattered/dissociated hadrons. Conventionally, this gap is quoted in units of rapidity. If the final state hadrons have masses M_1 and M_2, then they have a rapidity gap of $\Delta y = \ln(s/M_1 M_2)$. Since the size of the gap is usually $\Delta y \gtrsim 3$, there is a minimum C.o.M. energy $\sqrt{s} \gtrsim 4.5\,\mathrm{GeV}$ required for these events to occur. A related class of events, known as central diffraction events, show two large rapidity gaps separating centrally produced jets from forward/backward going hadrons (UA8 Collab., 1988). These can be interpreted as the interaction of two Pomerons, as shown in Fig. 3.7. They are of particular interest because the jet activity indicates the presence of a hard scale and the possibility to apply pQCD to their description.

At low energies the cross section for all these rapidity gap reactions equals approximately the elastic cross section, with the ratio of single to double diffractive events found to be $\approx 4:1$. As the C.o.M. energy dependence of the cross section for a fixed excited state is flat, the growth of the total dissociation cross section with $\sqrt{s}$ can be attributed to new excitation channels opening up. Experimentally the double diffractive dissociation cross section grows faster than the single diffractive dissociation cross section. This is in accord with the naïve expectation $\sigma_{\mathrm{DD}} \approx \sigma_{\mathrm{SD}}^2/\sigma_{\mathrm{tot}}$. The central diffraction cross section is a few per cent of the total dissociation cross section. At the LHC, a $\sqrt{s} = 14\,\mathrm{TeV}$, pp collider being built at CERN, predictions indicate that $\sigma_{\mathrm{tot}} \approx 105\,\mathrm{mb}$, $\sigma_{\mathrm{el}} \approx 25\,\mathrm{mb}$, $\sigma_{\mathrm{SD}} \approx 15\,\mathrm{mb}$ and $\sigma_{\mathrm{DD}} \approx 10\,\mathrm{mb}$ (Khoze et al., 2000; Block and Halzen, 2001).

The majority of the remainder of the total cross section is made up of what may be termed soft, inelastic collisions, see Fig. 3.8. These can be thought of as peripheral, or glancing, collisions which result in two fast, forward travelling fragments, which carry the quantum numbers of the incident hadrons and typically half of their energy, together with an intervening 'trail' of centrally produced soft particles. These central particles have exponentially damped transverse momenta, $\langle p_T \rangle = 350\,\mathrm{MeV}$, and are uniformly distributed in rapidity, $dN_{\mathrm{ch}}/dy \sim 2$. This implies that the multiplicity should grow logarithmically with the C.o.M. energy, $\langle N \rangle = A\ln(s/s_0) + B/\sqrt{s}$. The pion, kaon, and baryon composition is observed to be roughly 85%, 5% and 10% respectively (UA5 Collab., 1987). These soft particles also show short-range order characterized by positive correlations in rapidity. This structure is often interpreted as being due to the production and subsequent decay into stable hadrons of 'universal clusters'. The properties of the soft particles show only a weak dependence on the C.o.M. energy of the colliding hadrons.

The major components of total hadronic cross sections (elastic scattering, single/double diffractive dissociation and soft inelastic collisions) all feature 'small' transverse momentum transfers. This focuses our attention on scattering in the limit $s \to \infty$ whilst t is held relatively small. Here, a successful phenomenology has been developed based upon Regge theory. This pre-QCD theory treats the angular momentum in a scattering amplitude as a complex variable and proceeds to derive consequences from analyticity and crossing symmetries (Collins, 1977). A typical t-channel exchange amplitude for a two-to-two process takes the form

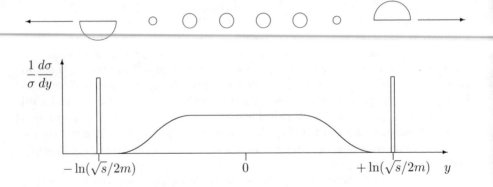

FIG. 3.8. A schematic diagram of a soft, inelastic collision and the associated rapidity distribution

$$\mathcal{M}(s,t) = (\mathrm{i} + \rho) \left(\frac{s}{M_i^2} \right)^{\alpha_i(t)} \mathrm{e}^{b_i t} \quad \text{with} \quad \rho(s,t) \equiv \frac{\mathcal{R}e[\mathcal{M}(s,t)]}{\mathcal{I}m[\mathcal{M}(s,t)]} \approx \rho(s,0) \,.$$
(3.66)

The index i in the above equation denotes the flavour quantum numbers exchanged in the interaction. As the practically t-independent ρ is $\mathcal{O}(0.1)$, the amplitudes are mainly imaginary. For the exchange of a particle of spin J the exponent would be $\alpha_i(t) = J$, but in eqn (3.66) this has been 'Reggeized' to include contributions from a whole family of particles lying on a linear Regge trajectory

$$\alpha_i(t) = \alpha_i(0) + \alpha_i' t + \mathcal{O}(t^2) \,.$$
(3.67)

The parameters of the trajectories can be found by fitting the spins and masses (s-channel poles) of real mesons and baryons using $J = \alpha_i(M^2)$. The slope is almost universal with $\alpha' \approx 1 \, \mathrm{GeV}^{-2}$, whilst the intercept depends on the flavour quantum numbers, i, being exchanged. One finds $\alpha_i(0) \approx 0.5$ for the leading (dominant) contribution from the non-strange vector mesons $\rho, \omega, a_2, f_2, \ldots$. The sub-leading pion trajectory has $\alpha_i(0) \approx 0$. The exponential t-dependence assumed in eqn (3.66) is empirical: it implies $b_{\mathrm{eff}} = 2[b_0 + \alpha' \ln(s/M_i^2)]$ and thus a 'shrinkage' of the t-distribution with increasing C.o.M. energy (Gribov, 1961), c.f. eqn (3.65). More formally it is a measure of the coupling strength between the scattering and exchanged particles. The mass M_i accounts for the dimensions and absorbs any numerical factors.

Using eqn (3.66) one can derive compact expressions for, for example, the total, elastic and singly diffractive cross sections as

$$\sigma_{\mathrm{tot}} = \frac{1}{M_i^2} \left(\frac{s}{M_i^2} \right)^{\alpha_i(0)-1}$$
(3.68)

$$\frac{\mathrm{d}\sigma_{\mathrm{el}}}{\mathrm{d}t} = \frac{(1+\rho^2)}{16\pi M_i^4} \left(\frac{s}{M_i^2} \right)^{2(\alpha_i(t)-1)} \mathrm{e}^{2b_{\mathrm{el}} t}$$
(3.69)

$$\frac{\mathrm{d}^2\sigma_\mathrm{D}}{\mathrm{d}t\,\mathrm{d}M_X^2} \sim \frac{1}{M_i^4}\frac{1}{M_X^2}\left(\frac{s}{M_X^2}\right)^{2(\alpha_i(t)-1)}\mathrm{e}^{2b_\mathrm{D}t} \tag{3.70}$$

For simplicity we have included only a single Reggeon exchange and omitted the electromagnetic contributions. Including the interference between the hadronic and, well known, electromagnetic amplitudes allows ρ to be measured experimentally. These formulae provide a very good description of reactions which involve the exchange of flavour. These typically fall as s^{-1}. To apply them to situations where no flavour is exchanged, where cross sections are constant or grow as $s \to \infty$, a new dominant contribution must be included. This is the Pomeron. It has the quantum numbers of the vacuum and the Regge parameters

$$\alpha_\mathbb{P}(0) \approx 1.08 \quad \text{and} \quad \alpha'_\mathbb{P} \approx 0.25 \ . \tag{3.71}$$

These values are derived from successful fits to a remarkably wide range of data (Donnachie and Landshoff 1992; 1994). This trajectory does not correspond to any presently known particles, though it has been conjectured that it is related to the predicted glueballs of QCD. Actually, since $\alpha_\mathbb{P}(0) > 1$, the Pomeron is 'super-critical' and, unless eqn (3.68) is modified, will lead to a violation of unitarity in the $s \to \infty$ limit. More apparent is the absurdity $\sigma_\mathrm{el}/\sigma_\mathrm{tot} \gtrsim (s/M_i^2)^{\alpha_\mathbb{P}(0)-1} > 1$ for s sufficiently large. The inclusion of the necessary multiple Pomeron exchanges and unitarization corrections leads to a more complex theory (Khoze et al., 2000).

It is important to remember that Regge theory has not been derived from QCD. One should therefore be wary of regarding it as doing anything more than providing an accurate and economical, phenomenological framework for describing data in the Regge limit. It also acts as a guide in framing the questions addressed in an experiment. That said, pQCD has been applied to the region $s \gg |t| > \Lambda_\mathrm{QCD}$, leading to the development of a hard Pomeron with an intercept significantly above one and a small slope. To distinguish it, the usual Pomeron is now often referred to as the soft Pomeron. This hard Pomeron is associated with the summation of leading logarithms of the form $\alpha_\mathrm{s}\ln(s/t)$ (Kuraev et al., 1977; Balitsky and Lipatov, 1978). The simplest model for such an object is the t-channel exchange of two gluons (Low, 1975; Nussinov, 1975) (or one 'Reggeized' gluon), which is suggestive of a glueball interpretation. The hard Pomeron also manifests itself in the small-x behaviour of structure functions where it sums leading $[\alpha_\mathrm{s}\ln(1/x)]^n$ logarithms. However, a word of caution should be sounded. As the hard Pomeron theory implies a rapid growth in the number of partons then non-perturbative methods will be required ultimately. The search for the predicted hard Pomeron is an active topic of research.

Finally, we turn to the rare, hard events which shall be our main focus of interest. By experimentally requiring an event to contain a large momentum scale we raise the possibility of applying pQCD to its description. Furthermore, the short-distance scales suggest working with the quark and gluon constituents rather than the colliding hadrons themselves. The situation is analogous to DIS and again a factorized formalism can be applied. This is illustrated in Fig. 3.9.

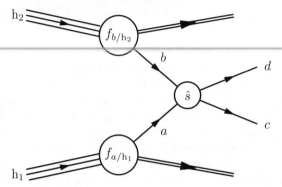

FIG. 3.9. A schematic diagram for the production of final state particles c and d in a hard collision of hadrons h_1 and h_2

The basic cross section formula for the collision of hadrons h_1 and h_2 to produce particles c and d is given by

$$d\sigma(h_1 h_2 \to cd) = \int_0^1 dx_1 dx_2 \sum_{a,b} f_{a/h_1}(x_1, \mu_F^2) f_{b/h_2}(x_2, \mu_F^2) d\hat\sigma^{(ab \to cd)}(Q^2, \mu_F^2) .$$

(3.72)

Here the f_{a/h_1} and f_{b/h_2} are the same p.d.f.s as arose in DIS, where the indices refer to partons $a, b \in \{q, \bar{q}, g\}$ in the interacting hadrons h_1 and h_2. Here there is a technical proviso that we are careful to use the same factorization scheme in the description of both processes. They are evaluated at the factorization scale μ_F, which is typically $\mathcal{O}(Q)$ — a hard scale characteristic of the scattering process. The use of the same p.d.f.s is possible because the presence of an incoming hadron does not cause the target hadron to modify its internal structure. This is the real significance of the factorization theorem and helps to make pQCD a predictive theory. In the matrix element for the hard subprocess the parton momenta are given by $p_a^\mu = x_1 p_{h_1}^\mu$ and $p_b^\mu = x_1 p_{h_2}^\mu$. In general, we do not expect $x_1 = x_2$ so that the hard subprocess will be boosted with $\beta = (x_1 - x_2)/(x_1 + x_2)$ with respect to the $h_1 h_2$ laboratory frame, resulting in the outgoing particles being thrown to one side or the other. The sum is over all partonic subprocesses which contribute to the production of c and d. For example, the production of a pair of heavy quarks receives contributions from $q\bar{q} \to Q\bar{Q}$ and $gg \to Q\bar{Q}$, whilst prompt photon production receives contributions from $qg \to q\gamma$ and $q\bar{q} \to g\gamma$. These two-to-two scatterings give the leading, $\mathcal{O}(\alpha_s^2)$ and $\mathcal{O}(\alpha_s \alpha_{em})$, contributions to the hard subprocess cross section. Beyond the leading order it is necessary to consider two-to-three, etc. processes, which gives rise to a perturbative expansion $\hat\sigma = C_{LO}\alpha_s^n + C_{NLO}\alpha_s^{n+1} + C_{NNLO}\alpha_s^{n+2} + \cdots$. A complication arises with the higher order corrections as they contain singularities when two incoming or outgoing partons become collinear. It is the factorization of these singularities, order by order, into the p.d.f.s and fragmentation functions which gives them their calculable μ_F^2 dependencies. This, logarithmically enhanced, near collinear radiation is manifested as the appearance of initial and final state jets associated with each of the incoming and outgoing partons.

The mix of hard subprocesses which contribute to eqn (3.72) depends non-trivially on the relative sizes of both the cross sections and the p.d.f.s. The latter are influenced by both the type and energy of the colliding beams and any requirements placed on the kinematics of the final state. For example, requiring the outgoing particles to be produced in a given rapidity range, perhaps corresponding to the geometry of a detector element, directly affects the x-ranges being sampled in the integral; see Ex. (3-13). To go further, we consider heavy quark production at the TEVATRON, a $\sqrt{s} = 1.8\,\text{TeV}$ (now $2\,\text{TeV}$) p$\bar{\text{p}}$ collider at FERMILAB. In the case of centrally ($y = 0$) produced bottom quarks one has $x_1 \approx x_2 \sim 2m_\text{b}/\sqrt{s} = 2 \times 5/1800 = 0.0056$, whilst for top quarks it is $x_1 \approx x_2 \sim 2 \times 175/1800 = 0.19$. At small x gluons dominate the p.d.f.s, whilst at large x only valence (anti)quarks are present; this is particularly true at the higher scale appropriate for top production, $Q \sim 2m_\text{Q}$. Thus, bottom quark production is dominated by gg $\rightarrow$ b$\bar{\text{b}}$ scattering, whilst top quark production is dominated by the annihilation process q$\bar{\text{q}} \rightarrow$ t$\bar{\text{t}}$. Here, we see that in a high mass 'annihilation process' it pays to have an antihadron in the initial state. In this result the larger cross section for gg $\rightarrow$ Q$\overline{\text{Q}}$ is overwhelmed by the p.d.f. contribution. As a second example we consider di-jet production. In the absence of any flavour determination the outgoing jets may be seeded by either a primary (anti)quark or gluon so that there are many contributing hard subprocesses: gg $\rightarrow$ gg, gq $\rightarrow$ gq, qq$'$ $\rightarrow$ qq$'$, etc. Loosely speaking, the relative hard subprocess cross sections are in the ratio $C_A^2 : C_A C_F : C_F^2$ etc., reflecting the colour charges of the colliding partons. This allows us to express the integrand in eqn (3.72) in terms of an effective p.d.f. (Combridge and Maxwell, 1984), see Ex. (3-14),

$$f_{\text{h}_1}^{\text{eff}}(x_1) f_{\text{h}_2}^{\text{eff}}(x_2) \text{d}\hat{\sigma}(\text{gg} \rightarrow \text{gg}) \quad \text{with} \quad f^{\text{eff}}(x) = g(x) + \frac{C_F}{C_A} \sum_{f=q,\bar{q}} f(x) . \quad (3.73)$$

Here $C_F/C_A = 4/9 \approx 1/2$. Thus at moderate transverse jet energies, equivalent to moderate x values, gg scattering will be dominant.

In addition to a hard subprocess such hadronic scatterings also involve an underlying event arising from the collision of the two beam remnants. In broad outline the underlying event is like a soft, inelastic collision between two hadrons of reduced C.o.M. energy squared $(1 - x_1 x_2)s$. Fortunately, the soft particles produced have limited transverse momentum and so do not unduly obscure the high transverse energy particles produced in the hard subprocess. Observationally there is an increased level of hadronic activity in hard events, even away from any jets, as compared to minimum bias events which are effectively equivalent to normal soft inelastic collisions. This is the so-called pedestal effect. Thus, more refined models build in an interplay between the hard subprocess and the underlying event (Sjöstrand and van Zijl, 1987). One possibility, which becomes more likely with increasing C.o.M. energy, is that a second hard scattering occurs between the partons in the beam remnants. By treating the two scatters as independent the rate of double scattering can be estimated as $\sigma_{12} = \sigma_1\sigma_2/\sigma_{\text{tot}}$. The assumption of independence is plausible provided all the momentum fractions remain small.

In an experiment it is necessary to supply a criterion to decide when to initiate the read-out of the detector. Typically, this trigger condition is based upon known/supposed features of the events which are of interest. This introduces inevitably a bias towards just such events. Therefore, it is also common to collect an 'unbiased' data sample based upon a minimal trigger condition such as the occurrence of a bunch crossing or the presence of an energy deposit somewhere in the detector. Given the relative cross sections for the hadron–hadron scatterings these minimum bias events coincide essentially with the soft, inelastic collision events. Since hadron–hadron colliders are often viewed as discovery machines searching for very rare events, there is a need to use high luminosities. Given large hadron number densities in the colliding bunches it becomes likely that more than one pair of hadrons from the colliding bunches may interact, most likely in soft, inelastic collisions. Thus, even when a hard trigger is satisfied it is quite possible that the detector is seeing an event of interest together with several soft, inelastic events. For example, at nominal luminosity at the planned LHC at CERN, each hard event is, on average, accompanied by $\mathcal{O}(10)$ simultaneous minimum bias events. Fortunately, these extra pile-up events produce mainly low transverse momentum particles, spread throughout longitudinal phase space, whilst the hard event must have high transverse momentum particles, typically restricted kinematically to the central ($y = 0$) region.

3.3 Born level calculations of QCD cross sections

In this section, we shall review the calculational techniques required to evaluate basic tree-level processes. We shall concentrate on the process $e^+e^- \rightarrow q\bar{q}$, which is a paradigm for several important processes, together with its lowest, $\mathcal{O}(\alpha_s)$, tree-level, QCD correction, $e^+e^- \rightarrow q\bar{q}g$, which we will use in our discussion of the QCD improved parton model. We will also look at the pure QCD process $q\bar{q} \rightarrow gg$ which will give us an insight into the nature of gauge invariance. We do assume some previous familiarity with Dirac spinors and working with Feynman diagrams. The interested reader can refresh their memory and find more details in any good text book, such as the one by Aitchison and Hey (1989) or by Peskin and Schroeder (1995).

3.3.1 e^+e^- annihilation to quarks at $\mathcal{O}(\alpha_s^0)$

The basic Feynman diagram for $e^+e^- \rightarrow q\bar{q}$ is given in Fig. 3.10(a). Strictly speaking, this lowest $\mathcal{O}(\alpha_s^0)$ process is more an electroweak than a QCD interaction. However, it remains of great importance in the description of e^+e^- annihilation to hadrons, and using crossing symmetry, also to deep inelastic scattering, Fig. 3.10(b) and the Drell–Yan process, Fig. 3.10(c). Furthermore, by replacing the lepton pair by a new quark pair ($q \neq q'$), we can learn about di-jet production

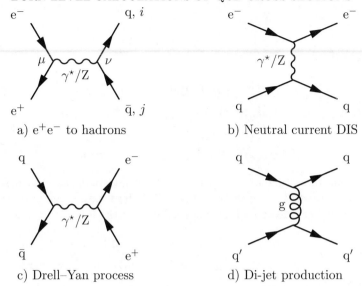

a) e^+e^- to hadrons   b) Neutral current DIS

c) Drell–Yan process   d) Di-jet production

FIG. 3.10. Examples of the basic processes contributing to hard lepton–lepton, lepton–hadron and hadron–hadron scattering. Our convention is such that the Feynman diagrams should be read from incoming states on the left to outgoing states on the right.

in hadron–hadron collisions, Fig. 3.10(d). It should also be noted that since the electroweak couplings are relatively small, $\alpha_{\rm em} \sim 10^{-2}$, diagrams involving single photon exchange should be sufficient for an accurate description of processes (a), (b) and (c). On the other hand, since the strong coupling is relatively large one might wonder about the size of the corrections to the single gluon exchange diagram (d). Asymptotic freedom will have something to say here.

The matrix element for $e^+e^- \to q\bar{q}$ is easily written down using the Feynman rules in Appendix B,

$$\mathcal{M} = \bar{v}(\ell^+) \cdot -\mathrm{i}\,e\gamma_\mu \cdot u(\ell^-) \times \frac{-\mathrm{i}\,\eta^{\mu\nu}}{Q^2} \times \bar{u}(q) \cdot -\mathrm{i}\,ee_{\rm q}\gamma_\nu\delta_{ij} \cdot v(\bar{q})$$

$$= e\bar{v}(\ell^+)\gamma_\mu u(\ell^-) \times (\mathrm{i}\,Q^{-2}) \times ee_{\rm q}\delta_{ij}\bar{u}(q)\gamma^\mu v(\bar{q}) \, . \qquad (3.74)$$

Here, we use the particle names to also represent their four-momenta and introduce $Q^\mu = (\ell^+ + \ell^-)^\mu$, the four-momentum transfer. For simplicity, only photon exchange is included, which is appropriate for $Q^2 \ll M_Z^2$, and we have chosen to work in the covariant Feynman gauge, $\xi = 1$. The quark colours are specified by the colour indices i and j which run from 1 to N_c. Note that we have explicitly included a colour conserving Kronecker δ-function at the quark–photon vertex which ensures that the $q\bar{q}$-pair forms a colour singlet. The order of the terms carrying spinor indices has been determined by working backwards along each fermion line. Next, we need to evaluate the matrix element squared: $|\mathcal{M}|^2 = \mathcal{M}\mathcal{M}^\star$. Now, whilst $\mathcal{M}$ is a complex number it is formed from a product of matrices, so that it is more convenient to use $\mathcal{M}^\dagger = \mathcal{M}^\star$ and rather evaluate $|\mathcal{M}|^2 = \mathcal{M}\mathcal{M}^\dagger$, namely,

$$
\begin{aligned}
|\mathcal{M}|^2 &= e^2 \cdot \bar{v}(\ell^+)\gamma_\mu u(\ell^-)\big[\bar{v}(\ell^+)\gamma_\nu u(\ell^-)\big]^\dagger \times Q^{-4} \\
&\quad \times (ee_q)^2 \cdot {}'\delta_{ij}\delta_{ji}{}' \cdot \bar{u}(q)\gamma^\mu v(\bar{q})\big[\bar{u}(q)\gamma^\nu v(\bar{q})\big]^\dagger \\
&= e^2 \cdot \bar{v}(\ell^+)\gamma_\mu u(\ell^-)\bar{u}(\ell^-)\gamma_\nu v(\ell^-) \times Q^{-4} \\
&\quad \times (ee_q)^2 \cdot N_c \cdot \bar{u}(q)\gamma^\mu v(\bar{q})\bar{v}(\bar{q})\gamma^\nu u(q) .
\end{aligned}
\tag{3.75}
$$

Here, care has been taken to keep the leptonic and hadronic terms separate and also to sum over the repeated indices. The Hermitian conjugated terms in eqn (3.75) have been dealt with as a special case of the following result.

$$
\begin{aligned}
[\bar{u}\Gamma_1\Gamma_2\cdots\Gamma_n v]^\dagger &= v^\dagger\Gamma_n^\dagger\cdots\Gamma_2^\dagger\Gamma_1^\dagger\gamma_0^\dagger u \\
&= \bar{v}(\gamma_0\Gamma_n^\dagger\gamma_0)\cdots(\gamma_0\Gamma_2^\dagger\gamma_0)(\gamma_0\Gamma_1^\dagger\gamma_0)u
\end{aligned}
$$

$$
\text{with} \quad \gamma_0\{\mathbf{1}, \gamma_5, \gamma_\mu, \gamma_\mu\gamma_5, \sigma_{\mu\nu}\}^\dagger\gamma_0 = \{+\mathbf{1}, -\gamma_5, +\gamma_\mu, +\gamma_\mu\gamma_5, +\sigma_{\mu\nu}\} \tag{3.76}
$$

Here Γ_i represents any one of the five basic 4×4 matrices.

At this point we pause to comment on the colour factor in eqn (3.75). Strictly speaking, the quark and antiquark come with colour polarization vectors, so that $\mathcal{M} \propto a^\star(q)_i\,\delta_{ij}\,a(\bar{q})_j$ in eqn (3.74). Then, when we sum $|\mathcal{M}|^2$ over these colour polarizations we must use the result

$$
\sum_{\text{col.pols}} a(q)_k a^\star(q)_i = \delta_{ki} \qquad \sum_{\text{col.pols}} a(g)_c a^\star(g)_a = \delta_{ca} , \tag{3.77}
$$

appropriate for unpolarized quarks. For completeness we have included the equivalent result for a gluon, where now the indices $\{a, c\} = 1\ldots N_c^2 - 1$. Thus,

$$
\begin{aligned}
|\mathcal{M}|^2 &\propto \sum a^\star(q)_i\,\delta_{ij}\,a(\bar{q})_j \times \big[a^\star(q)_k\,\delta_{kl}\,a(\bar{q})_l\big]^\star \\
&= \sum\Big(\sum a(q)_k a^\star(q)_i\Big)\delta_{ij}\Big(\sum a(\bar{q})_j a^\star(\bar{q})_l\Big)\delta_{lk} \\
&= \sum \delta_{ki}\delta_{ij}\delta_{jl}\delta_{lk} = \sum \delta_{ij}\delta_{ji} = N_c
\end{aligned}
\tag{3.78}
$$

Rather reassuringly, the reaction rate is found to be proportional to the number of quark colours, N_c. If a quark or gluon appears in the initial state then the corresponding colours should be averaged, as described in Appendix B. In practice, it is standard not to write out the colour polarization vectors and instead simply keep the same indices on the external particles in both $\mathcal{M}$ and $\mathcal{M}^\dagger$.

A similar result, originally due to van der Waerden, can be used to eliminate the spinor basis states still appearing in eqn (3.75). Denoting the spinor indices by $\{i, j, k, l\} = 1\ldots 4$, the following relations hold

$$
u(p)_i\bar{u}(p)_j = \frac{1}{2}\left[(\not{p} + m)(\mathbf{1} + \gamma_5\not{s})\right]_{ij}\bigg|_{s\|p} \approx \frac{1}{2}\left[(\not{p} + m)(\mathbf{1} \mp \gamma_5)\right]_{ij}
$$

$$v(\bar{p})_k \bar{v}(\bar{p})_l = \frac{1}{2} \left[(\not{p} - m)(1 + \gamma_5 \not{s}) \right]_{kl} \big|_{s \| p} \approx \frac{1}{2} \left[(\not{p} + m)(1 \mp \gamma_5) \right]_{kl} \quad (3.79)$$

The spin polarization state is specified by the space-like four-vector s which is orthogonal to p, $s \cdot p = 0$ and which for a pure state is normalized such that $s^2 = -1$. The approximate form is appropriate to the high energy limit, $m \ll E$, when the spin vector is parallel/antiparallel to the particle's direction of travel: the so-called helicity basis. Often the incoming particles in a collision are unpolarized, that is, they are an equal admixture of all possible polarizations. It is therefore conventional to include an average over the incoming particle spins; again see Appendix B. Concentrating on the hadronic part of eqn (3.75), making the spinor indices explicit and assuming no spin sensitive measurements are made on the outgoing quarks this result allows us to write:

$$\sum_{\text{spins}} \bar{u}(q)_i \, \gamma_{ij}^\mu \, v(\bar{q})_j \bar{v}(\bar{q})_j \, \gamma_{kl}^\nu \, u(q)_l = \left(\sum_{\text{spins}} u(q)_l \bar{u}(q)_i \right) \gamma_{ij}^\mu \left(\sum_{\text{spins}} v(\bar{q})_j \bar{v}(\bar{q})_k \right) \gamma_{kl}^\nu$$

$$= (\not{q} + m_{\text{q}})_{li} \cdot \gamma_{ij}^\mu \cdot (\not{\bar{q}} - m_{\text{q}})_{jk} \cdot \gamma_{kl}^\nu$$

$$= \text{Tr} \left\{ (\not{q} + m_{\text{q}}) \gamma^\mu (\not{\bar{q}} - m_{\text{q}}) \gamma^\nu \right\} . \quad (3.80)$$

In reaching this point we have been careful to make explicit the individual steps involved. Consequently, the derivation seems quite lengthy. However, with practice one can, in principle, go straight from $\mathcal{M}$ to the traces over propagators and vertices appearing in $|\mathcal{M}|^2$. One simply writes down a γ-matrix string from $\mathcal{M}$ followed by a second γ-matrix string from $\mathcal{M}^\dagger$ but with the order of the individual Γ-terms reversed, including minus signs for any γ_5 and $\gamma_\mu \gamma_5$ terms present, see eqn (3.76), and with spin-sums $(\not{p} \pm m)$, as appropriate for the external spinors, inserted between these strings.

To deal with such traces of γ-matrices we adopt the following strategy which is always guaranteed to work. First, expand out the brackets so that you have a sum of terms of the form $\text{Tr} \{ \gamma^{\mu_1} \gamma^{\mu_2} \cdots \gamma^{\mu_n} \}$. Second, set all terms where n is odd equal to zero; here remember that γ_5 is the product of an even number (four) of γ-matrices. Third, for traces of an even number of γ-matrices repeatedly use the following algorithm based on using the Clifford algebra, $\gamma^\mu \gamma^\nu = 2\eta^{\mu\nu} - \gamma^\nu \gamma^\mu$, to permute the first γ-matrix through the rest. We illustrate this for the case $n = 4$, where we need to iterate three times,

$$\text{Tr} \left\{ \gamma^\mu \gamma^\nu \gamma^\sigma \gamma^\tau \right\} = \text{Tr} \left\{ (2\eta^{\mu\nu} - \gamma^\nu \gamma^\mu) \gamma^\sigma \gamma^\tau \right\}$$

$$= 2\eta^{\mu\nu} \text{Tr} \left\{ \gamma^\sigma \gamma^\tau \right\} - \text{Tr} \left\{ \gamma^\nu \gamma^\mu \gamma^\sigma \gamma^\tau \right\}$$

$$= \cdots$$

$$= 2\eta^{\mu\nu} \text{Tr} \left\{ \gamma^\sigma \gamma^\tau \right\} - 2\eta^{\mu\sigma} \text{Tr} \left\{ \gamma^\nu \gamma^\tau \right\} + 2\eta^{\mu\tau} \text{Tr} \left\{ \gamma^\nu \gamma^\sigma \right\}$$

$$\qquad - \text{Tr} \left\{ \gamma^\nu \gamma^\sigma \gamma^\tau \gamma^\mu \right\}$$

$$\implies = \eta^{\mu\nu} \text{Tr} \left\{ \gamma^\sigma \gamma^\tau \right\} - \eta^{\mu\sigma} \text{Tr} \left\{ \gamma^\nu \gamma^\tau \right\} + \eta^{\mu\tau} \text{Tr} \left\{ \gamma^\nu \gamma^\sigma \right\} . \quad (3.81)$$

The last line follows because the final trace equals the original one by the cyclicity of traces. The algorithm reduces the number of γ-matrices in a trace by two each time it is applied. For $n = 2$ it gives $\mathrm{Tr}\{\gamma^\mu\gamma^\nu\} = \eta^{\mu\nu}\mathrm{Tr}\{\mathbf{1}\} = \eta^{\mu\nu}\cdot 4$. Thus the final result becomes

$$\mathrm{Tr}\{\gamma^\mu\gamma^\nu\gamma^\sigma\gamma^\tau\} = 4\left[\eta^{\mu\nu}\eta^{\sigma\tau} - \eta^{\mu\sigma}\eta^{\nu\tau} + \eta^{\mu\tau}\eta^{\nu\sigma}\right]. \tag{3.82}$$

Of course, in practical situations a number of tricks (short cuts) can often be used to speed up evaluations, for example,

$$
\begin{aligned}
\cdots\gamma_\mu\slashed{a}\slashed{a}\gamma_\nu\cdots &= +a^2 \times \cdots\gamma_\mu\gamma_\nu\cdots & \cdots\gamma_\mu\slashed{a}\slashed{b}\gamma^\mu\cdots &= 4a\cdot b \times \cdots\mathbf{1}\cdots \\
\cdots\gamma_\mu\slashed{a}\gamma^\mu\cdots &= -2 \times \cdots\slashed{a}\cdots & \cdots\gamma_\mu\slashed{a}\slashed{b}\slashed{c}\gamma^\mu\cdots &= -2 \times \cdots\slashed{c}\slashed{b}\slashed{a}\cdots,
\end{aligned}
\tag{3.83}
$$

where the dots are to remind us that the strings of consecutive γ-matrices may be embedded in a larger expression. Unfortunately, familiarity with these tricks only comes with practice. Given these trace results it is now straightforward to evaluate the hadron trace, eqn (3.80), to yield

$$
\begin{aligned}
\mathrm{Tr}\{(\slashed{q} + m_{\mathrm{q}})\gamma^\mu(\slashed{\bar{q}} - m_{\mathrm{q}})\gamma^\nu\} &= \mathrm{Tr}\{\slashed{q}\gamma^\mu\slashed{\bar{q}}\gamma^\nu\} - m_{\mathrm{q}}^2\mathrm{Tr}\{\gamma_\mu\gamma_\nu\} \\
&= 4\left[q^\mu\bar{q}^\nu - q\cdot\bar{q}\,\eta^{\mu\nu} + q^\nu\bar{q}^\mu - m_{\mathrm{q}}^2\eta^{\mu\nu}\right] \\
&= 4\left[q^\mu\bar{q}^\nu + \bar{q}^\mu q^\nu - (Q^2/2)\eta^{\mu\nu}\right].
\end{aligned}
\tag{3.84}
$$

The leptonic trace, coming from eqn (3.75), can be evaluated in the same way and gives essentially the same result, though with an extra factor $1/4$ reflecting the spin average in case of unpolarized beams. If we collect our results so far, we find

$$
\begin{aligned}
\overline{\sum}|\mathcal{M}|^2 &= \frac{1}{4Q^4}e^2\cdot\mathrm{Tr}\{(\slashed{\ell}^+ - m_\ell)\gamma_\mu(\slashed{\ell}^- + m_\ell)\gamma_\nu\} \\
&\quad \times (ee_{\mathrm{q}})^2 N_c\cdot\mathrm{Tr}\{(\slashed{q} + m_{\mathrm{q}})\gamma^\mu(\slashed{\bar{q}} - m_{\mathrm{q}})\gamma^\nu\} \\
&= \frac{1}{Q^4}e^2\cdot\left[\ell_\mu^+\ell_\nu^- + \ell_\mu^-\ell_\nu^+ - (Q^2/2)\eta_{\mu\nu}\right] \\
&\quad \times (ee_{\mathrm{q}})^2 N_c\cdot 4\left[q^\mu\bar{q}^\nu + \bar{q}^\mu q^\nu - (Q^2/2)\eta^{\mu\nu}\right] \\
&\equiv \frac{1}{Q^4}L_{\mu\nu}\cdot H^{\mu\nu}.
\end{aligned}
\tag{3.85}
$$

Here $\overline{\sum}$ is introduced to denote a sum over final state and average over initial state spins and colours. Also, for future reference we have introduced the lepton and hadron tensors $L_{\mu\nu}$ and $H^{\mu\nu}$. The Lorentz contractions are easily carried out to yield

$$
\begin{aligned}
\overline{\sum}|\mathcal{M}|^2 &= 4\frac{(e^2 e_{\mathrm{q}})^2 N_c}{Q^4}\left[2(\ell^+\cdot q)(\ell^-\cdot\bar{q}) + 2(\ell^-\cdot q)(\ell^+\cdot\bar{q}) + (m_\ell^2 + m_{\mathrm{q}}^2)Q^2\right] \\
&= (e^2 e_{\mathrm{q}})^2 N_c\frac{1}{2}\left(2 - \beta_{\mathrm{q}}^{\star 2} + \beta_{\mathrm{q}}^{\star 2}\cos^2\theta^\star\right).
\end{aligned}
\tag{3.86}
$$

In the second line the result is written in terms of the C.o.M. variables $\theta^\star$, the scattering angle between the incoming lepton and outgoing quark, and $\beta^\star$, the velocity of the final state quarks. In this C.o.M. frame, with massless leptons travelling along the z-direction and the scattering in the x-z plane, the four-momenta are given by

$$\ell^{\pm\mu} = \frac{\sqrt{Q^2}}{2}(1,0,0,\pm1) \tag{3.87}$$

and $\quad \overset{(-)\mu}{q} = \frac{\sqrt{Q^2}}{2}(1,\pm\beta_q^\star \sin\theta^\star,0,\pm\beta_q^\star \cos\theta^\star) \quad$ with $\quad \beta_q^\star = \sqrt{1 - \frac{4m_q^2}{Q^2}}$.

Note that as a Lorentz invariant quantity $|\mathcal{M}|^2$ could only depend on the particles' four-momenta via their invariants, for example, their scalar products, $p_i \cdot p_j$. In a two-to-two scattering, $ab \to cd$, it is common to introduce the Mandelstam variables

$$\begin{aligned} s &= (p_a + p_b)^2 & t &= (p_a - p_c)^2 & u &= (p_a - p_d)^2 \\ &= (p_c + p_d)^2 & &= (p_b - p_d)^2 & &= (p_b - p_c)^2 \,. \end{aligned} \tag{3.88}$$

Of these variables only two are independent since they are constrained to satisfy $s + t + u = m_a^2 + m_b^2 + m_c^2 + m_d^2$. Since $s > \max\{m_a^2 + m_b^2, m_c^2 + m_d^2\}$ is always positive, one of t and u, typically both, must be negative. Of course there is some freedom on which particle is labelled c or d. The motivation for a specific choice is to try and ensure that the Mandelstam variables naturally arise in the propagators: s in annihilation processes, for example, $e^+e^- \to q\bar{q}$, and t in scattering processes, for example, $\ell q \to \ell q$ DIS. One refers to s-, t- or u-channel contributions. In terms of the Mandelstam variables, with $t = (q - \ell^-)^2$, eqn (3.86) can be written

$$\sum|\mathcal{M}|^2 = (e^2 e_q)^2 N_c \frac{[t^2 + u^2 + 2(m_\ell^2 + m_q^2)(2s - m_\ell^2 - m_q^2)]}{s^2}. \tag{3.89}$$

According to eqn (B.4), to obtain a (differential) cross section we need to include a flux factor for the incoming particles and a (differential) phase space factor for the outgoing particles. In the massless-lepton limit the flux factor, eqn (B.5), is given by $1/(2s)$. The evaluation of two-body, Lorentz invariant phase space, $n = 2$ in eqn (B.6), is relatively straightforward and gives

$$\begin{aligned} d^2\Phi_2 &= \frac{1}{8\pi}\frac{|\boldsymbol{p}_{\text{out}}^\star|}{\sqrt{s}}d\cos\theta^\star\frac{d\phi}{2\pi} \\ &= \frac{1}{16\pi}\frac{dt}{|\boldsymbol{p}_{\text{in}}^\star|\sqrt{s}}\frac{d\phi}{2\pi}. \end{aligned} \tag{3.90}$$

The first expression for the two-body phase space element is appropriate for a description of the collision in the C.o.M. frame. In the second expression it is given in terms of the Mandelstam variable t. After integration over the azimuthal angle ϕ and introducing the frequently occuring fine structure constant α_{em},

$$\alpha_{em} = \frac{e^2}{4\pi} , \qquad (3.91)$$

the final result becomes

$$\frac{d\sigma}{d\cos\theta^\star} = \frac{\alpha_{em}^2 e_q^2 \pi N_c}{2s} (2 - \beta_q^{\star 2} + \beta_q^{\star 2} \cos^2\theta^\star)\beta_q^\star$$

$$\frac{d\sigma}{dt} = \frac{\alpha_{em}^2 e_q^2 \pi N_c}{s} \frac{[t^2 + u^2 - m_q^2(2s - m_q^2)]}{s^2} . \qquad (3.92)$$

If both lepton beams had a transverse polarization, then the matrix element eqn (3.86) would acquire a non-trivial ϕ dependence.

Equation (3.92) is easily integrated to give the lowest order expression for the total cross section for $e^+e^- \to q\bar{q}$,

$$\sigma_0 = \frac{2\pi\alpha_{em}^2}{s} e_q^2 N_c \left(1 - \frac{\beta_q^{\star 2}}{3}\right)\beta_q^\star \approx \frac{4\pi\alpha_{em}^2}{3s} e_q^2 N_c \approx \frac{86.8\,\text{nb}\,\text{GeV}^2}{s} e_q^2 N_c . \qquad (3.93)$$

The approximation holds well above threshold, $\sqrt{s} \gg 2m_q$, equivalent to $\beta_q^\star \to 1$. It should be noted that the fact that the total cross section depends only on s and m_q is a result of the quarks (and leptons) having no sub-structure, that is, they are pointlike. Furthermore, the polar angle dependence in eqn (3.86) is a direct consequence of the quarks (and leptons) having spin $1/2$. Given (pseudo-)vector boson exchange the lepton and quark spins like to align at the two vertices: a positive (negative) helicity particle with a negative (positive) helicity antiparticle. Thus, we have an initial spin-1 state annihilating into a final spin-1 state aligned at an angle $\theta^\star$ to the initial state. This $\theta^\star$ dependence in eqn (3.86) is usefully rewritten as

$$\frac{d\sigma}{d\cos\theta^\star} \propto (1 + \cos\theta^\star)^2 + (1 - \cos\theta^\star)^2 + 8\frac{m_q^2}{s}\sin^2\theta^\star . \qquad (3.94)$$

The first term corresponds to the contributions with a positive (negative) helicity lepton going to a positive (negative) helicity quark; angular momentum conservation then favours $\theta^\star \to 0$ over $\theta^\star \to \pi$. Likewise, the second term corresponds to a positive (negative) helicity lepton going to a negative (positive) helicity quark, which is favoured when the quark and lepton are antiparallel. The last term corresponds to a spin zero final state involving a spin-flip, which can only occur for massive quarks. That the first two terms contribute with equal weight is a consequence of QED (and QCD) being parity conserving. The photon couples with equal strength to the left- and right-handed fermions. The weak interaction violates parity conservation and the Z couples differently to left- and right-handed fermions. This changes the balance between the first two terms in eqn (3.94) and leads to a term linear in $\cos\theta^\star$ which induces a forward–backward asymmetry. A similar effect can be obtained by polarizing one or more of the fermions.

To obtain similar results for the processes $\ell^- q \to \ell^- q$ and $q\bar{q} \to \ell^+\ell^-$, shown in Fig. 3.10, one approach is to calculate the matrix element in exactly the same manner as above. An elegant alternative is to take the previous result for $\mathcal{M}(\ell^-\ell^+ \to q\bar{q})$ and use crossing symmetry. The basic idea is to swap particles between the initial and final states. An example is given by

$$\mathcal{M}^{a\bar{b}\to c\bar{d}}(p_a, p_{\bar{b}}, p_c, p_{\bar{d}}) \simeq \mathcal{M}^{ad\to bc}(p_a, -p_{\bar{d}}, -p_{\bar{b}}, p_c)$$
$$\implies |\mathcal{M}^{a\bar{b}\to c\bar{d}}(s,t,u)|^2 = |\mathcal{M}^{ad\to bc}(t,u,s)|^2 . \tag{3.95}$$

When fermions are involved the equality of the amplitudes is modulo an unobservable phase. Since the physical regions for the Mandelstam variables in the crossed and uncrossed process do not overlap, the arguments of the second amplitude have to be analytically continued. That this is possible places powerful constraints on the allowed form of the amplitude. The only real, though minor, complication is to remember that the spin and colour averages for the initial state may need to be changed as appropriate. Using these results and neglecting masses we quickly obtain

$$\overline{\sum}|\mathcal{M}(\ell^- q \to \ell^- q)|^2 = (e^2 e_q)^2 \frac{u^2 + s^2}{t^2}$$
$$\text{and } \overline{\sum}|\mathcal{M}(q\bar{q} \to \ell^-\ell^+)|^2 = (e^2 e_q)^2 \frac{1}{N_c} \frac{t^2 + u^2}{s^2} . \tag{3.96}$$

Observe that the result for $q\bar{q} \to \ell^-\ell^+$ is essentially the same as for $\ell^-\ell^+ \to q\bar{q}$, except for the extra factor $1/N_c^2$ due to the average over the colours of the incoming quarks. In situations where a number of processes are related to one another be crossing symmetries it is common practice to only quote one matrix element (squared) and expect the reader to derive the others using eqn (3.96).

Finally, before finishing our discussion of these processes, we return to a very important property of the lepton and hadron tensors defined by eqn (3.85). Suppose that we introduce a polarization vector for the exchanged (off mass-shell) photon and take $\epsilon(Q)^\mu \propto Q^\mu$, then it is easily verified that

$$Q^\mu L_{\mu\nu} = 0 = L_{\mu\nu}Q^\nu \quad \text{and} \quad Q^\mu H_{\mu\nu} = 0 = H_{\mu\nu}Q^\nu , \tag{3.97}$$

which is the embodiment of electromagnetic gauge invariance. As a consequence of this result, had we chosen a gauge in which $\xi \neq 1$, then the extra terms in the numerator of the photon propagator, which are proportional to Q^μ, would have given zero contribution. It was sufficient to use only $-\eta_{\mu\nu}$. This is a trivial example of a more general result which states that the sum of a gauge invariant set of amplitudes cannot depend on the arbitrary gauge parameter ξ. To see why a sum of amplitudes must vanish when a photon's polarization vector is replaced by its four-momentum consider the Fourier transform of a gauge transformation,

$$A^\mu \mapsto A^\mu + e^{-1}\partial^\mu\theta \quad \Longrightarrow \quad \epsilon(k)^\mu \mapsto \epsilon(k)^\mu + e^{-1}k^\mu\theta\,. \tag{3.98}$$

Since θ is arbitrary, the scalar product $\epsilon^\mu\mathcal{M}_\mu$ is only guaranteed to be gauge invariant if we require $k^\mu\mathcal{M}_\mu = 0$. A similar, though more delicate argument holds in the non-abelian QCD. For more explanations see eqn (3.118) and the discussion in Section 3.3.3.1. Constraints such as this significantly limit the possible tensor structures of amplitudes and provide very useful checks on the intermediate stages of a calculation.

3.3.2 e^+e^- annihilation to quarks at $\mathcal{O}(\alpha_s^1)$

We now consider the leading, tree-level QCD correction to the process $e^+e^- \rightarrow q\bar{q}$ in which a gluon is radiated from either the quark or the antiquark, $e^+e^- \rightarrow q\bar{q}g$. The two Feynman diagrams are shown in Fig. 3.11.

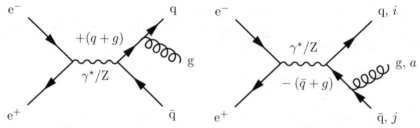

FIG. 3.11. The leading, tree level QCD corrections to $e^+e^- \rightarrow q\bar{q}$. The momentum shown at the internal quark flows in the direction of the fermion number, as indicated by the arrow.

Concentrating on the hadronic part of the amplitude, and again assuming only photon exchange, the matrix element is

$$\mathcal{M}_\mu = i\,ee_q g_s T^a_{ij} \bar{u}(q) \left[\gamma_\sigma \frac{(\slashed{q} + \slashed{g} + m_q)}{(q+g)^2 - m_q^2}\gamma_\mu + \gamma_\mu \frac{-(\slashed{\bar{q}} + \slashed{g}) + m_q}{(\bar{q}+g)^2 - m_q^2}\gamma_\sigma \right] v(\bar{q})\epsilon^*(g)^\sigma\,. \tag{3.99}$$

The minus sign in the antiquark propagator arises because the momentum is flowing in the opposite direction to the fermion number. Before proceeding to evaluate $|\mathcal{M}|^2$ it is instructive to verify electromagnetic gauge invariance. Using $Q^\mu = q^\mu + \bar{q}^\mu + g^\mu$ and neglecting the constant overall factors we have

$$\mathcal{M}_\mu Q^\mu \propto \bar{u}(q) \left[\slashed{\epsilon}^* \frac{1}{(\slashed{q} + \slashed{g} - m_q)}(\slashed{q} + \slashed{g} + \slashed{\bar{q}}) + (\slashed{q} + \slashed{g} + \slashed{\bar{q}}) \frac{1}{(-(\slashed{\bar{q}} + \slashed{g}) - m_q)}\slashed{\epsilon}^* \right] v(\bar{q})$$

$$= \bar{u}(q) \left[\slashed{\epsilon}^* \frac{1}{(\slashed{q} + \slashed{g} - m_q)}(\slashed{q} + \slashed{g} - m_q) - (m_q + \slashed{g} + \slashed{\bar{q}}) \frac{1}{(\slashed{\bar{q}} + \slashed{g} + m_q)}\slashed{\epsilon}^* \right] v(\bar{q})$$

$$= \bar{u}(q) \left[\not{\epsilon}^{*} \mathbf{1} - \mathbf{1} \not{\epsilon}^{*} \right] = 0 \, . \tag{3.100}$$

Here we wrote the propagators as inverses, using $(\not{p} - m)(\not{p} + m) = (p^2 - m^2)\mathbf{1}$, and exploited the Dirac equation

$$\begin{array}{ll} (\not{q} - m_{\mathrm{q}})u(q) = 0 & (\not{q} + m_{\mathrm{q}})v(\bar{q}) = 0 \\ \bar{u}(q)(\not{q} - m_{\mathrm{q}}) = 0 & \bar{v}(\bar{q})(\not{q} + m_{\mathrm{q}}) = 0 \, . \end{array} \tag{3.101}$$

Thus, we confirm that our expression for the amplitude is gauge invariant, provided we include the contributions from both diagrams.

At this point, we set the quark masses to zero, as they serve only to complicate our calculations. This is a good approximation for the light quarks. Since we already know the lepton tensor we only need to evaluate the hadron tensor,

$$\sum_{\mathrm{spins}} \mathcal{M}_{\mu\sigma} \mathcal{M}_{\nu\tau}^{\dagger} \epsilon^{*}(g)^{\sigma} \epsilon(g)^{\tau} = - \sum_{\mathrm{spins}} \mathcal{M}_{\mu\sigma} \mathcal{M}_{\nu}^{\dagger\sigma} \, . \tag{3.102}$$

Here, we have expressed $\mathcal{M}_{\mu}$ from eqn (3.99) as $\mathcal{M}_{\mu\sigma} \epsilon(g)^{*\sigma}$ and used

$$\sum_{\mathrm{pol}} \epsilon^{*}(g)^{\sigma} \epsilon(g)^{\tau} = -\eta^{\sigma\tau} \tag{3.103}$$

for the gluon's polarization tensor. We shall return to a consideration of this expression later. We start by considering the first diagram in Fig. 3.11, in which the gluon is emitted by the quark, which yields

$$(2q \cdot g)^2 \sum_{\mathrm{spins}} \mathcal{M}_{\mathrm{q}} \mathcal{M}_{\mathrm{q}}^{\dagger} \propto -\mathrm{Tr} \left\{ \not{q} \gamma_{\sigma}(\not{q} + \not{g}) \gamma_{\mu} \not{q} \gamma_{\nu} (\not{q} + \not{g}) \gamma^{\sigma} \right\}$$

$$= +2 \mathrm{Tr} \left\{ \not{q}(\not{q} + \not{g}) \gamma_{\mu} \not{q} \gamma_{\nu} (\not{q} + \not{g}) \right\}$$

$$= +2 \mathrm{Tr} \left\{ \not{q} \not{g} \gamma_{\mu} \not{q} \gamma_{\nu} \not{g} \right\}$$

$$= +2 \cdot 2(q \cdot g) \cdot 4 \left[\bar{q}_{\mu} g_{\nu} + g_{\mu} \bar{q}_{\nu} - (\bar{q} \cdot g) \eta_{\mu\nu} \right] . \tag{3.104}$$

Here, we used some tricks from eqn (3.83) to speed up the evaluation: $\gamma^{\sigma} \not{q} \gamma_{\sigma} = -2\not{q}$, $\not{q}\not{q} = q^2 = 0$ and $\not{q}\not{g} = 2q \cdot g - \not{g}\not{q}$ together with $g^2 = 0$. Note that scalars appearing in an expression which is a product of some γ-matrices have to be multiplied by a 4×4 unit matrix which is usually not written explicitly.

In eqn (3.104) we see that the factor $(q \cdot g)$ partly cancels the singularity from the propagator so that the contribution behaves as $(q \cdot g)^{-1}$ and not $(q \cdot g)^{-2}$ as might have been anticipated naïvely. This cancellation is typical of such calculations. In retrospect, this is not so surprising if we consider the $\mathrm{q} \to \mathrm{qg}$ branching in isolation. This can be achieved by expressing the numerator of the quark propagator, $\not{p}$, as a sum over bispinors, $u(p,s)\bar{u}(p,s)$, c.f. eqn (3.20), and picking out the term $\bar{u}(q)\not{\epsilon}^{*}(g)u(p)$. This expression has mass dimension one, and since the only relevant quantity carrying mass dimensions is the virtuality of the propagator, we have that it must be proportional to $\sqrt{2q \cdot g}$. The properties of such branchings will be studied later. They are important when the propagator becomes singular and they start to give the dominant contributions to a matrix element. These propagator singularities can be identified with two physical regions. Looking at the denominator

$$\left[(q+g)^2 - m_q^2\right] = 2q \cdot g = 2E_q E_g (1 - \beta_q \cos\theta_{qg}) , \qquad (3.105)$$

it is clear that this inverse propagator tends to zero for vanishing gluon energy, $E_g \to 0$, known as a soft singularity, and for a massless quark, $\beta_q = 1$, when the opening angle vanishes, $\theta_{qg} \to 0$, known as a collinear or mass singularity.

The evaluation of $\sum_{\text{spins}} \mathcal{M}_{\bar{q}} \mathcal{M}_{\bar{q}}^{\dagger}$ proceeds in the same manner. The result can be obtained from eqn (3.104) by exchanging q^μ and $\bar{q}^\mu$. The more cumbersome evaluation of the cross-term yields

$$(2q \cdot g)(2\bar{q} \cdot g) \sum_{\text{spins}} 2\mathcal{R}e \left\{ \mathcal{M}_q \mathcal{M}_{\bar{q}}^{\dagger} \right\}$$

$$\propto -16 \Big[(q \cdot \bar{q})(Q^2 \eta_{\mu\nu} - 2Q_\mu q_\nu - 2\bar{q}_\mu Q_\nu + 2\bar{q}_\mu q_\nu)$$

$$+ 2(\bar{q} \cdot Q) q_\mu q_\nu + 2(q \cdot Q) \bar{q}_\mu \bar{q}_\nu - Q^2 q_\mu \bar{q}_\nu \Big] . \quad (3.106)$$

Combining these results, restoring the constant factor and evaluating the sum over the final state quark colours, which is conveniently done using eqn (A.17), $\text{Tr}\{T^a T^a\} = C_F \delta_{ii} = C_F N_c$, gives the hadronic tensor

$$H_{\mu\nu} = \frac{4(ee_q)^2 g_s^2 C_F N_c}{(q \cdot g)(\bar{q} \cdot g)}$$

$$\times \Big\{ -Q^2 (q_\mu q_\nu + \bar{q}_\mu \bar{q}_\nu) - [(q \cdot Q)^2 + (\bar{q} \cdot Q)^2] \eta_{\mu\nu} \qquad (3.107)$$

$$+ (q \cdot Q)[q_\mu Q_\nu + Q_\mu q_\nu] + (\bar{q} \cdot Q)[\bar{q}_\mu Q_\nu + Q_\mu \bar{q}_\nu]$$

$$+ (q \cdot \bar{q})[Q_\mu (q - \bar{q})_\nu - (q - \bar{q})_\mu Q_\nu] + (q \cdot \bar{q} + Q^2/2)(q_\mu \bar{q}_\nu - \bar{q}_\mu q_\nu) \Big\} .$$

This can be contracted with the leptonic tensor, eqn (3.85), to yield the following expression for the matrix element squared:

$$L^{\mu\nu} H_{\mu\nu} = 8(e^2 e_q)^2 g_s^2 C_F N_c \frac{1}{Q^2} \frac{[(\ell^- \cdot q)^2 + (\ell^- \cdot \bar{q})^2 + (\ell^+ \cdot q)^2 + (\ell^+ \cdot \bar{q})^2]}{(q \cdot g)(\bar{q} \cdot g)}$$

$$(3.108)$$

Observe that only the first two terms in eqn (3.107) give non-zero contributions as the others are either proportional to Q^μ and vanish by gauge invariance or are antisymmetric under $\mu \leftrightarrow \nu$.

To obtain the differential cross section eqn (B.4) we need to include the flux factor, which is the same as for $e^+e^- \to q\bar{q}$, and the three-body phase space,

$n = 3$ in eqn (B.6). The evaluation of the phase space is best carried out in the C.o.M. frame and yields:

$$\mathrm{d}^5\Phi_3\big|_{\mathrm{CoM}} = \frac{1}{32\pi^3}\mathrm{d}E_1\mathrm{d}E_2\frac{\mathrm{d}\phi_{12}\mathrm{d}\Omega_1}{8\pi^2}$$

$$\mathrm{d}^2\Phi_3\big|_{\mathrm{CoM}} = \frac{1}{32\pi^3}\mathrm{d}E_1\mathrm{d}E_2 \tag{3.109}$$

$$= \frac{1}{128\pi^3 Q^2}\mathrm{d}m_{13}^2\mathrm{d}m_{23}^2$$

Here particles 1 and 2 can be any of the three final state particles; for the problem at hand 1=q, 2=$\bar{\mathrm{q}}$ and 3=g is the natural choice. The second expression follows after integrating over the decay plane's orientation and the third because, for example, $m_{13}^2 = Q^2 + m_2^2 - 2\sqrt{Q^2}E_2^{\mathrm{CoM}}$. Combining these results gives the fully differential cross section

$$\mathrm{d}^5\sigma = \frac{\alpha_{\mathrm{em}}^2 e_{\mathrm{q}}^2 N_c}{Q^2}\alpha_{\mathrm{s}}C_F\frac{8}{Q^2}\frac{[(\ell^- \cdot q)^2 + (\ell^- \cdot \bar{q})^2 + (\ell^+ \cdot q)^2 + (\ell^+ \cdot \bar{q})^2]}{(q \cdot g)(\bar{q} \cdot g)}$$

$$\times \mathrm{d}E_{\mathrm{q}}\mathrm{d}E_{\bar{\mathrm{q}}}\frac{\mathrm{d}\phi_{\mathrm{q}\bar{\mathrm{q}}}\mathrm{d}\Omega_{\mathrm{q}}}{8\pi^2} \ . \tag{3.110}$$

Often, in practice, we are not interested in the orientation of the decay plane. That the three particles lie in a plane, in the C.o.M. system, follows trivially from $q + \bar{q} + g = 0$. The integrations over $\phi_{\mathrm{q}\bar{\mathrm{q}}}$, ϕ_{q} and θ_{q} may be done explicitly in eqn (3.110) or done equivalently by replacing $L^{\mu\nu}$ with the orientation-averaged lepton tensor

$$\langle L^{\mu\nu}\rangle = e^2\frac{1}{3}\left(-\eta^{\mu\nu} + \frac{Q^\mu Q^\nu}{Q^2}\right)(Q^2 + 2m_\ell^2) \ . \tag{3.111}$$

Here m_ℓ^2 may be safely ignored. This tensor manifestly satisfies eqn (3.97) and may be thought of as the spin-averaged polarization tensor for the exchanged, off mass-shell, vector boson. Re-evaluating $L_{\mu\nu}H^{\mu\nu}$ then gives for the spin-averaged cross section for massless quarks

$$\mathrm{d}^2\sigma = \sigma_0\frac{\alpha_{\mathrm{s}}}{2\pi}C_F\frac{[(q \cdot Q)^2 + (\bar{q} \cdot Q)^2]}{(q \cdot g)(\bar{q} \cdot g)}\frac{4}{Q^2}\mathrm{d}E_{\mathrm{q}}\mathrm{d}E_{\bar{\mathrm{q}}}$$

$$\implies \frac{\mathrm{d}^2\sigma}{\mathrm{d}x_{\mathrm{q}}\mathrm{d}x_{\bar{\mathrm{q}}}} = \sigma_0\frac{\alpha_{\mathrm{s}}}{2\pi}C_F\frac{x_{\mathrm{q}}^2 + x_{\bar{\mathrm{q}}}^2}{(1 - x_{\bar{\mathrm{q}}})(1 - x_{\mathrm{q}})} \ . \tag{3.112}$$

In the second form we have introduced the variables x_i, equal to the energy fraction of the ith particle in the C.o.M. frame, defined by

$$x_i = 2\frac{p_i \cdot Q}{Q^2} = \frac{E_i^{\mathrm{CoM}}}{E_{\mathrm{beam}}} \ , \tag{3.113}$$

and such that $x_{\mathrm{q}} + x_{\bar{\mathrm{q}}} + x_{\mathrm{g}} = 2$. Projected onto the $x_{\mathrm{q}} - x_{\bar{\mathrm{q}}}$ plane (Fig. 3.12), a Dalitz plot, the allowed phase space lies in the upper-right triangle of the

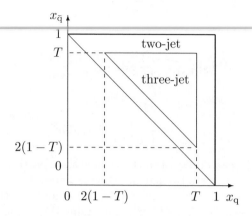

FIG. 3.12. A Dalitz plot showing the allowed region of the $x_q - x_{\bar{q}}$ plane for a $\gamma \to q\bar{q}g$ event with massless partons. The thick lines indicate the collinear singularities, $\theta_{\bar{q}g} \to 0 \Longrightarrow x_q = 1$ and $\theta_{qg} \to 0 \Longrightarrow x_{\bar{q}} = 1$, their intersection marks the position of the soft singularity, $x_g \to 0 \Longrightarrow x_q = 1 = x_{\bar{q}}$. Also shown is the boundary line separating the two- and three-jet regions based on the criterion $\max\{x_q, x_{\bar{q}}, x_g\} < T$.

$x_q = [0, 1]$ $x_{\bar{q}} = [0, 1]$ square, with lines of constant x_g running parallel to the diagonal edge, where $x_g = 1$. In eqn (3.112) we see very clearly the singularity structure for gluon radiation. If $\theta_{qg} \to 0$ then $2q \cdot g = (1 - x_{\bar{q}})Q^2 \to 0$, that is, $x_{\bar{q}} \to 1$, whilst if $E_g \to 0$ ($x_g \to 0$) then *both* x_q and $x_{\bar{q}} \to 1$.

3.3.3 $q\bar{q} \to gg$ *and the gauge invariant QCD Lagrangian*

We will now consider a pure QCD process, $q\bar{q} \to gg$, at leading order with, for convenience, massless quarks. Our approach will be to try to generalize the well understood theory of abelian QED to describe the non-abelian theory of QCD.

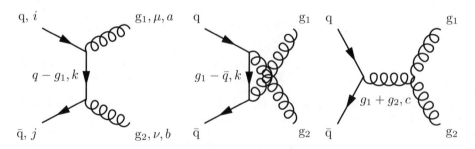

FIG. 3.13. The three leading order Feynman diagrams for $q\bar{q} \to gg$.

By direct analogy to the process $q\bar{q} \to \gamma\gamma$ we are quickly led to the first two Feynman diagrams shown in Fig. 3.13. Their matrix elements are given by

$$(\mathcal{M}_t + \mathcal{M}_u)(\epsilon^\star(g_1), \epsilon^\star(g_2))$$

$$= -i\,g_s^2 \bar{v}(\bar{q}) \left[T_{jk}^b T_{ki}^a \not{\epsilon}^\star(g_2) \frac{1}{\not{q} - \not{g}_1} \not{\epsilon}^\star(g_1) + T_{jk}^a T_{ki}^b \not{\epsilon}^\star(g_1) \frac{1}{\not{g}_1 - \not{\bar{q}}} \not{\epsilon}^\star(g_2) \right] u(q) \,.$$

$$(3.114)$$

Observe that we have included two non-commuting colour matrices at the quark–gluon vertices. This has significant consequences for gauge invariance. Replacing $\epsilon(g_2)$ by $g_2 = q + \bar{q} - g_1$, c.f. eqn (3.100), we easily find

$$(\mathcal{M}_t + \mathcal{M}_u)(\epsilon^\star(g_1), g_2) = -i\,g_s^2 (T^b T^a - T^a T^b)_{ji} \bar{v}(\bar{q}) \not{\epsilon}^\star(g_1) u(q)$$

$$= -g_s^2 f^{abc} T_{ji}^c \bar{v}(\bar{q}) \not{\epsilon}^\star(g_1) u(q) \,, \qquad (3.115)$$

so that, unlike the case of abelian QED, gauge invariance appears to be violated! The result is unaffected by including non-zero quark masses. Indeed this would be the case if there were no other diagrams contributing. However, since gluons carry colour charge, we might anticipate the existence of a triple-gluon vertex giving the third diagram in Fig. 3.13. In fact, looking at eqn (3.115), we see that the remainder has the form of a quark–gluon vertex proportional to $-i\,g_s T_{ji}^c$ times a new factor proportional to $+g_s f^{abc}$, that is, proportional to the appropriate colour matrix for an adjoint representation gluon. That the gluon should be placed in the adjoint representation makes sense from the group theoretical point of view, because when coupled to a particle in the representation R only the adjoint representation, which for SU(3) is an octet, is guaranteed to be contained within the tensor product $R \otimes \overline{R}$. Thus, only octet gluons can directly couple to particles from any other representation.

The properties of this new triple-gluon vertex $+g_s f^{abc} V_{\mu\nu\sigma}(g_1, g_2, g_3)$, with $g_1 + g_2 + g_3 = 0$, are easily established. Looking at eqn (3.115), and bearing in mind the s-channel propagator present in Fig. 3.13, it must have mass dimension one, be a Lorentz tensor and, since gluons are bosons, be totally symmetric under interchange of the labels on any pair of gluons. That is, since f^{abc} is completely antisymmetric in its indices, the same must hold for V. Thus, it must be constructed from terms of the form $\eta_{\mu\nu} g_{1\sigma}$, $\eta_{\mu\nu} g_{2\sigma}$, $\eta_{\mu\nu} g_{3\sigma}$ etc. and be antisymmetric under $\mu \leftrightarrow \nu$ and $g_1 \leftrightarrow g_2$ etc. A little experimentation shows that the only non-trivial possibility is

$$V_{\mu\nu\sigma}(g_1, g_2, g_3) = \left[\eta_{\mu\nu}(g_1 - g_2)_\sigma + \eta_{\nu\sigma}(g_2 - g_3)_\mu + \eta_{\sigma\mu}(g_3 - g_1)_\nu \right] \,, \qquad (3.116)$$

as also given by the Feynman rules in Appendix B. Taking over the form of the (Feynman gauge) photon propagator for the gluon we have

$$\mathcal{M}_s(\epsilon^\star(g_1), \epsilon^\star(g_2))$$

$$= \bar{v}(\bar{q}) \cdot -i g_s T^c_{ji} \gamma_\sigma \cdot u(q) \times \frac{-i \eta^{\sigma\tau}}{(g_1 + g_2)^2} \times + g_s f^{abc} \tag{3.117}$$

$$\times \left[\eta_{\mu\nu}(g_1 - g_2)_\tau + \eta_{\nu\tau}(g_1 + 2g_2)_\mu - \eta_{\sigma\mu}(2g_1 + g_2)_\nu \right] \cdot \epsilon^*(g_1)^\mu \epsilon^*(g_2)^\nu ,$$

and again replacing $\epsilon^*(g_2)^\nu$ by g_2^ν in this new contribution gives

$$\mathcal{M}_s\big(\epsilon^*(g_1), g_2\big) = + g_s^2 f^{abc} T^c_{ji} \bar{v}(\bar{q}) \left[\slashed{\epsilon}^*(g_1) + \slashed{g}_2 \frac{g_1 \cdot \epsilon^*(g_1)}{2 g_1 \cdot g_2} \right] u(q) . \tag{3.118}$$

Given the choice of unit numerical coefficient in eqn (3.116) this exactly restores gauge invariance, provided the first gluon is physical, that is, $g_1 \cdot \epsilon(g_1) = 0$.

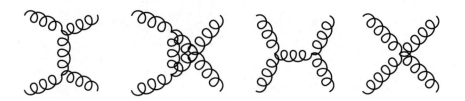

FIG. 3.14. The leading order Feynman diagrams contributing to gluon–gluon scattering

Having been led to introduce a triple-gluon coupling we ought to consider also the process gg → gg, which now can proceed via the first three diagrams in Fig. 3.14. If we test gauge invariance and replace $\epsilon^*(g_4)^\mu$ by g_4^μ, the sum of the first three diagrams in Fig. 3.14 gives the following contribution:

$$(\mathcal{M}_t + \mathcal{M}_u + \mathcal{M}_s)\big(\epsilon(g_1), \epsilon(g_2), \epsilon^*(g_3), g_4\big)$$

$$= -i g_s^2 \big[+ f^{abe} f^{cde} (\epsilon_1 \cdot \epsilon_3^* \, \epsilon_2 \cdot g_4 - \epsilon_1 \cdot g_4 \, \epsilon_2 \cdot \epsilon_3^*)$$

$$+ f^{ace} f^{dbe} (\epsilon_1 \cdot g_4 \, \epsilon_2 \cdot \epsilon_3^* - \epsilon_1 \cdot \epsilon_2 \, \epsilon_3^* \cdot g_4)$$

$$+ f^{ade} f^{bce} (\epsilon_1 \cdot \epsilon_2 \, \epsilon_3^* \cdot g_4 - \epsilon_1 \cdot \epsilon_3^* \, \epsilon_2 \cdot g_4) \big] \tag{3.119}$$

Here, we assume on mass-shell ($g_i^2 = 0$), physical ($g_i \cdot \epsilon(g_i) = 0$) external gluons and employ the Jacobi identity, eqn (A.12). Once more, we find problems with gauge invariance for this subset of diagrams which can be remedied by introducing a dimensionless, fully symmetric quartic-gluon vertex. The exact form can be read off directly from eqn (3.119) and can be found in Appendix B. This vertex has exactly the same structure as that derived from the gauge-kinetic term in the Yang–Mills (QCD) Lagrangian. What is more, these triple- and quartic-gluon vertices are sufficient to guarantee the gauge invariance of all QCD processes to all orders — no other gluon self-vertices are needed.

3.3.3.1 *Physical states and ghosts* In eqn (3.118) we demonstrated effectively the gauge invariance of the q$\bar{\text{q}}$ → gg amplitude provided that *both* gluons have physical polarizations. This raises the issue of how to treat the polarization tensor for the gluon: is the QED form, eqn (3.103), really adequate for QCD? For a vector particle of momentum k^μ its polarization vectors must satisfy

$$\epsilon(k; s) \cdot \epsilon^*(k; s') = -\delta_{ss'} \quad \text{and} \quad k \cdot \epsilon(k) = 0 . \tag{3.120}$$

The first equation imposes ortho-normality on the basis states whilst the second equation requires that they are orthogonal to the particle's direction of motion. Now, because the gluon (or photon) is massless it only has two physical polarization states whereas the second equation only imposes one constraint on the four components of a polarization vector. Thus, we require an additional constraint, which can be taken to be $n \cdot \epsilon = 0$, where n is any four-vector subject to $n \cdot k \neq 0$. As the physical polarization sum, $T_{\mu\nu}$, can be constructed only from $\eta_{\mu\nu}$, k_μ and n_ν, is required to satisfy $k^\mu T_{\mu\nu} = k^\nu T_{\mu\nu} = n^\mu T_{\mu\nu} = n^\nu T_{\mu\nu} = 0$ and have trace $T^\mu_{\ \mu} = -2$, it is straightforward to show that it must have the form

$$\sum_{\text{phys}} \epsilon(k)_\mu \epsilon^*(k)_\nu = -\eta_{\mu\nu} + \frac{(k_\mu n_\nu + n_\mu k_\nu)}{n \cdot k} - n^2 \frac{k_\mu k_\nu}{(n \cdot k)^2} ; \tag{3.121}$$

see Ex. (3-15). This is precisely the form of the numerator of the gluon propagator in a physical gauge. There is quite some freedom in the choice of n^μ which we can use to our advantage. In particular, having $n^2 = 0$ simplifies things.

Given the correct polarization tensor, eqn (3.121), you might ask why this expression is not used in QED calculations, since the arguments leading to its construction apply equally well to photons and gluons. The difference lies in the realization of gauge invariance. In QED we have

$$k^\mu \mathcal{M}_{\mu\nu\sigma...} = 0 \tag{3.122}$$

which for a physical photon is independent of the polarization of all other photons. In QCD all other gluons also had to be physical. Thus, the 'extra' terms in eqn (3.121) which subtract out the unphysical longitudinal and scalar polarizations and are proportional to k give vanishing contributions and can be safely dropped. In QED, we can use $-\eta_{\mu\nu}$ for the polarization sum. Likewise we can use $-\eta_{\mu\nu}$ in QCD, provided only a single external gluon is present. If two or more external gluons are present using $-\eta_{\mu\nu}$ will leave unwanted extra contributions. For example, using the full expression for the $q\bar{q} \to gg$ amplitude the additional contribution is given by:

$$\mathcal{M}_{\sigma\tau} \mathcal{M}^\dagger_{\sigma'\tau'} \left\{ [-\eta^{\sigma\sigma'}] \times [-\eta^{\tau\tau'}] - \left[-\eta^{\sigma\sigma'} + \frac{(g_1^\sigma n_1^{\sigma'} + n_1^\sigma g_1^{\sigma'})}{n_1 \cdot g_1} - n_1^2 \frac{g_1^\sigma g_1^{\sigma'}}{(n_1 \cdot g_1)^2} \right] \right.$$
$$\left. \times \left[-\eta^{\tau\tau'} + \frac{(g_2^\tau n_2^{\tau'} + n_2^\tau g_2^{\tau'})}{n_2 \cdot g_2} - n_2^2 \frac{g_2^\tau g_2^{\tau'}}{(n_2 \cdot g_2)^2} \right] \right\}$$

$$= \left| \mathrm{i} g_s^2 f^{abc} T_{ji}^c \frac{1}{2g_1 \cdot g_2} \bar{v}(\bar{q}) \not{g}_1 u(q) \right|^2 . \tag{3.123}$$

Here, we see that any dependence on the arbitrary four-vectors n_1 and n_2 vanishes, which is in fact required by gauge invariance. The final result also appears to be asymmetric. However, this is illusory: if we replace g_1 by $(q + \bar{q}) - g_2$, then the bracketed term vanishes and we are left with the same expression except that now $\not{g}_2$ appears, and f^{abc} becomes f^{bac}, that is, the result really is 1–2 symmetric.

The obvious way in which to avoid the contribution due to unphysical gluon polarizations, such as in eqn (3.123), is to always use eqn (3.121) for external gluons. However, this is cumbersome and an alternative is available. We retain eqn (3.103) for the polarization sum and add yet another diagram to those in Fig. 3.13 whose contribution is designed to cancel exactly the unwanted contribution. Looking at eqn (3.123) the structure is that of a $q\bar{q}g$ vertex, followed by a gluon propagator and finally a term proportional to $-g_s f^{abc} g_1^\mu$. This suggests the form of a new $\eta\bar{\eta}g$ vertex. We choose η to be a Lorentz scalar field, for simplicity, make it complex to give the vertex a 'directionality', so as to distinguish g_1 from g_2, and put it into a colour octet representation to justify the f^{abc} colour factor. Finally, we have this ghost field obey Fermi–Dirac statistics so that closed loops acquire an extra minus sign as do quarks. However, ghosts are scalars and this choice violates the spin-statistics theorem. As a consequence ghosts violate unitarity and give negative cross sections — which cancel precisely the unwanted term eqn (3.123).

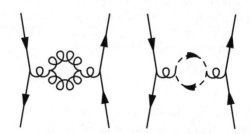

FIG. 3.15. Two of the diagrams contributing to the gluon self-energy

At this point the rôle of the unphysical ghost fields appears to be only to facilitate a trick intended to simplify the calculation of tree-level amplitudes. The situation is more subtle when loop diagrams occur, such as in Fig. 3.15. These diagrams can be viewed as either higher order contributions to the amplitude for $q\bar{q}$ scattering or, if the diagrams are 'cut' through the loop, as contributions to the amplitude squared, $\mathcal{M}$ (left-hand side) $\times \mathcal{M}^\dagger$ (right-hand side), for $q\bar{q} \to gg$. This dual interpretation can be made more formal. In the first interpretation there exists a region of the loop momentum integral (see Section 3.4 for an elaboration) in which both internal particles are real, that is, on their positive mass-shells, and generate an imaginary part in the complex amplitude. In the second interpretation this same integral corresponds to the phase-space integral appearing in the total cross section for the production of a pair of real gluons. (This non-linear relationship between S-matrix elements is in essence the optical theorem.) The significance of this equivalence is that we must restrict the internal gluons in any loop to only physical polarization states or violate unitarity. This can be achieved in two ways. Either, use a physical gauge, so called because only physical polarizations propagate for an on mass-shell gluon. Or use a covariant gauge and whenever a gluon loop occurs add the contribution from a ghost loop, remembering to include the extra minus sign, thereby ensuring the removal of any non-physical contributions.

3.3.4 The evaluation of colour factors

The evaluation of the colour factors which occur in the expressions we will encounter for matrix elements can always be found using essentially two results. For any particular sub-amplitude the colour factor consists of an ordered product of T^a and f^{abc} terms. The latter can be eliminated by applying the following identity, derived in Appendix A,

$$f^{abc} = -\frac{i}{T_F} \text{Tr} \left\{ T^a T^b T^c - T^c T^b T^a \right\}, \qquad (3.124)$$

to the fundamental representation. After applying eqn (3.124) we are left with a string of T^a matrices. Now all internal indices are already summed over and once the amplitude is squared and colour averaged so are all external indices. Thus, the matrices can be paired as $T^a_{ij} T^a_{kl} \, T^b_{pq} T^b_{rs} \, \cdots$, allowing the completeness relation to be substituted,

$$\sum_a T^a_{ij} T^a_{kl} = T_F \left[\delta_{il}\delta_{jk} - \frac{1}{N_c}\delta_{ij}\delta_{kl} \right]. \qquad (3.125)$$

Finally, by carefully contracting the colour δ-functions, a pure number, the colour factor, will remain.

We illustrate this approach using the process $q\bar{q} \to gg$. Schematically, the amplitude may be written

$$\mathcal{M} = T^a T^b \mathcal{M}_t + T^b T^a \mathcal{M}_u + i f^{abc} T^c \mathcal{M}_s$$
$$= T^a T^b (\mathcal{M}_t + \mathcal{M}_s) + T^b T^a (\mathcal{M}_u - \mathcal{M}_s). \qquad (3.126)$$

The first two terms are associated with the two orderings of the double bremsstrahlung contribution. The third term, which has been eliminated either directly using eqn (A.10) or less directly using eqn (3.124) and eqn (A.7), is associated with splitting of a radiated gluon via the triple-gluon vertex. The $T^a T^b$ and $T^b T^a$ terms in $\mathcal{M}\mathcal{M}^\dagger$, after colour averaging, each give rise to the colour factor

$$\text{Tr}\left\{ T^a T^b T^b T^a \right\} = \sum_{ab} \sum_{ijkl} T^a_{ij} T^b_{jk} T^b_{kl} T^a_{li}$$

$$= \sum_{ijkl} \sum_a T^a_{ij} T^a_{li} \times \sum_b T^b_{jk} T^b_{kl}$$

$$= \sum_{ijkl} T_F \left(\delta_{ii}\delta_{jl} - \frac{1}{N_c}\delta_{ij}\delta_{li} \right) \times T_F \left(\delta_{jl}\delta_{kk} - \frac{1}{N_c}\delta_{jk}\delta_{kl} \right)$$

$$= \sum_{jl} T_F \left(N_c - \frac{1}{N_c} \right) \delta_{jl} \times T_F \left(N_c - \frac{1}{N_c} \right) \delta_{jl}$$

$$= N_c C_F^2 \, . \tag{3.127}$$

After rather more work the cross-term can be evaluated similarly to obtain the colour factor

$$\text{Tr} \left\{ T^a T^b T^a T^b \right\} = N_c C_F \left(C_F - \frac{C_A}{2} \right) = -N_c C_F \frac{1}{2N_c} \, . \tag{3.128}$$

From these two results we can deduce that the colour factor associated with the $|\mathcal{M}_3|^2$ term is given by $N_c C_F C_A$. Thus, we learn that $|\mathcal{M}|^2$ is proportional to the number of quark colours, N_c; the radiation of a gluon off a(n anti)quark is proportional to C_F; whilst radiation off a gluon is proportional to C_A. Also, we see a generic behaviour, the interference between two colour flows, here $T^a T^b$ and $T^b T^a$, is suppressed by a typical factor N_c^{-2} compared to the direct contributions.

Again, in practice, a number of tricks may be applicable in special cases to speed up colour factor calculations. In the case of eqn (3.127) we could use eqn (A.17), $\sum_{aj} T^a_{ij} T^a_{jk} = C_F \delta_{ik}$, to immediately obtain the result. For eqn (3.128) one can substitute $T^a T^b = (T^b T^a + i f^{abc} T^c)$. The first term has just been evaluated. For the second term one can write

$$i f^{abc} \text{Tr} \left\{ T^c T^b T^a \right\} = i f^{abc} \text{Tr} \left\{ T^c T^b T^a - T^c T^a T^b \right\} / 2$$

$$= T_F f^{abc} f^{abc} / 2 \tag{3.129}$$

$$= T_F C_A (N_c^2 - 1) / 2 \, ,$$

where we used the antisymmetry of f^{abc}, eqn (3.124), and eqn (A.17). These two terms when combined give the quoted result. However, the approach discussed first is guaranteed to work in all cases and is ideally suited to being done by computer algebra techniques, thereby reducing the risk of error and the tedium.

3.4   Ultraviolet divergences and renormalization

When a Feynman diagram involves a closed loop, the momentum conserving δ-functions at the vertices prove insufficient to specify fully the momentum of the particles in the loop — an integral over a loop momentum remains. For example, in a self-energy diagram an integral of the following form may be encountered:

$$\int d^4k \frac{1}{[(k+p)^2 - m^2]\,k^2} \sim \int d\Omega_3 \int dk k^3 \frac{1}{k^2 \cdot k^2} \sim \int^\Lambda \frac{dk}{k} \sim \ln \Lambda \ . \quad (3.130)$$

Truncating the integral by an ultraviolet cut-off, Λ, we see that the integral is logarithmically divergent. Renormalization is the treatment of such divergences which are associated with the high frequency — short distance components of the fields. In essence, the procedure first involves using a regulator to artificially render all integrals finite, so that they can be safely manipulated. Second, new terms are introduced into the theory in such a way that all the divergences cancel between the contributions from the bare Lagrangian and the new counterterms when the regulator is removed. Significantly, these additional terms must have exactly the same structure as the original Lagrangian. Finally, the finite matrix elements can be compared to experimentally measured numbers and, after fixing any free parameters, higher order predictions can be made. We will demonstrate this procedure at one-loop. More details can be found in many of the standard field theory texts, such as (Ramond, 1990) or (Peskin and Schroeder, 1995).

Unfortunately, the crude cut-off used to make eqn (3.130) finite is not compatible with either Lorentz or gauge invariance. Technically speaking, the Ward identities are violated. However, it is easy to see that if the k-integral involves fewer dimensions, then the integral again becomes finite but importantly now no longer violates the two symmetries. Dimensional regularization is the preferred choice in pQCD calculations. The method of dimensional regularization and the standard procedures for dealing with (one) loop integrals are described in Appendix C.

3.4.1 Self-energy and vertex corrections

Armed with the knowledge of how to use dimensional regularization to calculate one-loop amplitudes, we now sketch the results for the quark and gluon self-energies and quark–gluon and triple-gluon vertices. These are the core corrections to the propagators and couplings in which all the external propagators are 'amputated'. We shall work in a covariant gauge with arbitrary gauge parameter $\xi \neq 1$. This will be followed by a discussion of how renormalization handles the divergences which we isolate.

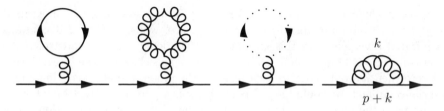

FIG. 3.16. The three tadpole-type diagrams and the quark–gluon loop diagram which contribute to the quark self-energy at leading non-trivial order, $\mathcal{O}(\alpha_s)$

At $\mathcal{O}(\alpha_s)$ four diagrams, shown in Fig. 3.16, appear to contribute to the quark self-energy. However, the three 'tadpole' diagrams give zero contributions. This is seen most easily by considering the diagram's colour factors, $\text{Tr}\{T^a\} = 0$ and $f_{abb} = 0$, for the quark and both the gluon and ghost tadpoles, respectively. This leaves only the fourth diagram, which apart from a group theoretical factor, is the same as in the abelian theory QED. The one-loop integral yields

$$-i\,\Sigma(p) = \int \frac{d^D k}{(2\pi)^D} \left\{ -i\,g_s\mu^\epsilon T^a_{ik}\gamma_\mu \times i\,\frac{(\not{k}+\not{p}+m)}{[(k+p)^2-m^2]} \times -i\,g_s\mu^\epsilon T^a_{kj}\gamma_\nu \right.$$
$$\left. \times \frac{i}{k^2}\left[-\eta^{\mu\nu} + (1-\xi)\frac{k^\mu k^\nu}{k^2}\right] \right\} \tag{3.131}$$
$$= -(g_s\mu^\epsilon)^2 C_F\delta_{ij} \int \frac{d^D k}{(2\pi)^D} \frac{\gamma_\mu(\not{k}+\not{p}+m)\gamma_\nu}{[(k+p)^2-m^2]k^2}\left[\eta^{\mu\nu} - (1-\xi)\frac{k^\mu k^\nu}{k^2}\right].$$

This expression should be familiar from the discussion in Appendix C and indeed coincides with eqn (C.1) for $\xi = 1$. The full result is

$$-i\,\Sigma(p) = i\,\frac{\alpha_s}{4\pi}C_F\delta_{ij}\left\{ \Delta_\epsilon\left[\xi\not{p} - (3+\xi)m\right] \right.$$
$$- \not{p}\left[1 + 2\int_0^1 d\alpha\left\{2\alpha\xi\ln\left(\frac{A(\alpha)}{\mu^2}\right) + (1-\xi)p^2\frac{\alpha^2(1-\alpha)}{A(\alpha)}\right\}\right]$$
$$\left. + m\left[2 + (3+\xi)\int_0^1 d\alpha\ln\left(\frac{A(\alpha)}{\mu^2}\right)\right]\right\}, \tag{3.132}$$

with $A(\alpha) = \alpha m^2 - \alpha(1-\alpha)p^2$.

The calculation of the gluon self-energy proceeds in a similar manner. At $\mathcal{O}(\alpha_s)$ the relevant Feynman diagrams are shown in Fig. 3.17. Again the four tadpole diagrams do not contribute. The first three diagrams are the same as for the quark propagator and vanish for the same reasons, the fourth diagram is more interesting as its colour factor is non-vanishing. In dimensional regularization it is proportional to a momentum integral which possesses no intrinsic scale,

$$\int \frac{d^D k}{(2\pi)^D}\frac{1}{k^2}. \tag{3.133}$$

As a result there is no value of D for which we can evaluate the integral. If $D \geq 2$ then it contains an ultraviolet divergence, whilst if $D \leq 2$ it contains an infrared divergence. In this circumstance we define such an integral to be identically zero (Collins, 1986). To help motivate this choice, if the integral were not zero then it would contribute a mass to the gluon, thereby violating gauge invariance. In fact, we could have also applied this argument to the loop integrals arising from the gluon and ghost tadpoles. The quark loop contribution, for a single quark flavour, is given by

$$-i\,\Pi(p)^{\mu\nu}_{q\bar{q}} = -\,(g_s\mu^\epsilon)^2\text{Tr}\{T^a T^b\}\int \frac{d^D k}{(2\pi)^D}\frac{\text{Tr}\{\gamma^\mu(\not{k}+\not{p}+m)\gamma^\nu(\not{k}+m)\}}{[(k+p)^2-m^2][k^2-m^2]}$$

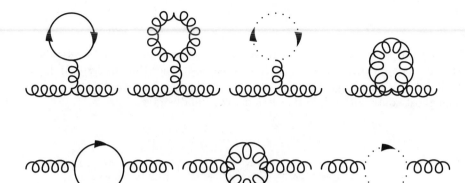

FIG. 3.17. The four tadpole-type diagrams and the quark, gluon and (in a covariant gauge) its associated ghost loop diagrams, which contribute to the gluon self-energy at leading non-trivial order, $\mathcal{O}(\alpha_s)$

$$= -\,\mathrm{i}\,\frac{\alpha_s}{4\pi}T_F\delta^{ab}\frac{4}{3}(p^2\eta^{\mu\nu}-p^\mu p^\nu)$$

$$\times\left[\Delta_\epsilon - 6\int_0^1 \mathrm{d}\alpha\,\alpha(1-\alpha)\ln\left(\frac{m^2-\alpha(1-\alpha)p^2}{\mu^2}\right)\right]. \qquad (3.134)$$

The only subtlety in this contribution to the gluon self-energy is the overall minus sign for the closed quark loop, otherwise the evaluation follows the earlier pattern and, if anything, is more straightforward. Modulo the coupling and colour factor the result is the familiar one from QED. In the non-abelian QCD, the new contributions come from the gluon and ghost loops. The gluon loop is given by

$$-\mathrm{i}\,\Pi(p)^{\mu\nu}_{\mathrm{gg}} = (g_s\mu^\epsilon)^2 f_{acd}f_{bcd}\frac{1}{2}\int\frac{\mathrm{d}^D k}{(2\pi)^D}$$

$$\times\left[\eta^\mu{}_\alpha(k+2p)_\beta - \eta_{\alpha\beta}(2k+p)^\mu + \eta_\beta{}^\mu(k-p)_\alpha\right]$$

$$\times\frac{1}{(k+p)^2}\left(\eta^{\alpha\gamma} - (1-\xi)\frac{(k+p)^\alpha(k+p)^\gamma}{(k+p)^2}\right)$$

$$\times\left[\eta^\nu{}_\delta(k-p)_\gamma - \eta_{\delta\gamma}(2k+p)^\nu + \eta_\gamma{}^\nu(k+2p)_\delta\right]$$

$$\times\frac{1}{k^2}\left(\eta^{\beta\delta} - (1-\xi)\frac{k^\beta k^\delta}{k^2}\right)$$

$$= \frac{\alpha_s}{4\pi}C_A\delta^{ab}\frac{1}{12}\left\{(19p^2\eta^{\mu\nu}-22p^\mu p^\nu)\left[\Delta_\epsilon - \ln\left(\frac{-p^2}{\mu^2}\right)\right]\right.$$

$$+\frac{116}{3}p^2\eta^{\mu\nu} - \frac{134}{3}p^\mu p^\nu$$

$$\left.+6(1-\xi)\left(\left[\Delta_\epsilon - \ln\left(\frac{-p^2}{\mu^2}\right)\right]-2\right)(p^2\eta^{\mu\nu}-p^\mu p^\nu)\right.$$

$$+ 3(1 - \xi)^2 (p^2 \eta^{\mu\nu} - p^\mu p^\nu) \bigg\} . \tag{3.135}$$

The overall factor of one half is to account for the diagram's symmetry factor. Despite its appearance the calculation is tedious rather than difficult and, given their relative simplicity, we have carried out the Feynman parameter integrals. The ghost contribution is rather less involved and is given by

$$-\mathrm{i}\,\Pi(p)^{\mu\nu}_{\eta\eta^\dagger} = -(g_s\mu^\epsilon)^2 f_{acd} f_{bcd} \int \frac{\mathrm{d}^D k}{(2\pi)^D} \frac{(k+p)^\mu k^\nu}{(k+p)^2 k^2}$$

$$= \mathrm{i}\frac{\alpha_s}{4\pi} C_A \delta^{ab} \frac{1}{12} \bigg\{ (p^2\eta^{\mu\nu} + 2p^\mu p^\nu) \bigg[\Delta_\epsilon - \ln\bigg(\frac{-p^2}{\mu^2}\bigg) \bigg]$$

$$+ \frac{8}{3} p^2 \eta^{\mu\nu} + \frac{5}{18} p^\mu p^\nu \bigg\} . \tag{3.136}$$

Note that there is an overall minus sign due to the fermionic nature of the ghost fields.

Referring back to the quark loop contribution to the gluon self-energy, given by eqn (3.134), we see that it has a transverse tensor structure, $p_\mu \Pi^{\mu\nu}_{q\bar{q}}(p) = 0 = p_\nu \Pi^{\mu\nu}_{q\bar{q}}(p)$, equivalent to $\Pi^{\mu\nu}_{q\bar{q}} \propto (p^2\eta^{\mu\nu} - p^\mu p^\nu)$. This structure is not shown by either the pure gluon, eqn (3.135), or ghost, eqn (3.136), contributions separately. However, they naturally form a set which when added together gives

$$- \mathrm{i}\,\Pi(p)^{\mu\nu}_{\text{gauge}} \equiv -\mathrm{i}\left(\Pi^{\mu\nu}_{gg} + \Pi^{\mu\nu}_{\eta\eta^\dagger} \right) \tag{3.137}$$

$$= \mathrm{i}\frac{\alpha_s}{4\pi} C_A \delta^{ab} \bigg\{ \frac{(13 - 3\xi)}{6} \bigg[\Delta_\epsilon - \ln\bigg(\frac{-p^2}{\mu^2}\bigg) \bigg] + \frac{115}{36} + \frac{1}{4}\xi^2 \bigg\} \times (p^2\eta^{\mu\nu} - p^\mu p^\nu) .$$

As we shall see, this is an important consequence of gauge invariance. A result all the more surprising since our Lagrangian contains an explicit gauge breaking term and the 'tree-level' propagator is not, in general, transverse.

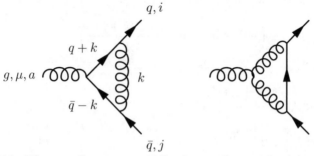

FIG. 3.18. The two diagrams contributing to the quark–gluon vertex at $\mathcal{O}(\alpha_s)$

The calculation of one-loop vertex corrections is more involved and results in rather complicated expressions which we choose not to give in full; details can

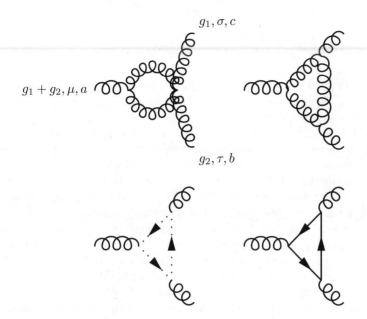

g_1, σ, c

$g_1 + g_2, \mu, a$

g_2, τ, b

FIG. 3.19. The four basic diagrams contributing to the triple-gluon vertex at $\mathcal{O}(\alpha_{\mathrm{s}})$. Note that two further permutations of the first diagram exist.

be found in the literature. Instead we only show, the relatively straightforward, divergent parts of the vertex corrections. Figure 3.18 shows the two diagrams which contribute to the quark–gluon vertex at $\mathcal{O}(\alpha_{\mathrm{s}})$. The first is essentially the same as in QED, whilst the second only arises in a non-abelian theory such as QCD. The result for the ultraviolet divergent part of the vertex correction is

$$\Gamma^a_{ij}(g, q)^\mu = -i\, g_s \mu^\epsilon T^a_{ji} \gamma^\mu \frac{\alpha_{\mathrm{s}}}{4\pi} \left\{ \left[\xi \left(C_F - \frac{C_A}{2} \right) + \frac{3(1+\xi)}{4} C_A \right] \Delta_\epsilon + \begin{array}{c} \mathrm{U.V.} \\ \mathrm{finite} \end{array} \right\} .$$

(3.138)

The first term comes from the 'QED-like' graph, proportional to $T^b T^a T^b$, whilst the second comes from the non-abelian graph, proportional to $T^b f_{abc} T^c$. Figure 3.19 shows the four basic diagrams which contribute to the triple-gluon vertex at $\mathcal{O}(\alpha_{\mathrm{s}})$. The result for the ultraviolet divergent part of the vertex is given by

$$\Gamma_{abc}(g_1, g_2)^{\mu\sigma\tau} = g_s \mu^\epsilon f_{abc} \left[\eta^{\sigma\tau}(g_1 - g_2)^\mu + \eta^{\tau\mu}(g_1 + 2g_2)^\sigma - \eta^{\mu\tau}(2g_1 + g_2)^\sigma \right]$$
$$\times \left\{ \frac{\alpha_{\mathrm{s}}}{4\pi} \left[\frac{4}{3} T_F n_f - \frac{(17 - 9\xi)}{12} C_A \right] \Delta_\epsilon + \begin{array}{c} \mathrm{U.V.} \\ \mathrm{finite} \end{array} \right\} .$$

(3.139)

The first term is given by the n_f fermions which contribute to the quark loop. In QED, this term would vanish according to Furry's theorem (Furry, 1937), which states that any quark loop joined to an odd number of photons vanishes. This is essentially because QED's charge conjugation symmetry guarantees that the contributions from a quark and an antiquark going round the loop in the opposite direction exactly cancel. Furry's theorem does not ensure the vanishing of this diagram because in QCD the order of the colour factors is important: $\text{Tr}\left\{T^a T^b T^c\right\} \neq \text{Tr}\left\{T^c T^b T^a\right\}$ so that the quark and antiquark contributions do not cancel. The second term in eqn (3.139) arises from the pure gauge field (gluon and ghost) diagrams.

As a result of these and other somewhat involved calculations we are able to write the one-loop corrections in terms of a divergent (in the $\epsilon \to 0$ limit) and a finite piece. Significantly, the coefficients of the divergent pieces follow the pattern established by the tree-level terms already present in the Lagrangian density — no new divergent interactions occur. This will prove to be of crucial importance in our approach to renormalization. It is also a very non-trivial statement. For example, a term of the form $\ln(p^2/\mu^2)/\epsilon$ appearing in a correction would certainly invalidate this statement. Fortunately, a theorem due to S. Weinberg states that the divergences may only be proportional to a polynomial in the external momenta and particle masses, of the order of the diagram's degree of divergence (Weinberg, 1960; 'tHooft, 1973; Weinberg, 1973b).

Before proceeding to consider the renormalization of the theory it is useful to collect our results so far. Beginning with the quark propagator, the sum of the tree and one-loop contributions may be written heuristically as

$$S_F + S_F(-i\Sigma)S_F + \mathcal{O}(\alpha_s^2)$$
$$= S_F + S_F(-i\Sigma)S_F + S_F(-i\Sigma)S_F(-i\Sigma)S_F + \cdots$$
$$= \left(S_F^{-1} + i\Sigma\right)^{-1}. \tag{3.140}$$

Here $S_F^{-1} = -i(\not{p}-m)$ is the inverse of the quark propagator and Σ is the quark self-energy. This should be calculated from one-particle irreducible diagrams, that is diagrams which cannot be split in two by cutting a single propagator, so as to avoid double counting in the series. Now the interpretation of the first expression is somewhat confused since its second term contains a double pole, $1/(p^2-m^2)^2$, due to the two quark propagators. In order to obtain an expression containing only a simple pole we follow Dyson and sum the infinite set of terms given in the second line. The expression in the third line gives the sum of this geometric series. Even though S_F is a 4×4 spinor matrix (and S_B, below, is a two-index Lorentz tensor), the usual trick of considering the difference of the geometric series and $S_F(-i\Sigma)$ times the series allows the sum to be calculated. The only proviso being that a little care is taken over the order of the non-commuting terms. Using the result eqn (3.132), the explicit expression for the summed (inverse) propagator is

$$i\left\{\not{p}\left(1+\frac{\alpha_s}{4\pi}C_F\left[\xi\Delta_\epsilon + \begin{smallmatrix}\text{U.V.}\\\text{finite}\end{smallmatrix}\right]\right) - m\left(1+\frac{\alpha_s}{4\pi}C_F\left[(3+\xi)\Delta_\epsilon + \begin{smallmatrix}\text{U.V.}\\\text{finite}\end{smallmatrix}\right]\right) + \mathcal{O}(\alpha_s^2)\right\}^{-1}. \tag{3.141}$$

Similar considerations apply to the gluon propagator which can be summed in the same fashion to give

$$S_B + S_B(-i\,\Pi)S_B + S_B(-i\,\Pi)S_B(-i\,\Pi)S_B + \cdots$$
$$= \left[(S_B^{-1} + i\,\Pi)^{-1}\right]^{\mu\nu}$$
$$= -i\left[(1 + \Pi_0)(p^2\eta^{\mu\nu} - p^\mu p^\nu) + \xi^{-1}p^\mu p^\nu\right]^{-1}$$
$$= \frac{i}{p^2}\left[\frac{1}{(1 + \Pi_0)}\left(-\eta^{\mu\nu} + \frac{p^\mu p^\nu}{p^2}\right) - \xi\frac{p^\mu p^\nu}{p^2}\right]. \tag{3.142}$$

Here S_B^{-1} is the inverse of the gluon propagator, as found in eqn (3.17), and $\Pi^{\mu\nu} = \Pi_0(p^2\eta^{\mu\nu} - p^\mu p^\nu)$ is the gluon self-energy. We do not show the diagonal colour matrix. Observe that the 'gauge fixing', longitudinal, term is unaffected by the corrections which remain orthogonal to it to all orders. Using the results of eqns (3.134) and (3.137) we have the explicit expression

$$1 + \Pi_0(p^2) = 1 + \frac{\alpha_s}{4\pi}\left(\left[\frac{4}{3}T_F n_f - \frac{(13 - 3\xi)}{6}C_A\right]\Delta_\epsilon + \frac{\text{U.V.}}{\text{finite}}\right) + \mathcal{O}(\alpha_s^2). \tag{3.143}$$

Here, we allow for n_f flavours of quarks in the loop. The vertex corrections are simpler to treat. The sum of the tree- and one-loop level contributions to the quark–gluon vertex is

$$-i\,g_s\mu^\epsilon T_{ji}^a\gamma^\mu\left\{1 + \frac{\alpha_s}{4\pi}\left(\left[\xi C_F + \frac{(3 + \xi)}{4}C_A\right]\Delta_\epsilon + \frac{\text{U.V.}}{\text{finite}}\right) + \mathcal{O}(\alpha_s^2)\right\}, \tag{3.144}$$

whilst the similar sum for the triple-gluon vertex is

$$i\,g_s\mu^\epsilon f_{abc}\left[\eta^{\sigma\tau}(g_1 - g_2)^\mu + \eta^{\tau\mu}(g_1 + 2g_2)^\sigma - \eta^{\mu\tau}(2g_1 + g_2)^\sigma\right] \tag{3.145}$$
$$\times\left\{1 + \frac{\alpha_s}{4\pi}\left(\left[-\frac{4}{3}T_F n_f + \frac{(17 - 9\xi)}{12}C_A\right]\Delta_\epsilon + \frac{\text{U.V.}}{\text{finite}}\right) + \mathcal{O}(\alpha_s^2)\right\}.$$

In the next section we discuss how to obtain physically meaningful results from these divergent expressions.

3.4.2 Renormalization

The basic analysis that led to eqn (3.132) can be applied to all the loop integrals encountered in pQCD. The loop integrals are first calculated in D dimensions, using analytic continuation, and then expanded in $\epsilon = (4 - D)/2$. The ultraviolet divergences manifest themselves as poles in $1/\epsilon$ (Speer, 1974; Breitenlohner and Maison, 1977). An N-loop amplitude has the Laurent expansion

$$I_N = \sum_{n=0}^{N}\left(\frac{\alpha_s}{4\pi}\right)^n\frac{C_n}{\epsilon^n} + \mathcal{O}(\epsilon). \tag{3.146}$$

The coefficients $\{C_n\}$ depend on combinations of the external momenta, typically arising via integrals over Feynman parameters, and an arbitrary mass μ. Clearly, the physical limit of an expression such as eqn (3.146) is not well defined. To make progress we must find a method of removing the $1/\epsilon$ poles, thus allowing the $D \to 4$ $(\epsilon \to 0)$ limit to be taken: this is renormalization.

The procedure we adopt is a pragmatic one. Using the original Lagrangian's Feynman rules, the divergent diagrams are identified and evaluated. Then knowing which interactions contain divergences, supplementary Feynman rules are added to the theory, one for each divergent interaction. These have coefficients that are carefully chosen so that they completely cancel the divergent $1/\epsilon$ poles generated by the original terms. In essence, the Lagrangian is supplemented by a counterterm Lagrangian which, treated as a perturbation, generates the new interactions necessary to render the theory finite. Schematically, this can be written as

$$\mathcal{L}_{\text{renorm}} = \mathcal{L} + \mathcal{L}_{\text{counter}}^{(1)} + \mathcal{L}_{\text{counter}}^{(2)} + \mathcal{L}_{\text{counter}}^{(3)} + \cdots$$
$$= \mathcal{L} + \mathcal{L}_{\text{counter}} . \tag{3.147}$$

The $\mathcal{O}(\alpha_s)$ term, $\mathcal{L}_{\text{counter}}^{(1)}$, is constructed to cancel the one-loop divergences generated by the original Lagrangian, $\mathcal{L}$. The $\mathcal{O}(\alpha_s^2)$ term, $\mathcal{L}_{\text{counter}}^{(2)}$, is constructed to cancel the 'two-loop' divergences generated by $\mathcal{L} + \mathcal{L}_{\text{counter}}^{(1)}$ etc.

Referring back to eqn (3.132) we see that two divergences occur in the quark propagator. One is associated with the quark's kinetic term, $\propto \not{p} \leftrightarrow \bar{\psi} i \not{\partial} \psi$, and one with its mass term, $\propto m \leftrightarrow m \bar{\psi} \psi$. Equations (3.134) and (3.137) show that another divergence is associated with the transverse part of the gluon's kinetic term, $\propto (p^2 \eta^{\mu\nu} - p^\mu p^\nu) \leftrightarrow (\partial^\mu A^{a\nu} - \partial^\nu A^{a\mu}) \partial_\mu A_\nu^a$; the longitudinal part of the gluon's kinetic energy term remains finite. Equation (3.138) contains a divergence of the same form as the quark–gluon vertex, $\propto T_{jk}^a \gamma^\mu \leftrightarrow T_{jk}^a \bar{\psi}_j A^{a\mu} \psi_k$, whilst eqn (3.139) contains a divergence of the same structure as the triple-gluon vertex, $\propto f_{abc}[\eta^{\sigma\tau}(g_1 - g_2)^\mu + \eta^{\tau\mu}(g_1 + 2g_2)^\sigma - \eta^{\mu\tau}(2g_1 + g_2)^\sigma] \leftrightarrow f_{abc}(\partial_\mu A_\nu^a) A^{b\mu} A^{c\nu}$. Proceeding in this way we find that the form for the $\mathcal{O}(\alpha_s)$ counterterm Lagrangian, in a general covariant gauge, is given by

$$\mathcal{L}_{\text{counter}}^{(1)} = \delta Z_\psi^{(1)} \, \bar{\psi}_i i \not{\partial} \psi_i - \delta Z_{m\bar{\psi}\psi}^{(1)} \, m \bar{\psi}_i \psi_i - \delta Z_A^{(1)} \frac{1}{2} (\partial_\mu A_\nu^a - \partial_\nu A_\mu^a) \partial^\mu A^{a\nu}$$

$$+ \delta Z_\eta^{(1)} \, (\partial^\mu \eta^{a\dagger})(\partial_\mu \eta^a) \tag{3.148}$$

$$- \delta Z_{A\bar{\psi}\psi}^{(1)} \, g_s \mu^\epsilon T_{jk}^a \bar{\psi}_j \not{A}^a \psi_k + \delta Z_{\eta^\dagger \eta A} \, g_s \mu^\epsilon f_{abc} (\partial_\mu \eta^{a\dagger}) \eta^b A^{c\mu}$$

$$+ \delta Z_{A^3}^{(1)} \, g_s \mu^\epsilon f_{abc} (\partial_\mu A_\nu^a) A^{b\mu} A^{c\nu} - \delta Z_{A^4}^{(1)} \, g_s^2 \mu^{2\epsilon} \frac{1}{4} f_{abc} f_{ade} A^{b\mu} A^{c\nu} A_\mu^d A_\nu^e .$$

Here, we have also added terms to cancel divergences which arise in the corrections to the ghost propagator, the quartic-gluon vertex and the ghost–gluon vertex. We have chosen not to include a term proportional to $(\partial^\mu A_\mu^a)^2$, as the longitudinal part of the gluon propagator receives no corrections.

As indicated earlier, in eqn (3.148) each term should be regarded as a perturbation to which new Feynman rules can be associated. The coefficients $\delta Z^{(1)}$ are chosen so that the sum of their individual contributions plus those of the corresponding one-loop contributions from the original Lagrangian are finite. Using the results from our earlier calculations, Section 3.4.1, and referring to the relationship between the quantum Lagrangian, eqn (3.21), and the Feynman rules derived from it, Appendix B, we can infer the values of six of the $\delta Z^{(1)}$ as

$$\delta Z_\psi^{(1)} = \frac{\alpha_s}{4\pi}\left[-\xi C_F \Delta_\epsilon + F_\psi\right]$$

$$\delta Z_{m\bar{\psi}\psi}^{(1)} = \frac{\alpha_s}{4\pi}\left[-(3+\xi)C_F \Delta_\epsilon + F_{m\bar{\psi}\psi}\right]$$

$$\delta Z_A^{(1)} = \frac{\alpha_s}{4\pi}\left[-\left(\frac{4}{3}T_F n_f - \frac{(13-3\xi)}{6}C_A\right)\Delta_\epsilon + F_A\right]$$

$$\delta Z_{A\bar{\psi}\psi}^{(1)} = \frac{\alpha_s}{4\pi}\left[-\left(\xi C_F + \frac{(3+\xi)}{4}C_A\right)\Delta_\epsilon + F_{A\bar{\psi}\psi}\right]$$

$$\delta Z_{A^3}^{(1)} = \frac{\alpha_s}{4\pi}\left[-\left(\frac{4}{3}T_F n_f - \frac{(17-9\xi)}{12}C_A\right)\Delta_\epsilon + F_{A^3}\right]. \quad (3.149)$$

The finite functions, F_ψ, $F_{m\bar{\psi}\psi}$, F_A, $F_{A\bar{\psi}\psi}$, F_{A^3} are arbitrary. Only the coefficients of the $1/\epsilon$ poles are prescribed. We shall return to the issue of how to choose the F_is later when we discuss renormalization schemes.

In eqn (3.148) we have the beginnings of a remarkable result which saves our approach from being merely *ad hoc*. Comparing eqn (3.21) and eqn (3.148) one is immediately struck by their similarity; a fact which has been proven to hold true to all orders in perturbation theory. We can make this similarity more manifest by rescaling, or if you will renormalizing, the fields, masses and couplings. To do this, introduce

$$Z = 1 + \delta Z \quad \text{where} \quad \delta Z = \delta Z^{(1)} + \delta Z^{(2)} + \delta Z^{(3)} + \cdots \quad (3.150)$$

and

$$\begin{aligned}
\psi_0 &= Z_\psi^{1/2}\psi & m_0 &= Z_{m\bar{\psi}\psi}Z_\psi^{-1}m \\
A_0^\mu &= Z_A^{1/2}A^\mu & &\equiv Z_m m \\
\eta_0 &= Z_\eta^{1/2}\eta & \xi_0 &= Z_A\xi \quad (Z_\xi = Z_A).
\end{aligned} \quad (3.151)$$

This allows us to write the sum of eqns (3.148) and (3.21), with $g_s \to g_s\mu^\epsilon$ as appropriate for $D = 4 - 2\epsilon$ dimensions, as

$$\mathcal{L}_{\text{renorm}} = \bar{\psi}_0(i\slashed{\partial} - m_0)\psi_0 - \frac{1}{2}\left[(\partial_\mu A_{0\nu}^a - \partial_\nu A_{0\mu}^a)\partial^\mu A_0^{a\nu} + \frac{1}{\xi_0}(\partial^\mu A_{0\mu}^a)^2\right]$$

$$+(\partial^\mu\eta_0^a)(\partial_\mu\eta_0^a) \quad (3.152)$$

$$-g_s\mu^\epsilon\frac{Z_{A\bar{\psi}\psi}}{Z_A^{1/2}Z_\psi}T_{jk}^a\bar{\psi}_{0j}\slashed{A}_0^a\psi_{0k} + g_s\mu^\epsilon\frac{Z_{A\eta^\dagger\eta}}{Z_A^{1/2}Z_\eta}f_{abc}(\partial_\mu\eta_0^{a\dagger})\eta_0^b A_0^{c\mu}$$

$$+g_s\mu^\epsilon\frac{Z_{A^3}}{Z_A^{3/2}}f_{abc}(\partial_\mu A_{0\nu}^a)A_0^{b\mu}A_0^{c\nu} - g_s^2\mu^{2\epsilon}\frac{Z_{A^4}}{Z_A^2}\frac{1}{4}f_{abc}f_{ade}A_0^{b\mu}A_0^{c\nu}A_{0\mu}^d A_{0\nu}^e.$$

In a renormalizable theory all the ultraviolet divergences arising in loop diagrams can be cancelled by counterterms corresponding to the finite number of

interactions of mass dimension four or less. By contrast, in a non-renormalizable
theory counterterms corresponding to new interactions must be added at each
new order in perturbation theory.

Unfortunately, looking at eqn (3.152), we appear to have lost gauge invariance
which, as we learnt in Section 3.3.3 for tree-level calculations, requires very particular relationships to hold between the various terms in the Lagrangian. This
is potentially a calamitous situation because the formal proof of renormalizability relies on the gauge symmetry to guarantee certain relationships amongst the
theory's Green's functions. Now, this apparent loss of gauge invariance already
appeared at the classical level because of the necessity to introduce a gauge fixing
(and ghost) term. Fortunately, Becchi, Rouet and Stora (BRS) (1974) have found
a rather unusual, but nonetheless exact, symmetry of the 'broken Lagrangian'.
Thus, a form of gauge invariance can be restored with the proviso that all the
gauge couplings are equal. This symmetry then allows a number of relationships,
known as Slavnov–Taylor (Taylor, 1971; Slavnov, 1972) identities (or Ward identities in the abelian QED case), to be established between the Green's functions
of the Yang–Mills theory. A consequence of these relationships is the requirement
for the following equations to hold:

$$\frac{Z_{A\bar{\psi}\psi}}{Z_A^{1/2} Z_\psi} = \frac{Z_{A\eta^\dagger\eta}}{Z_A^{1/2} Z_\eta} = \frac{Z_{A^3}}{Z_A^{3/2}} = \sqrt{\frac{Z_{A^4}}{Z_A^2}} \equiv Z_g = \frac{g_{s0}}{g_s \mu^\epsilon}. \tag{3.153}$$

This allows us to introduce a single renormalization factor Z_g and a unique gauge
coupling g_{s0} in the quark–gluon, triple-gluon, quartic-gluon and ghost–gluon
terms. Thus, the apparent proliferation of couplings in eqn (3.152) is illusory
provided that we choose the finite parts of the counterterms in eqn (3.149) so as
to respect eqn (3.153).

The renormalization prescription results in a finite Lagrangian, eqn (3.152),
whose form is exactly the same as that of the original Lagrangian, eqn (3.21),
written in terms of rescaled fields ψ_0, A_0 and η_0 and parameters g_{s0}, m_0 and ξ_0.
These are often referred to as the bare Lagrangian and bare fields and parameters.
In essence what we have are a remarkable series of cancellations, for example,

$$\psi = \psi_0 - (Z_\psi^{1/2} - 1)\psi$$
$$m = m_0 - (Z_m - 1)m \quad \text{etc.} \tag{3.154}$$

Here both the bare quantities and the corresponding counter terms are divergent
but their difference is finite.

We now return to the question of how to choose the finite terms, F_i, in
eqn (3.149). First, we should not be alarmed by their arbitrariness. This reflects
nothing more than the need to experimentally measure the actual masses and
couplings, something which would be true of any theory, irrespective of the need
for renormalization. Actually this cannot fix the wavefunction renormalizations,
Z_ψ, Z_A and Z_η, as they do not appear in physical quantities, though this also

means that we could live without them. The different choices for the F_i functions constitute different renormalization schemes. The most important point to remember with choosing a renormalization scheme is to use it consistently throughout the calculation so as not to spoil the critical relationships between Green's functions.

Perhaps the simplest scheme is the minimal subtraction, or MS, scheme ('tHooft, 1973) in which the counterterms only cancel the $1/\epsilon$ poles. A variant of this is the modified minimal subtraction, or $\overline{\text{MS}}$, scheme (Bardeen et al., 1978) in which the counterterms cancel the full $\Delta_\epsilon = 1/\epsilon + \ln(4\pi) - \gamma_{\text{E}}$ pieces. In eqn (3.149) this amounts to setting $F_i = 0$ in all the expressions. The difference between the two schemes is a difference in the finite parts proportional to $\ln(4\pi) - \gamma_{\text{E}} = 1.954$, equivalent to replacing the arbitrary μ^2 by $\overline{\mu}^2 = 4\pi\mu^2/e^{\gamma_{\text{E}}}$. Thus, in the $\overline{\text{MS}}$ scheme a potentially large contribution to radiative corrections is also removed, thereby aiding the convergence of the perturbative expansion. Despite its somewhat abstract nature the $\overline{\text{MS}}$ scheme's simplicity makes it the most popular one for pQCD calculations. A less manifest advantage of the scheme is the fact that the dimensionless Z_i do not depend on the combination m/μ. As we shall see, this mass independence simplifies the discussion of the renormalization group equations (RGEs). That this holds to all orders can be seen by the following heuristic argument. The counterterms are constructed to have just the bare bones necessary to remove the divergences which occur at high momentum. However, in this limit, we might expect any masses to be negligible so that they do not appear in the residues of the poles, and since $F_i = 0$, nor in the Z_i.

Other 'more physical' renormalization schemes may also be used. For example, when focusing on heavy quark properties the on mass-shell scheme may be used. Here F_ψ and F_m are adjusted so that the (real part) of the pole in the quark propagator occurs at the quark mass, $p^2 = m^2$, and has unit residue. In a similar fashion F_A and F_{A^3} may be adjusted so that the triple-gluon vertex, eqn (3.145), plus counterterm equals g_s at a particular external momentum configuration, typically chosen to be unphysical so as to avoid introducing extraneous singularities. In these schemes, it is common for a mass (m/μ) dependence to be introduced via the finite parts of the counterterms.

To finish this section we give the results of calculating the renormalization factors Z_i to two-loop approximation in pQCD. To be more specific, in a covariant gauge the $\overline{\text{MS}}$ prescription gives

$$Z_\psi = 1 - \frac{\alpha_s}{4\pi}\frac{1}{\epsilon}\xi C_F + \left(\frac{\alpha_s}{4\pi}\right)^2\left\{+\frac{1}{\epsilon}\left[T_F n_f + \frac{3}{4}C_F - \frac{(25 + 8\xi + \xi^2)}{8}C_A\right]\right.$$
$$\left. - \frac{1}{\epsilon^2}\left[\frac{\xi^2}{2}C_F + \frac{(3+\xi)\xi}{4}C_A\right]\right\}C_F$$

$$Z_A = 1 - \frac{\alpha_s}{4\pi}\frac{1}{\epsilon}\left[\frac{4}{3}T_F n_f - \frac{(13 - 3\xi)}{6}C_A\right]$$

$$+\left(\frac{\alpha_s}{4\pi}\right)^2\left\{-\frac{1}{\epsilon}\left[\left(2C_F+\frac{5}{2}C_A\right)T_Fn_f-\frac{(59-11\xi-2\xi^2)}{16}C_A^2\right]\right.$$

$$\left.+\frac{1}{\epsilon^2}\frac{(3+2\xi)}{3}\left[T_Fn_f-\frac{(13-3\xi)}{8}C_A\right]C_A\right\}$$

$$Z_\eta=1+\frac{\alpha_s}{4\pi}\frac{1}{\epsilon}\frac{(3-\xi)}{4}C_A+\left(\frac{\alpha_s}{4\pi}\right)^2\left\{-\frac{1}{\epsilon}\left[\frac{5}{12}T_Fn_f-\frac{(95+3\xi)}{96}C_A\right]\right.$$

$$\left.+\frac{1}{\epsilon^2}\left[\frac{1}{2}T_Fn_f-\frac{(35-3\xi^2)}{32}C_A\right]\right\}C_A$$

$$Z_m=1-\frac{\alpha_s}{4\pi}\frac{1}{\epsilon}3C_F+\left(\frac{\alpha_s}{4\pi}\right)^2\left\{+\frac{1}{\epsilon}\left[\frac{5}{3}T_Fn_f-\frac{3}{4}C_F-\frac{97}{12}C_A\right]\right.$$

$$\left.-\frac{1}{\epsilon^2}\left[2T_Fn_f-\frac{9}{2}C_F-\frac{11}{2}C_A\right]\right\}C_F$$

$$Z_g=1+\frac{\alpha_s}{4\pi}\frac{1}{\epsilon}\left[\frac{2}{3}T_Fn_f-\frac{11}{6}C_A\right]$$

$$+\left(\frac{\alpha_s}{4\pi}\right)^2\left\{\frac{1}{\epsilon}\left[\left(C_F+\frac{5}{3}C_A\right)T_Fn_f-\frac{17}{6}C_A^2\right]+\frac{1}{\epsilon^2}\frac{1}{24}\left[4T_Fn_f-11C_A\right]^2\right\}.$$

$$(3.155)$$

Here, we have done some work to derive the expression for Z_g, which cannot be obtained directly but is obtained indirectly using eqn (3.153). Typically, due to its relative simplicity, the correction to the ghost–gluon vertex is quoted to which one applies $Z_g=Z_{A\eta^\dagger\eta}/Z_A^{1/2}Z_\eta$; though any of the remaining three expressions in eqn (3.153) must lead to the same expression for Z_g. Also, you are reminded that the non-renormalization of the longitudinal part of the gluon propagator implies that $Z_\xi=Z_A$.

Before using these results we make a few observations. The choice of the (modified) minimal subtraction scheme leads to several simplifications which need not hold in other schemes. First, the leading terms in all the expressions are unity, $Z=1+\mathcal{O}(\alpha_s)$. Second, since all $F_i=0$, there is no dependence on any external momentum scales. All the coefficients of α_s^n are made up of pure numbers, the gauge fixing parameter and group factors. Third, there is also no dependence on the quark mass scale, m/μ. The MS prescriptions are the archetypical examples of mass independent renormalization schemes ('tHooft, 1973; Weinberg, 1973b). Fourth, both Z_m and Z_g are independent of ξ, the gauge fixing parameter (Caswell and Wilczek, 1974; Gross, 1976). A more general observation is that both δZ_ψ and δZ_m are proportional to the quark colour charge C_F, whilst δZ_η is proportional to C_A. All the above observations hold to all orders in α_s. We also see that certain choices of ξ give simplifications. In particular, in the Landau gauge, $\xi=0$, the $\mathcal{O}(\alpha_s)$ term in Z_ψ vanishes and the $\mathcal{O}(\alpha_s^2)$ term simplifies greatly.

3.4.3 *The renormalization group equations*

It should not have gone unnoticed that there appears to be a disturbingly large freedom in the application of the renormalization procedure. First is the freedom to select the finite parts of the counterterms, subject to respecting the Slavnov–Taylor identities. We have already exploited this freedom to absorb a numerically large coefficient in the change from the MS to $\overline{\text{MS}}$ schemes. Second is the freedom in the choice of the unit mass, μ, which sets a scale for the problem. However, physical quantities can not depend on any of these arbitrary choices. All prescriptions are ultimately equivalent. For example, in two schemes the quark masses are related as $Z_m(R)m_R = m_0 = Z_m(R')m_{R'}$ so that $m_R = [Z_m(R')/Z_m(R')]m_{R'}$, where the ratio is finite, because so is each m_R, even though neither Z_m is individually finite. In this way the invariance can be encapsulated in the group structure of the transformations connecting quantities $(g_s, m, \psi$, etc.) in different schemes.

The seemingly simple invariance of physical quantities under changes in μ leads to a very powerful differential equation conecting g_s, m, ψ , etc., defined at one scale μ to those at a second scale. To see how this arises, suppose we calculate the amplitude, that is amputated Green's function, for an operator describing the 'scattering' of n_ψ (anti)quarks and n_A gluons (we need not consider external ghosts). This amplitude can be written in terms of either the bare or renormalized quantities. Here, we have assumed that the counterterm for the interaction is proportional to itself so that the renormalization is multiplicative,[3] that is,

$$\Gamma_0(\alpha_{s0}, m_0, \xi_0, Q) = Z_\psi^{-\frac{n_\psi}{2}} Z_A^{-\frac{n_A}{2}} \Gamma(\mu, \alpha_s, m, \xi, Q) . \qquad (3.156)$$

Here, we use the single scale Q to characterize any external four-momenta present in the problem. For simplicity we only consider one quark mass, m. The left-hand side of eqn (3.156) is clearly independent of μ, as must be the right-hand side. Differentiating with respect to μ, using the chain rule, we obtain the following renormalization group equation (RGE)

$$0 = \mu \frac{\text{d}}{\text{d}\mu} \left[Z_\psi^{-\frac{n_\psi}{2}} Z_A^{-\frac{n_A}{2}} \Gamma(\mu, \alpha_s, m, \xi, Q) \right] \qquad (3.157)$$

$$\implies 0 = \left\{ \mu \frac{\partial}{\partial \mu} + \beta \frac{\partial}{\partial \alpha_s} + m\gamma_m \frac{\partial}{\partial m} + \xi \delta_\xi \frac{\partial}{\partial \xi} - n_\psi \gamma_\psi - n_A \gamma_A \right\} \Gamma .$$

The first term accounts for any explicit μ dependence, whilst the remainder takes care of any implicit dependences via $g_s(\mu)$, $m(\mu)$ and $\xi(\mu)$. Equation (3.157) serves to define the dimensionless coefficient functions

---

[3] In general the counterterm may involve other operators of the same mass dimension. An example is provided by the pQCD corrections to a weak decay. In the case of such operator mixing it is necessary to consider linear combinations of the operators which are diagonal.

$$\beta\left(\alpha_{\mathrm{s}},\frac{m}{\mu};\epsilon\right) = \mu\frac{\partial\alpha_{\mathrm{s}}}{\partial\mu} \qquad \gamma_\psi\left(\alpha_{\mathrm{s}},\frac{m}{\mu},\xi;\epsilon\right) = \frac{1}{2}\frac{\mu}{Z_\psi}\frac{\partial Z_\psi}{\partial\mu}$$

$$\gamma_m\left(\alpha_{\mathrm{s}},\frac{m}{\mu};\epsilon\right) = \frac{\mu}{m}\frac{\partial m}{\partial\mu} \qquad \gamma_A\left(\alpha_{\mathrm{s}},\frac{m}{\mu},\xi;\epsilon\right) = \frac{1}{2}\frac{\mu}{Z_A}\frac{\partial Z_A}{\partial\mu} \qquad (3.158)$$

$$\delta_\xi\left(\alpha_{\mathrm{s}},\frac{m}{\mu},\xi;\epsilon\right) = \frac{\mu}{\xi}\frac{\partial\xi}{\partial\mu}$$

which are all finite as $\epsilon \to 0$. You are warned that variants of these definitions, differing slightly in signs and normalizations, occur in the literature. Observe that β and γ_m, like Z_g and Z_m, are both independent of ξ.

A linear partial differential equation such as eqn (3.157) can be solved using the method of characteristics. To do this we introduce the functions $\bar\mu(t)$, $\bar\alpha_{\mathrm{s}}(t)$, $\bar m(t)$ and $\bar\xi(t)$ which satisfy the differential equations

$$\mathrm{d}t = \frac{\mathrm{d}\bar\mu}{\bar\mu} = \frac{\mathrm{d}\bar\alpha_{\mathrm{s}}}{\beta(\bar\alpha_{\mathrm{s}},\bar m/\bar\mu)} = \frac{\mathrm{d}\bar m}{\bar m}\frac{1}{\gamma_m(\bar\alpha_{\mathrm{s}},\bar m/\bar\mu)} = \frac{\mathrm{d}\bar\xi}{\bar\xi}\frac{1}{\delta_\xi(\bar\alpha_{\mathrm{s}},\bar m/\bar\mu,\bar\xi)} \qquad (3.159)$$

and pass through the point $\bar\mu(0) = \mu$, $\bar\alpha_{\mathrm{s}}(0) = \alpha_{\mathrm{s}}$, $\bar m(0) = m$ and $\bar\xi(0) = \xi$. These functions connect the parameters defined at the scale μ to those defined at a second scale $\bar\mu = \mu e^t$. The 'bar-notation' serves to highlight that we are now thinking of α_{s}, etc. as running parameters. Later on, except for $\bar m$, we will drop the special notation. The functions in eqn (3.159) define a characteristic parameterized by t. Since $\mathrm{d}\bar\xi \propto \bar\xi$, for $\delta_\xi \neq 0$, then $\bar\xi$ will remain identically zero if $\bar\xi(0) = 0$ and we can ignore any ξ-dependence in eqn (3.157): $\xi = 0$ is the Landau gauge which we adopt. The solution of these equations is straightforward if β, γ_m (and δ_ξ) do not depend on $\bar m/\bar\mu$. In a minimal subtraction scheme, or more generally a mass independent scheme, all the functions in eqn (3.158) are independent of m/μ. Adopting this further restriction we have

$$t = \ln\left(\frac{\bar\mu}{\mu}\right) = \int_{\alpha_{\mathrm{s}}}^{\bar\alpha_{\mathrm{s}}(t)}\frac{\mathrm{d}x}{\beta(x)} \quad \text{and} \quad \ln\left(\frac{\bar m(t)}{m}\right) = \int_{\alpha_{\mathrm{s}}}^{\bar\alpha_{\mathrm{s}}(t)}\mathrm{d}x\frac{\gamma_m(x)}{\beta(x)}. \qquad (3.160)$$

Explicit solutions for $\bar\alpha_{\mathrm{s}}$ and $\bar m$ require the actual expressions for β and γ_m. We shall derive these shortly but for the moment we assume that the solutions have been found. This allows us to rewrite eqn (3.157) as

$$\left\{\frac{\mathrm{d}}{\mathrm{d}t} - n_\psi\gamma_\psi(\bar\alpha_{\mathrm{s}}(t)) - n_A\gamma_A(\bar\alpha_{\mathrm{s}}(t))\right\}\Gamma(\bar\mu(t),\bar\alpha_{\mathrm{s}}(t),\bar m(t),Q) = 0. \qquad (3.161)$$

This ordinary differential equation is easily solved using an integrating factor:

$$\Gamma(\mu,\alpha_{\mathrm{s}},m,Q) = \exp\left(-n_\psi\int_{\alpha_{\mathrm{s}}}^{\bar\alpha_{\mathrm{s}}}\mathrm{d}x\frac{\gamma_\psi(x)}{\beta(x)} - n_A\int_{\alpha_{\mathrm{s}}}^{\bar\alpha_{\mathrm{s}}}\mathrm{d}x\frac{\gamma_A(x)}{\beta(x)}\right)\Gamma(\bar\mu,\bar\alpha_{\mathrm{s}},\bar m,Q).$$

$$(3.162)$$

The solution is a constant along our characteristic which we evaluate at $t = 0$. What it says is that the theory defined at $(\mu,\alpha_{\mathrm{s}},m)$ is equivalent to the theory defined at $\overline{\mu}$, provided that the coupling and mass are changed to take the effective values $\overline{\alpha}_s(\overline{\mu})$ and $\overline{m}(\overline{\mu})$ and the fields present are scaled appropriately.

It is useful to make explicit the dimensionality of the Green's function. The mass dimension of Γ is given by $d_\Gamma = D - n_\psi d_\psi - n_A d_A$, where again D is the dimensionality of space–time, and $d_\psi = (D-1)/2$ and $d_A = (D-2)/2$ are the mass dimensions of the quark and gluon fields. The significance of this dimension is that if $\{\mu, m, Q\}$ are all scaled by the same factor λ, then Γ scales as λ^{d_Γ}. This suggests rewriting eqn (3.162) as

$$\Gamma(\mu, \alpha_s, m, Q) \equiv \mu^{d_G} \overline{\Gamma}\left(\alpha_s, \frac{m}{\mu}, \frac{Q}{\mu}\right) \tag{3.163}$$

$$= \overline{\mu}^{d_\Gamma} \exp\left(-\int_{\alpha_s}^{\overline{\alpha}_s} dx \frac{n_\psi \gamma_\psi(x) + n_A \gamma_A(x)}{\beta(x)}\right) \overline{\Gamma}\left(\overline{\alpha}_s, \frac{m}{\overline{\mu}}, \frac{Q}{\overline{\mu}}\right)$$

$$= \exp\left(+\int_{\alpha_s}^{\overline{\alpha}_s} dx \frac{d_\Gamma - n_\psi \gamma_\psi(x) - n_A \gamma_A(x)}{\beta(x)}\right) \overline{\Gamma}\left(\overline{\alpha}_s, \frac{m}{\overline{\mu}}, \frac{Q}{\overline{\mu}}\right) .$$

In the second form, we have absorbed the Green's function's mass dimension into the exponential with the aid of eqn (3.160). This form suggests that the canonical mass dimension of the Green's function is modified in practice: for this reason the γ_i are called anomalous dimensions. Again eqn (3.163) tells us how to keep the physics the same should we choose to change the arbitrary unit mass μ. We can also use this equation to investigate the dependence on the scale of the external momenta, Q. To do this choose $\mu = 1$ and $\overline{\mu} = Q$ so that eqn (3.163) becomes

$$\overline{\Gamma}(\alpha_s, m, Q) = Q^{d_\Gamma} \exp\left(-\int_{\alpha_s}^{\overline{\alpha}_s} dx \frac{n_\psi \gamma_\psi(x) + n_A \gamma_A(x)}{\beta(x)}\right) \overline{\Gamma}\left(\overline{\alpha}_s, \frac{m}{Q}, 1\right) .$$

$$\tag{3.164}$$

Thus, the dependence on Q can be taken account of by simply scaling the Green's function, allowing for any anomalous dimensions, and evaluating it at an effective coupling, $\overline{\alpha}_s(Q)$, and an effective mass, $\overline{m}(Q)$. Note that even if $m = 0$, so that the classical theory was scale invariant, the anomalous dimensions and non-zero β-function would still imply a breaking of this naïve scale invariance in the quantum theory. This is possible because the renormalization procedure necessarily introduces a mass/momentum scale, here μ.

As the next section will show, in QCD both β and γ_m are negative. This means that the effective, or running, coupling, $\overline{\alpha}_s(Q)$, decreases logarithmically as the scale Q increases, eqn (3.22). This weakening of the strong force is the essence of asymptotic freedom (Gross and Wilczek, 1973; Politzer, 1974) and lies behind the success of perturbative QCD. The solution for $\overline{m}(Q)$ shows that it is also logarithmically suppressed; this is in addition to the factor $1/Q$ which appears in eqn (3.164). At this point it might be tempting to neglect quark masses, $m = 0$ in eqn (3.164), at high energies, $Q \gg \overline{m}(Q)$, so that all Q

dependence occurs via $\overline{\alpha}_{\mathrm{s}}(Q)$. However, this can lead to problems with low-energy or near collinear gluons so that for this approximation to make sense we must restrict ourselves to infrared safe quantities as discussed in Section 3.5.

Before moving on, we mention that a number of similar renormalization group equations have been derived in the literature. Foremost is the Callan–Symanzik equation (Symanzik, 1971; Callan, 1972) which is obtained by studying the Green's function's dependence on the physical mass m. The equation takes the form of eqn (3.157) with coefficient functions that only depend on g_s ($\gamma_m = 1$) and with the $\mu \partial/\partial\mu$ term replaced by an inhomogeneous term which may be neglected in the $m/Q \to 0$ limit.

3.4.4 Calculating the RGE coefficient functions

The values of the coefficient functions β, γ_m, γ_A, etc. defined in eqn (3.158) can be calculated as power series in α_s using our previous results. Here, we illustrate the method for the β-function. Consider the relationship, eqn (3.153), between the bare and renormalized couplings, $g_{s0} = \mu^\epsilon g_s Z_g$. Since the bare coupling g_{s0} can know nothing of the arbitrary scale μ, which was introduced only to facilitate renormalization, we must have

$$\mu \frac{\mathrm{d}g_{s0}}{\mathrm{d}\mu} = 0 \quad \Longrightarrow \quad 0 = \mu^\epsilon \left[\epsilon g_s Z_g + \beta_g \left(Z_g + g_s \frac{\partial Z_g}{\partial g_s} \right) \right] . \tag{3.165}$$

From this we can obtain $\beta = (g_s/2\pi)\beta_g$. In applying the chain rule we have made the simplifying assumption that we use a mass independent renormalization scheme such as $\overline{\mathrm{MS}}$. Thus Z_g only depends on the renormalized coupling g_s and not on $\overline{m}(\mu)/\mu$ (nor on ξ). Equation (3.165) determines how much $g_s(\mu)$ must change by when μ changes in order that g_{s0} stays constant. Referring to eqn (3.155) we can write Z_g as a Laurent series in inverse powers of ϵ,

$$Z_g = 1 + \sum_{n \geq 1} \frac{a_n(g_s)}{\epsilon^n} . \tag{3.166}$$

On the other hand, we want both g_s and β_g to be well defined in the limit $\epsilon \to 0$, that is to contain no poles in $1/\epsilon$. Thus, we write $\beta_g = A + B\epsilon$; all higher powers of ϵ must vanish. Substituting this into eqn (3.165) and collecting powers of ϵ gives

$$0 = (B + g_s)\epsilon + A + g_s B a_1' + \cdots \frac{\left[A a_{n-1} + (B + g_s)a_n + g_s A a_{n-1}' + g_s B a_n' \right]}{\epsilon^{n-1}} + \cdots , \tag{3.167}$$

where the prime indicates differentiation with respect to g_s. Setting the coefficients of ϵ^m, $m \leq 1$, to zero gives

$$\beta_g(g_s) = \lim_{\epsilon \to 0} \{ g_s^2 a_1' - \epsilon g_s \} \quad \text{and} \quad a_n' = a_1'(a_{n-1} + g_s a_{n-1}') , \quad n \geq 2$$

$$= g_s^2 a_1' \, . \tag{3.168}$$

At first sight, it may seem odd that we can calculate the β-function from just the residue of the $1/\epsilon$ pole. However, the conditions on the $a_{n \geq 2}$ ('tHooft, 1973), which ensure the absence of pole terms in β, together with the boundary conditions $a_n(0) = 0$ allow all the $a_{n \geq 2}$ to be calculated in terms of a_1.

Referring to eqn (3.155) the coefficient a_1 is easily read off, allowing us to infer the first two terms in the expansion of the β-function,

$$\beta = 2 \left\{ \frac{\alpha_s^2}{4\pi} \left[\frac{4}{3} T_F n_f - \frac{11}{3} C_A \right] + \frac{\alpha_s^3}{(4\pi)^2} \left[\left(4 C_F + \frac{20}{3} C_A \right) T_F n_f - \frac{34}{3} C_A^2 \right] \right.$$
$$+ \frac{\alpha_s^4}{(4\pi)^3} \left[\left(\frac{44}{9} C_F + \frac{158}{27} C_A \right) T_F^2 n_f^2 \right. \tag{3.169}$$
$$\left. \left. + \left(2 C_F^2 - \frac{205}{9} C_F C_A - \frac{1415}{27} C_A^2 \right) T_F n_f + \frac{2857}{54} C_A^3 \right] + \cdots \right\} \, .$$

The third term is given in (Tarasov $et\ al.$, 1980) for the $\overline{\text{MS}}$ scheme and the $\mathcal{O}(\alpha_s^5)$ (four loop) term is available in (Larin $et\ al.$, 1997), again for the $\overline{\text{MS}}$ scheme. Similar analyses to the above can be applied to $m_0 = m Z_m(g_s)$ to obtain γ_m, to obtain γ_A from $Z_A(g_s, \xi)$ and likewise γ_ψ and γ_η; see Ex. (3-20). Here we simply quote the results:

$$\gamma_m = -\frac{\alpha_s}{4\pi} 6 C_F + \left(\frac{\alpha_s}{4\pi} \right)^2 \left[\frac{20}{3} T_F n_f - 3 C_F - \frac{97}{3} C_A \right] C_F + \cdots$$

$$\gamma_\psi = +\frac{\alpha_s}{4\pi} \xi C_F - \left(\frac{\alpha_s}{4\pi} \right)^2 \left[2 T_F n_f + \frac{3}{2} C_F - \frac{(25 + 8\xi + \xi^2)}{4} \right] C_F + \cdots$$

$$\gamma_A = +\frac{\alpha_s}{4\pi} \left[\frac{4}{3} T_F n_f - \frac{(13 - 3\xi)}{6} C_A \right] \tag{3.170}$$
$$+ \left(\frac{\alpha_s}{4\pi} \right)^2 \left[(4 C_F + 5 C_A) T_F n_f - \frac{(59 - 11\xi - 2\xi^2)}{8} C_A^2 \right] + \cdots$$

$$\gamma_\eta = -\frac{\alpha_s}{4\pi} \frac{(3 - \xi)}{4} C_A + \left(\frac{\alpha_s}{4\pi} \right)^2 \left[\frac{5}{6} T_F n_f - \frac{(95 + 3\xi)}{48} C_A \right] C_A + \cdots \, .$$

Observe that both β and γ_m are independent of the gauge parameter ξ, a fact which can be traced to the ξ-independence of Z_g and Z_m in a mass independent renormalization scheme. This is not so for the wavefunction anomalous dimensions γ_ψ, γ_A and γ_η. This raises the issue of the scheme (in)dependence of our results. If we restrict ourselves to mass independent schemes (and neglect possible non-perturbative effects) then the first two terms in β and the first terms in γ_m, γ_ψ and γ_A are independent of the specific choices of counterterms. Beyond these leading terms the results are scheme dependent and, for example, the third and higher order terms in eqn (3.169) can be set equal to arbitrary values.

In eqn (3.165) we assumed a mass independent renormalization scheme had been used. If this is not the case, then the presence of m/μ dependences in the Z_i

significantly complicates the RGEs. First, the evaluation of β and the γ_i is made
~~harder by, for example, the extra term, $\frac{m}{\mu}(\gamma_m - 1)\partial Z_g/\partial(\frac{m}{\mu})$, in eqn (3.165).~~
Second, the solution of the coupled, linear, differential equations for the effective
coupling and mass, eqn (3.159), is made harder. Here, a possible approach to
avoiding these problems is to go to a regime where the mass(es) are negligible
compared to μ and all other scales (Q).

3.4.5   The running coupling and quark masses

A key lesson from our study of the RGEs is the need to express our results in
terms of the running, or effective, coupling and mass. If we use as argument Q^2,
the evolution equations in a mass independent scheme are

$$Q^2 \frac{\mathrm{d}\alpha_\mathrm{s}(Q^2)}{\mathrm{d}Q^2} = \beta(\alpha_\mathrm{s}(Q^2)) = -\alpha_\mathrm{s}^2(\beta_0 + \beta_1\alpha_\mathrm{s} + \beta_2\alpha_\mathrm{s}^2 + \cdots) \quad \text{and} \qquad (3.171)$$

$$\frac{Q^2}{\overline{m}(Q^2)} \frac{\mathrm{d}\overline{m}(Q^2)}{\mathrm{d}Q^2} = \gamma_m(\alpha_\mathrm{s}(Q^2)) = -\alpha_\mathrm{s}(\gamma_0 + \gamma_1\alpha_\mathrm{s} + \gamma_2\alpha_\mathrm{s}^2 + \cdots), \qquad (3.172)$$

c.f. eqn (3.159). Here, we have in mind that $\beta_0, \gamma_0 > 0$ so that $\beta, \gamma_m < 0$.
Taking into account that $Q^2\,\mathrm{d}/\mathrm{d}Q^2 = (1/2)Q\,\mathrm{d}/\mathrm{d}Q$, the coefficients in the series
expansions for the $\overline{\mathrm{MS}}$ scheme can be read off from eqns (3.169) and (3.170).

$$\begin{aligned}
\beta_0 &= \frac{11C_A - 4T_F n_f}{12\pi} && = \frac{(33 - 2n_f)}{12\pi} \\[2mm]
\beta_1 &= \frac{17C_A^2 - (6C_F + 10C_A)T_F n_f}{24\pi^2} && = \frac{(153 - 19n_f)}{24\pi^2} \\[2mm]
\gamma_0 &= \frac{3C_F}{4\pi} && = \frac{1}{\pi} \\[2mm]
\gamma_1 &= \frac{C_F(97C_A + 9C_F - 20T_F n_f)}{96\pi^2} && = \frac{(303 - 10n_f)}{72\pi^2}
\end{aligned} \qquad (3.173)$$

Here β_0, β_1 and γ_0 are common to any mass independent renormalization scheme,
whilst γ_1 is specific to the $\overline{\mathrm{MS}}$ scheme. The terms quoted above are given as
function of the colour factors C_F, C_A and T_F and thus are valid for any gauge
theory with an unbroken gauge symmetry. Note that T_F always appears in a
product with n_f, the number of active quark flavours. The coefficients on the
right-hand side apply for the case of colour SU(3), that is, they are specific to
QCD.

  Referring to eqn (3.160) and working to next-to-leading order the solution
for $\alpha_\mathrm{s}(Q^2)$ is given implicitly by

$$\ln\left(\frac{Q^2}{Q_0^2}\right) = + \int_{\alpha_\mathrm{s}(Q_0^2)}^{\alpha_\mathrm{s}(Q^2)} \frac{\mathrm{d}x}{\beta(x)}$$

$$= -\frac{1}{\beta_0^2} \int_{\alpha_\mathrm{s}(Q_0^2)}^{\alpha_\mathrm{s}(Q^2)} \mathrm{d}x \left[\frac{\beta_0^2}{x^2(\beta_0 + x\beta_1)} = \frac{\beta_0}{x^2} - \frac{\beta_1}{x} + \frac{\beta_1^2}{\beta_0 + x\beta_1}\right]$$

$$\Longrightarrow \quad \beta_0 \ln\left(\frac{Q^2}{Q_0^2}\right) = \frac{1}{\alpha_s(Q^2)} + \frac{\beta_1}{\beta_0} \ln\left(\frac{\alpha_s(Q^2)}{\beta_0 + \beta_1 \alpha_s(Q^2)}\right)\Bigg|_{Q_0^2}^{Q^2} . \tag{3.174}$$

Thus, given the value of $\alpha_s(Q_0^2)$ at one scale Q_0, it is possible to solve for $\alpha_s(Q^2)$ at a second scale Q; the one proviso being that we remain in the perturbative domain where eqn (3.171) is valid. An alternative and slightly simpler form of eqn (3.174) can be obtained using the boundary condition $\alpha_s(\Lambda_{\mathrm{QCD}}^2) = \infty$,

$$\beta_0 \ln\left(\frac{Q^2}{\Lambda_{\mathrm{QCD}}^2}\right) = \frac{1}{\alpha_s(Q^2)} + \frac{\beta_1}{\beta_0} \ln\left(\frac{\alpha_s(Q^2)}{\beta_0 + \beta_1 \alpha_s(Q^2)}\right) . \tag{3.175}$$

Here the parameter Λ_{QCD} is equivalent to giving the coupling $\alpha_s(Q^2)$ at a specific scale Q. At the quantum-level, QCD is specified by a dimensionful parameter. This is even true in the absence of any quark masses to set a classical-level scale. The appearance of such a scale at the quantum-level is known as dimensional transmutation. In eqn (3.175) $\alpha_s(Q^2)$ is given implicitly. By expanding in inverse powers of $\ln(Q^2/\Lambda_{\mathrm{QCD}}^2)$ an approximate explicit form can be derived,

$$\alpha_s(Q^2) = \frac{1}{\beta_0 \ln(Q^2/\Lambda_{\mathrm{QCD}}^2)} \left\{ 1 - \frac{\beta_1}{\beta_0^2} \frac{\ln\left[\ln(Q^2/\Lambda_{\mathrm{QCD}}^2)\right]}{\ln(Q^2/\Lambda_{\mathrm{QCD}}^2)} \right. \tag{3.176}$$

$$\left. + \frac{\beta_1^2}{\beta_0^4 \ln^2(Q^2/\Lambda_{\mathrm{QCD}}^2)} \left[\left(\ln\left[\ln(Q^2/\Lambda_{\mathrm{QCD}}^2)\right] - \frac{1}{2}\right)^2 - \frac{5}{4} \right] \right\} .$$

In practice the last $(-5/4)$ term in this expression is often neglected. This is equivalent to a redefinition of Λ_{QCD} by $\mathcal{O}(+10\%)$ (Buras $et\ al.$, 1977).

If we had worked at leading order, $\beta_1 = 0$, then eqn (3.174) can be solved to give

$$\alpha_s(Q^2) = \frac{\alpha_s(Q_0^2)}{1 + \beta_0 \alpha_s(Q_0^2) \ln(Q^2/Q_0^2)} \tag{3.177}$$

and the inverse of eqn (3.175) is given exactly by eqn (3.22) with $\Lambda_{\mathrm{QCD}}^2 = Q_0^2 \exp[-1/\beta_0 \alpha_s(Q_0^2)]$. As this expression for Λ_{QCD} suggests, changing the value of $\alpha_s(Q^2)$ for a fixed Q does not really alter the theory but gives the same theory with its unit of momentum rescaled. In this sense (massless) QCD is parameter free (Coleman and Weinberg, 1973). Equation (3.176) makes the asymptotic behaviour of $\alpha_s(Q^2)$ manifest — the QCD coupling decreases as $1/\ln(Q/\Lambda_{\mathrm{QCD}})$ for $Q \to \infty$. It is important to realize that this decrease in α_s justifies the use of pQCD, in particular the solution based on the first few terms in eqn (3.171). As Q decreases the converse is expected and indeed a strong growth of $\alpha_s(Q^2)$ is confirmed experimentally. However, the singularity at $Q = \Lambda_{\mathrm{QCD}}$ should not be taken too seriously as large values of α_s invalidate eqn (3.171) and any solutions based upon it. In this low-Q regime QCD is non-perturbative and no one knows yet how $\alpha_s(Q^2)$ behaves in reality, nor can it be claimed that this is a proof of confinement in QCD, though it does make it more plausible. It is safer to regard

$\Lambda_{\text{QCD}} \sim 200\,\text{MeV}$, roughly an inverse hadron size, as the scale at which non-perturbative physics becomes important. Finally, returning to eqn (3.176), if we substitute $\Lambda_{\text{QCD}}^2 = Q_0^2 \exp[-1/\beta_0 \alpha_s(Q_0^2)]$ we can derive an explicit expression relating the strong coupling at two scales,

$$\alpha_s(Q^2) = \frac{\alpha_s(Q_0^2)}{\omega}\left(1 - \frac{\beta_1}{\beta_0}\alpha_s(Q_0^2)\frac{\ln\omega}{\omega}\right) \tag{3.178}$$

$$\text{with} \quad \omega = 1 + \beta_0\alpha_s(Q_0^2)\ln\left(\frac{Q^2}{Q_0^2}\right),$$

which is accurate to next-to-leading order.

The crucial fact for asymptotic freedom is that β is negative, that is $\beta_0 > 0$. Referring to eqn (3.173) we see that quarks, and fermions in general, give a positive contribution, whilst non-abelian interactions amongst gluons, proportional to C_A, lead to an overall negative β, provided $n_f < 17$. In QED with abelian photons the β-function is positive and consequently electric charges grow as the scale of a measurement grows. How various particles contribute to the β-function has been extensively studied and it is now known that only theories containing non-abelian gauge bosons give negative contributions (Coleman and Gross, 1973). Since many extensions to the Standard Model have been proposed, it is interesting to see how a new particle would contribute to the β-function; see Ex. (3-22). At leading order only coloured particles can contribute to the QCD β-function, though in higher orders all particles contribute. The contributions to β_0 consist of two components related to the particles' colour and Poincaré group representations. The general expression is

$$\beta_0 = -\frac{1}{12\pi}\sum_{\substack{\text{coloured}\\\text{particles}}} D_i T_{R_i} . \tag{3.179}$$

Here D_i equals -11 for a vector boson, $+4$ for a Dirac fermion, $+2$ for a Weyl fermion, $+1$ for a complex scalar and $+1/2$ for a real scalar field, whilst T_R is a colour charge determined by the particle's SU(N_c) representation. For example, $T_F = 1/2$ (by convention) for the fundamental (triplet) representation, $T_A = 2N_c T_F$ for the adjoint (octet) representation, $(2N_c - 1)T_F$ for the sextet representation etc. For QCD with a colour SU(3) octet of vector bosons and n_f triplets of Dirac fermions this gives eqn (3.169).

The next-to-leading order solution for $\overline{m}(Q^2)$ follows similar lines to that for $\alpha_s(Q^2)$. One finds

$$\ln\left(\frac{\overline{m}(Q^2)}{\overline{m}(Q_0^2)}\right) = \int_{\alpha_s(Q_0^2)}^{\alpha_s(Q^2)} dx\, \frac{\gamma_m(x)}{\beta(x)}$$

$$= \frac{1}{\beta_0}\int_{\alpha_s(Q_0^2)}^{\alpha_s(Q^2)} dx\left[\frac{\beta_0(\gamma_0 + \gamma_1 x)}{x(\beta_0 + \beta_1 x)} = \frac{\gamma_0}{x} + \frac{\gamma_1\beta_0 - \beta_1\gamma_0}{\beta_0 + \beta_1 x}\right]$$

$$\implies \overline{m}(Q^2) = \overline{m}(Q_0^2) \left(\frac{\alpha_s(Q^2)}{\alpha_s(Q_0^2)}\right)^{\frac{\gamma_0}{\beta_0}} \left(\frac{\beta_0 + \beta_1 \alpha_s(Q^2)}{\beta_0 + \beta_1 \alpha_s(Q_0^2)}\right)^{\frac{(\beta_0 \gamma_1 - \beta_1 \gamma_0)}{\beta_0 \beta_1}}$$

$$\text{or} \quad \overline{m}(Q^2) = \overline{m}_0 \left(\alpha_s(Q^2)\right)^{\frac{\gamma_0}{\beta_0}} \left[1 + \frac{\beta_1}{\beta_0}\alpha_s(Q^2)\right]^{\frac{(\beta_0 \gamma_1 - \beta_1 \gamma_0)}{\beta_0 \beta_1}} . \tag{3.180}$$

Again, given $\overline{m}(Q_0^2)$ at one scale Q_0 we can calculate $\overline{m}(Q^2)$ at a second scale Q, provided that pQCD, and in particular eqn (3.172), remains valid. This offers a concise way of specifying a running quark mass as the mass when the scale equals its mass: $m = \overline{m}(m^2)$. In the second form of the solution $\overline{m}_0$ plays a similar rôle to Λ_{QCD}. Specializing to the leading order result, $\gamma_1 = 0 = \beta_1$, we have

$$\overline{m}(Q^2) = \overline{m}(Q_0^2) \left(\frac{\alpha_s(Q^2)}{\alpha_s(Q_0^2)}\right)^{\frac{\gamma_0}{\beta_0}} = \overline{m}(Q_0^2) \left(\frac{\ln(Q_0/\Lambda_{\text{QCD}})}{\ln(Q/\Lambda_{\text{QCD}})}\right)^{\frac{\gamma_0}{\beta_0}} . \tag{3.181}$$

Thus, we see that the quark mass falls as an inverse power of a logarithm as Q^2 increases. This quantum scaling violation, in addition to the classical $\overline{m}(Q^2)/Q$ suppression, adds justification to dropping light quark masses from our calculations.

Up to this point we have left unresolved the issue of how many quarks, n_f, to include in our calculations. The critical issue is the relative magnitude of a quark's mass to the overall scale Q. If the quark has $\overline{m}(Q^2) \gg Q$ then it can only make its presence felt via internal loops and it is possible to remove these contributions by suitable choices of the counterterms. This decoupling theorem (Symanzik, 1973; Appelquist and Carazzone, 1975) means that we can ignore a quark if $\overline{m}(Q^2) \gg Q$. On the other hand, if the quark has $\overline{m}(Q^2) \ll Q$, then we should include its contributions and infrared safe quantities can be evaluated using the approximation $m_q = 0$. The so-called light quarks, d, u and s, have $m_q < \Lambda_{\text{QCD}}$ so that in a pQCD calculation, characterized by $Q \gg \Lambda_{\text{QCD}}$, we always have $n_f \geq 3$. The issue is more delicate for the so-called heavy quarks, c, b and t, in situations where $\overline{m}_Q(Q^2) \sim Q$. Here we expect significant, process dependent, contributions from the quark mass which we must therefore include in our calculations. Furthermore, we have to decide how to cope with the change in the β-function above and below the quark mass threshold.

Well above $\overline{m}_Q$ we can use the 'full' theory containing $n_f + 1$ quarks and $\alpha_s^+(Q^2)$, whilst well below $\overline{m}_Q$ we can use an 'effective' theory with n_f light quarks and $\alpha_s^-(Q^2)$. At intermediate scales, $Q \sim \overline{m}_Q(Q^2)$, we must match the two versions of the theory so as to ensure that they give consistent results. This matching has been carried out to next-to-next-to-leading order for SU(3) in the $\overline{\text{MS}}$ scheme (Bernreuther, 1983) and results in a relationship between the two running couplings given by

$$\alpha_s^+(Q^2) = \alpha_s^-(Q^2) \left[1 + \frac{x}{6\pi}\alpha_s^-(Q^2) + \frac{(2x^2 + 33x - 11)}{72\pi^2}\left(\alpha_s^-(Q^2)\right)^2\right]$$

$$\text{with} \quad x = \ln\left(\frac{Q^2}{\overline{m}_Q^2(Q^2)}\right). \tag{3.182}$$

If we evaluate this expression at the point at which the scale equals the running quark mass, $m_Q = \overline{m}(m_Q^2)$, that is $x = 0$, then eqn (3.182) almost reduces to requiring α_s to be continuous at the scale m_Q (Marciano, 1984),

$$\alpha_s^+(m_Q^2) = \alpha_s^-(m_Q^2) - \frac{11}{72\pi^2}\left(\alpha_s^-(m_Q^2)\right)^3. \tag{3.183}$$

This explains why it is common to require α_s to be continuous at $Q = m_Q$ rather than at the production threshold $Q = 2m_Q$. Of course, imposing continuity on α_s implies a discontinuity in Λ_{QCD}, which subsequently becomes dependent on the number of active flavours, n_f. For example, at leading order it is easy to verify that the continuity of $\alpha_s(m_Q)$ as $n_f \to n_f + 1$ requires

$$\Lambda_{QCD}^{(n_f+1)} = \Lambda_{QCD}^{(n_f)}\left(\frac{\Lambda_{QCD}^{(n_f)}}{m_Q}\right)^{\frac{\beta_0^{(n_f)} - \beta_0^{(n_f+1)}}{\beta_0^{(n_f+1)}}} = \Lambda_{QCD}^{(n_f)}\left(\frac{\Lambda_{QCD}^{(n_f)}}{m_Q}\right)^{\frac{2}{31-2n_f}}. \tag{3.184}$$

Similar expressions can be derived at next-to-leading order given a specific equation for α_s, equivalent to a definition for Λ_{QCD}.

This raises the issue of how to quote a measurement of the running coupling, α_s. Two conventions in popular usage are to quote $\alpha_s(M_Z^2)$ or $\Lambda_{QCD}^{(5)}$. In both cases, this will typically involve having either to evolve α_s or to match Λ_{QCD} at flavour thresholds. In the case of Λ_{QCD} it is important to be specific as to which next-to-leading order equation is being used, for example, eqn (3.175) or (3.176) with or without the last term. The value of Λ_{QCD} also depends on the renormalization scheme, for example, $\Lambda_{\overline{MS}}^2 = 4\pi e^{-\gamma_E}\Lambda_{MS}^2$. Since there are more traps involved in specifying Λ_{QCD}, the preferred option has become to quote α_s at the scale of the Z mass.

An interesting aspect of converting a measurement at Q^2 to an α_s value at M_Z^2 is the effect on the measurement's error. By differentiating eqn (3.160) we find

$$\Delta\alpha_s(Q^2) = \frac{\beta(\alpha_s(Q^2))}{\beta(\alpha_s(M_Z^2))}\Delta\alpha_s(M_Z^2) \approx \left(\frac{\alpha_s(Q^2)}{\alpha_s(M_Z^2)}\right)^2\Delta\alpha_s(M_Z^2). \tag{3.185}$$

So that, if $Q^2 < M_Z^2$, then the error will shrink as we evolve from Q to M_Z. A second consequence of eqn (3.185) is that a change in $\alpha_s(M_Z^2)$ only causes an $\mathcal{O}(\alpha_s^2)$ change in $\alpha_s(Q^2)$. Thus, an experimentally determined quantity, $\sigma \pm \Delta\sigma$, at Q^2 must be compared to at least a next-to-leading order theoretical prediction in order to be able to meaningfully measure $\alpha_s(M_Z^2)$,

$$\sigma \pm \Delta\sigma = A\alpha_s^N\left[1 + B\alpha_s \pm \Delta C\alpha_s^2 \pm \frac{\Delta D}{Q^p}\right]. \tag{3.186}$$

That is, in addition to the leading $\mathcal{O}(\alpha_s^N)$ term we require the $\mathcal{O}(\alpha_s^{N+1})$ term to measure $\alpha_s(M_Z^2)$. In eqn (3.186) we also included an estimate of the next-to-next-to-leading order perturbative correction and a non-perturbative contribution that is parameterized as a power law correction. The measurement error on $\alpha_s(M_Z^2)$ can be estimated as

$$\frac{\Delta\alpha_s(M_Z^2)}{\alpha_s(M_Z^2)} = \frac{\alpha_s(M_Z^2)}{\alpha_s(Q^2)}\frac{1}{N}\left[\frac{\Delta\sigma}{\sigma} \pm \Delta C\alpha_s^2(Q^2) \pm \frac{\Delta D}{Q^p}\right]. \qquad (3.187)$$

Looking at the first term, the 'error telescoping effect' suggests using a small value of Q^2 together with an intrinsically higher order process, large N. However, the error associated with missing the second two terms favours using larger values of Q^2, where their contributions are smaller. It is also possible that $\Delta C, \Delta D \propto N$ so that there is no advantage to larger N.

3.4.6 *An explicit example*

In the previous sections we learnt that physical quantities are independent of the arbitrary renormalization scale μ if they are made functions of the running coupling and mass. We now repeat this rather formal analysis in a particular case. We focus on QCD corrections to the dimensionless R parameter defined in e^+e^- annihilation, eqn (2.30). Suppose we have calculated a perturbative series for R,

$$R\left(\frac{Q^2}{\mu^2}, \alpha_s(\mu^2)\right) = 1 + \sum_{n=1}^{\infty} r_n(Q^2/\mu^2)\alpha_s^n(\mu^2). \qquad (3.188)$$

We have removed an inessential factor $N_c e_q^2$ from eqn (3.188) with respect to the usual definition and assumed that all quark masses are zero. We comment shortly on the case of only a finite number of terms. Demanding that R, a physically measurable quantity, is independent of μ leads to a series of differential equations

$$\begin{aligned}
0 &= \mu^2\frac{d}{d\mu^2}R\left(\frac{Q^2}{\mu^2}, \alpha_s(\mu^2)\right) \\
&= \left[\mu^2\frac{\partial}{\partial\mu^2} + \beta(\alpha_s)\frac{\partial}{\partial\alpha_s}\right]R\left(\frac{Q^2}{\mu^2}, \alpha_s\right) \\
&= \sum_{n=1}\sum_{m=0}\left(\mu^2\frac{dr_n}{d\mu^2}\alpha_s^n - nr_n\beta_m\alpha_s^{n+m+1}\right). \qquad (3.189)
\end{aligned}$$

The first few terms are given explicitly by

$$0 = \mu^2\frac{dr_1}{d\mu^2}\alpha_s + \left(\mu^2\frac{dr_2}{d\mu^2} - 1r_1\beta_0\right)\alpha_s^2 + \left(\mu^2\frac{dr_3}{d\mu^2} - 2r_2\beta_0 - 1r_1\beta_1\right)\alpha_s^3 + \cdots. \qquad (3.190)$$

Since each coefficient must vanish individually we obtain a series of differential equations which are solved easily to give

$$r_1(t) = c_1$$

$$r_2(t) = c_2 + c_1\beta_0 t$$
$$r_3(t) = c_3 + (2c_2\beta_0 + c_1\beta_1)t + c_1\beta_0^2 t^2$$
$$r_n(t) = c_n + \cdots + c_1(\beta_0 t)^{n-1} . \tag{3.191}$$

Here the μ-dependence is via $t = \ln(\mu^2/Q^2)$ and the $\{c_i\}$ are numerical constants. In general, $r_n(t) \propto t^{n-1}$, so that the series contains terms of the form $(\alpha_s t)^n$. This raises a potentially embarrassing problem, for when $\alpha_s(\mu^2)\ln(\mu^2/Q^2) \geq 1$, which is inevitable for sufficiently large Q, the series appears not to be convergent. This problem is easily finessed if we rearrange eqn (3.191) to take account of these so-called leading logarithmic terms,

$$R(t, \alpha_s) = 1 + c_1[1 + \beta_0\alpha_s t + (\beta_0\alpha_s t)^2 + \cdots]\alpha_s$$
$$+ c_2\alpha_s^2 + [c_3 + (2c_2\beta_0 + c_1\beta_1)t]\alpha_s^3 + \cdots$$
$$= 1 + c_1 \frac{\alpha_s(\mu^2)}{1 + \beta_0\alpha_s(\mu^2)\ln(Q^2/\mu^2)} + \cdots . \tag{3.192}$$

It is then apparent that the leading logarithms can be summed by the use of the one-loop running coupling $\alpha_s(Q^2)$. You may wonder if this just means that the convergence problem has been shifted to the next-to-leading logarithmic terms $\propto (\alpha_s t)^n\alpha_s$. However, using the two-loop running coupling sums both the leading and next-to-leading logarithms and shows that the NLL terms are genuinely suppressed by α_s. In fact we already know the result of carrying this program to completion. It is given by eqn (3.162),

$$R\big(1, \alpha_s(Q^2)\big) = 1 + c_1\alpha_s(Q^2) + c_2'\alpha_s^2(Q^2) + c_3'\alpha_s^3(Q^2) \cdots , \tag{3.193}$$

a series whose convergence actually improves as $Q \to \infty$. The coefficients in eqn (3.193) have been calculated to $\mathcal{O}(\alpha_s^3)$ (Chetyrkin et al., 1996a). The coefficient c_1 is renormalization scheme independent, whereas c_n' for $n \geq 2$ depends on the scheme. We calculate the one-loop correction in Section 3.5.

It is useful to review the above calculation from a different perspective which gives an insight into how the RGEs work. Earlier, we encountered Weinberg's theorem when discussing the form of the counterterms needed to remove ultra-violet divergences (Weinberg, 1960). Once a diagram is rendered finite he went on to investigate its asymptotic behaviour as the scale of the external momenta becomes large. A typical behaviour is a dimensionful factor, Q^d, times a polynomial in $\ln(Q^2/\mu^2)$ (Mueller, 1981). Thus, we expect a typical cross section to have the form

$$\sigma = Q^d \sum_n \alpha_s^n S_n\big(\ln(Q^2/\mu^2)\big) \quad \text{with} \quad S_n(x) = a_{n0} + a_{n1}x + \cdots + a_{nm}x^m ,$$

$$\tag{3.194}$$

c.f. eqns (3.188) and (3.191). Now, because the terms in the RGE eqn (3.157) are of different orders in α_s, it interrelates S_n of different n and their coefficients a_{nm}. Indeed these constraints allow the S_n to be partially reconstructed. The resulting structure contains series of leading and next-to-leading logarithms etc., which correspond to expansions of the running coupling. Thus, the RGE enforces relationships between the coefficients such that all the large logarithms can be summed by using an effective coupling with a scale appropriate to the problem.

Before leaving eqn (3.188) we comment on two features of eqn (3.191). First is the seemingly trivial observation that $r_1(Q^2/\mu^2)$ is a constant, c_1. Since any μ dependence in r_n arises from the treatment of ultraviolet divergences, this means that at one-loop the QCD correction to the $\gamma^*q\bar{q}$ vertex is finite; a result which must in fact hold at all orders if QCD is not to spoil electric charge conservation.

Second is the effect of truncating the series eqn (3.188). As we have noted the coefficient of the $\mathcal{O}(\alpha_s^{n+1})$ term is related to coefficients of the $\mathcal{O}(\alpha_s^n)$, $\mathcal{O}(\alpha_s^{n-1})$, ..., $\mathcal{O}(\alpha_s^1)$ terms in such a way as to remove the μ-dependence to $\mathcal{O}(\alpha_s^n)$. For example, to two-loops one has

$$R^{(2)}\left(\frac{Q^2}{\mu^2}, \alpha_s\right) = 1 + c_1\alpha_s(\mu^2) + \left[c_2 - c_1\beta_0 \ln\left(\frac{Q^2}{\mu^2}\right)\right]\alpha_s^2(\mu^2) , \qquad (3.195)$$

which is μ-independent to $\mathcal{O}(\alpha_s)$ but μ-dependent at $\mathcal{O}(\alpha_s^2)$. Thus, a truncated series is μ-dependent and in practice we must decide what value(s) to use for μ. This is the scale setting problem. A conservative approach is to vary μ in a range centred on the characteristic scale $[Q/\lambda, Q\lambda]$. This should cover more specific prescriptions whilst, if λ is kept modest, avoid making the logarithm large and spoiling the validity of eqn (3.195). In this sense the measurement can be said to be of α_s at the scale Q^2. A word of caution: it is sometimes claimed that by varying μ one can estimate the size of the next (uncalculated) term in a series. For example, the α_s^2 term from the α_s term in eqn (3.195), but since c_2 is arbitrary (until calculated) this procedure is not without risk. Other more ambitious proposals for scale setting are available. The *principle of minimum sensitivity* (PMS) (P.M. Stevenson 1981) chooses μ so as to make the truncated series locally independent of μ. Applied to eqn (3.195) this yields

$$\mu^2 \frac{\mathrm{d}R^{(n)}}{\mathrm{d}\mu^2}(\mu^2)\bigg|_{\mu_{\mathrm{PMS}}^2} = 0 \implies \mu_{\mathrm{PMS}}^2 = Q^2 \exp\left(-\frac{c_2}{\beta_0 c_1} - \frac{\beta_1}{2\beta_0^2}\right) . \qquad (3.196)$$

The application is to eqn (3.195). *Fastest apparent convergence* (FAC) chooses μ so that the first non-trivial term gives the same result as the sum of the known terms. Applying this prescription to eqn (3.195) gives

$$R^{(1)}(\mu_{\mathrm{FAC}}^2) = R^{(n)}(\mu_{\mathrm{FAC}}^2) \implies \mu_{\mathrm{FAC}}^2 = Q^2 \exp\left(-\frac{c_2}{\beta_0 c_1}\right) . \qquad (3.197)$$

Again the application is to eqn (3.195). A third proposal by Brodsky, Lepage and Mackenzie (BLM) (1983) determines μ from the requirement that the n_f-dependence of the coefficients c_i vanishes. In all cases, by going to higher orders in α_s, the dependence on μ is reduced and the scale setting problem diminished.

3.5   Infrared safety

Ultraviolet divergences are not the only complication which arises in QCD, divergences also occur when real gluons are emitted with either very low energy or nearly collinear to the emitter. We already noticed this problem at tree level with the $e^+e^- \to q\bar{q}g$ calculation in Section 3.3.2. Throughout this section we will illustrate the basic methods and ideas used to deal with these infrared divergences using the important example of the $\mathcal{O}(\alpha_s)$ correction to electron–positron annihilation to hadrons. The underlying process is $e^+e^- \to \gamma^* \to q\bar{q}$ which, as we have seen, is also closely related to both deep inelastic scattering and the Drell–Yan process. The $\mathcal{O}(\alpha_s)$ correction is given by gluon emission off the final state quark or antiquark, though the following discussion applies with minimal modification for emission off a gluon. Recall the behaviour of the quark propagator prior to emission, eqn (3.105),

$$\frac{1}{(q+g)^2 - m_q^2} = \frac{1}{2E_g E_q (1 - \beta_q \cos\theta_{qg})} \quad \text{with} \quad \beta_q = \frac{|\boldsymbol{q}|}{E_q} = \sqrt{1 - \frac{m_q^2}{E_q^2}}. \tag{3.198}$$

Here we see that there are basically two singular regions, which may overlap:

$$(\text{Propagator})^{-1} \longrightarrow \begin{cases} 0 & E_g \to 0 \quad \text{soft} \\ 2E_g E_q (1 - \beta_q) \xrightarrow{m_q \to 0} 0 & \theta_{qg} \to 0 \quad \text{collinear}. \end{cases} \tag{3.199}$$

The collinear singularity is also known as a mass singularity since the propagator is strictly only divergent for gluon emission off a massless parton, quark or gluon. In both limits the virtuality of the emitter tends to zero, so that it travels a large space–time distance prior to the gluon emission. These divergences are therefore associated with the long distance, infrared behaviour of the theory. Similar infrared divergences occur also in virtual processes. This is because the integrals over loop momenta include phase space regions corresponding to the emission of both collinear and low-energy, real gluons for which the propagators are singular. This leads to very long-lived virtual fluctuations. It must now be admitted that we glossed over this issue in our earlier discussions of ultraviolet divergences. Since our calculations are perturbative, based on quarks and gluons, they will break down in these limits where non-perturbative contributions enter. In this section we shall consider how to cope with these divergences and under what circumstances they cancel.

3.5.1   Infrared cancellations and dimensional regularization

The key to treating infrared divergences lies in two observations. First, whilst real diagrams squared always give positive contributions to a cross section, interference involving virtual diagrams can give negative contributions. This opens up the possibility of arranging a cancellation between the singularities in the two sets of diagrams at the level of the amplitudes squared. Second, there is a striking similarity between the two singular configurations, soft and near collinear gluon emission, and the situation where no emission at all occurs. In practical situations a detector's energy resolution and granularity will not allow a sufficiently soft or collinear emission to be distinguished from no emission. If this is to be reflected in the corresponding theoretical calculation then the two contributions need to be added to give a useful result. In the case of electron–positron annihilation to hadrons the lowest order terms in the matrix element squared are given by

$$\mathcal{M} = \mathcal{M}_{q\bar{q}}^{(0)} + \sqrt{\alpha_s}\mathcal{M}_{q\bar{q}g}^{(0)} + \alpha_s\mathcal{M}_{q\bar{q}}^{(1)} + \cdots$$

$$|\mathcal{M}|^2 = \left|\mathcal{M}_{q\bar{q}}^{(0)}\right|^2 + \alpha_s\left(\left|\mathcal{M}_{q\bar{q}g}^{(0)}\right|^2 + 2\mathcal{R}e\left\{\mathcal{M}_{q\bar{q}}^{(0)}\mathcal{M}_{q\bar{q}}^{(1)\star}\right\}\right) + \cdots . \quad (3.200)$$

There is no $\sqrt{\alpha_s}$ cross-term in the squared result as there is no common final state: $|q\bar{q}\rangle \neq |q\bar{q}g\rangle$. At next-to-leading order we should take into account both the real process $e^+e^- \rightarrow q\bar{q}g$ and the interference between the tree-level and one-loop, virtual corrections to the process $e^+e^- \rightarrow q\bar{q}$. In eqn (3.200) we anticipate that the first and second terms at $\mathcal{O}(\alpha_s)$ will contain '$+\infty$' and '$-\infty$' infrared divergences and that these will cancel when added to leave a finite result.

As with the treatment of ultraviolet divergences, before we can manipulate any matrix elements we need to regulate any infrared divergences and make them finite. It is rather pleasing that dimensional regularization again provides a suitable method. To see how this works we shall consider the tree-level process $\gamma^\star \rightarrow q\bar{q}g$, for massless quarks. In the soft gluon limit the dominant contribution to the matrix element squared comes from the cross-term, eqn (3.106),

$$|\mathcal{M}|^2 \propto \frac{q \cdot \bar{q}}{(q \cdot g)(\bar{q} \cdot g)} \sim \frac{1}{E_g^2}\frac{1}{(1 - \cos\theta_{qg})} , \quad (3.201)$$

which behaves as E_g^{-2}. This expression, describing radiation off a colour–anticolour dipole, also contains collinear singularities for $g \rightarrow\parallel q$, where it behaves as θ_{qg}^{-2}, and for $g \rightarrow\parallel \bar{q}$, where it behaves as $\theta_{\bar{q}g}^{-2}$. As we shall learn in 3.6.7 and Sections 3.7, this simplification of a matrix element squared in the soft and collinear limits is generic. Thus the contribution to the cross section from a soft gluon emitted nearly parallel to the quark is given by the following D-dimensional phase space integral

$$\int \frac{\mathrm{d}^D g}{(2\pi)^D}\Theta^{(+)}(g^2)\frac{q \cdot \bar{q}}{(q \cdot g)(\bar{q} \cdot g)} \quad (3.202)$$

$$= \int \frac{\mathrm{d}^{D-3}\Omega}{(2\pi)^D}\int_0^\infty \frac{\mathrm{d}g_\parallel \mathrm{d}g_\perp g_\perp^{D-3}}{2E_g}\frac{1}{E_g\left[E_g - g_\parallel\right]f(\Omega)}\Bigg|_{E_g=\sqrt{g_\parallel^2+g_\perp^2}}$$

$$= \int_0^\infty \mathrm{d}E_g E_g^{D-5}\int_0^\pi \mathrm{d}\theta_{qg}\frac{\sin^{D-3}\theta_{qg}}{2(1 - \cos\theta_{qg})}\int \frac{\mathrm{d}^{D-3}\Omega}{(2\pi)^D}\frac{1}{f(\Omega)}$$

$$= \int_0^\infty dE_g E_g^{D-5} \int_0^\pi d\theta_{qg} \sin^{D-5}(\theta_{qg}/2) \cos^{D-3}(\theta_{qg}/2) \int \frac{d^{D-3}\Omega}{(2\pi)^D} \frac{2^{D-5}}{f(\Omega)} .$$

In the first line we isolate the component of the gluon's momentum parallel to the quark; $f(\Omega)$ describes the (non-singular) angular dependence of the dimensionless combination $(q \cdot \bar{q}/\bar{q} \cdot g) \times (E_g/E_q)$. In the second line we have introduced polar coordinates oriented along the direction q. As the third line makes clear, it is apparent that the soft, $E_g \to 0$, singularity is integrable provided $D > 4$, as is the collinear, $\theta_{qg} \to 0$, singularity. Thus to tame infrared divergences we work in $D = 4 - 2\epsilon$ dimensions, but now with $\epsilon < 0$ so that $D > 4$.

Readers may be aware that another method for regulating infrared divergences is to add a small mass to the gluon in intermediate calculations which can be removed in the final result. However, this is a 'short-sighted' solution as a gluon mass term violates gauge invariance at $\mathcal{O}(\alpha_s)$. As soon as two gluons are involved in a situation, a gluon mass cannot be used without destroying the basis for the theory. You may wonder if it is possible to extend the BRS symmetry even further to accommodate a gluon (or ghost) mass term as well as the gauge fixing term. Unfortunately, this is only known to be possible for an abelian theory such as QED.

3.5.2 e^+e^- annihilation to hadrons at NLO

We can now start to calculate the cross section for electron–positron annihilation to hadrons including our infrared regulator. This is given schematically as the product of a lepton tensor and a hadron tensor which is integrated over the final state phase space, c.f. eqn (3.85),

$$\sigma = \frac{1}{2Q^2} L_{\mu\nu} \frac{1}{Q^4} \int d\Phi H^{\mu\nu} . \tag{3.203}$$

Observe that in the photon propagator any terms proportional to $(1 - \xi)Q^\mu Q^\nu$ do not contribute thanks to electromagnetic gauge invariance which requires that both $Q^\mu L_{\mu\nu} = 0 = Q^\nu L_{\mu\nu}$ and $Q_\mu H^{\mu\nu} = 0 = Q_\nu H^{\mu\nu}$. Now, the only available objects that can carry the integrated hadron tensor's Lorentz indices are $\eta^{\mu\nu}$ and $Q^\mu Q^\nu$. Add to this the gauge invariance requirement, and we can restrict the integrated hadron tensor to the form

$$\int d\Phi H^{\mu\nu}(Q) = \frac{1}{(D-1)} \left(-\eta^{\mu\nu} + \frac{Q^\mu Q^\nu}{Q^2} \right) H(Q^2) \tag{3.204}$$

$$\text{with} \quad H(Q^2) = -\eta_{\mu\nu} \int d\Phi H^{\mu\nu}(Q^2) .$$

You may recognize the pre-factor as the appropriate form of the spin averaged polarization sum for an off mass-shell photon in D dimensions, c.f. eqn (3.111). Substituting eqn (3.204) into eqn (3.203) and using the standard form of the (massless) lepton tensor, eqn (3.85), then gives

$$\sigma = \frac{e^2}{4Q^4} \frac{(D-2)}{(D-1)} H(Q^2) . \tag{3.205}$$

In this way we have reduced the problem to the simpler one of calculating the contraction of the hadronic tensor with $-\eta_{\mu\nu}$ and integrating it over its phase space. To confirm eqn (3.205) we apply it to the underlying, lowest order subprocess for a single, massless quark flavour,

$$
\begin{aligned}
\sigma_0^{(\epsilon)} &= \frac{e^2}{4Q^4} \frac{2(1-\epsilon)}{(3-2\epsilon)} (ee_q\mu^\epsilon)^2 N_c \times -\eta_{\mu\nu} \int d\Phi_2 \mathrm{Tr}\{\not{q}\gamma^\mu \not{\bar{q}}\gamma^\nu\} \\
&= \frac{e^2}{4Q^4} \frac{2(1-\epsilon)}{(3-2\epsilon)} (ee_q\mu^\epsilon)^2 N_c \frac{1}{4\pi} \frac{1}{2} \left(\frac{4\pi}{Q^2}\right)^\epsilon \frac{\Gamma(1-\epsilon)}{\Gamma(2-2\epsilon)} 4(1-\epsilon)Q^2 \\
&= \frac{4\pi\alpha_{\mathrm{em}}^2}{Q^2} e_q^2 N_c \frac{(1-\epsilon)^2}{(3-2\epsilon)} \frac{\Gamma(1-\epsilon)}{\Gamma(2-2\epsilon)} \left(\frac{4\pi\mu^2}{Q^2}\right)^\epsilon .
\end{aligned}
\tag{3.206}
$$

The evaluation of the hadron tensor essentially follows the earlier treatment, Section 3.3.1, however, we now work in D dimensions. This means adding a factor μ^ϵ to the quark's coupling and remembering that $\eta_\mu{}^\mu = D = 4 - 2\epsilon$ when doing the γ-matrix algebra; see Ex. (3-19). We used eqn (C.22) for the phase space. The result coincides with eqn (3.93) in the limit $D \to 4$.

3.5.2.1 The real $\mathcal{O}(\alpha_s)$ contribution At $\mathcal{O}(\alpha_s)$ the most straightforward contributions to evaluate come from the real emission process $\gamma^* \to q\bar{q}g$; see Fig. 3.11 and the earlier discussion of Section 3.3.2. Working with massless quarks in D dimensions the projection of the $q\bar{q}g$ matrix element squared is

$$
\begin{aligned}
& \frac{-\eta_{\mu\nu} H_R^{\mu\nu}}{(ee_q g_s\mu^{2\epsilon})^2 C_F N_c} \\
&= 2(1-\epsilon)\mathrm{Tr}\{\mathbf{1}\} \left[(1-\epsilon)\left(\frac{g\cdot\bar{q}}{g\cdot q} + \frac{g\cdot q}{g\cdot\bar{q}}\right) + \frac{(q\cdot\bar{q})Q^2}{(g\cdot q)(g\cdot\bar{q})} - 2\epsilon\right] \\
&= 2(1-\epsilon)\mathrm{Tr}\{\mathbf{1}\} \left[(1-\epsilon)\left(\frac{(1-x_q)}{(1-x_{\bar{q}})} + \frac{(1-x_{\bar{q}})}{(1-x_q)}\right) + \frac{2(1-x_g)}{(1-x_q)(1-x_{\bar{q}})} - 2\epsilon\right] \\
&= 2(1-\epsilon)\mathrm{Tr}\{\mathbf{1}\} \left[\frac{x_q^2 + x_{\bar{q}}^2 - \epsilon x_g^2}{(1-x_q)(1-x_{\bar{q}})}\right] .
\end{aligned}
\tag{3.207}
$$

Here we employ the usual energy fractions, x_i, defined in eqn (3.113). In the first two lines it is easy to identify the individual contributions from the diagrams corresponding to gluon emission off the quark, emission off the antiquark and their interference. In this expression the collinear singularities appear as single poles in the limits $g\cdot q \to 0$ ($x_{\bar{q}} \to 1$) or $g\cdot\bar{q} \to 0$ ($x_q \to 1$). The soft singularity appears as poles in the limit $g\cdot q \to 0$ and $g\cdot\bar{q} \to 0$ ($x_q \to 1$ and $x_{\bar{q}} \to 1$). Observe that in this limit the interference term has a double pole. The appropriate D-dimensional, three-body phase space integral is given by eqn (C.23), so that $H_R(Q^2)$ is given by

$$
H_R(Q^2) = -\eta_{\mu\nu} \int d\Phi_3 H^{\mu\nu}(Q^2) = \alpha_{\mathrm{em}} e_q^2 \frac{\alpha_s}{2\pi} C_F N_c 2Q^2 \left(\frac{4\pi\mu^2}{Q^2}\right)^{2\epsilon} \frac{(1-\epsilon)}{\Gamma(2-2\epsilon)}
$$

$$\times \int_0^1 dx_q \int_{1-x_q}^1 dx_{\bar{q}} \frac{1}{[(1-x_q)(1-x_{\bar{q}})(x_q+x_{\bar{q}}-1)]^\epsilon}$$

$$\times \left[(1-\epsilon)\left(\frac{(1-x_q)}{(1-x_{\bar{q}})} + \frac{(1-x_{\bar{q}})}{(1-x_q)} \right) + \frac{2(x_q-x_{\bar{q}}-1)}{(1-x_q)(1-x_{\bar{q}})} - 2\epsilon \right]. \quad (3.208)$$

This phase space integral looks very daunting, but can in fact be rendered quite simple by means of a change of variables, $x_{\bar{q}} = 1 - vx_q$, and $x_q = x$,

$$\int_0^1 dx \int_0^1 dv \frac{x}{[x^2(1-x)v(1-v)]^\epsilon}$$

$$\times \left\{ (1-\epsilon)\left(\frac{(1-x)}{xv} + \frac{xv}{(1-x)} \right) + 2\frac{(1-v)}{(1-x)v} - 2\epsilon \right\}$$

$$= 2\left\{ (1-\epsilon)\frac{\Gamma(2-\epsilon)\Gamma(1-\epsilon)\Gamma(-\epsilon)}{\Gamma(3-3\epsilon)} + \frac{\Gamma(2-\epsilon)\Gamma^2(-\epsilon)}{\Gamma(2-3\epsilon)} - \epsilon\frac{\Gamma^3(1-\epsilon)}{\Gamma(3-3\epsilon)} \right\}$$

$$= \frac{\Gamma^3(1-\epsilon)}{\Gamma(1-3\epsilon)}\frac{1}{(1-3\epsilon)}\left(\frac{2}{\epsilon^2} - \frac{3}{\epsilon} + \frac{(1-4\epsilon)}{(2-3\epsilon)} \right)$$

$$= \frac{\Gamma^3(1-\epsilon)}{\Gamma(1-3\epsilon)}\left(\frac{2}{\epsilon^2} + \frac{3}{\epsilon} + \frac{19}{2} + \mathcal{O}(\epsilon) \right). \quad (3.209)$$

The x and v integrals are of the standard Euler β-function form, eqn (C.27), and the resulting Γ-functions have been manipulated using eqn (C.25). Here we see a double, $1/\epsilon^2$, pole which comes from the interference term and is associated with the soft gluon singularity. There is also a single, $1/\epsilon$, pole associated with the collinear/soft divergences. Combining eqns (3.209), (3.208) and (3.205) gives the real gluon emission cross section at $\mathcal{O}(\alpha_s)$,

$$\sigma_R^{(\epsilon)} = \sigma_0^{(\epsilon)}\frac{\alpha_s}{2\pi}C_F\left(\frac{4\pi\mu^2}{Q^2} \right)^\epsilon \frac{\Gamma^2(1-\epsilon)}{\Gamma(1-3\epsilon)}\left(\frac{2}{\epsilon^2} + \frac{3}{\epsilon} + \frac{19}{2} + \mathcal{O}(\epsilon) \right). \quad (3.210)$$

3.5.2.2 *The virtual $\mathcal{O}(\alpha_s)$ contribution* We now turn our attention to the $\mathcal{O}(\alpha_s)$ contribution coming from the interference of the virtual, one-loop, corrections, shown in Fig. 3.20, and the tree-level process $\gamma^* \to q\bar{q}$.

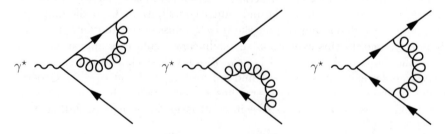

FIG. 3.20. The virtual, one-loop, QCD corrections to the $\gamma^* \to q\bar{q}$ vertex at $\mathcal{O}(\alpha_s)$

Recalling our earlier discussion of renormalization, Section 3.4.1, we anticipate that these virtual corrections contain both infrared and ultraviolet divergences. We begin by investigating the ultraviolet behaviour of the corrections.
The divergence in the quark self-energy can be read off from eqn (3.132). To
obtain the divergent part of the vertex correction we note its similarity to the
first, 'QED-like' graph in Fig. 3.18 which is identical upon making the replace-
ment $g_s(T^b T^a T^b)_{ij} \to ee_q(T^b T^b)_{ij} = ee_q C_F \delta_{ij}$. Thus, the divergence can be read
off from the first term in eqn (3.138). Temporarily abandoning our restriction to
massless, on mass-shell (anti)quarks, the one-loop contribution to the $\gamma q\bar{q}$ vertex
is given by

$$
-i\, ee_q \mu^\epsilon \delta_{ij} \frac{1}{2} \left[\Sigma(q) \frac{(\slashed{q}+m)}{q^2-m^2} \gamma^\mu + \gamma^\mu \frac{(-\slashed{q}+m)}{\bar{q}^2-m^2} \Sigma(-\bar{q}) \right] + \Gamma_{ij}(Q,q)^\mu
$$

$$
= -i\, ee_q \mu^\epsilon \delta_{ij} \frac{\alpha_s}{4\pi} C_F \left\{ \left(-\frac{1}{2} \left[\left(\xi - 3m \frac{(\slashed{q}+m)}{q^2-m^2} \right) \gamma^\mu \right. \right. \right.
$$

$$
\left. \left. \left. + \gamma^\mu \left(\xi + 3m \frac{(\slashed{\bar{q}}-m)}{\bar{q}^2-m^2} \right) \right] + \xi\gamma^\mu \right) \Delta_\epsilon + \text{U.V. finite} \right\}
$$

$$
= -i\, ee_q \delta_{ij} \frac{\alpha_s}{4\pi} C_F \times (\text{U.V. finite}) . \tag{3.211}
$$

Note that we only absorb half of the self-energy corrections into the renormalized
coupling appearing in the third line. The other half goes into the wavefunction
renormalization. We also use the Dirac equation acting on quark, $\bar{u}(q)$, and
antiquark, $v(\bar{q})$, basis states to eliminate the two remaining divergent terms.
The sum is then free of ultraviolet divergences. This result for the $\gamma q\bar{q}$ vertex
should be compared to that for the $gq\bar{q}$ vertex, Ex. (3-23).

The absence of ultraviolet divergent QCD corrections to the $\gamma q\bar{q}$ vertex is not
an accident but an important requirement for the acceptability of QCD. Its significance lies in the fact that QCD does not affect the renormalization of electric
charges or, if you wish, does not spoil electromagnetic gauge invariance. More
formally, the electric charge operator commutes with the QCD Hamiltonian. If
it did not, then electric charges would not be conserved. For example, consider
an antineutrino interacting with an electron to give hadrons, $\bar{\nu}_e e^- \to W^- \to d\bar{u}$.
The initial state has charge $-e$ and is unaffected by strong corrections, whereas
the final state is potentially affected by them and so might have a renormalized
charge $-e' \neq -e$ which is unacceptable. The $\gamma q\bar{q}$, $Zq\bar{q}$ and $Wq\bar{q}'$ vertices are free
of strong interaction divergences to all orders. This is the reason why we can cal-
culate the charge on a hadron from the sum of charges on its constituent quarks
without regard to any complex, non-perturbative, strong interaction dynamics.

Since we are focusing on the case of massless, on mass-shell (anti)quarks a
number of the virtual diagrams do not contribute. Recall eqn (3.131) for the
quark self-energy, which in the case of $p^2 = m^2 = 0$ reduces to

$$
-i\, \Sigma(p) = -(g_s \mu^\epsilon)^2 C_F \delta_{ij} \int \frac{d^D k}{(2\pi)^D} \frac{\gamma_\mu (\slashed{k}+\slashed{p})\gamma_\nu}{[k^2+2k\cdot p]k^2} \left[\eta^{\mu\nu} - (1-\xi)\frac{k^\mu k^\nu}{k^2} \right]
$$

$$= -(g_s\mu^\epsilon)^2 C_F \delta_{ij} \not{p} \int \frac{\mathrm{d}^D k}{(2\pi)^D} f(k) = 0 . \tag{3.212}$$

In the second line, we have made explicit the fact that the resulting loop integral is independent of any scale ($p^2 = 0$). This is the same situation that we encountered previously with the tadpole diagrams where the integral cannot be defined for any dimension D. In dimensional regularization these integrals are defined to be zero. Thus, the only one-loop diagram which can contribute is the vertex correction where the scale is set by $Q^2 = 2q \cdot \bar{q}$.

The contraction of the hadron tensor describing the interference between the vertex correction and tree-level diagrams is given by the real part of the following expression,

$$-\eta_{\mu\nu} H_V^{\mu\nu} = +\mathrm{i}\, 2(ee_q g_s \mu^{2\epsilon})^2 \mathrm{Tr}\left\{T^b T^b\right\} \int \frac{\mathrm{d}^D k}{(2\pi)^D} \frac{1}{k^2(k+q)^2(k-\bar{q})^2}$$

$$\times \left\{ \mathrm{Tr}\left\{\not{\bar{q}}\gamma_\sigma(\not{k}+\not{q})\gamma_\mu(\not{k}-\not{\bar{q}})\gamma^\sigma\not{q}\gamma^\mu\right\} - \frac{(1-\xi)}{k^2}\mathrm{Tr}\left\{\not{\bar{q}}\not{k}(\not{k}+\not{q})\gamma_\mu(\not{k}-\not{\bar{q}})\not{k}\not{q}\gamma^\mu\right\} \right\} . \tag{3.213}$$

Evaluating the two traces is a straightforward if tedious exercise,

$$\mathrm{Tr}\left\{\not{\bar{q}}\gamma_\sigma(\not{k}+\not{q})\gamma_\mu(\not{k}-\not{\bar{q}})\gamma^\sigma\not{q}\gamma^\mu\right\}$$
$$= +8(1-\epsilon)\left[Q^4 - 4(k\cdot q)(k\cdot\bar{q}) - 2k\cdot(q-\bar{q})Q^2 + \epsilon k^2 Q^2\right]$$
$$\mathrm{Tr}\left\{\not{\bar{q}}\not{k}(\not{k}+\not{q})\gamma_\mu(\not{k}-\not{\bar{q}})\not{k}\not{q}\gamma^\mu\right\}$$
$$= -4(1-\epsilon)(k+q)^2(k-\bar{q})^2 Q^2 . \tag{3.214}$$

The second result implies that the gauge dependent contribution is of the same type as the integral in eqn (3.212) and therefore vanishes. There is no ξ-dependence. In order to treat the remaining loop momentum integral we first combine the propagators using Feynman parameters,

$$\frac{1}{k^2(k+q)^2(k-\bar{q})^2} = \int_0^1 \mathrm{d}\alpha \int_0^{1-\alpha} \mathrm{d}\beta \frac{1}{[\alpha(k+q)^2 + \beta(k-\bar{q})^2 + (1-\alpha-\beta)k^2]^3}$$

$$= \int_0^1 \mathrm{d}\alpha \int_0^{1-\alpha} \mathrm{d}\beta \frac{1}{[(k+\alpha q - \beta\bar{q})^2 + \alpha\beta Q^2]^3} . \tag{3.215}$$

This suggests the change of variables $k^\mu \to k^\mu - \alpha q^\mu + \beta\bar{q}^\mu$ in the integral. Making this substitution in the integral's numerator the first trace, eqn (3.214), gives

$$\mathrm{Tr}\left\{\right\} = 8(1-\epsilon)\left\{Q^4 - 4\left[(k\cdot q)(k\cdot\bar{q}) - \alpha\beta(q\cdot\bar{q})^2\right]\right.$$
$$\left. -2\left[\beta q\cdot\bar{q} + \alpha q\cdot\bar{q}\right]Q^2 + \epsilon\left[k^2 - 2\alpha\beta q\cdot\bar{q}\right]Q^2\right\} \tag{3.216}$$
$$= 8(1-\epsilon)Q^2\left\{\left[1-\alpha-\beta + (1-\epsilon)\alpha\beta\right]Q^2 - (1-\epsilon)^2(2-\epsilon)^{-1}k^2\right\} .$$

In the first line we have discarded any terms that are linear in k and so give vanishing contributions in an isotropic integral. Whilst in the second line we have replaced $k^\mu k^\nu$ by $k^2 \eta^{\mu\nu}/D$, again by virtue of the integral's isotropy. When we substitute eqns (3.215) and (3.216) into (3.213) we see that we have an expression which only depends on Q^2. This makes the evaluation of the two-body phase space integral in eqn (3.204) trivial and we can go straight to eqn (3.205) for the result,

$$
\begin{aligned}
\sigma_V^{(\epsilon)} &= +i\sigma_0^{(\epsilon)} 4(g_s \mu^\epsilon)^2 C_F \int_0^1 d\alpha \int_0^{1-\alpha} d\beta \int \frac{d^D k}{(2\pi)^D} \frac{1}{[k^2 + \alpha\beta Q^2]^3} \\
&\quad \times \left\{ [1-\alpha-\beta+(1-\epsilon)\alpha\beta]Q^2 - (1-\epsilon)^2(2-\epsilon)^{-1}k^2 \right\} \\
&= -\sigma_0^{(\epsilon)} \frac{\alpha_s}{2\pi} C_F \left(\frac{4\pi\mu^2}{-Q^2} \right)^\epsilon \Gamma(1+\epsilon) \int_0^1 d\alpha \int_0^{1-\alpha} d\beta \frac{1}{(\alpha\beta)^\epsilon} \\
&\quad \times \left\{ \frac{[1-\alpha-\beta+(1-\epsilon)\alpha\beta]}{\alpha\beta} - \frac{(1-\epsilon)^2}{\epsilon} \right\}.
\end{aligned}
\tag{3.217}
$$

The loop momentum integrals are evaluated using eqn (C.11). The integral over the Feynman parameters is simplified by means of the change of variable $\beta = (1-\alpha)v$ which decouples the integrals and allows the use of eqn (C.27) to obtain

$$
\begin{aligned}
&\int_0^1 d\alpha \int_0^{1-\alpha} d\beta \frac{1}{(\alpha\beta)^\epsilon} \left\{ \frac{[1-\alpha-\beta+(1-\epsilon)\alpha\beta]}{\alpha\beta} - \frac{(1-\epsilon)^2}{\epsilon} \right\} \\
&= \int_0^1 d\alpha \int_0^1 dv \frac{(1-\alpha)}{[\alpha(1-\alpha)v]^\epsilon} \left\{ \frac{[(1-\alpha)(1-v)+(1-\epsilon)\alpha(1-\alpha)v]}{\alpha(1-\alpha)v} - \frac{(1-\epsilon)^2}{\epsilon} \right\} \\
&= \frac{\Gamma^2(1-\epsilon)}{\Gamma(2-2\epsilon)} \left[\frac{1}{\epsilon^2} + \frac{1}{2} - \frac{(1-\epsilon)}{2\epsilon} \right] \\
&= \frac{\Gamma^2(1-\epsilon)}{\Gamma(1-2\epsilon)} \frac{1}{2} \left[\frac{2}{\epsilon^2} + \frac{3}{\epsilon} + \frac{8}{(1-2\epsilon)} \right].
\end{aligned}
\tag{3.218}
$$

Substituting this result into eqn (3.217) gives us our final result,

$$
\begin{aligned}
\sigma_V^{(\epsilon)} &= -\sigma_0^{(\epsilon)} \frac{\alpha_s}{2\pi} C_F \left(\frac{4\pi\mu^2}{-Q^2} \right)^\epsilon \frac{\Gamma(1+\epsilon)\Gamma^2(1-\epsilon)}{\Gamma(1-2\epsilon)} \left(\frac{2}{\epsilon^2} + \frac{3}{\epsilon} + \frac{8}{(1-2\epsilon)} \right) \\
&= -\sigma_0^{(\epsilon)} \frac{\alpha_s}{2\pi} C_F \left(\frac{4\pi\mu^2}{-Q^2} \right)^\epsilon \frac{\Gamma(1+\epsilon)\Gamma^2(1-\epsilon)}{\Gamma(1-2\epsilon)} \left(\frac{2}{\epsilon^2} + \frac{3}{\epsilon} + 8 + \mathcal{O}(\epsilon) \right).
\end{aligned}
\tag{3.219}
$$

Here, it is important to remember that the real part is understood.

3.5.2.3 *The combined $\mathcal{O}(\alpha_s)$ contribution* After having calculated the real, eqn (3.210), and the virtual, eqn (3.219), gluon corrections to the lowest order expression, eqn (3.206), we can combine them to give the complete, $\mathcal{O}(\alpha_s)$ cross section for electron–positron annihilation to hadrons,

$$
\sigma = \sigma_0^{(\epsilon)} \left\{ 1 + \frac{\alpha_s}{2\pi} C_F \left(\frac{4\pi\mu^2}{Q^2} \right)^\epsilon \frac{\Gamma^2(1-\epsilon)}{\Gamma(1-3\epsilon)} \left[\left(\frac{2}{\epsilon^2} + \frac{3}{\epsilon} + \frac{19}{2} + \mathcal{O}(\epsilon) \right) \right. \right.
$$

$$+\mathcal{R}e\{(-1)^\epsilon\}\frac{\Gamma(1+\epsilon)\Gamma(1-3\epsilon)}{\Gamma(1-2\epsilon)}\left(-\frac{2}{\epsilon^2}-\frac{3}{\epsilon}-8+\mathcal{O}(\epsilon)\right)\right]\Bigg\}$$

$$= [\sigma_0 + \mathcal{O}(\epsilon)]\left\{1 + \frac{\alpha_s}{2\pi}C_F[1+\mathcal{O}(\epsilon)]\left[\left(\frac{2}{\epsilon^2}+\frac{3}{\epsilon}+\frac{19}{2}+\mathcal{O}(\epsilon)\right)\right.\right.$$

$$\left.\left.- [1+\mathcal{O}(\epsilon^4)]\left(\frac{2}{\epsilon^2}+\frac{3}{\epsilon}+8+\mathcal{O}(\epsilon)\right)\right]\right\}$$

$$= \sigma_0\left\{1 + \frac{3}{4}C_F\frac{\alpha_s}{\pi}+\mathcal{O}(\alpha_s^2)\right\}\ . \tag{3.220}$$

In the second line we made use of the expansion $\mathcal{R}e\{(-1)^\epsilon\} = \mathcal{R}e\{e^{i\pi\epsilon}\} = 1 - (\pi^2/2)\epsilon^2 + \mathcal{O}(\epsilon^4)$ and eqn (C.26). As a result, it becomes clear that both the $1/\epsilon^2$ and $1/\epsilon$ poles cancel to leave a finite result. As a consequence we can safely take $\epsilon \to 0$ and obtain the $D = 4$ limit. Thus, for a suitably inclusive definition of the hadronic cross section we avoid the potential infrared catastrophe associated with soft gluons.

Before going on to identify the general characteristics of infrared safe observables we mention a succinct way of organizing the above cancellation using cut-diagrams. The $\mathcal{O}(\alpha_s)$ cross section is represented by the diagrams shown in Fig. 3.21. The two basic diagrams have each been 'cut' in two ways. On the left-hand side of the cut we view it as the usual Feynman diagram corresponding to $\mathcal{M}$. On the right-hand side of the cut we view it as a 'reversed' Feynman diagram corresponding to $\mathcal{M}^\star$. Thus, the top-left diagram represents the interference between the two tree-level Feynman diagrams, whilst the bottom-left diagram represents the interference between the tree-level and one-loop correction to the basic $\gamma q \bar{q}$ vertex. You should notice that these are the same diagram cut in two different places. This one diagram encapsulates the contributions which must be taken into account to ensure the cancellation of infrared divergences.

3.5.3 Infrared safe observables

Our calculation of the cross section for electron–positron annihilation to hadrons demonstrates that it is free of infrared divergences to at least $\mathcal{O}(\alpha_s)$. In fact, the KLN-theorem (Kinoshita, 1962; Lee and Nauenberg, 1964), and its generalization to QCD (Poggio and Quinn, 1976; Sterman, 1976), guarantees that such a fully inclusive observable is infrared finite to all orders. This theorem can be extended so as to apply to other observables (Sterman and Weinberg, 1977; Dokshitzer *et al.*, 1980). The key requirement for the cancellation is that a quark and a quark accompanied by any number of soft gluons and/or collinear gluons and q$\bar{\text{q}}$-pairs are treated the same. Likewise, $|g\rangle$ and $\left|g + n_1 g_s + n_2 g_\parallel + n_3 (q\bar{q})_\parallel\right\rangle$ must give the same contribution to an observable. Now, in practice cross sections are often weighted by functions corresponding to physical measurements on the parton final states. In general a measurement is described by an expression of the form

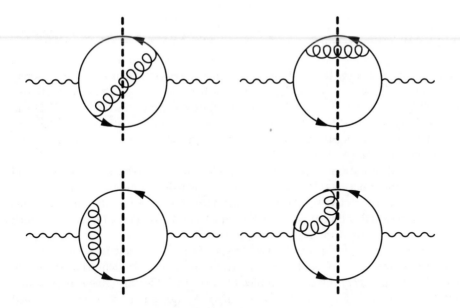

FIG. 3.21. The cut-diagrams which describe the contributions to the amplitude squared for $\gamma^* \to$ hadrons at $\mathcal{O}(\alpha_s)$

$$I = \frac{1}{\text{flux}} \sum_n \frac{1}{n!} \int d\Phi^n \overline{\sum} \left| \mathcal{M}^{(n)}(p_i) \right|^2 \rho_n(p_i) \qquad (3.221)$$

$$\text{with} \quad \begin{cases} \rho_n = 1 & I = \sigma & \text{(total cross section)} \\ \rho_n = \delta(X - \mathcal{X}_n(p_i)) & I = \dfrac{d\sigma}{dX} & \text{(differential cross section)} . \end{cases}$$

The sum includes the contributions from all n-parton final states and, on the assumption that quarks, antiquarks and gluons are not distinguished, we include the symmetrization factor $1/n!$. The weight functions of the n final state partons, $\rho_n(p_1, \ldots, p_n)$, define the measurement. For example, to obtain the total cross section, $I = \sigma$, we simply use $\rho_n(p_i) = 1$. Often, we are interested in the distribution of some variable X which is given by the function $\mathcal{X}_n(p_i)$ for n partons. Typical observables discussed later include the event's Thrust, see eqn (6.2), or a jet resolution parameter, see eqn (6.3) or (6.4). To obtain $I = d\sigma/dX$ we have to use $\rho_n = \delta(X - \mathcal{X}_n(p_i))$. However, in view of the cancellations required for infrared safety, the functions $\mathcal{X}_{n+1}$ and $\mathcal{X}_n$ must become equal in the soft and collinear limits,

$$\left. \begin{array}{l} \mathcal{X}_{n+1}(p_1, \ldots, \lambda p_n, (1 - \lambda)p_n) \\ \mathcal{X}_{n+1}(p_1, \ldots, p_n, 0) \end{array} \right\} = \mathcal{X}_n(p_1, \ldots, p_n) . \qquad (3.222)$$

Whilst this is a requirement imposed on theoretical grounds, you are reminded that it also has a basis in experimental reality. The results of a measurement

should be insensitive to changes in a detector's energy resolution or granularity.
~~As a result of this requirement any observable which is not a linear function of~~
the parton four-momenta will not be infrared safe and should not be used in
experiment or theory. If you do not heed this restriction then you will be sensitive to long-distance physics. This means either introducing a non-perturbative
model, upon which your results will depend, or accepting that using pQCD will
give infinity at NLO. Despite this established wisdom it is still too common to
see variables such as Sphericity (not to be confused with spherocity) or (some)
cone based jet finders which are infrared unsafe!

A special situation arises for observables such as structure functions and
fragmentation functions. By their nature the inclusive summation over initial or
final states is restricted by the requirement to contain a specific particle. This
results in residual collinear singularities which can be treated in a manner which
is reminiscent of renormalization: see Section 3.6.

The first example of an observable specifically designed to satisfy the infrared finiteness requirement is the Sterman–Weinberg jet definition (Sterman
and Weinberg, 1977). Other examples of early variables are given in (Basham
et al., 1978a) and (Fox and Wolfram, 1978). A hadronic e^+e^- event is defined
to be a two-jet event if a fraction $(1 - \epsilon)$ of the total C.o.M. energy is contained within two back-to-back cones of half angle δ. Whilst this clearly satisfies
eqn (3.222), in practice it is rarely used at e^+e^- colliders, though such 'cone
based' definitions are still used at hadron colliders. A more popular jet definition is based on the minimum invariant mass of particle pairs. At leading order
all events are two-jet events. At $\mathcal{O}(\alpha_s)$, an event is only classified as two-jet if one
of the three possible pairs of partons has $(p_i + p_j)^2/Q^2 < y$ for a given value of the
resolution parameter y. By construction we have $y < 1/3$. A two-jet event occurs
for two basic phase space configurations. Either, the gluon is radiated sufficiently
close in direction to the quark or with sufficiently low energy that $(q+g)^2/Q^2 < y$
and the qg-pair forms a jet recoiling against the $\bar{q}$ (or with q and $\bar{q}$ interchanged);
or, much less likely, the gluon has high energy and the $q\bar{q}$-pair forms a jet recoiling against the gluon. Using $(p_i + p_j)^2/Q^2 = 1 - x_k > y$ we see that the three-jet
region of phase space is confined to a triangular region, $2y < x_i < 1 - y$, in
the centre of the $x_q - x_{\bar{q}}$ plane away from the collinear and soft singularities,
see Fig. 3.12 where $y = 1 - T$. This means that calculating the $\mathcal{O}(\alpha_s)$, three-jet
cross section is relatively straightforward as, unlike the two-jet cross section, it
does not involve any infrared cancellations and therefore it does not require a
regulator. One can then use the known result $\sigma_2 + \sigma_3 = \sigma_0[1 + (3C_F\alpha_s/(4\pi)]$,
eqn (3.220), to infer the two-jet cross section. Rather than do this in one step
we first use eqn (3.221) with $\mathcal{X}_{n=3} = \min\{1 - x_1, 1 - x_2, 1 - x_3\}$, $X = y$ and the
matrix element squared given by eqn (3.112),

$$\frac{d\sigma_3}{dy} = \sigma_0 \frac{\alpha_s}{2\pi} C_F \int_{2y}^{1-y} dx \left\{ 2 \left. \frac{x^2 + \bar{x}^2}{(1-x)(1-\bar{x})} \right|_{\bar{x}=1-y} + \left. \frac{x^2 + \bar{x}^2}{(1-x)(1-\bar{x})} \right|_{\bar{x}=1+y-x} \right\}$$

$$= \sigma_0 \frac{\alpha_s}{2\pi} C_F \int_{2y}^{1-y} dx \left\{ 2 \frac{(1-y)^2 + 1 - (1-x)(1+x)}{y(1-x)} \right.$$

$$\left. + \frac{1}{1-y} \left(\frac{1}{1-x} + \frac{1}{x-y} \right) \left[1 + y^2 - 2(1-x)(x-y) \right] \right\}$$

$$= \sigma_0 \frac{\alpha_s}{2\pi} C_F \left[2 \frac{(3y^2 - 3y + 2)}{y(1-y)} \ln \left(\frac{1-2y}{y} \right) - 3 \frac{(1-3y)(1+y)}{y} \right]. \qquad (3.223)$$

This expression shows a characteristic logarithmic divergence as $y \to 0$, that is, as the singular regions are approached. In passing, we mention that this expression gives the Thrust distribution with $T = 1 - y$. Integrating this expression from y to the phase space limit $1/3$, we can calculate σ_3 and hence the n-jet rates defined by $f_n = \sigma_n/\sigma_{tot}$ as

$$f_3(y) = \frac{\alpha_s}{2\pi} C_F \left\{ \frac{5}{2} - \frac{\pi^2}{3} - 6y - \frac{9}{2}y^2 + (3 - 6y) \ln \left(\frac{y}{1-2y} \right) \right.$$

$$\left. + 4 \text{Li}_2 \left(\frac{y}{1-y} \right) + 2 \ln^2 \left(\frac{y}{1-y} \right) \right\}$$

and $\quad f_2(y) = 1 - f_3(y)$. $\qquad (3.224)$

Here, the dilogarithm or Spence function is defined by

$$\text{Li}_2(x) = - \int_0^x dz \frac{\ln(1-z)}{z}. \qquad (3.225)$$

It arises frequently in QCD calculations. Equation (3.224) shows that in the limit $y \to 0$, $f_3(y)$ diverges as $(\alpha_s C_F/\pi) \ln^2 y$. This can give the counter-intuitive result $f_2(y) < f_3(y)$ and worse $f_2(y) < 0$. Whilst this is unsettling, it is possible because $f_2(y)$ receives contributions from an interference term. In calculating $f_3(y)$ we restrict ourselves to a subregion of the real phase space so that the full cancellation of divergences which occurred in eqn (3.220) is now only partially complete and residual logarithms remain. At higher orders we may anticipate terms of the form $[\alpha_s C_F/(2\pi) \ln^2 y]^m$. Such large enhancements have obvious implications for the convergence of cross sections, and consequently a significant effort is dedicated to identifying and 'resumming' such contributions. For reference we also give the $\mathcal{O}(\alpha_s)$, three-jet rate for the Sterman–Weinberg jet definition,

$$f_3(\epsilon, \delta) = \frac{\alpha_s}{2\pi} C_F 8 \left\{ \ln \left(\frac{1}{\delta} \right) \left[\ln \left(\frac{1}{2\epsilon} - 1 \right) - \frac{3}{4} \right] + \frac{\pi^2}{12} - \frac{7}{16} + \mathcal{O}(\epsilon \ln \delta, \delta^2 \ln \epsilon) \right\}. \qquad (3.226)$$

Here, it is clear that the most singular term is associated with overlapping collinear, $\delta \to 0$, and soft, $\epsilon \to 0$, singularities.

3.6   The QCD improved parton model

The naïve (quark) parton model is independent of QCD as such and indeed was invented before QCD existed (Bjorken and Paschos, 1969; Feynman, 1972). It

began as a quasi-classical model for DIS, based upon the idea that a hadron
can be described as a collection of independent partons, with little transverse
momentum, off which a lepton can scatter via the exchange of a vector boson.
The constituent quark model supplies quantum numbers to these partons and
thereby suggests relationships between the various structure functions (Close,
1979). At this tree level all that pQCD supplies is support for treating the partons as independent, via asymptotic freedom, and candidates, the gluons, for
the electroweak neutral partons inferred from the apparent violation of the momentum sum rule (Llewellyn-Smith, 1972). This parton model picture of DIS is
easily generalized to hadron–hadron collisions.

The parton model comes to life when we add pQCD corrections (Altarelli,
1982). This brings to the fore quantum effects and changes our picture of the partons within a hadron. An essential feature of the parton model is the separation
of a cross section into hadron independent coefficient functions, which describe
the parton scatterings, and scattering independent p.d.f.s, which characterize the
hadrons; see, for example, eqn (3.227). In order to maintain this separation in the
presence of QCD corrections, we are obliged to make the p.d.f.s scale dependent,
that is, functions of both x and Q^2. This introduces the idea that a parton contains within it further 'daughter' partons and that these are revealed when the
Q^2 of the probing vector boson is increased. This scale dependence is governed
by the famous DGLAP equations and whilst it remains true that, in the absence
of suitable non-perturbative techniques for QCD, we cannot calculate the p.d.f.s
from first principles, we can deduce the p.d.f.s at one scale from a given set at
another scale. Introducing pQCD also forces us to give a precise meaning to
the idea of factorization, which has now been proved to hold in pQCD (Collins
and Soper, 1987). In doing so we give greater legitimacy to the QCD improved
parton model; so much so that the QCD improved parton model provides the
conventional framework for carrying out pQCD calculations.

Inevitably, when we add pQCD corrections to the naïve parton model, the
necessary mathematics becomes more involved. However, we believe that the underlying ideas are not that complicated. Therefore, after repeating the tree-level
treatment of DIS we give a heuristic development of the NLO pQCD corrections
to DIS and factorization. This is followed by the complete $\mathcal{O}(\alpha_s)$ calculation.
After this we switch attention to the DGLAP evolution equations and their generalizations. This is followed by a discussion of how these equations take account
of large logarithmic enhancements to a cross section. Finally, we show how the
factorization formalism is applied to the Drell–Yan process in hadron–hadron
collisions.

3.6.1 *DIS at the parton level*

The formal description of lepton–parton scattering follows that for lepton–hadron
scattering. The partonic cross section, $d\hat{\sigma}^{\ell f}$, is given by eqn (3.32) with two
modifications: the hadron momentum p^μ is replaced by the parton momentum
yp^μ and in the hadron tensor the state $|h\rangle$ is replaced by $|f\rangle$, $f = \{q, \bar{q}, g\}$, to

give $\hat{H}_{\mu\nu}^{(Vf)}$. Again gauge invariance ensures that this partonic tensor retains the form given in eqn (3.36) but with partonic structure functions. These partonic structure functions are functions of yp^μ and q_V^μ which, thanks to Bjorken scaling (Bjorken, 1969), occur in the combination $-q_V^2/(y2p \cdot q_V) = x/y$. Since we use the name of a particle to represent its four-momentum, we use q_V^μ for the four-momentum of the exchanged boson to avoid any confusion with the momentum of a quark, q^μ. The lepton tensor is, as before, given by eqn (3.33). This parton cross section is related to the hadron cross section by weighting it by the hadron's p.d.f.s,

$$d\sigma^{(\ell h)} = \sum_{f=q,\bar{q},g} \int_0^1 dy\, f_h(y) d\hat{\sigma}^{(\ell f)}\left(\frac{x}{y}\right). \tag{3.227}$$

This implies

$$H_{\mu\nu}^{(Vh)}(p, q_V) = \sum_{f=q,\bar{q},g} \int_0^1 \frac{dy}{y} f_h(y) \hat{H}_{\mu\nu}^{(Vf)}(yp, q_V), \tag{3.228}$$

where the factor $1/y$ can be traced to the scaling $p^\mu \to yp^\mu$ used to obtain the lepton–parton flux factor.

Rather than work with the full hadronic tensor, it is helpful to project out two combinations of structure functions,

$$H_{\Sigma}^{(Vh)} \equiv -\eta^{\mu\nu} H_{\mu\nu}^{(Vh)}$$
$$= (D-2)\frac{F_2}{2x}\left(1 + \frac{(2xM_h)^2}{Q^2}\right) - (D-1)\left[\frac{F_2}{2x}\left(1 + \frac{(2xM_h)^2}{Q^2}\right) - F_1\right]$$

$$H_L^{(Vh)} \equiv p^\mu p^\nu H_{\mu\nu}^{(Vh)}$$
$$= \frac{Q^2}{(2x)^2}\left[\frac{F_2}{2x}\left(1 + \frac{(2xM_h)^2}{Q^2}\right) - F_1\right]\left(1 + \frac{(2xM_h)^2}{Q^2}\right). \tag{3.229}$$

This will simplify the expressions with which we have to work. Here, with a view to future use, we have chosen to work in $D = 4 - 2\epsilon$ dimensions. Similar projections can be defined at the parton level. In the case of $\hat{H}_\Sigma$ this is straightforward but for $\hat{H}_L$ we have to use the parton momentum, yp, in the equivalent of eqn (3.229). Referring to eqn (3.228) we then have

$$H_\Sigma^{(Vh)} = \sum_{f=q,\bar{q},g} \int_0^1 \frac{dy}{y} f_h(y) \hat{H}_\Sigma^{(Vf)}(yp, q_\gamma) = \sum_{f=q,\bar{q},g} \int_x^1 \frac{dz}{z} f_h\left(\frac{x}{z}\right) \hat{H}_\Sigma^{(Vf)}(z)$$

$$H_L^{(Vh)} = \sum_{f=q,\bar{q},g} \int_0^1 \frac{dy}{y^3} f_h(y) \hat{H}_L^{(Vf)}(yp, q_\gamma) = \sum_{f=q,\bar{q},g} \frac{1}{x^2} \int_x^1 \frac{dz}{z} f_h\left(\frac{x}{z}\right) z^2 \hat{H}_L^{(Vf)}(z).$$
$$\tag{3.230}$$

Remember that scaling implies that the $\hat{H}_i^{(Vf)}$ are functions of $Q^2/(y2p \cdot q_\gamma) = x/y$. The advantage of the 'total' structure function, $\hat{H}_\Sigma^{(Vf)}$, is that it is essentially the matrix element squared for the vector boson–parton subprocess.

The 'longitudinal' structure function, $\hat{H}_L^{(Vf)}$, is particularly nice because many diagrams give vanishing contributions and those that do not vanish at $\mathcal{O}(\alpha_s)$ are free of infrared singularities. In eqn (3.229) we recognise the combination in square brackets as the longitudinal structure function; see Ex. (3-5). Once we have calculated H_Σ and H_L we can invert eqn (3.229) to give us the structure functions F_2 and F_1,

$$\frac{F_2(x)}{x} = \frac{1}{(1-\epsilon)}H_\Sigma^{(Vh)} + \frac{(3-2\epsilon)}{(1-\epsilon)}\frac{4x^2}{Q^2}H_L^{(Vh)}$$

$$= \sum_{f=q,\bar{q},g}\int_x^1\frac{dz}{z}f\left(\frac{x}{z}\right)\left[\frac{1}{(1-\epsilon)}\hat{H}_\Sigma^{(Vf)}(z) + \frac{(3-2\epsilon)}{(1-\epsilon)}\frac{4z^2}{Q^2}\hat{H}_L^{(Vf)}(z)\right]$$

$$F_1(x) - \frac{F_2(x)}{2x} = -\frac{4x^2}{Q^2}H_L^{(Vh)}$$

$$= -\sum_{f=q,\bar{q},g}\int_x^1\frac{dz}{z}f\left(\frac{x}{z}\right)\frac{4z^2}{Q^2}\hat{H}_L^{(Vf)}(z)\,. \tag{3.231}$$

Here we have made the simplifying assumption that the $2xM_h/Q$ terms are negligible. In what follows we shall also neglect all quark masses. This makes the algebra simpler and will cast into sharper relief any collinear singularities.

3.6.2 DIS at leading order

We now calculate the leading order contributions to DIS in the parton model. We will focus on electromagnetic exchange in which a photon couples to the electrically charged partons: quarks and antiquarks. The charge conjugation symmetry of QED and QCD ensures that quarks and antiquarks give the same contribution. Thus at $\mathcal{O}(\alpha_s^0)$ we need only consider the one tree-level subprocess $\gamma^\star q \to q'$. The treatment of Z and $W^\pm$ exchange involves only minor modifications.

To calculate $\hat{H}_\Sigma^{(\gamma q)}$ we first require the matrix element squared. This is easily evaluated in D dimensions,

$$\sum|\mathcal{M}(\gamma^\star q \to q')|^2 = e^2e_q^2N_c(1-\epsilon)\mathrm{Tr}\{\mathbf{1}\}Q^2\,. \tag{3.232}$$

Here $Q^2 = -(q'-q)^2 = 2q\cdot q' > 0$. Next, we average over the spin and colour polarizations of the incoming quark, $2N_c$, and include the one-body phase space integral, eqn (C.19), to obtain

$$\int d\Phi_1\overline{\sum}|\mathcal{M}(\gamma^\star q \to q')|^2 = 2e^2e_q^2(1-\epsilon)Q^2 \times 2\pi\delta(q'^2)\,. \tag{3.233}$$

Here we have used $\mathrm{Tr}\{\mathbf{1}\} = 4$. Since the struck quark carries a fraction y of the parent hadron's momentum, $q^\mu = yp^\mu$, the δ-function, which constrains the scattered quark to be on mass-shell, can be rewritten as

$$q'^2 = (yp + q_\gamma)^2 = y2p\cdot q^\gamma - Q^2 = 2p\cdot q^\gamma(y - x)$$

$$\implies \ \delta(q'^2) = \frac{1}{2p \cdot q^\gamma} \delta(y - x) . \tag{3.234}$$

Here x is the usual Bjorken-x, eqn (3.39). At this point we pause to observe that the origin of this $\delta(y-x)$ factor is purely kinematical and therefore we may anticipate that all the one-loop corrections to the $\gamma^\star q \to q'$ vertex will also be proportional to $\delta(y - x)$. Finally, following convention, we divide out a factor $4\pi e^2$ to obtain

$$\hat{H}_\Sigma^{(\gamma q)} \equiv \frac{1}{4\pi e^2} \int d\Phi_1 \overline{\sum} |\mathcal{M}(\gamma^\star q \to q')|^2 = e_q^2 (1 - \epsilon) \frac{Q^2}{2p \cdot q^\gamma} \delta(y - x)$$
$$= e_q^2 (1 - \epsilon) x \delta(y - x) . \tag{3.235}$$

The effect of the δ-function in eqn (3.235) is to select only those (anti)quarks with momentum fraction x. The presence of a δ-function also means that the partonic structure functions are formally distribution functions (in the mathematical sense) and so only have meaning when integrated with a sufficiently smooth ordinary function. The calculation of $\hat{H}_L^{(\gamma q)}$, eqn (3.229), is even easier as it vanishes. This follows because, assuming massless quarks so that $\slashed{q}u(q) = 0$, we have

$$q_\mu \mathcal{M}^\mu (\gamma^{\star\mu} q \to q') \propto \bar{u}(q') \slashed{q} u(q) = 0 . \tag{3.236}$$

Given $\hat{H}_\Sigma^{(\gamma q)}$, together with $\hat{H}_L^{(\gamma q)} = 0$, we use eqn (3.231) to obtain the lowest order electromagnetic structure functions

$$2x F_1^{(\gamma h)}(x) = F_2^{(\gamma h)}(x) = x \sum_{f=q,\bar{q}} e_f^2 f_h(x) . \tag{3.237}$$

This confirms the Callan–Gross relationship between F_1 and F_2 which holds at lowest order for scattering off a spin-1/2 parton. As we will see at $\mathcal{O}(\alpha_s)$ and beyond $F_2(x) \neq 2x F_1(x)$ and the two structure functions can no longer be regarded as equivalent. In Ex. (3-24) the same result is obtained using the explicit hadron tensor. This also demonstrates, as one might expect of the parity conserving QED, that $F_3^{(\gamma h)} = 0$.

This calculation has familiarized us with our notation and proved the Callan–Gross relationship between the structure functions appearing in the parton model at tree-level. We now wish to investigate how this picture changes with the inclusion of pQCD corrections.

3.6.3 A heuristic treatment of factorization

A number of processes contribute to the structure functions at $\mathcal{O}(\alpha_s)$. If the struck parton is a(n anti)quark we have the tree-level scattering $\gamma^\star q \to q'g$, the so-called QCD Compton process, shown in Fig. 3.22. To this must be added the interference between the tree-level and the one-loop corrections to the basic scattering $\gamma^\star q \to q'$. The situation here is similar to that encountered in the treatment of the pQCD corrections to the process $\gamma^\star \to q\bar{q}$; see Section 3.5.

There, both the ultraviolet and infrared singularities in the two sets of contributions cancelled to leave a finite result. Electroweak vector bosons do not directly couple to gluons. In order for gluons to contribute to the structure functions they must first split into a charged q$\bar{\text{q}}$-pair. Thus their contribution is at least $\mathcal{O}(\alpha_\text{s})$. The lowest order contribution comes from the tree-level process $\gamma^\star$g $\to$ q$\bar{\text{q}}$, the so-called boson–gluon fusion process, shown in Fig. 3.23.

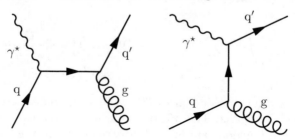

FIG. 3.22. The two diagrams contributing to the QCD Compton process, $\gamma^\star$q $\to$ q$'$g, at leading $\mathcal{O}(\alpha_\text{s})$ and, reading right to left, g$\bar{\text{q}}'$ $\to$ $\gamma^\star\bar{\text{q}}$

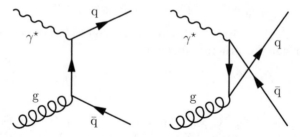

FIG. 3.23. The two diagrams contributing to the boson–gluon fusion process, $\gamma^\star$g $\to$ q$\bar{\text{q}}$, at leading $\mathcal{O}(\alpha_\text{s})$ and, reading right to left, q$\bar{\text{q}}$ $\to$ $\gamma^\star$g

We shall concentrate on the tree-level, $\mathcal{O}(\alpha_\text{s})$, process $\gamma^\star$q $\to$ q$'$g and in particular the partonic structure function $\hat{H}_\Sigma^{(\gamma\text{q})}$, since this contains examples of all the singularities with which we shall have to deal. As noted, $\hat{H}_\Sigma^{(\gamma\text{q})}$ is proportional to the matrix element squared for the hard subprocess. This can be obtained from that for the process $\gamma^\star \to$ q$\bar{\text{q}}$g, eqn (3.207), using crossing, eqn (3.95). That is, making the substitutions $q^\mu \to q'^\mu$, $\bar{q}^\mu \to -q^\mu$ and $Q^2 \to -Q^2$, reflecting the space-like nature of the photon's virtuality, one finds

$$\sum |\mathcal{M}(\gamma^\star\text{q} \to \text{q}'\text{g})|^2 = 8e^2 e_\text{q}^2 g_s^2 \text{Tr}\,\{T^a T^a\} \left[\frac{g\cdot q'}{g\cdot q} + \frac{g\cdot q}{g\cdot q'} + \frac{Q^2(q\cdot q')}{(g\cdot q)(g\cdot q')} \right] .$$

$$(3.238)$$

The three terms correspond to emission of the gluon off the outgoing quark, off the incoming quark and the interference between the two contributions. To obtain $\hat{H}_\Sigma^{(\gamma q)}$ we need to average over the incoming quark spins and colours, $2N_c$, divide by the conventional factor $4\pi e^2$ and integrate over the two-body phase space of the final state particles,

$$
\hat{H}_\Sigma^{(\gamma q)} = \frac{1}{4\pi e^2} \int d\Phi_2 \overline{\sum} |\mathcal{M}(\gamma^* q \to q' g)|^2
$$

$$
= 4e_q^2 \alpha_s C_F \int d\Phi_2 \left[\frac{g \cdot q'}{g \cdot q} + \frac{g \cdot q}{g \cdot q'} + \frac{Q^2 (q \cdot q')}{(g \cdot q)(g \cdot q')} \right] . \tag{3.239}
$$

Here we used $\mathrm{Tr}\{T^a T^a\} = C_F N_c$. Looking at the propagators, $\propto [E_g(1 - \cos\theta)]^{-1}$, we see that this expression has a number of singular regions; c.f. Section 3.3.2. There are collinear singularities when the gluon is emitted parallel to the incoming quark, $g \cdot q \to 0$, or the outgoing quark, $g \cdot q' \to 0$. There is also a soft singularity when the energy of the gluon vanishes. The collinear singularities are associated with the vanishing of the quark propagators just prior to or just after the interaction with the photon. Such a low-virtuality intermediate quark will travel a large distance, $x^\mu = k^\mu / k^2$, so that the gluon is emitted either well before or well after the hard subprocess. We may therefore anticipate that the initial state collinear singularity can be naturally associated with the incoming hadron and into which it might be absorbed. The final state collinear and soft gluon singularities both imply a zero mass particle in the final state,

$$
\hat{s} = (g + q')^2 = 2g \cdot q' \tag{3.240}
$$

$$
= (q + q_\gamma)^2 = y2p \cdot q_\gamma - Q^2 = Q^2 \left(\frac{y}{x} - 1 \right) = Q^2 \frac{(1 - z)}{z} .
$$

In the second form we let the incoming quark carry a momentum fraction y, that is, $q^\mu = yp^\mu$ and we introduced $z = x/y$ where x has its usual meaning, eqn (3.39). Thus, $\hat{s} = 2g \cdot q' \to 0$ is equivalent to $Q^2(y/x - 1) \to 0$ so that these singularities involve the kinematics of the lowest order hard scattering to which they must consequently be associated. They also raise the spectre of infrared singular cross sections.

Fortunately, for the analogous process $\gamma^* \to q\bar{q}g$ we learnt that including the contribution from the interference between the tree-level and one-loop corrections to the process $\gamma^* \to q\bar{q}$ ensured that all the singularities cancelled in the infrared safe total cross section, eqn (3.220); see Section 3.5. Hence we might expect that the singularities present in eqn (3.239) will cancel when we include the contribution coming from the interference between the tree-level and one-loop corrections to the process $\gamma^* q \to q'$. Unfortunately, in DIS the probing photon can tell the difference between the charged quark in a collinear qg-pair and a quark q with equal momentum. Thus, the equivalent cancellation for DIS is incomplete. In particular the final state collinear and soft gluon singularities cancel but the collinear singularity associated with the incoming quark remains. This removes the danger of a singular DIS cross section, provided we have a means of dealing with the remaining initial state singularity.

In view of the above discussion we will analyse the initial state, collinear singularity present in eqn (3.239). Equation (3.240) gives us one of the Lorentz invariants. To evaluate $g \cdot q$ and $q \cdot q'$ it is helpful to specialize to the C.o.M. frame. Here the momenta of the massless q, q' and g can be written as

$$
\begin{aligned}
q^\mu &= \hat{p}_{\text{in}}(1,0,0,1) & \hat{p}_{\text{in}} &= \frac{(\hat{s}+Q^2)}{2\sqrt{\hat{s}}} \\
q'^\mu &= \hat{p}_{\text{out}}(1,-\sin\theta^\star,0,-\cos\theta^\star) & & \\
g^\mu &= \hat{p}_{\text{out}}(1,+\sin\theta^\star,0,+\cos\theta^\star) & \hat{p}_{\text{out}} &= \frac{\sqrt{\hat{s}}}{2} .
\end{aligned}
\tag{3.241}
$$

These allow us to infer that

$$
\begin{aligned}
2g \cdot q &= 2\frac{\sqrt{\hat{s}}}{2}\frac{(\hat{s}+Q^2)}{2\sqrt{\hat{s}}}(1-\cos\theta^\star) & 2q \cdot q' &= 2\frac{(\hat{s}+Q^2)}{2\sqrt{\hat{s}}}\frac{\sqrt{\hat{s}}}{2}(1+\cos\theta^\star) \\
&= \frac{Q^2}{2z}(1-\cos\theta^\star) & &= \frac{Q^2}{2z}(1+\cos\theta^\star) .
\end{aligned}
\tag{3.242}
$$

The two-body phase space integral is given by eqn (C.21) with, for the moment, $\epsilon = 0$. In terms of these C.o.M. variables eqn (3.239) becomes

$$
\hat{H}_\Sigma^{(\gamma q)} = 4e_q^2\alpha_s C_F \frac{1}{8\pi}\frac{\hat{p}_{\text{out}}}{\sqrt{\hat{s}}}\int_{-1}^{+1} d\cos\theta^\star
$$
$$
\times \left[\frac{2(1-z)}{(1-\cos\theta^\star)} + \frac{(1-\cos\theta^\star)}{2(1-z)} + \frac{2z(1+\cos\theta^\star)}{(1-z)(1-\cos\theta^\star)}\right] . \tag{3.243}
$$

Referring to eqn (3.241) we see that the $\cos\theta^\star \to 1$ singularity in eqn (3.243) arises when the gluon direction approaches that of the incoming quark. The soft gluon and final state collinear singularities manifest themselves as the $z \to 1$ singularity, see eqn (3.240). Now, rather than work with $\cos\theta^\star$, we choose to use the transverse momentum of the gluon measured with respect to the incoming quark direction,

$$
\begin{aligned}
k_T^2 &= \hat{p}_{\text{out}}^2 \sin^2\theta^\star \\
&= \frac{Q^2}{4}\frac{(1-z)}{z}(1-\cos^2\theta^\star) \implies \frac{dk_T^2}{k_T^2} = -\frac{2\cos\theta^\star}{(1+\cos\theta^\star)}\frac{d\cos\theta^\star}{(1-\cos\theta^\star)} .
\end{aligned}
\tag{3.244}
$$

The limit $\cos\theta^\star \to 1$ now becomes $k_T^2 \to 0$. Making this change of variables in eqn (3.243) gives

$$
\hat{H}_\Sigma^{(\gamma q)} = \tag{3.245}
$$
$$
e_q^2\frac{\alpha_s}{2\pi}C_F \int_{\kappa^2}^{Q^2\frac{(1-z)}{4z}}\frac{dk_T^2}{k_T^2}\frac{2\cos\theta^\star}{(1+\cos\theta^\star)}\left[\frac{(1-z)^2+(1+\cos\theta^\star)z}{(1-z)} + \frac{1-\cos\theta^\star}{4(1-z)}\right] ,
$$

where $\cos\theta^\star$ is now implicitly given in terms of k_T. Notice that we have introduced a cut-off, κ^2, on the transverse momentum in order to regulate the collinear singularity. At small opening angles, $\cos\theta^\star \to 1$, the virtuality of the intermediate quark, $2g \cdot q$ in eqn (3.242), and the gluon's transverse momentum, eqn (3.244), are related by $2(g \cdot q) = k_T^2/(1-z)$. Thus, κ^2 is also a lower bound on the minimum virtuality of the intermediate quark, equivalent to an upper bound on the distance it travels. This *ad hoc* prescription will be replaced by dimensional regularization in Section 3.6.4. Focussing on the collinear region we obtain

$$\hat{H}_\Sigma^{(\gamma q)} = e_q^2 \frac{\alpha_s}{2\pi} C_F \left\{ \int_{\kappa^2}^{Q^2 \frac{(1-z)}{4z}} \frac{dk_T^2}{k_T^2} \left(\frac{1+z^2}{1-z} \right) + R''(z) + \mathcal{O}(\kappa^2/Q^2) \right\} . \quad (3.246)$$

The final state, $z \to 1$, singularity is still present in this result. Invoking the $\mathcal{O}(\alpha_s)$ corrections to the $\gamma^\star q \to q'$ vertex, which are proportional to $\delta(1-z)$, this singularity is removed. A more proper treatment would show us that the coefficient is actually a distribution,

$$\hat{H}_\Sigma^{(\gamma q)} = e_q^2 \frac{\alpha_s}{2\pi} C_F \left\{ \int_{\kappa^2}^{Q^2 \frac{(1-z)}{4z}} \frac{dk_T^2}{k_T^2} \left[\frac{1+z^2}{(1-z)_+} + C_{qq}\delta(1-z) \right] + R'(z) \right\}$$

$$\equiv e_q^2 \frac{\alpha_s}{2\pi} \left\{ P_{qq}^{(0)}(z) \ln\left(\frac{Q^2}{\kappa^2} \right) + R(z) \right\} . \quad (3.247)$$

Here, we introduced $1/(1-z)_+$ as a shorthand for $(1/(1-z))_+$, where the plus-prescription is defined by

$$F(z)_+ = F(z) - \delta(1-z) \int_0^1 dy\, F(y) . \quad (3.248)$$

Distributions only make sense when integrated with a suitable smooth function. We typically encounter the plus-prescription in a situation such as

$$\int_x^1 dz\, g(z) \left[\frac{f(z)}{(1-z)} \right]_+ \equiv \int_x^1 dz\, [g(z)-g(1)] \frac{f(z)}{(1-z)} - g(1) \int_0^x dz \frac{f(z)}{(1-z)} \quad (3.249)$$

which is free of divergences provided $g(z)$, and $f(z)$, are non-singular.

The full calculation would also have given us the form of the unspecified coefficients C_{qq} and $R(z)$. Later we shall use a physical argument to extract $C_{qq} = 3/2$. Given C_{qq}, then eqn (3.247) fully specifies the regularized, lowest order, Altarelli–Parisi splitting function $P_{qq}^{(0)}(z)$, eqn (3.50). Finally, given a similar result for $\hat{H}_L^{(\gamma q)}$, which is non-singular, we can use eqn (3.231) to derive F_2 and F_1. Including the lowest order contribution, eqn (3.237) and, for simplicity omitting the sum over quark flavours, gives

$$\frac{F_2(x, Q^2; \kappa)}{x e_q^2} = \int_x^1 \frac{dz}{z} q\left(\frac{x}{z}\right) \left\{ \delta(1-z) + \frac{\alpha_s}{2\pi} \left[P_{qq}^{(0)}(z) \ln\left(\frac{Q^2}{\kappa^2}\right) + R_{qq}(z) \right] \right\}$$

$$= q(x) + \int_x^1 \frac{dz}{z} q\left(\frac{x}{z}\right) \frac{\alpha_s}{2\pi} \left[P_{qq}^{(0)}(z) \ln\left(\frac{Q^2}{\kappa^2}\right) + R_{qq}(z) \right] . \quad (3.250)$$

In this expression we do not show explicitly any dependence on the renormalization scale μ_R which anyway does not enter at $\mathcal{O}(\alpha_s)$. Now, having identified and isolated the initial state singularity in F_2, we must decide how to deal with it.

Equation (3.250) exhibits a large logarithm coming from the collinear singularity which we have identified with long-distance physics. What we would like to do is to factorize eqn (3.250) in such a way that all long-distance physics is contained within the hadron specific p.d.f., whilst all short-distance physics is contained within a hard-subprocess specific coefficient function. In order to facilitate the separation of the long- and short-distance contributions we introduce a new factorization scale, μ_F, into eqn (3.250). Our aim is to move the logarithmic singularity into $q(x)$, but we are also free to move none, part or all of the finite term, R_{qq}, into $q(x)$. Reflecting this freedom we also introduce a finite, arbitrary function, $R_q^F(z)$, into eqn (3.250). The prescription for choosing R_q^F constitutes a factorization scheme. Introducing μ_F and R_q^F we rewrite eqn (3.250) as

$$\frac{F_2(x, Q^2; \kappa)}{x e_q^2} = q(x) + \int_x^1 \frac{dz}{z} q\left(\frac{x}{z}\right) \frac{\alpha_s}{2\pi} \left[P_{qq}^{(0)}(z) \ln\left(\frac{\mu_F^2}{\kappa^2}\right) + R_q^F(z) \right] \quad (3.251)$$

$$+ \int_x^1 \frac{dz}{z} q\left(\frac{x}{z}\right) \frac{\alpha_s}{2\pi} \left[P_{qq}^{(0)}(z) \ln\left(\frac{Q^2}{\mu_F^2}\right) + R_{qq}(z) - R_q^F(z) \right] .$$

This form suggests defining a factorization scale and scheme dependent p.d.f. which absorbs fully the collinear singularity

$$q^F(x, \mu_F^2, R_q^F; \kappa) = q(x) + \int_x^1 \frac{dz}{z} q\left(\frac{x}{z}\right) \frac{\alpha_s}{2\pi} \left[P_{qq}^{(0)}(z) \ln\left(\frac{\mu_F^2}{\kappa^2}\right) + R_q^F(z) \right] .$$
$$(3.252)$$

The second term on the right-hand side of eqn (3.252) is logarithmically divergent as $\kappa^2 \to 0$, but we expect the p.d.f. on the left-hand side to be finite. In an argument that is very reminiscent of renormalization we claim that the 'bare' p.d.f., $q(x)$, contains a compensating logarithmic divergence in κ^2 in just such a way that their sum is finite and independent of κ^2 in the $\kappa^2 \to 0$ limit,

$$q^F(x, \mu_F^2, R_q^F) = q(x; \kappa) + \int_x^1 \frac{dz}{z} q\left(\frac{x}{z}; \kappa\right) \frac{\alpha_s}{2\pi} \left[P_{qq}^{(0)}(z) \ln\left(\frac{\mu_F^2}{\kappa^2}\right) + R_q^F(z) \right] .$$
$$(3.253)$$

This 'physical' p.d.f. is now finite and so we may drop any reference to the κ regulator. In terms of eqn (3.253) we can rewrite eqn (3.250) to $\mathcal{O}(\alpha_s)$ as

$$\frac{F_2(x, Q^2)}{x e_q^2} = q^F(x, \mu_F^2, R_q^F)$$

$$+ \int_x^1 \frac{dz}{z} q^F\left(\frac{x}{z}, \mu_F, R_q^F\right) \frac{\alpha_s}{2\pi} \left[P_{qq}^{(0)}(z) \ln\left(\frac{Q^2}{\mu_F^2}\right) + R_{qq}(z) - R_q^F(z) \right]$$

$$= \int_x^1 \frac{dz}{z} q^F \left(\frac{x}{z}, \mu_F, R_q^F \right) \tag{3.254}$$

$$\times \left\{ \delta(1-z) + \frac{\alpha_s}{2\pi} \left[P_{qq}^{(0)}(z) \ln \left(\frac{Q^2}{\mu_F^2} \right) + R_{qq}(z) - R_q^F(z) \right] \right\} .$$

The right-hand side of these equations appears to depend on μ_F and R_q^F. However, all dependence on μ_F and R_q^F cancels to the calculated order, $\mathcal{O}(\alpha_s)$, and any dependence at $\mathcal{O}(\alpha_s^2)$ would also cancel if we included the neglected $\mathcal{O}(\alpha_s^2)$ terms in eqn (3.254). As the original expression, eqn (3.250), makes clear, the physical $F_2(x, Q^2)$ is independent of both the arbitrary factorization scale and scheme. What we have gained, as the second form in eqn (3.254) makes clear, is that all the long-distance behaviour is contained within the finite p.d.f. whilst the process dependent coefficient function only contains short-distance physics. The significance of μ_F is that it delimits the boundary between short, $Q > \mu_F$, and long, $Q < \mu_F$, distance physics.

In eqn (3.254) both the choice of μ_F and R_q^F are arbitrary. For the scale, choosing $\mu_F = Q$ is clearly advantageous, as it yields the simple expression

$$\frac{F_2(x, Q^2)}{xe_q^2} = \int_x^1 \frac{dz}{z} q^F \left(\frac{x}{z}, Q^2, R_q^F \right) \left\{ \delta(1-z) + \frac{\alpha_s}{2\pi} \left[R_{qq}(z) - R_q^F(z) \right] \right\} .$$
$$\tag{3.255}$$

The choice of which finite terms from R_{qq} in eqn (3.250) to absorb into R_q^F defines the factorization scheme. Two schemes are in popular usage. In the (modified) minimal subtraction scheme only the singular term is absorbed into the parton density function, that is, $R_q^{\overline{MS}} = 0$. In the DIS scheme all of the finite term, together with the singular term, are absorbed into the p.d.f., that is, $R_q^{DIS} = R_{qq}$. This scheme results in a particularly simple form for the structure function,

$$F_2(x, Q^2) = xe_q^2 q^{DIS}(x, Q^2) . \tag{3.256}$$

The above reasoning which leads to factorization applies equally well to $F_1(x, Q^2)$. One might therefore be tempted to define DIS p.d.f.s according to the equivalent of eqn (3.256) for F_1. However, you should be aware that at $\mathcal{O}(\alpha_s)$ $F_2 \neq 2xF_1$ and the two schemes will not be equivalent. Equation (3.256) is the conventional definition. It is significant that $F_L = F_2/(2x) - F_1$ is infrared finite and in particular contains no collinear, initial state singularities. This means that the same redefinition of the p.d.f.s used to render F_2 finite will also render F_1 finite. Although we will not demonstrate it, this is also true for F_3.

As the discussion of factorization schemes makes clear, the p.d.f.s should not be regarded as physical quantities, since they depend on the scheme used to define them. However, when convoluted with the appropriate coefficient function, eqn (3.254), they give rise to physical, measurable structure functions.

The crucial point to remember with regard to factorization schemes is that the same scheme must be used for both the p.d.f.s and the coefficient functions. If this is not the case then the cancellation implicit in eqn (3.254) will not occur and it will not be equivalent to eqn (3.250). Since the p.d.f.s in the $\overline{\text{MS}}$ scheme carry no information that is specific to lepton–hadron scattering, they are easier to use in applications to hadron–hadron scattering and thus often are the preferred choice. If DIS p.d.f.s were used to describe another process, then the new coefficient functions, describing that process's short-distance physics, would have to include a compensating factor of $R_{\text{q}}^{\text{DIS}}$ taken from the unrelated DIS process.

3.6.4 DIS at next-to-leading order

The above discussion of factorization avoided technical details so as to concentrate on the core ideas. We now explicitly carry out this process for the case of DIS. We will use dimensional regularization throughout to deal with all the singularities.

3.6.4.1 The process $\gamma^\star q \to q'$ at $\mathcal{O}(\alpha_s)$

There are two contributions to the process $\gamma^\star q \to q'$ which need to be considered at $\mathcal{O}(\alpha_s)$. The real, tree-level scattering $\gamma^\star q \to q'g$, Fig. 3.22, and the virtual, one-loop corrections to $\gamma^\star q \to q'$. These are both very similar to the pQCD corrections to the process $\gamma^\star \to q\bar{q}$ which we have already calculated; see Section 3.5. In fact, to obtain $-\eta_{\mu\nu}\hat{H}^{\mu\nu}$ we can use crossing, eqn (3.95), for the required matrix elements without any further calculation. The D-dimensional amplitude squared for the process $\gamma^\star \to q\bar{q}g$ is given by eqn (3.207). To obtain the amplitude squared for the process $\gamma^\star q \to q'g$ we need to make the substitutions $\bar{q}^\mu \to -q^\mu$, relabel the original q as q', replace Q^2 by $-Q^2$ and add an overall minus sign since we now have a closed quark loop. This gives

$$\sum |\mathcal{M}(\gamma^\star q \to q'g)|^2 = e^2 e_q^2 (g_s \mu^\epsilon)^2 C_F N_c 2\text{Tr}\{\mathbf{1}\}(1-\epsilon) \tag{3.257}$$

$$\times \left\{ (1-\epsilon)\left[\frac{g\cdot q}{g\cdot q'} + \frac{g\cdot q'}{g\cdot q} \right] + \frac{Q^2(q\cdot q')}{(g\cdot q)(g\cdot q')} + 2\epsilon \right\}.$$

Here we have replaced $\text{Tr}\{T^a T^a\}$ by $C_F N_c$. As before we chose to use the C.o.M. variables

$$2g\cdot q' = \frac{Q^2}{z}(1-z), \quad 2g\cdot q = \frac{Q^2}{z}v \quad \text{and} \quad 2q\cdot q' = \frac{Q^2}{z}(1-v), \tag{3.258}$$

which differ from eqn (3.240) and eqn (3.242) only in the replacement of $\cos\theta^\star$ by $v = (1+\cos\theta^\star)/2$. In terms of these variables eqn (3.257) becomes

$$\sum |\mathcal{M}(\gamma^\star q \to q'g)|^2 = e^2 e_q^2 (g_s \mu^\epsilon)^2 C_F N_c 2\text{Tr}\{\mathbf{1}\}(1-\epsilon) \tag{3.259}$$

$$\times \left\{ (1-\epsilon)\left[\frac{v}{(1-z)} + \frac{(1-z)}{v} \right] + \frac{2z}{(1-z)}\frac{(1-v)}{v} + 2\epsilon \right\}.$$

To this expression we should add an average over the spin and colour of the incoming quark, $2N_c$, divide by the conventional factor $4\pi e^2$ and include the two-body phase space integral, eqn (C.21). This gives

$\hat{H}_{\Sigma,R}^{(\gamma q)}$

$$= \frac{1}{4\pi e^2}\overline{\sum}|\mathcal{M}(\gamma^\star q \to q'g)|^2 \tag{3.260}$$

$$= e_q^2 \alpha_s C_F 4\frac{1}{4\pi}\frac{\hat{p}_{\text{out}}}{\sqrt{\hat{s}}}\left(\frac{\pi\mu^2}{\hat{p}_{\text{out}}^2}\right)^\epsilon \frac{(1-\epsilon)}{\Gamma(1-\epsilon)}$$

$$\times \int_0^1 dv\, v^{-\epsilon}(1-v)^{-\epsilon}\left\{(1-\epsilon)\left[\frac{v}{(1-z)}+\frac{(1-z)}{v}\right]+\frac{2z}{(1-z)}\frac{(1-v)}{v}+2\epsilon\right\}.$$

The v-integral is of the standard Euler β-function type, eqn (C.27), and is evaluated to yield

$$\int_0^1 dv\, v^{-\epsilon}(1-v)^{-\epsilon}\{\cdots\}$$

$$= \left\{(1-\epsilon)\left[\frac{1}{(1-z)}\frac{\Gamma(2-\epsilon)\Gamma(1-\epsilon)}{\Gamma(3-2\epsilon)}+(1-z)\frac{\Gamma(-\epsilon)\Gamma(1-\epsilon)}{\Gamma(1-2\epsilon)}\right]\right.$$

$$\left.+\frac{2z}{(1-z)}\frac{\Gamma(-\epsilon)\Gamma(2-\epsilon)}{\Gamma(2-2\epsilon)}+2\epsilon\frac{\Gamma^2(1-\epsilon)}{\Gamma(2-2\epsilon)}\right\}$$

$$= \frac{\Gamma^2(1-\epsilon)}{\Gamma(1-2\epsilon)}\left\{-\frac{(1-\epsilon)}{\epsilon}\left[(1-z)+\frac{1}{(1-2\epsilon)}\frac{2z}{(1-z)}\right]\right.$$

$$\left.+\frac{1}{2(1-z)}\frac{(1-\epsilon)}{(1-2\epsilon)}+\frac{2\epsilon}{(1-2\epsilon)}\right\}$$

$$= \frac{\Gamma^2(1-\epsilon)}{\Gamma(1-2\epsilon)}\left\{-\frac{1}{\epsilon}\frac{1+z^2}{(1-z)}-\frac{3}{2}\frac{1}{(1-z)}+3-z+\left(6-\frac{7}{2(1-z)}\right)\epsilon+\mathcal{O}(\epsilon^2)\right\}. \tag{3.261}$$

The simplifications in the second line have been achieved using eqn (C.25) whilst in the third line we have expanded out the expression in curly braces. Substituting eqn (3.261) into eqn (3.260) gives

$$\hat{H}_{\Sigma,R}^{(\gamma q)} = e_q^2\frac{\alpha_s}{2\pi}C_F\left(\frac{4\pi\mu^2}{Q^2}\frac{z}{(1-z)}\right)^\epsilon (1-\epsilon)\frac{\Gamma(1-\epsilon)}{\Gamma(1-2\epsilon)}$$

$$\times \left\{-\frac{1}{\epsilon}\frac{1+z^2}{(1-z)}-\frac{3}{2}\frac{1}{(1-z)}+3-z+\left(6-\frac{7}{2(1-z)}\right)\epsilon+\mathcal{O}(\epsilon^2)\right\}. \tag{3.262}$$

Identifying the $\epsilon \to 0$ limit in eqn (3.262) is a little tricky but using the identity

$$\frac{z^\epsilon}{(1-z)^{1+\epsilon}} = -\frac{1}{\epsilon}\delta(1-z)+\frac{1}{(1-z)_+}-\epsilon\left(\frac{\ln(1-z)}{1-z}\right)_+ +\epsilon\frac{\ln z}{1-z}, \tag{3.263}$$

see Ex. (3-26), we finally obtain

$$
\hat{H}^{(\gamma q)}_{\Sigma,R}
$$

$$
= e_q^2 \frac{\alpha_s}{2\pi} C_F \left(\frac{4\pi\mu^2}{Q^2}\right)^\epsilon (1-\epsilon) \frac{\Gamma(1-\epsilon)}{\Gamma(1-2\epsilon)} \left\{ \left(\frac{2}{\epsilon^2} + \frac{3}{2\epsilon} + \frac{7}{2}\right) \delta(1-z) - \frac{1}{\epsilon} \frac{1+z^2}{(1-z)_+} \right.
$$

$$
\left. +(1+z^2) \left(\frac{\ln(1-z)}{1-z}\right)_+ - \frac{1+z^2}{(1-z)} \ln z - \frac{3}{2} \frac{1}{(1-z)_+} + 3 - z + \mathcal{O}(\epsilon) \right\} .
$$

$$(3.264)$$

The double pole, $1/\epsilon^2$, is due to the soft gluon singularity.

A second contribution to the total structure function at $\mathcal{O}(\alpha_s)$ comes from the interference between the process $\gamma^* q \to q'$ at one-loop and at tree-level. The structures of the one-loop vertex and tree-level diagram are the same, so that we can combine them into an effective vertex

$$
i\Gamma^\mu = -i\,e e_q \gamma^\mu \left[1 - \frac{\alpha_s}{4\pi} C_F \left(\frac{4\pi\mu^2}{Q^2}\right)^\epsilon \frac{\Gamma(1+\epsilon)\Gamma^2(1-\epsilon)}{\Gamma(1-2\epsilon)} \left(\frac{2}{\epsilon^2} + \frac{3}{\epsilon} + 8\right)\right]
$$

$$
= -i\,e e_q \gamma^\mu \left[1 - \frac{\alpha_s}{4\pi} C_F \left(\frac{4\pi\mu^2}{Q^2}\right)^\epsilon \frac{\Gamma(1-\epsilon)}{\Gamma(1-2\epsilon)} \left(\frac{2}{\epsilon^2} + \frac{3}{\epsilon} + 8 + \frac{\pi^2}{3} + \mathcal{O}(\epsilon)\right)\right] .
$$

$$(3.265)$$

Here, the one-loop contribution has been inferred from eqn (3.219), the only difference being the absence of the $(-1)^\epsilon$ factor, reflecting the space-like nature of q_γ^μ in DIS. In the second line we used $\Gamma(1+\epsilon)\Gamma(1-\epsilon) = 1 + (\pi^2/6)\epsilon^2 + \mathcal{O}(\epsilon^4)$. Equation (3.265) has infrared singularities but is ultraviolet finite. The calculation of this additional contribution to $\hat{H}^{(\gamma q)}_\Sigma$ is straightforward, giving

$$
\hat{H}^{(\gamma q)}_{\Sigma,V} = e_q^2 (1-\epsilon)\delta(1-z)
$$

$$(3.266)$$

$$
\times \left\{ 1 - 2\frac{\alpha_s}{4\pi} C_F \left(\frac{4\pi\mu^2}{Q^2}\right)^\epsilon \frac{\Gamma(1-\epsilon)}{\Gamma(1-2\epsilon)} \left(\frac{2}{\epsilon^2} + \frac{3}{\epsilon} + 8 + \frac{\pi^2}{3} + \mathcal{O}(\epsilon)\right) \right\} .
$$

Adding eqn (3.264) and (3.266) together we see that the $1/\epsilon^2$ terms, the soft gluon pole, cancel,

$$
\hat{H}^{(\gamma q)}_\Sigma = e_q^2 \frac{\alpha_s}{2\pi} C_F (1-\epsilon) \left\{ -\left[\frac{1+z^2}{(1-z)_+} + \frac{3}{2}\delta(1-z)\right] \frac{1}{\epsilon} \frac{\Gamma(1-\epsilon)}{\Gamma(1-2\epsilon)} \left(\frac{4\pi\mu^2}{Q^2}\right)^\epsilon \right.
$$

$$
+(1+z^2)\left(\frac{\ln(1-z)}{1-z}\right)_+ - \frac{1+z^2}{1-z}\ln z - \frac{3}{2}\frac{1}{(1-z)_+}
$$

$$
\left. + 3 - z - \left(\frac{9}{2} + \frac{\pi^2}{3}\right)\delta(1-z) \right\} .
$$

$$(3.267)$$

The remaining $1/\epsilon$ pole is associated with the collinear singularity for gluon emission off the incoming quark, its coefficient is the regularized, one-loop, Altarelli–Parisi splitting function

$$P_{qq}^{(0)}(z) = C_F \left[\frac{1+z^2}{(1-z)_+} + \frac{3}{2}\delta(1-z) \right] = C_F \left(\frac{1+z^2}{1-z} \right)_+ . \qquad (3.268)$$

This calculation supplies us with the value of $C_{qq} = 3/2$ in eqn (3.247).

We also need to calculate the longitudinal part of the hadronic tensor. This is particularly easy to $\mathcal{O}(\alpha_s)$ since many of the potential contributions vanish in the massless quark limit. We have already seen, in eqn (3.236), that the tree-level diagram, and by virtue of eqn (3.265) its one-loop virtual correction, give no contribution. This implies that the longitudinal structure function, $F_L \propto H_L$, is at least $\mathcal{O}(\alpha_s)$. Turning to the $\mathcal{O}(\alpha_s)$ tree-level contribution and again assuming massless quarks, so that $\slashed{q}u(q) = 0$, we have

$$q_\mu \mathcal{M}(\gamma^{*\mu}q \to q'g) \propto \bar{u}(q') \left[\gamma_\sigma \frac{(\slashed{q}' + \slashed{g})}{(q'+g)^2}\slashed{q} + \slashed{q}\frac{(\slashed{q} - \slashed{g})}{(q-g)^2}\gamma_\sigma \right] u(q)\epsilon^\sigma(g)^\star$$

$$\propto \bar{u}(q')\frac{\slashed{q}\slashed{g}}{2q \cdot g}\gamma_\sigma u(q)\epsilon^\sigma(g)^\star . \qquad (3.269)$$

Thus, the diagram describing gluon radiation off the scattered quark gives no contribution, leaving only the diagram describing gluon radiation off the incoming quark. Squaring this diagram and summing over spins, where we can use $-\eta^{\sigma\sigma'}$ for the lone gluon's polarization tensor, gives

$$\sum |q_\mu \mathcal{M}(\gamma^{*\mu}q \to q'g)|^2 = -e^2 e_q^2 (g_s\mu^\epsilon)^2 C_F N_c \frac{1}{(2q \cdot g)^2}\text{Tr}\left\{ \slashed{q}'\slashed{q}\slashed{g}\gamma_\sigma\slashed{q}\gamma^\sigma\slashed{g}\slashed{q} \right\}$$

$$= e^2 e_q^2 (g_s\mu^\epsilon)^2 C_F N_c \frac{1}{(2q \cdot g)^2}2(1-\epsilon)\text{Tr}\left\{ \slashed{q}'\slashed{q}\slashed{g}\slashed{q}\slashed{g}\slashed{q} \right\}$$

$$= e^2 e_q^2 (g_s\mu^\epsilon)^2 C_F N_c \frac{1}{2q \cdot g}2(1-\epsilon)\text{Tr}\left\{ \slashed{q}'\slashed{q}\slashed{g}\slashed{q} \right\}$$

$$= e^2 e_q^2 (g_s\mu^\epsilon)^2 C_F N_c 2(1-\epsilon)\text{Tr}\left\{ \slashed{q}'\slashed{q} \right\}$$

$$= e^2 e_q^2 (g_s\mu^\epsilon)^2 C_F N_c (1-\epsilon)2q' \cdot q\,\text{Tr}\left\{ 1 \right\} . \qquad (3.270)$$

Here, we have used the trick in Ex. (3-19) and repeatedly used $\slashed{q}\slashed{g} = 2g \cdot q - \slashed{g}\slashed{q}$ together with $\slashed{q}\slashed{q} = q^2 = 0$. The result is non-zero and free of singularities and so we need not have used a regulator. Next, we average over the incoming quark's spin and colour polarizations, $2N_c$, use $\text{Tr}\{1\} = 4$, adopt the choice of variables in eqn (3.242) and include the integral over two-body phase space, eqn (C.20), to obtain

$$\hat{H}_L^{(\gamma q)} = \frac{1}{4\pi e^2}\int d\Phi_2 \overline{\sum} |q_\mu \mathcal{M}(\gamma^{*\mu}q \to q'g)|^2$$

$$= \frac{1}{2\pi}e_q^2 (g_s\mu^\epsilon)^2 C_F \frac{Q^2}{z}(1-\epsilon)\frac{1}{4\pi}\frac{\hat{p}_{out}}{\sqrt{\hat{s}}}\left(\frac{\pi}{\hat{p}_{out}^2} \right)^\epsilon \frac{1}{\Gamma(1-\epsilon)}\int_0^1 dv\, v^{-\epsilon}(1-v)^{1-\epsilon}$$

$$= \frac{1}{4}e_q^2 \frac{\alpha_s}{2\pi}C_F \frac{Q^2}{z}\left(\frac{4\pi\mu^2}{Q^2}\frac{z}{1-z} \right)^\epsilon \frac{\Gamma(2-\epsilon)}{\Gamma(2-2\epsilon)}$$

$$= \frac{1}{4}e_q^2 \frac{\alpha_s}{2\pi} C_F \frac{Q^2}{z} + \mathcal{O}(\epsilon) . \tag{3.271}$$

Given eqns (3.267) and (3.271) we can convolute them with the p.d.f. to reconstruct H_Σ and H_L and hence obtain the $\mathcal{O}(\alpha_s)$ (anti)quark's contribution to F_2 and F_1, c.f. eqn (3.231), as

$$
\begin{aligned}
\frac{F_2^{\gamma q}(x)}{x e_q^2} &= \frac{\alpha_s}{2\pi} \int_x^1 \frac{dz}{z} q\left(\frac{x}{z}\right) \left\{ -P_{qq}^{(0)}(z)\frac{1}{\epsilon}\frac{\Gamma(1-\epsilon)}{\Gamma(1-2\epsilon)}\left(\frac{4\pi\mu^2}{Q^2}\right)^\epsilon \right. \\
&\quad + C_F\left[(1+z^2)\left(\frac{\ln(1-z)}{1-z}\right)_+ - \frac{1+z^2}{(1-z)}\ln z \right. \\
&\quad \left.\left. -\frac{3}{2}\frac{1}{(1-z)_+} + 3 + 2z - \left(\frac{9}{2}+\frac{\pi^2}{2}\right)\delta(1-z)\right]\right\} \\
&= \frac{\alpha_s}{2\pi} \int_x^1 \frac{dz}{z} q\left(\frac{x}{z}\right) \left\{ -P_{qq}^{(0)}(z)\left[\frac{1}{\epsilon} - \gamma_E + \ln(4\pi) - \ln\left(\frac{Q^2}{\mu^2}\right)\right] \right. \\
&\quad \left. + C_F\left[\frac{1+z^2}{1-z}\left(\ln\left(\frac{1-z}{z}\right) - \frac{3}{4}\right) + \frac{5z+9}{4}\right]_+ \right\}
\end{aligned}
$$

$$F_1^{\gamma q}(x) = \frac{F_2^{\gamma q}(x)}{2x} - e_q^2 \frac{\alpha_s}{2\pi}\int_x^1 \frac{dz}{z} q\left(\frac{x}{z}\right) C_F z . \tag{3.272}$$

In the second expression for F_2 we have expanded out the coefficient of the splitting function and introduced a more compact form for the remainder term. The $1/\epsilon$ pole naturally arises in the combination Δ_ϵ, eqn (C.16). If we add in the leading order result, eqn (3.237), then eqn (3.272) takes the form of eqn (3.250) with ϵ acting as regulator. If the factorization procedure removes just the $1/\epsilon$ term we have the minimal subtraction scheme, if it removes the additional terms, Δ_ϵ, we have the modified minimal subtraction scheme, $\overline{\text{MS}}$. In the DIS scheme both the Δ_ϵ and finite terms are removed.

3.6.4.2 *The $\mathcal{O}(\alpha_s)$ process $\gamma^\star g \to q\bar{q}$* The calculation of the terms $-\eta_{\mu\nu}\hat{H}^{\mu\nu}$ and $g_\mu g_\nu \hat{H}^{\mu\nu}$ for the process $\gamma^\star g \to q\bar{q}$ follows the same lines as that for $\gamma^\star q \to q'g$ but is a little simpler in practice due to the lack of soft gluon singularities. Here, we just quote the results and leave their computation to the adventurous/diligent reader; see Ex. (3-27).

$$
\begin{aligned}
\hat{H}_\Sigma^{(\gamma g)} &\equiv -\eta_{\mu\nu}\hat{H}^{\mu\nu} = 2e_q^2 \frac{\alpha_s}{2\pi} T_F \left[z^2 + (1-z)^2\right] \\
&\quad \times \left\{ -\frac{1}{\epsilon}\left(\frac{4\pi\mu^2}{Q^2}\right)^\epsilon \frac{\Gamma(1-\epsilon)}{\Gamma(1-2\epsilon)} + \ln\left(\frac{1-z}{z}\right) + \mathcal{O}(\epsilon) \right\}
\end{aligned}
$$

$$\hat{H}_L^{(\gamma g)} \equiv g_\mu g_\nu \hat{H}^{\mu\nu} = e_q^2 \frac{\alpha_s}{2\pi} T_F Q^2 \frac{(1-z)}{z} + \mathcal{O}(\epsilon) \tag{3.273}$$

Using eqn (3.231) applied to the above results, which contain both the quark and antiquark terms, we obtain the $\mathcal{O}(\alpha_s)$ gluon's contribution to F_2 and F_1,

$$\frac{F_2^{\gamma g}}{x} = \frac{\alpha_s}{2\pi} T_F \sum_q e_q^2 \int_x^1 \frac{dz}{z} g\left(\frac{x}{z}\right)$$

$$\times \left\{ \frac{[z^2 + (1-z)^2]}{(1-\epsilon)} \left[-\frac{1}{\epsilon} \left(\frac{4\pi\mu^2}{Q^2}\right)^\epsilon \frac{\Gamma(1-\epsilon)}{\Gamma(1-2\epsilon)} + \ln\left(\frac{1-z}{z}\right) \right] + 6z(1-z) \right\}$$

$$= \frac{\alpha_s}{2\pi} \sum_q e_q^2 \int_x^1 \frac{dz}{z} g\left(\frac{x}{z}\right) \left\{ -P_{qg}^{(0)}(z) \left[\Delta_\epsilon - \ln\left(\frac{Q^2}{\mu^2}\right) \right] \right.$$

$$\left. + T_F \left[[z^2 + (1-z)^2] \ln\left(\frac{1-z}{z}\right) - 1 + 8z(1-z) \right] \right\}$$

$$F_1^{\gamma g} = \frac{F_2^{\gamma g}}{2x} - \frac{\alpha_s}{2\pi} T_F \sum_q e_q^2 \int_x^1 \frac{dz}{z} g\left(\frac{x}{z}\right) 4z(1-z) . \tag{3.274}$$

3.6.4.3   *The combined results for the $\mathcal{O}(\alpha_s)$ DIS structure functions*   The above results, eqns (3.272) and (3.274), can be combined to give the NLO formula for $F_1^{(\gamma h)}$ and $F_2^{(\gamma h)}$. In the modified minimal subtraction, $\overline{\text{MS}}$, scheme we have

$$\frac{F_{1,2}^{(Vh)}}{\frac{1}{2}, x}(x, Q^2) =$$

$$\int_x^1 \frac{dz}{z} \sum_{f=q,\bar{q}} g_{Vf}^2 \left\{ f^{\overline{\text{MS}}}\left(\frac{x}{z}, \mu_F^2\right) \left[\delta(1-z) + \frac{\alpha_s}{2\pi} \left(P_{qq}^{(0)}(z) \ln\frac{Q^2}{\mu_F^2} + C_{1,2}^{(Vq)}(z) \right) \right] \right.$$

$$\left. + g^{\overline{\text{MS}}}\left(\frac{x}{z}, \mu_F^2\right) \frac{\alpha_s}{2\pi} \left(P_{qg}^{(0)}(z) \ln\frac{Q^2}{\mu_F^2} + C_{1,2}^{(Vg)}(z) \right) \right\}$$

$$F_3^{(Vh)}(x, Q^2) =$$

$$\int_x^1 \frac{dz}{z} \sum_{f=q,\bar{q}} g_{Vf}^2 \left\{ f^{\overline{\text{MS}}}\left(\frac{x}{z}, \mu_F^2\right) \left[\delta(1-z) + \frac{\alpha_s}{2\pi} \left(P_{qq}^{(0)}(z) \ln\frac{Q^2}{\mu_F^2} + C_3^{(Vq)}(z) \right) \right] \right\}$$

$$\tag{3.275}$$

where we have also included the $\overline{\text{MS}}$ expression for $F_3^{(Vh)}$. In eqn (3.275) g_{Vf} gives the normalized strength of the exchanged gauge boson's coupling to the (anti)quark, for example $g_{\gamma q} = e_q$, whilst the coefficient functions are given by

$$C_1^{(Vq)} = \frac{1}{2} C_2^{(Vq)} - C_F z$$

$$C_2^{(Vq)} = C_F \frac{1}{2} \left[\frac{1+z^2}{1-z} \left(\ln\left(\frac{1-z}{z}\right) - \frac{3}{4} \right) + \frac{9+5z}{4} \right]_+$$

$$C_3^{(Vq)} = C_2^{(Vq)} - C_F(1+z) \tag{3.276}$$

$$C_1^{(Vg)} = \frac{1}{2} C_2^{(Vg)} - T_F 4z(1-z)$$

$$C_2^{(Vg)} = T_F z \left[[z^2 + (1-z)^2] \ln \frac{1-z}{z} - 1 + 8z(1-z) \right]$$

$$C_3^{(Vg)} = 0 .$$

Charge conjugation invariance implies $C_i^{(V\bar{q})}(z) = C_i^{(Vq)}(z)$. We can also infer the form of the quark p.d.f.s from our results for F_2. In the $\overline{\text{MS}}$ scheme the NLO (anti)quark, and for completeness the gluon, p.d.f.s are given by

$$q^{\overline{\text{MS}}}(x,\mu_F^2) = \int_x^1 \frac{dz}{z} \left\{ q\left(\frac{x}{z},\epsilon\right) \left(\delta(1-z) - \frac{\alpha_s}{2\pi} P_{qq}^{(0)}(z) \left[\Delta_\epsilon - \ln \frac{\mu_F^2}{\mu^2} \right] \right) \right.$$

$$\left. - g\left(\frac{x}{z},\epsilon\right) \frac{\alpha_s}{2\pi} P_{gq}^{(0)}(z) \left[\Delta_\epsilon - \ln \frac{\mu_F^2}{\mu^2} \right] \right\}$$

$$= \sum_{f=q,g} \int_x^1 \frac{dz}{z} f\left(\frac{x}{z},\epsilon\right) \left[\delta(1-z)\delta_{qf} - \frac{\alpha_s}{2\pi} P_{qf}^{(0)}(z) \frac{1}{\epsilon} \left(\frac{4\pi\mu^2}{\mu_F^2 e^{\gamma_E}} \right)^\epsilon \right]$$

$$g^{\overline{\text{MS}}}(x,\mu_F^2) = \sum_{f=q,\bar{q},g} \int_x^1 \frac{dz}{z} f\left(\frac{x}{z},\epsilon\right) \left[\delta(1-z)\delta_{gf} - \frac{\alpha_s}{2\pi} P_{gf}^{(0)}(z) \frac{1}{\epsilon} \left(\frac{4\pi\mu^2}{\mu_F^2 e^{\gamma_E}} \right)^\epsilon \right] .$$

$$(3.277)$$

In the deep inelastic scattering, DIS, scheme F_2 is given by

$$F_2^{(Vh)}(x,Q^2) =$$

$$\sum_{f=q,\bar{q}} g_{Vf}^2 \left\{ f^{\text{DIS}}(x,\mu_F^2) + \frac{\alpha_s}{2\pi} \int_0^1 \frac{dz}{z} f^{\text{DIS}}\left(\frac{x}{z},\mu_F^2\right) P_{qq}^{(0)}(z) \ln\left(\frac{Q^2}{\mu_F^2}\right) \right\} , \quad (3.278)$$

which is exact to all orders. At $\mathcal{O}(\alpha_s)$ F_1 and F_3 are given by eqn (3.275) with modified coefficient functions,

$$C_1^{(Vq)}(z) = -C_F z$$

$$C_3^{(Vq)}(z) = -C_F(1+z) \qquad\qquad (3.279)$$

$$C_1^{(Vg)}(z) = -T_F 4z(1-z) .$$

All other coefficient functions vanish to $\mathcal{O}(\alpha_s)$ in the DIS scheme. In order to maintain the same expression for F_2, modified (anti)quark p.d.f.s are required. The relation between DIS and $\overline{\text{MS}}$ scheme p.d.f.s is given by

$$q^{\text{DIS}}(x,\mu_F^2) = q^{\overline{\text{MS}}}(x,\mu_F^2) + \frac{\alpha_s}{2\pi} \int_x^1 \frac{dz}{z} \sum_{f=q,g} f^{\overline{\text{MS}}}\left(\frac{x}{z}\right) C_2^{(Vf)}(z)$$

$$g^{\text{DIS}}(x,\mu_F^2) = g^{\overline{\text{MS}}}(x,\mu_F^2) - \frac{\alpha_s}{2\pi} \int_x^1 \frac{dz}{z} \sum_{f=q,\bar{q},g} f^{\overline{\text{MS}}}\left(\frac{x}{z}\right) C_2^{(Vf)}(z) . \quad (3.280)$$

Here the expression relating g^{DIS} to the $\overline{\text{MS}}$ p.d.f.s is only the conventional one.

Comparing eqn (3.276) with (3.279) and eqn (3.277) with (3.280) we see the characteristic differences between the $\overline{\text{MS}}$ and DIS factorization schemes. In the $\overline{\text{MS}}$ scheme the coefficient functions are relatively complex whilst the p.d.f.s are very simple and contain no traces of any hard subprocess. By contrast, in the DIS scheme the coefficient functions describing DIS are very simple but the p.d.f.s are relatively complex and contain terms which are specific to the F_2 structure function of DIS. In the DIS scheme the simplicity of the coefficient functions only holds for DIS, whereas in the $\overline{\text{MS}}$ scheme the simplicity of the p.d.f.s holds for all processes.

Before moving on to discuss the scale dependence of the p.d.f.s it is worthwhile to remind ourselves of how this calculation proceeded. We began by deriving the corrections to the partonic scatterings $Vq, Vg \rightarrow X$. The result was then put into the form suggested by the factorization theorem eqn (3.47). This required the coefficient function, $\hat{F}_i^{Vf}(x, \mu_F^2)$, and the parton-to-parton p.d.f., $f_{f'}(x, \mu_F^2)$, to be defined. Now, the short-distance coefficient function is universal and can be equally well used to describe $Vh \rightarrow X$ scatterings using eqn (3.47) but now with hadron-to-parton p.d.f.s. By exploiting the separation of long- and short-distance physics we are able to finesse the need to deal directly with a non-perturbative hadron. Finally, given an experimental measurement of $F_i^{Vh}(x, Q^2)$, it is possible to extract a combination of the p.d.f.s $f_h(x, \mu_F^2)$ describing the hadron's constituents, which can then be used in the description of other processes.

3.6.5   The evolution of the parton density functions

The p.d.f. which appears in eqn (3.254) is not a perturbatively calculable quantity but one which must presently be extracted from experimental data within a particular factorization scheme. We also know that the left-hand side of eqn (3.254) is independent of the arbitrary factorization scale, μ_F. Indeed, $q^F(x, \mu_F^2)$ was constructed in eqn (3.252) to ensure that the right-hand side is μ_F-independent to $\mathcal{O}(\alpha_s)$. If we differentiate eqn (3.254), or (3.252), with respect to μ_F we obtain an equation for the scale dependence, setting $\mu = \mu_F$, of the p.d.f.,

$$\mu^2 \frac{\partial q(x, \mu^2)}{\partial \mu^2} = \int_x^1 \frac{dz}{z} \frac{\alpha_s}{2\pi} P_{qq}(z) q\left(\frac{x}{z}, \mu^2\right) . \tag{3.281}$$

This is the basic form of the DGLAP equation; see eqn (3.49) and the discussion in Section 3.2.2.

The explicit calculations of the previous section show that the evolution of a quark p.d.f. includes contributions from q $\rightarrow$ q(g) and also from g $\rightarrow$ q($\bar{\text{q}}$) splitting functions. Likewise, for an antiquark we should include contributions from $\bar{\text{q}}$ $\rightarrow$ $\bar{\text{q}}$(g) and g $\rightarrow$ $\bar{\text{q}}$(q) splitting functions. At leading $\mathcal{O}(\alpha_s)$ the gluon p.d.f. evolves according to a similar equation with contributions from g $\rightarrow$ g(g), q $\rightarrow$ g(q) and $\bar{\text{q}}$ $\rightarrow$ g($\bar{\text{q}}$) splitting functions. More generally we have to consider $a \rightarrow b(cd)$ and higher order vertices. This opens up the possibility of q $\rightarrow$ q'(X) and q $\rightarrow$ $\bar{\text{q}}$'(X) splitting functions etc. and leads us to the evolution equations

$$\mu^2 \frac{\partial q_i}{\partial \mu^2}(x,\mu^2) = \int_x^1 \frac{dz}{z} \frac{\alpha_s}{2\pi} \left[P_{q_i q_j}(z,\alpha_s) q_j\left(\frac{x}{z},\mu^2\right) + P_{q_i \bar{q}_j}(z,\alpha_s) \bar{q}_j\left(\frac{x}{z},\mu^2\right) \right.$$

$$\left. + P_{q_i g}(z,\alpha_s) g\left(\frac{x}{z},\mu^2\right) \right]$$

$$\mu^2 \frac{\partial \bar{q}_i}{\partial \mu^2}(x,\mu^2) = \int_x^1 \frac{dz}{z} \frac{\alpha_s}{2\pi} \left[P_{\bar{q}_i \bar{q}_j}(z,\alpha_s) \bar{q}_j\left(\frac{x}{z},\mu^2\right) + P_{\bar{q}_i q_j}(z,\alpha_s) q_j\left(\frac{x}{z},\mu^2\right) \right.$$

$$\left. + P_{\bar{q}_i g}(z,\alpha_s) g\left(\frac{x}{z},\mu^2\right) \right]$$

$$\mu^2 \frac{\partial g}{\partial \mu^2}(x,\mu^2) = \int_x^1 \frac{dz}{z} \frac{\alpha_s}{2\pi} \left[P_{gg}(z,\alpha_s) g\left(\frac{x}{z},\mu^2\right) + \sum_{f=q,\bar{q}} P_{gf}(z,\alpha_s) f\left(\frac{x}{z},\mu^2\right) \right] .$$

$$(3.282)$$

The kernel functions $P_{ab}(z,\alpha_s(\mu^2))$ are associated with the branchings $b \to a(X)$ and can be calculated as power series in α_s,

$$P_{ab}(z,\alpha_s) = P_{ab}^{(0)}(z) + \frac{\alpha_s}{2\pi} P_{ab}^{(1)}(z) + \cdots . \qquad (3.283)$$

In their general form, eqn (3.282), they look rather formidable but they actually simplify greatly since not all splitting functions are independent. The charge conjugation symmetry and the $SU(n_f)$ flavour symmetry of QCD, for equal mass quarks, imply the relationships

$$P_{q_j q_i} = P_{\bar{q}_j \bar{q}_i} \equiv \delta_{ij} P_{qq}^{NS} + P_{qq}^{S} \qquad P_{q_i g} = P_{\bar{q}_i g} \equiv P_{qg}$$
$$P_{\bar{q}_j q_i} = P_{q_j \bar{q}_i} \equiv \delta_{ij} P_{\bar{q}q}^{NS} + P_{\bar{q}q}^{S} \qquad P_{g q_i} = P_{g \bar{q}_i} \equiv P_{gq} . \qquad (3.284)$$

The $q \to q$ and $q \to \bar{q}$ splitting functions are usefully separated into flavour non-singlet (NS) and singlet (S) parts that are associated with the evolution of the valence and sea quarks, respectively. The splitting functions, P_{qq}^{S}, $P_{\bar{q}q}^{NS}$ and $P_{\bar{q}q}^{S}$ only start at $\mathcal{O}(\alpha_s^2)$, where $P_{qq}^{S(1)} = P_{\bar{q}q}^{S(1)}$. Thus, at leading order eqn (3.282) reduces to the simpler eqn (3.49). Since the virtualities involved in this initial state evolution are negative, these are the space-like splitting functions.

The treatment of radiative corrections for outgoing partons follows a similar pattern as that for incoming partons. It provides a similar factorization theorem for fragmentation functions and equations very much like eqn (3.282), which control the μ_F behaviour of the fragmentation functions (Owens, 1978; Uematsu, 1978). The leading order equations for the factorization scale dependence of the fragmentation functions are

$$\mu^2 \frac{\partial D_q^h}{\partial \mu^2}(x,\mu^2) = \int_x^1 \frac{dz}{z} \frac{\alpha_s}{2\pi} \left[P_{qq}(z,\alpha_s) D_q^h\left(\frac{x}{z},\mu^2\right) + P_{gq}(z,\alpha_s) D_g^h\left(\frac{x}{z},\mu^2\right) \right]$$

$$\mu^2 \frac{\partial D_{\bar{q}}^h}{\partial \mu^2}(x,\mu^2) = \int_x^1 \frac{dz}{z} \frac{\alpha_s}{2\pi} \left[P_{qq}(z,\alpha_s) D_{\bar{q}}^h\left(\frac{x}{z},\mu^2\right) + P_{gq}(z,\alpha_s) D_g^h\left(\frac{x}{z},\mu^2\right) \right]$$

$$\mu^2 \frac{\partial D_{\mathrm{g}}^{\mathrm{h}}}{\partial \mu^2}(x, \mu^2) = \int_x^1 \frac{\mathrm{d}z}{z} \frac{\alpha_{\mathrm{s}}}{2\pi} \left[P_{\mathrm{gg}}(z, \alpha_{\mathrm{s}}) D_{\mathrm{g}}^{\mathrm{h}}\left(\frac{x}{z}, \mu^2\right) + \sum_{f=q,\bar{q}} P_{\mathrm{qg}}(z, \alpha_{\mathrm{s}}) D_{\mathrm{f}}^{\mathrm{h}}\left(\frac{x}{z}, \mu^2\right) \right] .$$

$$(3.285)$$

Note the reversed order of the indices on the splitting functions compared to eqn (3.49). The structure and interpretation of these time-like equations are essentially the same as for the space-like equations and to $\mathcal{O}(\alpha_{\mathrm{s}})$ so are the splitting functions. However, beyond this leading order the space-like and time-like splitting functions differ. All the splitting functions are known to $\mathcal{O}(\alpha_{\mathrm{s}}^2)$ (Furmanski and Petronzio, 1982), see also (Hamberg and van Neerven, 1992), whilst partial results are becoming available at $\mathcal{O}(\alpha_{\mathrm{s}}^3)$. The $\mathcal{O}(\alpha_{\mathrm{s}}^2)$, time-like splitting functions can be found in Appendix E.

3.6.5.1 *Method of moments* Whilst the direct numerical solution of the space-like and time-like DGLAP equations is one option, semi-analytical approaches are also available. The convolution which occurs in the DGLAP equations can be separated using a Mellin transform into moment space:

$$\tilde{f}(n) = \int_0^1 \mathrm{d}x\, x^{n-1} f(x) \quad \Longleftrightarrow \quad f(x) = \frac{1}{2\pi\mathrm{i}} \int_{c-\mathrm{i}\infty}^{c+\mathrm{i}\infty} \mathrm{d}n\, x^{-n} \tilde{f}(n) . \quad (3.286)$$

The contour used in the inverse transformation must lie to the right of all singularities in the analytic continuation of $\tilde{f}(n)$. The moment space transformation of, for illustration only, a simplified DGLAP equation with $\mu_F^2 = Q^2$, is as follows:

$$Q^2 \frac{\partial f}{\partial Q^2}(x, Q^2) = \frac{\alpha_{\mathrm{s}}}{2\pi} \int_x^1 \frac{\mathrm{d}z}{z} P(z, \alpha_{\mathrm{s}}) f\left(\frac{x}{z}, Q^2\right)$$

$$\Longrightarrow \quad Q^2 \frac{\partial \tilde{f}}{\partial Q^2}(n, Q^2) = \frac{\alpha_{\mathrm{s}}}{2\pi} \gamma(n, \alpha_{\mathrm{s}}) \tilde{f}(n, Q^2) . \quad (3.287)$$

Here, the Mellin transform of the splitting function, $\gamma(n, \alpha_{\mathrm{s}}) \equiv \tilde{P}(n, \alpha_{\mathrm{s}})$, is known as the anomalous dimension. Equation (3.287) is now easily solved. Working to one-loop precision we find

$$\tilde{f}(n, Q^2) = \tilde{f}(n, Q_0^2) \left(\frac{Q^2}{Q_0^2}\right)^{\frac{\alpha_{\mathrm{s}}}{2\pi}\gamma^{(0)}(n)} = \tilde{f}(n, \mu^2) \left(\frac{\alpha_{\mathrm{s}}(Q_0^2)}{\alpha_{\mathrm{s}}(Q^2)}\right)^{\frac{\gamma^{(0)}(n)}{2\pi\beta_0}} . \quad (3.288)$$

The first solution assumes fixed α_{s} whilst the second assumes the use of the running coupling, eqn (3.22). As the first form makes clear the anomalous dimension is so called because it modifies the naïve Q-dependence and acts as an additional scaling power. All information on the nature of the solutions are contained within the anomalous dimensions, which are fully equivalent to the Altarelli–Parisi kernels. Given the solution eqn (3.288), one then applies the inverse Mellin transform, eqn (3.286), to go back from the moment to the x-space;

see examples Ex. (3-33) and (3-34). In the more general case the Mellin trans-
~~form of the DGLAP equations leads to matrix equations. These can be solved in~~
essentially the same way after they are first diagonalized.

We shall now use this moment space equation to determine the coefficient
C_{qq} in eqn (3.247) whilst avoiding the need to evaluate any virtual corrections.
Consider the leading order evolution equation for the difference of two quark
p.d.f.s, such as $(u - d)$ or $(u - \bar{u})$,

$$Q^2 \frac{\partial q^{\mathrm{NS}}}{\partial Q^2}(x, Q^2) = \int_x^1 \frac{dz}{z} \frac{\alpha_s}{2\pi} P_{qq}(z) q^{\mathrm{NS}}\left(\frac{x}{z}, Q^2\right). \tag{3.289}$$

All dependence on the gluon p.d.f. has cancelled. This is the evolution equation
for the non-singlet, in terms of its flavour SU(3) transformation properties, structure function. Exercise (3-8) investigates other useful combinations of p.d.f.s.
Now, $\int_0^1 dx\, q^{\mathrm{NS}}(x)$ is a constant by virtue of the conservation of flavour quantum
numbers within QCD. Referring to eqn (3.287) with $n = 1$, this implies that
$\gamma_{qq}(1, \alpha_s) = 0$ for the q $\to$ qg splitting function. That is,

$$
\begin{aligned}
0 &= \int_0^1 dz\, C_F \left\{ \frac{1 + z^2}{(1 - z)_+} + C_{qq}\delta(1 - z) \right\} \\
&= C_F \left\{ \int_0^1 dz \frac{(1 + z^2) - 2}{1 - z} + C_{qq} \right\} \\
&= C_F \left\{ -\frac{3}{2} + C_{qq} \right\},
\end{aligned}
\tag{3.290}
$$

which supplies us with the value of C_{qq} and completes the expression for $P_{qq}^{(0)}(z)$.
The same result viewed from an alternative perspective is discussed in Ex. (3-30)
and a similar approach based on using momentum conservation can be applied
to find C_{gg}, Ex. (3-29).

Of course these arguments rely on the physical interpretation of the p.d.f.s.
In Section 3.6.4 we proceeded by direct calculation to fully evaluate the splitting
functions. Given the full expressions for the splitting functions, eqn (3.50), we
can evaluate the anomalous dimensions appearing in eqn (3.287). At the lowest
order we find:

$$\gamma_{gg}^{(0)}(n) = 2C_A \left[\frac{1}{(n-1)n} + \frac{1}{(n+1)(n+2)} + \frac{11}{12} - \sum_{m=1}^n \frac{1}{m} \right] - \frac{2}{3} n_f T_F \tag{3.291}$$

$$\gamma_{qq}^{(0)}(n) = C_F \left[\frac{1}{n(n+1)} + \frac{3}{2} - 2\sum_{m=1}^n \frac{1}{m} \right] \tag{3.292}$$

$$\gamma_{qg}^{(0)}(n) = T_F \left[\frac{2 + n + n^2}{n(n+1)(n+2)} \right] \tag{3.293}$$

$$\gamma_{gq}^{(0)}(n) = C_F \left[\frac{2 + n + n^2}{(n-1)n(n+1)} \right] \tag{3.294}$$

These show that $\gamma_{\mathrm{qq}}^{(0)}(1) = 0$, thereby proving conservation of flavour. A number of other conservation laws are also implied by eqns (3.291)–(3.294).

$$0 = \int_0^1 dz \left\{ P_{\mathrm{qq}}^{\mathrm{NS}}(z) - P_{\bar{\mathrm{q}}\mathrm{q}}^{\mathrm{NS}}(z) + n_f \left[P_{\mathrm{qq}}^{\mathrm{S}}(z) - P_{\bar{\mathrm{q}}\mathrm{q}}^{\mathrm{S}}(z) \right] \right\} \qquad (3.295)$$

$$0 = \int_0^1 dz\, z \left\{ P_{\mathrm{gg}}(z) + 2 n_f P_{\mathrm{qg}}(z) \right\} \qquad (3.296)$$

$$0 = \int_0^1 dz\, z \left\{ P_{\mathrm{gq}}(z) + P_{\mathrm{qq}}^{\mathrm{NS}}(z) + P_{\bar{\mathrm{q}}\mathrm{q}}^{\mathrm{NS}}(z) + n_f \left[P_{\mathrm{qq}}^{\mathrm{S}}(z) + P_{\bar{\mathrm{q}}\mathrm{q}}^{\mathrm{S}}(z) \right] \right\} \qquad (3.297)$$

See Ex. (3-31) for further elaboration of how the momentum is shared within a hadron.

3.6.6 Leading logarithms

Our derivation of the DGLAP equation focused on treating a region of phase space which has a logarithmically enhanced cross section. Recall that introducing μ_F^2 to isolate the collinear singularity left behind a residual, large logarithm. There are two singular regions: the collinear region which gives logarithmic enhancements of the form $\alpha_{\mathrm{s}} \ln(Q^2/Q_0^2)$ and the soft region which gives logarithmic enhancements of the form $\alpha_{\mathrm{s}} \ln(1/x)$. These regions can overlap and give double logarithmic enhancements of the form $\alpha_{\mathrm{s}} \ln(Q^2/Q_0^2) \ln(1/x)$. Processes which involve multiple parton final states can have up to one $\ln(Q^2/Q_0^2)$ and one $\ln(1/x)$ factor for each power of α_{s}. The phase space regions which contribute these leading logarithmic enhancements are associated with configurations in which 'successive' partons have strongly ordered transverse, k_T, and/or longitudinal, $k_L (\equiv x)$, momenta:

$$\mathrm{LL}_Q\mathrm{A}: \quad \begin{cases} \alpha_{\mathrm{s}} L_Q \sim 1 \\ \alpha_{\mathrm{s}} L_x \ll 1 \end{cases} \quad Q^2 \gg k_{nT}^2 \gg \cdots \gg k_{1T}^2 \gg Q_0^2 \qquad (3.298)$$

$$\mathrm{DLLA}: \quad \begin{cases} \alpha_{\mathrm{s}} L_Q L_x \sim 1 \\ \alpha_{\mathrm{s}} L_Q \ll 1 \\ \alpha_{\mathrm{s}} L_x \ll 1 \end{cases} \quad \begin{cases} Q^2 \gg k_{nT}^2 \gg \cdots \gg k_{1T}^2 \gg Q_0^2 \\ x \ll x_n \ll \cdots \ll x_1 \ll x_0 \end{cases} \qquad (3.299)$$

$$\mathrm{LL}_x\mathrm{A}: \quad \begin{cases} \alpha_{\mathrm{s}} L_x \sim 1 \\ \alpha_{\mathrm{s}} L_Q \ll 1 \end{cases} \quad x \ll x_n \ll \cdots \ll x_1 \ll x_0 \qquad (3.300)$$

The solution of the DGLAP equation sums over all orders in α_{s} the contributions from the leading, single, collinear logarithms, $[\alpha_{\mathrm{s}} \ln(Q^2/Q_0^2)]^n$ and the leading, double logarithms $[\alpha_{\mathrm{s}} \ln(Q^2/Q_0^2) \ln(1/x)]^n$. This is the region of strongly ordered k_T and ordered x. It does not include the leading, single, soft singularities which are treated instead by the BFKL equation (Kuraev et al., 1977; Balitsky and Lipatov, 1978) which describes the x-evolution of p.d.f.s at fixed Q^2. Figure 3.24 shows the $\ln(Q^2)$–$\ln(1/x)$ plane and the regions which are described by the various *leading logarithmic* (LL) summations. Relaxing one of the strong ordering

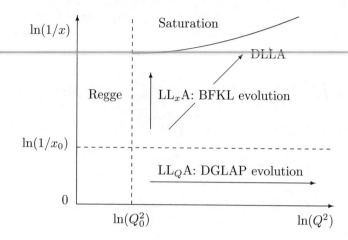

FIG. 3.24. The $\ln(Q^2)$–$\ln(1/x)$ plane showing the regions in which the LL_Q, LL_x, and DLL approximations hold. Also shown are the regions in which Regge phenomenology applies and where saturation/recombination effects have to be taken into account.

constraints in eqn (3.298) or eqn (3.300) gives rise to a *next-to-leading logarith-mic* (NLL) enhancement to the cross section. These are suppressed by a factor α_s with respect to the LL-enhancement. Including summed NLL-terms modi-fies the DGLAP or BFKL equations whilst maintaining their general structure. We now discuss the *double leading logarithmic* (DLL) approximation, the BFKL equation, the combined evolution equations which incorporate both DGLAP and BFKL evolution and the generalizations to include parton recombination. We shall make more explicit the relationship between these equations and the leading logarithms in the following Section 3.6.7.

3.6.6.1   *The double leading logarithmic approximation*   At small x and large Q^2 we must sum the leading $\alpha_s \ln(Q^2/Q_0^2) \ln(1/x)$ terms. This can be done directly from the DGLAP equations by keeping only the most singular $1/z$ terms in the splitting functions. At $\mathcal{O}(\alpha_s)$ only P_{gg} and P_{gq} have soft gluon singularities, but at $\mathcal{O}(\alpha_s^2)$ all splitting functions are singular as $z \to 0$. In this limit the lowest order parton distributions are given by (Rujula *et al.*, 1974)

$$F_2(x, Q^2) \sim xg(x, Q^2) \sim \tilde{g}(n_0, Q_0^2) \exp \sqrt{\frac{4C_A}{\pi\beta_0} \ln\left(\frac{\alpha_s(Q_0^2)}{\alpha_s(Q^2)}\right) \ln\left(\frac{1}{x}\right)} \quad (3.301)$$

$$\text{with} \quad n_0 = \sqrt{\frac{C_A \ln(\alpha_s(Q_0^2)/\alpha_s(Q^2))}{\pi\beta_0 \ln(1/x)}} \,.$$

See Ex. (3-33), which also gives sub-leading terms. This solution shows a strong growth in the small-x partons and hence the structure functions, in particular $F_2(x, Q^2)$ (Glück *et al.*, 1995). The dependence on the initial distribution is only via its n_0-th moment. If this initial distribution has too strong a small-x growth, then the above solution will not hold; for example, $xg(x, Q_0^2) \propto x^{-\lambda}$, $\lambda > 0$ leads to $xg(x, Q^2) \propto x^{-\lambda}$, independent of Q^2.

In what is known as 'double asymptotic scaling', in the limit $1/x$, $Q^2 \to \infty$ eqn (3.301) implies that for 'soft' $f(x, Q_0^2)$ then $\ln F_2(x, Q^2)$ depends linearly on $\sqrt{\ln(\alpha_s(Q_0^2)/\alpha_s(Q^2))} \times \ln(1/x)$ and is independent of the complementary combination $\sqrt{\ln(\alpha_s(Q_0^2)/\alpha_s(Q^2))} \div \ln(1/x)$ (Ball and Forte, 1994). Sub-leading terms only slightly complicate this statement.

3.6.6.2 *The BFKL equation* At small x and moderate $Q^2 > \Lambda_{\mathrm{QCD}}^2$, where gluons are dominant, we must sum the leading $\alpha_s \ln(1/x)$ terms whilst keeping the full Q^2-dependence. This means that we do not have strongly ordered k_T but instead integrate over the full range of k_T. This leads us to work with the unintegrated gluon p.d.f., $\mathcal{G}(x, k_T^2)$, which is related to the usual p.d.f. via

$$xg(x, Q^2) = \int^{Q^2} \frac{\mathrm{d}k_T^2}{k_T^2} \mathcal{G}(x, k_T^2) \,. \tag{3.302}$$

In phenomenological applications it is common to assume a narrow, Gaussian k_T distribution for the partons in a hadron, which is commensurate with confinement. Predictions for structure functions are then made using the so-called k_T factorization (Catani *et al.* 1990*a*; 1991*a*),

$$F_i(x, Q^2) = \int_0^x \frac{\mathrm{d}z}{z} \int \frac{\mathrm{d}k_T^2}{k_T^4} \hat{F}_i^{\mathrm{box}}(z, k_T^2, Q^2) \mathcal{G}\left(\frac{x}{z}, k_T^2\right) \,. \tag{3.303}$$

Here $\hat{F}_i^{\mathrm{box}}$ is derived from the quark box diagrams that describe virtual-photon virtual-gluon scattering, $\gamma^* g^* \to$ '$q\bar{q}$' $\to \gamma^* g^*$. The unintegrated gluon p.d.f. satisfies the BFKL equation (Kuraev *et al.*, 1977; Balitsky and Lipatov, 1978); see also (Mueller, 1994) for an alternative derivation in terms of colour dipoles. At leading order the BFKL equation is given by

$$\frac{\partial \mathcal{G}(x, k_T^2)}{\partial \ln(1/x)} = \frac{C_A \alpha_s}{\pi} k_T^2 \int_{k_0^2}^{\infty} \frac{\mathrm{d}q_T^2}{q_T^2} \left\{ \frac{\mathcal{G}(x, q_T^2) - \mathcal{G}(x, k_T^2)}{|q_T^2 - k_T^2|} + \frac{\mathcal{G}(x, k_T^2)}{\sqrt{4q_T^4 + k_T^4}} \right\} \,. \tag{3.304}$$

Given the unintegrated gluon p.d.f. at one value of x_0, this equation allows you to calculate its value at smaller values of x, that is, larger values of $\ln(1/x)$. If α_s is fixed, then the equation can be solved analytically. In the small-x limit this basically gives a power law behaviour in x,

$$\mathcal{G}(x, k_T^2) \approx \tilde{\mathcal{G}}\left(x_0, \frac{1}{2}\right)\left(\frac{x}{x_0}\right)^{-\lambda} \frac{\sqrt{k_T^2}}{\sqrt{2\pi[\lambda'' \ln(x_0/x) + A]}} \exp\left(\frac{-\ln^2(k_T^2/\bar{k}_T^2)}{2[\lambda'' \ln(x_0/x) + A]}\right) \,. \tag{3.305}$$

The solution follows by first applying a Mellin transform and then using the saddle point method to evaluate the inverse. Here

$$\tilde{\mathcal{G}}(x_0, \omega) \equiv \int_0^\infty \frac{dk_T^2}{k_T^2} \frac{\mathcal{G}(x_0, k_T^2)}{(k_T^2)^\omega}, \quad -\ln \bar{k}_T^2 = \frac{1}{\tilde{\mathcal{G}}} \frac{d\tilde{\mathcal{G}}}{d\omega}\left(x_0, \frac{1}{2}\right), \quad A = \frac{1}{\tilde{\mathcal{G}}} \frac{d^2\tilde{\mathcal{G}}}{d\omega^2}\left(x_0, \frac{1}{2}\right)$$

$$\tag{3.306}$$

$$\lambda = 4\ln 2 \frac{C_A \alpha_s}{\pi}\Bigg|_{\alpha_s = 0.2} \approx +0.5 \quad \text{and} \quad \lambda'' = 28\zeta(3)\frac{C_A \alpha_s}{\pi}. \tag{3.307}$$

The numerical value of the Riemann zeta-function is $\zeta(3) \approx 1.202\,056\,903\,2$. Due to eqn (3.303) the behaviour $\mathcal{G} \propto x^{-\lambda}$ feeds through to give $F_2 \propto x^{-\lambda}$. The k_T behaviour is typical of diffusion and reflects the lack of any k_T-ordering in BFKL dynamics; in essence there is a random walk in k_T as x decreases (Balitsky and Lipatov, 1978; Bartels and Lotter, 1993).

Actually, this observation highlights a problem. Given that the width of the Gaussian in $\ln(k_T^2/\bar{k}_T^2)$ is given by $\sqrt{[\lambda'' \ln(x_0/x) + A]}$, then for sufficiently small x there will be support for $\mathcal{G}(x, k_T^2)$ from the non-perturbative region in k_T^2. Thus, if we use a running coupling, $\alpha_s(k_T^2)$, then it is necessary to introduce infrared cut-offs, for example $k_0^2 > 0$ in eqn (3.304), and other possible refinements such as including momentum conservation (Collins and Landshoff, 1992; Bartels et al., 1996). Whilst numerical evaluations show that similar power law behaviour in x and diffusion in $\ln(k_T^2/\bar{k}_T^2)$ occurs (Askew et al., 1993), these are essentially misguided due to the inherent instablity of the BFKL equation with running coupling. The situation is made worse by the NLL$_x$ corrections to the BFKL kernel (Fadin and Lipatov, 1998), which gives

$$\lambda = 4\ln 2 \frac{C_A \alpha_s}{\pi}\left(1 - 6.3\frac{C_A \alpha_s}{\pi}\right)\Bigg|_{\alpha_s = 0.2} \approx -0.1. \tag{3.308}$$

Such a large, negative correction basically invalidates perturbation theory and, if taken seriously, leads to negative cross sections. The source of these large corrections has been traced to large $\ln(Q^2/Q_0^2)$ terms coming from phase space restrictions (Salam, 1998). There are a number of putative solutions to this situation, which include resummation (Ciafaloni et al. 1999a; 1999b), imposing momentum conservation (Altarelli et al., 2000) and imposing perturbative stability (Ball and Forte, 1999). A succinct review is provided by Ball and Landshoff (2000).

3.6.6.3 Combined evolution equations

The DGLAP and BFKL equations describe evolution in two complementary regions. A number of attempts have been made to give a combined description of both regions in a single equation. Amongst these are an attempt to include $\ln(1/x)$ terms into the usual collinear factorization by adding summed corrections into the P_{gg} kernel appearing in the DGLAP equations (Ellis et al., 1995; Ball and Forte, 1995). A second approach is given by the CCFM equation which uses angular ordering to describe both the x and the Q^2 evolution and has the DGLAP and the BFKL equations as limiting cases (Ciafaloni, 1988; Catani et al., 1990b), see also (Andersson et al., 1996a).

3.6.6.4 *Shadowing, gluon recombination and hot spots* If left unchecked, the rapid rise in the small-x gluon p.d.f. predicted by both the DGLAP and BFKL equations would violate unitarity. It also leads to a breakdown in the parton model picture of scattering off independent partons. At sufficiently high densities it becomes possible for a second parton to overlap in space with the first, so-called shadowing. The probability of this happening can be estimated as

$$\mathcal{P}_{\text{sat}} \sim \frac{\alpha_s(Q^2)/Q^2}{\pi R^2} N(x, Q) , \qquad (3.309)$$

where the partons, predominently gluons, are taken to have an effective area, given by a typical QCD cross section, $\sigma \sim \alpha_s(Q^2)/Q^2$ and number $N(x, Q)$. The denominator is taken to be of order the area of the hadron, with the radius $R \sim R_h = 1/M_h$. In general, $\mathcal{P}_{\text{sat}}$ is small but especially for small x it may become large. When it becomes $\mathcal{O}(1)$ the hadron is said to saturate and the usual DGLAP equation may need to be modified to account for parton recombination,

$$\mu^2 \frac{\partial g}{\partial \mu^2} = P_{gg} \otimes g + P_{gq} \otimes q - \frac{81\alpha_s^2}{16R^2\mu^2} \int_x^1 \frac{dy}{y} (yg)^2 . \qquad (3.310)$$

A similar modification can be applied to the BFKL equation. In this GLR equation (Gribov *et al.*, 1983) the familiar first two terms lead to a growth in $g(x, \mu^2)$ due to emission whilst the third involves a suppression due to recombination, $gg \to g$. The competition between these two terms ensures that the gluon p.d.f. equilibrates below the unitarity bound.

The validity of eqn (3.310) is not assured, but it appears reasonable to use it to estimate the onset of shadowing (Askew *et al.*, 1993). It has been derived at DLL accuracy (Mueller and Qiu, 1986); however, this neglects $1/N_c$ suppressed terms associated with pre-recombination interactions between the gluons (Bartels, 1993; Laenen and Levin, 1994). Its equivalent has also been derived for the BFKL equation in the colour dipole approach (Kovchegov, 1999). More significantly, it must be admitted that at saturation the high densities and field strengths occuring, $F^{\mu\nu} \sim 1/g_s$, imply that the perturbation theory is no longer valid. This has led to the development of a treatment in terms of a semi-classical, effective field theory (McLerran and Venugopalan, 1999), which also leads to parton recombination (Iancu *et al.*, 2000).

The choice $R = R_h$ in eqn (3.310) corresponds to a uniform distribution of the QCD fields across the hadron. However, it has been conjectured that this may not be the case and that partons inside the hadron may concentrate in dense hot spots centred on the valence quarks (Mueller, 1991). In this case one should use an $R < R_h$. Such a behaviour is predicted by the BFKL equation but not the DGLAP equation. It predicts the number of gluon jets per unit rapidity localized to a transverse region of size $\Delta x_T^2 \sim 1/\bar{k}_T^2$ as

$$\frac{dn}{d\ln(1/x)} = \frac{C_A \alpha_s}{\pi} \frac{x^{-\lambda}}{\sqrt{(\pi\lambda''/8)\ln(1/x)}} . \qquad (3.311)$$

3.6.7   The analysis of ladder diagrams

In our discussion of DIS, Section 3.6.3, we encountered the Altarelli–Parisi splitting function $P_{qq}(z)$, eqn (3.246), when investigating the limit of near collinear
emission. A point which may not yet have been appreciated is the universality of
this result. That is, whenever we have a process which contains a $q \rightarrow qg$ vertex,
then in the collinear limit its (azimuthally-averaged) contribution to the cross
section will be described by the same factor

$$\frac{\alpha_s}{2\pi} P_{qq}(z) dz \frac{dk_T^2}{k_T^2} . \qquad (3.312)$$

Similar expressions describe the collinear limits of $g \rightarrow q\bar{q}$ and $g \rightarrow gg$ vertices
with P_{qq} replaced by P_{qg} and P_{gg}, respectively. This factorization of the matrix
element squared then leads to much simpler expressions for a cross section in
the collinear limit. Furthermore, the collinear emission regions of phase space
are very important because they are responsible for one of the dominant, leading
logarithmic, contributions to the cross section. In the other dominant region
of phase space, the limit of soft gluon emission, we also have that the cross
section simplifies significantly; see Section 3.7. In this way we can use simplified
expressions to describe the bulk of a cross section. Of course, if our analysis
focuses attention on a region of phase space which involves hard, non-collinear
emission(s), then the approximate matrix elements may only be of limited use.

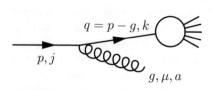

FIG. 3.25. The emission of a near collinear gluon off an incoming quark in an
$n + 1$ parton scattering

To see how this simplification occurs, consider the situation sketched in
Fig. 3.25 where a quark entering an n-particle scattering emits a real gluon,
which we shortly will take to be near collinear with the quark. The matrix element for this process is given by

$$\mathcal{M}_j^{(n+1)} = g_s T_{kj}^a \mathcal{M}_k^{(n)} \frac{(\not{p} - \not{g})}{(p-g)^2} \gamma_\mu \bar{u}(p) \epsilon^*(g)^\mu . \qquad (3.313)$$

Introducing a gauge vector n^μ with, for convenience, $n^2 = 0$ (and $g \cdot n \neq 0$),
we can use eqn (3.121) to sum over the gluon's physical, that is, transverse,
polarizations in the matrix element squared to obtain

$$\sum \left| \mathcal{M}^{(n+1)} \right|^2 \Big/ \left[g_s^2 T_{kj}^a T_{jk'}^a = g_s^2 C_F \delta_{kk'} \right]$$

$$= \mathrm{Tr} \left\{ \cdots \frac{(\not p - \not g)}{(p-g)^2} \gamma_\mu \not p \gamma^\nu \frac{(\not p - \not g)}{(p-g)^2} \cdots \right\} \left(-\eta^{\mu\nu} + \frac{[g^\mu n^\nu + n^\mu g^\nu]}{n \cdot g} \right)$$

$$= \mathrm{Tr} \left\{ \cdots \frac{(\not p - \not g)}{(p-g)^2} \left(-\gamma_\mu \not p \gamma^\mu + \frac{1}{n \cdot g} [\not g \not p \not n + \not n \not p \not g] \right) \frac{(\not p - \not g)}{(p-g)^2} \cdots \right\}.$$

$$= \mathrm{Tr} \left\{ \cdots \frac{(\not p - \not g)}{(p-g)^2} \left(2\not p + \frac{2}{n \cdot g} [(g \cdot p)\not n - (n \cdot g)\not p + (n \cdot p)\not g] \right) \frac{(\not p - \not g)}{(p-g)^2} \cdots \right\}$$

$$= \frac{1}{(2p \cdot g)^2} \frac{2}{(n \cdot g)} \mathrm{Tr} \left\{ \cdots (\not p - \not g) [(g \cdot p)\not n + (n \cdot p)\not g] (\not p - \not g) \cdots \right\}$$

$$= \frac{1}{(2p \cdot g)} \frac{2}{(n \cdot g)} \mathrm{Tr} \left\{ \cdots [(n \cdot (p-g))(\not p - \not g) + (p \cdot g)\not n + (n \cdot p)\not p] \cdots \right\}. \tag{3.314}$$

The ellipsis in these expressions represent the contributions from $\mathcal{M}_k^{(n)}$ and $\mathcal{M}_k^{(n)\star}$. In line two we used an identity based upon commuting γ-matrices to obtain line three and then again commuted γ-matrices in lines three and four, plus using $\not g \not g = g^2 = 0 = p^2 = \not p \not p$, to obtain an exact result. Now we wish to specialize to the near collinear limit. To do this we use a Sudakov decomposition of the quark and gluon momentum four-vectors (Sudakov, 1956),

$$q^\mu = zp^\mu + \beta n^\mu + k_\perp^\mu$$
$$p^\mu = q^\mu + g^\mu \implies g^\mu = (1-z)p^\mu - \beta n^\mu - k_\perp^\mu. \tag{3.315}$$

Here n^μ could have been any four-vector, subject to $n \cdot p \neq 0$, but it proves most useful to make this the gauge vector whilst $k_\perp^\mu$ is transverse to both p^μ and n^μ, $p \cdot k_\perp = 0 = n \cdot k_\perp$, and $k_\perp^2 \equiv -k_T^2 < 0$. The gluon's on mass-shell constraint, $g^2 = 0$, determines $\beta = -k_T^2/(2(1-z)n \cdot p)$. The (negative) virtuality of the intermediate quark is given by

$$q^2 = (p-g)^2 = -2p \cdot g = 2\beta n \cdot p = \frac{-k_T^2}{1-z}. \tag{3.316}$$

Adopting these variables eqn (3.314) becomes

$$g_s^2 C_F \delta_{kk'} \frac{(1-z)}{k_T^2} \frac{2}{(1-z)} \mathrm{Tr} \left\{ \cdots \left[(1+z^2)\not p + z\not k_T + \frac{k_T^2}{2(n \cdot p)}\not n \right] \cdots \right\}. \tag{3.317}$$

Now if we only wish to keep the leading term in the collinear limit, $k_T^2 \to 0$, then we can drop the second two terms in the square brackets to obtain

$$\sum \left| \mathcal{M}^{(n+1)} \right|^2 = 2\frac{(1-z)}{k_T^2} g_s^2 C_F \frac{1+z^2}{(1-z)} \delta_{kk'} \mathrm{Tr} \left\{ \cdots \not p \cdots \right\} + \mathcal{O}(1)$$

$$\approx 2\frac{(1-z)}{k_T^2} g_s^2 \hat{P}_{\mathrm{qq}}(z) \sum \left| \mathcal{M}^{(n)} \right|^2. \tag{3.318}$$

Thus, in the collinear limit the matrix element factorizes into the product of the unregularized, lowest order Altarelli–Parisi splitting function, $\hat{P}_{\mathrm{qq}}(z)$, and the matrix element squared for the process assuming no gluon emission took place. To obtain the cross section we must include the flux factor, an average over the initial spin and colour polarizations and the phase space, giving

$$\mathrm{d}^{n+1}\sigma = \frac{1}{16\pi^2}\frac{\mathrm{d}k_T^2}{(1-z)}\mathrm{d}z \times 2\frac{(1-z)}{k_T^2}g_s^2\hat{P}_{\mathrm{qq}}(z) \times \mathrm{d}^n\sigma$$

$$= \frac{\mathrm{d}k_T^2}{k_T^2}\mathrm{d}z\frac{\alpha_{\mathrm{s}}}{2\pi}\hat{P}_{\mathrm{qq}}(z) \times \mathrm{d}^n\sigma \ . \tag{3.319}$$

Here we have made explicit the phase space element for the near collinear gluon; see Ex. (3-35).

The collinear factorization of the matrix element squared, eqn (3.318), does not depend on the nature of the sub-matrix element, $\mathcal{M}_n$, for the other n particles involved in the scattering. In this sense the Altarelli–Parisi kernel $\hat{P}_{\mathrm{qq}}$ is universal. A similar analysis can be applied to the collinear limits of $\bar{\mathrm{q}} \to \bar{\mathrm{q}}\mathrm{g}$, $\mathrm{g} \to \mathrm{q}\bar{\mathrm{q}}$, see Ex. (3-36), and $\mathrm{g} \to \mathrm{gg}$.

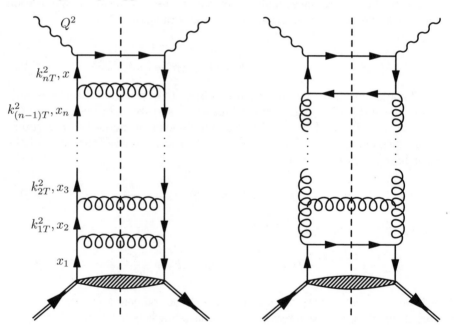

FIG. 3.26. On the left a ladder diagram dominated by $\mathrm{q} \to \mathrm{qg}$ branchings and on the right a ladder diagram dominated by $\mathrm{g} \to \mathrm{gg}$ branchings

Armed with collinear factorization we can give a less abstract and more physical interpretation of the relationship between the evolution equations and the leading logarithmic enhancements to cross sections. This traditional approach is based upon the analysis of Feynman diagrams (Dokshitzer, 1977). The aim is to identify those diagrams and the associated regions of phase space which give rise to the leading logarithms and then sum them. The treatment of the single collinear and double logarithms is a situation where the use of a physical gauge proves invaluable (Frenkel and Taylor, 1976; Dokshitzer et al., 1980). It can then be shown that the dominant contributions come from the so-called ladder graphs, such as Fig. 3.26, with strongly ordered branchings eqn (3.298) and (3.300). These ladders correspond to individual, tree-level Feynman diagrams squared. Diagrams involving the quartic gluon coupling give sub-leading contributions. Likewise, quantum interference between tree-level diagrams, which would give ladders with crossed rungs, only give sub-leading contributions. The strong ordering in k_T^2, which effectively implies ordering of the virtualities eqn (3.316), allows eqn (3.319) to be iteratively applied. In the case of an n-rung gluon ladder the cross section, $\sigma_n(x, Q^2)$, is given by:

$$\int_{Q_0^2}^{Q^2} \frac{dk_{nT}^2}{k_{nT}^2} \frac{\alpha_s(k_{nT}^2)}{2\pi} \cdots \int_{Q_0^2}^{k_{3T}^2} \frac{dk_{2T}^2}{k_{2T}^2} \frac{\alpha_s(k_{2T}^2)}{2\pi} \int_{Q_0^2}^{k_{2T}^2} \frac{dk_{1T}^2}{k_{1T}^2} \frac{\alpha_s(k_{1T}^2)}{2\pi}$$

$$\times \int_x^1 \frac{dx_n}{x_n} \hat{P}_{gg}\left(\frac{x}{x_n}\right) \cdots \int_{x_3}^1 \frac{dx_2}{x_2} \hat{P}_{gg}\left(\frac{x_3}{x_2}\right) \int_{x_2}^1 \frac{dx_1}{x_1} \hat{P}_{gg}\left(\frac{x_2}{x_1}\right) g(x_1, Q_0^2) .$$

$$(3.320)$$

Associated with each rung are a k_T and an x integral, both of which may contribute a large logarithm in the collinear or soft limits, respectively. Equation (3.320) embodies an almost classical picture of a parton shower in terms of successive branchings. It is this which lies behind our interpretation of the Altarelli–Parisi equations and which will be further exploited in the development of all-orders Monte Carlo event generators.

The usual collinear leading logarithmic approximation is characterized by strongly ordered transverse momenta, $Q^2 \gg k_{nT}^2 \gg \cdots \gg k_{2T}^2 \gg k_{1T}^2 \gg Q_0^2$. Using the one-loop expression for $\alpha_s(k_T^2)$ the nested transverse momentum integrals become:

$$\sigma_n(x, Q^2) \propto \int_{Q_0^2}^{Q^2} \frac{dk_{nT}^2}{k_{nT}^2} \frac{1}{2\pi} \frac{1}{\beta_0 \ln(k_{nT}^2/\Lambda^2)} \cdots \int_{Q_0^2}^{k_{2T}^2} \frac{dk_{1T}^2}{k_{1T}^2} \frac{1}{2\pi} \frac{1}{\beta_0 \ln(k_{1T}^2/\Lambda^2)}$$

$$= \frac{1}{(2\pi\beta_0)^n} \int_{Q_0^2}^{Q^2} d\ln\left[\frac{\ln(k_{nT}^2/\Lambda^2)}{\ln(Q_0^2/\Lambda^2)}\right] \cdots \int_{Q_0^2}^{k_{2T}^2} d\ln\left[\frac{\ln(k_{1T}^2/\Lambda^2)}{\ln(Q_0^2/\Lambda^2)}\right]$$

$$= \frac{1}{n!}\left[\frac{1}{2\pi\beta_0} \ln\left(\frac{\ln(Q^2/\Lambda^2)}{\ln(Q_0^2/\Lambda^2)}\right)\right]^n$$

$$= \frac{1}{n!}\left[\frac{1}{2\pi\beta_0} \ln\left(\frac{\alpha_s(Q_0^2)}{\alpha_s(Q^2)}\right)\right]^n .$$

$$(3.321)$$

In the second line, we have used a change of variables which makes the integrals simpler to evaluate. To do the momentum fraction integrals it is useful to work with the Mellin transform,

$$\tilde{\sigma}_n(m) \propto \int_0^1 dx \, x^{m-1} \int_x^1 \frac{dx_n}{x_n} \hat{P}_{gg}\left(\frac{x}{x_n}\right) \cdots \int_{x_2}^1 \frac{dx_1}{x_1} \hat{P}_{gg}\left(\frac{x_2}{x_1}\right) g(x_1, Q_0^2)$$

$$= \left[\tilde{P}_{gg}(m)\right]^n \tilde{g}(m, Q_0^2). \tag{3.322}$$

The result follows by repeatedly applying the fact that the Mellin transform of a convolution is the product of the Mellin transforms of the components. Combining eqns (3.321) and (3.322) gives

$$\sum_n \tilde{\sigma}_n(m, Q^2) = \tilde{g}(m, Q_0^2) \sum_n \frac{1}{n!} \left[\frac{\tilde{P}_{gg}^{(0)}(m)}{2\pi\beta_0} \ln\left(\frac{\alpha_s(Q_0^2)}{\alpha_s(Q^2)}\right)\right]^n$$

$$= \tilde{g}(m, Q_0^2) \left(\frac{\alpha_s(\mu_0^2)}{\alpha_s(Q^2)}\right)^{\frac{\tilde{P}_{gg}^{(0)}(m)}{2\pi\beta_0}}, \tag{3.323}$$

which coincides with eqn (3.288). This demonstrates that the DGLAP equation sums the leading $\alpha_s \ln(Q^2/Q_0^2)$ terms.

In the double leading logarithmic approximation we approximate $P_{gg}(z)$ by $2C_A/z$ and impose strong ordering on the longitudinal integrals, $x \ll x_n \ll \cdots \ll x_2 \ll x_1 \ll 1$. This is in addition to the strongly ordered transverse momentum integrals which we evaluated in eqn (3.321). The longitudinal integral becomes

$$x\sigma_n(x) \propto x \int_x^1 \frac{dx_n}{x_n} 2C_A \frac{x_n}{x} \cdots \int_{x_3}^1 \frac{dx_2}{x_2} 2C_A \frac{x_2}{x_3} \int_{x_2}^1 \frac{dx_1}{x_1} 2C_A \frac{x_1}{x_2} g(x_1, Q_0^2)$$

$$= (2C_A)^n \int_x^1 \frac{dx_n}{x_n} \cdots \int_{x_3}^1 \frac{dx_2}{x_2} \int_{x_2}^1 \frac{dx_1}{x_1} G_0(Q_0^2)$$

$$= \frac{1}{n!} \left[2C_A \log\left(\frac{1}{x}\right)\right]^n G_0(Q_0^2). \tag{3.324}$$

In the second line we have taken $xg(x, Q_0^2) = G_0(Q_0^2)$. Again the nested integrals are straightforward to evaluate and lead to a second $1/n!$ factor. Combining eqns (3.324) and (3.321) gives

$$\sum_n \sigma_n(x, Q^2) = G_0(Q_0^2) \sum_n \frac{1}{(n!)^2} \left[\frac{C_A}{\pi\beta_0} \ln\left(\frac{\alpha_s(Q_0^2)}{\alpha_s(Q^2)}\right) \ln\left(\frac{1}{x}\right)\right]^n$$

$$\sim G_0(Q_0^2) \exp\sqrt{\frac{4C_A}{\pi\beta_0} \ln\left(\frac{\alpha_s(Q_0^2)}{\alpha_s(Q^2)}\right) \ln\left(\frac{1}{x}\right)}. \tag{3.325}$$

Here, we have recognised the sum $\sum_n (y/2n!)^{2n}$ as the power series for the modified Bessel function $I_0(y)$ which has the asymptotic form $e^y/\sqrt{2\pi y}$ (Arfken and Weber, 1995). This result coincides with eqn (3.301).

Strictly speaking, the ladder diagrams, such as Fig. 3.26, are only schematic. For example, at one-loop they should be understood to also represent diagrams that include vertex and propagator corrections. This leads to a running coupling in eqn (3.320) which softens the Q^2 dependence in our results,

$$\int_{Q_0^2}^{Q^2} \frac{dk_T^2}{k_T^2} \alpha_s = \begin{cases} \alpha_s \ln(Q^2/Q_0^2) & \alpha_s \text{ fixed} \\ \beta_0^{-1} \ln(\alpha_s(Q_0^2)/\alpha_s(Q^2)) & \alpha_s \text{ running} . \end{cases} \tag{3.326}$$

Here, we have used the transverse momentum in the branching as the argument of α_s. An n-rung ladder diagram can also be used to represent the nth-order term, $[\alpha_s \ln(1/x)]^n$, in the solution of the BFKL equation (Gribov *et al.*, 1983). Here, each rung does not have a simple meaning but represents the sum of contributions from a set of real emission diagrams and interference with virtual diagrams. We also mention that the recombination term in eqn (3.310) can be represented by the merging of two ladders into a single ladder in what has been christened a 'fan diagram'.

3.6.8 *The Drell–Yan process*

The Drell–Yan process (Drell and Yan, 1971) is the production of high-mass lepton pairs from the decay of an electroweak boson produced in a hadron–hadron collision. Originally these were e^+e^- or $\mu^+\mu^-$ pairs coming from the decay of a virtual photon but, as collision energies have increased, it now includes the contribution from Z exchange and also $e\nu_e$ and $\mu\nu_\mu$ pairs coming from $W^\pm$ decays. Historically, the Drell–Yan process has proved very important; see the book by Cahn and Goldhaber (1989) for several original papers. It was pivotal in the discovery of heavy quarks, which manifested themselves as quarkonium resonances: charm and the $J(\psi)$ in 1974 and bottom and the Υ in 1978. It was also the process which in 1983 led to the discovery of the massive electroweak gauge bosons, the $W^\pm$ and the Z.

Theoretically, the Drell–Yan process is favoured because the final state particles are individually colourless and therefore are unaffected by the strong force. The high mass of the time-like photon, $Q^2 > 0$, ensures that small distance physics is probed and that pQCD is applicable. Within QCD the significance of the Drell–Yan process is due to its rôle as the prototype process within hadron–hadron collisions to be described using the same factorization approach that we used to treat DIS. Neglecting Z exchange, the underlying lowest order (tree-level) subprocess is quark–antiquark annihilation to a virtual photon, $h_1h_2 : q\bar{q} \to \gamma^*(Q^\mu) \to \ell^+\ell^-$. Therefore, the process offers a direct probe of a hadron's antiquark content; see Ex. (3-37). In the framework of eqn (3.72) the cross section is given by

$$\frac{d\sigma^{(0)}}{dQ^2}(h_1h_2 \to \gamma^*(Q) \to \ell^+\ell^-) = \tag{3.327}$$

$$\int_0^1 dx_1 \int_0^1 dx_2 \sum_q [q_{h_1}(x_1)\bar{q}_{h_2}(x_2) + \bar{q}_{h_1}(x_1)q_{h_2}(x_2)] \frac{d\hat{\sigma}^{(0)}}{dQ^2}(q\bar{q} \to \ell^+\ell^-)(Q,\hat{\tau}) .$$

Here, we introduce the following scaling variables

$$\hat{\tau} = \frac{Q^2}{\hat{s}} = \frac{\tau}{x_1 x_2} \quad \text{and} \quad \tau = \frac{Q^2}{s}, \tag{3.328}$$

with $\hat{s} = (x_1 p_1 + x_2 p_2)^2 = x_1 x_2 s$. Note that in eqn (3.327) we are careful to include both contributions coming from the quark (antiquark) being in hadron 1 (2) and *vice versa*. We have also temporarily suppressed the p.d.f.s' dependence on the scale Q^2. The differential cross section for the hard subprocess is given by:

$$\frac{d\hat{\sigma}}{dQ^2}(q\bar{q} \to \ell^+\ell^-) = \frac{4\pi\alpha_{\text{em}}^2}{3N_c Q^4} e_q^2 \delta(1 - \hat{\tau}) \equiv \frac{\sigma_{\text{DY}}^{(0)}}{Q^2} e_q^2 \delta(1 - \hat{\tau}). \tag{3.329}$$

This has been obtained from eqn (3.93) using crossing, allowing for the changed average over initial state colours and multiplying by unity in the form $1 = \int dQ^2 \delta(Q^2 - \hat{s})$. The factor $\sigma_{\text{DY}}^{(0)}$ sets the scale for the cross section and contains its dimensions. Combining eqns (3.327) and (3.329) gives

$$\frac{d\sigma}{dQ^2} = \frac{\sigma_{\text{DY}}^{(0)}}{Q^2} \tau \int_\tau^1 \frac{dx}{x} \left[q_{h_1}(x)\bar{q}_{h_2}\left(\frac{\tau}{x}\right) + \bar{q}_{h_1}(x)q_{h_2}\left(\frac{\tau}{x}\right) \right] e_q^2 \approx \frac{\tau F(\tau)}{Q^4}. \tag{3.330}$$

If the p.d.f.s are scale independent, that is, they do not depend on Q^2, as in the naïve parton model, then the differential cross section $d\sigma/dQ^2 \propto Q^{-4} \times$ a function of τ. This result follows on dimensional grounds and is the same scaling as we saw in DIS.

3.6.8.1 *The $\mathcal{O}(\alpha_s)$ corrections to the Drell–Yan process* The calculation of the $\mathcal{O}(\alpha_s)$ corrections to the Drell–Yan process follows very much the same procedures as those used for DIS (Altarelli *et al.*, 1979a; Kubar-André and Paige, 1979). Here we only outline the results using $\gamma^\star \to \ell^+\ell^-$ production for illustration; W and Z production follow the same lines whilst including the decay orientation of the lepton pair adds no new insights. A useful guide to the calculations is given by Willenbrock (1989).

There are basically two new contributions at $\mathcal{O}(\alpha_s)$; charge conjugation symmetry relates quark and antiquark initiated processes. There is the gluon bremsstrahlung correction to the lowest order process, $q\bar{q} \to \gamma^\star$. We expect this to show collinear singularities when the gluon becomes parallel to either the incoming quark or antiquark, but to be free of final state singularities after we include the one-loop, virtual corrections to $q\bar{q} \to \gamma^\star$. It is also free of ultraviolet singularities. There is also the gluon initiated process $gq \to \gamma^\star q'$. We expect this to contain a collinear singularity when the scattered quark lies antiparallel to the incoming quark in the C.o.M. frame. This is equivalent to the gluon undergoing a near collinear $g \to q\bar{q}$ branching in a fast moving frame. The appropriate generalization of eqn (3.327) is given by

$$\frac{d\sigma}{dQ^2}(h_1 h_2 \to \gamma^\star(Q) \to \ell^+\ell^-) = \int_0^1 dx_1 \int_0^1 dx_2$$

$$\times \left\{ \sum_q \left[q_{h_1}(x_1)\bar{q}_{h_2}(x_2) + \bar{q}_{h_1}(x_1)q_{h_2}(x_2) \right] \left(\frac{d\hat{\sigma}^{(0)}}{dQ^2}(q\bar{q} \to \gamma^\star) + \frac{d\hat{\sigma}^{(1)}}{dQ^2}(q\bar{q} \to \gamma^\star g) \right) \right.$$

$$\left. + \sum_{f=q,\bar{q}} \left[g_{h_1}(x_1)f_{h_2}(x_2) + f_{h_1}(x_1)g_{h_2}(x_2) \right] \frac{d\hat{\sigma}^{(1)}}{dQ^2}(gq \to \gamma^\star q') \right\}. \qquad (3.331)$$

The Feynman diagrams for the two new, hard subprocesses have been given in Figs. 3.23 and 3.22, but now are read from right to left. Crossing also allows us to obtain the matrix elements with minimal effort. However, we cannot directly take over the phase space integrals as they involve different regions. The result of these calculations are

$$\frac{d\hat{\sigma}^{(1)}}{dQ^2}(q\bar{q} \to \gamma^\star g) = \frac{\sigma_{\text{DY}}^{(0)}}{Q^2} e_q^2 \frac{\alpha_s}{2\pi} \left\{ -2P_{qq}^{(0)}(\hat{\tau}) \left[\Delta_\epsilon - \ln\left(\frac{Q^2}{\mu^2} \right) \right] + H_{q\bar{q}}(\hat{\tau}) \right\}$$

$$\frac{d\hat{\sigma}^{(1)}}{dQ^2}(gq \to \gamma^\star q') = \frac{\sigma_{\text{DY}}^{(0)}}{Q^2} e_q^2 \frac{\alpha_s}{2\pi} \left\{ -P_{qg}^{(0)}(\hat{\tau}) \left[\Delta_\epsilon - \ln\left(\frac{Q^2}{\mu^2} \right) \right] + H_{gq}(\hat{\tau}) \right\}, \qquad (3.332)$$

where for convenience we have introduced the coefficient functions

$$H_{q\bar{q}}(z) = C_F \left[4(1+z^2) \left(\frac{\ln(1-z)}{1-z} \right)_+ - 2\frac{1+z^2}{1-z}\ln z + \left(\frac{2\pi^2}{3} - 8 \right)\delta(1-z) \right]$$

$$H_{gq}(z) = T_F \left[2[z^2 + (1-z)^2]\ln\left(\frac{(1-z)^2}{z} \right) + 3 + 2z - 3z^2 \right]. \qquad (3.333)$$

The structure of eqns (3.331) and (3.332) is very similar to the corresponding expressions which arose in the NLO description of DIS, eqns (3.272) and (3.274). It is the power of factorization that essentially the same separation of the cross section into factorization scale dependent long-distance p.d.f.s and short-distance coefficient functions will treat the collinear singularities in the NLO description of the Drell–Yan process. Indeed, introducing the $\overline{\text{MS}}$ p.d.f.s, eqn (3.277), into eqn (3.331) gives

$$\frac{d\sigma}{dQ^2}(h_1 h_2 \to \gamma^\star(Q) \to \ell^+\ell^-) = \frac{\sigma_{\text{DY}}^{(0)}}{Q^2} \int_0^1 dx_1 \int_0^1 dx_2$$

$$\times \left\{ \sum_q \left[q_{h_1}^{\overline{\text{MS}}}(x_1, \mu_F^2)\bar{q}_{h_2}^{\overline{\text{MS}}}(x_2, \mu_F^2) + \bar{q}_{h_1}^{\overline{\text{MS}}}(x_1, \mu_F^2)q_{h_2}^{\overline{\text{MS}}}(x_2, \mu_F^2) \right] e_q^2 \right.$$

$$\times \left(\delta(1-\hat{\tau}) + \frac{\alpha_s}{2\pi} \left[2P_{qq}^{(0)}(\hat{\tau})\ln\left(\frac{Q^2}{\mu_F^2} \right) + H_{q\bar{q}}(\hat{\tau}) \right] \right)$$

$$\left. + \sum_{f=q,\bar{q}} \left[g_{h_1}^{\overline{\text{MS}}}(x_1, \mu_F^2)f_{h_2}^{\overline{\text{MS}}}(x_2, \mu_F^2) + f_{h_1}^{\overline{\text{MS}}}(x_1, \mu_F^2)g_{h_2}^{\overline{\text{MS}}}(x_2, \mu_F^2) \right] e_q^2 \right.$$

$$\times \frac{\alpha_s}{2\pi} \left[2P_{qg}^{(0)}(\hat{\tau}) \ln \left(\frac{Q^2}{\mu_F^2} \right) + H_{qg}(\hat{\tau}) \right] \Bigg\} , \qquad (3.334)$$

which is independent of the factorization scale to $\mathcal{O}(\alpha_s)$. The $\overline{\text{MS}}$ coefficient functions in this expression depend on the factorization scheme, that is, which finite terms are absorbed into the p.d.f.s and are therefore different in the DIS scheme; see Ex. (3-38). The $\mathcal{O}(\alpha_s^2)$ corrections to Drell–Yan have also been calculated (Zijlstra and van Neerven, 1992).

3.6.8.2 *Transverse momentum in Drell–Yan processes* The measurement of the W boson mass to high accuracy is very desirable as it facilitates tests of the Standard Model of electroweak interactions at the quantum (loop) level; see, for example, the report by Altarelli *et al.* (1989). Since the W decays to a charged lepton and a neutrino its mass reconstruction at hadron–hadron colliders must necessarily be indirect. The preferred method is based upon measuring the boost-invariant transverse momentum distribution of the charged lepton. Assuming that the W is produced with no transverse momentum and neglecting its width, one expects

$$\frac{1}{\sigma} \frac{d\sigma}{dp_{T\ell}^2} = \frac{3}{M_W^2} \left(1 - 2\frac{p_{T\ell}^2}{M_W^2} \right) \sqrt{1 - 4\frac{p_{T\ell}^2}{M_W^2}} . \qquad (3.335)$$

This strongly peaked distribution is very sensitive to M_W, but to be useful we must be confident that we understand the underlying transverse momentum distribution of the W boson.

At $\mathcal{O}(\alpha_s^0)$ the massive vector bosons produced in the Drell–Yan process have zero transverse momentum. At $\mathcal{O}(\alpha_s^1)$ the $q\bar{q} \to Vg$ and $gq \to Vq'$ ($g\bar{q} \to V\bar{q}'$) processes provide a good description of high-p_T vector boson production. Now, at order $\mathcal{O}(\alpha_s^n)$ the cross section behaves as

$$\frac{1}{\sigma} \frac{d\sigma}{dp_T^2} = \frac{1}{p_T^2} \left[\alpha_s^n A_{n,2n-1} \ln^{2n-1} \left(\frac{M_V^2}{p_T^2} \right) + \cdots \right] , \qquad (3.336)$$

so that care must be exercised in the low-p_T region, $M_V \gg p_T \gg \Lambda_{QCD}$. The need to sum these large logarithms was first recognized by Dokshitzer *et al.* (1980) and an impact parameter space formalism for summing them developed (Collins and Soper 1981; 1982; Collins *et al.* 1985). Impact parameters are the Fourier conjugate variables to p_T. In computing the effect of multiple gluon radiation it is important to impose momentum conservation, $\sum_n \boldsymbol{k}_{nT} = \boldsymbol{p}_T$, on the (soft) gluon bremsstrahlung (Parisi and Petronzio, 1979); this greatly reduces the possibility of obtaining $p_T = 0$. A numerical implementation of this formalism (Ladinsky and Yuan, 1994) proved the necessity to include a Gaussian smearing of the initial partons' impact parameter distribution in order to counter convergence and infrared problems. Analytic expressions for the coefficients in eqn (3.336) are available up to N^3LL accuracy (Kulesza and Stirling, 1999). In an alternative form of eqn (3.336) the right-hand side can be resummed by 'exponentiating' the leading logarithmic terms $\alpha_s^n \ln^{n+1}(M_V^2/p_T^2)$, which have been

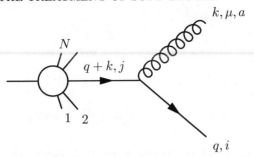

FIG. 3.27. The emission of a soft gluon from one of N hard, final state partons

calculated to NLLA accuracy (Frixione *et al.*, 1999). From the impact parameter space results it is then possible to derive expressions for the p_T-space. In order to describe the full p_T range, it is important to match these low-p_T summed calculations to the high-p_T fixed order calculations. This has been carried out (Ellis *et al.*, 1997; Ellis and Veseli, 1998), though there remain problems in ensuring a smooth cross-over at $p_T = M_V$. To compare with actual measurements, experimental data are available for the p_T-distribution of Z bosons (D0 Collab. 2000a, 2000b; CDF Collab. 2000) and W bosons (D0 Collab., 2001).

3.7 The treatment of soft gluons

We learnt in Section 3.5 that matrix elements become singular when soft gluons are emitted and that these singularities cancel in sufficiently inclusive observables such as a total cross section. When less inclusive measurements are made, for example, by observing any soft particles above an energy threshold, the cancellation is incomplete and there remain logarithmic enhancements to the emission probability. That said, the effects of these soft gluons are mitigated due to their lack of energy. In this section, first we show how matrix elements simplify in the limit of soft gluon emission and then we investigate their physical effects (Bassetto *et al.*, 1983).

As gluons become softer their wavelengths grow and they become sensitive only to an event's global structure. Specifically, the distribution of soft gluon radiation depends only on the momenta and colour connections of the hard, final state partons and not on any internal dynamics. This is known as Low's theorem for the case of soft photon emission. To see how this arises consider a hard process involving N final state partons with momenta $\{p_i^\mu\}$, all of whose relative transverse momenta are large compared to $\Lambda_{\rm QCD}$. This condition ensures that all internal particles are well off mass-shell, such that their large virtualities shield the collinear and soft singular regions. Thus, adding a soft gluon with momentum $k^\mu = \omega(1, \hat{\bm n})$, $\omega \ll E_i$, can give rise to soft singularities only if it is emitted from an external parton. A typical such Feynman diagram is illustrated in Fig. 3.27.

Suppose that the gluon is emitted by a final state quark, then the corresponding matrix element is given by

$$\mathcal{M}_i^{(N+1)}(q,k) = g_s T_{ij}^a \epsilon^\star(k)_\mu \bar{u}(q) \gamma^\mu \frac{(\slashed{q} + \slashed{k} + m)}{(q+k)^2 - m^2} \mathcal{M}_j^{(N)}(q+k)$$

$$= g_s T_{ij}^a \epsilon^\star(k)_\mu \bar{u}(q) \frac{[-(\slashed{q} - m)\gamma^\mu + 2q^\mu + \gamma^\mu \slashed{k}]}{2q \cdot k} \mathcal{M}_j^{(N)}(q+k)$$

$$= g_s T_{ij}^a \epsilon^\star(k)_\mu \frac{q^\mu}{q \cdot k} \bar{u}(q) \mathcal{M}_j^{(N)}(q) + \mathcal{O}(\omega) . \tag{3.337}$$

Here we have exploited, after a suitable re-arrangement, the Dirac equation satisfied by the basis spinor and neglected terms proportional to k compared to those proportional to q. This neglect of recoil effects is known as the eikonal approximation. The structure obtained is universal, that is, the same form would hold for emission off an antiquark, another gluon or even a coloured scalar; see Ex. (3-39). The reason is that the soft gluon has a correspondingly long wavelength and is unable to resolve the spin structure of the emitting particle. Adding up the contributions from all the hard partons we obtain

$$\mathcal{M}^{(N+1)} = g_s \epsilon^\star(k)_\mu \left(\sum_{i=1}^N \hat{T}_i^a \frac{p_i^\mu}{p_i \cdot k} \right) \times \mathcal{M}^{(N)} . \tag{3.338}$$

Here the colour charge operators, $\hat{T}_i^a$, generate the appropriate colour factors for a gluon, colour a, emitted by parton i. They act as follows:

$$\begin{aligned}
\hat{T}_q^a | \cdots ; q, i; \cdots \rangle &= +T_{ij}^a | \cdots ; q, j; \cdots \rangle \\
\hat{T}_{\bar{q}}^a | \cdots ; \bar{q}, i; \cdots \rangle &= -T_{ji}^a | \cdots ; \bar{q}, j; \cdots \rangle \\
\hat{T}_g^a | \cdots ; g, b; \cdots \rangle &= -i f_{abc} | \cdots ; g, c; \cdots \rangle .
\end{aligned} \tag{3.339}$$

To obtain the matrix element squared we need to use an expression for the gluon polarization sum, eqn (3.121), which allows us to write

$$\left| \mathcal{M}^{(N+1)} \right|^2 = -g_s^2 J \cdot J^\dagger \left| \mathcal{M}^{(N)} \right|^2 \quad \text{with} \quad J^{a\mu}(k; p_i) = \sum_i \hat{T}_i^a \left(\frac{p_i^\mu}{p_i \cdot k} - \frac{n^\mu}{n \cdot k} \right) . \tag{3.340}$$

Here J^μ is known as the insertion current. For a colour neutral system, such as in an e^+e^- event, we have $\sum_i \hat{T}_i = 0$, so that the second term, proportional to the gauge vector n^μ, can be safely dropped. It is also absent in the Feynman gauge. Henceforth we drop it. Including the phase space then gives the following form for the soft gluon emission cross section (Marchesini and Webber, 1990),

$$\begin{aligned}
&d\sigma^{(N+1)} \\
&= -\frac{\alpha_s}{2\pi} J(k) \cdot J^\dagger(k) \frac{d^3 k}{2\pi\omega} d^{(N)}\sigma \\
&= -\frac{\alpha_s}{2\pi} \sum_{i \neq j} \hat{T}_i \cdot \hat{T}_j \omega^2 \left[\frac{p_i \cdot p_j}{(p_i \cdot k)(p_j \cdot k)} - \frac{1}{2} \frac{p_i^2}{(p_i \cdot k)^2} - \frac{1}{2} \frac{p_j^2}{(p_j \cdot k)^2} \right] \frac{d\omega}{\omega} \frac{d\Omega_k}{2\pi} d^{(N)}\sigma
\end{aligned}$$

$$= -\frac{\alpha_s}{2\pi}\sum_{i\neq j}\hat{T}_i\cdot\hat{T}_j\,\omega^2\,\frac{p_i\cdot p_j}{(p_i\cdot k)(p_j\cdot k)}\,\frac{d\omega}{\omega}\,\frac{d\Omega_k}{2\pi}\,d\sigma^{(N)}\,. \tag{3.341}$$

Notice that we have manipulated the diagonal terms with the aid of the identity $\hat{T}_i = -\sum_{i\neq j}\hat{T}_j$; see Ex. (3-40). The last expression is appropriate to the case of massless hard partons. In this limit we see that the only contributions to the cross section come from quantum mechanical interference: the soft gluon can equally be associated with the hard partons i or j. In Section 4.2.5 we will discuss an approximation which allows us to recover a classical picture of soft gluon emission that is suitable for Monte Carlo simulation. The term in the sum is the classical expression for describing radiation off a dipole. In essence, the hard partons form an 'antenna' and the soft gluons follow the associated radiation pattern. The factor ω^2 ensures that it does not depend on the energy of the emitted soft gluon, only its direction. The energy spectrum is of the typical bremsstrahlung form $d\omega/\omega$. The cross section also contains collinear singularities whenever the direction of the gluon coincides with that of a hard external parton, $k\parallel p_i$.

We illustrate the use of eqn (3.341) with an historically important example. Namely, the comparison of the intra-jet energy flow in $\gamma^*/Z\to q\bar{q}\gamma$ and $\gamma^*/Z\to q\bar{q}g$ events which, modulo the couplings ($e_q^2\alpha_{\rm em}\leftrightarrow C_F\alpha_s$), are described by the same matrix element. The focus of our attention is the radiation pattern, given by J^2, which is straightforward to calculate. For simplicity we assume that the quarks are massless and obtain

$$J^\mu(k)_{q\bar{q}\gamma} = \hat{T}_q\left(\frac{q^\mu}{q\cdot k}-\frac{\bar{q}^\mu}{\bar{q}\cdot k}\right)\quad\Longrightarrow\quad -J^2(k)_{q\bar{q}\gamma} = 2C_F\frac{q\cdot\bar{q}}{(q\cdot k)(\bar{q}\cdot k)} \tag{3.342}$$

for the characteristics of soft gluon radiation in a $q\bar{q}\gamma$ event. Here, we employed colour conservation to substitute $\hat{T}_{\bar{q}} = -\hat{T}_q$ and that the quark's colour charge is given by $\hat{T}_q^2 = C_F$. The analogous calculation for gluon emission in a $q\bar{q}g$ event yields

$$J^\mu(k)_{q\bar{q}g} = \hat{T}_q\frac{q^\mu}{q\cdot k}+\hat{T}_{\bar{q}}\frac{\bar{q}^\mu}{\bar{q}\cdot k}+\hat{T}_g\frac{g^\mu}{g\cdot k}\quad\Longrightarrow$$

$$-J^2(k)_{q\bar{q}g} = C_A\left[\frac{q\cdot g}{(q\cdot k)(g\cdot k)}+\frac{\bar{q}\cdot g}{(\bar{q}\cdot k)(g\cdot k)}\right]-(C_A-2C_F)\frac{q\cdot\bar{q}}{(q\cdot k)(\bar{q}\cdot k)}\,. \tag{3.343}$$

Again, we have avoided detailed evaluation of the colour factors by making use of a trick,

$$2\hat{T}_q\cdot\hat{T}_g = (\hat{T}_q+\hat{T}_g)^2-\hat{T}_q^2-\hat{T}_g^2 \qquad\qquad 2\hat{T}_q\cdot\hat{T}_{\bar{q}} = (\hat{T}_q+\hat{T}_{\bar{q}})^2-\hat{T}_q^2-\hat{T}_{\bar{q}}^2$$
$$= (-\hat{T}_{\bar{q}})^2-\hat{T}_q^2-\hat{T}_g^2 \qquad\qquad\qquad\quad = (-\hat{T}_g)^2-\hat{T}_q^2-\hat{T}_{\bar{q}}^2 \tag{3.344}$$
$$= -C_A \qquad\qquad\qquad\qquad\qquad\qquad\quad = C_A-2C_F\,.$$

The two radiation patterns show markedly different behaviour in their inter-jet regions. In $q\bar{q}\gamma$ events the energy flow is concentrated between the coloured quark and antiquark. In qq̄g events the energy flow is concentrated between the colour connected q-g and q̄-g pairs, whilst it is suppressed as $(C_A - 2C_F)/C_A = 1/N_c$ between the qq̄-pair. In both cases, the soft radiation is predominantly in the plane of the event. The strong collinear enhancements in the directions of the coloured partons give rise to jets. Compared to the intra-jet energy flows the inter-jet energy flows are only modest. The resulting radiation patterns are illustrated in Fig. 3.28.

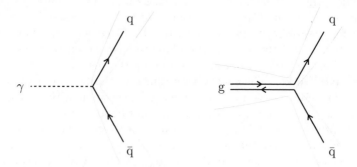

FIG. 3.28. The in-plane radiation patterns of qq̄γ (left) and qq̄g events (right) together with the underlying colour flows

Also shown in Fig. 3.28 are the associated colour flows based on representing the gluon as a pair of colour–anticolour lines. This is the planar approximation, equivalent to replacing $SU(N_c)$ by $U(N_c)$, which is correct to relative $\mathcal{O}(1/N_c^2)$. It is so-called because in this approximation it is always possible to draw any Feynman diagram such that no colour lines need cross. It suggests that a gluon should produce twice the amount of radiation as a quark by a factor of the ratio of their colour charges, $C_A/C_F = 2 + \mathcal{O}(1/N_c^2)$. We also see that we might view the qq̄g event as consisting of two colour dipoles in which the soft gluon radiation off one parton is restricted by the anticolour charge of its colour-connected partner. We shall investigate these ideas further in Section 4.2.5.

Looking further ahead we will see that the distribution of soft gluons predicted by pQCD coincides with the location of the string in the non-perturbative hadronization model of the same name (Section 3.8.3). Indeed, it was on the basis of this model that the 'string effect', the relative depletion of soft hadrons between the qq̄-pair in a qq̄γ event compared to a qq̄g event was predicted. Today the pQCD explanation is preferred. The experimental situation is discussed in Section 13.2.

A more sophisticated introduction to the treatment of soft gluons and pQCD in general can be found in the lectures by Dokshitzer (1996).

3.8 Hadronization models

The preceeding sections have equipped us with much of the technology needed to calculate properties of those multi-hadronic events that involve a hard scale, Q. The usual description follows a sequence of decreasing momentum transfers. First, a set of primary partons is produced in a hard subprocess that is described by a fixed-order matrix element. Second, a parton (or, equivalently, dipole) cascade evolves these primary partons from the hard scale down to a variable number of final state partons at a fixed cut-off scale $Q_0 \approx 1\,\text{GeV}$. The leading logarithmic behaviour of these cascades is governed by the same pQCD evolution equations that were developed to describe the scale dependence of p.d.f.s and fragmentation functions. Third, the final state quarks and, predominantly, gluons arrange themselves into colour-neutral hadrons. Given the low scales inherent in this hadronization process, and the subsequently large strong coupling, this process is non-perturbative and so, at present, beyond our means to calculate it. Fourth, pretabulated decay tables (PDG, 2000) are used to sequentially decay any unstable primary particles into the stable hadrons, leptons and photons that are seen directly in the detector. As an illustration, the fragmentation function for finding a hadron h in a jet initiated by an outgoing primary parton a of maximum scale Q, such that it carries a fraction x of a's momentum, is given schematically by

$$D_a^{\text{h}}(x, Q^2) = (\text{pQCD evolution: } Q^2 \to Q_0^2) \otimes (\text{model: } b \to \text{H})\Big|_{Q_0^2} \qquad (3.345)$$
$$\otimes (\text{tables: } \text{H} \to \text{h, h}', \dots)\,.$$

The convolution runs over all partons b created in the pQCD evolution, which then hadronize into all kinds of possible intermediate unstable hadrons H that in turn decay into the final state particles seen in an experiment. The hadronization process is believed to be essentially independent of Q, up to power corrections, but sensitive to the arbitrary cut-off Q_0. In combination with the pQCD evolution the net result should, of course, be independent of the actual choice Q_0.

Now, it is important to remember that our present understanding of the hadronization process, and hence its effects, is based largely on models and not QCD as such. Fortunately, it is expected to be both local in nature and not to involve any large momentum transfers. This helps to justify our belief that it is the first two calculable stages of an event, characterized by relatively large momentum transfers, $Q \gg \Lambda_{\text{QCD}}$, that are dominant in determining its global features, such as energy dependences, event shapes, multiplicities etc. This makes possible meaningful tests of pQCD. However, this is not to say that the effects of hadronization are completely insignificant. For example, in e^+e^- collisions, since pQCD predicts that soft gluon radiation in a three-jet event is preferentially in the plane defined by the inital quark–antiquark pair and the first hard gluon, it does have some influence on event shape variables which are sensitive to 'out-of-plane' activity. It also determines the rates of production of identified particles and their correlations.

In this section, after setting the scene with a review of the space–time development of a multi-hadronic event, we discuss the three most popular types of model for the parton-to-hadron transition. These models play an essential rôle in the numerical simulation of complete multi-hadronic events. It is therefore important to understand these models in order to both constrain their influence on calculable event properties and to gain insights into the non-perturbative physics which might underly them.

3.8.1 *Space–time structure of multi-hadron events*

When we calculate a hard scattering process using Feynman diagrams we work usually in momentum space. This is the Fourier transform of configuration space and, loosely, we can transform between the two descriptions using the uncertainty relation, $\Delta q \sim (\Delta x)^{-1}$. Here, we will develop an intuitive space–time picture of a hard scattering event which, for simplicity, we take to be e^+e^- annihilation to hadrons in the C.o.M. frame. The discussion is based on arguments presented by Dokshitzer *et al.* (1989, 1991). The essence of the hard subprocess, $e^+e^- \to q\bar{q}$, is the production of a highly virtual photon of momentum Q^μ, which impulsively kicks a virtual $q\bar{q}$-pair out of the vacuum. This happens on a short time scale, $t_{\mathrm{hard}} \sim 1/\sqrt{Q^2}$. These primary partons would be surrounded ordinarily by a cloud of virtual particles. However, in this sudden process they shake off most of their cloud, so that any structure they possess is on a scale below $1/\sqrt{Q^2}$, the wavelength of the hard probe. Effectively they behave as bare, or more properly half dressed, colour charges until the gluon field has had time to regenerate out to a typical light hadron size $R_0 \sim m_{\mathrm{had}}^{-1} \sim \Lambda_{\mathrm{QCD}}^{-1} \sim 1$ fm. In the rest frame of the hadron this takes a time R_0. After allowing for the boost to the laboratory, which in our case coincides with the C.o.M. frame, the typical hadronization time becomes

$$t_{\mathrm{had}} \approx \gamma \cdot \tau = \frac{E}{m} \cdot R_0 \sim \frac{Q}{R_0^{-1}} \cdot R_0 = QR_0^2 . \tag{3.346}$$

The picture so far is of a fast, hard subprocess followed by the relatively slow, $t_{\mathrm{had}} \gg t_{\mathrm{hard}}$, hadronization of a set of primary partons. The potential for the production of hadrons at wide, space-like separations raises immediately the issue of how the various charges are conserved. Of course, the decelerating partons will radiate gluons and here two new time scales enter the discussion. Consider Fig. 3.29 in which a relatively soft gluon with energy k is emitted at a small opening angle θ from a parton of energy $E \gg k$, which is the favoured region for gluon emission. We assume these partons to be massless and so moving at the speed of light. The first time scale is the formation time of the gluon, defined to be the lifetime of the virtual emitter which is determined by its off mass-shellness

$$q^2 = 2Ek(1 - \cos\theta) \approx Ek\theta^2 = \frac{E}{k}k_\perp^2 . \tag{3.347}$$

Here we have introduced $k_\perp = k\theta$, the transverse momentum of the gluon. Using the uncertainty relation and applying a boost, c.f. Ex. (3-2), this gives

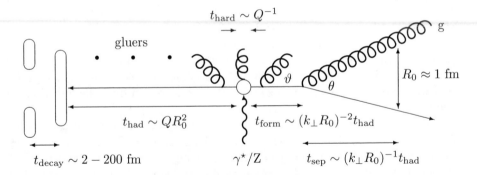

FIG. 3.29. A schematic diagram showing the time scales involved in a $\gamma^\star \to q\bar{q}g$ event. The first gluers form almost immediately and result in two colour neutral systems of partons. Figure from Knowles and Lafferty(1997).

$$t_{\text{form}} \sim \frac{E+k}{\sqrt{q^2}} \cdot \frac{1}{\sqrt{q^2}} = \frac{E+k}{E} \frac{k}{k_\perp^2} \approx \frac{k}{k_\perp^2} \, . \tag{3.348}$$

The second time scale is the time taken for the gluon to reach a transverse separation of R_0 and become independent of the emitter. It is given by

$$t_{\text{sep}} \sim \frac{R_0}{\theta} = \frac{R_0 k}{k_\perp} \, . \tag{3.349}$$

Comparing these formation, separation and hadronization times

$$t_{\text{form}} : t_{\text{sep}} : t_{\text{had}} = \frac{k}{k_\perp^2} : \frac{R_0 k}{k_\perp} : k R_0^2 = 1 : (k_\perp R_0) : (k_\perp R_0)^2 \, , \tag{3.350}$$

we easily see that for the time sequence in this picture to make sense, we require the gluon to be emitted above a minimum transverse momentum, $k_\perp > R_0^{-1}$. This is natural as it implies that $\alpha_s(k_\perp^2) < 1$, so that perturbation theory is applicable and our picture is justified. If $k_\perp < R_0^{-1}$ we can say nothing. On the borderline are quanta with $k_\perp = \theta k = R_0^{-1}$. These quanta feel the strong force and are responsible for holding hadrons together. We distinguish them from the essentially free, perturbative gluons by the name gluers. The first gluers form at time R_0 and have $k \sim k_\perp \sim R_0^{-1}$. Thus, gluers form immediately in the wake of the primary partons providing a means to blanch their colour field and leave two fast separating colour neutral systems. The slowest hadrons form first, close to the interaction point, in what may be called an 'inside-out' pattern (Bjorken, 1973). On quite general grounds the distribution of these gluers can be estimated using QCD,

$$d N_{\text{gluers}} \sim C_F \left[\int_{k_\perp \sim R_0^{-1}} \frac{dk_\perp^2}{k_\perp^2} \frac{\alpha_s(k_\perp^2)}{\pi} \mathcal{P}(k) \right] \times \frac{dk}{k} \propto \frac{dk}{k} = dy \, . \tag{3.351}$$

Here the logarithmic $k_\perp^2$ dependence reflects the fact that the coupling is dimensionless. The logarithmic k dependence follows from the gluon being massless.
Only the kernel function, $\mathcal{P}$, which is of order unity, depends on details of the
parton–gluon vertex. The distribution is thus governed by longitudinal phase
space. It predicts hadron production that is uniformly distributed in rapidity
and with typical transverse momenta $\sim R_0^{-1}$ as measured from the directions of
the primary partons.

Now, not all quanta are emitted at low $k_\perp$ and in a significant range, $R_0^{-1} \ll$
$k_\perp < k \ll \sqrt{Q^2}$, a shower of perturbative gluons is possible. The low $k_\perp$ gluers
do not shower. A hard gluon, g, emitted at an angle θ_g becomes an independent
colour source at $t_{\text{sep}} \sim R_0/\theta_g$, again raising the issue of charge conservation.
Fortunately, a gluer emitted at the same angle θ_g would appear just in the
right place and at the right time to blanch the tail of the separating gluon's
colour field. This gluon then acts as a new source of gluers, restricted such that
$k \gtrsim (\theta_g R_0)^{-1}$. It is equivalent to a quark jet of a reduced scale $Q = g_\perp$, boosted
with $\gamma = 1/\theta_g$, and with the substitution $C_F \mapsto C_A$ for the parton's charge in
eqn (3.351) (Bjorken, 1992). It is important to note that gluers emitted at larger
angles, $\vartheta > \theta_g$, have $R_0/\vartheta < R_0/\theta_g$, the separation time of the hard gluon, and
should therefore be associated with the parent parton–gluon system. This obser-
vation lies behind the strong angular ordering condition, $\vartheta \ll \theta$, which defines
the dominant phase space region in a time-like parton shower. The coherence of
soft emissions simply reflects colour charge conservation and is fundamental to
a gauge theory such as QCD. We shall investigate further this colour coherence
in Section 4.2.5; see also Ex. (4-3). If we consider two gluers emitted at the same
angle, ϑ, from the two hard partons, then when they hadronize simultaneously
their transverse separation is given by

$$d_\perp = \frac{R_0}{\vartheta}\theta_g \gg R_0 . \tag{3.352}$$

This large inter-gluer separation suggests that the hadronizing gluers form a
rather low density system in configuration space, thus limiting the influence one
hadronizing parton system can have on a neighbouring parton system.

Finally, the time scales for the decay of the primary particles are set by
the reciprocal of their mass-widths Γ. For the strong resonances typical values
of $\Gamma \sim 1$–$100\,\text{MeV}$ give $t_{\text{decay}} \sim 200$–$2\,\text{fm}$. Clearly, these time scales are commensurate with those for primary hadron production, so that any distinction
between seperate hadronization and resonance decay phases may be only se-
mantics. The time scale for the weak decays of s, c and b hadrons is of order
$t_{\text{decay}} \sim (M_W/m_Q)^2 k/m_Q$, where k/m_Q is the hadron's γ-factor. The flight distances are typically $\mathcal{O}(10^{12}\,\text{fm})$ for c and b and much longer for s, and thus these
decays may safely be treated as separate subprocesses. The case of the very heavy
top quark is special. Since the intermediate $W^\pm$ appearing in the decay may be
real, the width behaves instead as $\Gamma_t \propto m_t^2/M_W \approx 2.5\,\text{GeV}$. The decay is very
fast and should properly be treated as part of the hard subprocess.

In the resulting picture, at the end of the parton shower we have a low density system of quarks and, predominantly, gluons, each dressed in a uniform sheath of strongly interacting gluers. What happens next is essentially a mystery but as the parton virtualities drop below a critical value, $Q_0 \sim 1\,\mathrm{GeV}$, a phase transition occurs and the relevant degrees of freedom cease to be partons, q, q̄ and g, and become mesons and baryons, that is, qq̄ and qqq bound states. Now, due to the strength of the strong force at low momentum transfers, it is energetically favourable for the vacuum to be populated with virtual qq̄-(Cooper)-pairs of opposite flavour, momentum and helicity but non-zero net chirality. It is common to all the hadronization models considered here that the energy stored in the gluer field is used to promote these virtual qq̄ and perhaps qq'-q̄q̄' pairs into real particles, which then arrange themselves into colour singlet hadrons. In practice, what the models must supply are rules that can be applied iteratively to determine which quark flavours are produced, which primary hadrons subsequently form and how much momentum these hadrons carry away.

The space–time picture developed above suggests a simple 'tube' model for describing hadronization (Feynman, 1972). This is suitable for semi-quantitative evaluations of the effects of hadronization on event shape variables etc. However, for detailed studies of whole events it is necessary to specify every hadron produced in an event. This requires more sophisticated models as described below and in the literature, for example, the review by Knowles and Lafferty (1997).

3.8.2 Independent hadronization

Independent hadronization is the oldest and simplest realistic model for hadronization. Today, it is synonymous with the work of Field and Feynman (1978). The idea is to iterate a sequence of universal branchings, $q_1 \to q_2 + h$, based on the excitation of new quark, $q_2\bar{q}_2$, or diquark pairs. Gluons are treated by replacing them by a light qq̄-pair. The basic 'unit cell' is illustrated in Fig. 3.30.

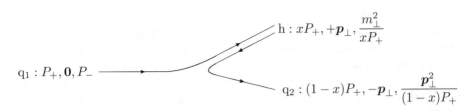

$$h : xP_+, +\boldsymbol{p}_\perp, \frac{m_\perp^2}{xP_+}$$

$$q_1 : P_+, 0, P_-$$

$$q_2 : (1-x)P_+, -\boldsymbol{p}_\perp, \frac{\boldsymbol{p}_\perp^2}{(1-x)P_+}$$

FIG. 3.30. The basic 'unit cell', showing the light-cone momenta, which is iterated in independent hadronization models

Unfortunately, the model has no strong theoretical underpinning so that it is rather arbitrary in its details. The flavours of the excited (di)quark pair are selected in fixed ratios. Empirically, it is found necessary to suppress both strange quarks and diquarks, $d : u : s = 1 : 1 : r_s$, with $r_s \approx 1/3$, and $q : qq' = 1 : r_{qq}$, with $r_{qq} \approx 1/9$. Further rules are required to choose between the various low lying hadrons, $h : (q_1 \bar{q}_2)$, of a specified flavour. The hadronization process is described using light-cone coordinates, $p_\pm \equiv (E \pm p_z)$ and $p_\perp$, with the z-axis chosen to be the direction of the parent parton. The mass-shell constraint is given in terms of the transverse mass by $p_+ p_- \equiv m_\perp^2 = m^2 + p_\perp^2$. The light-cone momentum fraction of the hadron, $x = p_+^h / p_+^{q_1}$, is specified by an arbitrary, longitudinal fragmentation function. A typical parametric form is

$$f(x) = 1 - a + a(1 + b)(1 - x)^b , \qquad (3.353)$$

though more sophisticated forms are not uncommon when heavy quarks (c,b) are present. The transverse momentum is selected from a Gaussian distribution, $\exp(-\boldsymbol{p}_\perp^2 / 2\sigma^2)$. As a result of this branching, the primary quark, q_1, acquires a mass. Thus the iteration may be terminated either at the point just before the sum of the parton masses becomes greater than the available energy or just before the remnant quark, q_2, starts to travel backwards, $p_z < 0$. Since each parton is treated in isolation, this requires an *ad hoc* prescription to ensure overall momentum and quantum number conservation; only in the unit cell are these conserved automatically. Enforcing four-momentum conservation proves particularly troublesome, since in $e^+ e^-$ collisions event shape variables and hence α_s determinations, see Chapter 8, are sensitive to the nature of the chosen solution (Bengtsson *et al.*, 1986).

More sophisticated variants of this basic scheme are still employed in the COJETS (Odorico, 1990) and ISAJET (Paige and Protopopescu, 1986) event generators. However, independent hadronization coupled with their use of incoherent parton showers, typically with a rather large cut-off parameter, $Q_0 \sim 3\,\text{GeV}$, does not give the best description of today's exacting data. This is particularly true in $e^+ e^-$ annihilation, where these programs are no longer widely used.

3.8.3 *String hadronization*

String hadronization is often assumed to mean the 'Lund' model (Andersson *et al.*, 1983*a*). However, there are really several models, each based on a common starting point. When an oppositely coloured quark–antiquark pair moves apart, as in $e^+ e^- \to q\bar{q}$, it is believed that the self-interacting colour field between them collapses into a narrow 'flux tube' of uniform energy density per unit length, or string tension κ, estimated to be $\kappa \approx 1\,\text{GeV/fm}$. This model therefore corresponds to a linear confining potential.

The transverse size of a string, $\langle r_\perp^2 \rangle = \pi/(2\kappa)$, is small compared to its typical length. It is therefore plausible that its dynamics can be described by a massless, one-dimensional, relativistic string possessing no transverse excitations. The classical equations of motion follow from a covariant action equal to the space–time area swept out by the string. In the solution for a massless $q\bar{q}$-pair, seen from the string's rest frame, the end-point quarks oscillate repeatedly outwards and inwards at the speed of light, passing through one another

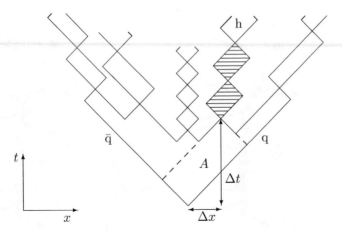

FIG. 3.31. The space–time evolution of a string as it breaks up into hadrons. Highlighted is a slowly right-moving string fragment, a hadron, exhibiting yo-yo modes. Observe that the slowest hadrons form at the earliest times near the centre of the string and the fast hadrons form latest near the ends of the string.

and transferring energy to and from the string in what are termed yo-yo modes (Artru, 1983), seen as diamonds in Fig. 3.31. Solutions also exist that exhibit energy–momentum carrying 'kinks', which have been identified successfully with hard gluons (Andersson and Gustafson, 1980). At the end of the perturbative shower, colourless string segments form between neighbouring partons, each segment terminating on a quark and an antiquark. Figure 3.32 shows such a string network. The full three-dimensional evolution of such networks is complicated, but robust treatments exist (Sjöstrand, 1984; Morris, 1987). Further complications arise for massive end-point quarks when the classical equations become non-linear (Chodos and Thorn, 1974; Bardeen *et al.*, 1976).

The motion of a string is described classically. Adding quantum mechanics introduces the possibility of a string breaking up, *à la* the snapping of a magnet, via the creation of intermediate $q\bar{q}$-pairs from the field energy in the string. The probability for this to occur is given by Wilson's exponential area decay law (Wilson, 1974)

$$\frac{\mathrm{d}\mathcal{P}}{\mathrm{d}A} = P_0 \mathrm{e}^{-P_0 A} . \tag{3.354}$$

Here A is the space–time area within the backward light-cone of the point at which the $q\bar{q}$-pair is created, see Fig. 3.31, and P_0 is a constant, reflecting the uniformness of strings. It is important to note that the position of the break-up point and the momentum of the fragment are linearly related, $E = \kappa \Delta t$ and $p = \kappa \Delta x$. Thus, for example, the area decay law is equivalent to a mass-squared decay law, $\mathrm{d}\mathcal{P}/\mathrm{d}m^2 = b\exp(-bm^2)$ with $b = P_0/2\kappa^2$. This implies that the average string fragment has $\langle \tau_{\mathrm{form}}^2 = t^2 - x^2 \rangle = 1/P_0$ and $\langle m_{\mathrm{string}}^2 \rangle = 2\kappa^2/P_0$

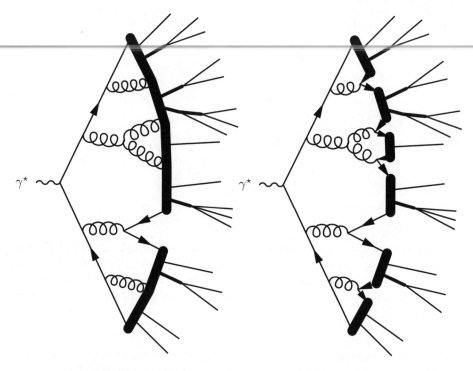

FIG. 3.32. The hadronization of a $\gamma^\star \to q\bar{q}$ event into resonances and long–lived hadrons, as pictured in the string model, left, and cluster model, right. The parton cascade contains a perturbative $g \to q\bar{q}$ vertex causing two strings to form, into which gluons introduce 'kinks'. In the cluster model these final state gluons are forcibly split into $q\bar{q}$-pairs just prior to cluster formation.

so that the slowest moving fragments typically form first, near the centre of the string assuring an inside-out pattern (Bjorken, 1973). Since the break-up points are space-like separated this is true in any frame.

In practice, rather than working in terms of the time and place of the string's break-up point, (t, x), it is more convenient to work in terms of the equivalent (transverse) mass and light-cone momentum fraction of the fragment, $(m_\perp^2 = m^2 + \boldsymbol{p}_\perp^2, z)$. Given rules for choosing the mass of the fragment and a fragmentation function, the break-up can proceed iteratively. In order to be more specific it is necessary to identify the string fragments. These can either be continuous-mass substrings, called clusters, which then decay into hadrons (Artru and Mennessier, 1974; Gottschalk, 1984), or actual fixed-mass hadrons (Andersson *et al.*, 1979*a*). The former choice gives the original Artru–Mennessier (1974) and later CalTech-II (Gottschalk and Morris, 1987) schemes, whilst the latter choice gives the Lund (Andersson *et al.*, 1983*a*) and UCLA (Chun and Buchanan, 1993) schemes.

The pure area decay law is equivalent to the joint $m_\perp^2 - z$ distribution

$$\frac{\mathrm{d}^2 \mathcal{P}}{\mathrm{d}z \mathrm{d}m_\perp^2} = \frac{b}{z} \exp\left(-b\frac{m_\perp^2}{z}\right) \quad \Longrightarrow \quad \frac{\mathrm{d}\mathcal{P}}{\mathrm{d}m_\perp^2} = bE_1(bm_\perp^2) \overset{m_\perp^2 \to 0}{\sim} -b\ln(bm_\perp^2) .$$

(3.355)

Here E_1 is the first exponential integral function (Arfken and Weber, 1995). Unfortunately, the resulting $m_\perp^2$ distribution diverges as the mass of the daughter cluster tends to zero. Thus, in the string-cluster schemes, unless a mass regulator is introduced, this will result in a sequence of ever lighter clusters, precluding any physical interpretation (Artru and Mennessier, 1974; Bowler, 1981). In the more fully developed CalTech-II model (Gottschalk and Morris, 1987) this is resolved by introducing a probability for a cluster to undergo further fission which vanishes as its mass approaches a lower threshold. The clusters which are prevented from further fissions then undergo a comparatively complex sequence of decays, resulting in hadrons whose flavours and momenta are determined by phase space considerations (see also Section 3.8.4). Since the model's success in confronting data has been mixed, so that it has fallen out of favour and is no longer being actively developed, we do not discuss it any further.

In the string-hadron schemes, first, a fixed mass hadron and transverse momentum are selected according to rules discussed below. Second, a light-cone momentum fraction is chosen according to a fragmentation function compatible with the area decay law eqn (3.354), of which there are many (Sjöstrand, 1994). This choice is restricted in the Lund scheme by requiring that a set of plausible assumptions hold (Andersson et al., 1983b; Bowler, 1984). These are that the same results are obtained on average, independent of whether the hadronization proceeds from either the left or right end of the string, that a central rapidity plateau occurs, that the dynamics are those of a massless relativistic string with no transverse excitations and that edge effects are negligible. This gives the Lund symmetric fragmentation function (LSFF),

$$f(z) = \frac{N_{\alpha\beta}}{z} z^{a_\alpha} \left(\frac{1-z}{z}\right)^{a_\beta} \exp\left(-b\frac{m_\perp^2}{z}\right)$$

$$= \frac{N}{z}(1-z)^a \exp\left(-b\frac{m_\perp^2}{z}\right) .$$

(3.356)

Here, α labels the parent quark flavour and β the daughter quark flavour. In practice, normally only quarks, a_{q}, and diquarks, a_{qq}, are treated as separate flavours. Observe that when $m_\perp$ is large in eqn (3.356), it favours larger values of z, effectively stiffening the fragmentation function of heavy hadrons. As the hadronization proceeds simultaneously from both ends of the string, it is necessary to have a special treatment of the central fragment so as to guarantee quantum number conservation.

In the Lund scheme, one takes seriously the picture of q$\bar{\mathrm{q}}$-pairs forming in the strong colour field which is the string. The choice of which hadrons form as the string breaks up is based on an attempt to model, via flavour and spin selection rules, the supposed dynamics of the process (Andersson *et al.*, 1980*a*). Since no transverse string excitations are allowed, the quarks have equal and opposite transverse momenta, $\pm \boldsymbol{p}_\perp$, and, according to their flavour, mass m_q. In order to supply the energy to produce a q$\bar{\text{q}}$-pair of transverse mass $2m_\perp$, it is necessary to 'eat' a length, $2m_\perp/\kappa$, of string. Assuming that the quarks in a q$\bar{\text{q}}$-pair are created at a point, then they must tunnel quantum-mechanically to the classically required separation. This probability is estimated to be

$$\exp\left(-\frac{\pi}{\kappa}(m_q^2 + \boldsymbol{p}_\perp^2)\right) = \exp\left(-\frac{\pi}{\kappa}m_q^2\right) \times \exp\left(-\frac{\pi}{\kappa}\boldsymbol{p}_\perp^2\right) . \tag{3.357}$$

This expression is only regarded as an approximation to the 'true' formula and is essentially used as a guide. Indeed, in two-dimensional QED, the exact expression for the production of a single q$\bar{\text{q}}$-pair is known to be $\sum_n n^2 \exp(-n\pi m_q^2/\kappa)$, for which the first term is a poor approximation to the full sum (Casher *et al.*, 1979). For example, a Gaussian transverse momentum distribution, which is independent of the flavour of the produced quark, is found to be compatible with data but only if the width is increased from the 'predicted' value, $\langle \boldsymbol{p}_\perp \rangle = \sqrt{\kappa/\pi} = 0.25\,\text{GeV}$ to $\langle \boldsymbol{p}_\perp \rangle \approx 0.40\,\text{GeV}$ (ALEPH Collab., 1998*a*; DELPHI Collab., 1996*a*; OPAL Collab., 1996*a*; Doyle *et al.*, 1999). Likewise, whilst eqn (3.357) implies that the production of heavy c- and b-quarks in the string is negligible, it is unable to quantify γ_s, the degree to which strange quarks are suppressed relative to the light u- and d-quarks. In practice, several free parameters are needed to describe quark and diquark production. Finally, having selected the flavour and transverse momentum of the q$\bar{\text{q}}$-pair and hence the flavour of the string fragment, it is necessary to decide which hadron it forms. In addition to the phase space factor $2J + 1$, J being the hadron's spin, consideration of a hadron's wavefunction suggests a suppression of heavy hadrons proportional to $1/m_\perp^2$ (Andersson *et al.*, 1983*a*). In practice, the Lund scheme makes further free parameters available.

The predictive power of the Lund scheme is somewhat undermined by uncertainties in the properties of the (di)quarks. For example, what masses m_u, m_s, m_{uu}, m_{us} and m_{ss} are to be used in eqn (3.357). The UCLA scheme (Chun and Buchanan, 1993) attempts to finesse this difficulty by only working in terms of the known properties of hadrons. The essence of the difference is a reinterpretation of eqn (3.356) (Chun and Buchanan, 1987). In the Lund scheme, eqn (3.356) is used to choose z, given a fixed transverse mass hadron, so that the normalization is given by

$$N^{-1} = \int_0^1 dz \frac{(1-z)^a}{z} \exp\left(-b\frac{m_\perp^2}{z}\right) \equiv F(m_\perp^2) . \tag{3.358}$$

In the UCLA scheme eqn (3.356) is used to choose both z and $m_\perp$, so that the normalization becomes

$$N^{-1} = \sum_h (CG)^2 \int d\boldsymbol{p}_\perp^2 F(m_h^2 + \boldsymbol{p}_\perp^2) , \tag{3.359}$$

where the sum is over all hadrons containing the end-point quark and CG represents the corresponding constituent quark model, SU(6), Clebsch–Gordan coefficients. Since N is now a common constant, the transverse mass dependence appearing in the exponential term in eqn (3.356) immediately implies a suppression of heavy or large transverse momentum hadrons.

The actual UCLA implementation (Chun and Buchanan, 1993) is slightly more sophisticated than described above, as it tries to 'look ahead'. For example, if the first hadron leaves behind a u- or s-quark, then the next hadron is most likely to be a pion or a kaon, respectively, and the latter choice is (doubly) suppressed due to the larger masses of strange hadrons. This is really an attempt to mimic the quantum mechanical projection of a whole string on to a set of hadrons. Even in its more sophisticated implementation the scheme has remarkably few free parameters.

The relationship between the various string schemes discussed above is illustrated by the family tree shown in Fig. 3.33.

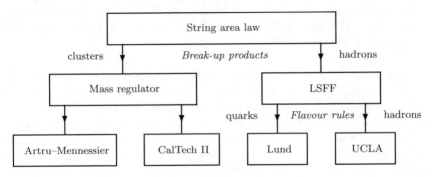

FIG. 3.33. A family tree for string models. Figure from Knowles and Lafferty(1997).

An interesting aspect of the string models are the inferences that can be drawn from their associated space–time pictures. We mention two here. At a string break the q and q̄ are formed with equal and opposite transverse momenta, $|\boldsymbol{p}_\perp| \approx \sqrt{\kappa/2\pi}$, a distance $2m_\perp/\kappa$ apart, so that an orbital angular momentum, $L = 2|\boldsymbol{p}_\perp|m_\perp/\kappa$, is necessarily introduced. Since total angular momentum, $J = L \otimes S_{\mathrm{q}} \otimes S_{\bar{\mathrm{q}}}$, must be conserved and $\langle L \rangle \approx 1\hbar$, the qq̄-pair must typically form in a $^3P_0$ state, particularly so at higher values of $|\boldsymbol{p}_\perp|$. This implies correlations between $\boldsymbol{p}_\perp$ and spin for neighbouring hadrons (Andersson et al., 1979b). The second consequence concerns identical particles (Andersson and Hofmann, 1986; Artru and Bowler, 1988). Provided that they do not have exactly the same momentum, there is an ambiguity in the order in which they may appear along the string. These two string configurations enclose different areas with $\Delta A \propto (p_1 - p_2)^2$. Now, if the string dynamics and area decay law derive from the modulus squared of an amplitude, $\mathcal{M} = \exp[iA(\kappa+iP_0/2)]$, c.f. eqn (3.354), then quantum interference will occur and the joint production probability becomes

$$\left(\mathcal{P}_{12} + \mathcal{P}_{21}\right) \times \left(1 + \frac{\cos(\kappa \Delta A)}{\cosh(P_0 \Delta A/2)}\right),\tag{3.360}$$

which is clearly different from the naïve sum of weights. This Bose–Einstein correlation effect is largest for small ΔA, that is, as $(p_1 - p_2)^2 \to 0$. This approach has close parallels to the more geometric approach based on the Fourier transform of the source distribution (Gyulassy *et al.*, 1979); see Section 13.3.3.

FIG. 3.34. A schematic diagram of baryon production in the diquark model (left) and popcorn model (right) leading to $MB\overline{B}$ and $BM\overline{B}$ configurations, respectively (M=meson, B=baryon). The lines show the flavour correlations. Figure from Knowles and Lafferty(1997).

The production of baryons is not so well understood as the above discussion implies. The diquark mechanism (Andersson *et al.*, 1982; Meyer, 1982) based on the production of $qq'\bar{q}\bar{q}'$ pairs, in a straightforward generalization of the meson production mechanism, is not unique. Indeed, the first model proposed and developed was based on a stepwise, multiple $q\bar{q}$-pair production mechanism (Casher *et al.*, 1979), which has since developed into the popcorn model (Andersson *et al.*, 1985). Recall that when a $q\bar{q}$- or $qq'\bar{q}\bar{q}'$-pair forms in a string's colour field with the same colour as the end-point quarks, it precipitates a string break: this is the diquark mechanism. However, if the $q\bar{q}$-pair has a different colour, then an anti-aligned, colour triplet, gluon field remains between them. This introduces the possibility for further $q\bar{q}$-pairs to form, allowing the string to break before the first (virtual) $q\bar{q}$-pair recombines. If the string does break, then the end fragment, containing three quarks, will have non-zero baryon number. The main difference with the diquark mechanism is that now the baryons need not be nearest neighbours on the string and intervening mesons may occur when more then one $q\bar{q}$-pair forms in the internal colour field. The situation is illustrated in Fig. 3.34. At some level, the two mechanisms reflect the ambiguity in whether a baryon should be regarded as three quarks or as a quark–diquark bound state (Anselmino *et al.*, 1993).

3.8.4 *Cluster hadronization*

Cluster-like structures have a long history in models of non-perturbative physics. They occur as intermediate states in string models (Artru and Mennessier, 1974), in the statistical bootstrap model (Hagedorn, 1965; Frautschi, 1971), in multi-peripheral models (Berger and Fox, 1973; Hamer and Peierls, 1973) and cluster

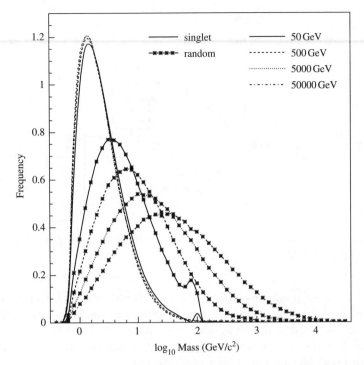

FIG. 3.35. The mass spectrum of colour singlet and random $q_1\bar{q}_2$ clusters in $e^+e^- \to u\bar{u}$ events at four beam energies. In the random sample $q\bar{q}$-pairs originating from the same gluon splitting are excluded as the final state gluon has a unique mass. Figure from Knowles and Lafferty(1997).

hadronization (Gottschalk, 1984). Today, they are best known through the hadronization model implemented in the HERWIG event generator (Webber, 1984; Corcella *et al.*, 2001). Here, the motivation for the use of clusters is based on the pre-confinement property proved in pQCD (Amati and Veneziano, 1979; Bassetto *et al.*, 1980; Marchesini *et al.*, 1981). At the end of a parton shower, the mass and spatial distributions of colour singlet clusters spanned by quark–antiquark pairs have a universal distribution. In practice, gluons remaining at the cut-off scale, Q_0, are forcibly split into light $q\bar{q}$-pairs, the Wolfram ansatz (Field and Wolfram, 1983), so that in the large N_c, or planar, approximation ('tHooft, 1974) neighbouring $q\bar{q}'$-pairs form colour singlets. This procedure is illustrated in Fig. 3.32.

Figure 3.35 shows the resultant cluster mass spectra in $e^+e^- \to u\bar{u}$ events, generated with HERWIG, for four beam energies. The four colour singlet mass spectra are almost indistinguishable. Each has a mean mass $\langle m \rangle \approx \mathcal{O}(Q_0)$ and a tail which, whilst growing with beam energy, is known to fall faster than any

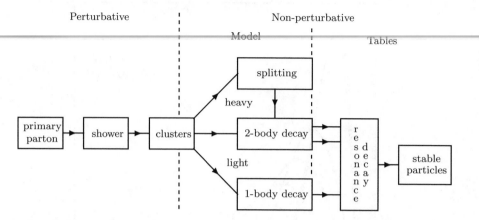

FIG. 3.36. A schematic diagram for cluster hadronization. Figure from Knowles and Lafferty(1997).

power of mass for a colour-coherent cascade (Bertolini and Marchesini, 1982). This is in marked contrast to the behaviour of the random, non-singlet mass spectra which broaden markedly with the energy of the primary quark. This behaviour holds also for down-type quark and gluon–initiated jets. Similar observations apply to the spatial sizes of clusters.

The universal properties of these colour singlet clusters, which arise naturally at the end of a coherent parton shower, has led to their use as the basis for a hadronization model (Gottschalk, 1984; Webber, 1984). A typical mass of a few GeV suggests associating clusters with 'super-resonances' that decay, independent of one another, into the familiar hadron resonances. Here, simplicity is used as a guide to describing this process. Two-body decays are assumed for the majority of clusters, $Cl(q_1\bar{q}_2) \rightarrow h_1 + h_2$. The hadrons h_1 and h_2 are subject only to flavour conservation. If $|h_1\rangle = |q_1\bar{x}\rangle$, then we have $|h_2\rangle = |x\bar{q}_2\rangle$ with $x = q_3$ or q_3q_4; the field energy of the cluster being used to create either one or two $q\bar{q}$-pairs. The probability of a given pair of hadrons being selected is made proportional to its available phase space,

$$\mathcal{P}(Cl \rightarrow h_1 + h_2) \propto (2J_{h_1} + 1)(2J_{h_2} + 1)p^\star(m_{Cl}, m_{h_1}, m_{h_2})\Theta(m_{Cl} - m_{h_1} - m_{h_2}) .$$
$$(3.361)$$

Here $p^\star$ is the C.o.M. frame three-momentum in the two-body decay. The Θ-function ensures that the putative decay is allowed physically. The decay is taken to be isotropic in the cluster's rest frame.

Referring to Fig. 3.35, two special cases suggest themselves. Some clusters will be too light to undergo any two-body decay. Such clusters undergo one-body decays with excess momenta being redistributed amongst neighbouring clusters. More significant are the limited number of rather heavy clusters found in the tail of the distribution which occur when there is little perturbative parton showering. Here, two-body, isotropic cluster decays would seem implausible, given the large momenta involved. By using the (repeated) device of introducing light $q\bar{q}$-pairs, these clusters are split forcibly into lighter daughter clusters whose directions of motion are aligned along the original $q_1 - \bar{q}_2$ axis. This appearance of a preferred, colour field, axis is reminiscent of the string model. The situation is summarized in Fig. 3.36.

As described above, cluster hadronization represents a well-founded attempt to go as far as possible with as little as possible. The original scheme is simple, compact and predictive, though some special cases require *ad hoc* solutions. Since typical clusters are light, $p^\star$ in eqn (3.361) is small and only limited transverse momentum can be generated during hadronization. Likewise, $p^\star$ is reduced for heavier hadrons, leading to a suppression of baryons and strange hadrons. Furthermore, the spin ratios for iso-flavour hadrons are predicted partly from the $(2J + 1)$ factor in eqn (3.361) and partly from the heavier mass of higher spin states.

3.8.5 *A comparison of the main hadronization models*

All the main hadronization models reduce to a set of rules that are used in recursively applied branchings. Their motivations range from the QCD inspired, complex dynamics of strings, through the minimalism of clusters, to the simple expediency of independent hadronization. Table 3.1 compares their more important features.

Table 3.1 *A comparison of the main hadronization model approaches*

| | Hadronization Model | | | |
| | | | String | |
Feature	Cluster	Independent	Lund	UCLA
Principle	Very simple	Simple	Complex	Less complex
Lorentz invariant	Yes	No	Yes	
Flavour, charge etc., conservation	Automatic	*ad hoc*	Automatic	
Mass dep. via	Hadrons	Quarks	Quarks	Hadrons
Strangeness supp.	Predicted	Free param.	Restricted params.	Predicted
Baryon supp.	Predicted	Free param.	Restricted params.	Predicted
J^P ratios	Predicted	Free params.	Restricted params.	Predicted
Limited $p_\perp$	Natural	Built in	Built in	Natural
Frag. func.	—	Free	Restricted by L–R symmetry	
Cut-off (Q_0) dep.	Significant	Very strong	Modest	
Stability	Infrared prob.	Collinear prob.	Stable	
Limitations	Massive clusters treated like 'strings'	Requires large Q_0	Light strings treated as clusters	

Reflecting its lack of a strong theoretical underpinning, the rather arbitrary nature of the independent hadronization model is immediately apparent. A par-

ticular problem is its lack of Lorentz invariance and *ad hoc* quantum number
conservation. At first sight, in the case of back-to-back $q\bar{q}$-pairs, string models
appear to be just like an independent hadronization model. This comparison is
most apparent with the Lund scheme. However, its belief in an underlying model
does place significant restrictions on the many free parameters, most notably the
fragmentation function. That the model is not more predictive appears to be a
direct consequence of its working at the quark level. This difficulty has been
finessed in the UCLA scheme, which is consequently far more predictive than
the Lund scheme — it has few free parameters. The advantage of working at
the hadron level is equally apparent in the cluster model, which would be fully
determined were it not for the need to treat exceptional-mass clusters. That
said, it is critical for a cluster model to be reliable that the distribution of any
perturbative soft gluons be accurate.

In practice string-based models provide the best descriptions of experimental
data. Whether this reflects the flexibility provided by their large numbers of
parameters or some underlying physics remains to be seen. Cluster models in
their simplest form are less successful and it is likely that the use of isotropic
phase space decays will need to be refined. Independent hadronization models,
combined with incoherent parton showers, provide inadequate descriptions of the
data.

In comparing the different hadronization models to experimental data one
employs Monte Carlo event generators. Unfortunately, these inextricably couple
the perturbative parton shower and non-perturbative hadronization model. Furthermore, both components contain free parameters which can be simultaneously
tuned to the data. As a consequence, it is often difficult to correlate the inherent properties of the hadronization model with the quality of a Monte Carlo's
description of the data. As a pragmatic approach to assessing the uncertainties
associated with hadronization, it is common practice to use the spread of results
based on different models.

Exercises for Chapter 3

3–1 Confirm explicitly the transformation law eqn (3.8) using the defi-
    nition eqn (3.7) and the transformation law for A^μ, eqn (3.5). You
    will need the identity $\partial_\mu U^{-1} = -U^{-1}(\partial_\mu U)U^{-1}$ which follows from
    differentiating $\mathbf{1} = UU^{-1}$.

3–2 Use the uncertainty principle to relate the four-momentum of a vir-
    tual particle to the space–time distance that it travels.

3–3 For inclusive, unpolarized lepton–hadron scattering, write down the
    most general form of the hadronic tensor $H^{\mu\nu}(q,p)$. Verify that the
    current conservation constraints, $q_\mu \cdot H^{\mu\nu} = 0 = H^{\mu\nu} \cdot q_\nu$, imply that
    the general tensor reduces to eqn (3.36).

3–4 Several choices of variable are used to describe lepton–hadron colli-
    sions. Verify the following identities in the limit of vanishing lepton
    masses,

$$\int_\phi \frac{1}{2\pi} \frac{\mathrm{d}^3\ell'}{E_{\ell'}} = \frac{M_\mathrm{h}}{(s - M_\mathrm{h}^2)} \mathrm{d}\nu \mathrm{d}Q^2 = \frac{Q^2}{2x^2(s - M_\mathrm{h}^2)} \mathrm{d}x \mathrm{d}Q^2 = M_\mathrm{h}\nu \mathrm{d}x \mathrm{d}y \,,$$

where $\int_\phi$ denotes integration over the azimuthal angle of the momentum vector of the final state lepton.

3–5 For lepton–hadron scattering in the target rest frame, $p^\mu = (M_\mathrm{h}, \mathbf{0})$, with axes orientated such that $q^\mu = (\nu, 0, 0, \sqrt{\nu^2 + Q^2})$, the exchanged vector boson has the following helicity ± 1 (transverse) and 0 (longitudinal) polarization vectors

$$\epsilon_\pm^\mu = \frac{1}{\sqrt{2}}(1, 0, \mp i, 0) \quad \text{and} \quad \epsilon_0^\mu = \frac{1}{Q}(\sqrt{\nu^2 + Q^2}, 0, 0, \nu) \,.$$

Calculate the structure functions $H_\lambda \equiv \epsilon_\lambda^\mu \cdot H^{\mu\nu} \cdot \epsilon_\lambda^{*\nu}$ that describe the absorbtion of vector bosons with specific polarizations. For an off mass-shell boson there is also a scalar polarization, $\epsilon_s^\mu = q^\mu/Q$. Why does this not give any contribution and what can be inferred about H_0 from this observation?

3–6 Show that, if charm production is not kinematically allowed, then the Adler sum rule

$$\frac{1}{2} \int_0^1 \frac{\mathrm{d}x}{x} \left[F_2^{\bar\nu \mathrm{h}}(x) - F_2^{\nu \mathrm{h}}(x) \right] = 2T_3(\mathrm{h}) + \sin^2\theta_c \left[3Q(\mathrm{h}) - 4T_3(\mathrm{h}) \right]$$

holds for a charged current neutrino–hadron interaction. Here θ_c is the Cabbibo mixing angle, and T_3 and Q the third component of the weak isospin and the electric charge of the hadron, respectively.

3–7 The Altarelli–Parisi splitting functions, eqn (3.50), satisfy a number of inter-relationships. Removing the colour factors by writing $P_{qq} \equiv C_F V_{qq}$ etc., verify the following symmetries for $z \neq 1$, at leading order.

Parton exchange $(a \to bb')$: $\qquad V_{b'a}(z) = V_{ba}(1 - z)$

Drell–Levy–Yan crossing: $\qquad V_{ab}(z) = (-1)^{2S_a + 2S_b - 1} z V_{ba}\left(\dfrac{1}{z}\right)$

Conformal invariance: $\qquad \left(z\dfrac{\mathrm{d}}{\mathrm{d}z} - 2\right) V_{qg}(z) = \left(z\dfrac{\mathrm{d}}{\mathrm{d}z} + 1\right) V_{gq}(z)$

Supersymmetry: $\qquad V_{qq}(z) + V_{gq}(z) = V_{qg}(z) + V_{gg}(z)$

The third relationship is the simplest of several due to the conformal invariance of QCD. The fourth relationship is based on an associated supersymmetric field theory in which the (Majorana) 'quarks' and 'gluons' lie in the same adjoint representation of the colour group so that $C_A = C_F = T_F$ and 'quark' and 'gluon' decays become equally likely (Dokshitzer, 1977).

3–8 As a precursor to solving the DGLAP equations it is helpful to try to decouple them as much as possible. Using eqns (3.282) and (3.284), derive the equations satisfied by the flavour non-singlet combinations $q_i^{NS} = q_i - \bar{q}_i$ and $q_{ij}^{NS} = q_i + \bar{q}_i - q_j - \bar{q}_j$ ($i \neq j$), and the flavour singlets g and $\Sigma = \sum_i q_i$.

3–9 Use the leading order form of the DGLAP equations, eqn (3.49), and the definition eqn (3.44) to derive an evolution equation for $F_2^\gamma(x, \mu^2)$ in terms of itself and $G(x, \mu^2) = xg(x, \mu^2)$. Also, give the result in the limit $x \to 0$, where gluons dominate the evolution.

3–10 Confirm the approximations given in eqn (3.59) for $t \ll Q^2 + M_X^2$ and $M_h^2 \ll W^2 + Q^2$. Show also that for fixed values of Q^2 and M_X^2 the minimum value of $-t$ is given approximately by $-t_{min} = M_h^2(Q^2 + M_X^2)^2/W^4$.

3–11 Use the optical theorem, $\mathcal{I}m[\mathcal{M}_{el}(s, t = 0)] = \lambda^{1/2}(s; m_a, m_b)\sigma_{tot}$, to derive a relationship between the forward, elastic scattering cross section, $d\sigma_{el}/dt(s, t)|_{t=0} = |\mathcal{M}|^2/(16\pi\lambda)$, and the total cross section. ($\lambda = [s - (m_a + m_b)^2][s - (m_a - m_b)^2]$ is the kinematic function.) Indicate how, in practice, you might use this expression to measure the total cross section.

3–12 Verify the relationship between rapidity and pseudorapidity given in eqn (3.55).

3–13 Consider a hadron–hadron collision in which two partons carrying momenta $x_a p_a$ and $x_b p_b$ collide head-on to produce two outgoing particles of mass, rapidity and transverse momenta $(m_1, y_1, \boldsymbol{p}_T)$ and $(m_2, y_2, -\boldsymbol{p}_T)$, measured with respect to the beam axis. Show that

$$x_a = \frac{1}{2p_a}(m_{T_1}e^{+y_1} + m_{T_2}e^{+y_2}) \qquad x_b = \frac{1}{2p_b}(m_{T_1}e^{-y_1} + m_{T_2}e^{-y_2})$$

where $m_T = \sqrt{m^2 + \boldsymbol{p}_T^2}$ is the transverse mass.

3–14 In a hadron–hadron scattering there are many subprocesses which contribute to di-jet production. Below are the leading order, $\mathcal{O}(\alpha_s^2)$, matrix elements for the two-to-two QCD scatterings of massless partons (Combridge et al., 1977). They are given as a function of the Mandelstam invariants $\hat{s} = x_1 x_2 s$, $\hat{t} = -\hat{s}(1 - \cos\theta^\star)/2$ and $\hat{u} = -\hat{s}(1 + \cos\theta^\star)/2 = -(\hat{s} + \hat{t})$, where $\theta^\star$ is the partons' C.o.M. frame scattering angle. The matrix elements are averaged over the initial state, and summed over the final state colours and spins, a factor g_s^4 has been extracted.

$$gg \to gg: \qquad \frac{9}{2}\left(3 - \frac{\hat{t}\hat{u}}{\hat{s}^2} - \frac{\hat{u}\hat{s}}{\hat{t}^2} - \frac{\hat{s}\hat{t}}{\hat{u}^2}\right)$$

$$gg \to q\bar{q}: \qquad (\hat{t}^2 + \hat{u}^2)\left(\frac{1}{6}\frac{1}{\hat{t}\hat{u}} - \frac{3}{8}\frac{1}{\hat{s}^2}\right)$$

$$\left.\begin{array}{r} \text{gq} \rightarrow \text{gq} \\ \text{g}\bar{\text{q}} \rightarrow \text{g}\bar{\text{q}} \end{array}\right\} : \quad (\hat{s}^2 + \hat{u}^2)\left(-\frac{1}{6}\frac{1}{\hat{s}\hat{u}} + \frac{3}{8}\frac{1}{\hat{t}^2}\right)\frac{8}{3}$$

$$\text{q}\bar{\text{q}} \rightarrow \text{gg} : \quad (\hat{t}^2 + \hat{u}^2)\left(\frac{1}{6}\frac{1}{\hat{t}\hat{u}} - \frac{3}{8}\frac{1}{\hat{s}^2}\right)\frac{64}{9}$$

$$\left.\begin{array}{r} \text{qq} \rightarrow \text{qq} \\ \bar{\text{q}}\bar{\text{q}} \rightarrow \bar{\text{q}}\bar{\text{q}} \end{array}\right\} : \quad \frac{4}{9}\left(\frac{\hat{s}^2 + \hat{u}^2}{\hat{t}^2} + \frac{\hat{s}^2 + \hat{t}^2}{\hat{u}^2}\right) - \frac{8}{27}\frac{\hat{s}^2}{\hat{t}\hat{u}}$$

$$\text{q}\bar{\text{q}}' \rightarrow \text{q}\bar{\text{q}}' : \quad \frac{4}{9}\frac{\hat{s}^2 + \hat{u}^2}{\hat{t}^2}$$

$$\left.\begin{array}{r} \text{qq}' \rightarrow \text{qq}' \\ \bar{\text{q}}\bar{\text{q}}' \rightarrow \bar{\text{q}}\bar{\text{q}}' \end{array}\right\} : \quad \frac{4}{9}\frac{\hat{s}^2 + \hat{u}^2}{\hat{t}^2}$$

$$\text{q}\bar{\text{q}} \rightarrow \text{q}\bar{\text{q}} : \quad \frac{4}{9}\left(\frac{\hat{u}^2 + \hat{s}^2}{\hat{t}^2} + \frac{\hat{u}^2 + \hat{t}^2}{\hat{s}^2}\right) - \frac{8\hat{u}^2}{27\hat{t}\hat{s}}$$

$$\text{q}\bar{\text{q}} \rightarrow \text{q}'\bar{\text{q}}' : \quad \frac{4}{9}\frac{\hat{t}^2 + \hat{u}^2}{\hat{s}^2}$$

Give the parton model expression, eqn (3.72), for di-jet production in terms of these two-to-two scatterings. Using the matrix elements show that their ratios, relative to that for the dominant gg $\rightarrow$ gg scattering, are nearly constant as a function of $\cos\theta^\star$ and evaluate them in the dominant phase space region, $\cos\theta^\star \rightarrow 1$. Hence, show how to write the di-jet cross section in terms of the effective p.d.f. defined in eqn (3.73).

3–15 Show that the polarization sum $T_{\mu\nu}$, eqn (3.121), for an on mass-shell gluon with four-momentum k^μ, follows unambiguously from the conditions $k^\mu T_{\mu\nu} = k^\nu T_{\mu\nu} = n^\mu T_{\mu\nu} = n^\nu T_{\mu\nu} = 0$ and $T^\mu{}_\mu = -2$. Here n is an auxiliary four-vector which satisfies $n \cdot k \neq 0$.

3–16 Calculate the three-jet cross section $e^+e^- \rightarrow q\bar{q}g$ for a scalar gluon. Assume massless quarks and that the orientation of the event plane is averaged over. (★★)

3–17 By considering the Lagrangian eqn (3.21) in D dimensions determine the mass dimensions of the fields and show that consistency requires the replacement $g_s \rightarrow g_s\mu^{(4-D)/2}$, where μ has mass dimension one.

3–18 Derive the expression for the $(D-1)$-dimensional angular integral, eqn (C.8), by evaluating the expression

$$\int_{-\infty}^{+\infty} d^D x \exp(-x^2/2)$$

in both Cartesian and polar coordinates.

3–19 Derive the equivalents of the identities in eqn (3.83) for D dimensions.

3–20 By considering $m_0 = mZ_m$ derive an expression for γ_m, defined in eqn (3.158), in terms of the Laurent expansion of Z_m. Assume a mass-independent renormalization scheme, so that $Z_m = Z_m(g_s)$. What constraints are placed on the coefficients in the expansions? Using eqn (3.155) evaluate your results. $^{(\star)}$

3–21 Using the next-to-leading order expression for Λ_{QCD} in terms of $\alpha_s(Q^2)$ and Q^2, eqn (3.174), show that for fixed n_f it is independent of Q^2.

3–22 Supersymmetry is a popular extension of the Standard Model in which gluons are partnered by a set of Weyl spinors in the adjoint representation, the gluinos $\tilde{g}$, whilst the left- and right-handed quarks are partnered by complex scalars in the fundamental representation, the squarks $\tilde{q}_L$ and $\tilde{q}_R$. What is the β-function in supersymmetric QCD assuming complete supersymmetric families?

3–23 Following the approach used for the $\gamma q \bar{q}$ vertex, eqn (3.211), determine the ultraviolet behaviour of the $\mathcal{O}(\alpha_s)$ corrections to the $gq\bar{q}$ vertex.

3–24 Show explicitly that the hadronic tensor describing $\gamma^* q \to q'$ is of the form of eqn (3.34) and extract the structure functions $F_{1,2,3}$.

3–25 Confirm the plus-prescription identity in eqn (3.268).

3–26 By considering the integral $\int_0^1 \mathrm{d}x\, f(x)/(1-x)^{1+\epsilon}$ for arbitrary smooth $f(x)$ show that

$$\lim_{\epsilon \to 0} \frac{1}{(1-x)^{1+\epsilon}} = -\frac{\delta(1-x)}{\epsilon} + \left(\frac{1}{1-x} \right)_+ .$$

In the derivation, to make the integral convergent work with $\epsilon < 0$ and substitute $f(x) \to [f(x) - f(1)] + f(1)$. By a similar means derive eqn (3.263). $^{(\star)}$

3–27 Following the same approach as that for $\gamma^* q \to q'g$ in Section 3.6.4.1, calculate the contributions to the structure functions from the subprocess $\gamma^* g \to q\bar{q}$. $^{(\star\star)}$

3–28 Using eqn (3.275) and the identity $\int_0^1 \mathrm{d}x\, f_+(x) = 0$, calculate the $\mathcal{O}(\alpha_s)$ correction to the Gross–Llewellyn-Smith sum rule, defined by $\int_0^1 \mathrm{d}x\, F_3(x, Q^2)$.

3–29 Overall momentum conservation implies that the following combination of parton density functions is independent of the scale μ^2:

$$\int_0^1 \mathrm{d}x\, x \left\{ g(x, \mu^2) + \sum_{f=q,\bar{q}} f(x, \mu^2) \right\} \tag{3.362}$$

By considering the μ^2 evolution equation for this quantity derive a constraint on the integral of the $g \to gg$ splitting function and hence determine C_{gg}, the coefficient of the $\delta(1 - z)$ term. $^{(\star)}$

3–30 Verify the result for $\gamma_{qq}^{(0)}$, eqn (3.292). Next, working at one-loop precision, carry out the Mellin transform of the DGLAP evolution equation for a non-singlet quark distribution, eqn (F.2), and solve the resulting equation. What is the physical significance of the case $n = 1$? $^{(\bigstar)}$

3–31 At one-loop level the Mellin transform of the coupled singlet-gluon equations, eqn (F.4), are given by

$$\mu^2 \frac{\partial}{\partial \mu^2} \begin{pmatrix} \tilde{\Sigma}(n, \mu^2) \\ \tilde{g}(n, \mu^2) \end{pmatrix} = \frac{\alpha_s}{2\pi} \begin{pmatrix} \gamma_{qq}^{(0)} \, , & 2n_f \gamma_{qg}^{(0)} \\ \gamma_{gq}^{(0)} \, , & \gamma_{gg}^{(0)} \end{pmatrix} \begin{pmatrix} \tilde{\Sigma}(n, \mu^2) \\ \tilde{g}(n, \mu^2) \end{pmatrix} .$$

Solve these equations for the case $n = 2$ and give the physical interpretation of the solution. $^{(\bigstar)}$

3–32 Show that in the DLLA the DGLAP equation for gluons can be written in the form

$$\frac{\partial^2 [xg(x, Q^2)]}{\partial \ln(\ln(Q^2/\Lambda^2)) \partial \ln(1/x)} = \frac{C_A}{\pi \beta_0} \, xg(x, Q^2) ,$$

and that eqn (3.301) satisfies this equation.

3–33 In the Mellin transform method of solution, the small-x behaviour of the p.d.f. $g(x, \mu^2)$ is dominated by $\gamma_{gg}(n \approx 1)$, assuming a sufficiently non-singular $g(x, \mu_0^2)$. Consider a gluon-only theory and apply the saddle-point method (Arfken and Weber, 1995) to evaluate $xg(x, \mu^2)$. What assumptions have to be made about the position of the poles in $\tilde{g}(n, \mu_0^2)$? $^{(\bigstar\bigstar)}$

3–34 At small x, when colour coherence is taken into account, the Mellin transform of the fragmentation function, $\tilde{D}(n, Q^2)$, behaves as

$$\exp \left(\frac{1}{\beta_0} \sqrt{\frac{2C_A}{\pi \alpha_s(Q^2)}} - \frac{(n-1)}{4\beta_0 \alpha_s(Q^2)} + \frac{(n-1)^2}{48\beta_0} \sqrt{\frac{2\pi}{C_A \alpha_s^3(Q^2)}} + \cdots \right) .$$

How does the multiplicity of a jet behave as a function of Q and, using the saddle-point method, what is the shape of the fragmentation function $xD(x, Q^2)$?

3–35 Confirm the expression for the gluon's phase space element used in eqn (3.319).

3–36 Derive an expression for the square of a matrix element in which an initial state gluon undergoes a near collinear branching to a $q\bar{q}$-pair.

3–37 In a Drell–Yan event suppose the mass and rapidity of the lepton pair are measured. Derive a leading-order expression for the double differential cross section in terms of these variables.

3–38 Confirm that eqn (3.334) equals the μ_F^2-independent eqn (3.331) to $\mathcal{O}(\alpha_s^2)$ and find the coefficient functions in the DIS scheme analogue of eqn (3.334).

3–39 Repeat the derivation leading to eqn (3.337) for soft gluon emission off an antiquark and adapt the method to treat emission off a hard(er) gluon. To do this treat the parent gluon as almost real and decompose the numerator of its progagator as a sum over transverse polarization vectors, eqn (3.121).

3–40 Confirm that $\hat{T}_q^2 = C_F \mathbf{1} = \hat{T}_{\bar{q}}^2$ and $\hat{T}_g^2 = C_A \mathbf{1}$. Also confirm that $(\hat{T}_q + \hat{T}_{\bar{q}})|q, i;\ \bar{q}, j\rangle = 0$, $(\hat{T}_q + \hat{T}_{\bar{q}} + \hat{T}_g)|q, i;\ \bar{q}, j;\ g, b\rangle = 0$ and $(\hat{T}_{g_1} + \hat{T}_{g_2} + \hat{T}_{g_3})|g_1, b;\ g_2, c;\ g_3, d\rangle = 0$ for colour singlet states.

3–41 Check that, in the soft gluon limit, the matrix element squared for $\gamma^\star \to q\bar{q}g$, eqn (3.107), factorizes as

$$|\mathcal{M}(q\bar{q}g)|^2 \xrightarrow{\ g^\mu \to 0\ } C_F g_s^2 |\mathcal{M}(q\bar{q})|^2 \frac{q \cdot \bar{q}}{(q \cdot g)(\bar{q} \cdot g)} \ .$$

4

MONTE CARLO MODELS

A Monte Carlo event generator is a computer program which numerically implements the predictions of a cross section calculation. Their use has become ubiquitous in modern particle physics. Two factors have fuelled this rise to prominence. First, in order to compare with real experimental results, cross section formulae often have to be integrated over complex regions of phase space. Whilst this may prove practically impossible to do in an analytical approach, it is usually trivial to implement the required cuts in a numerical approach. Second, there is a need for realistic simulated events in order to do everything from designing an experiment to analysing its data.

Here, we discuss the theory behind event generation, whilst avoiding the many technical aspects. In addition to the various program manuals, details of the specialized numerical techniques can be found in the report by James (1980) and another approach to the theory can be found in the review by Webber (1986). Also, when a new experimental facility is being designed or about to begin operation, it is common for an accompanying physics study to be published. These comprehensive works often dedicate significant space to the available event generators and provide a valuable source of information on the state of the art. Useful sources include: for LEP1 (Altarelli *et al.*, 1989), vol. 3, for LEP2 (Altarelli *et al.*, 1996) vol. 2, for HERA (Doyle *et al.*, 1999) and for the LHC (Catani *et al.*, 2000).

There are two basic types of Monte Carlo which are distinguished both by their approaches to approximating the full cross section calculation and by the extent to which they describe a complete event.

4.1 Fixed-order Monte Carlos

Fixed-order Monte Carlos take the most straightforward approach. They generate sets of partons, distributed according to an appropriate cross section calculated to some fixed, finite order in perturbation theory. If the calculation is at tree-level, then all that is required are a set of phase space cuts designed to isolate the collinear and soft regions, where the matrix element is singular. The matrix element squared is positive-definite and can be used as a multi-dimensional probability density function for parton configurations. This can then be used in one of two ways. In a weighted Monte Carlo, an 'event' is selected from within the allowed phase space and this is given a weight equal to its matrix element squared. The results of any measurements made on this event must then be given this weight. Of course, to generate events efficiently, avoiding those with small weights, is not so simple, but several approaches to this importance sampling problem are available. In an unweighted Monte Carlo the phase space points are selected so as to be distributed as the matrix element squared. These events are designed to occur with the same frequency as would occur in nature and can be treated as 'real events' in any analysis. Again, techniques, such as von Neumann's accept/reject algorithm, are available to ensure that the events are distributed correctly throughout the allowed phase space with the predicted distribution.

   If the calculation of the matrix element is at one-loop, or higher, then matters become more complicated. This is due to the need to implement numerically the cancellations between the $(N + 1)$-parton contribution and the higher order N-parton contribution to the matrix element squared; see Section 3.5. We illustrate the two standard approaches to this problem with a schematic example. Suppose we have used dimensional regularization to isolate a matrix element's singularity as $x \to 0$, where x might be a gluon energy or opening angle. Also, suppose that $O(x)$ is an infrared safe observable, that is, $O(x)$ tends smoothly to $O(0)$ as $x \to 0$. A measurement of O on the final state partons is given by an expression of the form

$$\langle O \rangle = \int_0^1 \frac{\mathrm{d}x}{x^{1+\epsilon}} O(x) + \frac{1}{\epsilon} O(0) \,, \qquad (4.1)$$

where the first term represents the real contribution and the second the virtual contribution. In principle, the two $\epsilon \to 0$ divergences in eqn (4.1) cancel. However, if we have to resort to numerical integration, then we must re-arrange eqn (4.1) in such a way that we can take the physical, $\epsilon \to 0$, limit before carrying out the integration. In the subtraction method (Frixione et al., 1996; Catani and Seymour, 1997; Phaf and Weinzierl, 2001) we project the $(N + 1)$-parton phase space on to the N-parton phase space and evaluate our observable using a simplified matrix element, which matches the full expression in (all) the singular limits. Adding and subtracting this approximate expression gives the exact result

$$\langle O \rangle = \int_0^1 \frac{\mathrm{d}x}{x^{1+\epsilon}} O(x) - \int_0^1 \frac{\mathrm{d}x}{x^{1+\epsilon}} O(0) + \int_0^1 \frac{\mathrm{d}x}{x^{1+\epsilon}} O(0) + \frac{1}{\epsilon} O(0)$$

$$= \int_0^1 \frac{\mathrm{d}x}{x} \left[O(x) - O(0) \right] \,, \qquad (4.2)$$

which is finite; c.f. Ex. (3-26). In the phase space slicing method (Giele et al., 1993; Kilgore and Giele, 1997; Keller and Laenen, 1999), we introduce an arbitrary parameter, δ, which is used to isolate the singular region(s). Simplified matrix elements then allow the necessary integrals to be carried out analytically in the singular region(s),

$$\langle O \rangle = \int_0^\delta \frac{\mathrm{d}x}{x^{1+\epsilon}} O(x) + \int_\delta^1 \frac{\mathrm{d}x}{x^{1+\epsilon}} O(x)$$

$$= \int_\delta^1 \frac{\mathrm{d}x}{x} O(x) + O(0) \ln \delta + \mathcal{O}\big(O'(0)\delta\big) \,. \qquad (4.3)$$

This result is only exact in the limit $\delta \to 0$. In practice, numerical values become independent of δ when it falls well below any of the physical scales associated with the problem. A second, more significant issue with one-loop calculations is that the cross section may be negative for particular phase space configurations. This means that only a weighted Monte Carlo is possible without resorting to rather artificial means.

The principle advantage of a fixed-order Monte Carlo is that it gives the exact result to a given order in perturbation theory. They are suited ideally to describing well separated, hard jets, which correspond to parton configurations away from the soft and collinear regions. In these regions we know that large logarithmic enhancements imply the need to use resummed calculations. This is the domain of the all-orders Monte Carlos. The main disadvantage of a fixed-order Monte Carlo is that it has partonic final states rather than realistic hadronic final states. Again, this is the domain of all-orders Monte Carlos. Of course, it is in principle possible to generate a hard subprocess using a fixed-order Monte Carlo and then pass the final state partons to an all-orders Monte Carlo for showering and hadronization.

4.2   All-orders Monte Carlos

All-orders Monte Carlos aim to simulate events in their entirety, that is, to the point of generating a fully specified set of final state hadrons. As such they are rather complicated programs, which must take account of both perturbative and non-perturbative QCD. If they are to be useful, they must also conserve exactly four-momentum, flavour etc. This means that the programs must deal with all of an event's 'messy details', something which is often neglected at the level of accuracy of pQCD calculations.

The basic structure of an event generator reflects the factorized form of pQCD cross sections. First, a hard subprocess is generated according to the standard, fully differential, leading-order matrix elements and, if required, the p.d.f.s. Typically, the user specifies a set of subprocesses of interest. The methods used are essentially the same as those for a fixed-order Monte Carlo, where the art lies in improving efficiency. Second, the 'primary' partons emerging from or entering into the hard subprocess are showered to produce a set of 'final state' partons. The generation of the incoming, space-like, and outgoing, time-like, parton showers is based on the use of modified forms of the DGLAP equations for p.d.f.s and fragmentation functions. These stages are perturbative. Third, a non-perturbative hadronization model is used to convert the final state partons into 'primary hadrons'. In a scattering that involves initial state hadron beam(s) there is a need to treat the beam remnant(s) that are left behind when a parton is removed for the hard subprocess. These remnants generate a soft underlying event, the model for which is often closely related to the hadronization model. Finally, many of the primary hadrons are actually highly unstable, strong resonances or electromagnetically/weakly decaying particles whose decays must also be modelled. On the whole, these chains of decaying particles are gen-

erated using simple matrix elements and tables derived in part from measured
branching ratios (PDG, 2000) and in part from per/in-spiration. A common exception are the lightest B hadrons, whose decays are often treated as secondary
hard subprocesses, followed by potential showering and hadronization. In some
circumstances both these approaches may prove inadequate to describe certain
particles, for example, taus and Bs. In this case, the particle should be set stable
in the Monte Carlo and passed on to a specialized decay package.

4.2.1   *The parton evolution equations*

The parton showers convert highly virtual, primary partons into low virtuality,
final state partons. These final state partons are either positive virtuality partons
just prior to hadronization, or negative virtuality partons just emerging from any
incoming beam hadrons, one per beam hadron. We have learnt in Sections 3.2.2
and 3.6.5 that the Q^2-evolution of parton-within-parton distribution functions
are described by time-like and space-like DGLAP equations, eqns (3.285) and
(3.49). These equations sum the leading effects of repeated parton branchings.
What we would like to do is to generate explicitly these parton branchings and
in the process create a parton shower.

To be a little more explicit, in Section 3.6.7 we learnt that in an axial gauge
the DGLAP equations correspond to a sum of ladder diagrams (Frenkel and
Taylor, 1976; Dokshitzer *et al.*, 1980), with interference or crossed-rung diagrams
giving sub-leading contributions. This ensures that the parton model language
is appropriate and facilitates a description of the leading contributions via a
classical Markov process. Basically, each parton is assigned a set of probabilities
for its possible branchings, or no branching at all, and is randomly evolved. The
procedure is iterated for any daughters. This jet calculus (Konishi *et al.*, 1979;
Bassetto *et al.*, 1983) led to the development of Monte Carlo event generators
(Fox and Wolfram, 1980; Field and Wolfram, 1983).

Unfortunately, in the form given originally, the DGLAP equations are not
suitable for a numerical treatment, as the splitting kernels, eqn (3.50), are singular. However, we can reformulate them in an equivalent, finite form,

$$t\frac{\partial}{\partial t}f_{a/h}(x,t) = \int_x^{1-\epsilon_{a'}^s(t)}\frac{dz}{z}\frac{\alpha_s(t,z)}{2\pi}\bar{P}_{ab}(z)f_{b/h}\left(\frac{x}{z},t\right)$$

$$-f_{a/h}(x,t)\sum_{a\to cc'}\int_{\bar{\epsilon}_c(t)}^{1-\epsilon_{c'}^s(t)}dz\frac{\alpha_s(t,z)}{2\pi}\bar{P}_{ca}(z) \qquad (4.4)$$

$$t\frac{\partial}{\partial t}D_a^h(x,t) = \int_x^{1-\epsilon_{b'}(t)}\frac{dz}{z}\frac{\alpha_s(t,z)}{2\pi}\bar{P}_{ba}(z)D_b^h\left(\frac{x}{z},t\right)$$

$$-D_a^h(x,t)\sum_{a\to cc'}\int_{\epsilon_c(t)}^{1-\epsilon_{c'}(t)}dz\frac{\alpha_s(t,z)}{2\pi}\bar{P}_{ca}(z)\,, \qquad (4.5)$$

where h now represents a parent/daughter parton. Equation (4.4) pertains to
p.d.f.s, that is, a space-like evolution, while eqn (4.5) deals with the time-like

evolution of fragmentation functions. We use the $\mathcal{O}(\alpha_s)$ expressions for the split-ting kernels, $\bar{P}_{ba}$, which are associated with one-to-two branchings, $a \to bb'$, but omit the plus-prescriptions and δ-functions in the diagonal $a \to ag$ kernels. In Section 3.6 we used a physical argument, based on the need to conserve quan-tum numbers and momentum, to determine the singular terms. However, in event generators these quantities are tracked explicitly and conserved trivially. By in-troducing suitable cut-offs on the soft gluon momenta, the distribution function regulators can be replaced by the equivalent subtraction terms found in eqns (4.4) and (4.5). Their rôle is to provide correctly normalized branching probabilities. Formally, the subtraction terms follow from unitarity and the infrared finiteness of inclusive observables and are given to leading logarithmic accuracy by virtual diagrams of the same order as the real emission diagrams. Strictly, both the vir-tual and real emission diagrams are divergent with only their sum being finite. In eqns (4.4) and (4.5) the divergent part of the real emission is isolated below the z cut-offs. Upon adding this 'unresolved', real radiation to the virtual diagrams, the finite subtraction term is obtained.

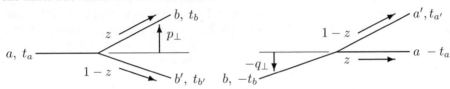

FIG. 4.1. A unit in a parton cascade, left, for a 'forward evolving' time-like branching $a \to b + b'$, and, right, a 'backward evolving' space-like branching $a + a' \to b$

As with the original DGLAP equations, both eqns (4.4) and (4.5) possess simple classical interpretations, which prove very helpful in formulating the par-ton shower algorithms. We illustrate this for the case of initial state radiation. First, recall that the p.d.f., $f_{b/h}(x,t)\mathrm{d}x$, is interpreted as the number of partons of type b with momentum fraction in the interval $[x, x + \mathrm{d}x]$, within h at the scale t. Second, note that the splitting kernels give the probability that in evolv-ing from the scale $-t$ to $-(t + \delta t)$ parton b will undergo a resolved branching into partons a and a', each carrying fractions z and $(1 - z)$, respectively, of b's momentum,

$$\mathcal{P}\bigg(b \to a(z) + a'(1 - z) \text{ when } t \to t + \delta t\bigg) = \frac{\delta t}{t} \frac{\alpha_s(t, z)}{2\pi} \bar{P}_{ab}(z) \,. \qquad (4.6)$$

This basic unit is illustrated in Fig. 4.1. Thus, in eqn (4.4) the first (source) term represents the possibility that a parton b present at a slightly lower scale, t, with a momentum fraction larger than x, underwent a resolved branching, $b \to aa'$, yielding the specified parton a. The second (sink) term, which is summed over all the available decay channels $a \to cc'$, represents the possibility that the already present parton a is removed from the specified momentum interval by a resolved branching.

4.2.2   Branching kinematics

As a preliminary to obtaining solutions to eqn (4.4) and eqn (4.5) it is necessary to give the exact definitions of the evolution variable, t, and splitting fraction, z; see Fig. 4.1. These, in turn, determine the form of the z limits, $\epsilon_b(t)$, appropriate to each branching. In practice the definitions depend on the specific Monte Carlo implementation, as does the choice of argument for α_s. Typical examples are $t = p_a^2$ or $p_\perp^2$ and $z = E_b/E_a$ or $(E+p_z)_b/(E+|p|)_a$, which become equivalent in the limit $E_a \to \infty$. As a matter of convention, for initial state space-like radiation where virtualities are negative, we take the evolution variable to be $-t$. Thus, for both initial and final state evolution we have $t \sim Q^2$ at the hard scattering, with decreasing values away from it until the respective cut-offs t_s and t_0 are reached.

To see how the kinematic limits, $\epsilon_b(t)$, depend on the precise meanings of t and z consider a time-like branching with t equal to the mother's virtuality and z the light-cone momentum fraction of b. Momentum (p_-) conservation implies

$$p_\perp^2 = z(1-z)t_a - (1-z)t_b - zt_{b'} \geq 0 . \tag{4.7}$$

Now, each parton species has a cut-off placed on its virtuality, below which pQCD will no longer be valid and the shower evolution is terminated. For example, $t_b > t_0^b \equiv (m_b + m_0)^2$, where m_b is the parton's mass and m_0 is a free parameter. Applying these flavour–dependent limits on the daughters' virtualities in eqn (4.7) implies the following limits on the allowed z-range:

$$\left. \begin{array}{c} z_{max} \\ z_{min} \end{array} \right\} = \frac{1}{2}\left[1 + \frac{(t_0^b - t_0^{b'})}{t_a} \pm \sqrt{1 - 2\frac{(t_0^b + t_0^{b'})}{t_a} + \frac{(t_0^b - t_0^{b'})^2}{t_a^2}} \right] \approx \left\{ \begin{array}{c} 1 - \dfrac{t_0^{b'}}{t_a} \\ \\ \dfrac{t_0^b}{t_a} \end{array} \right. \tag{4.8}$$

To reach the required level of accuracy (in the Sudakov form factor eqn (4.11)) it is actually sufficient to use the simpler limits, $z \in [\epsilon_b(t_a), 1 - \epsilon_{b'}(t_a)]$, with

$$\epsilon_b(t_a) = \frac{t_0^b}{t_a} . \tag{4.9}$$

Finally, if a branching does occur, then eqn (4.7) implies the upper bounds

$$t_b < zt_a , \qquad t_{b'} < (1-z)t_a \tag{4.10}$$

on the daughter virtualities. The space-like case is left as an exercise, Ex. (4-1).

4.2.3 Time-like Monte Carlo algorithm

Proceeding formally, a set of Sudakov form factors (Sudakov, 1956) is introduced, defined by

$$\ln \Delta_a(t, t_0^a) = - \sum_{a \to cc'} \int_{t_0^a}^{t} \frac{dt'}{t'} \int_{\epsilon_c(t')}^{1-\epsilon_{c'}(t')} dz \frac{\alpha_s(t, z)}{2\pi} \bar{P}_{ca}(z) \qquad (4.11)$$

$$\Delta_a(t, t_0^a) \equiv \prod_{a \to cc'} \Delta_{a \to cc'}(t, t_0^a) \,,$$

where t_0^a is a cut-off associated with the time-like parton a. This allows the second term in eqn (4.5) to be removed and the equations to be rewritten as

$$t \frac{\partial}{\partial t} \left(\frac{D_a^h(x,t)}{\Delta_a(t, t_0^a)} \right) = \int_x^{1-\epsilon_{b'}(t)} \frac{dz}{z} \frac{\alpha_s(z,t)}{2\pi} \bar{P}_{ba}(z) \frac{1}{\Delta_a(t, t_0^a)} D_b^h\left(\frac{x}{z}, t \right)$$

$$\implies \quad D_a^h(x,t) = \Delta_a(t, t_0^a) D_a^h(x, t_0^a) \qquad (4.12)$$

$$+ \int_{t_0^a}^{t} \frac{dt'}{t'} \int_x^{1-\epsilon_{b'}(t')} \frac{dz}{z} \Delta_a(t, t') \frac{\alpha_s(t', z)}{2\pi} \bar{P}_{ba}(z) D_b^h\left(\frac{x}{z}, t' \right) \,.$$

In the integrated form we used $\Delta_a(t_0^a, t_0^a) = 1$ and $\Delta_a(t, t_0^a)/\Delta_a(t', t_0^a) = \Delta_a(t, t')$. This set of (inhomogeneous, Voltera) integral equations is solved easily by repeated back substitution to obtain the Neumann series solution.

$$D_a^h(x,t) = \Delta_a(t, t_0^a) D_a^h(x, t_0^a)$$

$$+ \int_{t_0^a}^{t} \frac{dt_1}{t_1} \int_x^{1-\epsilon_{b'}(t_1)} \frac{dz_1}{z_1} \Delta_a(t, t_1) \frac{\alpha_s(t_1, z_1)}{2\pi} \bar{P}_{ba}(z_1) \Delta_b(t_1, t_0^b) D_b^h\left(\frac{x}{z_1}, t_0^b \right)$$

$$+ \int_{t_0^a}^{t} \frac{dt_1}{t_1} \int_x^{1-\epsilon_{b'}(t_1)} \frac{dz_1}{z_1} \int_{t_0^b}^{t_1} \frac{dt_2}{t_2} \int_{x/z_1}^{1-\epsilon_{c'}(t_2)} \frac{dz_2}{z_2}$$

$$\times \Delta_a(t, t_1) \frac{\alpha_s(t_1, z_1)}{2\pi} \bar{P}_{ba}(z_1) \Delta_b(t_1, t_2) \frac{\alpha_s(t_2, z_2)}{2\pi} \bar{P}_{cb}(z_2) \Delta_c(t_2, t_0^c) D_c^h\left(\frac{x}{z_1 z_2}, t_0^c \right)$$

$$+ \cdots . \qquad (4.13)$$

This expansion forms the basis of the Monte Carlo implementation of final state, time-like parton cascades.

To give eqn (4.13) a physical interpretation we need to understand the significance of the Sudakov form factor $\Delta_a(t, t_0^a)$. It is the probability for parton a to evolve down from the scale t to t_0^a, its cut-off, without any resolvable radiation. This is most easily seen from the fact that Δ_a satisfies the differential equation implied by

$$P(t + \delta t \to t_0 \mid \text{no res. rad.}) = \left[1 - \sum_{a \to cc'} \int_{\epsilon_c(t)}^{1-\epsilon_{c'}(t)} dz \frac{\alpha_s(t, z)}{2\pi} \bar{P}_{ca}(z) \frac{\delta t}{t} + \mathcal{O}(\delta t^2) \right]$$

$$\times \mathcal{P}\big(t \to t_0 \mid \text{no res. rad.}\big) , \qquad\qquad (4.14)$$

which is essentially a radioactive decay law. In eqn (4.14) the summation over all resolved branchings which remove the parton $a \to cc'$ gives rise to the sub-probabilities $\Delta_{a \to cc'}$ indicated in eqn (4.11). Armed with this interpretation, we see that the first term in eqn (4.13) represents the possibility that the highly off mass-shell parton a first evolves down to the low scale t_0^a, without any radiation being resolved. Then it undergoes a, by now non-perturbative, transition producing h with a fraction x of its momentum. Similarly, the second term represents the possibility that initially a evolves down to the intermediate scale, t_1, at which point a resolved branching, $a \to bb'$, occurs. Parton b, which receives a large fraction $z_1 > x$ of a's momentum, then evolves without radiation to its cut-off scale t_0^b at which point it fragments to give a parton of the desired properties. Similar interpretations apply for all the terms in eqn (4.13). Their sum gives a solution of the DGLAP equation based on explicit branchings.

Before evaluating the Sudakov form factor, eqn (4.11), it is important to specify the argument of α_s. This choice is guided by the need to resum the hard infrared $\ln(1 - z)$ terms which arise when soft gluon radiation is inhibited. At NLO, in the limit $z \to 1$, the q $\to$ qg splitting function contains $\ln(1 - z)$ terms, whilst in the case of the g $\to$ gg splitting function there are also singular $\ln z$ terms. By making the local transverse momentum, $p_\perp^2 \approx z(1 - z)Q^2$, the argument of α_s these $\ln(1 - z)$ and $\ln z$ corrections are accounted for automatically (Amati $et\ al.$, 1980); see Ex. (4-6). This choice does introduce spurious singularities into P_{gq} and P_{qg}, as $z \to 0$ and $0, 1$, respectively, but these are inconsequential because these splitting functions are finite in these limits. Using either $\alpha_s(t = Q^2)$ or $\alpha_s(p_\perp^2)$ in eqn (4.11) we find that $\Delta_a(t, t_0)$ is a heavily damped function,

$$\Delta_q(t, t_0)\Big|_{\alpha_s(t)} \sim \left[\frac{t_0}{t} \left(\frac{\alpha_s(t)}{\alpha_s(t_0)} \right)^{\ln(t_0/\Lambda^2)} \right]^{\frac{C_F}{\pi\beta_0}}$$

$$\Delta_q(t, t_0)\Big|_{\alpha_s(p_\perp^2)} \sim \left(\frac{\alpha_s(t)}{\alpha_s(t_0)} \right)^{\frac{C_F}{\pi\beta_0} \ln(t/\Lambda^2)} . \qquad (4.15)$$

For $\alpha_s(t)$ it decreases as an inverse power of t, whilst for $\alpha_s[p_\perp^2 = z(1 - z)Q^2]$ it falls faster than any inverse power. The difference is important for the pre-confinement property of pQCD (Bertolini and Marchesini, 1982). In both cases, eqn (4.15) shows that it becomes increasingly unlikely, as t increases, to evolve from $t \sim Q^2$ to t_0 without any resolvable radiation.

The numerical implementation of eqn (4.13) is as a Markov process that proceeds in a series of iterated steps. Given an off mass-shell parton, a, of maximum scale $t > t_0^a$, the channel-by-channel Sudakov form factors, $\Delta_{a \to cc'}(t, t_0^a)$, are used to select a possible intermediate scale at which a specific branching can occur. Of course, it is possible that the putative intermediate scale is below t_0^a, in which case the parton is placed on 'mass-shell', $p_a^2 = t_0^a$. If a valid branching does occur, then the momentum fraction is selected with distribution $\alpha_s(t, z)P_{ca}(z)$, and the daughter particles are given new maximum scales, t_c and $t_{c'}$, derived from t and z. This procedure is repeated until all partons are placed on their 'mass-shell'.

It is worth remembering that the branching algorithms are designed primarily to give good approximations to the full matrix element in the dominant phase space regions. In non-soft, non-collinear regions, for example, $e^+e^- \to q\bar{q}g$ with the gluon recoiling against a near-aligned $q\bar{q}$-pair, they may offer a poor approximation. It is therefore common to correct the first branching to the full, known matrix element. It may also be necessary to apply corrections to populate regions of the full phase space which are not covered by the branching's (t, z) phase space.

4.2.4 *Space-like Monte Carlo algorithm*

In view of the structural similarity between eqns (4.5) and (4.4) we can employ exactly the same method to obtain the following formal series solution to eqn (4.4),

$$
f_{a/h}(x, t) = \Delta_a(t, t_s^a) f_{a/h}(x, t_s^a)
$$
$$
+ \int_{t_s^a}^t \frac{dt_1}{t_1} \int_x^{1 - \epsilon_{b'}^s(t_1)} \frac{dz_1}{z_1} \Delta_a(t, t_1) \frac{\alpha_s(t_1, z_1)}{2\pi} \bar{P}_{ab}(z_1) \Delta_b(t_1, t_s^b) f_{b/h}\left(\frac{x}{z_1}, t_s^b\right)
$$
$$
+ \int_{t_s^a}^t \frac{dt_1}{t_1} \int_x^{1 - \epsilon_{b'}^s(t_1)} \frac{dz_1}{z_1} \int_{t_s^b}^{t_1} \frac{dt_2}{t_2} \int_{x/z_1}^{1 - \epsilon_{c'}^s(t_2)} \frac{dz_2}{z_2}
$$
$$
\times \Delta_a(t, t_1) \frac{\alpha_s(t_1, z_1)}{2\pi} \bar{P}_{ab}(z_1) \Delta_b(t_1, t_2) \frac{\alpha_s(t_2, z_2)}{2\pi} \bar{P}_{bc}(z_2) \Delta_c(t_2, t_s^c) f_{c/h}\left(\frac{x}{z_1 z_2}, t_s^c\right)
$$
$$
+ \cdots . \tag{4.16}
$$

Again, a simple physical interpretation is possible. The presence of parton a, inside h at the scale t, for example virtuality, may be accounted for in an infinite number of ways. Reading from right to left, term two, for example, represents the possibility that at the cut-off t_s^b parton b is found in h with a large momentum fraction $x/z_1 > x$. Parton b then evolves without resolvable radiation to the scale $t_1 \in [t, t_s^b]$ at which the branching $b \to aa'$ occurs, leaving parton a with energy fraction x. Parton a then evolves from t_1 to the scale t, again without resolvable radiation. Similar interpretations exist for all the other terms.

Unfortunately, this forward evolution scheme, eqn (4.16), for the initial state is no longer generally regarded as suitable for Monte Carlo implementation. Its fatal flaw is that it is not possible to guarantee that the generated parton cascade will produce a parton compatible with that required for the pre-selected hard scattering. Despite attempts to improve this situation using pretabulation techniques (Odorico, 1984), it remains a grossly inefficient algorithm. Fortunately, the problem has been finessed by a reformulation as a backward evolution scheme, in which the known p.d.f.s, already used to choose the hard subprocess, are employed to guide the evolution of partons from the hard-scattering to the low-scale incoming hadron (Sjöstrand, 1985; Gottschalk, 1986).

By judiciously inserting 1 written as a ratio of p.d.f.s, eqn (4.16) can be manipulated to obtain an equivalent form,

$$
\begin{aligned}
1 = {}& \prod_a(t, t_s; x) \\
&+ \int_{t_s^a}^{t} \frac{\mathrm{d}t_1}{t_1} \int_{x}^{1-\epsilon_{b'}^s(t_1)} \frac{\mathrm{d}z_1}{z_1} \prod_a(t, t_1; x) \frac{\alpha_s(t_1, z_1)}{2\pi} \bar{P}_{ab}(z_1) \frac{f_{b/h}(\frac{x}{z_1}, t_1)}{f_{a/h}(x, t_1)} \prod_b\left(t_1, t_s; \frac{x}{z_1}\right) \\
&+ \int_{t_s^a}^{t} \frac{\mathrm{d}t_1}{t_1} \int_{x}^{1-\epsilon_{b'}^s(t_1)} \frac{\mathrm{d}z_1}{z_1} \int_{t_s^b}^{t_1} \frac{\mathrm{d}t_2}{t_2} \int_{x/z_1}^{1-\epsilon_{c'}^s(t_2)} \frac{\mathrm{d}z_2}{z_2} \\
&\times \prod_a(t, t_1; x) \frac{\alpha_s(t_1, z_1)}{2\pi} \bar{P}_{ab}(z_1) \frac{f_{b/h}(\frac{x}{z_1}, t_1)}{f_{a/h}(x, t_1)} \\
&\times \prod_b\left(t_1, t_2; \frac{x}{z_1}\right) \frac{\alpha_s(t_2, z_2)}{2\pi} \bar{P}_{bc}(z_2) \frac{f_{c/h}(\frac{x}{z_1 z_2}, t_2)}{f_{b/h}(\frac{x}{z_1}, t_2)} \prod_c\left(t_2, t_s^c; \frac{x}{z_1 z_2}\right) \\
&+ \cdots .
\end{aligned}
$$

(4.17)

Here the usual time-like Sudakov form factor, eqn (4.11), has been replaced by

$$
\prod_a(t, t_s^a, x) = \Delta_a(t, t_s^a) \frac{f_{a/h}(x, t_s^a)}{f_{a/h}(x, t)} .
$$

(4.18)

The interpretation of $\prod_a(t, t_s^a, x)$ is as the probability that the parton a, present in h, evolves backwards from the scale t and momentum fraction x to the scale $t_s^a < t$ and unchanged momentum fraction, x, without any resolved emissions. To see this, let $\mathcal{F}_a(t'; t, x)\mathrm{d}t'$ be the fraction of partons of type a present at t with momentum fraction x, which came from a resolved branching in the interval $[t', t' + \mathrm{d}t']$. Then one has

$$
\begin{aligned}
\prod_a(t, t_s^a, x) &= 1 - \int_{t_s^a}^{t} \mathrm{d}t' \mathcal{F}_a(t'; t, x) \\
&= 1 - \int_{t_s^a}^{t} \mathrm{d}t' \frac{1}{f_{a/h}(x, t)} \int_{x}^{1-\epsilon_{a'}^s(t)} \frac{\mathrm{d}z}{z} \Delta_a(t, t') \frac{\alpha_s(x, t')}{2\pi t'} \bar{P}_{ab}(z) f_{b/h}\left(\frac{x}{z}, t'\right) \\
&= 1 - \frac{1}{f_{a/h}(x, t)} \int_{t_s^a}^{t} \mathrm{d}t' \frac{\partial}{\partial t'} \left(\frac{f_{a/h}(x, t')}{\Delta_a(t', t)} \right) \\
&= \Delta_a(t, t_s^a) \frac{f_{a/h}(x, t_s^a)}{f_{a/h}(x, t)} .
\end{aligned}
$$

(4.19)

In the third line, we have used the DGLAP equation, eqn (4.4), in a form equivalent to eqn (4.12) and when required used $\Delta_a(t', t) = \Delta_a^{-1}(t, t')$. Recalling that the DGLAP evolution suppresses high-x partons, $f_{a/h}(x, t) < f_{a/h}(x, t_s^a)$, we see that the effect of the p.d.f. ratio in eqn (4.18) compared to eqn (4.11) is to make $\Pi > \Delta$. The converse is true at small x. This reflects the observation that a high-x parton is less likely to have undergone a branching than a small-x parton.

A second modification in eqn (4.17) are the p.d.f. factors accompanying the splitting function, $\bar{P}_{ab}(z)f_{b/\mathrm{h}}(x/z,t)/f_{a/\mathrm{h}}(x,t)$. The presence of $f_{b/\mathrm{h}}$ helps to guide the evolution to the correct parton content, whilst $f_{a/\mathrm{h}}$ ensures the correct normalization. Strictly speaking, these p.d.f.s and those appearing in eqn (4.18), are the solutions to eqn (4.4). However, in practice, the p.d.f.s used are those from the standard fits; see Chapter 7.

We are now able to interpret eqn (4.17) as a correctly normalized sum of probabilities for all the chains of branchings that take a parton a, present at the scale t, back to an initial 'parent' parton at the scale t_s^a. Reading from left to right, the second term, for example, represents the probability that the desired parton a, present at t, had evolved, without resolvable radiation, from the scale t_1. At t_1 it had been produced in the branching of a parton b of larger momentum fraction x/z_1, which had originally come from the scale t_s^b without resolvable emission.

The backwards evolution scheme implicit in eqn (4.17) is solved numerically using the following iterative algorithm, which is similar to that for the time-like case. Given a space-like parton a of maximum scale t with momentum fraction x, the modified Sudakov form factor $\Pi_a(t, t_s^a, x)$ is used to select a putative branching scale. If this new scale is below the space-like cut-off t_s^a, or leaves insufficient phase space for any branching, then no resolved branchings are taken to have occurred and the parton is placed on 'mass-shell', $p_a^2 = t_s^a$. If a branching does occur then the type, $b \to aa'$, and momentum fraction, z, are selected according to $\alpha_{\mathrm{s}}(t,z)\bar{P}_{ab}(z)f_{b/\mathrm{h}}(x/z,t)$. New maximum virtualities are constructed for the daughter partons and the above procedure repeated for b, whilst a' is treated using the time-like algorithm. Remember that only those partons lying directly on the chain linking the hard subprocess to the incoming hadron have negative evolution variables (virtualities) and only these partons are governed by eqn (4.4). All partons on the side branches have positive evolution variables and are described by the same time-like equations as above.

A final issue is how to match the low-scale parton, at the start of the initial state shower, to the quark content of the incoming hadron. In the case of gluons and sea quarks flavour conservation requires a minimum of up to two 'forced' branchings. For example, the simplest solution to the worst case of a $\bar{s}$ emerging from a proton requires a branching sequence such as p : (uu)d $\to$ dg, g $\to$ s$\bar{s}$. Since we are now in the non-perturbative regime, these branchings need not follow the same prescription as in the cascade, though momentum etc. conservation should still be respected. In view of the parton's Fermi motion within the hadron it is common to give the initiator parton, d in the above example, and remnant uu, a Gaussian transverse momentum.

4.2.5 *Soft gluon logarithms*

Logarithmic enhancements to a multi-parton cross section occur not only in the collinear limit but also in the limit of soft gluon emission. It is therefore important that soft gluon effects are included properly in Monte Carlo programs. In Section 3.7 we learnt that the cross section for the emission of a soft gluon is dominated by interference between Feynman diagrams. Thus in order to permit a Monte Carlo implementation we need to impose first a classical structure. In essence, we must find a way to associate the emission of a soft gluon with a particular hard parton, ensure positive emission probabilities and maintain the correct total cross section (Bassetto *et al.*, 1983). This will lead us to a simple modification of the DGLAP equations and the so-called *modified leading logarithmic approximation* (MLLA).

The basic expression for describing soft gluon emission is given by eqn (3.341). This contains the dipole radiation terms

$$W_{ij}(k) \equiv \omega^2 \left[\frac{p_i \cdot p_j}{(p_i \cdot k)(p_j \cdot k)} - \frac{1}{2} \frac{p_i^2}{(p_i \cdot k)^2} - \frac{1}{2} \frac{p_j^2}{(p_j \cdot k)^2} \right]$$

$$= \frac{1}{2} \left[\frac{2\zeta_{ij}}{\zeta_{ik}\zeta_{jk}} - \frac{(1-\beta_i^2)}{\zeta_{ik}^2} - \frac{(1-\beta_j^2)}{\zeta_{jk}^2} \right] . \tag{4.20}$$

In the second line we have introduced the angular variables

$$\zeta_{ij} \equiv \frac{p_i \cdot p_j}{E_i E_j} = 1 - \beta_i \beta_j \cos\theta_{ij} , \tag{4.21}$$

which range from $(1 - \beta_i\beta_j) \approx 0$ to $(1 + \beta_i\beta_j) \approx 2$ as θ_{ij} goes from 0 to π. In the massless limit, $\beta \to 1$, eqn (4.20) contains collinear singularities as either $\theta_{ik} \to 0$ or $\theta_{jk} \to 0$. This suggests the following decomposition, based upon adding and subtracting $(1/\zeta_{ik} - 1/\zeta_{jk})$,

$$W_{ij}(k) = \frac{1}{2} \left[\frac{1}{\zeta_{ik}} \left\{ 1 - \frac{(1-\beta_i^2)}{\zeta_{ik}} + \frac{(\zeta_{ij} - \zeta_{ik})}{\zeta_{jk}} \right\} + (i \leftrightarrow j) \right]$$

$$\equiv W_{ij}^{(i)}(k) + W_{ij}^{(j)}(k) . \tag{4.22}$$

The first piece, $W_{ij}^{(i)}(k)$, has only a collinear singularity for $\boldsymbol{k}$ parallel to $\boldsymbol{p}_i$, where $W_{ij}^{(i)}(k) \sim (1 - \cos\theta_{ij})^{-1}$. It is finite for $\boldsymbol{k}$ parallel to $\boldsymbol{p}_j$. Thus, it may naturally be associated with emission off parton i.

In the cross section formula, eqn (3.341), we think of the directions of the hard partons as fixed and the direction of the soft gluon as variable. Therefore, let us consider $W_{ij}^{(i)}(k)$ for fixed θ_{ij} and study it as a function of θ_{ik} and ϕ_i, the azimuthal angle w.r.t. $\boldsymbol{p}_i$. In terms of these variables we have $\cos\theta_{jk} = \cos\theta_{ij}\cos\theta_{ik} + \sin\theta_{ij}\sin\theta_{ik}\cos\phi_i$ and thus

$$W_{ij}^{(i)}(k) = \frac{\beta_i}{2\zeta_{ik}} \left[\frac{\beta_i - \cos\theta_{ik}}{1 - \beta_i \cos\theta_{ik}} + \frac{\cos\theta_{ik} - \beta_j \cos\theta_{ij}}{1 - \beta_j[\cos\theta_{ij}\cos\theta_{ik} + \sin\theta_{ij}\sin\theta_{ik}\cos\phi_i]} \right].$$

(4.23)

Figure 4.2 shows the angles and $W_{ij}^{(i)}$ as function of ϕ for various θ_{ij} and θ_{ik}.

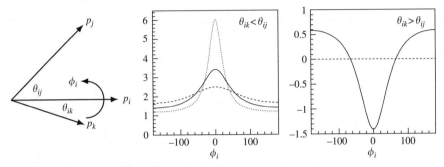

FIG. 4.2. The angles between the soft gluon k, the hard parton i and its interference partner j. The radiation function $2\zeta_{ik} \times W_{ij}^{(i)}(k)$ as a function of ϕ_i. Left, $\pi/2 = \theta_{ij} > \theta_{ik}$ and $\theta_{ik} = \pi\{1, 2, 3\}/8$ for the dashed, solid and dotted lines, respectively. Right, $\pi/4 = \theta_{ij} < \theta_{ik} = \pi/2$. In all cases $\beta_i = 1 = \beta_j$.

A significant feature of eqn (4.23), as shown in Fig. 4.2, is that it is everywhere positive-definite for $\theta_{ik} < \theta_{ij}$. In contrast to this, for $\theta_{ik} > \theta_{ij}$ the distribution goes negative, that is, there is destructive interference. In fact, azimuthal averaging gives

$$\left\langle W_{ij}^{(i)} \right\rangle_{\phi_i} = H_{ij}^{(i)}(\theta_{ik}) \frac{\beta_i}{\zeta_{ik}}$$

(4.24)

with

$$H_{ij}^{(i)}(\theta_{ik}) = \frac{1}{2} \left(\frac{\beta_i - \cos\theta_{ik}}{(1 - \beta_i \cos\theta_{ik})} + \frac{\cos\theta_{ik} - \beta_j \cos\theta_{ij}}{\sqrt{|\cos\theta_{ik} - \beta_j \cos\theta_{ij}|^2 + (1 - \beta_j^2)\sin^2\theta_{ik}}} \right)$$

$$\implies \quad \Theta(\theta_{ij} - \theta_{ik}) \quad \text{for} \quad \beta_{i,j} \to 1 ;$$

(4.25)

see Ex. (4-4). This rather complex looking expression for $H_{ij}^{(i)}(\theta_{ik})$ reduces to the Heaviside step-function for massless, hard partons. This requires the opening angles to be nested, $\theta_{ik} < \theta_{ij}$ ($\equiv \zeta_{ik} < \zeta_{ij}$), in order have any net soft radiation. This restriction is the basis of the angular ordering prescription. The form of $H_{ij}^{(i)}(\theta_{ik})$ for massive particles is shown in Fig. 4.3. It follows the same form as the step-function but the inclusion of particle masses softens its shape. For wide-angle emission the first term in eqn (4.25) is positive, whilst the second goes negative for $\cos\theta_{ik} < \beta_j \cos\theta_{ij}$ and the function falls away quickly. A new feature is the vanishing of radiation at small opening angles, where the collinear singularity is shielded by the particle mass: the inverse propagator behaves as

$2(1 - \beta_i \cos \theta) \sim (m/E)^2 + \theta^2$. Inspection of eqn (4.25) shows that the destructive interference begins when $\cos \theta_{ik} > \beta_i \ (\equiv \theta_{ik} < m_i/E_i)$. This is the so-called 'dead cone' for emission close to a heavy particle's direction of travel. In effect, soft radiation is restricted to the screened cone $R_{ij}^{(i)}$ defined via

$$R_{ij}^{(i)}: \quad \beta_j \cos \theta_{ij} < \cos \theta_{ik} < \beta_i \quad \Longleftrightarrow \quad \frac{m_i^2}{E_i^2} < \zeta_{ik} \lesssim \zeta_{ij} . \tag{4.26}$$

This dominant region of phase space is illustrated in Fig. 4.3.

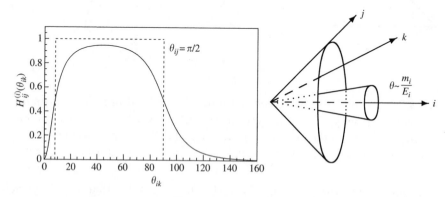

FIG. 4.3. The screened cone $R_{ij}^{(i)}$ for the emission of the soft gluon k associated with parton i, whose interference partner is j. Here we assumed $\beta_i = 0.99$, which implies a dead cone up to $\theta_{ik} \approx 8°$ and $\beta_j = 0.95$, which softens the cut-off at $\theta_{ik} = \theta_{ij} = \pi/2$.

Using these results we can rewrite eqn (3.341) in a form which is everywhere positive-definite,

$$d\sigma^{(N+1)} = -\frac{\alpha_s}{2\pi} \sum_{i \neq j} \hat{T}_i \cdot \hat{T}_j \left\{ \Theta(\zeta_{ij} - \zeta_{ik}) W_{ij}^{(i)}(\phi_i) \frac{d\phi_i}{2\pi} d\zeta_{ik} + (i \leftrightarrow j) \right\} \frac{d\omega}{\omega} d\sigma^{(N)} .$$
$$\tag{4.27}$$

This version is for massless partons. It is exact in the angular ordered region and outside this region, where the soft gluon radiation is prohibited, correct on the azimuthal average. To include hard parton masses, replace $d\zeta_{ik}$ by $d\zeta_{ik}/\beta_i$ and $\Theta(\zeta_{ij} - \zeta_{ik})$ by $\Theta(R_{ik}^{(i)})$, so that the radiation is restricted to the screened cone. One might also include a multiplicative factor to compensate for the neglected tails of the $H_{ij}^{(i)}$ distribution which lie outside the screened cone.

The above reasoning applies equally well to QED, where it is known as the Chudakov effect (Chudakov, 1955) in recognition of the observation of a lack of wide-angle soft photons emitted by the e^+e^- pairs found in cosmic ray showers. Heuristically, if the soft photon is emitted at too wide an angle to the hard leptons, it will have insufficient, transverse, resolving power to distinguish the separated charges and will only see their total charge, which is zero: see Ex. (4-3). In QCD the argument is the same but the total charge need not be zero. To understand the situation better we will consider how to treat successive soft gluon emission in a parton shower.

Consider a colour neutral system of hard partons and let us focus on partons i and j, collecting the remainder together in $\sum_l$. The square of the insertion current is given by

$$-\frac{1}{2}J \cdot J^\dagger = \hat{T}_i \cdot \left(\hat{T}_i + \sum_l \hat{T}_l\right) W_{ij}^{(i)} + \left(\hat{T}_j + \sum_l \hat{T}_l\right) \cdot \hat{T}_j W_{ij}^{(j)}$$
$$- \sum_l \hat{T}_l \cdot \hat{T}_i W_{il} - \sum_l \hat{T}_l \cdot \hat{T}_j W_{jl} - \sum_{l \neq l'} \hat{T}_l \cdot \hat{T}_{l'} W_{ll'} . \quad (4.28)$$

In the first line, we have decomposed W_{ij} using eqn (4.22) and made use of the identity $\hat{T}_i + \hat{T}_j + \sum_l \hat{T}_l = 0$. Re-arranging this expression gives

$$-\frac{1}{2}J \cdot J^\dagger = C_i W_{ij}^{(i)} + C_j W_{ij}^{(j)} - \sum_l \hat{T}_l \cdot \hat{T}_i \left[\left(W_{il}^{(i)} - W_{ij}^{(i)}\right) + W_{il}^{(l)}\right]$$
$$- \sum_l \hat{T}_l \cdot \hat{T}_j \left[\left(W_{jl}^{(j)} - W_{ij}^{(j)}\right) + W_{jl}^{(l)}\right] - \sum_{l \neq l'} \hat{T}_l \cdot \hat{T}_{l'} W_{ll'} , \quad (4.29)$$

where we have used $\hat{T}_i^2 = C_i$, the quadratic Casimir, colour charge. In view of the third and fourth terms we introduce a new function,

$$\tilde{W}_{lj}^{(i)} \equiv W_{il}^{(i)} - W_{ij}^{(i)} \quad \text{such that} \quad \left\langle \tilde{W}_{lj}^{(i)} \right\rangle_{\phi_i} = \begin{cases} \left\langle W_{il}^{(i)} \right\rangle_{\phi_i} & \zeta_{ij} < \zeta_{ik} < \zeta_{il} \\ 0 & \text{otherwise} \end{cases}$$
$$(4.30)$$

where we took $\zeta_{ij} < \zeta_{il}$. In $\tilde{W}_{lj}^{(i)}(p_k)$ the collinear singularity for p_k parallel to p_i has cancelled and in fact the function is only non-zero on average when the soft gluon is radiated at an angle greater than θ_{ij}, but less than θ_{il}. Now, we make the assumption that i and j are close in angle so that in the non-singular W-functions, which are scale independent, p_i and p_j can be replaced by their sum $p_s = p_i + p_j$. In this approximation eqn (4.29) becomes

$$-\frac{1}{2}J \cdot J^\dagger = C_i W_{ij}^{(i)} + C_j W_{ij}^{(j)} - \sum_l \hat{T}_l \cdot \hat{T}_s \left[\tilde{W}_{ls}^{(s)} + W_{ls}^{(l)}\right] - \sum_{l \neq l'} \hat{T}_l \cdot \hat{T}_{l'} W_{ll'} , \quad (4.31)$$

where $\hat{T}_s = \hat{T}_i + \hat{T}_j$. The first two terms represent emission off either i or j which is limited by their common opening angle and with weights C_i. Then, if we think of s as the parent of i and j the remaining terms represent the square of the insertion current for the $s + \sum l$ system prior to the '$s \to i + j$ branching' with the restriction that 'radiation from s' must be at an angle greater than θ_{ij}. Of course the parent s is only a theoretical construct which helps us to encapsulate the effect of colour coherence in this small-angle limit, but the concept proves very useful in developing a Monte Carlo implementation.

By iterating the procedure which led to eqn (4.31) we see that we can capture the leading contribution from successive soft gluon emissions by generating a parton cascade in which successive opening angles decrease. The cross section for generating successive branchings is given by

$$d\sigma^{(N+1)} = \frac{\alpha_s}{2\pi} \sum_i 2C_i \Theta(R_{ii'}^{(i)}) \left\langle W_{ii'}^{(i)} \right\rangle_{\phi_i} (\zeta_{ik}) \frac{d\phi_i}{2\pi} d\zeta_{ik} \frac{d\omega}{\omega} d\sigma^{(N)} , \qquad (4.32)$$

where the interference partner, i', is the parton produced in the same branching as i. This raises the obvious question: how does this expression relate to eqn (4.6) which describes the collinear singularities? Now, the diagonal splitting functions, $P_{aa}(1-z)$ with a=q or g, are both singular in the limit of soft gluon emission $z \to 0$. Indeed, in this limit the two branching probabilities coincide,

$$\lim_{z\to 0} P_{aa}(1-z)dz\frac{dt}{t} = 2C_a \frac{dz}{z}\frac{dt}{t} = 2C_a \frac{d\omega}{\omega}\frac{d\zeta_{ag}}{\zeta_{ag}} , \qquad (4.33)$$

so that, neglecting azimuthal correlations, the distribution of soft gluon radiation is correctly described by the usual Altarelli–Parisi kernel with an angular ordering constraint $\Theta(\zeta_{ag} - \zeta_{aa'})$ imposed. A natural way to achieve this is to use the evolution variable $t_a = E_a^2(1 - \cos\theta_{bc})$ and the orderings $t_b < z^2 t_a$ and $t_c < (1-z)^2 t_a$. Including the distribution of the azimuthal angle requires two sets of quantum mechanical correlations to be taken into account, those which are due to soft gluon interference, eqn (4.23), and those due to the spin polarization of hard gluons in a cascade (Knowles, 1990; Richardson, 2001).

The analysis of soft gluon coherence given above applies to the final state. A similar approach holds for soft gluons in the initial state, though the analysis is more complicated due to the asymmetric kinematics of space-like branchings. That said, the effect of colour coherence also can be included by placing a restriction on phase space in terms of an angle-like variable which decreases away from the hard scattering, towards the incoming hadron (Catani $et\ al.$, 1986). The precise variable is defined in terms of the incoming hadron momentum as

$$t_i = E_i\sqrt{\zeta_i} \approx E_i\varphi_i \quad \text{with} \quad \zeta_i = \frac{p_{i'} \cdot P_h}{E_{i'} E_h} \approx 1 - \cos\varphi_i , \qquad (4.34)$$

where i' is the time-like daughter at i's branching. Observe that in the $x = z_1 \cdots z_n \to 1$ limit ($z_i \sim 1$), ordering in t_i becomes equivalent to ordering in φ_i. Due to destructive interference this is more restrictive than the earlier ordering in virtuality, though disordered transverse momenta are permitted by t_i ordering. In the opposite $x \to 0$ limit ($z_i \sim 0.5$), ordering in t_i is equivalent to ordering in the branching's transverse momenta. Actually, in this small-x region the DGLAP equation is superseded by the BFKL equation; see Section 3.6. This leads to strict ordering in φ, which is less restrictive than ordering in transverse momentum, and the need for a new, non-local, non-Sudakov form factor, whose effect is to dampen the $1/z$ singularity in $P_{gg}(z)$ (Catani *et al.*, 1991*b*). In practice, only very specialized programs treat coherence fully in the initial state.

To summarize, so far we have seen that the effect of soft gluon coherence within parton showers can be included by imposing angular ordering. It remains to specify the initial opening angles. In order to take account of inter-jet coherence these should be determined by the colour flow in the hard subprocess. Sometimes this is unique, as is the case $\gamma^*/Z \to q\bar{q}g$, shown in Fig. 3.28. Here the appropriate initial angles are those between the colour-connected pairs θ_{qg} and $\theta_{\bar{q}g}$. Unfortunately, the colour flow is usually not unique, as in the case of $\gamma^*/Z \to q\bar{q}gg$, where two colour flows contribute. This requires weights for the competing colour flows, so that on an event-by-event basis one can be selected at random. The least ambiguous prescription (Odagiri, 1998) is to decompose the amplitude for the hard subprocess into the separate colour flow contributions and assign weights according to the square of these sub-amplitudes. This implies the neglect of the colour interference terms when assigning the weights. However, this is not to say that they should be neglected when calculating the cross section for the hard subprocess!

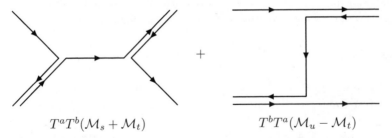

FIG. 4.4. The two colour flows contributing to the amplitude for quark–gluon scattering

To see how this works in practice consider $qg \to gq$ scattering. At leading order three Feynman diagrams contribute to the amplitude: s-channel quark exchange with colour factor $T^a T^b$, u-channel quark exchange with colour factor $T^b T^a$ and t-channel gluon exchange with colour factor $if_{abc}T^c = T^a T^b - T^b T^a$. The two colour flows for this process are illustrated in Fig. 4.4. Referring back to Section 3.3.4, the amplitude squared is given by

$$\left| T^a T^b (\mathcal{M}_s + \mathcal{M}_t) + T^b T^a (\mathcal{M}_u - \mathcal{M}_t) \right|^2$$
$$\propto C_F N_c \frac{(u^2 + s^2)}{t^2} \left[C_F \frac{-u}{s} + \left(\frac{C_A}{2} - C_F \right) + C_F \frac{s}{-u} \right]. \qquad (4.35)$$

The first and third terms correspond to the squares of the sub-amplitudes, which are used as weights when selecting a colour flow. The second term corresponds to the neglected interference term. It is both colour suppressed, being of relative magnitude $\mathcal{O}(1/N_c^2)$, and dynamically suppressed as it has neither a $1/s$ nor a $1/u$ singularity.

4.2.6 *The colour dipole model*

The formulation of pQCD evolution in terms of initial and final state parton showers is not unique. A complementary, or dual, approach is possible in terms of colour dipole cascades (Gustafson, 1986; Gustafson and Pettersson, 1988). This is equivalent, to MLLA accuracy, to a parton shower with angular ordering automatically incorporated (Azimov *et al.*, 1985*a*). The basic observation is that gluon emission from, say, a q$\bar{\text{q}}$-pair is described, in direct analogy with electromagnetism, by the same formula as gluon emission by a colour dipole. To $\mathcal{O}(1/N_c^2)$ subsequent gluon emission may then be described by radiation from two independent dipoles. The resulting picture is of a colour-ordered chain in which gluons join dipoles and dipoles join partons. A close correspondence to the string model, in which gluons cause 'kinks' along a string, makes this formulation seem rather appealing. Colour dipole cascades are implemented in the program ARIADNE (Lönnblad, 1992).

Gluon emission from one of the three basic dipole types, $\gamma^\star \to \text{q}\bar{\text{q}} \to \text{q}_1\text{g}_2\bar{\text{q}}_3$ with aligned quark spins, $\text{q} \to \text{qg} \to \text{q}_1\text{g}_2\text{g}_3$, $\bar{\text{q}} \to \bar{\text{q}}\text{g} \to \bar{\text{q}}_1\text{g}_2\text{g}_3$ or $\text{g} \to \text{gg} \to \text{g}_1\text{g}_2\text{g}_3$, is described by the cross section formulae for massless partons

$$d\sigma = C_{nm}\frac{\alpha_s(k_\perp^2)}{2\pi}\frac{x_1^m + x_3^n}{(1-x_1)(1-x_3)}dx_1dx_3 \qquad m,n = \begin{cases} 2 \text{ for } \text{q},\bar{\text{q}} \\ 3 \text{ for } \text{g}. \end{cases} \qquad (4.36)$$

Here $x_{1,3}$ represent twice the energy fractions of the two emitting partons in the dipole's rest frame, so that the gluon has $x_2 = 2 - x_1 - x_3 = 2E_\text{g}/\sqrt{s_\text{dip}} \in [0,1]$. Here s_dip is the square of the dipole's invariant mass. The coefficient is given by $C_{22} = 2/3$, $C_{23} = 3/4 = C_{33}$ for the three vertices, respectively. This expression is designed to be fully equivalent to the conventional approach, see Ex. (4-7). The splitting $\text{g} \to \text{q}\bar{\text{q}}$ does not arise naturally in the above formalism for gluon emission. Its treatment is inspired by appeal to the full $\gamma^\star \to \text{qq}'\bar{\text{q}}'\bar{\text{q}}$ cross section (Andersson *et al.*, 1990). In the limit of a low $k_\perp^2$, low mass, intermediate gluon this factorizes, after azimuthal averaging, as $\sigma(\text{q}\bar{\text{q}}\text{g})P_{\text{q}'\text{g}}(z)$, which can be separated into two equal parts associated with the q-g and g-$\bar{\text{q}}$ dipoles. This prescription for the dipole formalism enhances effectively $\text{g} \to \text{q}\bar{\text{q}}$ splitting compared to the parton formalism (Seymour, 1995).

The dipole formalism also uses a different description of the phase space. In the limit of soft gluon emission, off massless partons, the following 'rapidity' and 'transverse momentum' variables, both with respect to the mother dipole's axis, suggest themselves,

$$y = \frac{1}{2}\ln\left(\frac{1-x_1}{1-x_3}\right) \quad \text{and} \quad \kappa = \frac{1}{2}\ln\left(\frac{k_\perp^2}{\Lambda^2}\right)$$

$$\text{with} \quad k_\perp^2 = s_\text{dip}(1-x_1)(1-x_3). \qquad (4.37)$$

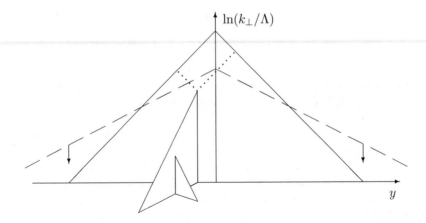

FIG. 4.5. The fractal phase space in the colour dipole model. The dotted lines
show the allowed region neglecting the $k_\perp$ ordering constraint. The dashed
lines indicate the additional constraint when extended sources are present.

Here Λ is of the order of the QCD scale $\Lambda_{\rm QCD}$. Parton mass effects can be
included by replacing $(1 - x_1)(1 - x_3)$ in eqns (4.36) and (4.37) by $(1 - x_1 +
(m_1^2 - m_3^2)/s_{\rm dip})(1 - x_3 + (m_3^2 - m_1^2)/s_{\rm dip})$. In light-cone coordinates the gluon
has momentum $(k_+, k_-) = \sqrt{k_\perp^2}(e^{+y}, e^{-y})$. Adopting these variables eqn (4.36)
becomes

$$\mathrm{d}\sigma = \frac{C_{nm}}{2\pi\beta_0}\left[\left(1 - e^{\kappa+y}\frac{\Lambda}{\sqrt{s_{\rm dip}}}\right)^n + \left(1 - e^{\kappa-y}\frac{\Lambda}{\sqrt{s_{\rm dip}}}\right)^m\right]\frac{\mathrm{d}\kappa}{\kappa}\mathrm{d}y \sim \frac{\mathrm{d}\kappa}{\kappa}\mathrm{d}y\,.$$

$$(4.38)$$

Using four-momentum conservation we find the following constraint on y for a
given $k_\perp$,

$$x_2\sqrt{s_{\rm dip}} = 2k_\perp\cosh y \quad\Longrightarrow\quad \ln\left(\frac{\sqrt{s_{\rm dip}}}{k_\perp}\right) \geq \ln(2\cosh y) \approx |y|\,. \qquad (4.39)$$

This is an approximately triangular region in the $y - \ln(k_\perp/\Lambda)$ phase space.
When a gluon is emitted, the phase space for subsequent branchings is increased
as indicated by the projecting region in Fig. 4.5. Repeated gluon emission leads
to a fractal-like structure (Andersson *et al.*, 1989a; Gustafson and Nilsson, 1991).
When a g $\to$ q$\bar{\rm q}$ branching occurs, the phase space is effectively cut along the
middle of the projecting region.

   Given a branching at $(k_{1\perp}, y_1)$, two new dipoles are formed and, as shown
in Fig. 4.5, the phase space forms two 'overlapping triangular regions' separated
by the gluon. Reflecting the gluon's double colour charge, in the planar approximation, radiation is permitted on either side of the projecting region. Two
constraints are placed on the phase space for a second, subsequent branching,

$$k_{2\perp} \begin{cases} \leq \left(\frac{1}{2}\right) e^{|y_1|} \sqrt{k_{1\perp}} \sqrt{s_{\mathrm{dip}}} & \text{kinematics} \\ \ll k_{1\perp} & \text{dynamics .} \end{cases} \tag{4.40}$$

The kinematic condition follows from the masses of the daughter dipoles, see Ex. (4-8), and can allow $k_{2\perp} > k_{1\perp}$. The dynamic condition is required for the validity of the dipole cascade approximation. Applying the stricter $k_\perp$ ordering compromises the dipoles' independence but guarantees angular ordering.

The cascade implementation of the dipole branching probabilities is similar to that of a conventional partonic cascade. Again, there is a divergence as $k_\perp \to 0$ which is tamed by a Sudakov form factor. This includes the virtual corrections and thereby provides a finite probability distribution for the $k_\perp$ of gluon emissions. For example, the probability distribution for the branching qg $\to$ qgg, first occurring at transverse momentum $k_\perp$, is given by

$$\mathcal{P}_{\mathrm{qg} \to \mathrm{qgg}}(k_\perp) = f_{\mathrm{qg} \to \mathrm{qgg}}(k_\perp^2) \tag{4.41}$$

$$\times \exp\left\{ -\int_{k_\perp^2}^{s_{\mathrm{dip}}/4} \mathrm{d}p_\perp^2 \left[f_{\mathrm{qg} \to \mathrm{qgg}}(p_\perp^2) + \sum_{q'} f_{\mathrm{qg} \to \mathrm{qq'\bar{q}'}}(p_\perp^2) \right] \right\}.$$

Here $f_{...}(k_\perp^2)$ is the branching probability of eqn (4.38) integrated over the rapidity interval eqn (4.39). This expression is applied to each possible branching in turn and the one with the largest $k_\perp$ value above the cut-off, $k_\perp^{\min}$, is accepted. Given a particular type of branching, a value of y can then be chosen according to the appropriate function $f_{...}(k_\perp^2)$, subject to the boundary conditions, eqn (4.39).

Given values of $k_\perp$ and y, it is possible to reconstruct the opening angles between the partons. However, this leaves unspecified the two degrees of freedom, polar angle θ and azimuthal angle ϕ, associated with the orientation of the branching plane with respect to the original dipole's axis. These determine the distribution of transverse momentum recoils. The ϕ-distribution is taken to be uniform. The θ-distribution depends on the type of branching. In the case of a q-$\bar{q}$ dipole, spin considerations show that, working in the C.o.M. frame, parton 1 or 3 should retain its direction with a probability proportional to x_1^2 and x_3^2, respectively (Kleiss, 1986). In the remaining cases, no such prescription exists and it is postulated that the recoils are chosen so as to minimize the disturbance to the neighbouring dipoles' colour flows (Gustafson and Pettersson, 1988). Specifically, the gluon retains its direction in a g-q(g-$\bar{q}$) dipole whilst for a g-g dipole $p_\perp^2(\mathrm{g}_1)+$ $p_\perp^2(\mathrm{g}_3)$ is minimized.

The above formalism can be applied also to lepton–hadron and hadron–hadron collisions in much the same way as for e^+e^- annihilation. Figure 4.6 shows an initial configuration for the three dipoles associated with the hard subprocess gg $\to$ q$\bar{q}$ in a hadron–hadron collision. These dipoles are then evolved as described above, but with the proviso that the transverse momentum in the hard subprocess is used as an upper bound on the $k_\perp$ in any branchings. This is to avoid confusion over what constitutes the hard subprocess, also referred to as 'double counting'. Observe that this dipole evolution gives a naturally unified,

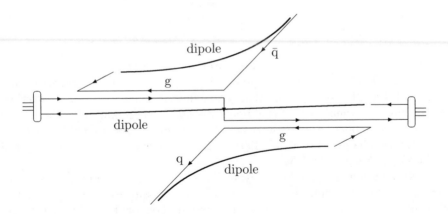

FIG. 4.6. One of the two colour flows associated with the hard subprocess
gg → qq̄. Also indicated are the initial three dipoles associated with this
colour flow. This is the same colour flow that determines the initial opening
angles in a coherent parton shower.

gauge invariant treatment of bremsstrahlung, without the need to separate it
into initial and final state radiation.

There is one complication that necessitates a refinement of the colour dipole
model in hadronic collisions. When a remnant forms at the end of a dipole, it
must be remembered that it should not be regarded as pointlike. In the soft
radiation model (Andersson *et al.*, 1989*b*) it is noted that a coherent gluon of
transverse momentum $k_\perp$ is only aware of a portion of an extended emitter
$\lambda_\perp/R_0 \sim \mu/k_\perp$, where μ corresponds to the source's inverse size. Thus, it can
only obtain its light-cone momentum from a restricted portion of the emitter's
momentum. This reduces the available phase space to

$$k_\pm < \left(\frac{\mu}{k_\perp}\right)^\alpha P_\pm^{\text{rem}} = \frac{\mu_0}{k_\perp} P_\pm . \tag{4.42}$$

Here $P_\pm$ is the relevant light-cone momentum of the dipole end's source, for a
beam remnant $P_\pm^{\text{rem}} = (1-x)P_\pm$, and α reflects its dimensionality. Experimental
data favour $\alpha \sim 1$ and $\mu \sim 1\,\text{GeV}$. Actually, it is assumed that rather than
the dipole's (inverse) size, its rest frame energy density is a constant, so that
$\mu = \mu_0/(1 - x)$ should be used in the case of a beam remnant. This gives the
second expression in eqn (4.42), which is independent of both x and Q^2, although
$k_+ < P_+^{\text{rem}}$ is still required. Equation (4.42) can be rewritten to give the phase
space constraint

$$\kappa \leq \ln\left(\frac{\sqrt{\mu_0 P_\pm}}{\Lambda}\right) \mp \frac{y}{2} \tag{4.43}$$

which is indicted by the dashed lines in Fig. 4.5. We also note that since only
a part of the remnant participates in any radiation, a careful treatment of the

remnant and recoils is required (Andersson et al., 1989b).

The basic colour dipole model has been extended to incorporate further physics. This includes photon radiation from final-state electromagnetic dipoles (Lönnblad and Kniehl, 1992), the production of heavy quarkonia (Ernström and Lönnblad, 1997; Ernström et al., 1997) and a description of diffractive scattering. An alternative formulation of the dipole cascade for DIS, known as the Linked Dipole Chain model (Andersson et al. 1996a, 1996b) is also available as an option (Kharraziha and Lönnblad, 1998). This is equivalent to using the CCFM equation; see Section 3.6.6.3.

Two situations do not fit easily into the dipole model. First, it does not incorporate boson–gluon fusion in DIS and this necessitates the use of a matrix element matching scheme for the first emission (Lönnblad, 1996a). Second, in Drell–Yan the dipole joins the two beam remnants, so that without a special scheme to transfer momentum, the vector boson would acquire no transverse momentum, which is at odds with experimental observations (Lönnblad, 1996b).

4.2.7 The soft underlying event model

In events that involve one or two incoming hadrons we must treat the beam remnant(s), that is the particles which are left in the incoming hadron(s) after extracting the parton shower/hard scattering initiators. Essentially by definition, the relevant physics may only involve relatively small momentum transfers and thus necessarily probe non-perturbative processes. Therefore, once again programs rely on models rather than calculations from first principle. A number of approaches are available, including a parameterization of existing data, a straight generalization of the hadronization model, a multiple scattering mini-jet model or Pomeron inspired models.

At this point it is useful to be aware that, whilst the basic physics of the soft underlying event (SUE) is assumed to be rather similar to that of the soft hadronic collisions which dominate minimum bias data, significant differences occur when a high-Q^2 hard scattering is present. In particular, the associated particle and energy flow in high-Q^2 events, for example on the shoulders of high-$p_\perp$ jets, is significantly larger than in a minimum bias event with equivalent $\sqrt{s}$. This 'pedestal effect' has been seen for jet events (UA2 Collab., 1985a; UA1 Collab., 1988), W/Z production (UA2 Collab., 1987) and Drell–Yan pairs (UA2 Collab., 1985c). Typically, an enhancement in the activity of between 1.5 and 4 times that in a minimum bias event is required. Both experimentally (UA2 Collab., 1985a) and theoretically (Gaisser et al., 1986; Marchesini and Webber, 1988) a two-component picture is favoured. Initial state QCD bremsstrahlung provides a small component that grows with Q^2 and soft physics provides a contribution that appears to saturate for sufficiently large Q^2. The latter is what we mean by the SUE.

Approaches based on ad hoc parameterizations of experimental data are exemplified by HERWIG. This takes UA5's GENCL Monte Carlo (UA5 Collab., 1987), fitted to the then existing data, and adapts it to use its own cluster hadronization scheme. Clusters are generated assuming a flat, central rapidity plateau with Gaussian tails and limited transverse momenta. The resulting hadrons are then required to conform to a negative binomial multiplicity distribution, whose parameters are s-dependent. A special prescription exists for the leading hadrons. No supporting theory is attempted, so that energy extrapolations are open to question. It is interesting to note that, in order to reproduce the pedestal seen in the hard scattering data for hadron–hadron collisions, no enhancement of the underlying event was found necessary. Perturbative soft gluon radiation proved to be adequate (Marchesini and Webber, 1988). A characteristic asymmetry in the average multiplicity in a jet's two shoulders is anticipated. A similar approach is used by COJETS (Odorico, 1990), though it works at the hadron level and assumes a KNO-scaling form (Koba et al., 1972) for the multiplicity distribution.

PYTHIA (Sjöstrand et al., 2001) provides two basic models for the SUE. The first simply follows the colour flows through an event, enabling the coloured beam remnant(s) to be connected to the rest of the event using a string, which is then hadronized as usual. A significantly more ambitious approach, based on perturbative estimates for multiple, semi-hard parton scatterings, is also available (Sjöstrand and van Zijl, 1987). The mini-jet cross section, as a function of a $p_\perp^{min}$ cut-off and normalized to the fitted, inelastic, non-diffractive cross section, is used as a mean scattering probability. Multiple scatters above $p_\perp^{min}$ are then generated from a Poisson type distribution and the resulting event connected using simplified string drawings. As an option, a double Gaussian spatial distribution for the partons within a hadron can be folded into the probabilities above. If a hard scattering occurs, then a large hadronic overlap is more likely and multiple scatterings can be anticipated, which naturally gives an enhanced underlying event. A particular feature to be expected in such a multiple scattering model are correlations in azimuth, and to a lesser extent rapidity, for mini-jet observables (Wang, 1993). Variants of these mini-jet/string models can also be found in the FRITIOF (Pi, 1992) and HIJING (Gyulassy and Wang, 1991) Monte Carlo programs, which concentrate on unified descriptions of soft and hard QCD events in hadron–hadron and heavy ion collisions.

A final class of models for the SUE is inspired by 'Pomeron physics' descriptions of soft and hard collisions. ISAJET (Paige and Protopopescu, 1986) uses the AKG cutting rules (Abramovskii et al., 1972). The basic unit is a cut Pomeron, which gives rise to a chain of hadrons uniformly distributed in rapidity and with a Poissonian multiplicity distribution. The number of such Pomerons and the mean hadron $p_\perp$ are adjusted separately for soft and hard scattering events so as to reproduce the data. After separately assigning leading baryons, the Pomerons are given rescaled longitudinal momentum fractions from a uniform distribution. Finally, each Pomeron is fragmented in its own C.o.M. frame, using an independent fragmentation function made energy dependent to reproduce the observed rise in dN/dy with s. The program DTUJET (Aurenche et al., 1994) uses the more elaborate Dual Topological Unitarization model which again involves cut Pomerons.

4.3   Multi-purpose event generators

Whilst there are a large number of Monte Carlo programs available the majority
of these either treat specialized event types or supply partonic final states using
fixed-order matrix element calculations. There are relatively few multi-purpose
event generators which offer complete and unified descriptions of many types
of events. Here we outline briefly the main features of four of the more widely
used programs and refer the interested reader to the detailed descriptions in the
program manuals and our earlier discussion.

PYTHIA (Sjöstrand *et al.*, 2001), which subsumes JETSET, is a general
purpose event generator that has grown out of the well developed Lund string
model, which provides the default hadronization scheme. It contains a wide range
of hard subprocesses and relatively elaborate models for soft physics. These typically come with many options and parameters. The basic parton cascade uses
virtuality ordering with colour coherence imposed in the time-like cascades via
a veto on opening angles. In addition, the nearest neighbour, intrajet, spin correlations are included.

ARIADNE (Lönnblad, 1992) only provides pQCD cascades, using the colour
dipole model; see Section 4.2.6. This gives an automatically coherent and unified
treatment of initial and final state cascades which merge naturally with string
hadronization. It must be interfaced to another generator, such as PYTHIA, to
handle the hard subprocess, hadronization and particle decays.

HERWIG (Corcella *et al.*, 2001) is a general purpose event generator which
places its emphasis on the perturbative description of an event. It uses comparatively sophisticated parton showers that build in colour coherence automatically
via ordering of suitable evolution variables. The first branching matches the exact
LO result and in the semi-inclusive, $x \to 1$, limit the algorithms are accurate to
NLO, which allows Λ_{MC} to be related to $\Lambda_{\overline{MS}}$; see Ex. (4-6). Angular correlations
are also fully included. Hadronization uses a cluster model.

ISAJET (Paige and Protopopescu, 1986) is a more basic event generator. It
employs relatively primitive incoherent parton showers and independent hadronization. It allows fast studies using its wide range of hard subprocesses.

4.3.1   *Using event generators*

Monte Carlo event generators for the simulation of hadronic events are based
on QCD. Therefore they should, in principle, have only seven free parameters:
the strength of the coupling and the six quark masses. However, the need to
use perturbative approximations and non-perturbative models introduces many
more parameters. This raises the questions of how to select these input values
and what systematic errors to put on the subsequent predictions.

Input parameters are usually selected by either accepting the default values, which may well be mere 'guestimates' or, better, by tuning the program to
experimental data. Tuning a Monte Carlo is both something of an art and com-
putationally expensive, so it is not done lightly. The basic method is to first select
a set of data to fit and related parameters to tune. For example, the data points

may include the fraction of hadronic Z decays with Thrust T (see Section 6.2) in the range $T \in [0.90, 0.95]$, or the number of K$^+$ particles with momentum fraction in the range $x_p \in [0.01, 0.02]$ etc. Next, by estimating starting values and rough ranges, one can create a grid of values in parameter space. Second, at each grid point the Monte Carlo is run and a measure for the difference between simulation and experimental data, usually the χ^2, is evaluated for all the data points; sufficient events should be generated so that the statistical error on the Monte Carlo prediction is negligible compared to the uncertainties of the data. For a single data point, the χ^2 is defined as the normalized square of the differences between the data and prediction, $(D - T)^2/\sigma^2$. It measures the quality of the fit in units of the 'error' squared (Cowan, 1998). Given the total χ^2 value on the grid, an interpolating function can be constructed to enable a Monte Carlo prediction to be made for arbitrary parameter values. Third, the total χ^2 is minimized as a function of the parameters. This should result in a set of 'best-fit' parameter values and their one-sigma errors.

Lest the above procedure seem too straightforward, you are reminded of the large size of the data samples, the large number of experimentally measured distributions and the size of the parameter space. To this should be added potential complications arising from correlations between parameters and possible instabilities in the minimization. Part of the art is to know which subset of parameters to focus attention on and which subset of the data to fit.

To date, the most complete parameter tunings have been based on LEP data, where typically a two-tier approach has been taken in order to make the task tractable (ALEPH Collab., 1998a; DELPHI Collab., 1996a; OPAL Collab., 1996a). A first tuning concentrates on the 'major' parameters, such as Λ_{MC}, the shower cut-off(s) and the principle hadronization model parameters, a, b and σ (JETSET), or CLMAX and CLPOW (HERWIG). Experience shows that these parameters control semi-inclusive observables, such as event shape variables, jet rates and the momentum spectrum. This largely fixes the partonic phase of the fragmentation and an event's global structure. A second tuning is then carried out for the 'minor' parameters, mainly associated with the hadronization model. Exclusive measurements of identified particle production rates and spectra are added to the fit and an enlarged tuning repeated. In this way, remarkably good descriptions of the LEP1 data have proved possible (Altarelli et $al.$, 1996). Tunings based on the HERA data are also available (Doyle et $al.$, 1999).

Two standard procedures for assigning errors to a Monte Carlo prediction are either to vary some input parameters between 'reasonable limits', or to quote the difference between two independent programs. In both cases, the Monte Carlo's predictions are only considered acceptable when they lie within one or two standard deviations of the experimental data. These procedures are somewhat ad hoc and have proved insufficient, such as for example, in the LEP1 electroweak precision measurement of the hadronic forward-backward asymmetry, determined using jet charges (ALEPH Collab., 1996a). The forward-backward asymmetry is the difference in probabilities for the quark in an e$^+$e$^-$ → q$\bar{\text{q}}$ event to be

produced in the same or opposite hemisphere as the incoming electron. It is di-
~~rectly related to the electroweak mixing angle $\sin^2 \theta_W$. A jet charge is essentially~~
a momentum weighted sum of the track charges in a jet. By using a Monte Carlo
model, the jet charges can be correlated to the direction of travel of the primary
$q\bar{q}$-pair coming from the Z decay. To quantify this correlation the following more
sophisticated approach was used. Distributions were identified that involved the
same physics and Monte Carlo model assumptions as affected the jet charges:
$\pi^{\pm}$, $K^{\pm}$, p, Λ, ρ, η, K^*, ... momentum spectra, baryon-antibaryon correlations,
etc. The Monte Carlo parameters were then varied, subject to still describing
the 'constraint distributions', and the allowed range of values of the correlation
measured. By using an understanding of the measurement and the Monte Carlo's
workings, attention was focussed upon relevant parameters, whose values were
then constrained by demanding accurate descriptions of related measurements.
Of course, if the Monte Carlo proves incapable of describing the data and the
constraint distributions, it would be foolish to use it to make any inferences from
the data.

We leave the reader with a very important caution.

Warning

*Monte Carlo event generators are complicated programs that will
almost inevitably contain bugs, incorrect assumptions and ill-
chosen parameters. It is therefore vital that a user does not take
any results at face value. As a minimum at least two completely
independent programs should be used in any physics study.*

Exercises for Chapter 4

4–1 For a space-like branching, $p_a + p_{a'} \leftarrow p_b$, see Fig. 4.1, show that

$$0 \leq \frac{z}{(1-z)} q_\perp^2 = -t_b + \frac{t_a}{z} - \frac{t_{a'}}{(1-z)} \,,$$

where t is the parton virtuality and z is the light-cone momentum fraction $(E + p_z)_a = z(E + p_z)_b$. If limits are placed on the
daughter virtualities, $t_{a'} > t_0^{a'}$ and $t_b > t_s^b$, associated with the
non-perturbative physics of the p.d.f.s, what is the allowed range
$z \in [z_{\min}, z_{\max}]$? If a branching occurs, what are the upper limits on
the virtualities t_b and $t_{a'}$?

4–2 An alternative form for the space-like Sudakov form factor is given
by (Sjöstrand, 1985):

$$\Pi'_a(x, t, t_s) = \exp\left\{ -\int_{t_s}^t \frac{\mathrm{d}t'}{t'} \int_x^{1-\epsilon_{a'}(t')} \frac{\mathrm{d}z}{z} \bar{P}_{ab}(z) \frac{f_{b/h}(x/z, t')}{f_{a/h}(x, t')} \right\} \,.$$

By comparing derivatives, show that $\prod'$ is equivalent to eqn (4.18). What is its physical interpretation?

4–3 Consider two colour-connected hard partons, i and j, separated by an angle θ_{ij} and suppose parton i emits a soft gluon k, at an angle θ_{ik}. By considering the wavelength of the gluon compared to the distance between the hard partons at the time of emission, show that the gluon will not resolve the individual charges in the (ij)-system unless $\theta_{ik} < \theta_{ij}$.

4–4 Derive the expression for $\cos\theta_{jk}$ and hence confirm eqn (4.23) and the result for its azimuthal average, eqn (4.25). You may find it helpful to transform an integral of the form $\int d\phi (A - B\cos\phi)^{-1}$ into a contour integral by means of the substitution $z = e^{i\phi}$.$(\bigstar)$

4–5 Repeat the derivation of eqn (4.31) for the colour neutral three-parton system $(ij)l$.

4–6 In the limit $z \to 1$ the two-loop splitting function for $q \to qg$ takes the form

$$\frac{\alpha_s}{2\pi} P_{qq}^{(2)}(z) \approx \frac{\alpha_s}{2\pi} \frac{2C_F}{1-z} \left\{ 1 + \frac{\alpha_s}{2\pi} \left[-\frac{(11C_A - 4T_F n_f)}{6} \ln(1-z) \right.\right.$$
$$\left.\left. + C_A \left(\frac{67}{18} - \frac{\pi^2}{6} \right) - T_F n_f \frac{10}{9} \right] \right\}.$$

By changing the argument of $\alpha_s(Q^2)$ and rescaling Λ show how the leading order term can also reproduce the next-to-leading order term in this expression.

4–7 Show that in the limit $x_1 \to 1$ and with $x_3 = z$, eqn (4.36) yields the conventional expression for a branching in terms of Altarelli–Parisi splitting kernels.

4–8 Calculate the masses of the two daughters in a dipole branching in terms of $k_{1\perp}$ and y_1 and hence the bound on $k_{2\perp}$. Show that these masses correspond to the apexes of the two triangles shown in Fig. 4.5.

4–9 In the colour dipole model branchings are ordered in $k_\perp^2$, that is vertically in Fig. 4.5. How does ordering in virtuality, Q^2, or $t = E^2\zeta$, see the discussion below eqn (4.33), proceed in Fig. 4.5?

5

EXPERIMENTAL SET-UP

In order to study QCD or in general the interactions between the fundamental constituents of matter, one has to go to the highest possible energies. From what has been discussed before, this may be evident for the case of perturbative QCD, where the colour charge, as for example in deep inelastic scattering, becomes only visible when probing the nucleon at a scale significantly below 1 fm. It equally holds for weak interactions, where the fundamental scale as set by the masses of the gauge bosons is in the 100 GeV range, as well as for electromagnetic interactions. In the latter case high energy processes allow sensitive tests of the theory at the level of quantum corrections and to probe the structure of the electroweak unification, which is very difficult at lower energies. In this chapter, we will discuss some aspects of the technological challenges involved in doing experiments at high energies, both from the accelerator and the detector point of view.

5.1 Accelerators

In collisions between fundamental particles the entire C.o.M. energy is available to study the interactions between the basic constituents of matter. Ideally, one thus would like to accelerate particles which have no internal structure and make them collide.

This may sound obvious, but it is certainly worthwhile to think a bit more about the concept, since from an operational point of view it does not matter whether a particle is really fundamental or built from more basic constituents, provided the binding energy per constituent is significantly larger than the kinetic energy.

To illustrate the point, let's start with the case of a proton, built out of three valence quarks which carry about half of the momentum of the particle. Since the binding energy of those quarks is on the order of hundreds of MeV, a 20 MeV proton, which provides a convenient probe in nuclear physics, can be viewed as a fundamental particle. This is no longer the case for 315 GeV protons stored and collided with antiprotons of the same energy in the CERN *Super Proton Synchrotron* (SPS). Instead of 630 GeV the effective C.o.M. energy in most collisions between constituents of the proton and the antiproton is much lower, on average even below 100 GeV. Still, it was sufficient for the discovery of the W and Z bosons and at the time the only technologically feasible way to reach the required C.o.M. energies in collisions between fundamental particles.

The apparent drawback that in hadron colliders the beam energy is shared between the partons, thereby leading to a reduced effective C.o.M. energy, can also
be turned into an asset. High luminosity and a wide spectrum of effective collision energies make hadron colliders ideal 'discovery machines'. Once the energy
scale for a particular phenomenon is known, one would of course prefer a well
defined C.o.M. energy in order to be able to perform precision measurements.
Here e^+e^- colliders are usually the machines of choice.

Another instructive example is the case of an ordinary golf ball. With a typical
mass around 45 g and a density close to that of water, one can estimate that
it is a compound of something like 10^{24} individual molecules, held together by
intermolecular forces. Hit with an initial velocity of 60 m/s the ball has a kinetic
energy of $E = mv^2/2 \approx 81$ J or $5 \cdot 10^{20}$ eV. The kinetic energy per molecule thus
is only 0.5 meV, which would be available when colliding two such balls head-on.
Since at room temperature the binding energy per molecule must be larger than
the thermal energy of $kT = 25$ meV, it follows that under the conditions of a
game of golf, the ball will indeed behave like a fundamental object.

Continuing this type of experiments and hitting the ball ever harder, one
would eventually realize that it has more fundamental building blocks. It is made
from molecules, and what appeared to be a high energy interaction between
two golf balls was in fact a low energy interaction at the level of the molecules.
Repeating these steps would in turn reveal that the molecules are made of atoms,
that the atoms have an electron cloud and a nucleus, and that the nucleus is a
bound system of protons and neutrons which in turn consist of quarks. With
electrons and quarks, according to today's knowledge, the fundamental level is
reached, that is, even the highest energy interactions show no evidence for a
substructure in those particles.

Whether one uses composite or pointlike particles depends on the scope of the
experiment. Taking into account that an efficient acceleration requires charged
stable particles, and that free quarks do not exist, the obvious candidates are
protons, electrons and their antiparticles. Also heavy ions are of interest for
studies of matter under extreme conditions, such as for the case of the quark–
gluon plasma. At extremely high energies also muons live sufficiently long to be
of interest. A muon collider, however, is still beyond the capabilities of today's
technology, although numerous studies are under way.

The standard procedure for the study of fundamental interactions is thus
to accelerate the most elementary particles which are available to the highest
possible energies. In addition, the interaction rate has to match the requirements
of an experiment. The relevant quantities in a collision of elementary particles
are the C.o.M. energy, that is, that part of the total energy of a collision which
does not go into the kinetic energy of the entire system and the luminosity $\mathcal{L}$,
defined as the proportionality factor between cross section and event rate,

$$\frac{\mathrm{d}}{\mathrm{d}t} N = \mathcal{L}\sigma .$$
(5.1)

5.1.1 *Accelerator systems*

The most efficient way to maximize the C.o.M. energy is the use of colliding beams rather than fixed target operations. In case antiparticles can be produced abundantly, an elegant technology to achieve this is via particle–antiparticle collisions inside a storage ring. Because particles and antiparticles have the same mass and opposite charge, they can travel in opposite direction through the same accelerator structure. This concept was realized for example in the TEVATRON at FERMILAB, the CERN SPS collider or the *Large-Electron-Positron* collider (LEP). The technical realization in all cases requires a rather intricate system of accelerators, which shall be illustrated here at the example of the CERN set-up.

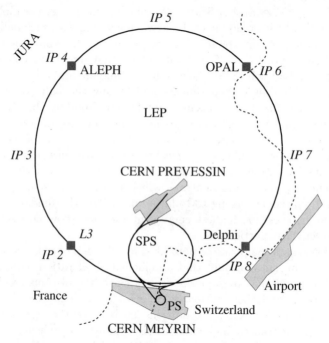

FIG. 5.1. Simplified view of the CERN accelerator complex. The LEP ring has a circumference of approximately 27 km. Figure from Müller(2000).

A simplified view of this system is sketched in Fig. 5.1, showing the *Proton Synchrotron* (PS), the SPS and the LEP ring. The PS as the oldest ring was originally built to accelerate protons to an energy of typically 25 GeV. In order to achieve higher energies the SPS was constructed to go up to 450 GeV beam energy, and is still in use to support a fixed target physics program. It can also be operated as a proton–antiproton collider and served as a pre-accelerator for the LEP, where four big experiments were located to study the physics of e^+e^- annihilations at C.o.M. energies from the scale of the Z-resonance to $\sqrt{s} >$

200 GeV. In addition, both the PS and the SPS are used to supply test beams for research and development work on future projects. The whole system is rather flexible, supporting physics with a large variety of beams and a wide range of energies. One has the opportunity to work either directly with the particles from an accelerator or, alternatively, with secondary particle beams which are created in fixed target interactions of the primary beams. It is clearly beyond the scope of this book to describe all these possibilities in detail. Below we will focus on the general steps needed to reach the highest energies. Some key parameters of LEP, as given in Table 5.1, should convey an impression of what is technologically feasible.

Table 5.1 *Characteristic parameters of the* LEP *storage ring*

Parameter	Value
Circumference	26658.88 m
Magnetic radius	3096 m
Revolution frequency	11245.5 Hz
RF frequency	352 MHz
Injection energy	approx. 20 GeV
Achieved peak energy per beam	104.5 GeV
Achieved peak luminosity	4 pb^{-1}/day
Number of bunches	4, 8 or 12
Typical current/bunch	0.75 mA

At the beginning of any accelerator complex is a source of particles which are to be accelerated. This is simple in the case of electrons or protons which can be obtained from either a heated filament or from ionized hydrogen, respectively. More recent technologies are based on laser ionization or *radio-frequency* (RF) guns, which also allow to prepare polarized beams. The first acceleration step is usually electrostatic, like in a TV set with a cathode ray tube. The particles pass through some high-voltage difference which brings them up to a speed where an RF-system can take over by which the beam is accelerated through the electric field inside a cavity. Both standing wave and travelling wave devices are used. Since the energy transfer can be efficient only if the phase velocity of the wave is similar to the speed of the particle, this principle works best for relativistic beams.

So far only particles are involved. For a proton–antiproton or an e$^+$e$^-$ collider, however, one also needs the corresponding antiparticles. These can be created by directing a primary beam of sufficient energy on a fixed target. Amongst the reaction products there will also be the respective antiparticles. Although by no means trivial, it is today a well understood technology to collect positrons or antiprotons from such a reaction and feed them into an accumulator ring which collects the antiparticles and reduces their large initial momentum spread to the

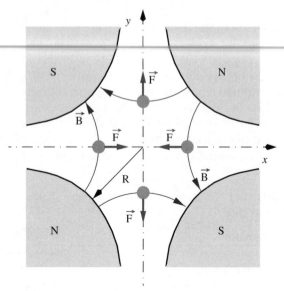

FIG. 5.2. Schematic view of a quadrupole magnet. Also shown are the forces on
    positive particles in the region between the pole faces moving into the plane
    of the drawing. Figure from Müller(2000).

level where they form a well collimated beam. When enough current has been
accumulated, the beams are injected into further accelerator stages and finally
filled into the same storage ring as the particles.

Large storage rings such as the LEP or the SPS operate as a synchrotron,
which allows not only to store particles but also to accelerate them further
and feed back energy losses. For the LEP this continuous acceleration is vital, since electrons and positrons on a circular orbit emit synchrotron radiation. In natural units the power radiated by a relativistic particle is given by
$dE/dt = \alpha_{em}2p^4/3m^4\rho^2$, where p is the momentum of the particle, m its mass
and ρ the magnetic radius of the ring. In more familiar units the energy loss per
turn of an electron thus becomes $\Delta E = 0.0885(E/\text{GeV})^4/(\rho/\text{m})$ MeV. At the
highest LEP energies this corresponds to an energy loss of over 3 GeV per turn,
which has to be fed back by the RF-system. Without an accelerating structure
beams of light particles would thus be quickly lost. As a side effect, the RF-
acceleration in a synchrotron leads to a bunched beam, which needs to be taken
into account when designing an experiment. The obvious advantage is that the
timing of possible interactions is precisely known, a disadvantage may be that
in case of high luminosity running one has to cope with simultaneous multiple
interactions. High luminosities require large beam currents and small transverse
dimensions in the interaction points, which are achieved by the layout of the
elements which guide the beam through the machine. Some basics of these beam
optics are discussed in the next section.

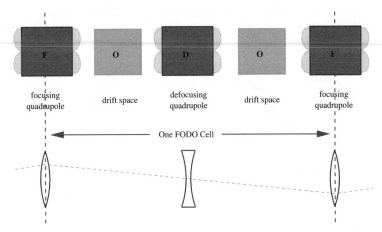

5.1.2 *Beam optics*

The beams are kept on their orbit by dipole magnets. Since the bending radius
is proportional to the momentum, it follows that only particles with an energy
close to the nominal energy would have a stable orbit in a ring consisting entirely
of dipole magnets. All others would be lost because they are deflected either too
little or too much by the bending field. Because of synchrotron radiation and
other imperfections there is always some energy spread in the beam. One thus
has to deal with particles which may have a substantial momentum deviation,
that is, a focusing mechanism is needed to bend the particle back into the acceptance of the accelerator ring. This is achieved by quadrupole magnets (Fig. 5.2),
where the magnetic flux density B grows proportional to the distance from the
centre. Particles on the nominal orbit do not see a field, but particles with an
offset are deflected towards the centre or away from it. In such a system particles
with a momentum offset start oscillating around the orbit of a nominal particle.
Unlike an optical lens, which usually is focusing or defocusing horizontally and
vertically, a quadrupole field which is focusing in the horizontal plane is defocusing in the vertical one, and *vice versa*. Still, combining a focusing and defocusing
quadrupole with a drift space in between, it is possible to achieve a net focusing. Intuitively this can be understood by the fact that a particle which first
has a large offset in the focusing quadrupole sees a strong field which deflects it
towards the centre. Therefore the defocusing magnet is traversed closer to the
centre where the field is weaker. The defocusing is thus not as large and the net
effect is a focusing of the beam. A similar argument holds if the first quadrupole
acts as a defocusing lens. The beam then is directed outward and enters the
focusing quadrupole at a larger distance from the centre. The focusing is again
stronger than the defocusing and the whole system performs a focusing of the

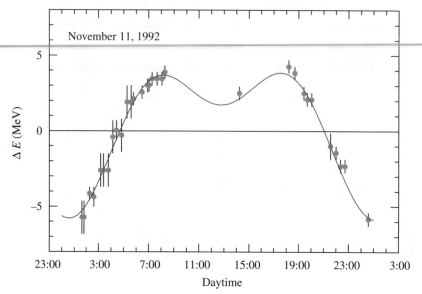

FIG. 5.4. Effect of terrestrial tides on the LEP beam energy. Figure from
Müller(2000).

beam. Figure 5.3 shows the schematic layout of such a FODO-structure, where
F stands for 'Focusing', D for 'Defocusing' and O for the drift space between two
magnetic lenses.

Accelerators and storage rings are complex devices, which must be able to
control a particle beam with high precision over large distances. For the LEP one
is faced with the situation that the beam energy is very sensitive to the effective
radius of the ring. As a consequence it turned out, somewhat surprisingly at
the time, that the LEP beam energy is affected by such effects as terrestrial
tides, the level of the water in lake Geneva or the time-table of the French–Swiss
railroad network. The first two effects act through a slight deformation of the ring
through the elastic forces on the geological environment of the region. Because
of these deformations, the average LEP beam deviates from the nominal orbit
and sees additional bending fields from the quadrupoles. The so-called *phase-
focusing* (see, for example, Wiedemann 1993,1995) of the accelerator structure
then automatically re-adjusts the beam energy to match the integrated bending
field. An example of how tidal forces created by the sun and the moon affect
the beam energy is shown in Fig. 5.4. The measurements are compared to an
absolute prediction based on the known geological properties of the ground.

Another example for external influences on the LEP beam energy is shown in
Fig. 5.5. On top of a smooth behaviour there are many erratic spikes which in the
beginning were not understood. Also the drift of the beam energy was kind of a
mystery; but let's start with the spikes. A first hint is obtained from the fact that

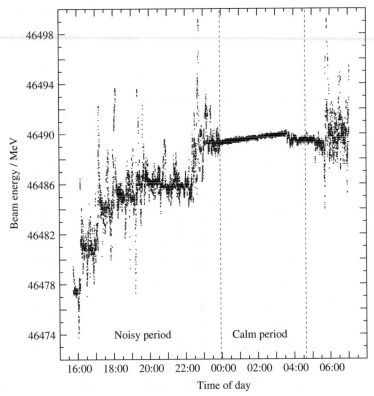

FIG. 5.5. Typical evolution of the LEP beam energy during the night. Figure from Müller(2000).

the situation is very quiet between midnight and 6 AM, that is, it somehow seems to be related to human activities. The solution finally turned out to be related to the high speed train, TGV, from Geneva to Paris, which causes vagabond currents that also reach the LEP tunnel and flow along the beam pipe, thereby inducing additional magnetic fields which affect the beam energy. The individual spikes in Fig. 5.5 could in fact be correlated with the arrival and departure times of the TGV at some regional train stations. Residual magnetizations caused by the currents associated to the spikes finally explain at least part of the drift in the average beam energy.

5.2 Detectors at high energy colliders

A detector for high energy interactions ideally should be able to record all particles including their properties which are created in the reaction. Many excellent books (Blum and Rolandi, 1994; Grupen, 1996; Kleinknecht, 1999) exist which cover the state of the art of this subject in detail. Below, only a brief overview of the most important techniques can be given.

Table 5.2 *Typical masses and lifetimes of elementary particles*

Particle	Mass/GeV	Decay	Approximate lifetimes
$\rho^{0,\pm}$	0.770	Strong	4.4×10^{-24} s
$K^{*0,\pm}$	0.892	Strong	1.3×10^{-23} s
Δ	1.232	Strong	5.5×10^{-24} s
π^0	0.135	Electromagnetic	8.4×10^{-17} s
$\pi^\pm$	0.140	Weak	2.6×10^{-8} s
Λ	1.116	Weak	2.6×10^{-10} s
$K^\pm$	0.494	Weak	1.2×10^{-8} s
K^0_S	0.498	Weak	8.9×10^{-11} s
K^0_L	0.498	Weak	5.2×10^{-8} s
D^0	1.865	Weak	4.2×10^{-13} s
$D^\pm$	1.869	Weak	1.0×10^{-12} s
B^0	5.279	Weak	1.6×10^{-12} s
$B^\pm$	5.279	Weak	1.7×10^{-12} s
$\mu^\pm$	0.106	Weak	2.2×10^{-6} s
$\tau^\pm$	1.777	Weak	2.9×10^{-13} s
n	0.940	Weak	8.9×10^2 s
$e^\pm$	0.0005	Stable	$> 4.3 \times 10^{23}$ years
$p^\pm$	0.938	Stable	$> 1.6 \times 10^{25}$ years
γ	0	Stable	∞

Particles created in high energy interactions can be classified as short-lived resonances, which decay via strong interaction processes on a time scale of 10^{-23} s, particles which decay through electromagnetic forces with typical lifetimes of 10^{-16} s and long-lived particles with weak decays. Table 5.2 shows masses and lifetimes for a selection of particles, which illustrate the hierarchies of lifetimes observed in nature.

Experimentally strong and electromagnetic decays usually cannot be resolved from the creation vertex of the mother particles. Weak decays of relativistic particles, on the other hand, occur after macroscopic flight distances. For heavy flavour particles with lifetimes around 10^{-12} s these are typically in the range of a few hundred micron to a few millimetres, and particles with lifetimes of 10^{-8} s or larger can be considered as absolutely stable since they live long enough to travel distances of many metres before decaying.

An event from a high energy interaction will ultimately consist of long-lived final state particles which traverse the entire detector. For a full reconstruction one has to measure as precisely as possible these particles, both charged and neutral ones. The primary quantities to be determined are charge, momentum and mass of those particles. Knowledge about the precise direction close to the primary interaction vertex then allows to reconstruct heavy flavour decays or decay chains in general, which is important for detailed studies of production processes. The determination of the electric charge is greatly simplified by the fact that all charged long-lived particles carry either plus or minus one positron charge, that is, the direction of curvature in a magnetic field already allows to infer the charge of a particle.

In general, a detector consists of the so-called *tracking devices* which determine momentum and point of origin of charged particles, *calorimeters* to register also neutral particles and measure their energies, and particle identification devices. We will discuss them in turn.

5.2.1 *Tracking detectors*

Tracking devices exploit the fact that charged particles passing through matter create an ionization signal along the trajectory of the particle. It is basically the same principle for cloud chambers, bubble chambers, emulsions or all kinds of electronic tracking detectors. The primary particle passes through the medium and loses some energy which goes into the creation of electron–ion pairs. In a cloud chamber containing supersaturated vapour, the free charges trigger condensation; in a bubble chamber with superheated fluid they create bubbles from local boiling of the fluid. In both cases, the trajectory becomes visible and can be analysed. In photographic emulsions the charges liberated by the passage of a high energy particle lead to photo-chemical reactions that are made permanent by the development process.

The ionization from a charged particle can also be recorded directly by electronic detectors. A typical example is a *multi-wire proportional chamber* (MWPC), where the primary ionization in a gas volume is amplified in the strong electric field around a wire. The amplification starts when electrons approaching the wire are accelerated to the point where they create secondary ionization. Since the secondary charges as well are accelerated in the field around the wire, an avalanche develops which, for not too high fields, is proportional to the primary ionization and large enough to be recorded electronically. Since the primary charge drifts to the closest wire, a signal on a wire yields a measurement of one coordinate on a particle track, with a resolution proportional to the distance of the wires. An ensemble of such chambers allows to sample the entire trajectory and, placed inside a magnetic field, to determine simultaneously the particle's charge and momentum vector.

A modification of the scheme is applied in the so-called *time projection chamber* (TPC). Here, an entire ionization track created inside a drift volume, typically a cylinder, drifts in parallel electric and magnetic fields to an end plate where it is recorded by a MWPC. As before, the signal is enhanced by gas amplification at the wires. To improve the measurement, additional electrodes, so-called *pads*, are placed behind the wires. If a primary charge creates an avalanche on a wire, a signal is also induced on the closest pads. Reading out the pads thus yields a picture of the two-dimensional projection of the track. Recording also the drift time, that is, the time between creation of the primary ionization by a particle track from an interaction and its arrival at the end plate, gives the third coordinate. A TPC thus performs three-dimensional imaging of charged tracks inside a magnetic field. The magnetic field actually serves a double purpose. Evidently, the bending of the tracks inside the field provides the basis for the momentum measurement. In addition, a B-field parallel to the electric drift field reduces transverse diffusion of the electrons, which in strong fields tend to follow the B-field lines, thereby ensuring a high resolution image of the track. As a final point it is worth mentioning that the electric field is adjusted such that the drift velocity for the electrons is maximal. This again has a double benefit since it minimizes the readout time and at the same time renders the drift velocity insensitive against small local variations of the drift field. In other words, setting the drift field to the maximum of the drift velocity guarantees optimal resolution in the longitudinal direction.

While MWPCs can be read out very fast, the event rate of a TPC is limited by the drift velocity of charges. Both types of tracking detectors have their preferred applications, and it is the experimental environment which drives the choice of technology. For e^+e^- physics TPCs are excellent detectors, at high luminosity hadron machines the high event rate would render them useless. They also cannot be used to provide information for fast trigger decisions. For these kinds of application conventional wire chambers or similar technologies have to be used.

The types of tracking detectors discussed so far are usually large devices which have to follow a track over a long distance in order to perform precise measurements of the deflection of high energy particles inside magnetic fields. Another task which is vital for the reconstruction of secondary vertices, for example from heavy flavour decays, is the precise determination of the track direction and position close to the primary vertex. The high resolution needed for this task exceeds that of conventional wire chambers by a factor of up to 100. It is reached with silicon strip or pixel detectors. Again, a charged particle passing through the detector ionizes the material. In case of silicon, electron-hole pairs are created, and diode structures implanted on the wafer collect the produced charges very much like the readout wires of a MWPC. Since the energy loss in solid state material is much larger than in gases, even in relatively thin detectors of typically 300 μm thickness a minimum ionizing track generates sufficient charge for direct readout by a charge-sensitive amplifier. No further amplification is required.

5.2.2 Calorimeters

Neutral particles have to be detected by a different technology called *calorimetry*. The basic idea is to put a sufficient amount of absorber material for the particles to interact with the detector such that they deposit their entire energy. The energy deposit then is converted into a proportional signal which can be read out electronically. With proper calibration a calorimeter thus is able to measure the energy of an incident particle. Arranged in projective segments pointing towards the interaction point, it also measures the direction of a neutral particle and thus allows to infer its momentum vector.

Two types of calorimeters are commonly employed: electromagnetic and hadronic calorimeters. Although both follow the same basic principle, the actual realizations differ considerably. An electromagnetic calorimeter is designed to absorb photons. At high energies the energy deposition mainly starts with pair creation in the strong electric field around a heavy nucleus, where the energy of the incident photon is converted into an e^+e^- pair. These particles in turn radiate when passing close to another nucleus and the bremsstrahlung photons emitted in the process can again be converted into new e^+e^- pairs. A so-called *electromagnetic shower* develops, where the initial energy is successively distributed over a large number of secondary particles. The shower development stops once the energy per particle falls below the threshold for e^+e^- pair-creation. The characteristic length describing the shower development is the so-called *radiation length*, X_0, defined as the mean distance over which a high energy electron loses all but a fraction $1/e \approx 0.368$ of its energy by bremsstrahlung. The radiation length as an atomic quantity is usually given in units of g/cm^2. Dividing by the density of the actual medium yields X_0 in units of cm. A rough estimate, which is correct to better than 5% for all elements, is given by

$$X_0 = \frac{1433A}{Z(Z+1)(11.4 - \ln(Z))} \frac{\mathrm{g}}{\mathrm{cm}^2}. \tag{5.2}$$

Here A is the atomic number and Z the charge of the nucleus. As a consequence of the electromagnetic nature the dominant quantity for X_0 is the charge Z. The most common choice is lead, with a radiation length $X_0 = 6.37$ g/cm^2 or, after dividing by the density, $X_0 = 0.56$ cm.

The total number of particles in the shower is proportional to the energy of the incident particle. Since the number of particles in the shower grows exponentially with the length, it follows that the length of a shower is proportional to the logarithm of the energy of the primary particle.

In a calorimeter the measurement of the energy of an incident particle is transformed to a measurement of the number of secondary particles created in the shower. This can be achieved in a variety of ways. One possibility is to use wire chambers and measure the ionization signal created by the particles in the shower. Another technique is to use scintillator materials which emit light on the passage of a charged particle and record the light signal by means of photomultipliers. In both cases the calorimeter is built in a sandwich-like structure with alternating layers of absorber and detector material. Such a design always is a compromise between high density, which is needed for the shower evolution, and sensitivity, since particles produced inside the shower should not be reabsorbed by the passive material before having passed through some sensitive areas. A more favourable solution would be to use homogeneous calorimeters, such as lead-glass blocks or high-Z scintillator crystals. In both cases the entire volume acts as shower medium. In lead-glass light is created in the form of Cherenkov radiation from the secondaries, crystals such as BaF_2, BGO, NaI(Tl),

Table 5.3 *Characteristics of some materials commonly used in calorimeters*

Element	Z	A	$\rho/(\mathrm{g/cm^3})$	X_0/cm	λ_I/cm
Liquid Argon	18	39.9	1.40	14.0	75.7
Iron	26	55.8	7.87	1.76	16.8
Lead	82	207.2	11.35	0.56	17.1
Uranium	92	238.0	18.95	0.32	10.5

CsI(Tl), PbWO$_4$ emit scintillation light which can be detected with photomultipliers. Another kind of homogeneous calorimeters uses liquid Argon or Krypton as absorber material. The noble gases are particularly well suited for calorimeters because the filled electron shells do not absorb electrons created in the shower. Electrodes immersed in the liquid are used to read out the ionization signal. While a liquid Argon calorimeter still requires lead plates to get a sufficiently short radiation length, Krypton already has sufficiently high Z to act as an effective radiator. It is thus possible to sample a shower longitudinally, which makes it a very fast detector with excellent energy and position resolution. The typical energy resolution that can be achieved with large electromagnetic calorimeters is in the region around $\sigma(E)/E = 0.10/\sqrt{E/\mathrm{GeV}}$.

The above discussion was mainly dealing with electromagnetic calorimeters, but the basic principles of course are equally valid for hadronic calorimeters. The main difference is the type of interaction which gives rise to the shower development. Electromagnetic calorimeters for the detection of photons or electrons, which after the first step develop an identical shower, are based on a cascade of bremsstrahlung and pair creation. Bremsstrahlung is favourably produced by light particles in strong electric fields, which is the reason why electromagnetic calorimeters are built preferably from high-Z radiator material.

Heavier particles, such as muons, pions, kaons or protons have a very small cross section for bremsstrahlung and thus hardly shower in a calorimeter which readily absorbs a high energy photon or electron. Hadrons, however, can start a shower based on nuclear interactions. The length scale governing the evolution of a hadronic shower is the interaction length λ_I, which corresponds to the mean free path of a hadron before an inelastic interaction with a nucleus. A rough approximation for λ_I is given by

$$\lambda_I = 35 A^{1/3} \frac{\mathrm{g}}{\mathrm{cm^2}} \,. \tag{5.3}$$

As strong interactions are insensitive to the charge of the nucleus, the interaction length only depends on the number of nucleons for a given element and scales with the radius of the nucleus, that is, proportional to the distance travelled inside nuclear matter. The above definition of λ_I again defines an atomic quantity. Dividing by the actual density of the medium yields the interaction length in units of cm.

The characteristics of some materials which are commonly used in calorimeters are listed in Table 5.3. It is evident that lead is one of the preferred materials for electromagnetic calorimeters. Interestingly, for hadronic calorimeters iron is better suited than lead. It is also less expensive and in addition easier to handle when building large devices. An electromagnetic calorimeter with a thickness of ten radiation lengths amounts to 5.6 cm lead, which corresponds to only 30% of an interaction length, that is, a hadronic shower at most would just start inside such a calorimeter, while a 100 GeV electromagnetic shower would almost be completely absorbed.

Another interesting material for building calorimeters is Uranium. Because it is difficult to handle, while having only a marginally smaller radiation length than lead, it is not very attractive for electromagnetic calorimeters. It is, however, a good option for hadronic calorimeters. In contrast to electromagnetic showers, hadronic showers also deposit a sizeable amount of energy into the creation of slow neutrons or other nuclear fragments, which usually do not create a signal that can be read out. This invisible energy and the larger fluctuations in the shower evolution result in a rather bad energy resolution. There is no way to reduce the fluctuations in the shower, but it is possible to recover some of the invisible energy. One possibility would be to use active materials that also are sensitive to neutrons, like hydrogen-rich scintillator materials. Fast neutrons then could hit protons from hydrogen nuclei which in turn would produce scintillation light. In Uranium calorimeters, on the other hand, slow neutrons can induce fission reactions that liberate extra energy and thus compensate for the components that otherwise would be lost. While conventional hadronic calorimeters yield energy resolutions around $\sigma(E)/E = 0.7/\sqrt{E/\text{GeV}}$, compensating devices using Uranium were shown to be up to a factor of two more accurate.

So far we have discussed electromagnetic calorimeters for photons or electrons and hadronic calorimeters for the detection of all kinds of charged and neutral hadrons. While the charged particles are also seen by tracking detectors, calorimeters are the only efficient way to measure neutral particles. As such, they are also employed as neutrino detectors. Because of the small cross section for neutrinos, really massive devices are required even if one is content to see only a tiny fraction of the total flux. A good example for such a detector was the CDHS-experiment (see, for example, (CDHS Collab., 1982)) which consisted of magnetized iron toroids interleaved with scintillation counters and drift chambers having a total mass of 1400 tons. The iron served a double purpose as target material for the neutrinos and as showering medium for hadronic showers from remnants of a struck nucleus. The scintillators then allowed to measure the energy in the shower. For neutral current events where the neutrino–nucleon interaction proceeds through the exchange of a Z boson, an isolated hadronic shower inside the detector is all that is seen. In case of charged current reactions, where an incident muon-neutrino is transformed into a muon, one observes a final state muon in addition. The muon is too heavy to initiate an electromagnetic shower. Since it does not feel the strong nuclear force either, it also cannnot create a hadronic

shower and simply penetrates the detector as a minimum ionizing particle. Because of the high energy of the incident neutrino, such a muon can travel many
meters even through a massive detector, where its momentum is determined from
the track curvature in the magnetic field as measured by MWPCs.

5.2.3 *Passage of particles through matter*

Many of the methods to determine the momentum of particles created in high
energy interactions exploit that charged particles ionize the medium they are
traversing. What we have neglected so far is that the passage through matter
also affects the particle.

Evidently, the particle must lose energy due to the creation of electron–ion
pairs. As a rule of thumb one can assume an excitation energy

$$I \approx (10 \pm 1) \times Z \text{ eV} , \tag{5.4}$$

where Z is the charge of the respective element. The average energy loss per unit
length is described by the Bethe–Bloch formula

$$\frac{\mathrm{d}E}{\mathrm{d}x} = 0.3071 z^2 \frac{Z}{A} \frac{1}{\beta^2} \left[\ln \frac{2m_e c^2 \beta^2 \gamma^2}{I} - \beta^2 - \frac{\delta}{2} \right] \frac{\text{MeV cm}^2}{\text{g}} . \tag{5.5}$$

In this expression $\mathrm{d}x$ is measured in mass per unit area. To convert to centimetres
one has to multiply with the density of the medium. The quantity z is the charge
of the incident particle in units of the proton charge, $\beta = v/c$ and $\gamma = 1/\sqrt{1 - \beta^2}$
are the usual relativistic variables.

Equation (5.5) shows that for a given particle the energy loss is only a function
of the particle velocity. For small velocities it is dominated by the $1/\beta^2$ term,
which takes into account that the energy loss is larger if more time is available
for an interaction between the particle and the medium. With increasing energy
the specific energy loss drops until it reaches a minimum before starting to rise
logarithmically proportional to $\ln \gamma \beta$. This so-called *relativistic rise* is caused by
the transformation of the electric field of the incident particle, which flattens
and extends as the particle becomes relativistic, so that the contributions from
distant collisions increasingly contribute to the energy loss. Since real media
become polarized, the extension of the field is attenuated, which is described by
δ, the so-called *density effect*. At very high energies one has

$$\frac{\delta}{2} \simeq \ln \frac{28.8 \text{eV} \sqrt{\rho(Z/A)}}{I} + \ln \beta \gamma - \frac{1}{2} \tag{5.6}$$

with the density ρ of the medium in units of g/cm^3.

An important observation is that the energy loss for charged particles in
matter has a minimum. Depending on the material the value is typically in the
range $\mathrm{d}E/\mathrm{d}x_{\min} \sim 2$ MeV cm^2/g. Since the rise is only logarithmic with the
particle energy, this minimum constitutes the typical energy loss for secondary
particles from high energy interactions.

Another effect is elastic scattering of charged particles off atoms in matter. Here the main effect comes from the electric field of the nucleus, which reaches much further than the strong nuclear force and therefore dominates all multiple scattering. Like in the case of electromagnetic showers the atomic quantity to parameterize multiple scattering is the radiation length X_0. Small angle scattering in a plane is well described by a Gaussian distribution of width Θ_0,

$$\Theta_0 = z \frac{13.6\,\text{MeV}}{\beta c p} \sqrt{\frac{x}{X_0}} \left[1 + 0.038 \ln \left(\frac{x}{X_0} \right) \right] . \tag{5.7}$$

As before z is the charge of the particle in units of the proton charge, βc and p are the velocity and momentum of the particle and x/X_0 the thickness of the scattering medium in units of radiation lengths. Scattering angles greater than a few Θ_0 have to be described by Rutherford scattering, which enhances the tails of the distribution compared to a Gaussian.

5.2.4 *Particle identification*

In addition to the determination of the momenta of all final state particles one also would like to determine the particle type. For all practical purposes this is equivalent to a measurement of the particle mass. The direct methods to achieve this are based on different ways to measure the velocity of a particle. The combined information from velocity and momentum then allows to extract the particle mass.

The most straightforward way to determine the velocity is via a simple *time-of-flight* (TOF) measurement. The method is limited by the fact that the velocity of all relativistic particles approaches the velocity of light, c. In typical applications the relevant observable thus is the difference to c rather than the velocity, which puts high demands on the time resolution of a TOF-system.

Another way to determine the velocity is from the specific energy loss of a charged particle in matter. From eqn (5.5) one sees that for a given charge dE/dx is only a function of the velocity. Particle identification via dE/dx is an attractive possibility since it also works for highly relativistic particles by exploiting the relativistic rise in the specific energy loss. It can be performed simultaneously with the momentum determination in a tracking system by measuring the ionization charge together with the position. A potential problem of the method is related to the fact that ionization energy loss is subject to large fluctuations. One therefore needs many samples along a track in order to get a reliable estimate for the average energy loss dE/dx. Particle identification via dE/dx therefore is preferentially done in detectors with large drift chambers and TPCs.

The velocity of a particle can also be measured by means of Cherenkov radiation, which is emitted when a charged particle travels faster than the speed of light inside the medium. A charged particle in matter polarizes the medium. For a slowly moving particle the polarization built up in front of the particle compensates the relaxation behind it. However, if the velocity is larger than the speed of light in the medium, then there is no polarization in front of the particle, simply because the presence of a charge cannot be communicated ahead and the relaxation of the polarization created by the passage of the particle leads to the emission of real photons. The photons are emitted as spherical waves along the trajectory and propagate with the speed of light in the medium. For a straight track they thus are emitted as a cone with opening angle Θ_c,

$$\cos \Theta_c = \frac{c}{n}\frac{1}{v} = \frac{1}{n\beta} , \qquad (5.8)$$

where n is the index of refraction in the medium. The number of photons emitted per length interval dx and energy interval dE of photon energy is given by

$$\frac{d^2N}{dxdE} = \frac{\alpha z^2}{\hbar c}\left(1 - \frac{1}{\beta^2 n^2(E)}\right) \approx 370z^2 \sin^2 \Theta_c(E)\frac{1}{\text{eV cm}} \qquad (5.9)$$

or equivalently

$$\frac{d^2N}{dxd\lambda} = \frac{2\pi\alpha z^2}{\lambda^2}\left(1 - \frac{1}{\beta^2 n^2(\lambda)}\right) . \qquad (5.10)$$

As in the case of the specific ionization, also the number of Cherenkov photons is proportional to the square of the particle charge. The total energy emitted into Cherenkov radiation can be obtained by integrating eqn (5.9) over the entire radiator length x and the photon energy spectrum, which is finite because only those regions contribute where $n(E) > \beta$.

The simplest way to exploit Cherenkov radiation for particle identification is realized by a threshold counter. Particles which are faster than c/n emit light, the others don't. This simple scheme is often already sufficient to discriminate for example between pions and heavier particles. A modification of this approach is a set of Cherenkov counters with different indices of refraction n. With two indices $n_1 < n_2$ one can then distinguish between pions, kaons and protons over certain ranges in momentum. Pions give light in both radiators, kaons only emit light in n_2, while protons emit no Cherenkov radiation at all.

A more elegant way to determine the velocity of any particle which emits Cherenkov radiation is to meaure the opening angle of the light cone. This is done with the so-called RICH, *Ring-Imaging-CHerenkov*, counters. The basic principle is that a focusing mirror or lens images the light from the Cherenkov cone on a ring, whose radius is proportional to the opening angle of the cone. The position of the centre of the ring contains the information about the track direction, which allows to match a RICH ring to a particle registered by a tracking system.

Note that the methods for particle identification discussed so far are usually applied under the assumption that the magnitude of the charge of the particle is the fundamental charge e. This is no problem when dealing with the identification of known particles, but must be taken into account when looking, for example, for free quarks which are expected to carry fractional charges. Just as a reminder, a tracking system determines z/p, a dE/dx measurement yields $z^2 f(v)$ and a TOF or a RICH system gives v. The number of photons recorded in a RICH also is proportional to z^2. Taking information from different systems thus allows a simultaneous determination of p, m and z.

In addition to the active methods of particle identification by means of dedicated measurements, identification is also possible by combining information from different devices. Electrons, for example, are usually identified as charged particles which deposit their entire energy in the electromagnetic calorimeter, charged hadrons are absorbed in hadronic calorimeters and muons are recognized as charged particles with extraordinary penetrating power. They are too heavy to create electromagnetic showers and as leptons do not interact via the strong interaction, and thus even pass through calorimeters just leaving an ionization track.

5.2.5   ALEPH: an example of a LEP detector

A typical detector exploiting the technologies discussed above is shown in Fig. 5.6. It hermetically surrounded one interaction region at LEP, in order to record all particles emanating from an e^+e^- annihilation. Only particles emitted along the beam pipe (1) and non-interacting particles such as neutrinos escape. The inner regions are instrumented with tracking detectors, then come the electromagnetic and the hadronic calorimeter and finally again two layers of tracking chambers to identify high energy muons. The inner tracking system consists of a silicon vertex detector (2) for high precision measurements of secondary vertices from heavy flavour decays. The next layer is a small cylindrical MWPC (3) which provides fast tracking information for triggering. In addition, it serves for extrapolation from the main tracking chamber, a large TPC (5), to the vertex detector. The TPC with a radius of 1.8 m is immersed in a homogeneous solenoidal B-field of 1.5 T created by a superconducting coil (7) which allows precision measurements of the momentum of charged particles. In addition it performs particle identification from dE/dx by sampling the specific ionization up to 180 times along a track. The next detector outside the TPC is the electromagnetic calorimeter (6), which is a lead-MWPC sandwich structure. This calorimeter is still inside the magnet coil (7), which minimizes the number of radiation lengths in front of the calorimeter and thus takes full advantage of the excellent position resolution of the MWPCs. Additional small calorimeters (4) close to the beam pipe are used to measure Bhabha scattering for precision measurements of the luminosity. Outside the electromagnetic calorimeter comes the superconducting coil and then the hadron calorimeter (8). The hadron calorimeter is an iron streamer-tube sandwich structure. The streamer tubes, like MWPCs, utilize gas amplification to detect ionization signals, but in contrast to the former they are not operated in proportional mode but at higher voltage where the signal saturates. This results in large pulses which can easily be read out via induction pads, but yields only digital information. The energy of a hadronic shower thus is proportional to the number of pads above a noise threshold. The iron in the hadron calorimeter

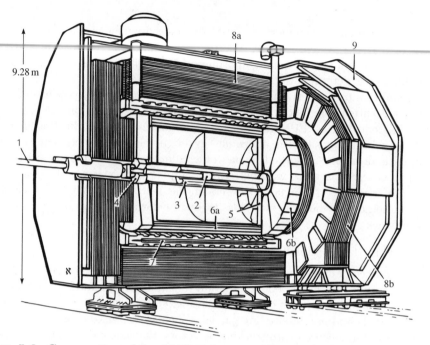

FIG. 5.6. Cross section of the ALEPH detector. The numbers denote the different sub-detectors or parts of ALEPH: (1) beam-pipe, (2) silicon vertex detector VDET, (3) inner tracking chamber ITC, (4) luminosity monitor LCAL,
(5) time projection chamber TPC, (6a) electromagnetic calorimeter ECAL
(barrel part), (6b) ECAL (end-cap), (7) superconducting coil, (8a) hadronic
calorimeter HCAL (barrel part), (8b) HCAL (end-cap), (9) muon chambers.
Figure from Müller(2000).

not only serves as shower medium, but also as mechanical support of the entire
detector and as return yoke for the magnetic flux. It absorbs all remaining particles from the primary interaction except high energy muons, which thus are
identified by the signals they leave in the tracking chambers (9) mounted on the
outside of the detector.

The example of the ALEPH detector illustrates how different detector technologies can be combined to cover the requirements of a typical high energy
physics experiment. It has to be emphasized that not only are the technologies
important, but also the granularity that is needed in order to record reliably
all final state particles from a primary interaction. Each of the major devices
described above has between 10^4 and 10^5 readout channels, and this number is
going to increase for future detectors. The large number of channels matches the
overall complexity of the events that are to be recorded, and it should not come
as a surprise that this number has a tendency to approach the level of biological
sensors, such as, for example, the human eye, which also have to deal with highly

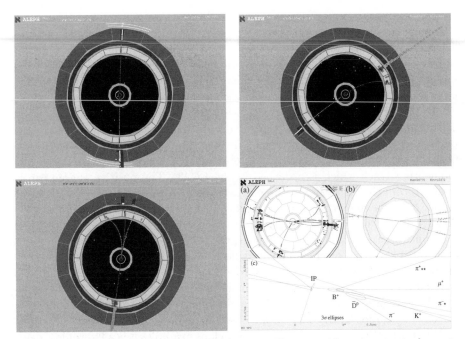

FIG. 5.7. The working of the ALEPH detector illustrated by some typical events as recorded by the experiment.

complex patterns.

The working of the ALEPH detector is illustrated in Fig. 5.7 by *event displays*, which present the information recorded by the detector in a graphical way. The view is along the beam direction. Going from the centre to the outside, one sees the vertex detector and inner tracker, followed by the TPC, ECAL, magnet coil, HCAL and muon chambers. Hits in any of the tracking detectors are marked and connected by the reconstructed tracks. The amount of energy seen in the calorimeters is shown in the form of histograms, where the histogram area is proportional to the energy deposit.

The upper left frame of Fig. 5.7 shows an e^+e^- annihilation into a pair of tau leptons, which both decayed into a muon and two neutrinos that are not seen. Both charged final state particles pass as minimum ionizing particles through the entire system, leaving also hits in the muon chambers outside of the magnet yoke. The event topology is the same as for an e^+e^- annihilation into a $\mu^+\mu^-$ pair, except that in the latter case both muons would have the beam energy. For the event shown, analysis of the track curvature in the magnetic field reveals that the muons are less energetic, which leads to the conclusion that one has actually observed the reaction $e^+e^- \rightarrow \tau^+\tau^-$.

In the upper right an $e^+e^- \rightarrow \tau^+\tau^-$ event is shown, where the τ^+ decays into a positron and two neutrinos and the τ^- into a muon and two neutrinos.

The neutrinos are not seen, the muon again passes through the entire detector. The positron emits most of its energy into a bremsstrahlung photon and thus suffers a strong deflection in the magnetic field, the photon travels in a straight line until it is absorbed in the electromagnetic calorimeter, where it deposits a large amount of energy. The positron of course is also stopped in the ECAL.

The lower left shows another $e^+e^- \to \tau^+\tau^-$ event. Here the τ^+ decays into a positron and two neutrinos; the τ^- decays hadronically into three charged pions and a neutrino. The neutrinos do not interact and escape unseen. The positron as an electromagnetically interacting particle is recognized as a high momentum track which is absorbed in the ECAL. The positive charge can be inferred from the curvature in the magnetic field. Although barely visible in the figure, it can be reliably measured in the TPC which has a momentum resolution $\Delta p/p < 10^{-3}p/(\text{GeV}/c)$. From the three pions from the τ^- decay the two with the same direction of curvature are π^-s, the third one is a π^+. One of the π^-s is of so low energy that after significant deflection in the TPC it is absorbed already in the ECAL. The other two pass through the ECAL and are only stopped in the hadron calorimeter, where they finally deposit their energy.

The lower right frame of Fig. 5.7 finally illustrates the importance of a good vertex detector for the reconstruction of secondary vertices and decays of beauty- and charm-particles. In contrast to the previous examples it is a hadronic Z decay with a so-called *two-jet structure*. Zooming in to the primary vertex, one sees how it is possible to reconstruct the entire decay chain of a primary B-meson, using the precise vertexing, and the good momentum measurement and particle identification capabilities of the detector.

Exercises for Chapter 5

5–1　What is the required energy for a proton beam on a hydrogen target in order to reach the same C.o.M. energy as a 7 TeV on 7 TeV proton–proton collider? What would be the relativistic γ-factor of the C.o.M. system in the fixed target scenario?

5–2　The motion of a particle in an accelerator structure is parameterized by its transverse position (x, y) with respect to the nominal orbit and the direction $(\mathrm{d}x/\mathrm{d}s, \mathrm{d}y/\mathrm{d}s)$, where s is the longitudinal coordinate. Considering only x, the effect of an optical element can be described by a so-called *transfer matrix* M, which transforms the state vectors $(x, x')^T$ of the beam, that is,

$$\begin{pmatrix} x \\ x' \end{pmatrix}_{\text{new}} = M \begin{pmatrix} x \\ x' \end{pmatrix}.$$

Construct the transfer matrix of a drift volume of length L and that of a quadrupole with focal length f. For the quadrupole employ the thin-lens approximation, that is, ignore the length of the device. Finally, by multiplying the transfer matrices of the individual components, calculate the combined effect of a focusing and a defocusing quadrupole separated by a drift space and derive a condition which has to be fulfilled if the entire assembly is to have a net focusing effect. (★★)

5–3 Show that in the absence of multiple scattering the momentum resolution of a magnetic spectrometer behaves like $dp/p \sim p$. Assume that the spectrometer measures the radius of curvature for the particle track from three equidistant points on a circular arc, and that the individual position measurements have Gaussian errors. (★★)

5–4 A charged particle travelling along the x-axis enters a magnetic field pointing in the z-direction. The field integral seen by the particle is $\int B ds = 1.5$ Tm. Behind the magnet the track direction is measured by two MWPCs spaced 1 m apart. In the absence of multiple scattering, what position resolution is needed in the chambers if this spectrometer has to be able to determine the charge of a pion with $p = 100$ GeV/c with a significance of three standard deviations? How much material, in units of radiation lengths, can be tolerated in front of the second chamber if for a particle with $p = 5$ GeV/c the error from multiple scattering and chamber resolution are allowed to be of equal size? (★)

5–5 Show that the energy resolution of a calorimeter follows $dE/E \sim 1/\sqrt{E}$.

5–6 What is the required time resolution of a TOF-system built from two scintillators with a separation of 2 m that has to distinguish between a $p = 1$ GeV/c pion and kaon with a significance of three standard deviations?

5–7 Show that the Cherenkov light from a particle moving along a straight line is mapped into a ring at a distance f behind a lens with focal length f. (★)

6

QCD ANALYSES

6.1 General concepts

The property of confinement implies that we never observe quarks or gluons in our experiments, but only hadrons and their decay products, which can be of a hadronic or leptonic nature. This means that we are not able to observe, directly, the fundamental particles of QCD, the theory we want to test, since the QCD Lagrangian is built out of quark and gluon fields. In addition, until now nobody has been successful in computing the transition from partons to hadrons from first principles, a problem which cannot be solved by the well known methods of perturbation theory. Only phenomenological models are available, as has been described in Chapter 3.

This obviously constitutes a major complication for the experimentalist, who has to find correlations between the properties of the observed hadronic final state and the properties of the partonic state, that is, the quarks and gluons. This is also in sharp contrast to experiments which test theories with leptonic final states, such as *quantum electrodynamics* (QED). There the asymptotic states of the theory are directly observable in the experimental apparatus. Furthermore calculations have been carried out up to high orders in perturbation theory, which altogether leads to very precise measurements, with measurement errors well below the per mille level. Such an accuracy is beyond reach for any test of the strong interactions. There, the most precise measurements rather have errors of a few per cent, in particular, the measurements of the strong coupling constant. A major theoretical breakthrough in perturbative as well as non-perturbative methods would be absolutely necessary in order to push the precision into a new regime.

So, the challenge for the experimentalist is first to measure as precisely as possible the properties of the final state hadrons, and second to get information about the partons. Of course, the latter task is not required if the experiment only aims at studies of the properties of hadrons.

In order to perform the former task, we have to build an apparatus which is able to measure the momentum and energy of charged and neutral particles, as well as the decay vertices and decay products of short-lived states, and furthermore to give estimates for the particle type. A description of modern detectors is given in Chapter 5.

Now consider, for example, e^+e^- annihilations at LEP at a C.o.M. energy of 91 GeV, where 20 charged hadrons and a similar number of neutral particles are observed on average. Figure 6.1 shows an event display of a hadronic event

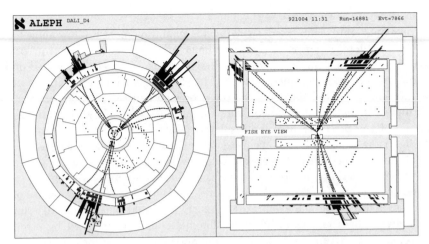

FIG. 6.1. Event display of a hadronic event, observed with the ALEPH detector at the LEP collider.

at LEP1, measured with the ALEPH detector. Most of the charged particles are pions, with in addition some protons, kaons and leptons (electrons and muons) from decays due to weak interactions or from photon conversions into lepton pairs. The stable neutral particles are mostly photons, which stem from the decays of neutral pions. Short-lived particles include neutral kaons and the Λ baryon. Finally, via reconstructing the invariant masses from the momenta of the stable decay products, even very short-lived hadronic resonances such as the ρ meson can be identified.

Particle identification is obtained from measurements of the specific ionization loss when the particles traverse the detector material, from measurements of the Cherenkov radiation or from the time of flight. Flavour separation, that is, the distinction between hadrons which contain light (up, down, strange), charm or beauty quarks, is possible since the decays of hadrons with heavy quarks can be tagged either by measuring displaced vertices with respect to the main event vertex, or by looking for high momentum leptons from the weak decay.

The information on the hadronic final state can finally be used to reconstruct, for example, *jets* and to study their properties. The precise meaning of the concept of jets will be explained later.

In summary, the detector requirements are: excellent tracking devices combined with good particle identification capabilities, and good energy measurements from calorimeters with high resolution in energy and angles.

For deep inelastic scattering experiments, in particular at the electron–proton collider HERA at DESY, the precise measurement of the four-momentum of the scattered electron is most important, since all the relevant kinematic quantities of the reaction can be deduced from the electron's scattering angle and energy. This again requires excellent calorimetry as well as precise tracking. At HERA also the hadronic final state is analysed, since its energy and momentum flow allow for an independent determination of the basic kinematic variables for electron–proton scattering. In addition, the hadronic final state is analysed in further detail, looking for jets and identified particles. In charged current interactions this is the only source of information, since the produced neutrino escapes undetected.

Finally, at hadron–hadron colliders the primary task of QCD studies is the analysis of the production of jets, isolated photons and di-lepton pairs. Good calorimetry is required for an accurate determination of the jet energies. In particular, good knowledge of the absolute energy scale as well as the energy resolution is mandatory, and in order to have a reliable energy measurement up to the highest jet energies ($\mathcal{O}(100\,\mathrm{GeV})$ and more), good linearity of the response of the calorimeter has to be achieved. Furthermore, it should be possible to separate well between hadronic jets and electromagnetic showers induced by electrons and photons. Good hermeticity of the apparatus will allow for a measurement of missing transverse momentum, which, for example, is induced by neutrino production. Last but not least, precise tracking enables flavour tagging as in the case of e^+e^- colliders.

After this brief overview we will now describe in some more detail the individual steps of a QCD analysis and give some concrete examples at the end of the chapter.

6.1.1 *Event selection*

Event selection deals with the decision about which events from the reaction between two colliding beams are of interest for a particular analysis. It is clear that not all the events observed in the detector are exactly of the kind we want to measure, therefore some filtering has to be applied. The filtering should be such that the efficiency for the detection of interesting events is large, that is, only a few of the 'good' events are lost. At the same time, a high purity of the selected event sample shall be achieved, which means that only a few of the accepted events are actually from processes we did not want to select. In the following section, both of these aspects will be discussed for the various collider types.

6.1.1.1 *Electron–positron annihilation* In QCD studies at LEP we are looking for decays of the Z boson into quark–antiquark pairs, where the Z boson stems from the annihilation of the incoming electron and positron. Of course, also highly virtual photons can be exchanged between the initial and final state. However, at the peak of the Z resonance this process is suppressed with respect to Z exchange. In any case, for the pure QCD point of view it does not really matter if quarks of a certain flavour are produced from a Z or a photon, as long as the event orientation is not considered.

Triggering on hadronic Z decays is rather easy. Typically, one looks for an energy deposit in the calorimeter above some threshold, of the order of several GeV, or for a coincidence of track segments with hits in outer calorimeter modules, where the track origin should be near the interaction point. These very simple requirements lead already to an efficiency greater than 99.9% for selecting hadronic events.

The next step is the rejection of all possible backgrounds. There are back-grounds which are not related to the scattering of the electron and the positron, such as beam–gas interactions. These are already suppressed at the level of the first trigger decision, mainly by requiring a certain energy deposit in the central calorimeter. More serious backgrounds would be the leptonic decays of the Z boson, or photon–photon scattering, where either photon is radiated from one of the incoming beams.

The leptonic Z decays distinguish themselves by their low multiplicity. The decay into electron or muon pairs at first order gives two back-to-back tracks, with the muons possibly identified by some muon detection system. This simple picture is altered by bremsstrahlung and/or subsequent conversion of the brems-strahlung photon into lepton pairs. Tau production can lead to higher track multiplicities, since the tau lepton can decay hadronically. However, in the tau decay also neutrinos are produced, which escape the apparatus undetected and thus lead to a reduced total measured energy.

Photon–photon interactions are characterized by a reduced effective C.o.M. energy, and in contrast to the genuine electron–positron interactions the momenta along the beam axis are not balanced, which causes the final state to be boosted along the beam direction. Therefore many final state particles are lost undetected into the beam pipe.

After the triggering stage, the analysis tries to combine all the information of the tracking and calorimeter devices into a measurement of the energy flow, with a given track multiplicity and total energy or invariant mass computed from the reconstructed four-momenta. By requiring, for example, at least five charged tracks in the event, and a minimum amount of total energy or invariant mass, all the backgrounds mentioned above are rejected very efficiently. This is illustrated in Fig. 6.2, where the measured distributions of the two last mentioned quantities are shown. Also the contributing processes and the applied selection cuts are indicated there. Already such a simple selection reduces the backgrounds to the per mille level. The small remainder stems mainly from hadronic tau decays.

It should be pointed out that, when going to C.o.M. energies well above the Z resonance, in particular above the threshold for W pair production at 161 GeV or the threshold for Z pair production at 182 GeV, additional backgrounds from the hadronic decays of the $W^\pm$ and the Z arise. The rejection of this *four-fermion background* is somewhat more difficult, and one ends up with reduced efficiencies and purities. In addition, the problem of the so-called *radiative return* has to be dealt with (c.f. Section 3.2.1). Namely, there is a high probability that one of the incoming electrons or positrons radiates a very energetic photon such that the effective C.o.M. energy of the e^+e^- scattering is reduced to the mass of the Z. So, in order to be able to study hadronic Z decays at or very close to the real C.o.M. energy, $\approx 200\,\text{GeV}$ at the final LEP2 stage, it has to be checked that no very energetic photon is found in the detector, or that the hadronic system has no strong boost along the beam line, which would be a signal for the radiative

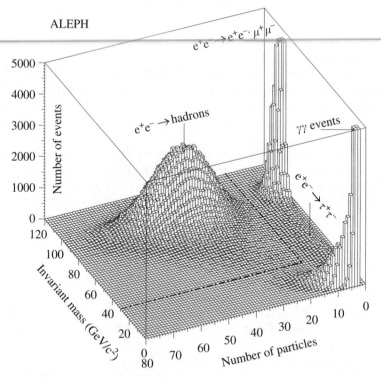

FIG. 6.2. Distribution of the invariant mass vs. the track multiplicity of events measured by the ALEPH detector. Figure from ALEPH Collab.(1995 *a*).

photon to have escaped into the beam pipe. Figure 6.3 shows the distribution of $\sqrt{s'}$, which is the effective C.o.M. energy reconstructed in the detector, for a total LEP energy of 161 GeV. A clear two-peak structure can be observed, where the peak close to 161 GeV stems from events where only a small amount of energy has been lost because of initial state photon radiation, and the peak around the Z mass is due to events with very hard photon radiation. If one is only interested in events with the highest effective energy, one will usually cut at large values of $\sqrt{s'}$, typically about 90% of the total LEP energy. It is clear from the plot that in this case only a small fraction of the collected events remains for further analysis, which is the main reason that many of the QCD analyses at LEP2 energies suffer from limited statistical precision. The bulk of the most precise QCD measurements comes from LEP1, with about four million hadronic Z decays registered by each of the four experiments ALEPH, DELPHI, L3 and OPAL. The LEP1 results at the Z resonance are complemented by important contributions from SLD at the SLAC linear collider SLC, based on a smaller dataset obtained from highly polarized beams.

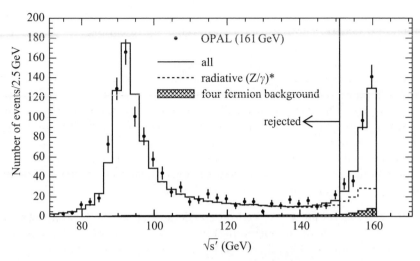

FIG. 6.3. Distribution of the measured effective C.o.M. energy $\sqrt{s'}$ of the hadronic system, for e^+e^- collisions at 161 GeV collected by the OPAL detector. Figure from OPAL Collab.(1997a).

6.1.1.2 *Deep inelastic lepton–nucleon scattering* In deep inelastic scattering experiments, the basic process is the scattering of a lepton (electron, positron, muon or neutrino) off a constituent (quark or gluon) of a hadron, which for example is given by a proton at HERA. The scattering should occur with large momentum transfer from the lepton to the hadron. The reaction leads to a breakup of the hadron into a more complicated hadronic final state. Out of this general sample of events those with particular kinematic configurations are selected further, for example, events with extremely large momentum transfer, or events where the hadronic final state is such that a large angular region between the outgoing hadrons is completely depleted of any hadronic activity, so-called *rapidity gap* or *diffractive* events, c.f. Section 3.2.2.

The basic event selection criteria at HERA, where electrons or positrons of about 30 GeV collide with protons of roughly 820 GeV or 920 GeV energy, are briefly described now. The trigger looks for energy deposits of at least $\approx 5\,\text{GeV}$ in the barrel or the backward calorimeter, which should stem from the electron scattered at large momentum transfer. Here 'backward' refers to the part of the detector being in the flight direction of the electron, that is, backwards to the direction of the incoming proton and most of the produced hadronic final state. In order to strengthen the hypothesis, some isolation criteria in terms of energy deposits around the electron candidate are imposed. The offline selection then applies many more cuts in order to ensure high purity. The main criteria are that the energy of the scattered electron exceeds a threshold of around 10 GeV, and that there is good momentum balance between the electron and the hadronic system, in order to remove events with initial state radiation.

Backgrounds arise from interactions outside the detector, for example, beam–gas scattering, which can be pinned down by looking at timing information, and mainly from events where the electron has escaped undetected into the beam pipe, but a photon or a hadron actually fakes the presence of an energetic lepton in the detector. The latter is studied by measuring the number of electrons at very small angles with dedicated detectors far downstream with respect to the main detector, or by analysing simulated events.

Whereas the trigger efficiency is very high, close to 100%, the variety of the additional offline analysis cuts reduces the overall efficiency quite considerably. The remaining backgrounds amount to a few per cent.

6.1.1.3 *Hadron colliders* At hadron colliders, such as the proton–antiproton collider TEVATRON or the future proton–proton collider LHC, the most interesting reaction for QCD studies is the hard scattering between two partons, one from each of the hadrons, where 'hard' indicates that the outgoing partons are produced with large transverse momentum of at least several tens or even hundreds of GeV. This requirement is necessary in order to distinguish the hard scattering process from the underlying event, that is, the system formed by the partons which do not participate in the hard scattering. The underlying event is characterized by high track multiplicities. However, the transverse momenta are very small, typically below 1 GeV, and the main activity occurs at low angles with respect to the beam pipe. Thus, looking for large energy deposits in the more central region of the calorimeter is an efficient trigger for a hard interaction. A further reason for studying the highest momenta is the fact that it gives a handle for probing for new physics, such as substructure of quarks. Finally, the total cross section for the production of any kind of hadronic final states is enormous. Thus, by selecting hard scatterings the event rates are reduced to a manageable level.

Figure 6.4 shows a sample, which of course is not exhaustive, of possible processes to occur, and which are interesting to select. The incoming partons have to be thought as being resolved within the hadrons, and the outgoing final state has to be produced with large transverse momentum. In the upper part various scatterings between quarks and gluons are indicated, which lead to the production of multi-jet events. The outgoing quarks are mostly light, but can also be very heavy, as is the case for top quark production. In the lower part first one diagram for *direct photon production* is shown, which is a means for measuring the gluon distribution inside the proton. The photon line can be replaced by a Z or a $W^\pm$ boson, which are tagged by their leptonic decays. Finally, a Drell–Yan process is drawn, where again only the purely leptonic final state is tagged. This process is useful for the measurement of the quark distributions inside the proton, c.f. Section 7.4.1.

In principle, triggering on this variety of processes is rather straightforward. It suffices to look for one or more very energetic clusters in the calorimeters to

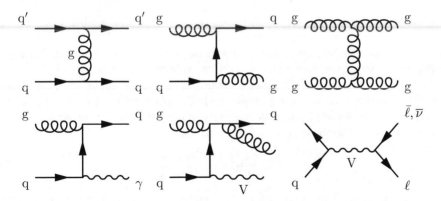

FIG. 6.4. A sample of leading order Feynman diagrams of possible processes to occur in hadron–hadron collisions (V= Z or $W^{\pm}$)

tag multi-jet events. Purely electromagnetic energy deposits, which in addition are well isolated, are a trigger for photons or leptonic decays of the Z and the $W^{\pm}$ boson, where the lepton was an electron or a positron. Well identified muons with high transverse momentum are a further indication for Z and/or $W^{\pm}$ production. In the case of the W also missing transverse energy should be looked for, caused by the escaping neutrino. These triggers are extremely efficient, and in the case of multi-jet events, where the event rates are very high, the trigger has even to be pre-scaled, if the energy threshold is low. Pre-scaling by a factor of n means that only every nth trigger is actually recorded.

Possible backgrounds are the following: photons and electrons can be faked by hadronic jets; leptons with large transverse momentum are also produced in semi-leptonic decays of heavy quarks; imperfections or measurement errors in the calorimeter can lead to missing transverse energy. A general problem for the quantitative understanding of all processes is the underlying event. Careful studies are needed in order to understand the amount of energy and momentum produced in the underlying event and the extent to which it is mis-assigned to the hard-scattering final state at large transverse momentum.

6.2   Observables

Observables which should be calculable within the framework of perturbative QCD have to fulfill a basic requirement: They have to be *infrared safe*. This means that, given any final state composed of quarks and gluons, the value of the observable does not change if a very soft parton is added to this final state, or if the added parton has its momentum parallel to one of the other partons. If this is true, then the cancellation of soft and collinear singularities is ensured. These singularities appear when calculating radiative corrections via real and virtual parton emission. A more detailed discussion of this very important aspect has been given in Section 3.5.

Now some examples for observables are given. A first rather obvious case is the total hadronic cross section in e^+e^- annihilation. Here, the number of actually observed hadronic events has to be counted, and this counting is not affected by the appearance of very soft or collinear particles in the final state. The very soft particles might simply not be resolved at all, and two collinear ones would be counted as a single particle with the momentum given by the sum of their momenta. In e^+e^- annihilation, rather than looking at the absolute total hadronic cross section, it is preferred to measure the ratio

$$R_l = \frac{\sigma(e^+e^- \rightarrow \text{hadrons})}{\sigma(e^+e^- \rightarrow \text{leptons})} \ . \tag{6.1}$$

The electroweak cross section for e^+e^- annihilation into a quark or a lepton pair is the same, modulo the different electric charges and the number of colour charges. It thus cancels in the ratio, and what remains is sensitive to these charges and to radiative corrections from QCD. Also, from the experimental point of view, it is a good observable, since systematic errors such as the error on the luminosity measurement cancel out in the ratio. The latter is needed in order to translate the measured number of events into a cross section. A more detailed discussion of the observable R_l will follow in Chapter 8.

In deep inelastic lepton–nucleon scattering the total inclusive cross section for the scattering of the lepton with some momentum transfer Q is the basic observable, since it is directly related to the nucleon structure function. Totally inclusive means that the hadronic final state is not analysed further. At the end of this chapter we will give an example for such a measurement, and in Chapter 7 the present knowledge on the structure functions is summarized.

Of course we would like to scrutinize the hadronic final state, and not just count the overall number of events. In e^+e^- annihilation there exists a class of observables, the so-called *event shape* variables, which allow to characterize the basic properties of the final state. As an example for such an event shape variable, which also fulfills the criterion of being infrared safe, we will now discuss the Thrust variable T (Farhi, 1977). It is defined as

$$T = \max_{\boldsymbol{n}} \frac{\sum\limits_{i=1}^{m} |\boldsymbol{p}_i \cdot \boldsymbol{n}|}{\sum\limits_{i=1}^{m} |\boldsymbol{p}_i|} \ . \tag{6.2}$$

The sum runs over the three-momenta of all m final state particles. The maximization is performed with respect to a unit vector $\boldsymbol{n}$, that is, one looks for the direction for which the sum of longitudinal momentum components is maximal. The vector $\boldsymbol{n}_T$, for which this maximum is obtained, is called the *Thrust axis* of the event. It is a rather simple exercise to show that the Thrust variable is infrared safe (Ex. (6-1)). For the simplest case of a final state with only two back-to-back particles we have $T = 1$. For the other extreme case, namely a perfectly spherical m-particle configuration with $m \rightarrow \infty$, the Thrust is $T = 1/2$.

So, for every event, a Thrust value between these two extreme cases will be measured. A nice aspect of event shape variables is the fact that the shape of their distribution is rather sensitive to radiative corrections, such as gluon radiation off quarks. Therefore, they offer a handle for measuring the strong coupling constant, as will be described in Chapter 8. The basic idea is that the larger the coupling is, the more gluon radiation will occur from the initial quark–antiquark system. Part of this radiation will occur at large angles and with high energies, leading to more spherical rather than pencil-like event configurations, and thus enhance the Thrust distribution at low Thrust values.

Another class of very useful observables are quantities related to jet production, which means that the final state particles are not distributed uniformly over phase space, but rather collimated within bundles of particles, the so-called *jets*. On the perturbative level the reason can be found in the combined effect of the particular bremsstrahlung spectrum which enhances logarithmically the probability to observe low-angle radiation, and the behaviour of the running coupling, which increases logarithmically with decreasing energy scales, giving an additional effective enhancement of multiple low-angle and low-energetic radiation. The jet picture on the parton level is not altered drastically by hadronization effects, provided the overall energy scale is sufficiently high, let's say above 30 GeV or so. This is because the transverse momenta, which are produced during hadronization and transferred to the hadrons, can not be much larger than the typical overall hadronization scale, which is of the order of $\Lambda \sim 200 - 300$ MeV. These jets are indeed observed at colliders. Figure 6.1 shows an event display where the jet structure of the outgoing particles is evident.

Now the goal is to give a definition of a jet in the form of an algorithmic prescription, which can be implemented in the theoretical calculations as well as in computer programs used to analyse the experimental final states. Two classes of algorithms exist. At e^+e^- colliders jet algorithms of the *Jade* type are usually employed, whereas at proton machines the *cone based* jet finders are preferred.

The JADE collaboration (1986, 1988a) proposed the following scheme: First, one defines a resolution criterion or metric for two particles based on their four-vectors,

$$y_{ij}^{\mathrm{J}} = \frac{2E_i E_j (1 - \cos\theta_{ij})}{E_{\mathrm{vis}}^2} , \qquad (6.3)$$

where E_i, E_j are the particles' energies, $\cos\theta_{ij}$ is the angle between them, and E_{vis} is the energy sum over all particles in the final state. Note that for massless particles y_{ij}^{J} corresponds to the invariant mass squared of the two particles, normalized to the total energy squared. Then we compute the distances y_{ij}^{J} for all particle pairs (i, j) in the final state. Taking the smallest one, we next compare it to some predefined cut-off parameter y_{cut}. If $y_{i_{\min} j_{\min}}^{\mathrm{J}} > y_{\mathrm{cut}}$, then the algorithm stops, and all the particles are defined as jets. However, if $y_{i_{\min} j_{\min}}^{\mathrm{J}} < y_{\mathrm{cut}}$, then the two particles $i_{\min}$ and $j_{\min}$ are combined or *clustered* to form a new pseudo-particle. There exist several prescriptions to perform the combination of the two

four-momenta. For example, in the so-called *E-scheme*, the four-momenta are
simply added, whereas in the E_0-scheme first the four-momenta are added, but
then the new three-momentum is rescaled in order to have a massless pseudo-
particle. After this clustering step the number of (pseudo)particles in the event
is reduced by one, and the algorithm starts again by recomputing all $y_{ij}^{\rm J}$ of the
newly defined final state. The iteration continues until the stopping criterion
$y_{i_{\min}j_{\min}}^{\rm J} > y_{\rm cut}$ is reached.

At LEP, a slightly modified version of this algorithm has been widely used,
the so-called *Durham* clustering algorithm (Dokshitzer, 1990). The difference can
be found in the definition of the resolution criterion, which here is

$$y_{ij}^{\rm D} = \frac{2\min(E_i^2, E_j^2)(1 - \cos\theta_{ij})}{E_{\rm vis}^2}. \tag{6.4}$$

For small angles this corresponds to the transverse momentum of the less energetic particle with respect to the higher energetic one. The Durham scheme
is preferred from the theoretical viewpoint, since using this prescription it is
possible to resum leading logarithmic terms to all orders of perturbation theory,
something which is not possible with the Jade scheme (Catani, 1991).

The idea of the cone jet finder arises rather naturally from the characteristics of hadron collider experiments, where fine-grained calorimetry plays a major
role and where one has the problem of the underlying event. Remember that the
energy flow of the underlying event is distributed uniformly in rapidity and in
azimuth, Section 3.2.3. Therefore, looking for jets means looking for enhancements in the energy flow above this background in certain intervals of rapidity
and azimuth. The calorimeters are usually composed of projective towers seg-
mented in azimuth, $\Delta\Phi$, and pseudorapidity, $\Delta\eta$, where η is related to the polar
angle θ by $\eta = \ln(\cot(\theta/2))$. First, preclusters (seeds) are found by summing
the transverse energy in contiguous towers, requiring a certain minimum energy
per tower, typically about $1\,{\rm GeV}$ at the TEVATRON experiments. The transverse energy E_T is defined by $E_T = E\sin\theta$, where E is the tower energy. For
each precluster having a total transverse energy larger than a fixed threshold, for
example, $2\,{\rm GeV}$, a centroid is computed by transverse-energy weighting. Then,
a cone of radius $R = \sqrt{(\Delta\Phi)^2 + (\Delta\eta)^2}$ is formed around the centroid of each
precluster. Here $\Delta\eta$ ($\Delta\Phi$) refer to the difference in pseudorapidity (azimuth) between a tower location and the centroid. A typical values is $R = 0.7$. All towers
above a third threshold, for example, $0.2\,{\rm GeV}$, in transverse energy inside the
cone form a cluster; the centroid is then recalculated, and a new cone is formed.
The process is repeated until the list of towers inside the cone remains unchanged
in successive iterations. Finally, if two clusters overlap, they are merged if either
cluster shares more than 50% of its energy with the other; if not, they remain
separate and towers common to both are assigned to the nearest cluster. The
jet transverse energy is obtained as the scalar sum of transverse energies of the
towers belonging to the jet. The jet's rapidity and azimuth are computed as

weighted sums of tower rapidities and azimuths, with the transverse energy of the tower as weight.

In contrast to the Jade-type algorithms, in the cone jet finder not all particles (towers) are necessarily clustered into a jet. The energy thresholds help to reduce the contribution from the underlying event. It is worth noting that the cone algorithm in its basic definition actually is not infrared safe in all orders of perturbation theory. Additional algorithmic steps have to be introduced in order to avoid this problem. A detailed discussion can be found in the literature (Seymour, 1998).

Having finally defined our jets, we can construct many interesting observables: at e^+e^- colliders we look at the two-, three-, four-, five- and more jet rates as function of the resolution parameter y_{cut}. The smaller it is chosen, the more jets will be resolved, until we enter the non-perturbative regime where we start to resolve single hadrons. Furthermore, jet properties such as particle or subjet production within the jet are studied, and one looks for jet variables which differentiate between a jet coming from a quark, and a jet originating from a gluon.

At hadron colliders one of the most important measurements is the determination of the inclusive transverse energy spectrum of jets. The cross section has been measured over many orders of magnitude, and the tails of this distribution, where jets with very large transverse energy are produced, actually allow to pose rather stringent limits on potential deviations from the Standard Model, such as quark substructure. Other typical observables are di-jet rates and the angular distribution of the jets, as well as the ratio of the cross sections for the production of a $W^{\pm}$ together with one jet, and the cross section for a $W^{\pm}$ plus two jets, which is another means for measuring the strong coupling constant.

Studies of event shape distributions and jet production are useful in order to test perturbative QCD, but of course allow also to investigate the hadronization phase. Here one mainly concentrates on the tests of the phenomenological models available, such as, for example, the Lund string model, c.f. Section 3.8. Typical observables are inclusive charged particle distributions such as the multiplicity or the momentum distribution, or the production rates of identified hadrons, for example neutral pions or vector mesons.

It is worth noting that many observables studied in e^+e^- experiments are also being investigated in deep inelastic scattering at HERA. If the final state is analysed in the *Breit frame of reference*, see Ex. (6-3), and if only the particles from the current hemisphere are considered, as indicated in Fig. F.5, then there is a very close correspondence to a single hemisphere in e^+e^- annihilation. The current hemisphere is the hemisphere of the struck quark or gluon, whereas the target hemisphere refers to the region where the proton remnant can be found. A Breit frame analysis in deep inelastic scattering has the advantage that the available energy Q is variable, whereas in e^+e^- annihilation it is fixed. So within a single experiment the energy dependence of the observables mentioned above can be looked at over a wide energy range.

6.3   Corrections

Performing a measurement involves a rather intricate chain of detectors and
analysis steps. As a consequence the result will be a somewhat distorted image
of the physical truth, and usually considerable effort goes into the understanding
and undoing of these distortions.

6.3.1   Detector corrections

Every detector has finite resolution and acceptance. This means that we measure
the momenta and energies within some uncertainty range, and we lose particles
because the detector is not able to detect them, as is the case for neutrinos, or
because it is not perfectly hermetic. At collider experiments, where the sensitive
region only starts above some minimum polar angle, there are unavoidable losses
of tracks into the beam pipe, and there are losses because of regions in the
detector which are not equipped with sensitive devices, such as transition regions
between central and forward detector components. However, if we know our
detector very well, that is, if we know which regions of the detector cannot
measure anything, and if we know what the resolution of our measuring device
is, then we can correct for these detector effects.

As a simple example, consider an experiment where we would like to measure
the decay rate of some radioactive source. Imagine we would like to measure
the number of α decays per second. Now suppose that we were able to build
a perfectly spherical detector which is able to record every α particle passing
through it, and we put the source at the centre of the detector. Unfortunately,
because of bad manufacturing, a quarter of the detector stops functioning at the
early stage of the experiment. This means that from now on we have to correct for
its finite acceptance, that is, in order to get the total number of decays N_{true}, the
actually measured number N_{meas} has to be multiplied by a factor $4/3$, since only
$3/4$ of the solid angle are giving signals. As an example of finite resolution one
could imagine that our detector has some finite timing resolution, that is, it can
not resolve two decays which occur within some small time interval t_{resol}. If this
is the case, we will miss some counts because they are too close to one-another
and we have to correct for this when giving our final result.

Of course, collider experiments are much more complex, but the basic ideas
remain the same. Very elaborate computer models of the detector are built, which
then deliver (in simple words) the fraction of the solid angle which is equipped
with sensitive detectors, and the resolution of the energy, momenta and timing
measurements. For example, in e^+e^- annihilations first hadronic final states
are generated by some Monte Carlo model, and the distribution D_i^{gen} of some
variable is computed from this final state. Technically, one usually approximates
the distribution by a histogram, and the index i introduced above refers to the ith
bin of the histogram. Then all particles of these final states are passed through
the detector simulation. At the end, one obtains a set of measurements which
should resemble a real set of measurements, and from these new final states a
distribution D_i^{sim} is obtained. For the sake of simplicity, assume that the same

binning as before is used. Now, the so-called bin-by-bin correction factors are computed according to

$$C_i^{\text{det}} = \frac{D_i^{\text{gen}}}{D_i^{\text{sim}}} \ . \tag{6.5}$$

Knowing these correction factors, the measured distribution D_i^{meas} can be corrected for detector effects in order to obtain the *true* distribution D_i^{corr},

$$D_i^{\text{corr}} = C_i^{\text{det}} \, D_i^{\text{meas}} \ . \tag{6.6}$$

This final distribution can be compared to theoretical calculations or to results from other experiments. Since those usually have different resolution and acceptance, it would be meaningless to compare directly measured distributions D_i^{meas}.

In this simple bin-by-bin approach it is assumed that there is only a weak dependence on the Monte Carlo generator used to obtain D_i^{gen}. One must also take care that the binning is in accordance with the detector resolution, that is, detector effects should not induce too much migration from bin i to bin j, $i \neq j$. If migration effects are important, then a more sophisticated solution is needed, based on the construction of a transition matrix M_{ij}^{det}, which relates D_j^{gen} to D_i^{sim} according to $D_i^{\text{sim}} = \sum_j M_{ij}^{\text{det}} D_j^{\text{gen}}$. The problem is that, in order to get D_i^{corr}, a simple inversion of M_{ij}^{det} often leads to unstable results, so that more advanced methods have to be applied in order to unfold the detector effects. We cannot go into the discussion of this topic here, but refer the reader to the literature (Cowan, 1998; Schmelling, 1994).

6.3.2   *Hadronization corrections*

If the distribution D_i^{gen} introduced above is obtained with some Monte Carlo model which generates hadronic final states, our corrected distribution D_i^{corr} corresponds to the measured distribution of some quantity as obtained from a final state built of hadrons. Usually the experiments publish their results in this form, that is, corrected to the *hadron level*, as is the jargon. However, in order to test predictions from perturbative QCD, or to determine quantities which enter the perturbative predictions, first these predictions have to be corrected for the transition from partons to hadrons, called *hadronization* or *fragmentation*, and only then they can be compared to D_i^{corr}. The size of the hadronization effects depends very much on the observable and on the overall energy scale Q. Generally speaking, the size of the corrections decreases according to some inverse power of Q. Very well resolved and energetic jets, as well as the multi-jet regions of event shape variables, are less affected than for example close-by and low-energetic jets. Some quantities can change drastically, such as total particle multiplicities.

The solution of this problem is again obtained by running Monte Carlo programs which first generate partonic final states according to exact or approximate perturbative QCD, and then simulate the transition to hadrons according to some phenomenological models. The basic ideas of such programs have been

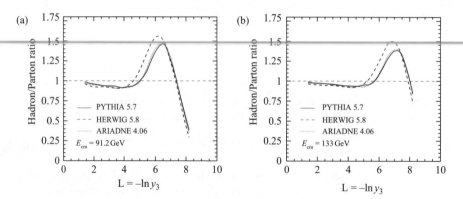

FIG. 6.5. Hadronization corrections for the event shape variable y_3, measured in e^+e^- annihilations at two different C.o.M. energies. Figures from
ALEPH Collab.(1997*a*).

described in Chapter 4. Then the approach is very similar to the detector correction procedure described above. Analysing the parton (hadron) level, we obtain
distributions D_i^{par} (D_i^{had}), and correction factors $C_i^{\text{had}} = D_i^{\text{had}}/D_i^{\text{par}}$, or some
transition matrix M_{ij}^{had}. Now we are able to correct predictions from perturba-
tive QCD, D_i^{QCD}, which might depend on one or more free parameters, for the
hadronization step, and compare them to measured distributions,

$$D_i^{\text{QCD,corr}} = C_i^{\text{had}} D_i^{\text{QCD}} \quad \text{or} \quad D_i^{\text{QCD,corr}} = \sum_j M_{ij}^{\text{had}} D_j^{\text{QCD}} . \qquad (6.7)$$

Care has to be taken if the corrections deviate too much from unity. It is clear
that an observable which has hadronization corrections of, for example, 50% will
depend a lot on the actual hadronization model applied, and it gets more and
more difficult to extract reliably the properties of the partonic system.

As an example, at LEP it is typically checked that the hadronization corrections do not exceed the 10% level, if precise measurements of basic quantities
such as α_s are to be obtained. Figure 6.5 shows the correction factor C_i^{had} for the
event shape variable y_3. This variable is found by applying the Durham clustering algorithm to the final state (partons or hadrons), until only three jets are left.
Then y_3 is given by $y_3 = \min y_{ij}^{\text{D}}$, as described in Section 6.2. It determines the
resolution value for which this final state would transit from a three- to a two-
jet configuration. The region of small values is populated by two-jet like events,
whereas the region of large y_3, or equivalently small values of $-\ln y_3$, is filled by
multi-jet events, and often used to measure the strong coupling constant. From
the plots it is clearly visible that the corrections, computed with three different
Monte Carlo models, are fairly small in this multi-jet region, but they increase
when going to the two-jet region, where two out of the three jets tend to be close

by. It can also be observed that the corrections slightly decrease when going to a larger C.o.M. energy.

6.4 Systematic uncertainties

The topic of systematic uncertainties is a very difficult one, and since there is some lack of general and objective rules, it is not always free of controversy. The sources of systematic uncertainties depend very much on the analysis, and since it is practically impossible to give an exhaustive discussion here, we will rather try to outline some basic aspects to be considered.

Generally, we can distinguish between two large classes of uncertainties, the experimental systematic uncertainties and errors related to the theory. Systematic uncertainties because of the experimental procedure arise because in fact we never understand perfectly well our detector, as we have assumed in Section 6.3. This means that the Monte Carlo simulation of the detector, which we use to estimate acceptance corrections or the resolution, is not perfect. An easy check of the stability of a result is obtained by applying different cuts in order to select events and/or particles. If the detector simulation would be exact, then every set of cuts should lead to the same final result, since the change in the event sample would be tracked by a change in the correction factors. However, if the final result is not invariant, then we have a hint that some of the quantities, where we apply cuts on, are not described correctly by the simulation. As a trivial example, assume that in the Monte Carlo model the sensitive region of the detector starts at a polar angle θ_{MC}, whereas in the real detector it starts only at $\theta_{DE} > \theta_{MC}$. If we then use some overall event axis, such as the Thrust axis, to define the orientation of an event, and apply an acceptance cut on the polar angle θ_{TH} of this axis to ensure that the event is well contained within the detector, our acceptance calculation will be correct as long as $\theta_{TH} > \theta_{DE}$. However, if $\theta_{TH} < \theta_{DE}$, our Monte Carlo simulation will predict a wrong acceptance, and the final result will change.

Another critical aspect of an analysis is the good knowledge of the detector resolution, such as the jet energy resolution, and the overall energy scale. A good understanding of the latter can be particularly important for processes with cross sections that fall steeply as a function of (transverse) jet energy. Often independent processes with respect to the one actually under study can be employed to estimate the energy scale and resolution, if they offer some physical constraint such as energy balance or mass constraints. The final systematic uncertainty on the original measurement might only be limited by the statistical uncertainty of the cross check analysis.

Theoretical uncertainties simply arise because of limited theoretical knowledge. As an example, in Fig. 6.5 the hadronization corrections for an event shape variable are plotted, as predicted by three different models for the hadronization process. Since we are still not able to compute this mechanism from first principles, every model which is able to give a good overall description of the properties of hadronic final states has to be considered for the computation of hadronizat-

ion corrections. However, as we can see from Fig. 6.5, there are some differences
in the predicted corrections, which in the end can lead to differences in some
measured quantities such as α_s. This kind of uncertainty is called *hadronization
uncertainty*.

Another typical error source is a limitation of calculations within the framework of perturbative QCD. Every calculation has to stop at some order in the
expansion parameter, which usually is given by α_s. The problem is that this
truncated perturbation series depends on the unphysical renormalization scale,
c.f. Section 3.4, and in principle every choice for this scale is equally well justified. Different choices will lead to different final answers, such as different values
for α_s. As systematic uncertainty one usually defines the variation of the final
results under variations of the renormalization scale over some reasonable range.
What is 'reasonable' is a question very much under debate.

As a final warning it should be mentioned that usually the systematic uncertainties are treated as if they had the same meaning as a statistical error.
However, because of the arbitrariness of some of the evaluation procedures, or
simply because of the intrinsic non-probabilistic nature of certain systematic uncertainties, the interpretation of specific results should be taken with care. This
is particularly important when deviations from the expectation are observed, and
if the significance of these deviations is measured with respect to the quadratic
sum of statistical and systematic uncertainties.

6.5   Examples

In the following sections we will give some concrete examples of QCD measurements. We will describe in more detail the main steps to be taken, the problems
which are typically encountered and how they are solved. It should be clear to
the reader that only a rough sketch of an analysis can be given, and that a detailed description of all the aspects, such as detector specific topics, analyses of
systematic uncertainties or subtleties of the theoretical framework, would go far
beyond the scope of this book.

6.5.1   *Structure function measurement at* HERA

In the first example, we will discuss the measurement of the structure function
$F_2(x, Q^2)$ from deep inelastic electron–proton scattering at the HERA collider.
An in-depth discussion of deep inelastic scattering can be found in Section 3.2.2.
In Fig. 2.1 we have depicted the basic Feynman diagram for the deep inelastic
scattering process. What is shown is only the exchange of a virtual photon, since
we will delimit our discussion to momentum transfers $Q^2 = -(l - l')^2 \ll M_Z^2$
so that the exchange of a Z boson can be safely neglected. In this case, the
differential cross section is given in terms of two structure functions $F_2(x, Q^2)$
and $F_L(x, Q^2)$,

$$\frac{\mathrm{d}^2\sigma}{\mathrm{d}x\mathrm{d}Q^2} = \frac{2\pi\alpha_{\mathrm{em}}^2}{xQ^4} \left[\left(1 + (1-y)^2\right) F_2(x, Q^2) - y^2 F_L(x, Q^2)\right] , \qquad (6.8)$$

where α_{em} is the fine structure constant, and $F_L = F_2 - 2xF_1$ is the longitudinal structure function. As a reminder, the kinematic variables x, Bjorken's scaling variable, and y are defined as

$$x = \frac{Q^2}{2p \cdot q} , \quad y = \frac{p \cdot q}{p \cdot l} , \tag{6.9}$$

$0 < x, y < 1$. When neglecting particle masses we find $s = (p + l)^2 = 2p \cdot l$ for the C.o.M. energy squared and $Q^2 = sxy$.

Now let us concentrate on the kinematical situation at HERA. There protons with an energy of $E_p = 820\,\text{GeV}$ (later $920\,\text{GeV}$) collide with electrons or positrons of $E_e = 27.5\,\text{GeV}$. Let's choose a frame such that $p = (E_p, 0, 0, E_p)$ and $l = (E_e, 0, 0, -E_e)$, that is, the protons are moving in the positive z-direction. The C.o.M. energy is given by $\sqrt{s} = \sqrt{4E_pE_e} \approx 300\,\text{GeV}$. Since the momenta are highly unbalanced, the C.o.M. system is moving with respect to the laboratory system in positive z-direction with a velocity $\beta = v/c = |\boldsymbol{p}_{tot}|/E_{tot} = (E_p - E_e)/(E_p + E_e) \approx 0.935$. This has an important impact on the detector design. In fact, the detector layout is asymmetric with respect to z. In the forward (positive z) direction the detector is more densely instrumented than in the backward direction, since most of the time a big fraction of the outgoing hadronic system will move into the forward hemisphere under rather small angles with respect to the beam pipe. In the backward direction a dedicated calorimeter is installed in order to measure precisely the scattered electron or positron. This most often is found at large angle, $\theta_e > 150°$, with respect to the incoming proton, that is, low angle with respect to the beam line in negative z-direction.

The kinematic variables can be reconstructed in terms of the measured energy and direction of the outgoing electron, without analysing the hadronic system. This follows from

$$Q^2 = 2l \cdot l' = 2E_e E'_e (1 + \cos\theta_e) = Q_e^2 , \tag{6.10}$$

$$y = 1 - \frac{p \cdot l'}{p \cdot l} = 1 - \frac{E'_e}{2E_e} (1 - \cos\theta_e) = y_e , \tag{6.11}$$

when neglecting particle masses. The index e indicates that the variables are reconstructed with the so-called *electron method*. When expressing E_e in terms of y_e we find

$$Q_e^2 = \frac{E'^2_e \sin^2\theta_e}{1 - y_e} , \quad x_e = \frac{Q_e^2}{sy_e} = \frac{E'^2_e \sin^2\theta_e}{sy_e(1 - y_e)} . \tag{6.12}$$

From these expressions it is clear that, before applying any event selection cuts, the accessible range in Q^2, x and y is limited by the angular acceptance of the detector and the energy thresholds.

In order to get a feeling for the numbers involved, let's take a few examples. To start with, assume that the electron is found in the backward calorimeter at $\theta_e = 160°$, with $E'_e = 1\,\text{GeV}$. Then $y_e = 0.965$, $Q_e^2 = 3.32\,\text{GeV}^2$ and $x_e = 3.8 \times 10^{-5}$.

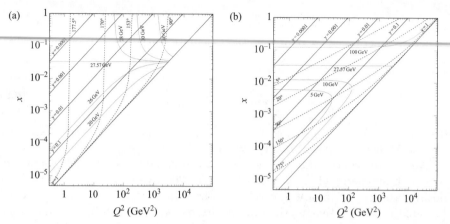

FIG. 6.6. Kinematic domain in the (Q^2, x) plane : (a) lines of constant E'_e, θ_e and y; (b) lines of constant E_q, θ_q and y, where E_q and θ_q are the energy and angle of the scattered quark in the parton model. Figures from Clerbaux(1998).

When going to a higher energy, say $E'_e = 10\,\text{GeV}$, we find $y_e = 0.647$, $Q^2_e = 33.2\,\text{GeV}^2$ and $x_e = 5.7 \times 10^{-4}$. Finally, when $E'_e \approx E_e$, the values $y_e = 0.03$, $Q^2_e = 91.2\,\text{GeV}^2$ and $x_e = 0.034$ are found. In Fig. 6.6(a) the kinematic domain in the (Q^2, x) plane is shown (Clerbaux, 1998). It is evident that for $y \lesssim 0.1$ lines of constant Q^2 correspond to constant θ_e, and y is basically determined by the energy of the scattered electron for $E'_e < 27\,\text{GeV}$.

An important feature of deep inelastic scattering within a hermetic collider experiment is that also the hadronic final state can be used for the determination of the kinematic variables. However, since a complete 4π angular coverage is impossible to achieve, and since hadronic activity often occurs at rather low angle, it is preferable to find variables which are rather insensitive to losses into the beam pipe along the proton direction. This is achieved, for example, by the variable

$$\Sigma = \sum_a (E_a - p_{z,a}) = \sum_a E_a (1 - \cos\theta_a) \ , \qquad (6.13)$$

where the sum runs over all hadrons a found in the final state. Hadron masses are neglected for the second equality. It is evident that this variable is rather insensitive to particles lost at very low angles with respect to the proton direction, since these anyway would not contribute to the sum. It was first proposed by Jacquet and Blondel (1979) to express y and Q^2 in terms of

$$y = \frac{\Sigma}{2E_e} = y_h \ , \quad Q^2 = \frac{P^2_{\perp,h}}{1 - y_h} = Q^2_h \ , \qquad (6.14)$$

where $P_{\perp,h} = \sum_a E_a \sin\theta_a$ is the transverse momentum of the hadronic system, and the subscript h indicates that the observables are constructed from hadronic variables. The derivation of these formulas is discussed in Ex. (6-4). There it is also shown that the hadron angle defined as

$$\cos \theta_h = \frac{P_{\perp,h}^2 - \Sigma^2}{P_{\perp,h}^2 + \Sigma^2} \tag{6.15}$$

corresponds to the angle of the struck quark in the parton model. In Fig. 6.6(b) again the kinematic domain in Q^2 and x is shown, but now for lines of constant θ_h and constant energy of the struck quark as derived in Ex. (6-4).

A third method for the kinematic reconstruction can be employed in order to avoid the following problem. From eqn (6.14) we see that y_h depends on E_e. Using the beam energy for E_e is correct only as long as the incoming electron does not radiate photons before undergoing the actual hard interaction with the proton. If instead there is a large amount of initial state photon radiation (ISR), the relevant electron energy can be reduced substantially below the beam energy, and we would make a large mistake in the determination of y_h. This effect can be largely reduced by using the *Sigma method* (Bassler and Bernardi, 1995) for the reconstruction, where

$$y_\Sigma = \frac{\Sigma}{\Sigma + E_e'\,(1 - \cos \theta_e)}\;, \qquad Q_\Sigma^2 = \frac{E_e'^2 \sin^2 \theta_e}{1 - y_\Sigma}\;. \tag{6.16}$$

Again we refer to Ex. (6-4) for a derivation of these formulae.

So far we have assumed that we can determine the kinematic quantities of the event with arbitrary precision. Of course this is not the case. In fact, for the choice of the final event selection cuts the detector resolution plays an important role. Take for example the measurement of y. In the electron and the Jacquet–Blondel methods we find for the resolution of y, for fixed θ_e,

$$\frac{\delta y_e}{y_e} = \frac{1 - y_e}{y_e} \frac{\delta E_e'}{E_e'}\;, \qquad \frac{\delta y_h}{y_h} = \frac{\delta \Sigma}{\Sigma}\;. \tag{6.17}$$

In the first method, the resolution on y_e is determined by the resolution of the energy measurement for the scattered electron and the actual value of y_e. At large y_e the resolution is very good, however, at low y_e it diverges. The second method does not suffer from this problem. Here, only the resolution of the hadronic energy measurement is of importance. These facts are illustrated in Fig. 6.7, where a Monte Carlo simulation of the H1 detector has been used (Glazov, 1998) in order to show the transition from a true (generated) y to a measured y_e (a) or y_h (b) because of detector resolution effects. The degradation of the resolution on y_e for low y is clearly visible, whereas it is rather constant over y when taking y_h. At large y the second method is worse because of the intrinsic lower resolution on the hadronic energy compared to the electron energy. These plots suggest to combine the two methods for a best possible determination of the kinematic variables over a large area in the (Q^2, x) plane. Namely, take the electron method for large y,

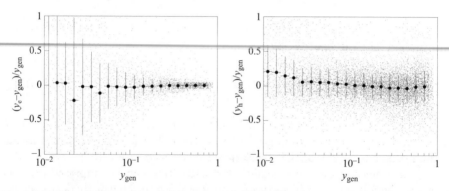

FIG. 6.7. Resolution of y_e (left) and y_h (right). Here y_{gen} is the true simulated value of y. Figures from Glazov(1998).

say $y_e > 0.3$, the Jacquet–Blondel or Sigma method for the low y region, and restrict the overall event selection to regions with acceptable resolution. Finally the (Q^2, x) plane is divided into bins, the sizes of which are chosen such that the bin-to-bin migration from true to reconstructed values because of the detector resolution is kept small.

After a basic event selection such as described in Section 6.1.1.2, the determination of the kinematic variables, further event selection refinements and the choice of the range and binning in Q^2 and x, we can proceed to the final cross section measurement. For every bin (i, j) in (Q^2, x) we have

$$\left.\frac{\mathrm{d}^2\sigma}{\mathrm{d}x\mathrm{d}Q^2}\right|_{i,j} = \frac{N_{i,j}^{\text{data}} - N_{i,j}^{\text{bkg}}}{C_{i,j}\,\mathcal{L}}\,\delta_{\text{add}}\,. \tag{6.18}$$

Here $N_{i,j}^{\text{data}}$ is the number of measured events in the relevant bin, and $N_{i,j}^{\text{bkg}}$ is the number of expected background events, usually determined from Monte Carlo simulations. Possible backgrounds have been described in Section 6.1.1.2. The factor $C_{i,j}$ stands for the acceptance correction and is also determined from a Monte Carlo simulation. It is important here that the simulation reproduces well the detector acceptance and resolution. Furthermore, it has to be checked that the correction factors do not depend on the choice of a priori structure functions needed for the Monte Carlo generation of simulated events.

The luminosity $\mathcal{L}$ is determined in the following manner. A process is taken for which the cross section is very well known theoretically, and the equation

$$\sigma = (N^{\text{data}} - N^{\text{bkg}})\,\epsilon\,\mathcal{L}\,, \tag{6.19}$$

is inverted with respect to $\mathcal{L}$. Again $N^{\text{data(bkg)}}$ is the number of observed (expected background) events, and ϵ the efficiency to collect such events. The two latter numbers are taken from simulations and/or auxiliary measurements. At HERA a suitable process is given by the Bethe–Heitler Bremsstrahlung process

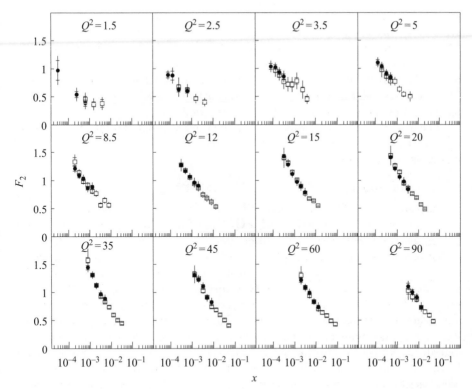

FIG. 6.8. Measurement of the F_2 structure function by H1 (1996) using the electron (closed circles) and the sigma method (open squares). The inner error bar is the statistical error, the full error bar includes also systematic uncertainties.

$ep \rightarrow ep\gamma$, where the outgoing electron and photon(s) are measured in coincidence at very low angles, far downstream of the detector.

Finally, δ_{add} represents all kinds of possible additional corrections, such as higher order radiative corrections to the Born cross section, or bin centre corrections which can be of importance in regions of a fast varying cross section.

Having determined the cross section, eqn (6.8) can be inverted in order to extract $F_2(x, Q^2)$. This is done in two steps. At low $y < 0.1$ the contribution from F_L is negligible, therefore the extraction of F_2 is straightforward. At high $y > 0.6$ the contribution of F_L becomes more important. Here the previous F_2 measurement or another independent determination of F_2 is evolved to the relevant kinematic range using QCD evolution equations and inserted in eqn (6.8). Then F_L is extracted. Of course this method is based on the assumption that F_2 at large y is in accordance with the predictions of perturbative QCD.

Figure 6.8 shows a measurement of the structure function F_2 in bins of x and Q^2 (H1 Collab., 1996). The error bars in Fig. 6.8 include statistical and

systematic uncertainties. The statistical errors range between 1% and 6%. The
systematic uncertainties are of the order of 5% to 10%, depending on the region
in y. Many sources of systematic uncertainties contribute, such as uncertainties
on the luminosity measurement, on the electron and hadron energy scales in
the respective calorimeters, on various reconstruction efficiencies, as well as on
background expectations and radiative corrections.

The measured structure functions can be used as inputs for QCD fits in order
to extract parton distribution functions in the proton. Analyses of this kind will
be discussed in detail in Chapter 7.

6.5.2 *Inclusive jet production at the* TEVATRON

The measurements of the cross section for inclusive jet production in p$\bar{\text{p}}$ collisions
are an important test for perturbative QCD, and allow to put constraints on the
parton distribution functions as will be described in Section 7.4.4. Furthermore,
deviations of the data from the predicted cross sections at the highest jet energies
could be indications for new physics such as substructure of quarks, since with
jet transverse energies of about $400\,\text{GeV}$ distance scales as small as 10^{-19}m are
probed, c.f. eqn (7.1). The production of jets is understood as the result of a hard
scattering of partons from the incoming hadrons, giving outgoing partons with
large transverse momenta which fragment into jets of particles. The theoretical
ingredients for the description of the process have been outlined in Section 3.2.3.
In this section, we will rather concentrate on the experimental aspects of the
cross section measurements. In particular we will describe the measurement of
the cross section for p$\bar{\text{p}} \to \text{jet} + X$, that is, the probability to observe a jet with
a certain transverse energy E_T at a pseudo-rapidity η, without considering the
remaining system X. Measurements of this type have been performed at the ISR
pp collider at a C.o.M. energy of $\sqrt{s} = 63\,\text{GeV}$ (AFS Collab., 1983), the CERN
p$\bar{\text{p}}$ collider at 546 and $630\,\text{GeV}$ (UA2 Collab. 1985b, 1991; UA1 Collab. 1986),
and the FERMILAB TEVATRON p$\bar{\text{p}}$ collider at $1800\,\text{GeV}$ (CDF Collab. 1989,
1992, 1996; D0 Collab. 1999a). A recent review can be found in the article by
Blazey and Flaugher (1999).

The selection of events proceeds via several hardware trigger stages and a
final offline software selection. The first trigger looks for an inelastic p$\bar{\text{p}}$ collision
indicated by signals from detectors placed very closely to the beam pipe at both
sides of the interaction region. Next, transverse energy above a threshold of a
few GeV has to be measured in calorimeter cells dedicated to the trigger. At
the following stage jets are reconstructed with a cone algorithm, described in
Section 6.2. As an example, in the D0 experiment at the TEVATRON the events
are recorded and classified if at least one jet with energy above a threshold of
30, 50, 85 or $115\,\text{GeV}$ is found. With these requirements integrated luminosities
of up to $100\,\text{pb}^{-1}$ have been collected at the TEVATRON Run 1A and Run 1B.
Finally, at the offline analysis level the cone jet finder is applied again, with a
typical cone size of $R = 0.7$. The transverse energy of a jet is computed according
to

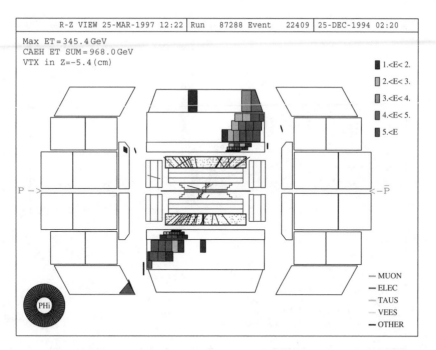

| R-Z VIEW 25-MAR-1997 12:22 | Run   87288 Event   22409 | 25-DEC-1994 02:20 |

Max ET = 345.4 GeV
CAEH ET SUM = 968.0 GeV
VTX in Z=-5.4 (cm)

■ 1.<E< 2.
□ 2.<E< 3.
▨ 3.<E< 4.
▨ 4.<E< 5.
▨ 5.<E

P —>   <— P̄

— MUON
— ELEC
— TAUS
— VEES
— OTHER

PHi

FIG. 6.9. Event display of a di-jet event in the D0 detector. Figure from Blazey and Flaugher(1999).

$$E_T^{\text{jet}} = \sum_i (E_T)_i = \sum_i E_i \sin\theta_i \, , \qquad (6.20)$$

where the sum goes over all calorimeter towers within the jet cone, with E_i the measured tower energy and θ_i the polar angle of the tower. Note that a jet typically illuminates around 20 towers. The jet direction is determined by the energy-weighted average over angular tower positions and the location of the main event vertex. The latter is determined from charged tracks measured in the tracking system.

Figure 6.9 shows an event display of a di-jet event in the D0 detector. An alternative way to illustrate jet events is given in Fig. 6.10 for the case of a multi-jet event recorded by the CDF detector.

Usually the cross section is integrated over some region in η, such as $|\eta| < 0.5$ (D0 Collab., 1999a). This central detector region has the best energy resolution and uniformity. The bin-averaged double differential cross section over a certain angular range, $\langle \mathrm{d}^2\sigma/(\mathrm{d}E_T\mathrm{d}\eta)\rangle$, is determined by simply counting the number N of jets within a specific bin ΔE_T, normalized to the bin width and the integrated luminosity,

$$\left\langle \frac{\mathrm{d}^2\sigma}{\mathrm{d}E_T\mathrm{d}\eta} \right\rangle = \frac{N}{\Delta E_T \Delta\eta\, \epsilon\, \mathcal{L}} \, . \qquad (6.21)$$

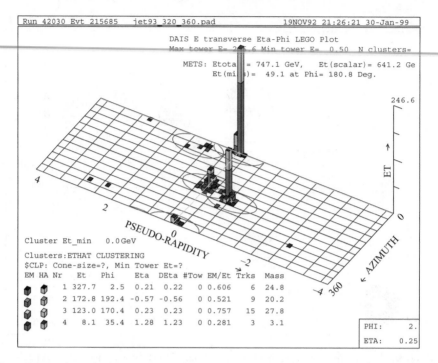

| Run 42030 Evt 215685 | jet93_320_360.pad | 19NOV92 21:26:21 30-Jan-99 |

DAIS E transverse Eta-Phi LEGO Plot
Max tower E= 2?.6 Min tower E= 0.50 N clusters=

METS: Etotal = 747.1 GeV, Et(scalar)= 641.2 Ge
Et(miss)= 49.1 at Phi= 180.8 Deg.

246.6

ET

0

Cluster Et_min   0.0 GeV

Clusters:ETHAT CLUSTERING
$CLP: Cone-size=?, Min Tower Et=?

EM HA Nr	Et	Phi	Eta	DEta	#Tow	EM/Et	Trks	Mass
1	327.7	2.5	0.21	0.22	0	0.606	6	24.8
2	172.8	192.4	-0.57	-0.56	0	0.521	9	20.2
3	123.0	170.4	0.23	0.23	0	0.757	15	27.8
4	8.1	35.4	1.28	1.23	0	0.281	3	3.1

| PHI: | 2. |
| ETA: | 0.25 |

FIG. 6.10. A multi-jet event in the CDF detector, shown in the Lego-plot representation on the η-ϕ plane. The height of the towers indicates the amount of transverse jet energy. A jet clustering cone of radius 0.7 is shown around each jet. Figure from Blazey and Flaugher(1999).

The jet selection efficiency is $\epsilon > 95\%$. The luminosity is monitored by dedicated detectors close to the beam pipe which are sensitive to almost the total p$\bar{\text{p}}$ cross section. The energy dependence of the total cross section has been given in eqn (3.63). D0 estimates an uncertainty on the luminosity measurement of 6%.

Backgrounds which are potentially dangerous are noisy calorimeter cells or jets induced by beam losses. These are rejected by imposing cuts on the quality of the jets, such as the expected jet shape. The contamination because of backgrounds is reduced to below 1%, checked by simulation studies and even visual scanning of the events with the highest energetic jets.

The most critical experimental aspect is the understanding of the jet energy measurement. From Fig. 6.11 we see that the energy spectrum is falling very steeply, with $\sigma \propto E_T^{-n}$, $n \gg 1$. This means that a small error δE_T on the energy measurement of a jet will lead to a large error on the cross section estimate, $|\delta\sigma/\sigma| \propto n |\delta E_T/E_T|$. A precise measurement is only obtained if the absolute energy scale of the jets is well known, and if the jet energy resolution and consequently the smearing of the jet's E_T is well under control.

The absolute energy calibration of the calorimeters is initially taken from test

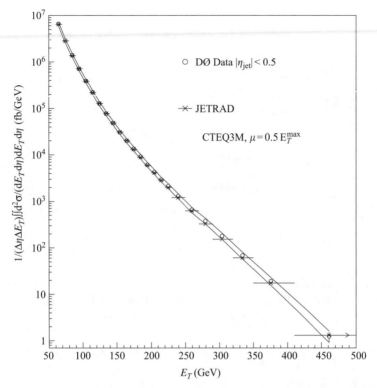

FIG. 6.11. The inclusive jet cross section for $|\eta| < 0.5$, measured by D0 and compared to a NLO QCD prediction. The error band indicates the experimental systematic uncertainty; statistical errors are invisible. Figure from Blazey and Flaugher(1999).

beam measurements. More importantly, it is monitored *in-situ* during the actual running of the experiment by looking for physics processes with well defined constraints on the energy deposits. For the electromagnetic calorimeter these are Z decays into electron–positron pairs or photon pairs from π^0 decays. The hadronic response is checked with photon–jet events, where the precisely measured photon transverse energy has to be balanced by the jet. Other contributions to the absolute jet energy scale come from instrumental effects, pile-up from previous beam–beam crossings, and energy deposits from the underlying event, which is theoretically not so well understood. It is studied in events with no hard scattering at all or from the energy deposits in random cones orthogonal to the jets. It amounts to about 1 GeV for jets with cone-size $R = 0.7$.

The finite resolution of the jet energy measurement distorts the steeply falling energy spectrum. In D0 this resolution is measured (from balancing E_T in jet events) to be of Gaussian shape with a width of about 7 GeV for $E_T = 100$ GeV.

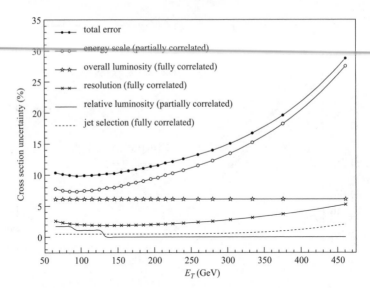

FIG. 6.12. Contributions to the systematic uncertainty of the cross section measurement shown in Fig. 6.11. Figure from Blazey and Flaugher(1999).

The distortion is corrected by parameterizing the shape of a test spectrum,
such as $(aE_T^b)(1 - 2E_T/\sqrt{s})^c$, smearing this spectrum with the known resolution
function, and determining the parameters from a comparison of the smeared
spectrum to the measured distribution. From the ratio of the thus found test and
smeared distributions bin-to-bin correction factors are computed and applied to
the data. This resolution correction reduces the cross section observed by D0 by
$(13 \pm 3)\%$ at $E_T = 60\,\text{GeV}$ and $(8 \pm 2)\%$ at $400\,\text{GeV}$. In CDF both the energy
scale and resolution corrections are determined in a single step by computing
response functions with a Monte Carlo simulation. This simulation is tuned to
describe well electron and hadron test beam results as well as jet fragmentation
such as charged track multiplicities and momenta. They obtain correction factors
for the jet energy between 1.0 and 1.2.

The result of the inclusive jet cross section measurement by D0 (1999a) is
displayed in Fig. 6.11. The jet E_T spectrum is measured for energies between 60
and 450 GeV, with a cross section falling by seven orders of magnitude. The good
agreement within error bands of a next-to-leading order QCD calculation over
the full spectrum is remarkable and a success for the theory. These data give no
indication for new physics. A further discussion of theory-data comparisons for
the inclusive jet cross section will follow in Section 7.4.4.

Figure 6.12 shows the systematic uncertainties of the D0 measurement as
a function of E_T. The total error varies from 10% at the lowest energies up to
about 25% at the upper limit of the spectrum. It is dominated by the uncertainty
on the energy scale.

6.5.3 *Jet rates at* LEP

In our last example we will discuss the measurement of jet rates in e^+e^- annihilations at LEP. We will restrict ourselves to data taken at the Z resonance. The n-jet rate as a function of the resolution parameter y_{cut} is defined as the number of events where exactly n jets are found for a fixed value of y_{cut}, normalized to the total number of hadronic events,

$$R_n(y_{\text{cut}}) = \frac{N_n(y_{\text{cut}})}{N_{\text{had}}} . \tag{6.22}$$

Jet finding algorithms and the concept of resolution parameter have been introduced in Section 6.2. The measurement of jet rates is a basic test for the description of gluon radiation off quarks and allows for a determination of the strong coupling constant.

The standard procedure of triggering for hadronic events has already been described in Section 6.1.1.1. At the offline analysis level further quality cuts are imposed. Here we will summarize typical cuts as used in the ALEPH experiment. For every event classified online as a hadronic event first a sample of 'good' charged tracks is selected. These tracks must have at least four well measured coordinates from the Time Projection Chamber, a polar angle of $|\cos\theta| < 0.95$, a momentum in excess of 200 MeV, and originate from within a cylinder of radius 2 cm and length 10 cm centred around the interaction point. These requirements select tracks which are well within the acceptance of the detector and reject low momentum particles which are possibly related to beam-induced background or nuclear interactions in the detector material. The event is then retained for further analysis if at least five such good tracks are found, if the sum of the charged energy exceeds half of the C.o.M. energy, and if the polar angle of the Thrust axis, computed with good charged tracks only, satisfies $|\cos\theta_{\text{Thrust}}| < 0.9$. The last cut ensures that the event is well contained within the detector, whereas the previous cuts reject background from $Z \to \tau^+\tau^- \to$ hadrons $+ X$ and $\gamma^\star\gamma^\star \to$ hadrons to below a few per mille.

Once an event is selected, the jet clustering algorithm of the Durham- or Jade-type is applied to a list of reconstructed tracks and neutral particles. The latter are identified as energy deposits in the calorimeters above a threshold of several hundred MeV. The algorithm is applied for a fixed number of y_{cut} values, ranging for example from 0.001 to 0.1. The number of jets found defines the classification of the event for a specific y_{cut}. After having analysed all selected hadronic events, which amount to several million at LEP, the n-jet rates are computed according to eqn (6.22). This defines the raw data distribution D^{raw} of jet rates.

Exactly the same analysis is applied to a sample of simulated events, where a Monte Carlo model for quark fragmentation and hadron production, such as described in Section 4, is combined with a detailed simulation of the detector response, giving D^{MC}. Furthermore, the analysis is also applied to the set of simulated hadrons before any detector simulation, which defines the 'true' distribution D^{true}. The distribution of jet rates as a function of y_{cut}, D^{corr}, corrected

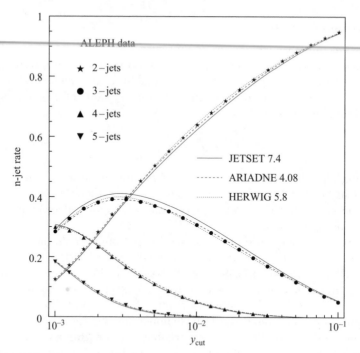

FIG. 6.13. Distribution of jet rates measured by ALEPH as a function of the resolution parameter y_{cut}. Figure from ALEPH Collab.(1998a).

for detector acceptance and resolution effects, is then obtained by computing for every y_{cut} value

$$D^{\mathrm{corr}} = \frac{D^{\mathrm{true}}}{D^{\mathrm{MC}}} D^{\mathrm{raw}} \qquad (6.23)$$

as outlined in Section 6.3.1. To first approximation the correction factor does not depend on the particular Monte Carlo model used. A residual dependence is checked for by computing it with a different model. This leads to systematic uncertainties of the order of 1%. Further systematic checks are performed by repeating the above procedure, but changing the selection cuts on track and event quality. A good test of the reliability of the Monte Carlo simulation is obtained by repeating the analysis using charged tracks only. In the case of a perfect simulation every analysis variation should lead to the same corrected distribution. If there are differences they are taken as estimates of the systematic uncertainty. Because of the huge data sample at LEP the statistical errors are very small, and the total uncertainty is dominated by the systematic uncertainties of a few per cent.

The result of such an analysis by ALEPH is shown in Fig. 6.13 (ALEPH Collab., 1998a). We see that the number of jets found increases with decreasing y_{cut}, since a smaller resolution parameter resolves more and more gluon radiation. The

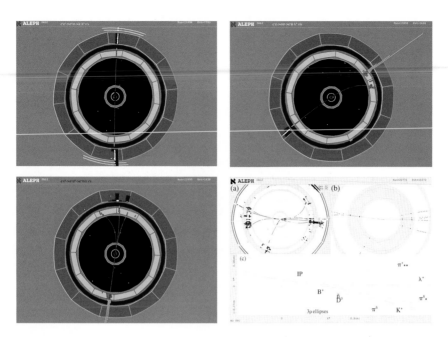

FIG. 5.7. The working of the ALEPH detector illustrated by some typical events as recorded by the experiment. (See p. 225.)

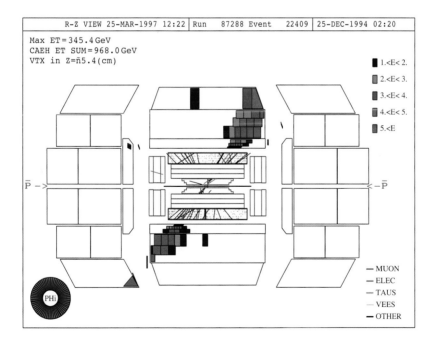

FIG. 6.9. Event display of a di-jet event in the D0 detector. Figure from Blazey and Flaugher (1999). (See p. 251.)

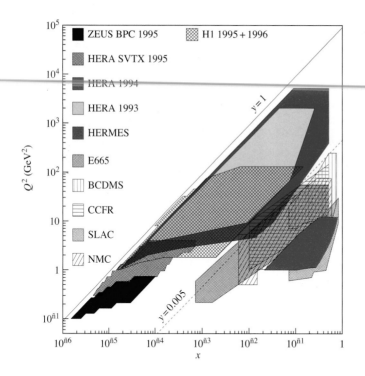

FIG. 7.1. Coverage of the (x, Q^2) plane by various DIS experiments. Figure from Garcia Canal and Sassot (2000). (See p. 259.)

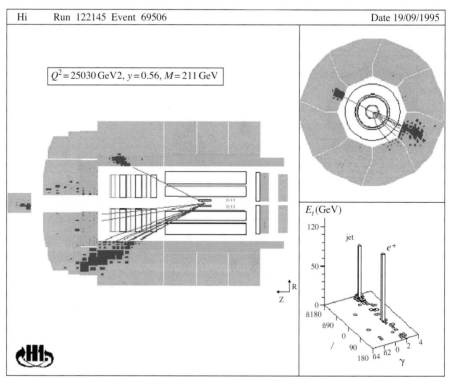

FIG. 7.2. Event display of a neutral current event, observed with the H1 detector at the HERA collider. (See p. 261.)

data are compared to predictions of standard Monte Carlo models described in Section 4. They give a quite satisfactory description of the data, considering the fact that these n-jet rates have not been used for the adjustment of the model parameters.

Exercises for Chapter 6

6–1 Show that the Thrust variable is indeed infrared safe.

6–2 In hadron–hadron collisions the actual hard scattering takes place between two partons coming from the two hadrons, and these two partons carry some fractions x_1 and x_2 ($x_{1,2} \geq 0$) of the longitudinal momentum of the hadrons, leading to a longitudinal momentum imbalance, $x_1 \neq x_2$. Therefore the final state will be boosted along the direction of the beam axis, usually defined as the z-axis. Only transverse to this axis is momentum balance obtained, since the small (and different) transverse momenta of the partons can be neglected in many circumstances. Show that differences of the rapidity y of two jets, $\Delta y = y_{j_1} - y_{j_2}$, are invariant under Lorentz boosts along the z-axis.

6–3 The Breit frame of reference is defined as the Lorentz frame in which the photon, which is exchanged between the incoming lepton and a parton from the probed hadron, is purely spatial ($E = 0$) and collides head on with the hadron. For a particular choice of the z-axis the photon's four-momentum takes on the form $Q = (0, 0, 0, -Q)$, that is, it has zero energy and only its z-momentum component is non-zero. Show that a transformation from the laboratory frame into the Breit frame exists, and discuss the phase space for the incoming and outgoing scattered partons. $^{(\bigstar\bigstar)}$

6–4 Express the kinematic variables Q^2 and y in terms of energies and production angles of the particles in the hadronic system of a deep inelastic scattering event at HERA, assuming the particles to be massless. Exploit the energy-momentum constraint to derive the expressions of the Jacquet–Blondel and of the Sigma method. Assuming the parton model, derive an expression for the angle of the struck parton in terms of the previously derived quantities. $^{(\bigstar)}$

7

STRUCTURE FUNCTIONS AND PARTON DISTRIBUTIONS

Deep inelastic scattering (DIS) of leptons on hadrons is the most fundamental experimental tool for the study of the structure of hadrons. In the experiments either charged (electrons, muons) or neutral (neutrinos) leptons and antileptons are used. The scattering occurs on single free protons (HERA collider) or on bound protons and neutrons within a nucleus (fixed-target experiments). At large momentum transfer the inelastic scattering is described as the incoherent sum of elastic scattering off the hadron constituents, the partons, which are assumed to be pointlike. These constituents are u and d quarks in the simplest quark parton model, whereas QCD predicts the existence of further partons within the hadrons, such as heavier quarks, antiquarks and gluons.

The lepton doesn't feel strong interactions, therefore in principle it can only probe hadron constituents which carry electric or weak charges. Gluons carry neither of these charges, but their existence is probed by the sensitivity of the measurements to radiative corrections as predicted by QCD.

The kinematics of the DIS process has already been discussed in detail in previous chapters. As a reminder, the most important quantities are the momentum transfer $Q^2 = -q^2$ with $q = l - l'$, where $l(l')$ is the momentum of the incoming (outgoing) lepton, and the Bjorken scaling variable $x = Q^2/(2p \cdot q)$, where p is the hadron's momentum. In the parton model x corresponds to the fraction of the hadron momentum carried by the struck parton. Finally, the quantity $y = (p \cdot q)/(p \cdot l) = Q^2/(sx)$ corresponds to the relative energy loss of the lepton in a Lorentz frame where the hadron is at rest. The second equality is true when hadron and lepton masses are neglected, since then the squared C.o.M. energy is simply given by $s = 2p \cdot l$. In Fig. 7.1, the accessible kinematic domain in the (x, Q^2) plane for different experiments is shown. Experiments at HERA (H1, ZEUS) cover a very extended range in $Q^2 \approx 0.1 - 10^4 \, \text{GeV}^2$ as well as in $x \approx 10^{-5} - 10^{-1}$. In particular the very low x and very high Q^2 regions can not be covered by fixed-target experiments, which instead probe the regions at intermediate Q^2 and larger x in a nicely complementary way. It can also be observed that the lower C.o.M. energy available in the fixed-target experiments compared to HERA effectively corresponds to a rescaling to lower values in y. It is worth noting that the momentum transfer Q is a measure of the 'magnifying power' of the lepton probe, since the probed distance scale d is given by the simple relation

$$d \approx \frac{\hbar c}{Q} \approx 0.2 \frac{\text{GeV fm}}{Q} \, .\tag{7.1}$$

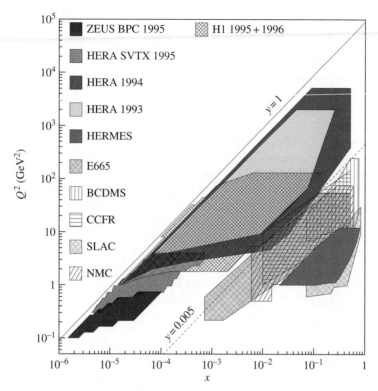

FIG. 7.1. Coverage of the (x, Q^2) plane by various DIS experiments. Figure from Garcia Canal and Sassot(2000).

So we find that $Q^2 = 4 \, \text{GeV}^2$ corresponds to $d \approx 10^{-14} \, \text{cm}$, and correspondingly $Q^2 = 40000 \, \text{GeV}^2$ to $d \approx 10^{-16} \, \text{cm}$.

The theoretical ingredients for the description of DIS experiments are also discussed in previous chapters. There it is explained how the relevant cross sections are expressed in terms of structure functions $F(x, Q^2)$, and the relation between these structure functions and the parton density functions, p.d.f.s, (or parton distributions), which parameterize the sharing of the hadron momentum between the partons, is given. Remember that the structure functions and cross sections are physical (measurable) quantities in the sense that they do not depend on the details of the calculations, such as the number of computed perturbative terms or the factorization scheme, whereas the parton distributions do. Therefore the parton distributions can be viewed as effective quantities such as a running coupling. Once determined, they can be used to predict other structure functions and/or cross sections computed within the same theoretical scheme and order of perturbative expansion.

Why are the measurements of the hadron structure so important? First, they constitute a basic test of the theory, such as the confirmation of the predicted deviations from the simple quark parton model over a large kinematic domain. Second, a detailed understanding of the proton structure, such as the knowledge of which constituents carry which fraction of the proton's momentum, is a necessary ingredient for the computation of cross sections in proton–(anti)proton collisions. Therefore, the future experiments at the TEVATRON and LHC colliders rely heavily on the past, present and future measurements of structure functions and extractions of parton distributions.

In this chapter, we will start with a description of deep inelastic charged lepton–nucleon scattering, with a major focus on experiments performed at the HERA collider at DESY. Then we will discuss neutral lepton–nucleon scattering which is studied in experiments with neutrino beams. Tests of QCD predictions and constraints on parton distributions are obtained from measurements of sum rules. Parton distributions are further constrained by measurements of Drell–Yan, direct-photon and inclusive jet cross sections in hadron–hadron collisions. Finally, all the information is gathered in order to perform so-called global QCD analyses, where the p.d.f.s are determined from combined fits of next-to-leading order (NLO) perturbative QCD predictions to the large data sets from many experiments.

7.1   Charged lepton–nucleon scattering

In this first section, we will concentrate on the reaction $\ell N \to \ell' X$, where the incoming lepton ℓ is charged, such as an electron, positron or muon. N stands for a free nucleon (proton/neutron), or for protons or neutrons within some nucleus, such as deuterium. In the final state we can have any kind of hadronic system, denoted by X. The type of the outgoing lepton ℓ' depends on the type of interaction which occured, that is, on the type of the exchanged particle.

7.1.1   *Neutral current interactions*

If the exchanged particle is a virtual photon $\gamma^\star$ or a Z we talk about a *neutral current* (NC) interaction. An example which is studied in detail at HERA is $e^+ p \to e^+ X$. Up to small electroweak corrections the double differential cross section for NC events is given by

$$\frac{d\sigma_{\mathrm{NC}}}{dx\,dQ^2} = \frac{2\pi\alpha_{\mathrm{em}}^2}{x}\left(\frac{1}{Q^2}\right)^2 \phi_{\mathrm{NC}}(x, Q^2)\,, \tag{7.2}$$

where

$$\phi_{\mathrm{NC}}(x, Q^2) = Y_+ \tilde{F}_2(x, Q^2) - Y_- x\tilde{F}_3(x, Q^2) - y^2 \tilde{F}_L(x, Q^2)\,. \tag{7.3}$$

Here $\alpha_{\mathrm{em}} = 1/137$ is the fine structure constant, and $Y_\pm = 1 \pm (1 - y)^2$. The structure functions $\tilde{F}_{2,3}$ can be expressed as

$$\begin{aligned}
\tilde{F}_2 &= F_2^\gamma - v_e P(Q^2) F_2^{\gamma Z} + (v_e^2 + a_e^2)\, P^2(Q^2)\, F_2^Z \\
\tilde{F}_3 &= - a_e P(Q^2) F_3^{\gamma Z} + (2v_e a_e)\, P^2(Q^2)\, F_3^Z
\end{aligned} \tag{7.4}$$

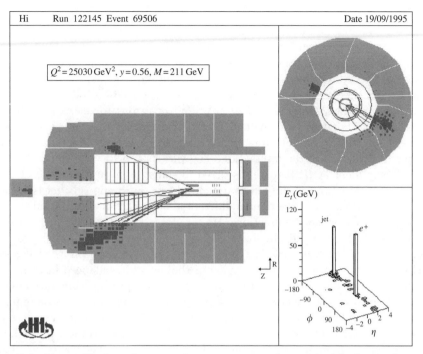

$Q^2 = 25030\,\text{GeV}^2,\ y = 0.56,\ M = 211\,\text{GeV}$

$E_t(\text{GeV})$

120—

jet

e^+

50—

0—

−180

−90

0

ϕ 90

180

−4

−2

0

2

4

η

FIG. 7.2. Event display of a neutral current event, observed with the H1 detector at the HERA collider.

with $P(Q^2) = \kappa^2 Q^2/(Q^2 + M_Z^2)$. M_Z is the Z mass, $\kappa^2 = 1/(4\sin^2\theta_{\rm w}\cos^2\theta_{\rm w})$ is a function of the weak mixing angle $\theta_{\rm w}$, and $v_{\rm e} = -1/2 + 2\sin^2\theta_{\rm w}$ and $a_{\rm e} = -1/2$ are the vector and axial vector couplings of the positron to the Z. F_2^γ is the electromagnetic structure function, often denoted as $F_2^{\rm em}$ or simply as F_2, which originates from photon exchange only. The other structure functions describe the contributions from Z exchange and from Zγ interference. The latter become only relevant for very large momentum transfers, when Q^2 is of the same order or larger than M_Z^2. In particular, the contribution from $x\tilde{F}_3$ becomes visible, leading to a significant reduction in the e$^+$p cross section, only well above $Q^2 = 5000\,\text{GeV}^2$. The contribution from the longitudinal structure function $\tilde{F}_L$ amounts to a few per cent correction to the cross section for $y \geq 0.65$ and $Q^2 \leq 1500\,\text{GeV}^2$, and is negligible for $y\lesssim 0.4$. In leading order QCD we have $\tilde{F}_L = 0$. At this order, and for small enough momentum transfers where contributions from Z exchange can be ignored, the cross section is determined by the electromagnetic structure function $F_2(x, Q^2) \approx \frac{1}{Y_+}\phi_{\rm NC}^{\rm LO}$, which is related to the parton distributions as

$$F_2(x, Q^2) = x\left[\frac{4}{9}\left(u(x) + c(x) + \bar{u}(x) + \bar{c}(x)\right) + \frac{1}{9}\left(d(x) + s(x) + \bar{d}(x) + \bar{s}(x)\right)\right].$$

$$(7.5)$$

In this Q^2 regime and at large x the cross section is mainly determined by the valence u-quark distribution $u_v(x) = u(x) - \bar{u}(x) \approx u(x)$.

For the description of the measurements we will mainly concentrate on experiments at HERA, which cover the largest kinematic range. This is because of the large C.o.M. energy of $\sqrt{s} \approx 300\,\text{GeV}$, achieved by a head-on collision of protons with $E_p = 820\,\text{GeV}$, later even $920\,\text{GeV}$, and positrons (or electrons) close to $E_e = 28\,\text{GeV}$. Furthermore, the collider experiments H1 and ZEUS are almost hermetic multi-purpose detectors, which allow to measure to high accuracy the complete final state of the interaction. The various aspects of a structure function measurement at HERA are discussed in some detail in Section 6.5.1, such as the reconstruction of the kinematic variables, the luminosity measurement and the measurement of the double differential cross section and the subsequent extraction of the structure functions.

In Fig. 7.2 an event display of a NC event in the H1 detector is shown. The proton beam enters from the right side. The positron arrives from the left and is backscattered into the upper left region of the detector, balanced in transverse momentum by a hadronic jet in the lower region. The proton remnant manifests itself as hadronic activity close to the beam pipe in the proton direction of flight.

The event selection of NC events is based on the identification of a scattered positron with large transverse momentum $p_T \gtrsim 10\,\text{GeV}$ with respect to the beam line, as, for example, clearly visible in Fig. 7.2, and an inelasticity y smaller than about 0.9 in order to ensure a precise kinematic reconstruction. Several methods as described in Section 6.5.1 are used for this latter purpose, since in NC events there is redundant information from the simultaneous reconstruction of the scattered positron and the hadronic final state. Depending on the kinematic region the method with the best resolution is adopted. Measurements have been performed up to the highest values of $Q^2 \approx 40000\,\text{GeV}^2$ by H1 (2000) and ZEUS (1999a). With integrated luminosities around $40\,\text{pb}^{-1}$ from the data taking period 1994–97, the experiments have collected up to 75000 NC events in the range $Q^2 \approx 150 - 40000\,\text{GeV}^2$. Dedicated measurements of the low-Q^2 region will be discussed in a later subsection. The cross sections are measured with a few per cent statistical accuracy and systematic errors of about 4%, dominated by uncertainties in the energy scale and identification efficiency of the scattered positron as well as by uncertainties on the energy scale of the hadronic final state. The luminosity and thus the overall normalization is known within a precision of 1.5%. For $Q^2 \lesssim 5000\,\text{GeV}^2$ the systematic uncertainties dominate, whereas at larger Q^2 the statistical uncertainties increase strongly because of the steeply falling cross section.

The electromagnetic structure function $F_2(x, Q^2)$ is extracted from the measured double differential cross section by correcting for electroweak radiative effects, the contributions from Z exchange and the contributions from F_3 and F_L as predicted by the NLO DGLAP evolution equations. Examples of measurements as a function of x and Q^2 are shown in Figs. 7.3 and 7.4. Nice consistency with fixed-target data at smaller Q^2 by NMC (1995) is found. Measurements at

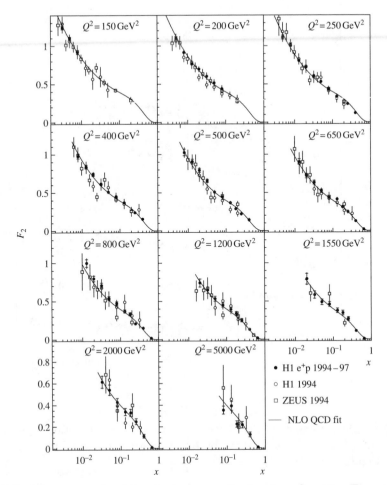

FIG. 7.3. Measurements of the electromagnetic structure function F_2 as a function of x for various Q^2 values, by H1 and ZEUS, compared to a NLO QCD fit. Figure from H1 Collab.(2000).

lower Q^2 are shown in Fig. 6.8. All measurements are well described by the Q^2 evolution of F_2 as predicted by the NLO DGLAP equations from $Q^2 \approx 1\,\text{GeV}^2$ up to the highest measured Q^2. For further discussion see also Sections 3.2.2, 3.6 and 7.5. A positive slope as a function of Q^2 is visible in Fig. 7.4 for the low-x data points and this slope decreases with increasing x as expected from QCD. This is because for low x and increasing Q^2 the exchanged photon can resolve more and more partons within the proton, arising from parton cascades like gluon radiation and gluon splitting into $q\bar{q}$-pairs. At larger x the probability decreases to find gluons or sea quarks. There the structure function is dominated by the valence quark distribution, which varies more slowly with Q^2. In Fig. 7.3

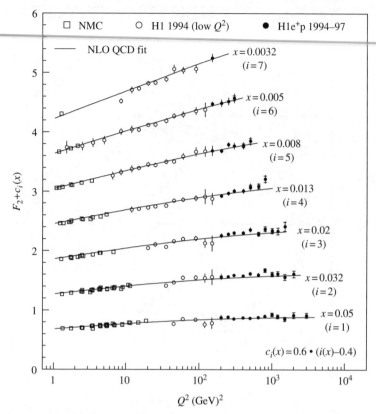

Fig. 7.4. Measurements of the electromagnetic structure function F_2 as a function of Q^2, for various x values, by H1 and NMC, compared to a NLO QCD fit. Figure from H1 Collab.(2000).

a strong rise of the structure function at low x can be seen, which is interpreted as the strong rise of the contribution from gluons in this kinematic regime. In contrast, at the largest x values the structure function decreases rapidly, due to a rapid decrease of the valence quark contribution.

7.1.2 *Charged current interactions*

If the exchanged particle in the deep inelastic scattering is a W boson, we talk about a *charged current* (CC) interaction. An example which is studied at HERA is $e^+ p \rightarrow \bar{\nu}_e X$. Up to small electroweak corrections the double differential cross section for CC events is given by

$$\frac{d\sigma_{CC}}{dx\,dQ^2} = (1 - P_e) \frac{G_F^2}{2\pi x} \left(\frac{M_W^2}{Q^2 + M_W^2} \right)^2 \phi_{CC}(x, Q^2) . \qquad (7.6)$$

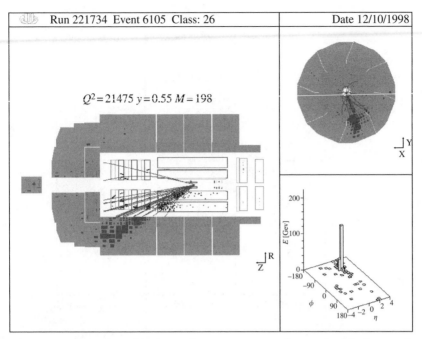

Run 221734 Event 6105 Class: 26   Date 12/10/1998

$Q^2 = 21475 \; y = 0.55 \; M = 198$

FIG. 7.5. Event display of a charged current event, observed with the H1 detector at the HERA collider.

Here $P_e = (N_L - N_R)/(N_L + N_R)$ is the degree of polarization of the lepton beam, where $N_{L(R)}$ is the number of left(right)-handed positrons. We see that the CC cross section has a structure very similar to the NC cross section, eqn (7.2), the only difference being that the fine structure constant α_{em} is replaced by the Fermi coupling constant G_F and the photon propagator term $1/Q^4$ is replaced by the corresponding W propagator. This propagator structure tells us immediately that the CC cross section is much smaller than the NC one, and relevant contributions can only be expected for $Q^2 \gtrsim M_W^2$. The term $\phi_{CC}(x, Q^2)$ can be decomposed into structure functions in exactly the same way as $\phi_{NC}(x, Q^2)$. However, let us write it down immediately in terms of p.d.f.s:

$$\phi_{CC}^{LO} = x \left[(\bar{u}(x) + \bar{c}(x)) + (1 - y)^2 \, (d(x) + s(x)) \right] . \qquad (7.7)$$

This expression is exact in LO QCD. In order to take into account quark mixing the individual terms would have to be weighted by the relevant squared matrix elements of the CKM matrix. Since for an incoming positron the exhanged W boson has positive charge, the cross section is sensitive to down-type quarks and up-type antiquarks. In the case of an incoming electron the expression would be changed to $\left[(u + c) + (1 - y)^2 (\bar{d} + \bar{s}) \right]$. In addition, the polarization factor is changed to $(1 + P_e)$. Thus we see that CC interactions can distinguish be-

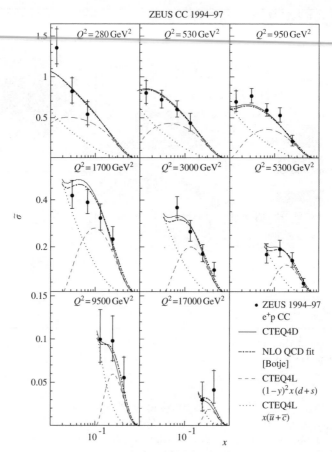

FIG. 7.6. Measurement of the reduced charged current cross section by ZEUS as a function of x in different Q^2 bins. Also indicated are NLO QCD fits and several parameterizations of p.d.f.s. Figure from ZEUS Collab.(2000a).

tween quarks and antiquarks, which is not possible for photon exchange in NC interactions.

The experiments at HERA have observed this type of inelastic scattering. The event selection of CC events is based on the identification of large missing transverse momentum $p_T^{\text{miss}} \gtrsim 10$ GeV with respect to the beam line. An example of a CC event candidate observed with the H1 detector is displayed in Fig. 7.5. In the plane orthogonal to the beam line, the xy-plane, we see very clearly the momentum imbalance, since the antineutrino has escaped undetected. There is hadronic activity with large transverse momentum in the lower region of the detector, and some energy deposits are observed very close to the beam pipe in the proton direction stemming from the proton remnant. A further event selection criterion is to restrict the inelasticity y to a region, for example, $0.03 -$ 0.85, where a good kinematic reconstruction is possible. In this case the kinematic reconstruction has to be performed using information from the hadronic final state only, as described in Section 6.5.1. The backgrounds are more difficult to reject in this case. Transverse momentum imbalance can be caused by NC events where the scattered positron is not identified, by photo-production events with $Q^2 \approx 0$ where the positron escapes undetected into the beam pipe, or by events which are not induced by e^+p collisions, such as cosmic ray interactions. Because of the much lower cross section and the tighter selection cuts that are needed in order to reject backgrounds, the number of selected CC events for a luminosity of about $40\,\mathrm{pb}^{-1}$ is of the order of 1000 only, to be compared to 75000 NC events. This leads to larger statistical uncertainties. Also the systematic uncertainties are a factor of two larger than for analyses of NC interactions. Recent measurements are summarized in publications by the HERA experiments (H1 Collab., 2000; ZEUS Collab., 2000a).

Figure 7.6 shows the reduced CC cross section $\tilde{\sigma}_{CC}$, which up to electroweak corrections is equal to ϕ_{CC}, measured by ZEUS as a function of x in Q^2 bins from $Q^2 = 280\,\mathrm{GeV}^2$ to $Q^2 = 17000\,\mathrm{GeV}^2$. Good agreement is seen with a NLO QCD fit. Also shown are parameterizations of the p.d.f.s extracted by various groups, c.f. Section 7.5. As anticipated, the cross section at large x is dominated by the scattering off a d-quark. For lower x the cross section rises because of the increasing contribution from $\bar{u}$ and $\bar{c}$ quarks from the sea. At the same time d- and s-quarks get suppressed because of the $(1-y)^2$ factor. Remember that low x corresponds to large y values.

An interesting aspect to study in NC and CC deep inelastic scattering is the helicity dependence of the cross sections at large x. The experiments at HERA are sensitive to the contributions from Z and W exchange, and thus to electroweak effects. In particular, only left(right)-handed (anti)quarks participate in the weak part of the interaction. Since certain spin configurations are forbidden by angular momentum conservation, as shown in Ex. (7-1), an asymmetry in the positron scattering angle $\theta_e^\star$ defined in the positron-quark C.o.M. system appears. The weighting factor $(1-y)^2$ in ϕ_{CC} for down-type quarks can be understood because of the relation $\cos^2(\theta_e^\star/2) = 1 - y$. In Fig. 7.7 the H1 measurement of the cross section term ϕ_{CC} is shown as a function of $(1-y)^2$, for various bins in the large x region. In leading order we expect a dependence proportional to $(1-y)^2$ from positron–quark (d,s) scattering, and an isotropic distribution from positron–antiquark ($\bar{u},\bar{c}$) scattering. In fact we observe an almost linear dependence of ϕ_{CC}, with a finite offset, which decreases with increasing x. Therefore these measurements can help to constrain strongly the various quark contributions in the large x region.

Finally, the measurements of the Q^2 dependence of the NC and CC cross sections up to the highest values of Q^2 allow for a beautiful visualization of the unification of electromagnetic and weak interactions. This is illustrated in Fig. 7.8. At low Q^2 virtual photon exchange dominates the NC interactions, and

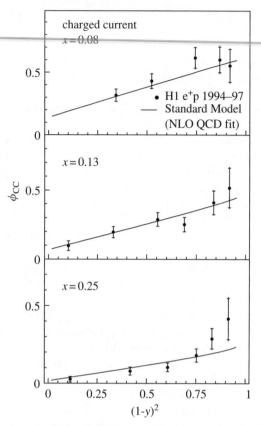

FIG. 7.7. Measurement of the CC cross section term ϕ_{CC} by H1 as a function of $(1-y)^2$, together with a NLO QCD fit. Figure from H1 Collab.(2000).

CC events are suppressed by many orders of magnitude. However, with increasing Q^2 both cross sections approach each other, showing that the electromagnetic and weak contributions become of similar size. Note the fact that for large Q^2 the CC cross section for electrons is higher than for positrons. This is because for electrons a W^- is exchanged, which couples to up-type quarks. Those are more abundant in a proton than down-type quarks. In addition, for positron scattering the helicity structure of the interaction leads to an additional suppression in certain phase-space regions, as discussed above. In the highest Q^2 region also the NC cross section is larger for electrons than for positrons. In this region the interference between photon and Z exchange becomes relevant, which explains the observed asymmetry.

Any deviations of the measurements from the Standard Model prediction at the largest Q^2 values would indicate the possible appearance of new physics, such as the exchange of a lepto-quark instead of a photon. Lepto-quarks are

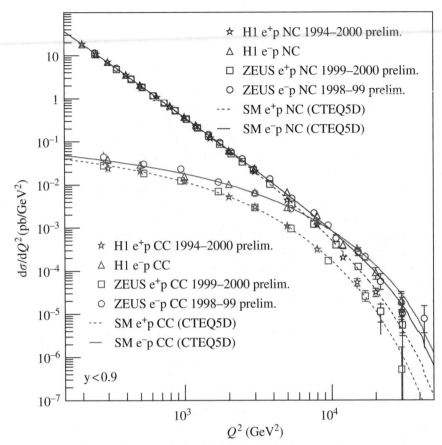

FIG. 7.8. Measurements of the neutral and charged current cross sections as a function of Q^2 compared to the Standard Model (SM) prediction. Figure from Long *et al.*(2001).

hypothetical particles which carry lepton and baryon quantum numbers. So far no indications for such particles have been found.

7.1.3 *The low-x and low-Q^2 region*

So far we have learned about structure function measurements in the medium and large Q^2 regime at HERA. We have seen that there is a strong rise in the electromagnetic structure function $F_2(x, Q^2)$ when going to small x, $x \lesssim 0.01$. In this regime the contribution from valence quarks is small, whereas gluons and quarks from the sea dominate. In fact, the strong rise in $F_2(x, Q^2)$, which is of the form $F_2 \propto x^{-\lambda}$, $\lambda > 0$, is attributed to a strong rise in the gluon and sea quark distributions. This rise is also generated by the DGLAP evolution equations, if restricted to the perturbative regime $Q^2 \gtrsim 1\,\text{GeV}^2$. In the small-$x$ region the evolution is governed by

$$\frac{\mathrm{d}F_2(x, Q^2)}{\mathrm{d}\ln Q^2} \propto \alpha_\mathrm{s}(Q^2)\, P_{\mathrm{qg}} \otimes g(x)\,, \tag{7.8}$$

where P_{qg} is the Altarelli–Parisi splitting function as defined in Section 3.2.2, and $g(x)$ is the gluon distribution. We use here the short-hand notation $\otimes$ for the convolution integral as explained in eqn (3.49). In order to solve the equation, $F_2(x, Q^2)$ has to be parameterized as a function of x at some starting scale Q_0^2, since in contrast to the evolution the absolute value of $F_2(x, Q_0^2)$ can not be predicted by perturbative QCD.

At this stage several interesting questions arise. When approaching the very low Q^2 region, $Q^2 \approx 1\,\mathrm{GeV}^2$ and below, thus entering the non-perturbative regime, does this rise in F_2 for low x persist? On the other hand, when staying in a regime where perturbative QCD is applicable, do the DGLAP evolutions equations give the correct prediction for the extremely small $x \lesssim 10^{-3}$ region, where $\ln(1/x)$ terms might become important? Finally, what is the asymptotic behaviour of F_2 for $x \to 0$? Some theoretical introduction to these topics is already given in Section 3.6. Here we look at the subject also from the experimental point of view.

In order to tackle the first question, we definitely have to advocate non-perturbative models. Deep inelastic lepton–proton scattering at $Q^2 \lesssim 1\,\mathrm{GeV}^2$ is related to the cross section for the scattering of a virtual photon $\gamma^\star$ and a proton, $\sigma^{\gamma^\star \mathrm{p}}$, which is proportional to F_2. A possible model for the description of $\gamma^\star$p scattering is based on the so-called *Generalized Vector Meson Dominance* (GVMD) model (Sakurai and Schildknecht, 1972) combined with ideas from Regge theory introduced in Section 3.2.3. At very low Q^2 the virtual photon fluctuates into vector meson states, for example, ρ, ω, ϕ, which then interact with the proton. Regge theory predicts the energy dependence of the total cross section for hadron–hadron scattering,

$$\sigma^{\mathrm{h_1,h_2}} \propto A\, s^{\alpha_{\mathbb{R}} - 1} + B\, s^{\alpha_{\mathbb{P}} - 1}\,, \tag{7.9}$$

where s is the energy squared of the hadron–hadron interaction, $\alpha_{\mathbb{R}} = 0.5$ is the *Reggeon* contribution, describing the exchange of hadronic states with definite quantum numbers such as vector mesons, and $\alpha_{\mathbb{P}} \approx 1.1$ is due to *Pomeron* exchange, an apparently colour-less multi-gluon state with the quantum numbers of the vacuum. The relation to deep inelastic scattering at low x is given by the fact that the energy squared W^2 for the $\gamma^\star$p (or vector-meson–proton) scattering can be written as $W^2 \approx Q^2/x$, eqn (2.4). Putting all these ingredients together, a behaviour for F_2 such as

$$F_2(x, Q^2) \propto \left(\frac{M_0^2}{M_0^2 + Q^2}\right) \left[A\left(\frac{Q^2}{x}\right)^{\alpha_{\mathbb{R}} - 1} + B\left(\frac{Q^2}{x}\right)^{\alpha_{\mathbb{P}} - 1} \right] \tag{7.10}$$

is expected, where the first term in brackets is motivated by GVMD, and the second by Regge theory. We see that $F_2(x \to 0) \propto x^{-\lambda_{\mathrm{eff}}}$, with a rather small

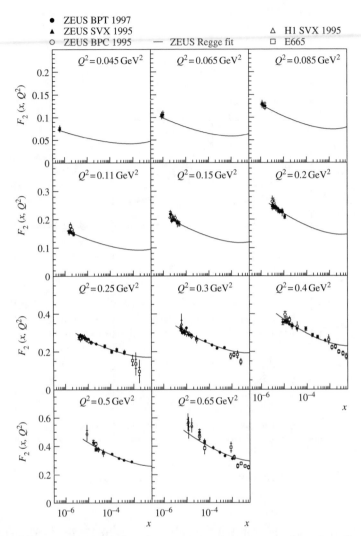

ZEUS BPT 1997
ZEUS SVX 1995   △ H1 SVX 1995
ZEUS BPC 1995   — ZEUS Regge fit   □ E665

FIG. 7.9. Measurements of the electromagnetic structure function F_2 for very
    small Q^2 by ZEUS, H1 and E665. Superimposed are predictions based
    on Regge theory by Donnachie and Landshoff (1992, 1994). Figure from
    ZEUS Collab.(2000b).

$\lambda_{\mathrm{eff}} \approx 0.1$. So a very soft and Q^2-independent x-behaviour is predicted, without
any strong rise.

In order to test these predictions experimentally, the experiments at HERA
have upgraded their detectors and used dedicated HERA runs. Very small Q^2 values have to be measured, which is equivalent to detecting the outgoing positron or
electron at very small angles. Special detectors very close to the beam pipe have

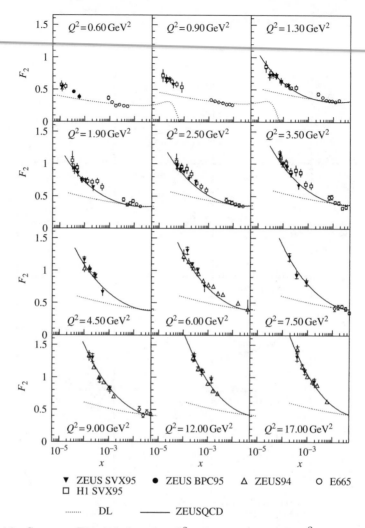

FIG. 7.10. Same as Fig. 7.9, but for Q^2 values up to $17\,\mathrm{GeV}^2$. Also shown are the predictions from NLO perturbative QCD (full lines) and Regge theory by Donnachi and Landshoff (1992, 1994) (dotted lines). Figure from ZEUS Collab.(1999b).

been installed, such as the *Beam Pipe Calorimeter* (BPC) and the *Beam Pipe Tracker* (BPT) in the case of ZEUS. These detectors are positioned at a distance of about 3 m from the main interaction vertex, in the positron-beam direction. Another way of accessing smaller scattering angles is to shift the interaction vertex by about 70 cm in the proton-beam direction, as was done for a short period in 1995. Previously unexplored regions such as $0.045\,\mathrm{GeV}^2 < Q^2 < 0.65\,\mathrm{GeV}^2$ and

$6 \cdot 10^{-7} < x < 10^{-3}$ have thus been studied (ZEUS Collab. 1999b, 2000b; H1 Collab. 1997a). The main background for these measurements is photo-production, where the positron escapes undetected into the beam pipe, and a signal is faked in the low angle detectors by electromagnetic showers induced mainly by π^0s. The extraction of the structure function after the measurement of the double differential cross section follows mainly the lines described in Section 6.5.1.

In Figs. 7.9 and 7.10 the measurements of $F_2(x, Q^2)$ are displayed, from $Q^2 = 17\,\mathrm{GeV}^2$ down to $Q^2 = 0.045\,\mathrm{GeV}^2$, as obtained by the HERA experiments and by the fixed-target experiment E665 (1996) at FERMILAB. A decrease of the rise for low x is very nicely visible when going to smaller and smaller Q^2 values. In the perturbative region $Q^2 \gtrsim 1\,\mathrm{GeV}^2$ a NLO QCD fit describes the data well, whereas the Regge model predicts a too soft x-dependence. However, the latter model, eqn (7.10), has been fitted to the data (ZEUS Collab., 2000b) at $Q^2 < 1\,\mathrm{GeV}^2$, and indeed a good fit is obtained, with $A = 147.8 \pm 4.6\mu\mathrm{b}$, $\alpha_{\mathrm{IR}} = 0.5$, $B = 62.0 \pm 2.3\mu\mathrm{b}$, $\alpha_{\mathrm{IP}} = 1.102 \pm 0.007$ and $M_0^2 = 0.52 \pm 0.04\,\mathrm{GeV}^2$. The quoted uncertainties combine statistical and systematic errors.

The transition from the perturbative to the Regge behaviour for F_2 has been studied by measurements of $\lambda_{\mathrm{eff}} = \mathrm{d}\ln F_2/\mathrm{d}\ln(1/x)$, obtained by fitting $F_2 = ax^{-\lambda_{\mathrm{eff}}}$ to ZEUS and E665 data with $x < 0.01$ (ZEUS Collab., 1999b). For $Q^2 < 1\,\mathrm{GeV}^2$ the data indicate a Q^2-independent parameter $\lambda_{\mathrm{eff}} \approx 0.1$, as suggested by Regge theory. At larger Q^2 values λ_{eff} increases with increasing Q^2 to values around $\lambda_{\mathrm{eff}} \approx 0.3$, and the data are nicely described by a NLO QCD fit based on the DGLAP evolution equations.

It is worth noting that, for very small x, terms of the form $(\alpha_s \ln(1/x))^n$ become relevant for every order in the perturbative series. The DGLAP evolution equations only resum logarithmic terms $(\alpha_s \ln Q^2)^n$. A resummation of the logarithms $(\ln(1/x))^n$ is achieved by the BFKL equations, see Section 3.6, which in leading order predict $\lambda_{\mathrm{eff}} \approx 0.4 - 0.5$. However, studies of NLO contributions to this prediction indicate rather large corrections, so that further development in this area is to be awaited before final conclusions can be drawn.

At this stage the third question about the asymptotic behaviour is to be faced. At large Q^2 the rise in the gluon distribution is strong, and at lower Q^2 the sea increases steadily. However, the rise cannot persist for $x \to 0$, as we hit basic physical boundaries. If the number of partons within the proton becomes very large, they cannot be regarded as free particles any more, as assumed in the parton model. Dynamical mechanisms such as gluon recombination should lead to parton saturation, c.f. Section 3.6.6.4. This is a developing field of QCD, and future precise HERA data should help to constrain further the various models which are on the market.

7.1.4 *The gluon density in the proton*

A standard method to infer the gluon distribution in the proton from DIS data is to perform a global analysis of scaling violations in the structure function $F_2(x, Q^2)$. This will be described in detail in Section 7.5. However, this method

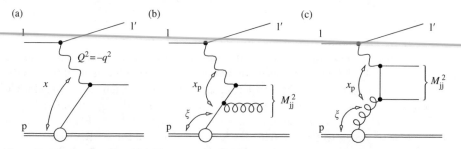

FIG. 7.11. Diagrams of different processes in deep inelastic lepton–proton scattering: (a) Born process, (b) QCD Compton scattering and (c) the boson–gluon fusion. Figures from H1 Collab.(2001).

is indirect in the sense that in DIS the exchanged boson probes valence quarks at large x, and sea quarks at low x, and the dependency on the gluon distribution enters only via the QCD evolution equations as indicated in eqn (3.49) or (7.8).

A more direct determination of the gluon density can be obtained by the reconstruction of the interacting partons in gluon-induced processes. Such methods are more sensitive to local variations of the gluon distribution. However, they are still limited in statistics. Nevertheless they constitute an independent test of perturbative QCD. Actual examples are the measurements of production rates for jets or heavy quark mesons.

Let's first discuss jet rate measurements. The hadronic final state of NC DIS events is analysed by looking for more than one highly energetic jet in addition to the proton remnant. The production of such events occurs via the diagrams depicted in Fig. 7.11. In leading order (diagram (a)) only one jet with large transverse momentum with respect to the beam line can be produced, and no sensitivity to the gluon distribution is obtained. When going to NLO ((b) and (c)), two or more jets can be produced, via gluon radiation from a quark line or via the boson–gluon fusion process, which then gives direct sensitivity to the gluon density. The multi-jet final state can be characterized by variables such as the invariant mass M_{jj} or the scaling variables

$$\xi = x \left(1 + \frac{M_{jj}^2}{Q^2} \right) \quad \text{and} \quad x_p = \frac{x}{\xi} , \qquad (7.11)$$

where $x = Q^2/(2p \cdot q)$ is the standard Bjorken scaling variable. In the leading order picture ξ is the momentum fraction of the struck parton, whereas x_p specifies the fractional momentum of the incoming parton seen by the boson. In the hadronic final state jets have to be defined according to some jet algorithm, such as described in Section 6.2. A good jet algorithm should approximate well the kinematics of the underlying partonic reaction.

H1 (2001) has performed a measurement of jet cross sections in the kinematic region $0.2 < y < 0.6$ and $5 < Q^2 < 15000 \,\text{GeV}^2$. The cross sections to produce

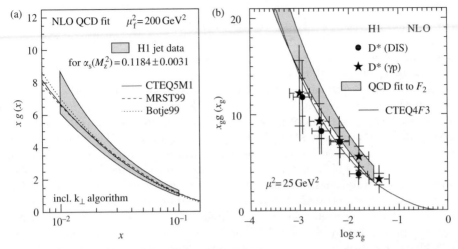

FIG. 7.12. Direct measurements of the gluon density in the proton by H1, based on jet cross sections (a) and from an analysis of D* production (b). A comparison is given to the results from global analyses of scaling violations. Figures from H1 Collab.(2001) (a) and H1 Collab.(1999a) (b).

two or more jets within the angular acceptance have been measured as a function of variables such as the jet transverse energy. The fraction of di-jet events in NC DIS varies strongly with Q^2, between 1% at $Q^2 = 5\,\mathrm{GeV}^2$ to $\approx 20\%$ at $Q^2 = 5000\,\mathrm{GeV}^2$. When requiring an approximate relative transverse momentum of $k_T > 10\,\mathrm{GeV}$ between jets, using a Durham type jet finder, H1 finds about 11000 (2900) di-jet events in the low (high) Q^2 sample. The rates are measured within systematic uncertainties of about 10%, mainly dominated by the uncertainties on the calorimeter energy scales.

The measured rates are well described by Monte Carlo simulations based on perturbative QCD calculations. This gives confidence for the extraction of the p.d.f.s $q^{\mathrm{P}}(x)$ in the proton according to $\sigma_{\mathrm{jet}} = \hat{\sigma}_{\mathrm{q}} \otimes q^{\mathrm{P}}$, where σ_{jet} is any kind of jet cross section, obtained from the convolution of the related partonic cross section $\hat{\sigma}_{\mathrm{q}}$ involving a parton q with the respective p.d.f. For jet production in NC DIS the relevant distributions are the sum over quark and antiquark distributions, weighted by the square of the quark charges e_{q}^2, $x\Sigma(x) = x\sum e_{\mathrm{q}}^2(q^{\mathrm{P}}(x) + \bar{q}^{\mathrm{P}}(x))$, and the gluon distribution $g(x)$. A measurement of the inclusive DIS cross section is mainly sensitive to Σ, that is, $\sigma_{\mathrm{inc,DIS}} \propto \Sigma$, which can then be used as constraint in the extraction of the gluon density, since $\sigma_{\mathrm{jet}} \propto \alpha_{\mathrm{s}}(c_{\mathrm{g}}g + c_{\Sigma}\Sigma)$. The coefficients c_{g} and c_{Σ} are calculable in perturbative QCD. The only remaining unknowns are α_{s} and $g(x)$. Taking the world average value of α_{s}, a direct determination of $g(x)$ is possible. The H1 data allow for such a determination within the range $0.01 < x < 0.1$. The result obtained for a factorization scale $\mu_F^2 = 200\,\mathrm{GeV}^2$ is shown in Fig. 7.12 (a), where the shaded band reflects experimental and theoretical uncertainties. Good agreement is observed with the results obtained from global analyses of scaling violations. By integrating $xg(x)$ it is found that at a scale of $200\,\mathrm{GeV}^2$ gluons carry about 23% of the proton momentum over the measured x-interval.

Another direct test of the gluon distribution is obtained from the measurement of the production rate of heavy quarks. Heavy quark production proceeds almost exclusively via boson–gluon fusion, as shown in Fig. 7.11 (c), where the two quark lines now represent a c- or b-quark. Additional theoretical input is needed in order to describe the transition of a heavy quark to a heavy meson, such as c $\to$ D*, which is detected in the experiment. For this purpose phenomenological fragmentation functions are employed, which model the fraction of the quark's energy transferred to the heavy meson. An example is given by the Peterson et al. fragmentation function (1983). The fraction x_g of the proton's momentum carried by the struck gluon is again computed from the standard DIS kinematic variables and the charm quark kinematics. Compared to the dijet case smaller invariant masses of the partonic subprocess can be accessed in charm production. Therefore it is possible to extend the gluon density determination towards smaller momentum fractons x_g. The heavy quark analyses suffer from low statistics, but are less sensitive to background induced by such diagrams as (a) and (b) in Fig. 7.11.

Charm quarks are tagged by searching for D* mesons in the hadronic final state. A search for the decay chain D$^{\star+}$ $\to$ D$^0\pi_{\mathrm{slow}}^+$ $\to$ $(\mathrm{K}^-\pi^+)\pi_{\mathrm{slow}}^+$ and its charged conjugate is performed, for which the overall branching fraction is 2.62%. Here π_{slow}^+ indicates a pion track with very low momentum, below a few GeV. This decay chain is subject to tight kinematic conditions, and subsequently D$^{\star+}$ production is found as a distinct enhancement in the distribution of the mass difference $\Delta M = M(\mathrm{K}^-\pi^+\pi_{\mathrm{slow}}^+) - M(\mathrm{K}^-\pi^+)$ around the expected value of 145.4 MeV.

The H1 Collaboration (1999a) has performed such a measurement for NC DIS and photoproduction events. In a data sample of $\mathcal{L} = 9.7\,\mathrm{pb}^{-1}$ they find about 580 events, with an estimated efficiency for reconstructing the scattered electron and a D* meson of around 42%. The gluon distribution is extracted according to

$$\sigma(x_g^{\mathrm{obs}}) = \int dx_g \left[g(x_g, \mu^2)\hat{\sigma}(x_g, \mu^2) A(x_g^{\mathrm{obs}}, x_g, \mu^2) \right] + \sigma_q(x_g^{\mathrm{obs}}) , \qquad (7.12)$$

where $\sigma(x_g^{\mathrm{obs}})$ is the measured D* cross section as a function of the reconstructed gluon momentum fraction, x_g is the true momentum fraction, $g(x_g, \mu^2)$ is the gluon distribution at a given scale μ^2, $\hat{\sigma}(x_g, \mu^2)$ is the partonic cross section for charm quark production from boson–gluon fusion, and $\sigma_q(x_g^{\mathrm{obs}})$ is the background from quark-induced processes. The resolution matrix $A(x_g^{\mathrm{obs}}, x_g, \mu^2)$ gives the probability to observe a value x_g^{obs} when the actual true value was x_g. This matrix is obtained from Monte Carlo simulations. The average scale μ^2 relevant for the H1 measurement is determined to be $25\,\mathrm{GeV}^2$. The results of this analysis are displayed in Fig. 7.12 (b) for DIS and photoproduction (γp). Within the statistical and systematic uncertainties good agreement is found with the indirect determinations from scaling violation analyses.

7.2 Neutrino–nucleon scattering

After the discussion of charged lepton–nucleon scattering, we now focus on neutrino–nucleon or equivalently neutral lepton–nucleon scattering. The reaction of interest is $\nu_\ell(\bar{\nu}_\ell)\mathrm{N} \to \ell(\bar{\ell})X$, with an incoming (anti)neutrino scattering on a nucleon N, which may actually be inside a nucleus such as iron, and the corresponding lepton as well as a hadronic system X in the final state. Thus, the same situation as depicted in Fig. 2.1 is given, except that the incoming charged lepton is replaced by a neutrino, and the exchanged virtual photon by a W boson. Note also that by interchanging the order of the neutrino and the lepton, the configuration for charged current DIS is restored.

The particular interest in neutrino–nucleon scattering is understood from the following observation. The neutrino couples to other particles only through weak interactions, via the exchange of a W boson, which can probe different quantum numbers of the proton's constituents compared to a virtual photon. In particular, it is possible to distinguish between particles and antiparticles. In this section we will concentrate on charged current interactions. However, neutral current processes where a Z is exchanged with consequently a neutrino instead of a lepton in the final state, can occur and are of interest for precision tests of the Standard Model of electroweak interactions. As an example, the weak mixing angle $\sin^2 \theta_\mathrm{w}$ is measured in such experiments. A very comprehensive review on measurements with high energy neutrino beams can be found in the literature (Conrad *et al.*, 1998), which summarizes experimental tests of QCD as well as electroweak theory.

Let us first summarize briefly the kinematics as well as the relevant cross sections and structure functions before going to a description of the experimental results. The relevant Lorentz-invariant quantities are the same as for charged lepton DIS, namely the momentum transfer $Q^2 = -q^2$, the Bjorken scaling variable $x = Q^2/(2p \cdot q)$, the inelasticity $y = (p \cdot q)/(p \cdot l)$ and the energy transfer $\nu = (p \cdot q)/M$, where M is the proton mass. All neutrino experiments are fixed-target experiments, where the nucleon is at rest to a very good approximation. Therefore, in the laboratory frame the kinematic variables can be expressed in terms of the measured energy E_μ and angle θ_μ of the outgoing lepton (typically a muon) as well as of the energy E_H of the hadronic system:

$$\nu = E_\mathrm{H} \ , \ y = \frac{E_\mathrm{H}}{E_\mathrm{H} + E_\mu} \ , \ Q^2 = (E_\mathrm{H} + E_\mu)E_\mu\theta_\mu^2 \ , \ x = \frac{(E_\mathrm{H} + E_\mu)E_\mu\theta_\mu^2}{2ME_\mathrm{H}} \ , \quad (7.13)$$

where we have used a small-angle approximation and $E_\nu = E_\mu + E_\mathrm{H}$, neglecting the proton rest energy in the initial state. The C.o.M. energy squared is $s = (2E_\nu M + M^2) \approx 2E_\nu M$.

The cross section for neutrino–nucleon scattering is again expressed in terms of structure functions,

$$\frac{d^2\sigma^{\nu(\bar{\nu})N}}{dx\,dQ^2} = \frac{G_F^2}{2\pi x}\left(\frac{M_W^2}{Q^2+M_W^2}\right)^2\left[F_2^{\nu(\bar{\nu})N}(x,Q^2)\left(\frac{y^2+(2Mxy/Q)^2}{2+2R_L^{\nu(\bar{\nu})N}(x,Q^2)}\right)\right.$$
$$\left.+1-y-\left(\frac{Mxy}{Q}\right)^2\right) \pm xF_3^{\nu(\bar{\nu})N}(x,Q^2)\,y\left(1-\frac{y}{2}\right)\right], \quad (7.14)$$

where the $\pm$ is $+(-)$ for $\nu(\bar{\nu})$ scattering. Instead of $F_1(x,Q^2)$ here we have used the structure function R_L, which can be interpreted as the ratio of the longitudinal and transverse virtual boson absorption cross section,

$$R_L(x,Q^2) = \frac{\sigma_L}{\sigma_T} = \frac{F_2(x,Q^2)(1+(4M^2x^2)/Q^2)-2xF_1(x,Q^2)}{2xF_1(x,Q^2)}. \quad (7.15)$$

In leading order QCD and neglecting the proton mass we have $R_L = 0$ because of the Callan–Gross relation $F_2 = 2xF_1$. In Ex. (7-2) the cross section eqn (7.14) is compared to the cross section for charged current lepton–nucleon scattering as discussed in Section 7.1.2.

In the naive quark-parton model the neutrino structure functions are simple functions of x, and can be expressed in terms of p.d.f.s,

$$F_2^{\nu p}(x) = 2x\left[d(x)+s(x)+\bar{u}(x)+\bar{c}(x)\right] \quad (7.16)$$
$$xF_3^{\nu p}(x) = 2x\left[d(x)+s(x)-\bar{u}(x)-\bar{c}(x)\right]. \quad (7.17)$$

These are valid for neutrino–proton scattering. We observe that only negatively charged partons contribute to the interaction. This is because in neutrino scattering a W^+ is exchanged. By isospin symmetry the structure functions for neutrino–neutron scattering are simply obtained by replacing in the above expressions the d- (ū-) quarks by the u- (d̄-) quarks. The sea quark distributions, $s(x), \bar{c}(x)$, are assumed to be the same in protons and neutrons. The structure functions for an *isoscalar* target are obtained by taking the average of the proton and neutron structure functions, since an isoscalar target has an equal number of protons and neutrons. The result is

$$F_2^{\nu N}(x) = x\left[u(x)+d(x)+2s(x)+\bar{u}(x)+\bar{d}(x)+2\bar{c}(x)\right] \quad (7.18)$$
$$xF_3^{\nu N}(x) = x\left[u(x)+d(x)+2s(x)-\bar{u}(x)-\bar{d}(x)-2\bar{c}(x)\right]. \quad (7.19)$$

Antineutrino scattering occurs via the exchange of a W^-, therefore only positively charged partons can contribute to the process. If we assume $s(x) = \bar{s}(x)$ and $c(x) = \bar{c}(x)$, we find $F_2^{\bar{\nu}N} = F_2^{\nu N}$ and $xF_3^{\bar{\nu}N} = xF_3^{\nu N} - 4x\left[s(x)-c(x)\right]$.

Looking at the above expressions we can summarize:

- $F_2^{\nu N}$ and $F_2^{\bar{\nu}N}$ measure the sum of quark and antiquark distributions.

- $xF_3 = 1/2(xF_3^{\nu N} + xF_3^{\bar{\nu} N})$ measures the sum of valence quark distributions $u_v(x) = u(x) - \bar{u}(x)$ and $d_v(x) = d(x) - \bar{d}(x)$.
- The difference $\Delta(xF_3) = xF_3^{\nu N} - xF_3^{\bar{\nu} N}$ is sensitive to the strange and charm content of the nucleon.
- Comparing $F_2^{\nu N}$ and $F_2^{e(\mu)N}$ constitutes a test of the fractional charge assignment to quarks and gives sensitivity to the strange quark distribution (see Ex. 7-3).

In addition, the search for two muons in the final state gives a direct handle to test the strange quark contribution. This measurement is described in more detail below.

Of course, the structure functions above are modified because of strong interactions, which generate scaling violations, that is, the parton distributions are functions of x and Q^2 and sensitivity to the gluon distribution is obtained from their Q^2 dependence. An often used notation is

$$F_2(x, Q^2) = x \sum_{i=u,d,\dots} q_i(x, Q^2) + \bar{q}_i(x, Q^2) \qquad (7.20)$$

$$xF_3(x, Q^2) = x \sum_{i=u,d,\dots} q_i(x, Q^2) - \bar{q}_i(x, Q^2) \quad , \qquad (7.21)$$

where $q_i(x, Q^2)$ stands for a quark distribution. Note that in the DIS factorization scheme (Altarelli *et al.*, 1978), Section 3.6, this expression remains valid also beyond leading order in α_s.

7.2.1 *Experimental issues*

Neutrino beams are generated by firing a high intensity proton beam into a target, such as beryllium. In the interactions with the target pions and kaons are produced, a large fraction of which decay semileptonically into muons and muon-neutrinos, ν_μ. As an example, the neutrino beam for the CCFR experiment at FERMILAB consisted in about 86.4% ν_μ, 11.3% $\bar{\nu}_\mu$, and 2.3% ν_e and $\bar{\nu}_e$. Depending on whether there are sign- and momentum selecting magnets behind the target, the neutrino beam can cover a narrow or wide energy range, such as from 30 to 300 GeV for the CCFR experiment. The x and Q^2 ranges typically covered are $0.01 < x < 0.8$ and $0.1\,\text{GeV}^2 < Q^2 < 100\,\text{GeV}^2$, as shown in Fig. 7.1. The accessible range is set by the beam energies and by the experimental acceptance.

The effect of the small interaction cross section, $\sigma^{\nu\text{Fe}}/E_\nu \approx 0.7 \times 10^{-38}$ cm^2/GeV, is overcome by modern experiments through the use of high-intensity beams coupled to massive detectors, resulting in luminosities of about 10^{36} cm^{-2} s^{-1} and correspondingly in collected data samples of up to a million events. A short description of a typical neutrino fixed-target detector has already been given in Section 5. An important aspect of the detector is the absorption material, often also employed as absorber for the calorimeter. Just to mention a few, iron is used by CCFR (now NuTeV) at FERMILAB and CDHSW at CERN, marble

(CaCO$_3$) by CHARM, glass by CHARMII and neon or deuterium by BEBC, all located at the CERN SPS neutrino beam. Whereas marble and deuterium are isoscalar targets, iron is not. Therefore the structure functions obtained from iron first have to be corrected before being compared to parameterizations such as in eqns (7.18) and (7.19).

Furthermore, because of the scattering on heavy targets, nuclear effects have to be taken into account, in particular when extracting structure functions. An example is gluon recombination, which can occur in a large nucleus between partons of neighbouring nucleons, leading to an A (=atomic number) dependent depletion of low-x partons (Nikolaev and Zakharov, 1975; Mueller and Qiu, 1986). Other effects are Fermi motion of nucleons in the nucleus, or the EMC effect, which is a suppression of structure functions from high-A targets compared to deuterium for $0.2 < x < 0.7$. A discussion of these effects and others such as target and 'higher-twist' effects can be found in the review by Conrad and collaborators (1998) and references therein.

An important experimental issue is the precise determination of the neutrino fluxes, in order to get the normalization for the cross section measurements right. Of course, a direct monitoring of neutrino beams is difficult, therefore various indirect methods are employed. One method consists in monitoring the produced secondary pion and kaon beams, whereas another method is based on the measurement of the y dependence of the cross section for $y \to 0$, or equivalently $E_H \to 0$, as discussed in Ex. (7-4). Finally, a Monte Carlo simulation of the neutrino beam and its interactions in the target can be iterated by varying the beam conditions until good agreement is obtained between data and calculations. The flux measurements contribute a large fraction to the final overall systematic uncertainties on the cross sections.

7.2.2 Measurements of F_2 and xF_3

Several experimental techniques exist in order to extract the structure functions from cross section measurements. For example, the CCFR Collaboration (1997) used the following method, which in the end led to rather small systematic uncertainties of the order of 2%, allowing for precision tests of NLO QCD. After having determined the neutrino flux $\Phi(E_\nu)$, the number of observed events N^{obs} is compared to the predicted cross section for every x and Q^2 bin according to

$$N^{\mathrm{obs}} = \rho\, L\, N_A \int_{x\,\mathrm{bin}} \mathrm{d}x \int_{Q^2\,\mathrm{bin}} \mathrm{d}Q^2 \int_{\mathrm{all\,energies}} \mathrm{d}E_\nu \Phi(E_\nu) \frac{\mathrm{d}^2\sigma}{\mathrm{d}x\,\mathrm{d}Q^2}\,, \qquad (7.22)$$

where ρ is the target density, L is the target length and N_A is the Avogadro number. The structure functions are extracted by an iterative procedure, where input structure functions for the Monte Carlo prediction of the cross section are varied until good agreement in eqn (7.22) is found.

The results are shown in Fig. 7.13 for F_2 and Fig. 7.14 for xF_3. Very good agreement with a NLO QCD fit based on the DGLAP evolution equations is found, that is, scaling violations are clearly observed. It should be noted that

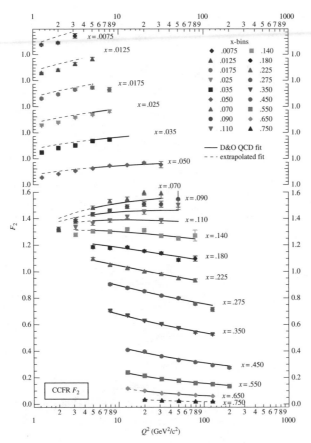

FIG. 7.13. Measurements of the neutrino structure function F_2 by CCFR. The errors are statistical only. Figure from CCFR Collab.(1997).

for fixed Q^2 F_2 increases with decreasing x, whereas xF_3 first increases and then decreases again. This is seen more clearly in Fig. 7.15, where the results of several experiments are shown as a function of x. Because of the averaging over Q^2, the results are not expected to agree perfectly, since the experiments measured different Q^2 ranges. The low-x behaviour of F_2 is understood by the increasing importance of the gluon and sea quark contributions. For xF_3 we have seen that it is sensitive to the valence quark distributions, $xF_3 \propto x[q(x) - \bar{q}(x)] = xq_{\mathrm{v}}(x)$. In the simplest quark parton model without strong interactions we would expect this distribution to have a narrow peak around 1/3 with some smearing due to Fermi motion, since the three valence quarks should each carry about one-third of the proton momentum. Because of QCD effects the distribution is shifted to lower values and smeared out much further, as seen in Fig. 7.15.

An important aspect of the measurements of F_2 and xF_3 are the assumptions

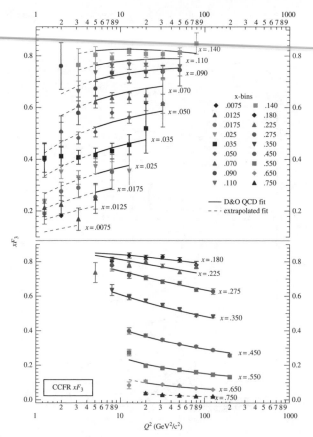

FIG. 7.14. Measurements of the neutrino structure function xF_3 by CCFR. The errors are statistical only. Figure from CCFR Collab.(1997).

made about R_L, which enters in the cross section prediction eqn (7.14). Often it is simply set to zero, or its QCD prediction is used. However, it can be measured by fitting the function

$$F(x, Q^2, \epsilon) = \frac{\pi}{G_F^2} \frac{2x(1-\epsilon)}{yQ^2} \left(\frac{\mathrm{d}^2\sigma^\nu}{\mathrm{d}x\,\mathrm{d}y} + \frac{\mathrm{d}^2\sigma^{\bar{\nu}}}{\mathrm{d}x\,\mathrm{d}y} \right) = 2xF_1(x, Q^2) \left[1 + \epsilon R_L(x, Q^2) \right],$$
(7.23)

which can be derived from eqns (7.14) and (7.15) for $Q, E_\nu \gg M$, $Q^2 \ll M_W^2$ and $xF_3^\nu = xF_3^{\bar{\nu}}$. Because of the last assumption a correction for $\Delta(xF_3)$ has to be applied. The term $\epsilon = 2(1-y)/[1+(1-y)^2]$ is the polarization of the virtual W boson. The ratio R_L is extracted from linear fits in ϵ to F in fixed x and Q^2 bins. Results are shown in Fig. 7.16, together with measurements from charged lepton DIS. Within the still limited precision of these measurements it is seen that $R_L \to 0$ for increasing Q^2. Perturbative QCD predicts $R_L \propto \alpha_s(Q^2)$,

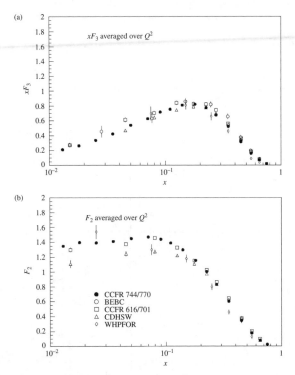

FIG. 7.15. Neutrino structure functions F_2 and xF_3 as a function of x, averaged over Q^2, obtained by several experiments. Figure from Conrad *et al.*(1998).

so that the Q^2 behaviour of R_L can be understood from the decrease of $\alpha_s(Q^2)$ with increasing Q^2.

In order to study ν scattering on protons and neutrons separately, experiments at CERN have used the Big European Bubble Chamber BEBC (1985, 1994) filled with deuterium. A ν or $\bar{\nu}$ interaction was identified as coming from a neutron if it had either an even number of charged tracks, assuming conservation of electric charge which is zero in the initial state, or an odd number of charged tracks together with a low momentum proton, interpreted as spectator in the reaction ν-neutron within the deuterium. All other events with an odd number of charged tracks and thus a net total charge were classified as ν-proton interactions. Corrections were applied for mis-identifications by studying Monte Carlo simulations. From the measured structure functions $F_2^{\nu p}$ and $F_2^{\nu n}$ it is observed that $F_2^{\nu n} \approx 2 F_2^{\nu p}$ over almost the full kinematic range, and in particular at larger x where valence quarks dominate. This can be understood since the virtual W^+ from a ν scattering must interact with a negatively charged quark, which at large x is with high probability a d-quark. In a neutron there are twice as much d valence quarks as u-quarks, which explains the larger neutron structure function. So these data clearly indicate the flavour sensitivity of neutrino

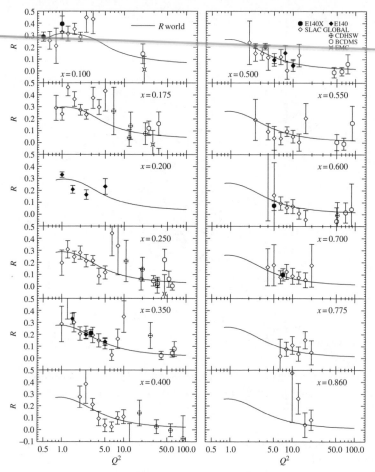

Fig. 7.16. Measurements of R_L by neutrino and charged lepton DIS experiments. Figure from Conrad *et al.*(1998).

scattering.

An interesting test of the universality of parton distributions is obtained by comparing the results of F_2 measurements from charged lepton to neutrino deep inelastic scattering. In Ex. (7-3) the ratio of structure functions is derived as

$$\frac{F_2^{\mu N}}{F_2^{\nu N}} = \frac{5}{18}\left[1 - \frac{3(s+\bar{s})}{5(q+\bar{q})}\right],\qquad(7.24)$$

where $q+\bar{q}$ represents the sum over all quark flavours. This ratio is determined by the electric charge assignment to quarks, and the expression above stays valid at all orders of perturbation theory if defined in the DIS factorization scheme. When going to large x, where the sea contribution is small, we expect

$F_2^{\mu N}/F_2^{\nu N} \to 5/18 = 0.2778$. Indeed, comparing neutrino data from CCFR and CDHSW (1989) to muon data from BCDMS (1987) at high x yields a ratio of 0.278 ± 0.010, in excellent agreement with the expectation.

7.2.3 The gluon distribution

A simultaneous analysis of scaling violations in the neutrino structure functions F_2 and xF_3 allows for a determination of the gluon distribution in the medium and large x domain. In particular, the behaviour of $xF_3 \propto x(q - \bar{q})$ is exploited. We know from Ex. (3-8) that the evolution equation for non-singlet quark distributions $q_i^{NS} = q_i - \bar{q}_i$ only depends on α_s and the non-singlet distribution itself, whereas the evolution of flavour singlet distributions q_i^S also depends on the gluon distribution $g(x)$,

$$\frac{dq^{NS}}{dQ^2} = \alpha_s P_{qq} \otimes q^{NS} , \quad \frac{dq^S}{dQ^2} = \alpha_s \left(P_{qq} \otimes q^S + P_{qg} \otimes g \right) . \tag{7.25}$$

This allows for a precise determination of α_s without a detailed knowledge of the gluon distribution, or equivalently the gluon distribution can be determined largely uncorrelated with α_s, which is in contrast to the situation in charged lepton DIS described previously. An analysis by CCFR (1997) yields

$$xg(x, Q^2 = 5\,\text{GeV}^2) = (2.22 \pm 0.34)\,(1 - x)^{4.65 \pm 0.68} \tag{7.26}$$

in the region $0.04 < x < 0.7$. This result is then evolved to $Q^2 = 32\,\text{GeV}^2$ and compared to HERA results (H1 Collab., 1995a), as shown in Fig. 7.17. Good agreement between the various measurements and the fits from global analyses of parton distributions is found. It is worth noting that the HERA results discussed in Section 7.1.4 are more recent than those of Fig. 7.17.

7.2.4 The strange quark distribution

Charm production via neutrino scattering off strange quarks can be used to isolate the strange quark distribution $s(x)$ in the nucleon. The distinct experimental signature for charm production are two oppositely signed muons in the final state, which is understood to occur via

$$\nu_\mu + N \to \mu^- + c + X$$
$$\ \ \ \hookrightarrow s + \mu^+ + \nu_\mu .$$

Indeed, charm production is a substantial fraction of the total cross section, as large as 10% at high energies, $E_\nu \approx 300\,\text{GeV}$. The cross section for di-muon production is the product of the cross section for charm production, a fragmentation function $D(z)$, with $z = p_D/p_D^{max}$, which describes the transition of a c-quark to a charmed D meson, and the branching ratio B_c for the subsequent semileptonic decay of the D meson,

$$\frac{d^3\sigma(\nu_\mu N \to \mu^- \mu^+ X)}{d\xi\,dy\,dz} = \frac{d^2\sigma(\nu_\mu N \to c\,\mu^- X)}{d\xi\,dy}\,D(z)\,B_c(c \to \mu^+ X) . \tag{7.27}$$

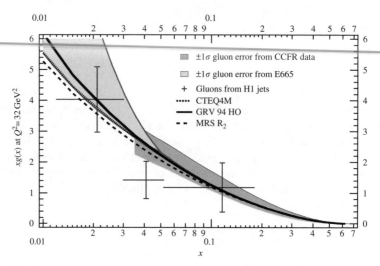

FIG. 7.17. The gluon distribution measured by CCFR. Figure from Conrad *et al.*(1998).

Here instead of the Bjorken scaling variable x the variable ξ appears, which is approximately given by $\xi = x(1 + m_c^2/Q^2)(1 - x^2M^2/Q^2)$. This so-called *slow rescaling* (Barnett, 1976; Georgi and Politzer, 1976) takes effects of the charm quark mass m_c and of the target mass M into account. The leading order cross section for charm production is

$$\sigma^{\mathrm{LO}}(\nu_\mu N \to cX) \propto \left[\left(\xi u(\xi, \mu^2) + \xi d(\xi, \mu^2)\right)|V_{cd}|^2 + 2\xi s(\xi, \mu^2)|V_{cs}|^2\right] \ . \quad (7.28)$$

Note that this expression holds for an isoscalar target, where also the u-distribution appears because of isospin invariance between protons and neutrons, $u(x) = u^p(x) = d^n(x)$. We see that the cross section depends on the CKM matrix elements. The contribution from scattering off d- or d̄-quarks is Cabbibo suppressed, $|V_{cd}| \ll |V_{cs}|$. The sensitivity to $s(x)$ becomes obvious from eqn (7.28). In the case of ν scattering about 50% of the charm production is due to scattering off s-quarks, whereas for $\bar{\nu}$ scattering 90% comes from s̄, since only d̄ from the sea can contribute in this case.

The largest data samples are available from the CCFR (1995) and CDHS (1982) Collaborations. An important experimental issue is the control of the backgrounds, such as muons from pion and kaon decays. Fortunately, the dense calorimeters with short interaction lengths minimize these contributions. Furthermore, a control of the contributions from ν versus $\bar{\nu}$ is of importance. Systematic uncertainties arise from the limited knowledge of the charm fragmentation function $D(z)$.

The size and shape of the strange parton distribution are determined in the following way: The measurements of F_2 and xF_3 are used to constrain the sum of antiquark distributions,

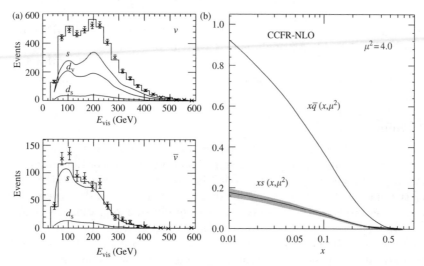

FIG. 7.18. (a) Distribution of the visible energy for the CCFR ν- and $\bar{\nu}$-induced
di-muon events. Indicated are also the contributions from the strange sea
quarks, s, and the valence and sea d-quarks, $d_{\rm v}$ and $d_{\rm s}$. (b) Measured quark
and strange sea distributions at $\mu^2 = 4\,{\rm GeV}^2$. The grey band around the
strange quark distribution indicates the $\pm 1\sigma$ uncertainty. Figures from Conrad *et al.*(1998).

$$\frac{1}{x}F_2 - F_3 \approx q^{\rm S} - q^{\rm NS} = 2\bar{q} = 2(\bar{u} + \bar{d} + \bar{s}) = 2(\bar{u} + \bar{d}) + (s + \bar{s}) . \qquad (7.29)$$

Thus, assuming $s(x) = \bar{s}(x)$, the strange distribution can be related to the non-
strange distributions. A possible parameterization is

$$\kappa = \frac{\int_0^1 {\rm d}x\,[xs(x) + x\bar{s}(x)]}{\int_0^1 {\rm d}x\,[x\bar{u}(x) + x\bar{d}(x)]} \quad , \quad xs(x) = A_{\rm s}(1 - x)^\alpha \frac{1}{2}[x\bar{u}(x) + x\bar{d}(x)] , \qquad (7.30)$$

where the parameters κ and α are obtained from a fit of predictions from Monte
Carlo simulations, based on eqn (7.27), to distributions measured in the final
state of charm production events, such as $E_{\rm vis} = E_{\rm H} + E_\mu$, Fig. 7.18(a). There the
various contributions from different parton distributions are indicated, showing
in particular the importance of the strange quarks for $\bar{\nu}$ scattering. Taking the
world average values for the CKM matrix elements, the parameters turn out to
be $\kappa = 0.477 \pm 0.051$ and $\alpha = -0.02 \pm 0.63$ (CCFR Collab., 1995), which are
obtained together with measurements of $B_{\rm c}$ and $m_{\rm c}$. This means that within the
current experimental precision the shape of the strange quark distribution cannot
be distinguished from the non-strange component. However, $\kappa < 1$ demonstrates
a violation of SU(3) flavour symmetry in the quark sea. Intuitively, we could
explain this by the larger s-quark mass with respect to u and d, which would
suppress the gluon splitting into strange quark pairs. However, non-perturbative

phenomena could be important as well. The extracted parton distributions are shown in Fig. 7.18(b). An analysis by CCFR in order to disentangle $s(x)$ from $\bar{s}(x)$ was not yet conclusive. Another handle on the strange quark distribution could be obtained from the approximate relation $xs + x\bar{s} = 5/3F_2^{\nu N} - 6F_2^{\mu N}$ (Ex. 7-3).

7.3   Sum rules

Sum rules are integrals over structure functions or parton distributions, express-ing usually the conservation law for some quantum number of the proton. Basic sum rules have already been mentioned in Section 3.2.2, and a further discus-sion is found in Ex. (7-5). QCD predictions exist for a wide variety of sum rules. These predictions are tested experimentally by integrating the measured structure functions $F_{2,3}(x, Q^2)$ over the full x range at a given Q^2. A basic prob-lem arises from the fact that only a subrange $[x_{min}, x_{max}] \subset [0, 1]$ is accessible experimentally. A possible solution is to combine measurements from different experiments which cover different kinematic ranges. Otherwise various structure function measurements at different Q^2 scales can be evolved to a single scale Q_0^2, using QCD evolution equations or approximations such as simple power laws. The result should cover a wide enough range in x for the integral to be computed reliably. Nevertheless, usually some assumptions about the behaviour at $x \to 0$ and $x \to 1$ have still to be made in order to complete the integration range. These assumptions lead to systematic uncertainties on the sum rule measurements.

   Although it is quite common to call all these integrals over structure functions simply *sum rules*, one should be aware that actually there exist different classes of sum rules. We might call a sum rule an *exact QCD sum rule* if its result found within the context of the parton model is not altered by any radiative or non-perturbative correction. An example is given by the Adler sum rule. Next there are sum rules where the parton model result is modified by radiative corrections and maybe by power-suppressed terms, such as the Gross–Llewellyn-Smith or the Bjorken sum rules. Finally, we have sum rules which are strongly affected by non-perturbative physics, thus deviations of the measurements from the predictions of perturbative QCD might not be a surprise. Examples are the Gottfried and the Ellis–Jaffe sum rules. A last particular case is the momentum sum rule. When going beyond leading order QCD, it should be seen as a convenient constraint on the definition of the p.d.f.s rather than a basic QCD sum rule.

7.3.1   *The Adler sum rule*

The Adler sum rule (Adler, 1963) is defined as

$$I_A = \int_0^1 \frac{dx}{x} \left(F_2^{\nu n}(x, Q^2) - F_2^{\nu p}(x, Q^2) \right) = 2 \int_0^1 dx \, (u_v(x) - d_v(x)) = 2 .$$

(7.31)

This result is obtained by inserting the parton model expressions for the structure functions, $F_2^{\nu p} = 2x(d(x) + s(x) + \bar{u}(x) + \bar{c}(x))$ and $F_2^{\nu n} = 2x(u(x) + s(x) + $

$\bar{d}(x) + \bar{c}(x))$. The sum rule expresses the difference of valence quark distributions in neutrons and protons. It is not modified by perturbative or non-perturbative QCD corrections because of the conservation of the charged weak current. The same result can be obtained from

$$I_A = \int_0^1 \frac{dx}{x} \left(F_2^{\bar{\nu}p}(x, Q^2) - F_2^{\nu p}(x, Q^2) \right) = 2 , \qquad (7.32)$$

using the constraint $\int_0^1 dx(c - \bar{c}) = \int_0^1 dx(s - \bar{s}) = 0$ (*cf.* Ex. 3-6).

An experimental problem arises from the fact that usually isoscalar targets are employed. For such targets we have $F_2^{\bar{\nu}N} = F_2^{\nu N}$, and the contributions from proton or neutron scattering are difficult to disentangle. The only measurement in the literature is from the BEBC Collaboration (1984, 1985). We have described previously how they used neutrino and antineutrino scattering on a deuterium target in order to disentangle $F_2^{\nu p}$ from $F_2^{\nu n}$. Their result for the Adler sum rule is $I_A = 2.02 \pm 0.40$, in agreement with the theoretical expectation $I_A = 2$. A re-evaluation of the BEBC data (Conrad *et al.*, 1998) using two different approximations in order to cover the unmeasured low-x region gave $I_A = 1.87 \pm 0.15$ and $I_A = 2.05 \pm 0.15$, depending on the extrapolation method. Again both results agree with the expectation.

7.3.2   The Gross–Llewellyn Smith sum rule

The Gross–Llewellyn-Smith (GLS) sum rule (Gross and Llewellyn Smith, 1969) is defined as

$$I_{GLS} = \int_0^1 \frac{dx}{x} x F_3(x, Q^2) = \int_0^1 dx \frac{1}{2} \left(F_3^\nu(x, Q^2) + F_3^{\bar{\nu}}(x, Q^2) \right) . \qquad (7.33)$$

In the simple parton model this sum rule is the integral over the valence quark distributions, with an expectation of 3 for protons or neutrons, that is,

$$I_{GLS} = \int_0^1 dx \left(u_v(x) + d_v(x) \right) = 3 . \qquad (7.34)$$

This leading order expectation is modified by radiative corrections from QCD discussed in Ex. (3-28) and Section 8.3.3. There it is shown that the measurement of the GLS sum rule can be used in order to obtain a measurement of α_s.

Measurements of xF_3 by several neutrino experiments have been combined (Conrad *et al.*, 1998), and the GLS sum rule is determined to be $I_{GLS} = 2.64 \pm 0.06$. In a recent analysis of data from CCFR (1998) and other neutrino experiments the GLS integral has been evaluated at various average Q^2 values, giving for example

$$I_{GLS} = 2.49 \pm 0.08_{stat} \pm 0.10_{syst} \quad \text{at } \langle Q^2 \rangle = 2 \, \text{GeV}^2$$
$$I_{GLS} = 2.78 \pm 0.06_{stat} \pm 0.19_{syst} \quad \text{at } \langle Q^2 \rangle = 5 \, \text{GeV}^2$$

$$I_{\mathrm{GLS}} = 2.80 \pm 0.13_{\mathrm{stat}} \pm 0.18_{\mathrm{syst}} \quad \text{at } \langle Q^2 \rangle = 12.6\,\mathrm{GeV}^2 \ .$$

It is clearly seen that the results satisfy $I_{\mathrm{GLS}} < 3$, as expected from eqn (8.10), since the QCD corrections are negative. Furthermore, we can see that the integral approaches the naive expectation of $I_{\mathrm{GLS}} = 3$ when going to larger Q^2, since at larger scales the strong coupling constant and thus the radiative corrections decrease.

7.3.3 The Gottfried sum rule

The Adler sum rule is measured in deep inelastic neutrino scattering experiments. A comparable sum rule for charged lepton scattering is given by the Gottfried sum rule (Gottfried, 1967),

$$I_{\mathrm{G}} = \int_0^1 \frac{\mathrm{d}x}{x} \left(F_2^{\mu\mathrm{p}}(x, Q^2) - F_2^{\mu\mathrm{n}}(x, Q^2) \right) = \frac{1}{3} + \frac{2}{3} \int_0^1 \mathrm{d}x \left(\bar{u}(x) - \bar{d}(x) \right) ,$$

(7.35)

which can be obtained by inserting eqn (7.5) and using isospin symmetry, that is, $u(x) = u^{\mathrm{p}}(x) = d^{\mathrm{n}}(x)$, $d(x) = d^{\mathrm{p}}(x) = u^{\mathrm{n}}(x)$. Starting from the assumption that the sea quark distributions $\bar{u}(x)$ and $\bar{d}(x)$ in the nucleon are the same, we expect $I_{\mathrm{G}} = 1/3$. However, the NMC experiment measured $I_{\mathrm{G}}(x = 0.004 - 0.8) = 0.227 \pm 0.007_{\mathrm{stat}} \pm 0.014_{\mathrm{syst}}$ at $Q^2 = 4\,\mathrm{GeV}^2$, and $I_{\mathrm{G}} = 0.240 \pm 0.016$ when extrapolated to the full x range (NMC Collab., 1991). This result is significantly different from $1/3$. Although many possible experimental biases have been considered, no explanation could be found except a difference in the sea quark distributions of protons and neutrons, or equivalently that there is a light quark flavour asymmetry in the sea, $\int_0^1 \mathrm{d}x(\bar{u}(x) - \bar{d}(x)) = -0.140 \pm 0.024 \neq 0$. Lately this observation has been confirmed by the E866 experiment, as will be discussed later. Non-perturbative processes are thought to be at the origin of this asymmetry. For a more in-depth discussion we refer to the literature (Kumano and Londergan, 1991; E866/NuSEA Collab., 1998b).

7.3.4 The momentum sum rule

As will be described in Section 7.5, a global QCD analysis of deep inelastic scattering experiments results in a set of p.d.f.s for quarks and gluons. It is shown (Botje, 2000) that the integral over these distributions gives the following values when evaluated at $Q^2 = 4\,\mathrm{GeV}^2$,

$$I_{\mathrm{q}} = \int_0^1 \mathrm{d}x\, x \sum_q (q(x) + \bar{q}(x)) = 0.594 \pm 0.018 ,$$

$$I_{\mathrm{g}} = \int_0^1 \mathrm{d}x\, x\, g(x) = 0.394 \pm 0.018 .$$

The errors combine statistical and systematic uncertainties in quadrature. We see that the quarks carry only about 60% of the nucleon momentum. This was a big surprise when observed for the first time, and suggested the presence of further proton constituents, c.f. Section 2.1. Indeed, we now know that the gluons carry the very large missing fraction of the proton momentum.

7.3.5 Sum rules for polarized structure functions

So far we have only discussed structure function measurements which are extracted from spin-averaged cross sections. However, if the probing beam and the target are polarized, then similar experiments can be performed in order to measure polarized structure functions. We will only give a very short overview here. The interested reader is referred to the literature for a more detailed discussion (Ellis *et al.*, 1996*b*), or a general review (Windmolders, 1999) on recent results in spin physics.

The basic observable in polarized scattering is the spin asymmetry, defined as the asymmetry of the cross sections $d\sigma^{\uparrow\downarrow}$ $(d\sigma^{\uparrow\uparrow})$ with the lepton and nucleon having antiparallel (parallel) polarization, normalized to the spin-averaged case,

$$A(x, Q^2) = \frac{d\sigma^{\uparrow\downarrow}(x, Q^2) - d\sigma^{\uparrow\uparrow}(x, Q^2)}{d\sigma^{\uparrow\downarrow}(x, Q^2) + d\sigma^{\uparrow\uparrow}(x, Q^2)} = \frac{A_{\mathrm{meas}}(x, Q^2)}{P_B P_T f} . \tag{7.36}$$

The second equality shows the relation between the observable and the measured asymmetry A_{meas}. The factors $P_{B(T)}$ give the degree of polarization of the beam (target), which for modern experiments such as HERMES at HERA range from 40% up to 90%. The dilution factor f accounts for the fraction of polarizable nucleons in the target. The uncertainties on these three factors are important sources of systematic uncertainties on the spin asymmetry A.

For the kinematic range relevant for most of the fixed-target experiments (where only photon exchange needs to be considered) the spin asymmetry is given by $A(x, Q^2) \approx g_1(x, Q^2)/F_1(x, Q^2)$, where g_1 is the polarized equivalent of the structure function F_1 from unpolarized scattering. Thus a measurement of A and F_1 will determine g_1. In the parton model approximation g_1 is defined as

$$g_1(x) = \frac{1}{2} \sum_q e_q^2 [\Delta q(x) + \Delta \bar{q}(x)] ; \quad \Delta q = q^{\uparrow} - q^{\downarrow} , \tag{7.37}$$

with e_q^2 the electric charges of the quarks. The parton distributions relevant for spin-averaged scattering are $q = q^{\uparrow} + q^{\downarrow}$. The simple picture is modified by strong interactions, and evolution equations can be found which describe the Q^2 dependence of polarized parton distributions.

An important observable and test for the theory is the Bjorken sum rule (Bjorken 1966, 1970) which measures the difference between the integrated polarized structure function g_1 in proton and neutron,

$$I_B = \int_0^1 dx \, (g_1^{\mathrm{p}} - g_1^{\mathrm{n}}) = \frac{1}{6} \left| \frac{g_A}{g_V} \right| = 0.209 \pm 0.001 . \tag{7.38}$$

Here g_A/g_V is the ratio of axial-vector over vector couplings measured from the neutron β-decay. The sum rule is modified by radiative corrections, expressed as a power series in the strong coupling constant, and hence was also employed to measure α_s, as discussed in Section 8.3.3.

Similarly, when considering only the proton, the Ellis–Jaffe sum rule is obtained (Ellis and Jaffe, 1974),

$$I_{\mathrm{EJ}} = \int_0^1 \mathrm{d}x\, g_1^{\mathrm{p}} = \frac{1}{12} \left| \frac{g_A}{g_V} \right| (1 + C) = 0.185 \pm 0.003 \,, \qquad (7.39)$$

where C is a correction not computable within perturbative QCD. It can be obtained, for example, from an analysis of n $\rightarrow$ p and $\Sigma^- \rightarrow$ n decays (Ellis *et al.*, 1996*b*; Close and Roberts, 1993). A comparison with experimental results, such as $I_{\mathrm{EJ}} = 0.142 \pm 0.008_{\mathrm{stat}} \pm 0.011_{\mathrm{syst}}$ at $Q^2 = 10\,\mathrm{GeV}^2$ (SMC Collab., 1994), reveals a large discrepancy, which can not be compensated by radiative corrections. A basic assumption which enters the theoretical prediction eqn (7.39) is that the strange quark sea is unpolarized. Therefore the discrepancy could be an indication for polarization of the strange sea quarks. An additional important contribution should come from the gluons. This contribution cancels out in the Bjorken sum rule. The measurements of these sum rules as well as of the shape of the polarized parton distributions are a field of active research at existing and future experiments, such as COMPASS at CERN.

7.4   Hadron–hadron scattering

Information on the p.d.f.s can be obtained not only from structure functions measured in DIS experiments, but also from analyses of hadron–hadron scattering. In order to reduce the complexity of these reactions, and to allow for cross sections to be calculable, the final state is required to have particular properties. For example, one looks for lepton-pair or photon production, both final states being of non-hadronic nature. However, even hadronic final states are studied, since jets produced in hadron–hadron collisions with very large transverse momentum can be understood in terms of a simple underlying hard parton–parton scattering. The sensitivity to the parton distributions becomes clear when looking at the general expression for a cross section, as given in eqn (3.72), Section 3.2.3. There we also have discussed various possible parton–parton processes relevant for different final states observed in the experiment. In the following we will briefly review several experiments which help to constrain particular p.d.f.s.

7.4.1   *The Drell–Yan process*

A basic discussion of the hard subprocess, relevant for Drell–Yan production of lepton pairs in nucleon–nucleon scattering, has been given in Section 3.6.8. We look for the experimentally very clean signature of two oppositely charged leptons, typically muons. The lepton pair is produced by the annihilation of a $q\bar{q}$-pair, the quarks being constituents of either of the two hadrons. They annihilate into a photon or a Z boson, which subsequently decays into the two leptons (c.f. Fig. 6.4, lower rightmost Feynman diagram). The invariant mass of the

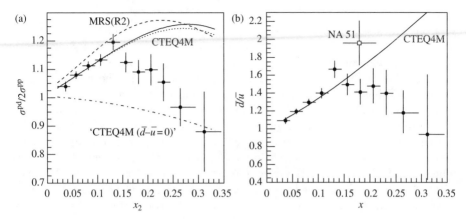

FIG. 7.19. (a) The ratio $\sigma^{\mathrm{pd}}/2\sigma^{\mathrm{pp}}$ of Drell–Yan cross sections measured by E866 and compared to various parton distributions. (b) The ratio of $\bar{d}/\bar{u}$ as a function of x in the proton, determined by E866. Also shown is the result by NA51 and a comparison to a parton distribution obtained from a global QCD analysis. Figures from E866/NuSea Collab.(1998a).

lepton pair is given by $M^2 = s x_1 x_2$, where s is the overall C.o.M. energy and $x_{1(2)}$ is the hadron's momentum fraction carried by the first (second) quark. Testing therefore different regions of the invariant mass distribution allows to probe different x regions of the parton distributions. Since for the annihilation a q$\bar{\mathrm{q}}$-pair is required, the Drell–Yan process is particularly useful to test the sea (anti)quark distributions. Experimentally, it has been studied at fixed-target experiments and colliders, such as the TEVATRON. Here we will concentrate on a recent FERMILAB fixed-target experiment, E866, which has confirmed the flavour asymmetry of the light quark sea in the proton, as indicated in Section 7.3.3.

This experiment measures the muon-pair yield from the Drell–Yan process in a 800 GeV proton bombardement of liquid deuterium and hydrogen targets. Using different targets they are able to test the sea quark distributions in protons and neutrons, and ultimately determine the $\bar{d}/\bar{u}$ and $\bar{d} - \bar{u}$ distributions in the proton over the range $0.020 < x < 0.345$. The observable is the ratio of Drell–Yan cross sections for deuterium and hydrogen, $\sigma^{\mathrm{pd}}/2\sigma^{\mathrm{pp}}$, which they have measured from 140000 Drell–Yan pair events with a total systematic uncertainty of less than 1% (E866/NuSea Collab., 1998a). For a particular kinematic region this ratio can be approximated as

$$\left.\frac{\sigma^{\mathrm{pd}}}{2\sigma^{\mathrm{pp}}}\right|_{x_1 \gg x_2} \approx \frac{1}{2} \frac{\left(1 + \frac{1}{4}\frac{d_1}{u_1}\right)}{\left(1 + \frac{1}{4}\frac{d_1}{u_1}\frac{\bar{d}_2}{\bar{u}_2}\right)} \left(1 + \frac{\bar{d}_2}{\bar{u}_2}\right) , \tag{7.40}$$

where the indices 1(2) refer to the partons belonging to the incoming (target) hadron. The sensitivity to $\bar{d}/\bar{u}$ is nicely seen in this ratio, which would be unity if the light quark flavour sea were symmetric. However, the data show a clear deviation from unity (Fig. 7.19(a)), which indicates an excess of $\bar{d}$ over $\bar{u}$ sea quarks. This is further illustrated in Fig. 7.19(b), where the measured cross section ratio has been translated into a distribution of $(\bar{d}/\bar{u})(x)$. Several parton distributions, which are obtained from global QCD analyses, assuming a non-symmetric light flavour sea, are able to reproduce the low-x rise of $(\bar{d}/\bar{u})(x)$, but not the fall-off towards unity above $x \approx 0.2$. Note that a very recent global QCD analysis by the CTEQ Collaboration leads to a set of p.d.f.s, called CTEQ5M, which are able to describe the ratio $\bar{d}/\bar{u}$ over the entire x range (CTEQ Collab., 2000). As already mentioned previously, this asymmetry in the sea quarks must have its origin in non-perturbative phenomena of strong interactions.

7.4.2 The W rapidity asymmetry

Similar to the Drell–Yan process, $W^{\pm}$ bosons are produced in p$\bar{\text{p}}$ collisions mainly by the annihilation of u- (d-) quarks in the proton with a $\bar{\text{d}}$- ($\bar{\text{u}}$-) quark from the antiproton. This is particularly true at TEVATRON energies of about 1.8 TeV. As global QCD analyses show (Section 7.5, Fig. 7.24), u-quarks carry on average more momentum than d-quarks. Therefore the W^{+} bosons tend to follow the direction of the incoming proton and the W^{-} bosons that of the antiproton, which leads to a charge asymmetry in the angular (rapidity) distribution of the produced W bosons. For $Q^2 \approx M_W^2$ this W rapidity asymmetry is sensitive to the u- and d-quark distributions in the proton, in particular to d/u, as can be seen from the following approximate relationship,

$$A_W(y) = \frac{d\sigma(W^+)/dy - d\sigma(W^-)/dy}{d\sigma(W^+)/dy + d\sigma(W^-)/dy}$$
$$\approx \frac{u(x_1)d(x_2) - d(x_1)u(x_2)}{u(x_1)d(x_2) + d(x_1)u(x_2)} = \frac{R(x_2) - R(x_1)}{R(x_2) + R(x_1)}$$

(7.41)

with $R(x) = d(x)/u(x)$, $x_{1(2)} = x_0 \exp(\pm y)$, $x_0 = M_W/\sqrt{s}$, y being the rapidity and $\sqrt{s}$ the C.o.M. energy. This relation is derived under the assumption of $u^P(x) = \bar{u}^{\bar{P}}(x)$ and $d^P(x) = \bar{d}^{\bar{P}}(x)$. In fact, the W asymmetry puts the most stringent constraints on quark distributions in the proton of all measurements at the TEVATRON.

The production of W bosons is tagged by their semileptonic decays, W $\rightarrow$ $\mu\nu_\mu$ or W $\rightarrow$ eν_e, where the leptons must have a large transverse momentum $\gtrsim$25 GeV, and also the missing transverse momentum due to the neutrino has to be of similar size. Hadronic decays of the W are not looked for because of the enormous background from QCD jet production. Unfortunately, the semileptonic decays give rise to an experimental problem. The longitudinal momentum of the neutrinos can not be measured, and therefore the W rapidity is not directly accessible. It is rather inferred from the lepton rapidity y_l. However, because of the V-A structure of the semileptonic decay, which gives rise to an opposite charge

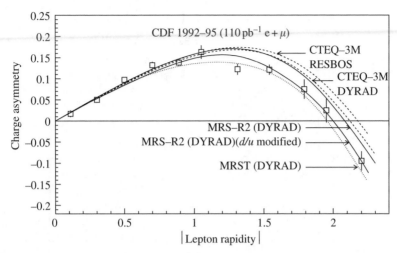

FIG. 7.20. The lepton charge asymmetry as a function of rapidity, measured by CDF, with statistical and systematic uncertainties added in quadrature. The data are compared to NLO QCD predictions based on various parton distribution sets. Figure from CDF Collab.(1998).

asymmetry effect, the original W rapidity asymmetry is diluted. Nevertheless, the TEVATRON data still have enough sensitivity to constrain the parton distributions from the measurement of the lepton rapidity asymmetry.

The CDF Collaboration has analysed the data taken during 1992–95, corresponding to an integrated luminosity of $110\,\mathrm{pb}^{-1}$ or about 90000 events after all selection cuts (CDF Collab., 1998). They have looked for muons and electrons over a rapidity range $0 < |y_l| < 2.5$, which allows to constrain the quark momentum distributions in the proton for $0.006 < x < 0.34$ at $Q^2 \approx M_{\mathrm{W}}^2$. The result is shown in Fig. 7.20, where the data points are compared to different NLO QCD predictions, based on various parton distribution sets. The data are combined for positive and negative lepton rapidity. Recent global QCD analyses (CTEQ Collab., 2000; Martin *et al.*, 1998) which include these CDF data give parton distributions, in particular $(d/u)(x)$, which reproduce the asymmetry well over the full y_l range.

7.4.3 Direct-photon production

Direct-photon production in hadron–hadron collisions is not only a useful tool for testing perturbative QCD, but in particular to constrain the gluon distribution in the proton at medium and large x values, $x \approx 0.2 - 0.6$. There DIS experiments have less sensitivity to the gluon contribution. Direct-photon production occurs mainly via QCD compton scattering $gq \to \gamma q$ and annihilation processes $q\bar{q} \to \gamma g$, that is, at $\mathcal{O}(\alpha_{\mathrm{em}}\alpha_{\mathrm{s}})$. The relative importance of the two contributions depends on the type of hadrons colliding and the $x \approx x_T = 2p_T^\gamma/\sqrt{s}$ range. For

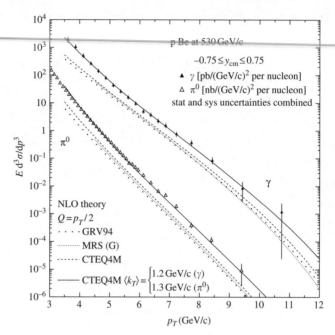

FIG. 7.21. The inclusive π^0 and direct-photon cross sections as functions of the transverse momentum p_T, measured by the E706 experiment. The data are compared to NLO QCD predictions with different parton distributions, in one case supplemented with intrinsic transverse momentum for the incident partons. Figure from E706 Collab.(1998).

example, the QCD Compton process dominates in pp collisions and at large x_T, where the antiquark sea is suppressed. Therefore fixed-target experiments are particularly suited for testing the gluon distribution. In p$\bar{\text{p}}$ collisions the annihilation contribution can be significant, because the $\bar{u}$ and $\bar{d}$ distributions in the antiproton are the same as u and d in the proton.

In the final state of this process we find a photon with large transverse momentum p_T, for example, $3.5\,\text{GeV} < p_T < 12\,\text{GeV}$ for the fixed-target experiment E706 at FERMILAB, balanced by a recoiling jet on the opposite side of the experiment. The experimental advantage of detecting photons is that their energy and momentum can be measured more precisely than for jets. Photons deposit all their energy within a few cells of the electromagnetic calorimeter, whereas jets are more spread out and deposit a large fraction of their energy also in the hadronic calorimeter. Overall this leads to a better energy resolution for photons. In addition, jets have to be defined by some algorithm, and an ambiguity arises from the assignment of particles belonging to the underlying event or to the jet.

The experimental disadvantage is the relatively low rate of single photons compared to jet production, namely $\mathcal{O}(\alpha_{\text{em}}\alpha_{\text{s}})/\mathcal{O}(\alpha_{\text{s}}^2) = \mathcal{O}(\alpha_{\text{em}}/\alpha_{\text{s}}) \approx 10^{-3}$.

Furthermore, direct-photon production suffers from a very large background of π^0 decays into two photons which often form overlapping clusters in the calorimeter. They can be identified by studying the profile of the energy deposits in the calorimeter, which is different for single or two nearby photons. Also isolation criteria are applied, since π^0s are usually accompanied by other high energy hadrons, whereas direct photons are not.

An interesting result has been obtained in a recent measurement by the E706 experiment (1998). There the inclusive π^0 and direct-photon cross sections are measured for 530 and 800 GeV proton beams and a 515 GeV π^- beam incident on beryllium targets. This fixed-target experiment features a large lead and liquid argon electromagnetic calorimeter and a charged particle spectrometer. When comparing the E706 data to NLO calculations rather large discrepancies of factors of two and more are found over almost the entire p_T range, Fig. 7.21. These discrepancies can not be reduced by using different parton distribution sets or changing the renormalization and/or factorization scales. A possible solution is to assume that the incident partons have Gaussian transverse momentum distributions with on average $\langle k_T \rangle \approx 1 - 1.5\,\mathrm{GeV}$, which could be due to multiple soft gluon radiation.

However, it is worth noting that results from other measurements, for example by ISR experiments, or by the WA70 and UA6 Collaborations, do not lead to the same conclusions. There the data are in better agreement with the NLO QCD predictions. Since there appear to be even inconsistencies between the various experimental data, many issues concerning the phenomenology of direct-photon production remain to be clarified. For further reading we refer to a recent discussion of this topic by Aurenche *et al.* (1999, 2000).

7.4.4 *Inclusive jet production*

The measurements of the inclusive cross section for jet production as a function of the jet transverse energy at the TEVATRON during recent years have triggered some excitement and stimulated interests for studying parton distributions, in particular the gluon distribution, at large x. A basic description of the process as well as a detailed discussion of the experimental issues can be found in previous chapters. The origin of the excitement is understood by looking at Fig. 7.22. There a recent measurement by CDF is compared to a NLO QCD prediction convoluted with a standard parton distribution set, CTEQ5M (CTEQ Collab., 2000). From the comparison of the cross sections and even more clearly from the ratio data/theory an excess of the data over the prediction is found for energies above 250 GeV. Such an excess at the largest energy scales could be due to new physics, such as substructure of quarks. The discrepancy can not be reduced by changing the renormalization as well as factorization scales within reasonable ranges, since this changes the predictions above $E_T \approx 100\,\mathrm{GeV}$ only by $2 - 9\%$ (Blazey and Flaugher, 1999). However, it is found that the theoretical uncertainty from the parton distributions can give rise to variations up to 30%. It is further pointed out that the measurements from CDF and D0 agree well

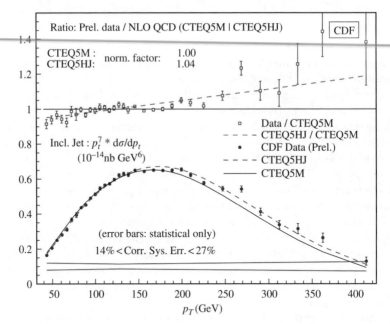

FIG. 7.22. Comparison of the inclusive jet cross section from CDF with two different parton distribution sets by the CTEQ group. The upper plot shows the ratio data/theory. Figure from CTEQ Collab.(2000).

within the experimental systematic uncertainties.

Since the inclusive jet cross section at the largest jet energies is very sensitive to the large-x regime, theoretical effort has been invested in order to scrutinize this region further. The precise data which are dominated by systematic uncertainties with known correlations, together with data from DIS, help to constrain the parton distributions and in particular the gluon contribution in the range $0.05 < x < 0.25$. The jet data probe the gluon at a much larger energy scale than for example the direct-photon measurements, and in addition they are not plagued by multiple soft-gluon radiation effects. The CTEQ group has performed a dedicated global QCD analysis (CTEQ Collab., 2000) where they have adjusted the parameterization of the gluon distribution in order to generate a significant enhancement at large x. This parameterization, called CTEQ5HJ, leads to a good description of the full set of DIS and Drell–Yan data, with only a marginally worse fit result compared to the standard parton distribution set. However, in addition the CTEQ5HJ distributions are able to restore good agreement between theory and high-E_T jet data, as seen in Fig. 7.22.

Similar studies have been performed based on measurements of the di-jet cross section at the TEVATRON. Since in leading order the invariant mass of the di-jet pair is given by the geometric average of the momentum fractions of the scattering partons times the C.o.M. energy, $M_{\mathrm{jj}} = \sqrt{x_1 x_2 s}$, a measurement of the differential cross section as a function of the di-jet mass is sensitive to the parton distributions. The same conclusions as above can be drawn from these studies (Blazey and Flaugher, 1999). Using standard parton distribution sets, again an excess of events at the largest di-jet masses is found. However, the theory can accomodate also this excess by introducing an enhanced gluon contribution at large x.

7.5   Global QCD analyses: parton distribution fits

We have learned previously that lepton–hadron or hadron–hadron cross sections can be expressed as a convolution of partonic cross sections and parton momentum distributions. The former are calculable within perturbative QCD, whereas the latter are not, since they describe non-perturbative long-distance phenomena of strong interactions. However, they are universal, that is, process-independent. So they can be determined from experiment and subsequently be used to predict cross sections for different processes and/or other energy scales. Similarly, when going beyond leading order, also the structure functions are expressed as convolutions of perturbatively calculable functions and parton distributions. Measurements of structure functions are therefore particularly useful to determine parton distributions. A precise determination of parton distribution sets is of paramount importance not only for tests of QCD, but also for precision measurements of Standard Model parameters such as the W mass, as well as the determination of signals and backgrounds for new physics searches at hadron colliders.

In order to obtain such a determination one has to use a large set of data which together cover a large range in x and Q^2 and put stringent constraints on the various parton types within the proton. Throughout this chapter we have summarized many measurements which are used for global QCD analyses. We have seen that the structure function measurements in charged lepton and neutrino DIS experiments constrain the quark contributions, and indirectly, via scaling violations, also the gluon distribution. Sea quark distributions are constrained by neutrino data (strange quarks), by Drell–Yan data $(\bar{d}, \bar{u})$ and heavy quark production (charm). The latter also gives a handle on the gluon distribution, together with direct-photon and jet production in hadron–hadron collisions. These global QCD analyses have been performed by several groups (CTEQ (1997, 2000), MRST (Martin *et al.*, 1998), GRV (Glück *et al.*, 1998)). Another recent determination was performed by Botje(2000).

We will now summarize the steps to be taken in such an analysis, avoiding however a discussion of details which can be found in the references mentioned above. The first step consists in a choice of experimental data, with possible restrictions on the x and Q^2 range in order to avoid experimentally and/or theoretically critical regions. Then an ansatz for the functional form of the parton distributions at a fixed scale Q_0^2 is chosen, for example, $Q_0^2 = 4\,\text{GeV}^2$ (Botje, 2000) or $1\,\text{GeV}^2$ (CTEQ Collab., 2000). As an example (Botje, 2000) the following parameterizations for the gluon distribution (xg), the sea quark distribution

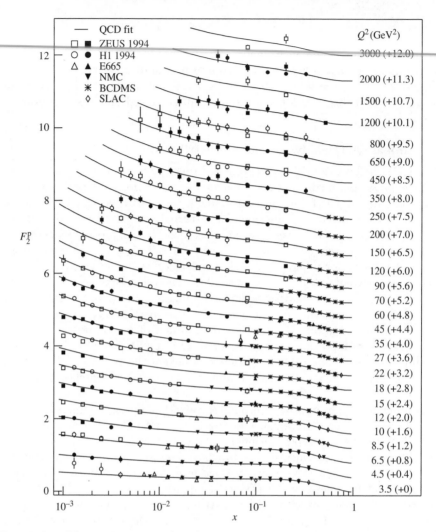

FIG. 7.23. The proton structure function F_2 versus x at fixed values of Q^2 from data included in the QCD analysis of Ref. (Botje, 2000). The full line indicates the result of the QCD fit. The constants in brackets are added to F_2.

xS, the difference of down and up antiquarks ($x\bar{\Delta}$) and the valence quark distributions ($xu_{\mathrm{v}}, xd_{\mathrm{v}}$) are used:

$$xg(x, Q_0^2) = A_{\mathrm{g}} x^{\delta_{\mathrm{g}}} (1-x)^{\eta_{\mathrm{g}}} (1 + \gamma_{\mathrm{g}} x) \,, \tag{7.42}$$

$$xS(x, Q_0^2) = 2x(\bar{u} + \bar{d} + \bar{s}) = A_{\mathrm{s}} x^{\delta_{\mathrm{s}}} (1-x)^{\eta_{\mathrm{s}}} (1 + \gamma_{\mathrm{s}} x) \,, \tag{7.43}$$

$$x\bar{\Delta}(x, Q_0^2) = x(\bar{d} - \bar{u}) = A_{\Delta} x^{\delta_{\Delta}} (1-x)^{\eta_{\Delta}} \,, \tag{7.44}$$

Table 7.1 *The values of the parameters obtained from a global QCD fit. The parameters are defined in the text. The errors are statistical only.*

Parameter	xg	xS	$x\bar{\Delta}$	xu_v	xd_v
A	1.32	0.81 ± 0.03	0.31 ± 0.10	2.72	1.98
δ	-0.26 ± 0.03	-0.15 ± 0.01	0.57 ± 0.09	0.62 ± 0.02	0.65 ± 0.02
η	5.19 ± 1.53	3.96 ± 0.19	7.47 ± 1.00	3.89 ± 0.02	3.07 ± 0.18
γ	-0.52 ± 1.42	-1.32 ± 0.06		2.79 ± 0.31	-0.82 ± 0.12

$$xu_v(x, Q_0^2) = x(u - \bar{u}) = A_u x^{\delta_u} (1 - x)^{\eta_u} (1 + \gamma_u x) , \tag{7.45}$$
$$xd_v(x, Q_0^2) = x(d - \bar{d}) = A_d x^{\delta_d} (1 - x)^{\eta_d} (1 + \gamma_d x) . \tag{7.46}$$

Thus the parameters δ_* (η_*) govern the large (small) x behaviour. The strange quark distribution $x(s+\bar{s}) = 2x\bar{s}$ is fixed to be a fraction κ of the non-strange sea as found by neutrino experiments (Section 7.2.4). The normalization parameters A_* are fixed by the momentum sum rule $\int_0^1 (xg + xS + xu_v + xd_v)\, dx = 1$ and valence quark counting rules $\int_0^1 dx\, u_v(x) = 2$ and $\int_0^1 dx\, d_v(x) = 1$. The remaining free parameters have to be determined from the fits to the data.

Once the parameterization as well as the theoretical scheme, that is, renormalization and factorization scheme, such as $\overline{\text{MS}}$, are fixed, the DGLAP evolution equations at NLO perturbative QCD are employed to compute the parton distributions and subsequently cross sections and structure functions at the scales relevant for the various experiments. The predictions are then fitted to the data by a least-squares method. In Botje(2000) 1578 data points are fitted, resulting in an excellent fit with a χ^2/point of 0.97. The fit to F_2 structure function measurements is shown in Fig. 7.23, and the resulting fit parameters are given in Table 7.1. Similarly, the CTEQ Collaboration uses about 1000 data points from DIS, 140 of Drell–Yan and 57 of jet experiments, with equally good fit results. Plots of the fitted parton distributions are given in Fig. 7.24 for (a) $Q^2 = 10\,\text{GeV}^2$ (Botje, 2000) and (b) $Q^2 = 25\,\text{GeV}^2$ (CTEQ Collab., 2000). It is nicely seen that the valence quark distributions peak around $x = 10^{-1}$ and vanish at small x, whereas the gluon and sea quark distributions increase with decreasing x. It should be noted that the gluon distribution has to be scaled down in order to be shown on the plot, which demonstrates the dominance of gluons in the proton at small x values.

These parameters can be used to compute other predictions within the same order of perturbation theory and the same theoretical scheme. They are implemented in computer programs to be used together with analytical calculations or Monte Carlo simulations. Because of practical reasons the energy dependence of the parameters is often parameterized by simple power laws, which avoids the need for computing the full DGLAP evolution every time a new prediction is needed.

There are many subjective choices which have to be made within a global analysis. This unavoidably leads to differences in the results from different groups

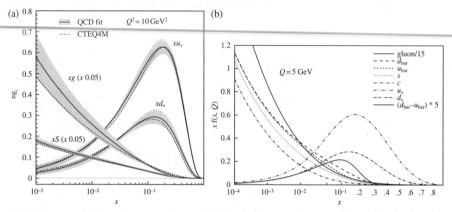

FIG. 7.24. The parton momentum distributions in the proton as extracted by Botje (2000) (a) and the CTEQ group (2000) (b). The shaded areas indicate the bands of uncertainty.

and consequently in intrinsic systematic uncertainties. These choices cover

- the selection of data sets and kinematic boundaries. As an example, the use of direct-photon data is particularly debated because of the discrepancies between data and NLO predictions (c.f. Section 7.4.3).

- the theoretical scheme used to compute the perturbative predictions. In particular the treatment of heavy quarks (threshold and mass effects) is not free of ambiguities.

- the choice of the strong coupling constant α_s, which is strongly correlated with the gluon distribution. Recently the preferred approach has become to fix the coupling to $\alpha_s(M_Z^2) = 0.118$ and to study the effect on the parton distributions when varying it within a certain range, for example, by ± 0.005.

- the correction for non-perturbative, so-called higher-twist contributions. One ansatz is $F_2^{\mathrm{HT}} = F_2^{\mathrm{QCD}}(1 + H(x)/Q^2)$, where $H(x)$ is a polynomial function. Its parameters are determined from the global fit.

- corrections for target mass and nuclear effects of data from fixed target experiments. Only phenomenological models exist for such corrections, and the groups differ in their choice of data to be corrected or not.

- the propagation of experimental and theoretical uncertainties into uncertainties of the parton distributions and ultimately the uncertainties of cross section predictions. Currently this is a field of active studies and discussions.

The user of a particular set of parton distributions should be aware of these problems and carefully evaluate their impact on the resulting predictions.

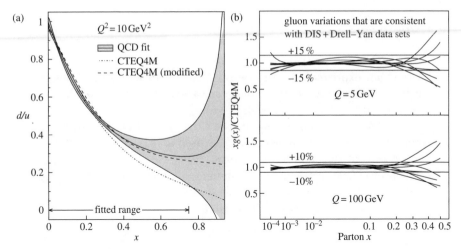

FIG. 7.25. (a) The ratio $(d+\bar{d})/(u+\bar{u})$ versus x at $Q^2 = 10\,\text{GeV}^2$ from the QCD fits of (Botje, 2000) and (CTEQ Collab., 1997). (b) Study of the uncertainty on the gluon distribution (Huston *et al.*, 1998). Shown is the ratio to the distribution obtained in (CTEQ Collab., 1997) at two different Q values.

After many years of global QCD analyses our knowledge of the proton structure is quite rich and precise. The valence quark distributions are rather precisely known. The gluon distribution is well constrained over the range $0.05 \lesssim x \lesssim 0.25$, however, the very low-$x$ behaviour is much less well known, and also at large $x \gtrsim 0.25$ it suffers from large uncertainties (Fig. 7.25(b)). In the large-x region also the d/u ratio is very much unconstrained, as can be seen from Fig. 7.25(a). The large-x domain could definitely gain from a better theoretical control of direct-photon production in hadron–hadron collisions. Concerning sea quarks, the fact that $\bar{d} \neq \bar{u}$ is well established by now. Future neutrino experiments should help to understand if $s = \bar{s}$. The charm quark sea still suffers from theoretical ambiguities in handling heavy quark effects. Finally, the question of uncertainties on the parton distributions and their propagation into uncertainties on the cross sections has to be addressed further in the near future.

Exercises for Chapter 7

7–1 Show that in a deep inelastic lepton–proton interaction the lepton scattering angle $\theta_e^\star$ is given by $\cos^2(\theta_e^\star/2) = 1 - y$, when computed in the C.o.M. frame defined by the momenta of the lepton and the struck quark. Here y is the inelasticity. Discuss various helicity configurations in the case of a W exchange, as is the case for a charged current interaction (Hint: c.f. the short discussion in Section 3.3.1 around eqn (3.94)).

7-2 Write down the expressions for charged current $\nu_e p$ and $e^+ p$ scattering in terms of the target hadrons p.d.f.s. Comment on the helicity structure of the contributing hard subprocesses.

7-3 Derive an expression for the ratio of the F_2 structure functions in lepton–nucleon and neutrino–nucleon scattering, in terms of the nucleon's p.d.f.s.

7-4 Show that the y dependence in the limit $y \to 0$ of the neutrino–nucleon scattering cross section can be used to determine the relative fluxes of neutrinos and antineutrinos.

7-5 The proton has electric charge 1 (in units of e), baryon number 1 and zero strangeness and charm. Use these constraints in order to derive sum rules for the valence quark distributions, such as $\int_0^1 dx\, u_v(x) = 2$.

8

THE STRONG COUPLING CONSTANT

If one accepts that QCD is described by an unbroken SU(3) gauge symmetry, then, apart from quark masses, the theory of strong interactions contains only one free parameter, the strong coupling constant α_s. Of course one has to verify experimentally that SU(3) really is the underlying gauge symmetry of the theory, a topic which is addressed in a later chapter. Here we will take for granted the arguments which motivated to build QCD on the gauge group SU(3) and describe the various approaches to measure the coupling strength. The rationale behind using as many methods as possible to determine α_s is to demonstrate that QCD really is the correct theory of strong interactions by showing that one universal coupling constant describes all strong interactions phenomena.

The processes discussed below cover a large range of both space-like and time-like momentum transfers. Reactions include neutrino and charged lepton nucleon scattering, proton-(anti)proton collisions, e^+e^- annihilation and decays of bound states of heavy quarks. Observables are, for example, cross section measurements, scaling violations, branching ratios, global event shape variables and the production rates of hadron jets.

To leading order all these processes can be described as interactions between spin-1/2 fermions and spin-1 gauge bosons. While the involved spins will determine the general structure of angular distributions and correlations between initial and final state particles, the rates can always be adjusted by introducing effective coupling constants. The real test of QCD comes with controlling higher order effects to the level where precise quantitative predictions can be made as function of the fundamental coupling of the theory. If a universal coupling exists, then all data within the respective combined experimental and theoretical uncertainties must be consistent with those calculations.

Over the past years significant progress has been made both experimentally and in theoretical calculations. As explained in detail in previous chapters, *next-to-leading order* (NLO) predictions are generally available. For some inclusive quantities also the *next-to-next-to-leading order* (NNLO) corrections have been calculated and estimates of the next higher terms exist. In addition, all-order resummations of *leading-logarithmic* (LL) and *next-to-leading-logarithmic* (NLL) corrections have been performed in some cases. Power law corrections are treated either by heuristic methods which introduce new free parameters, or more rigorously in the framework of the *operator product expansion* (OPE) or the resummation of renormalon chains.

8.1   Theoretical predictions

The QCD prediction for a cross section $\sigma(Q)$ at an energy scale Q can be expressed as sum of perturbative terms δ_{PT} varying logarithmically with energy, and non-perturbative power law corrections δ_{NP}. In next-to-leading order the perturbative part is of the form

$$\delta_{\mathrm{PT}} = \alpha_{\mathrm{s}}(\mu^2)A + \alpha_{\mathrm{s}}^2(\mu^2)\left(B + A\beta_0 \ln\frac{\mu^2}{Q^2}\right). \tag{8.1}$$

Here $\mu = \mu_R$ is the arbitrary renormalization scale of the calculation and β_0 the leading order coefficient of the QCD β-function. Setting $\mu^2 = Q^2$ the above expression eqn (8.1) simplifies to a power series in $\alpha_{\mathrm{s}}(Q^2)$ with energy independent coefficient functions for the LO and NLO contributions, respectively. The functions A and B are determined by the Lorentz structure of the contributing Feynman diagrams and thus essentially by the spins of the interacting fields. QCD only enters to the extent that it defines the relative weights for the different kinds of couplings. For $\mu^2 = Q^2$ the energy dependence of eqn (8.1) comes only from the running of the coupling $\alpha_{\mathrm{s}}(Q^2)$.

While the choice $\mu^2 = Q^2$ may appear to be the natural one, there is a priori no reason to single out a specific renormalization scale since the full theory is invariant with respect to μ. For the perturbative expansion things are not so obvious. If the renormalization scale and the scale of the physics process are very different, $|\ln(\mu^2/Q^2)| \gg 1$, then the perturbative expansion is no longer in the strong coupling but in an effective expansion parameter $\alpha_{\mathrm{s}}(\mu^2)\ln(\mu^2/Q^2)$ which can be larger than unity or even negative. While always being formally a valid expansion of the prediction for a physical observable, the consequences for a numerical evaluation of the series can be disastrous. A convergent series in $\alpha_{\mathrm{s}}(Q^2)$ can become divergent in $\alpha_{\mathrm{s}}(\mu^2)$, that is, while the full theory does not depend on the choice of μ, the truncated perturbative expansion certainly does. One also sees that the renormalization scale should be chosen not too different from the natural scale, which at least guarantees convergence of the perturbative series for asymptotic energies where α_{s} goes to zero.

The above discussion shows that varying the renormalization scale is a way to shift around higher order contributions. Scale variations thus are a convenient way to probe the sensitivity of the prediction to uncalculated higher orders and thereby to assess theoretical uncertainties. One must, however, emphasize that a small scale dependence at the calculated order does not mean that higher order terms are negligible. In general, a sound assessment of theoretical uncertainties will have to take other ways of estimating theoretical errors into account as well, such as dropping the highest order term which is known, variation of the so-called *matching scheme* which is used to merge a fixed order calculation and a resummed prediction, or using Padè approximations to rewrite the perturbation series. Without going into details it is fair to say that all these methods can be interpreted as attempts to guess and resum the uncalculated higher orders. Unfortunately, to date no generally established way to assess theoretical errors exists. The methods mentioned above are used in various combinations, which makes it very difficult to compare theoretical uncertainties. Nevertheless, and even if it is not possible to assign confidence levels in a strict mathematical sense to theoretical errors, they are best estimates of the actual uncertainties constructed such that it is reasonable to interpret them like conventional 68% confidence level intervals.

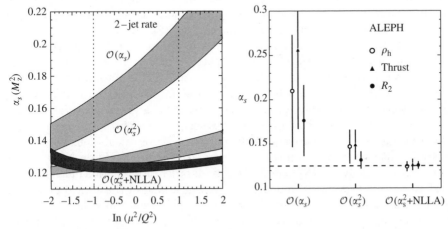

FIG. 8.1. Estimate of theoretical uncertainties for a measurement of the strong coupling from global event shape variables. A detailed discussion is given in the text.

This is illustrated in Fig. 8.1 by means of some measurement of the strong coupling constant performed on global event shape variables by the ALEPH collaboration. The variables will be described later. The left plot shows error bands in measurements of $\alpha_s(M_Z^2)$ based on the LO, NLO and NLO+NLLA predictions for the two-jet rate R_2 as function of $\ln(\mu^2/Q^2)$. The widths of the bands indicate what happens when switching from the perturbative prediction of R_2 to that of $\ln R_2$. The theoretical error was taken to be the range of values covered by the projection of the respective bands over $-1 < \ln \mu^2/Q^2 < 1$ on the abscissa. The right figure shows how the central values and errors obtained this way for three different shape variables converge with improvements in the theory. That this procedure yields reasonable error estimates is demonstrated by the fact that for a fixed level of theoretical precision the errors cover the scatter between the different variables, and that they also match the convergence observed when using better predictions.

8.2 Comparison and combination of results

To compare measurements of the strong coupling which were performed at different scales, one has to take into account that α_s is energy dependent. Measurements of α_s can be compared, either by evolving backwards to the point Λ_{QCD}, that is, $\Lambda_{\overline{MS}}$ when working in the $\overline{MS}$-scheme, where α_s diverges, or by evolving to a common reference scale, which in recent years has become the Z mass. Details about how to perform the evolution including the proper treatment of flavour thresholds can be found in Appendix D.1. In addition it is interesting to plot a measurement of the strong coupling for the scale where it has been measured in order to see the running. In cases where a reaction between pointlike particles happens at a fixed C.o.M. energy, this energy is quoted as the scale Q of the measurement. If the data cover a range $[Q_{min}, Q_{max}]$, then either the central scale quoted by the authors is given or, since α_s varies with the logarithm of the energy, the scale $Q = \sqrt{Q_{min}Q_{max}}$ which satisfies $\ln Q = (\ln Q_{min} + \ln Q_{max})/2$.

For an overview over various α_s-measurements one is not only interested in the final result, but also in a breakdown of the uncertainties into experimental and theoretical ones. While the former are fixed for a given measurement, the latter in principle could be reduced by further progress on the theoretical side. One thus also should keep track of the current level of calculations used in the analysis. All measurements presented below at least are based on NLO predictions, many already on NNLO or NLO plus all-order resummation of leading and next-to-leading logarithms. Independent of which kind of theoretical prediction has been used as a basis for a particular analysis, all results are comparable since the theoretical errors should account for the effect of missing higher order terms.

When combining results one has to keep in mind that the precision of many measurements of the strong coupling constant is limited by theoretical uncertainties. This leads to a situation where measurements from different experiments, while being statistically independent, are correlated through the uncertainties in the theoretical prediction. A global average has to take this into account. Unfortunately it is practically impossible to quantify these correlations on a case by case basis since the methods that have been employed in the determination of the theoretical uncertainties vary considerably between different measurements. These correlations usually do not affect the value of a weighted average, but they have to be taken into account in order to derive a realistic error estimate. In situations where they cannot be quantified reliably, the best one can do is to treat hidden correlations in an effective way. Methods to do this in a systematic way have been proposed by one of the authors (Schmelling 1995b, 2000), where an estimate for the size of the correlations between measurements is derived from the typical scatter of the results around a common average compared to the quoted errors of the measurements. The basic idea is that if the average scatter is smaller than the average errors, then the errors are correlated.

In the following sections we will describe the most important methods that are used to measure α_s and the status of our knowledge reached at the turn of the millenium. The field is still rapidly evolving, and the reader is also referred to the literature for further information. Good reviews are found for example in (Bethke, 2000; Hinchliffe and Manohar, 2000). Results are collected in Table 8.3 and, unless mentioned otherwise in the text, quoted from (Bethke, 2000).

8.3 Inclusive measurements

Inclusive measurements of α_s depend only on one energy scale characterizing the process, and are theoretically best understood. The perturbative prediction is known to $\mathcal{O}(\alpha_s^3)$, non-perturbative effects can be treated in the framework of the OPE (Shifman *et al.*, 1979). Even the results of higher order calculations are reasonably compact. The main ingredients needed for the theoretical prediction of R_γ, R_l and R_τ are collected in Appendix D.

8.3.1 *The ratios R_γ and R_l*

The quantities R_γ and R_l are defined by the ratios of the hadronic to the leptonic branching fractions of the virtual photon or the Z boson. In both cases the hadronic system is formed through the electroweak coupling of a vector boson (photon or Z) to a primary quark–antiquark pair. The sensitivity to the strong coupling comes from gluon radiation off the primary quarks, as depicted in Fig. 8.2. This radiation opens up new final states for the hadronic system and thus increases the hadronic width with respect to the purely electroweak expectation. Since the amount of gluon radiation in leading order is proportional to the value of the strong coupling constant, a measurement of the hadronic cross section allows to determine α_s.

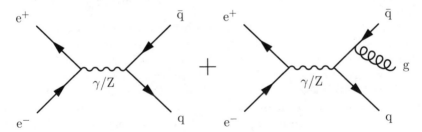

FIG. 8.2. Electroweak and leading order QCD contribution to the hadronic cross section in e$^+$e$^-$ annihilation.

In practical applications one usually determines the cross section ratio

$$R = \frac{\sigma(e^+e^- \to \text{hadrons})}{\sigma_{\text{born}}(e^+e^- \to \text{leptons})} , \qquad (8.2)$$

where the final state lepton pair is assumed to be massless and different from the inital state e$^+$e$^-$ pair. Experimentally, this means that one determines the number of electron–positron annihilations into hadrons compared to the number of annihilations into muon pairs, where the muon rate has to be corrected for higher order electroweak effects, the finite mass of the muons and in general also for a detector acceptance which is slightly different from the one for hadronic events. In the theoretical prediction for R phase-space factors cancel, making it a dimensionless function of the relevant couplings. It can be written in the form

$$R = R_{\mathrm{EW}} \left(1 + \delta_{\mathrm{QCD}} + \delta_{\mathrm{m}} + \delta_{\mathrm{NP}}\right) , \qquad (8.3)$$

where the overall factor R_{EW} depends on the electroweak couplings of the quarks. The corrections are dominated by the perturbative QCD correction δ_{QCD}. The other terms δ_{m} and δ_{NP} take into account the finite quark masses and the non-perturbative corrections. Since the leading non-perturbative effects are $\mathcal{O}(1/Q^4)$ they are negligible for high energy e^+e^- annihilation processes. Mass effects, in particular from the large top-bottom mass splitting, have been calculated (Chetyrkin et al., 1996b) to $\mathcal{O}(\alpha_s^2)$.

Ignoring mass corrections and non-perturbative effects, the theoretical prediction for R_γ becomes

$$R_\gamma = 3 \sum_q e_q^2 \left(1 + \frac{\alpha_s}{\pi} + 1.41 \frac{\alpha_s^2}{\pi^2} - 12.8 \frac{\alpha_s^3}{\pi^3} \cdots \right) , \qquad (8.4)$$

where the strong coupling has to be evaluated at the scale of the C.o.M. energy $\sqrt{s}$ of the experiment. Various determinations of the strong coupling $\alpha_s(Q^2)$ based on eqn (8.4) have been performed, spanning the energy range from just above the $b\bar{b}$-threshold (CLEO Collab., 1998), $Q = 10.52$ GeV, until $Q = 34$ GeV (Bethke, 2000) and $Q = 42.4$ GeV (Haidt, 1995). The results are given in Table 8.3.

Also the theoretical prediction for R_l, where an electron and a positron annihilate into a Z boson, is known to third order in α_s. Because of the simultaneous contribution of vector and axial currents in the Z–fermion couplings and the different nature of mass corrections in the presence of axial currents, the coefficients in the theoretical prediction are slightly different from those for R_γ. An explicit description to calculate the theoretical prediction including the leading order corrections is given in Appendix D.3. A more complete calculation leads to the following parameterization (Tournefier, 1998) as function of $\alpha_s(M_Z^2)$,

$$R_l = 19.934 \left(1 + 1.045 \frac{\alpha_s}{\pi} + 0.94 \frac{\alpha_s^2}{\pi^2} - 15 \frac{\alpha_s^3}{\pi^3}\right) . \qquad (8.5)$$

Using the combined result from all four LEP experiments (Abbaneo et al., 2000), $R_l = 20.768 \pm 0.024$, one finds $\alpha_s(M_Z^2) = 0.124 \pm 0.004$, where the error is purely statistical. The full result including theoretical uncertainties from scale variations and the uncertainty of the Higgs mass is given in Table 8.3.

8.3.2 Measurement of α_s from R_τ

An α_s-measurement can also be obtained from $R_\tau = B_{\mathrm{hadr}}/B_{\mathrm{e}}$, the ratio of the hadronic to the electronic branching ratio of the tau lepton. Here QCD radiative corrections affect the hadronic final state from a W decay, the leading order diagram of which is shown in Fig. 8.3.

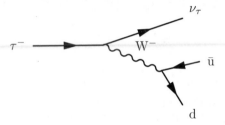

FIG. 8.3. Leading order diagram for the decay of a τ lepton into hadrons. QCD corrections manifest themselves as gluon radiation off the quark lines.

In contrast to α_s-determinations from R_γ or R_l, the mass of the hadronic system is not fixed in τ decays, that is, energy scales from the mass of the pion to the mass of the τ lepton contribute. For the theoretical prediction of R_τ one thus essentially has to integrate the QCD correction to R_γ over the contributing mass range, weighted with the mass spectrum of the hadronic system. Compared to R_γ and R_l this makes the quantity R_τ doubly inclusive, that is, integrated over all hadronic final states at a given mass and integrated over all masses between M_π and M_τ. Expansion in powers of $\alpha_s(M_\tau^2)$ yields the following expression:

$$R_\tau = 3.0582 \left(1 + \frac{\alpha_s}{\pi} + 5.2023 \frac{\alpha_s^2}{\pi^2} + 23.366 \frac{\alpha_s^3}{\pi^3} + \cdots \right) . \qquad (8.6)$$

Despite the low energy scale the perturbative expansion still seems to be well behaved. Indeed, there is no reason to re-expand the integration over the mass spectrum which resums certain terms of the perturbative prediction to all orders and thus results in an even more convergent expression. The evaluation of this resummed perturbative correction to R_τ is described in Appendix D.4.

However, working at a low mass scale one has to worry about the non-perturbative corrections δ_{NP}. The Shifman–Vainstain–Zhakarov (SVZ) ansatz (Shifman *et al.*, 1979) controls non-perturbative corrections in the framework of the OPE,

$$\delta_{NP} = \sum_{n=2}^{\infty} C_n \frac{\langle 0| O_n |0 \rangle}{M_\tau^{2n}} , \qquad (8.7)$$

with coefficients C_n that can be determined in perturbation theory, and universal vacuum expectation values of operators O_n, so-called condensates. Values for those condensates obtained from phenomenological analyses can be found in the literature, for example (Braaten *et al.*, 1992). Alternatively, they can be extracted together with the strong coupling from higher moments of the mass spectrum of hadronic τ decays (Le Diberder and Pich, 1992*a*). Assuming that the SVZ ansatz holds, it turns out that the non-perturbative corrections to R_τ are surprisingly small, below 1%. This, together with the finding (Girone and Neubert, 1996) that perturbative QCD may be applicable down to mass scales below 1 GeV, implies that an α_s-measurement based on R_τ is potentially very accurate.

Experimentally R_τ can be determined in a variety of ways, and it is quite interesting to see how it is inferred from at first glance completely unrelated inputs. The obvious way to proceed is of course to measure the hadronic and the electronic branching fractions, B_{hadr} and B_{e}, and take the ratio. Assuming lepton universality and the completeness relation $B_{\mathrm{hadr}} + B_{\mathrm{e}} + B_\mu = 1$, R_τ can also be obtained from B_{e} and B_μ alone. Note that the Standard Model predicts a slight difference between B_{e} and B_μ, $B_\mu/B_{\mathrm{e}} = f_\mu = 0.97256$, as consequence of the larger mass of the muon. One obtains

$$R_\tau = \frac{B_{\mathrm{hadr}}}{B_{\mathrm{e}}} = \frac{1 - B_{\mathrm{e}} - B_\mu}{B_{\mathrm{e}}} = \frac{1}{B_{\mathrm{e}}} - 1 - f_\mu = \frac{f_\mu}{B_\mu} - 1 - f_\mu . \quad (8.8)$$

Within the Standard Model it is also possible to infer B_{e} and thus R_τ comparing mass and lifetime of the tau lepton to that of the muon. Since the partial width of weak decays is proportional to the fifth power of the mass, and assuming again lepton universality, one has

$$\frac{\Gamma_e(\mu)}{\Gamma_e(\tau)} = \frac{\Gamma_{\mathrm{tot}}(\mu)}{B_{\mathrm{e}}\Gamma_{\mathrm{tot}}(\tau)} = \left(\frac{M_\mu}{M_\tau}\right)^5 \implies B_{\mathrm{e}} = \frac{\tau_\tau}{\tau_\mu}\left(\frac{M_\tau}{M_\mu}\right)^5 . \quad (8.9)$$

Table 8.1 shows the averages for some basic parameters of the tau lepton as they are given by Pich (2000). The value for R_τ is the weighted average of the three independent measurements that can be extracted from those numbers. Using the formalism described in Appendix D.4, this value translates into a measurement of the strong coupling $\alpha_s(M_\tau^2) = 0.347 \pm 0.012$ or $\alpha_s(M_Z^2) = 0.121 \pm 0.002$. Different approaches in the theoretical prediction (Braaten $et\ al.$, 1992; Le Diberder and Pich, 1992a; Ball $et\ al.$, 1995; Girone and Neubert, 1996) indicate that the theoretical uncertainties are somewhat larger, which is taken into account for the values and errors given in Table 8.3.

Table 8.1 $Averages\ for\ some\ parameters\ of\ the$
$tau\ lepton$

Quantity	Value
M_τ	$(1777.05^{+0.29}_{-0.26})$ MeV$/c^2$
τ_τ	(290.77 ± 0.99) fs
B_{e}	$(17.791 \pm 0.054)\%$
B_μ	$(17.333 \pm 0.054)\%$
R_τ	3.643 ± 0.010

8.3.3 α_s from sum rules

Sum rules are discussed in detail in Section 7.3. Determinations of the strong coupling constant based on sum rules are fully inclusive measurements at very low Q^2. In the quark parton model of non-interacting constituents sum rules have a very intuitive interpretation. The Gross–Llewellyn-Smith sum rule (GLS-SR), for example, counts the number of valence quarks in the nucleon, while the Bjorken sum rule (Bj-SR) extracted from polarized lepton nucleon scattering experiments is directly proportional to the ratio of axial to vector coupling in neutron β-decay.

Strong interactions between the quarks modify the predictions. Also here the QCD corrections (Larin and Vermaseren, 1991) are known to $\mathcal{O}(\alpha_\mathrm{s}^3)$, and estimates for the size of the non-perturbative effects exist (Braun and Kolesnichenko, 1987; Stein et al., 1995; Mueller, 1993). The QCD corrections for the GLS-SR and the Bj-SR are very closely related,

$$\int_0^1 dx F_3(x) = 3 \left(1 - \frac{\alpha_\mathrm{s}}{\pi} - 3.5833 \frac{\alpha_\mathrm{s}^2}{\pi^2} - 18.976 \frac{\alpha_\mathrm{s}^3}{\pi^3} \cdots \right) , \qquad (8.10)$$

and

$$\int_0^1 dx (g_1^\mathrm{p}(x) - g_1^\mathrm{n}(x)) = \frac{1}{6} \left| \frac{g_\mathrm{A}}{g_\mathrm{V}} \right| \left(1 - \frac{\alpha_\mathrm{s}}{\pi} - 3.5833 \frac{\alpha_\mathrm{s}^2}{\pi^2} - 20.852 \frac{\alpha_\mathrm{s}^3}{\pi^3} \cdots \right) . \quad (8.11)$$

A measurement of the strong coupling constant based on the GLS-SR, taking into account also non-perturbative terms (Braun and Kolesnichenko, 1987), has been performed by the CCFR/NuTeV Collaboration (CCFR Collab., 1998), α_s-measurements from the Bj-SR were done by Ellis et al. (1996a). Both analyses determined $\alpha_\mathrm{s}(Q^2)$ at a scale $Q = \sqrt{3}$ GeV.

It is interesting to note that despite the low energy scale the result (Ellis et al., 1996a) is dominated by experimental errors. The small size $\delta\alpha_\mathrm{s}(Q^2) = 0.016$ of the theoretical uncertainties is inferred from the stability of the analysis with respect to various ways of estimating missing higher order terms using Padé approximants (PA). The PA $[N/M]$ of a function is the ratio of two polynomials of order N and M, respectively, which to order $N + M$ has the same Taylor-expansion as the original function. Padé approximants thus offer a systematic way to guess how a perturbative expansion resums, by rewriting a perturbative series as a ratio of two polynomials. Compared to the original expression, the PA introduces poles into the prediction in a similar fashion as expected from renormalons. This may explain why PAs seem to be able to approximately resum the perturbative series, a finding corroborated, for example, by the observation (Burrows et al., 1996) that measurements of the strong coupling from global event shape variables become much more consistent when the conventional second order prediction is replaced by its PA $[1/1]$.

8.4 Measurements of α_s from heavy flavours

Heavy flavour physics not only allows to study quantum flavour dynamics, but in addition provides unique opportunities to determine the strong coupling constant. Information about α_s can be obtained from decay rates by comparing, for

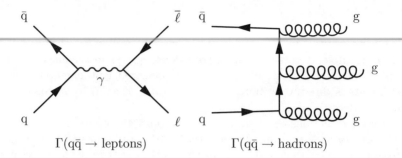

FIG. 8.4. Born level amplitudes for the decays of a spin-1 heavy flavour system into leptonic and hadronic final states

example, hadronic to leptonic final states, or from the level splittings between different orbital excitations of the bound system, which probe the QCD potential between quark and antiquark. Since the latter approach requires a good quantitative understanding of the non-perturbative inter-quark potential, it is pursued in the context of lattice gauge theories.

8.4.1 *Decays of heavy quarkonia*

A precise α_s-measurement is obtained from the ratio of the hadronic to the leptonic width of a spin-1 bound state of a heavy quark–antiquark pair. The dominant Born level amplitudes for the decay into leptons and hadrons are shown in Fig. 8.4. While the leptonic channel can proceed via an intermediate photon, colour conservation prevents the analog process through a single gluon since the two quarks form a colour singlet state, while the gluon is a colour octet. As far as colour is concerned, a decay into two gluons would be allowed, but that again is forbidden by angular momentum conservation because the gluons are massless and thus cannot form a spin-1 system. Finally, since the strong coupling is much larger than the electromagnetic coupling, the annihilation of the heavy quark pair into light quarks is only a small correction to the three gluon decay. The leading order diagram for hadronic decays thus becomes the decay into three gluons which subsequently hadronize.

It follows that to leading order the ratio of the hadronic to leptonic width is proportional to α_s^3/α_{em}^2. The perturbative prediction is known to NLO. Considering the ratio has the additional advantage that the bound state wave function at the origin cancels, making a measurement of the strong coupling independent of precise knowledge of the QCD potential between the quarks. One has, however, to take into account relativistic corrections which are proportional to the average $\langle v^2/c^2 \rangle$ of the quarks. Here we will present a result (Kobel, 1992) based on the following parameterization

$$\frac{\Gamma(q\bar{q} \to \text{hadrons})}{\Gamma(q\bar{q} \to e^+e^-)} = \frac{\alpha_s^3}{\alpha_{em}^2} A \left(1 - B\alpha_s\right) \left(1 + D \left\langle \frac{v^2}{c^2} \right\rangle\right) \qquad (8.12)$$

for the NLO description of the decay of a colour singlet $q\bar{q}$-state. The coefficients A and B are known from perturbation theory, D is a free parameter of the model. Assuming D to be a universal constant, it was extracted (Kobel, 1992) together with α_s in a combined analysis of Υ and the J/Ψ decays. Theoretical uncertainties in the perturbative sector were studied by introducing ad hoc NNLO terms, renormalization scale variation and by Padé-like rewriting $(1 - B\alpha_s)$ as $1/(1 + B\alpha_s)$. The result at a scale $Q = 10$ GeV given in Table 8.3 is dominated by theoretical uncertainties.

The CLEO Collaboration (1996) has performed a similar analysis using the ratio $\Gamma(\Upsilon \to \mathrm{gg}\gamma)/\Gamma(\Upsilon \to \mathrm{ggg})$, which to leading order is proportional to α_{em}/α_s. The result, determined with the help of a Monte Carlo model for the hard photon spectrum at a scale $Q = 9.7$ GeV, is also given in Table 8.3. Here, the systematic errors of the measurement, reflecting the difficulty of discriminating against the photon background in hadronic Upsilon decays, and theoretical uncertainties contribute equally to the total error.

8.4.2   Lattice calculations

Precise determinations of α_s were also performed in the analysis of level splittings between the S- and the P-states in the Υ-system (Davies et al., 1996; El-Khadra, 1996) by means of lattice calculations. The physical scale of these determinations of the strong coupling is $Q = 10$ GeV. The lattice approach tries to reproduce the observed energy levels in bound state systems by ab-initio calculations based on the full QCD Lagrangian. Technically this is done by letting QCD fields evolve on a discretized space–time lattice using a Monte Carlo method. While the quantization of the space–time continuum makes the problem amenable to numerical simulations it also introduces so-called lattice artefacts which vanish only in the continuum limit. A good understanding of how to extrapolate the numerical results to the continuum limit is essential for a precision measurement of the strong coupling. In the past years significant progress has been made in reducing lattice spacing errors, in the conversion from the lattice to the continuum $\overline{\mathrm{MS}}$ coupling constant and in the treatment of fermions in loop corrections. Due to the anticommuting nature of the fermion fields a numerical treatment on the lattice is highly non-trivial. Calculations exist with $n_f = 0$ and $n_f = 2$ dynamic fermions, which give only marginally different results and thus allow a safe extrapolation to the physical case of $n_f = 3$ light flavours. Two early results (Flynn, 1996) are $\alpha_s(M_Z^2) = 0.118 \pm 0.003$ (NRQCD Collaboration) and $\alpha_s(M_Z^2) = 0.116 \pm 0.003$ (FNAL Collaboration). A more recent calculation using a different numerical technique (Spitz et al., 1999) as before derived a lower value $\alpha_s(M_Z^2) = 0.112 \pm 0.002$. The errors are purely theoretical. In Table 8.3 the unweighted mean and standard deviation are quoted as the nominal result and error for a measurement of the strong coupling based on lattice gauge theories.

8.5 Scaling violations

Scaling violations are an immediate consequence of the fact that quarks are strongly interacting particles and thus provide a rather direct and theoretically clean way to measure the strong coupling constant. In DIS processes with space-like momentum transfer scaling violations are observed as a softening in the nucleon structure functions, that is, the average momentum fraction per parton becomes smaller with increasing momentum transfer. In e^+e^- annihilation into hadrons, which proceeds via time-like momentum transfer, they manifest themselves as a softening of the fragmentation functions, which means that the average momentum fraction per final state particle decreases with increasing Q^2. Neither structure functions nor fragmentation functions can be calculated within perturbative QCD, but given at a certain scale, QCD predicts the energy evolution as a softening with increasing Q^2, $d \ln F / d \ln Q^2 \propto \alpha_s(Q^2)$, which is described by the DGLAP evolution equations (Gribov and Lipatov, 1972; Altarelli and Parisi, 1977; Dokshitzer, 1977), c.f. Section 3.6.

For structure functions the softening comes about because higher momentum transfers resolve more partons from vacuum fluctuations in the nucleon, for fragmentation functions the growing phase space permits additional gluon radiation and thus particle production in jets. The theory (Curci *et al.*, 1980; Furmanski and Petronzio, 1980; Altarelli *et al.*, 1979b; Nason and Webber, 1994) is known to next-to-leading order accuracy. In a determination of α_s the non-perturbative effects can be disentangled by their energy dependence. Here DIS processes are favoured because non-perturbative effects decrease rapidly with $1/Q^2$. In e^+e^- annihilation they decay only proportional to $1/Q$, which renders the measurements of α_s from fragmentation functions less precise than the ones from DIS, even if they are performed at much larger momentum transfers.

The analysis of scaling violations in DIS is covered in Sections 3.6 and 7.5. Here we will now complement that discussion by describing how an α_s measurement is performed by analysing scaling violations in fragmentation functions. The theoretical framework as well as the phenomenological inputs are discussed in detail in order to illustrate the limitations of such an analysis. Explicit expressions for the functions entering the analysis are collected in Appendix E.

8.5.1 *Scaling violations in fragmentation functions*

The single inclusive particle spectrum produced in the process $e^+e^- \rightarrow$ hadrons can be written as a sum of a 'transverse' (T) and a 'longitudinal' (L) cross section:

$$\frac{1}{\sigma_{tot}(s)} \frac{d\sigma(s)}{dx} = \frac{1}{\sigma_{tot}(s)} \frac{d\sigma^T(s)}{dx} + \frac{1}{\sigma_{tot}(s)} \frac{d\sigma^L(s)}{dx}, \qquad (8.13)$$

where the transverse component has an angular distribution proportional to $1 + \cos^2 \theta$ and the longitudinal one a distribution proportional to $\sin^2 \theta$. Here 'transverse' and 'longitudinal' refer to the polarization states of the intermediate photon or Z boson. The angle θ is measured with respect to the incoming beams. The Lorentz invariant variable x is defined through $x = 2(k \cdot Q)/(Q \cdot Q)$ where

k is the four-momentum of the produced hadron and Q the four-momentum of the virtual photon or Z. In the C.o.M. frame of the collision, which generally coincides with the laboratory frame of an e^+e^- collider as long as initial state radiation can be neglected, it reduces to $x = E_{hadron}/E_{beam}$.

The transverse component is the dominant contribution from the annihilation of an e^+e^- pair into a pair of spin-1/2 fermions. The longitudinal cross section gets contributions from higher order corrections and is suppressed by a factor α_s with respect to $\sigma^T(s)$. Disentangling the two contributions at a fixed energy thus would also allow to measure α_s. In the following we will, however, not pursue this possibility and rather focus on a measurement based on the energy dependence of $d\sigma(s)/dx$.

In QCD, the observable hadron spectra are related to a set of fragmentation functions $D_p(x, s)$. Here the D_p are defined as x-weighted particle spectra resulting from the fragmentation of a parton of type p. The weighting with x results in a less singular behaviour of the fragmentation functions for $x \to 0$ and results also in simpler evolution equations. The functions $D_p(x, s)$ cannot be calculated perturbatively, but once given at a fixed C.o.M. energy $\sqrt{s_0}$, the evolution with s is predicted by QCD. To leading order the fragmentation functions can directly be associated with the observable hadron spectra. In general the connection between fragmentation functions and observable distributions is given by

$$\frac{x}{\sigma_{tot}(s)}\frac{d\sigma^{T,L}(s)}{dx} = 2\frac{\sigma_0(s)}{\sigma_{tot}(s)}\int_x^1 dz \sum_p w_p(s)P_p^{T,L}(z, s, \mu_F)D_p(\frac{x}{z}, \mu_F) , \quad (8.14)$$

where the sum adds contributions from all quark flavours and the gluon. The weights $w_p(s)$ are the s-dependent relative electroweak production cross sections for the different flavours. The weight for gluons is $w_g(s) = 1$. The factorization scale μ_F appearing in this expression relates the fragmentation functions at the energy scale μ_F to the cross section at a C.o.M. energy $\sqrt{s}$. Note that for the treatment of scaling violations we have to be explicit about the renormalization scale μ_R and the factorization scale μ_F. The integral kernels $P_p^{T,L}$ are calculable in perturbation theory. The overall factor of 2 accounts for the fact that two primary quarks fragment in a multi-hadron event. Note that the functions $P_p^{T,L}$ appearing here must not be confused with the Altarelli–Parisi splitting kernels appearing in the theory of scaling violations in deep inelastic scattering, even though the two sets of functions are related via crossing symmetry.

The evolution of the fragmentation functions with energy is described by the DGLAP evolution equations. It is most conveniently done by decomposing the quark fragmentation functions into flavour *non-singlet* parts $N_i(x, s)$ which evolve independently and a flavour *singlet* part $S(x, s)$ which is coupled to the evolution of the gluon fragmentation function $G(x, s) \equiv D_g(x, s)$,

$$S(x, s) = \frac{1}{n_f}\sum_{i=u,d,s,c,b} D_i(x, s) \quad \text{and} \quad N_i(x, s) = D_i(x, s) - S(x, s) . \quad (8.15)$$

Input	Measurement at $\sqrt{s_i}$
$D(x, \mu_F^2(i)) \quad \rightarrow \quad \text{QCD}(\mu_F^2(i), s_i) \quad \rightarrow$	$\dfrac{d\sigma(s_i)}{dx} \quad \rightarrow \quad \dfrac{d\sigma(s_i)}{dx_e}$
$\downarrow$	
$\text{QCD}(\mu_F, \mu_R)$	
$\downarrow$	Measurement at $\sqrt{s_f}$
$D(x, \mu_F^2(f)) \quad \rightarrow \quad \text{QCD}(\mu_F^2(f), s_f) \quad \rightarrow$	$\dfrac{d\sigma(s_f)}{dx} \quad \rightarrow \quad \dfrac{d\sigma(s_f)}{dx_e}$

FIG. 8.5. Schematic of an analysis of scaling violations. The variable x is the momentum fraction in the perturbative sector, x_e the experimentally observed quantity which includes non-perturbative contributions.

One then obtains for the non-singlet parts

$$s\frac{d}{ds}N_i(x, s) = \int_x^1 dz\, P^{\text{NS}}(z, s, \mu_R) N_i\left(\frac{x}{z}, s\right) . \tag{8.16}$$

The evolution of the singlet components is described by a coupled system:

$$s\frac{d}{ds}G(x, s) = \int_x^1 dz\, \left[P_{GG}^{\text{S}}(z, s, \mu_R) G\left(\frac{x}{z}, s\right) + P_{GQ}^{\text{S}}(z, s, \mu_R) S\left(\frac{x}{z}, s\right)\right] \tag{8.17}$$

$$s\frac{d}{ds}S(x, s) = \int_x^1 dz\, \left[P_{QG}^{\text{S}}(z, s, \mu_R) G\left(\frac{x}{z}, s\right) + P_{QQ}^{\text{S}}(z, s, \mu_R) S\left(\frac{x}{z}, s\right)\right] \tag{8.18}$$

Again the kernels P^{NS} and P_{pp}^{S}, with $p = Q$ or G are calculable in perturbation theory. The renormalization scale μ_R relates the splitting kernels at the C.o.M. energy $\sqrt{s}$ to the strong coupling constant at the energy scale μ_R.

Non-perturbative corrections which are expected to show up as corrections proportional to $1/\sqrt{s}$ to the logarithmic scaling violations from perturbative QCD are discussed in depth by Nason and Webber (1994). A simple way of parameterizing non-perturbative effects is a change of variables, relating the perturbative variable x to the measured quantity x_e through $x = g(x_e, s) = x_e(1+h/\sqrt{s})$. The connection between the perturbative prediction $d\sigma/dx$ and the experimentally observed cross section $d\sigma/dx_e$ can then be derived from the constraint of energy conservation before and after the transformation $\int dx\, x\, d\sigma/dx = \int dx_e\, x_e\, d\sigma/dx_e$, which gives

$$\frac{d\sigma}{dx_e} = \frac{g(x_e)g'(x_e)}{x_e}\frac{d\sigma}{dx}. \tag{8.19}$$

At this point the set of ingredients needed for an α_s-measurement from scaling violations in fragmentation functions is complete. The explicit expressions of the functions needed to perform the analysis are given in Appendix E. Schematically a QCD test based on measurements of inclusive cross sections at different C.o.M. energies is sketched in Fig. 8.5. The input is a set of fragmentation functions $D(x, \mu_F^2(i))$ at an initial factorization scale $\mu_F(i)$. Perturbative QCD then relates those fragmentation functions to an observable cross section which, after inclusion of the non-perturbative corrections, can be compared with experimental data at an initial C.o.M. energy $\sqrt{s_i}$ (horizontal arrows). The energy evolution of the fragmentation functions is described by perturbative QCD (vertical arrows), where the energy variation at the scale μ is expressed as function of μ and the renormalization point μ_R. The evolution equations yield the fragmentation functions at a new factorization scale μ_F where they again can be related to observable cross sections at a new C.o.M. energy $\sqrt{s_f}$. The test of QCD consists in showing that all available data are described consistently with one set of fragmentation functions, a universal coupling constant and a global parameter h to account for non-perturbative effects. In an actual analysis the fragmentation functions α_s and h are adjusted in a global fit to the available data.

For the parameterization of the fragmentation functions some guidance is provided by perturbative QCD in the framework of the *modified leading-log approximation* (MLLA), which predicts that the momentum spectrum of final state particles should exhibit an approximately Gaussian peak in $\ln x$ (Fong and Webber, 1991). From this one would infer a functional form for the fragmentation function

$$\frac{d\sigma}{d\ln x} \sim \exp\left(-c(d - \ln x)^2\right) \qquad \Longleftrightarrow \qquad \frac{d\sigma}{dx} \sim \exp\left(-c\ln^2 x\right) x^{2cd-1} . \quad (8.20)$$

Combined with the expectation that the momentum spectrum falls off with some power of $(1 - x)$ for $x \to 1$, then finally yields the ansatz

$$D(x) = N \exp(-c\ln^2 x)x^b(1 - x)^a, \quad (8.21)$$

where N is a normalization constant and a, b and c are free parameters which have to be determined from the data. The parameters a and b depend on the quark mass, that is, they are different for light quarks, c-quarks, b-quarks and gluons, whereas c should be flavour independent.

A priori the scales μ_F and μ_R are unconstrained. When calculating to all orders, any dependence on the choice of the scales vanishes. In finite order perturbation theory a residual scale dependence is related to uncalculated higher order terms. In order to avoid large logarithms in the theoretical predictions the natural choice of scales is $\mu_F^2(i)/s_i = \mu_F^2(f)/s_f = \mu_R^2/\mu_F^2 = 1$. Varying the scales is one possible way to estimate the theoretical uncertainties of the prediction. A common choice to parameterize the scale dependence is through a common factor $f = \mu_F^2(i)/s_i = \mu_F^2(f)/s_f$ to probe the factorization scale dependence and another factor $f_r = \mu_R^2/\mu_F^2$ for the renormalization scale dependence.

The full NLO theoretical framework has been used in two determinations
of the strong coupling constant from scaling violations in fragmentation functions (ALEPH Collab., 1995b; DELPHI Collab., 1997a). Both are based on inclusive distributions and flavour enriched data samples from hadronic Z decays,
that is, momentum spectra for light quarks, c-quarks, b-quarks and gluons, combined with measurements from lower C.o.M. energies down to $\sqrt{s} = 14$ GeV.
The strong coupling constant was determined together with parameterizations
for the fragmentation functions of the different parton types and a power law correction describing non-perturbative effects. The latter was found to be small for
the energy range under consideration. The combined result is given in Table 8.3.

8.5.2 *Scaling violations in structure functions*

The study of scaling violations in structure functions is one of the classical approaches for the determination of the strong coupling constant. Measurements
were performed in DIS with neutrino beams and charged leptons on targets of
heavy and light nuclei. The quantity extracted from these measurements was
the QCD scale $\Lambda_{\overline{MS}}^{(4)}$ for four active flavours. Since those experiments covered
Q^2-ranges between 0.5 GeV^2 and 260 GeV^2, with a typical central scale around
$\mu = 7.1$ GeV, the contributions from heavier quarks are negligible. The result
from a combined analysis (Virchaux and Milsztajn, 1992) of SLAC-BCDMS measurements of the structure function $F_2(x)$, based on a NLO theoretical prediction
is $\Lambda_{\overline{MS}}^{(4)} = 263 \pm 42$ MeV. The error is the combined experimental and theoretical
uncertainty. In this analysis also the gluon distribution and non-perturbative contributions have been fitted. In a similar analysis (Kataev *et al.*, 2000) of CCFR
data (1997), using NNLO theoretical predictions, the strong coupling constant
was extracted from the evolution of the flavour singlet distribution $F_3(x)$ as measured in neutrino–nucleon scattering. Finally we should also mention a NNLO
analysis (Santiago and Yndurain, 1999) based on the evolution of moments of
F_2, where in addition to fixed target measurements from electron and muon scattering also HERA data in the Q^2-interval from 2.5 GeV^2 to 230 GeV^2 were used.
The respective results are summarized in Table 8.3.

8.6   Measurements at hadron colliders

As we have learned, high energy hadron–hadron collisions with large momentum
transfer arise from the scattering of strongly interacting partons. It follows that
the cross section is sensitive to α_s. In general, however, determinations of the
strong coupling in hadronic interactions are less precise, on one hand because
of uncertainties in the parton density functions, and on the other due to the
underlying event and other soft contributions from the spectator partons which
do not participate in the hard scattering. In addition, we have to keep in mind
that the value of α_s also enters in the determination of the parton density functions. This introduces an additional, although higher order, uncertainty, in any
measurement of the strong coupling which requires the knowledge of the parton

density functions as input. Nevertheless, competitive measurements were performed from direct-photon production processes, measurements of the $b\bar{b}$ cross sections and the inclusive jet cross section in high energy nucleon–(anti)nucleon collisions.

Direct-photon production in hard parton–parton scattering is a Compton-like process of $\mathcal{O}(\alpha_s \alpha_{em})$. The NLO QCD corrections are known, so that a reliable measurement of the strong coupling is possible. In the cross section difference $\sigma(p\bar{p} \to \gamma X) - \sigma(pp \to \gamma X)$ the sea quark and gluon parton density functions of the nucleon cancel, that is, only the well known valence quark distributions are needed as external input for an α_s-determination. Still it is a difficult measurement to disentangle prompt photons from background caused by π^0 and η decays into pairs of photons. The result from the UA6 Collaboration (1999), obtained at the CERN SPS collider is given in Table 8.3. Despite a $p\bar{p}$ C.o.M. energy of 630 GeV the typical scale of the hard parton–parton scattering for this measurement was only $Q = 24.3$ GeV. The theoretical error of this measurement covers uncertainties from the choice of the renormalization scale and the variations of the parton density functions.

Heavy quarks which are not present in the initial state are produced by quark–antiquark annihilation or gluon–gluon fusion processes which are of $\mathcal{O}(\alpha_s^2)$. The theory is developed to NLO. Experimentally those reactions can be tagged by the decay characteristics of the heavy hadrons. Table 8.3 lists one such measurement (UA1 Collab., 1996) from an analysis of $b\bar{b}$+jets production in $p\bar{p}$ collisions.

A measurement of the strong coupling (Giele *et al.*, 1996) is based on the jet cross sections $d\sigma/dE_T$, with E_T the transverse energy of a jet, measured by the CDF Collaboration (1992) at a C.o.M. energy $\sqrt{s} = 1.8$ TeV. The analysis was performed in bins of E_T using a Monte Carlo model based on the NLO QCD matrix elements to give $\alpha_s(E_T)$ in the range 30 GeV$< E_T < 400$ GeV.

8.7 Global event shape variables

Many determinations of the strong coupling constant that have been performed in the past years are based on global event shape variables, that is, observables designed to describe the structure of hadronic final states in e^+e^- annihilation by one characteristic number or distribution which in perturbation theory can be related to the strong coupling constant. A large number of these variables has been invented in the past years, and it is beyond the scope of this book to deal with all of them. Instead, we will focus on the most important ones, discussing various approaches to describe the properties of multihadron events in a quantitative way.

8.7.1 *Theoretical predictions*

The theoretical prediction for all global event shape variables is known to NLO $\mathcal{O}(\alpha_s^2)$, based on a numerical integration (Kunszt *et al.*, 1989) of the ERT matrix elements (Ellis *et al.*, 1981). For some variables also leading-logarithmic and next-to-leading logarithmic corrections, which dominate the cross section in the two-jet limit, have been resummed to all orders. In the latter case, an improved theoretical prediction is obtained by combining the resummed prediction with the second order matrix elements, which then is exact to $\mathcal{O}(\alpha_s^2)$ over the whole phase space and contains the dominant terms for two-jet like configurations to all orders. There is a certain freedom in performing the matching (Catani *et al.*, 1993) which gives rise to differences at $\mathcal{O}(\alpha_s^3)$. This freedom can be employed as an alternative to the variation of the renormalization scale to probe the sensitivity of an α_s-measurement to unknown higher order perturbative corrections.

The non-perturbative transition from partons to hadrons gives rise to power law corrections, which in contrast to the case of inclusive variables cannot be handled in the framework of the OPE. That these non-perturbative effects are important can be seen from the following simple argument. The observed cross section can be written as a sum of perturbative terms known to $\mathcal{O}(\alpha_s^2)$ and non-perturbative contributions of the order $\mathcal{O}(1/Q)$. At the scale of the Z mass with $Q \approx 100$ GeV and $\alpha_s(Q^2) \approx 0.1$ one finds

$$\sigma = \sigma_{\mathrm{PT}} + \sigma_{\mathrm{NP}} = \mathcal{O}(\alpha_s^1) + \mathcal{O}(\alpha_s^2) + \mathcal{O}(1/Q) \sim 10^{-1} + 10^{-2} + 10^{-2} \ .$$

At LEP one thus has to deal with non-perturbative effects in the same order of magnitude as the NLO perturbative contribution. It follows that a precision measurement of the strong coupling is only possible if one can control the non-perturbative contributions.

The conventional approach to handle this kind of corrections is based on Monte Carlo simulations as described in Section 4 which explicitly model the hadronization process. A conceptual drawback of this approach is the fact that the models contain both perturbative QCD and a parameterization for the hadronization phase, where phenomenological parameters are adjusted such that the models describe the experimental data as closely as possible. Since the perturbative phase can only be handled to some approximation, it follows that for a well adjusted model the hadronization part corresponds to a mixture of non-perturbative and higher order perturbative effects. The best one can do in order to safeguard against significant biases from this approach is to use different models for the analysis of non-perturbative effects, and quote the spread of the results as a systematic uncertainty.

8.7.2 *Event shape variables*

The most intuitive example for an event shape variable is the ratio R_3/R_2 of the number of three-jet events to the number of two-jet events observed in the final state. Even without going into details of the jet definition and just identifying a collimated bundle of particles with a distinct primary parton fragmenting into observable hadrons, it is evident that a two-jet event would correspond to the process $e^+e^- \to q\bar{q}$, while a three-jet event would originate from $e^+e^- \to q\bar{q}g$, that is, a process where one of the primary quarks has radiated a hard gluon. Since the probability of gluon radiation is proportional to α_s, the ratio R_3/R_2

is a measure of the strong coupling constant. However, in order to perform a measurement of α_s one needs to know the proportionality constant between the ratio R_3/R_2 and α_s which depends on such details as the jet finding algorithm and the selected kinematics. It is also affected by higher order corrections, which because of the large value of the strong coupling constant usually are not negligible. Turning the simple concept of global event shape variables into a precision measurement of α_s thus is a rather complex business.

In order to be useful for a measurement of the strong coupling, an event shape variable must be calculable in perturbation theory, that is, the soft and the collinear singularities must cancel. The variable must be 'infrared' or 'soft and collinear safe', c.f. Sections 3.5 and 6.1. To illustrate the point, we will first discuss some variables which at first glance may appear as possible quantities to extract α_s, but which do not satisfy those criteria. After that we will turn to describe 'safe' observables.

An example for a variable which is neither soft nor collinear safe is the final state multiplicity, which is increased by one unit if an additional low energy particle is emitted or if a given momentum is split into two collinear ones. Although the number of particles is expected to grow with α_s, a perturbative calculation and thus a measurement of α_s is not possible. Another event shape variable which is widely used for event classification is the Sphericity S, defined as

$$S = \frac{3}{2}\min \frac{\sum_{\boldsymbol{p}} p_{\mathrm{T}}^2}{\sum_{\boldsymbol{p}} \boldsymbol{p}^2} \tag{8.22}$$

where the sum runs over the three-momenta of all particles in the event. The determination of S involves finding the direction, with respect to which the sum of the squares of the transverse momenta p_T is minimized. For a perfect two-jet event one has $S = 0$, for isotropic particle flow $S = 1$. Since gluon radiation tends to make a two-jet event more spherical, one might suspect that it can be used for a determination of α_s. However, since S is quadratic in the momenta it is not collinear safe. Like in the case of the total multiplicity, a measurement of α_s using Sphericity would have to be based on an exact solution of QCD, for example, in the framework of lattice gauge theories.

One of the earliest infrared safe variables was Thrust (Farhi, 1977), discussed in Section 6.2. It follows the same rationale as Sphericity but is amenable to a perturbative treatment. Although Thrust is a conceptually very simple variable, its calculation in a multihadron event is a complicated combinatorial task. This motivated the definition of an alternative quantity, the C-parameter (Parisi, 1978; Donoghue et al., 1979; Ellis et al., 1981), defined through the second order invariant of the infrared safe linear momentum tensor

$$\Theta_{ij} = \frac{1}{\sum_{\boldsymbol{p}} |\boldsymbol{p}|} \cdot \sum_{\boldsymbol{p}} \frac{p_i p_j}{|\boldsymbol{p}|}, \tag{8.23}$$

as

$$C = 3\left(\Theta_{11}\Theta_{22} + \Theta_{22}\Theta_{33} + \Theta_{33}\Theta_{11} - \Theta_{12}\Theta_{12} - \Theta_{23}\Theta_{23} - \Theta_{31}\Theta_{31}\right) . \quad (8.24)$$

In the definition of Θ_{ij} the sum is again over all final state momenta $\boldsymbol{p}$, p_i denotes the ith component, $i = 1, 2, 3$, of the momentum vector $\boldsymbol{p}$. For an ideal two-jet event of two back-to-back systems of particles with vanishing transverse momenta one has $C = 0$. An isotropic flow yields $C = 1/3$. Equation (8.24) is equivalent to a definition of the C-parameter based on the eigenvalues $\{\lambda_1, \lambda_2, \lambda_3\}$ of Θ_{ij} which is usually found in the literature: $C = 3(\lambda_1\lambda_2 + \lambda_2\lambda_3 + \lambda_3\lambda_1)$.

Another set of event shape variables is based on the division of the event into two hemispheres, a and b, by a plane perpendicular to the Thrust axis $\boldsymbol{n}_T$. Despite the numerical problems in the determination of the Thrust axis in a multihadron event, the hemisphere definition based on Thrust is a very natural one when looking at the pQCD prediction for distributions of event shape variables. The leading order QCD correction to the two-jet cross section is the emission of one gluon from either of the primary quarks. Because of momentum conservation the three-particle final state is a planar configuration, with the Thrust axis coinciding with the highest energy parton. The two hemispheres a and b thus have one particle along the event axis recoiling against a system of two particles. The probability of a large angle gluon emission grows with increasing α_s. Since large angle emissions also mean large invariant masses of the two-particle system, one is led to a definition of event shape variables based on hemisphere masses. With M_a and M_b the invariant masses of the two hemispheres, the normalized heavy jet mass ρ_H is defined as

$$\rho_H = \frac{1}{E_{\mathrm{vis}}^2} \max(M_a^2, M_b^2) , \quad (8.25)$$

where the quantity E_{vis} is the total visible energy in the event. The sum goes over all particles which contribute in the evaluation of the event shape variable. The normalization to E_{vis}^2 serves two purposes. First, it normalizes the range of values for the shape variable, making distributions roughly independent of the C.o.M. energy of the reaction, second it reduces the size of systematic uncertainties in an α_s-determination since part of the efficiency and acceptance corrections of the detector cancel.

The event shape variables discussed so far characterize the structure of a multihadron event by just one number. Alternatively one can also measure the structure of the momentum flow by means of a correlation function which yields a distribution for each event. The classic example for this kind of event shape measures is the *energy–energy-correlation function* (EEC) (Basham *et al.* 1978a, 1978b), defined as the distribution of the weighted opening angles χ_{ij} between all pairs i, j of final state particles with energies E_i and E_j,

$$\mathrm{EEC}(\chi) = \frac{1}{N\Delta\chi} \sum_N \sum_{i,j} \frac{E_i E_j}{E_{\mathrm{vis}}^2} \ \Theta\left(\frac{\Delta\chi}{2} - |\chi - \chi_{ij}|\right) . \quad (8.26)$$

Here N is the total number of events to be analysed. The EEC is defined as the continuous average of the discrete distributions measured from the individual events. The weights are given by the product of the energy fraction carried by the individual particles, which makes the EEC independent of the final state multiplicity. The normalization E_{vis}^2 makes sure that the EEC is a normalized probability density function. The Θ-function in eqn (8.26) is just the formal expression for the fact that in an actual measurement the EEC-distribution is represented by a histogram, where all opening angles χ_{ij} in the range $\chi - \Delta\chi/2 < \chi_{ij} < \chi + \Delta\chi/2$ are filled in the bin centred at χ with a total bin width of $\Delta\chi$.

For a perfect two-jet event the EEC consists of two spikes, one at $\chi = 0°$ if particles from the same jet are combined, and one at $\chi = 180°$ when looking at the correlation between the two jets. Gluon radiation which widens the jets and produces also large transverse momenta with respect to the initial event axis then fills the distribution in between the two extremes. The level of the distribution again is a measure of the strong coupling constant.

A derived quantity from the EEC is the *energy-energy-correlation-asymmetry* (AEEC),

$$\text{AEEC}(\chi) = \text{EEC}(180° - \chi) - \text{EEC}(\chi) \tag{8.27}$$

where the two-jet component of the distribution cancels and which thus can be expected to yield a cleaner measurement of the strong coupling constant.

Finally, we will now turn to jet rates. Intuitively, a jet is a bundle of particles associated with a high energy parton from the hard QCD process. This simple picture, however, holds only approximately, since the creation of a multihadron final state proceeds via a parton showering mechanism. Jets are formed because of the angular ordering in the parton cascade. If an initial high energy parton from the early stages of the cascade radiates soft particles the resulting final state hadrons will be more or less aligned with the original parton. If a high energy parton is emitted at a large angle, it will produce its own jet. These considerations show that an event with a completely obvious jet structure will be the exception rather than the rule, since it requires a showering pattern with a few hard partons that in turn radiate only soft particles. To classify an arbitrary event thus requires a jet finding algorithm, which selects jets according to some formal criterion. The criterion is arbitrary to some extent. It must only be constructed such that it is applicable also to few parton states and that it can be applied in an identical way for the theoretical prediction. Different jet finding algorithms were already introduced in Section 6.2, so we will not repeat them here.

Event shape variables which can be extracted from such a jet-clustering process and which are sensitive to the strong coupling constant are, for example, the fraction of two-jet events R_2 as function of the resolution parameter y_{cut} or the distribution of the resolution parameter y_3 where a given event makes the transition from a three-jet to a two-jet event. The relation between the two is given by

$$R_2(y) = -\int_y^\infty dy_3 \, \frac{1}{\sigma_{tot}} \frac{d\sigma(y_3)}{dy_3}, \tag{8.28}$$

that is, both quantities contain essentially the same information. From the experimental point of view the y_3-distribution is to be preferred since here relative frequencies at different values of y_3 are statistically independent.

Although at first glance dealing with the same information, different jet algorithms or generally event shape variables have different uncalculated higher order corrections in the theoretical prediction and different sensitivities to hadronization and detector effects. Studying a large set of observables thus gives a feeling for the systematic limitations of an α_s-measurement.

Many measurements of the strong coupling based on global event shape variables have been performed in e^+e^- annihilation reactions in the energy range between 22 GeV up to LEP2 energies. Historically, the first set of precise measurements was based on NLO theoretical predictions. Later improved calculations became available, where leading- and next-to-leading-logarithmic contributions could be resummed to all orders. As measurements based on these improved predictions turned out to be more precise, and since also older data have been re-analysed in view of those theoretical improvements, we now are in the lucky situation to have measurements over the range from 22 GeV to around 200 GeV based on a homogeneous theoretical framework. Measurements from many authors (JADE Collab. 1998, 1999; TOPAZ Collab. 1993; ALEPH Collab. 1998a, 1999; DELPHI Collab. 1999a; L3 Collab. 1997a; OPAL Collab. 1993a, 1996b, 1997a, 2000a; SLD Collab. 1995) contribute to the values listed in Table 8.3.

8.8 Analytical approaches to power law corrections

Although it is currently not possible to calculate non-perturbative effects for global event shape variables from first principles, there exists at least a phenomenological ansatz, which allows to relate non-perturbative corrections for different shape variables to a few measurable parameters. The basic argument (Dokshitzer and Webber, 1995; Webber, 1995) goes as follows.

Consider a dimensionless quantity $F(Q)$, for example, the average Thrust measured at a C.o.M. energy Q. This average will have contributions from gluon radiation spanning a range from very small momenta, which are essentially non-perturbative, up to the hard scale Q. Formally one can write

$$F(Q) = \int_0^Q dk\, f(k) \quad \text{with} \quad f(k) = a_F \alpha_s(k^2) \frac{k^p}{Q^{p+1}} + \mathcal{O}(\alpha_s^2) \text{ for } k \to 0 \tag{8.29}$$

where the leading order terms of $f(k)$ can be determined in perturbation theory. Note that the running coupling $\alpha_s(k^2)$ effectively resums a part of the higher order perturbative contributions. The purely perturbative prediction for $F(Q)$ is of the form

$$F(Q) = F_1 \alpha_s(\mu_R^2) + \left(F_2 + \beta_0 \ln \frac{Q^2}{\mu_R^2} \right) \alpha_s^2(\mu_R^2) + \mathcal{O}(\alpha_s^3), \tag{8.30}$$

that is, a power series in the strong coupling constant evaluated at a renormalization scale μ_R. Equation (8.29) can now be used to examine the low energy contributions to $F(Q)$, up to a matching scale μ_I which should be somewhere in the transition region between the perturbative and the non-perturbative regime, $\Lambda_{\overline{\text{MS}}} \ll \mu_I \ll Q$. A typical value would be $\mu_I = 2$ GeV. One then can introduce parameters $\bar{\alpha}_p(\mu_I)$,

$$\int_0^{\mu_I} dk\, \alpha_s(k^2) k^p = \frac{\mu_I^{p+1}}{p+1} \bar{\alpha}_p(\mu_I^2)\,, \tag{8.31}$$

which are moments of the low-energy behaviour of the strong coupling constant. These moments are the key element of this approach. In the absence of detailed knowledge of the low energy behaviour of $\alpha_s(k^2)$, one assumes that the true physical function is sufficiently regular for the moments to exist. As such they are an effective way to parameterize low energy QCD, including both perturbative and non-perturbative contributions. Although the moments currently cannot be calculated, they are universal parameters of QCD, which in principle can be measured. They also offer a convenient way to treat non-perturbative effects in α_s-measurements based on event shape variables. With eqn (8.31) the total contribution to $F(Q)$ up to the matching scale μ_I is given by

$$\int_0^{\mu_I} dk\, a_F \alpha_s(k^2) \frac{k^p}{Q^{p+1}} = \frac{a_F}{p+1} \left(\frac{\mu_I}{Q}\right)^{p+1} \bar{\alpha}_p(\mu_I^2) = I_{\text{PT}} + I_{\text{NP}}. \tag{8.32}$$

The perturbative part in this expression can be determined by substituting the perturbative running of the strong coupling,

$$\alpha_s(k^2) = \alpha_s(\mu_R^2) - \beta_0 \alpha_s^2(\mu_R^2) \ln \frac{k^2}{\mu_R^2} + \mathcal{O}(\alpha_s^3)\,, \tag{8.33}$$

to yield

$$I_{\text{PT}} = \frac{a_F}{p+1} \left(\frac{\mu_I}{Q}\right)^{p+1} \left\{ \alpha_s(\mu_R^2) + \beta_0 \alpha_s^2(\mu_R^2) \left(\ln \frac{\mu_R^2}{\mu_I^2} + \frac{1}{p+1} \right) \right\}. \tag{8.34}$$

Subtracting the perturbative part which is already contained in eqn (8.30) from eqn (8.32) thus allows to extract the non-perturbative contribution. It depends on the coefficient a_F which can be determined analytically for a given event shape variable F, and the moment $\bar{\alpha}_0(\mu_I)$, which is a universal parameter of QCD. To give a few examples, for the mean value of $1 - T$ one has $a_T = 4C_F/\pi$, for the mean value of ρ_H it is $a_\rho = 2C_F/\pi$ and for the mean value of the C-parameter the result is $a_C = 6C_F$. Apparently, non-perturbative effects are more pronounced for the C-parameter than for Thrust or heavy jet mass.

The formalism has been applied to the mean values of $1 - T$ and ρ_H by the DELPHI Collaboration (1999a), covering the energy range between 14 GeV and 183 GeV. The result is shown in Fig. 8.6. One clearly sees the importance of non-perturbative corrections up to the highest LEP energies, but one also observes an impressive agreement between the theoretical prediction and the

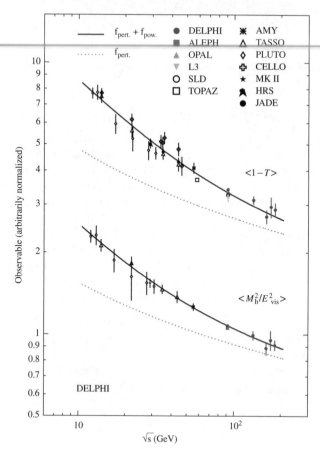

FIG. 8.6. Measurement of α_s based on mean values of global event shape variables measured between $\sqrt{s} = 14$ GeV and $\sqrt{s} = 183$ GeV. The experimental data are compared to the purely perturbative QCD prediction and an extended theory including power law corrections. Figure from DELPHI Collab.(1999a).

two-parameter fit to the experimental data determining simultaneously $\alpha_s(M_Z^2)$ and the effective non-perturbative coupling $\alpha_0(4\,\mathrm{GeV}^2)$. The fit results are given in Table 8.2. Although the overall agreement between data and theory is very satisfactory, the quality of the fit as expressed by the χ^2/ndf-values points to some internal problems with the data. To account for this the value quoted in Table 8.3 was taken to be the result from $1 - T$, with the errors scaled by $\sqrt{\chi^2/\mathrm{ndf}}$.

Finally, it has to be mentioned that power law corrections can not only be calculated for mean values or other scalar observables, but also for distributions of event shape variables. Here the correction to the purely perturbative

Table 8.2 *Results from a simultaneous fit of $\alpha_s(M_Z^2)$ and the non-perturbative effective coupling $\alpha_0(4\,\mathrm{GeV}^2)$ to the energy evolution of global event shape variables*

Observable	$\alpha_0(4\,\mathrm{GeV}^2)$	$\alpha_s(M_Z^2)$	χ^2/ndf
$\langle 1-T \rangle$	$0.493 \pm 0.009 \pm 0.004$	$0.1191 \pm 0.0015 \pm 0.0051$	$50.3/26$
$\langle M_h^2/E_{\mathrm{vis}}^2 \rangle$	$0.550 \pm 0.024 \pm 0.013$	$0.1192 \pm 0.0022 \pm 0.0037$	$2.65/15$

prediction can be parameterized as a shift in the argument of the distribution function (Dokshitzer, 1999), for example,

$$\frac{1}{\sigma}\frac{\mathrm{d}}{\mathrm{d}C}\sigma(C) = \frac{1}{\sigma_{\mathrm{PT}}}\frac{\mathrm{d}}{\mathrm{d}C}\sigma_{\mathrm{PT}}\left(C - D_C/Q\right) . \tag{8.35}$$

8.9 Jets in deep inelastic scattering

Also in deep inelastic lepton–nucleon scattering it is possible to analyse distributions of event shape variables for the final-state hadronic system and extract a measurement of the strong coupling constant. So far the most precise measurements are from the study of jet production rates. One contribution to multi-jet production in ep-collisions is, for example, gluon radiation off a quark scattered by a large Q^2 virtual photon. Quark and gluon emerge as two jets in addition to the jet from the proton remnant. The production rate R_{2+1} of those (2+1)-jet final states is known to next-to-leading order $\mathcal{O}(\alpha_s^2)$ (Körner *et al.*, 1989; Brodkorb *et al.*, 1989). A very appealing feature of this kind of analyses is that kinematic selection of the scattered electron allows to control the Q^2 of the process, thereby making it possible to establish the running of the strong coupling constant within a single experiment. Measurements from the ZEUS Collaboration (1995) and the H1 Collaboration (1995*b*, 1998, 1999*b*) covering an energy range from 6 GeV to 100 GeV were combined in (Bethke, 2000) and quoted in Table 8.3.

8.10 Summary of α_s measurements

A summary of the various types of measurements discussed above to determine the value of the strong coupling constant is given in Table 8.3. The same information is displayed graphically in Fig. 8.7 and Fig. 8.8. The results are sorted according to the energy scale that is probed by the different methods. The table contains the value of α_s at the scale of the measurement and, for a global comparison, the value at the scale of the Z mass. In both cases, the total uncertainty of the measurement is split into an experimental and a theoretical error. The last column indicates the theoretical framework employed in the measurement. The precision is at least NLO. If in addition leading and next-to-leading logarithmic contributions have been resummed this is specified by 'resum'. The term 'LGT' refers to calculations based on lattice gauge theories.

Table 8.3 *Summary of measurements of the strong coupling constant*

Observable	Q/GeV	$\alpha_s(Q)$	$\alpha_s(M_Z^2)$	Theory
Bj-Sumrule	1.73	$0.320^{+0.031}_{-0.053}\pm0.016$	$0.118^{+0.004}_{-0.007}\pm0.002$	NNLO
GLS-Sumrule	1.73	$0.280\pm0.061^{+0.035}_{-0.030}$	$0.112^{+0.008}_{-0.011}\pm0.005$	NNLO
R_τ	1.78	$0.323\pm0.005\pm0.030$	$0.118\pm0.001\pm0.003$	NNLO
$F_3(x)$	5.0	$0.214^{+0.018}_{-0.016}\pm0.010$	$0.118\pm0.005\pm0.003$	NNLO
$\int dx\, x^k F_2(x)$	5.0	$0.208\pm0.007\pm0.007$	$0.117\pm0.002\pm0.002$	NNLO
$F_2(x)$	7.1	$0.177\pm0.008^{+0.011}_{-0.010}$	$0.113\pm0.003\pm0.004$	NLO
Υ Decays	9.7	$0.163\pm0.009\pm0.010$	$0.111\pm0.004\pm0.005$	NLO
$J/\Psi,\Upsilon$ Decays	10.0	$0.167\pm0.002^{+0.015}_{-0.011}$	$0.113\pm0.001^{+0.007}_{-0.005}$	NLO
$Q\bar{Q}$ bound states	10.0	$0.171\pm0.000\pm0.007$	$0.115\pm0.000\pm0.003$	LGT
R_γ	10.5	$0.200\pm0.060^{+0.008}_{-0.005}$	$0.130^{+0.021}_{-0.029}{}^{+0.003}_{-0.002}$	NNLO
$p\bar{p}\to b\bar{b}$+jets	20.0	$0.145^{+0.012}_{-0.010}{}^{+0.013}_{-0.016}$	$0.113^{+0.007}_{-0.006}{}^{+0.008}_{-0.009}$	NLO
e^+e^- Jets/Shapes	22.0	$0.161\pm0.009^{+0.014}_{-0.006}$	$0.124\pm0.005^{+0.008}_{-0.003}$	resum
$p\bar{p},\,pp\to\gamma+X$	24.3	$0.135\pm0.006^{+0.011}_{-0.005}$	$0.110\pm0.004^{+0.007}_{-0.003}$	NLO
ep DIS/Jets	24.5	$0.148\pm0.013\pm0.011$	$0.118\pm0.008\pm0.007$	NLO
R_γ	34.0	$0.160\pm0.019^{+0.005}_{-0.003}$	$0.133\pm0.013^{+0.003}_{-0.002}$	NNLO
e^+e^- Jets/Shapes	35.0	$0.145\pm0.002^{+0.012}_{-0.007}$	$0.123\pm0.002^{+0.008}_{-0.005}$	resum
$e^+e^- D(x)$	36.0	$0.147^{+0.009}_{-0.010}\pm0.013$	$0.125^{+0.006}_{-0.007}\pm0.009$	NLO
R_γ	42.4	$0.144\pm0.029^{+0.004}_{-0.003}$	$0.126\pm0.022^{+0.003}_{-0.002}$	NNLO
e^+e^- Jets/Shapes	44.0	$0.139\pm0.004^{+0.010}_{-0.007}$	$0.123\pm0.003^{+0.007}_{-0.005}$	resum
$e^+e^-\langle1-T\rangle$	50.6	$0.131\pm0.003\pm0.008$	$0.119\pm0.002\pm0.007$	NLO
e^+e^- Jets/Shapes	58.0	$0.132\pm0.003\pm0.008$	$0.123\pm0.003\pm0.007$	resum
e^+e^- Jets/Shapes	91.2	$0.121\pm0.001\pm0.006$	$0.121\pm0.001\pm0.006$	resum
R_l	91.2	$0.124\pm0.004^{+0.003}_{-0.002}$	$0.124\pm0.004^{+0.003}_{-0.002}$	NNLO
$\sigma(p\bar{p}\to\text{jets}(E_T))$	110.	$0.118\pm0.008\pm0.005$	$0.121\pm0.008\pm0.005$	NLO
e^+e^- Jets/Shapes	133.	$0.113\pm0.003\pm0.006$	$0.120\pm0.003\pm0.006$	resum
e^+e^- Jets/Shapes	161.	$0.109\pm0.004\pm0.005$	$0.118\pm0.005\pm0.006$	resum
e^+e^- Jets/Shapes	172.	$0.104\pm0.004\pm0.005$	$0.114\pm0.005\pm0.006$	resum
e^+e^- Jets/Shapes	183.	$0.109\pm0.002\pm0.004$	$0.121\pm0.002\pm0.005$	resum
e^+e^- Jets/Shapes	189.	$0.110\pm0.001\pm0.004$	$0.123\pm0.001\pm0.005$	resum

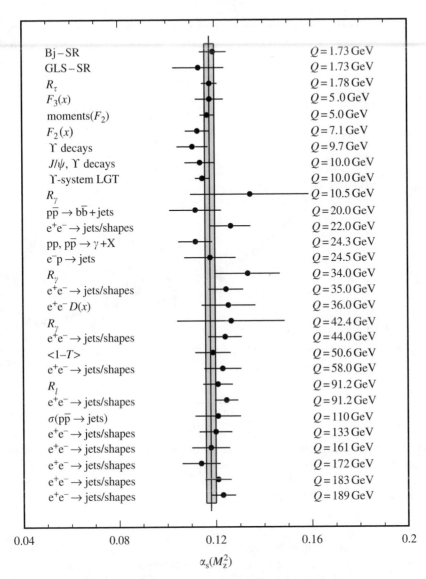

FIG. 8.7. Measurements of the strong coupling constant evolved to the scale of the Z mass. Also quoted is the energy scale of the actual measurement. The data are compared to the global average given by the PDG (2000).

All numbers given in Table 8.3 are rounded to three decimal digits. This is still perfectly adequate in most cases, although in some places the intrinsic precision has reached a point where four digits would be preferable.

At low energies, sum rules allow a measurement at the scale close to the

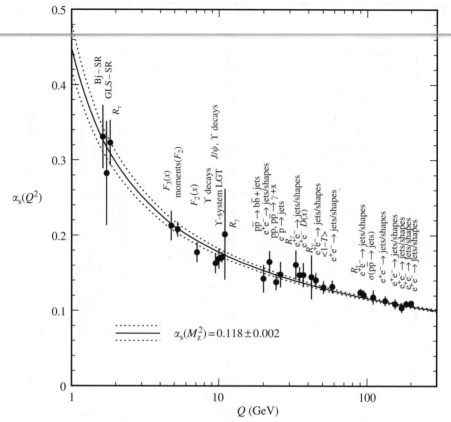

FIG. 8.8. Measurements of the strong coupling constant plotted at the scale of the respective measurements. The running is clearly visible. The experimental data are compared to the QCD expectation for the world average value of α_s as quoted by the PDG (2000).

nucleon mass. Also the hadronic branching ratio of the tau lepton is very sensitive to the strong coupling. Another group of measurements with a still fairly low scale comes from the measurement of scaling violations in structure functions as seen in deep inelastic lepton–nucleon interactions. Here two measurements are based on the evolution of the structure functions $F_2(x)$ and $F_3(x)$, respectively, a third result comes from the evolution of the moments of F_2. Also done at comparatively low scales are measurements using decays or level splittings in heavy quarkonia.

Going to higher energies most measurements are derived from the study of final-state jet topologies or in general from event shape variables. This covers jets in deep inelastic lepton–nucleon scattering, in proton–antiproton collisions or in e^+e^- annihilation. In addition there are measurements based on the total hadronic cross sections, R_γ and R_l, heavy flavour or direct-photon production in proton–antiproton interactions or scaling violations in fragmentation functions.

The overall agreement of the results is rather impressive, giving convincing evidence that QCD really is the universally correct theory of strong interactions. The measurements cover space-like and time-like momentum transfers, energy scales between the mass of the nucleon and the highest LEP energies and are based on lepton–lepton, lepton–hadron and hadron–hadron interactions. The observables are sum rules, branching ratios, level splittings in bound state systems, scaling violations, cross section ratios or event shape variables. Within their respective experimental and theoretical uncertainties all results are consistent.

How to merge all available measurements into a single combined result is not entirely obvious (Schmelling 1996, 2000). Applying the method proposed in the latter reference to all data quoted in Table 8.3 yields a global average of $\alpha_s(M_Z^2) = 0.1180 \pm 0.0043$. One has to keep in mind that the error obtained by this method (Schmelling, 2000) tends to be overestimated if some of the input data are quoted with overly conservative errors. This kind of bias can be reduced by a careful selection of the data which are used in the average. Examples can be found in the literature (Bethke, 2000; PDG, 2000). Here, Bethke arrives at a global average $\alpha_s(M_Z^2) = 0.1184 \pm 0.0031$, while the Partice Data Group quotes $\alpha_s(M_Z^2) = 0.1181 \pm 0.0020$. This average is compared to the individual measurements in Figs. 8.7 and 8.8. Note that the estimates for the central value of the global average essentially agree. The error estimates differ by as much as a factor of two, reflecting different assumptions about intrinsic precision and correlations between individual results. At the time of writing this book we thus can conclude that the value of the strong coupling is $\alpha_s(M_Z^2) = 0.118$ with a total uncertainty of $(3 \pm 1)\%$.

Exercises for Chapter 8

8–1 Show how the Sphericity of an event can be calculated from the eigenvalues of the Sphericity-tensor

$$W_{ij} = \frac{\sum_{\boldsymbol{p}} p_i p_j}{\sum_{\boldsymbol{p}} p^2} = \frac{\sum_{\boldsymbol{p}} \boldsymbol{p}\,\boldsymbol{p}^T}{\sum_{\boldsymbol{p}} p^2}$$

where p_i is the ith component, $i = 1, 2, 3$, of the momentum vector $\boldsymbol{p}$. (Hint: It may be convenient to work with the form of W where the tensor products are expressed in matrix notation, using column vectors for the individual momenta.)$(\bigstar)$

8–2 Show that Sphericity is not collinear safe.

8–3 Show that the C-parameter is collinear safe.

8–4 Show that the C-parameter has a value between $C = 0$ for ideal two-jet events and $C = 1/3$ for events with isotropic momentum flow.

9

TESTS OF THE STRUCTURE OF QCD

When discussing tests of QCD, so far the main emphasis has been on the determination of the strong coupling constant, which, apart from the quark masses, is the only free parameter of the theory. If the theory is correct, then the value for $\alpha_s(M_Z^2)$ must be the same for all types of strong interaction processes. However, in order to go further and check the entire theory of QCD, one also has to verify that the quantum numbers of the interacting fields, such as spin and colour charge, are what they are assumed to be in the QCD Lagrangian. Here we will study the spin of the quarks and the gluon and test the flavour independence of α_s, that is, the assumption that all quarks carry the same colour charge. The ratio of the colour charge between quarks and gluons will be examined in the following chapter.

9.1 Parton spins

One finding from the phenomenology of DIS processes was that quarks should be spin-1/2 particles, c.f. Section 2. Using a Yang–Mills gauge theory to describe strong interactions between quarks by gluon exchange leads to the prediction that gluons are spin-1 fields. Both assumptions can be tested experimentally in e^+e^- annihilations.

9.1.1 *The quark spin*

The most straightforward way to probe the quark spin in e^+e^- annihilation is by looking at the angular distribution of the event axis, which coincides with the direction of the primary $q\bar{q}$-pair. For spin-1/2 fermions annihilating into a vector boson, conservation of angular momentum predicts a distribution

$$\frac{d\sigma}{d|\cos\Theta|} \sim 1 + \cos^2\Theta \tag{9.1}$$

if the final state particles have spin-1/2 and

$$\frac{d\sigma}{d|\cos\Theta|} \sim 1 - \cos^2\Theta \tag{9.2}$$

for scalar particles. Here Θ is the polar angle with respect to the direction of the incoming particles. If parity is conserved, that is, when the angular distribution does not have terms proportional to $\cos\Theta$, the absolute values can be omitted. For e^+e^- annihilation into hadronic final states these simple predictions are modified by higher order corrections and non-perturbative effects.

The first measurements (Schwitters *et al.*, 1975) of this type were performed at rather low energies. Because of the running of the strong coupling, higher order corrections are smaller at LEP energies. In addition power law corrections and non-perturbative effects in general are smaller and the Thrust axis to a very good approximation aligns with the direction of the primary quarks. The resulting angular distribution found in (ALEPH Collab., 1998*a*) is shown in Fig. 9.1. Since no attempt was made to distinguish q and q̄ and in order to average over the forward-backward asymmetry in the coupling of the Z to the primary qq̄-pair, the angle Θ between the incoming beam and the direction of the final state quarks is always taken in the range $0 \leq \Theta \leq 90°$.

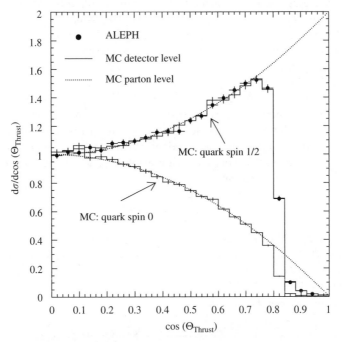

FIG. 9.1. Angular distribution of the Thrust axis in hadronic Z decays. Figure from ALEPH Collab.(1998*a*).

The experimental data are compared to a Monte Carlo calculation which includes the full detector simulation. The data are in perfect agreement with the spin-1/2 assignment for the quarks. The sensitivity to the quark spin is illustrated by comparing the measurements to the expectation from a spin-0 assignment, which is clearly excluded. For large polar angles the shape is perfectly described by the simple expectation $1 + \cos^2 \Theta$. The sharp drop in the distribution around $\cos \Theta = 0.8$ is due to the event selection cuts and finite detector acceptance and is also well reproduced by the Monte Carlo simulation.

9.1.2   *The gluon spin*

While the quark spin can be determined already by looking at two-jet events, the study of the spin of the gluon requires at least three jets, that is, events with at least one hard gluon in the final state. Taking for granted that quarks are spin-1/2 particles, information about the spin of the gluon can be obtained from internal correlations in three-jet events, or the angular distribution of the event plane as function of the Thrust. Measurements have been published by many experiments (TASSO Collab., 1980; PLUTO Collab., 1980a; CELLO Collab., 1982; L3 Collab., 1991a; OPAL Collab., 1991a; DELPHI Collab., 1992a; ALEPH Collab., 1998a). The sensitivity of the differential qq̄g cross section to the spin of the gluon arises from the fact that the coupling of the vector gluon to the quarks is helicity conserving, while a scalar Higgs-like gluon would induce a spin flip. Comparing measurements to the predictions for both hypotheses thus allows to gauge the sensitivity of an observable to the spin of the gluon and check the spin-1 assignment.

The kinematics of a three-jet event is that of a three-particle Z decay into a quark q, antiquark q̄ and gluon g. In total there are nine degrees of freedom. After imposing the constraint of energy and momentum conservation, only five are independent, three of which describe the overall orientation of the event. Thus, there remain only two independent variables to study the topology of three-jet events beyond its angular orientation. Let x_i denote the jet energies normalized to the beam energy,

$$x_i = \frac{2E_i}{\sqrt{s}} \quad i = \text{q}, \bar{\text{q}}, \text{g} , \tag{9.3}$$

with $x_\text{q} + x_{\bar{\text{q}}} + x_\text{g} = 2$ by energy conservation. The leading order cross sections for the vector and the scalar gluon hypothesis and massless partons, normalized to the Born level cross section σ_0, are given in the literature (Laermann *et al.* 1980, 1991 OPAL Collab. 1991a). Ignoring parton masses, which are small compared to the jet energies, one finds

$$\frac{1}{\sigma_0} \frac{\text{d}^2\sigma}{\text{d}x_\text{q}\text{d}x_{\bar{\text{q}}}} = \frac{\alpha_s C_F}{2\pi} \frac{1}{\sum v_\text{q}^2 + a_\text{q}^2} \left(\sum v_\text{q}^2 S_v + \sum a_\text{q}^2 S_a \right) , \tag{9.4}$$

where the sums run over all contributing quark flavours, with v_q and a_q the vector and axial-vector couplings of the quarks to the intermediate photon or Z boson. For e^+e^- annihilation via a photon only the vector coupling contributes, on the Z resonance both terms have to be taken into account. The kinematical functions S_v and S_a depend on the spin of the gluon. For a vector gluon they are

$$S_v^V = \frac{x_\text{q}^2 + x_{\bar{\text{q}}}^2}{(1 - x_\text{q})(1 - x_{\bar{\text{q}}})} \quad \text{and} \quad S_a^V = S_v^V. \tag{9.5}$$

Thus, the normalized differential three-jet cross section is the same on and off the Z resonance. In case of a scalar gluon the situation is different. Here the kinematical functions are

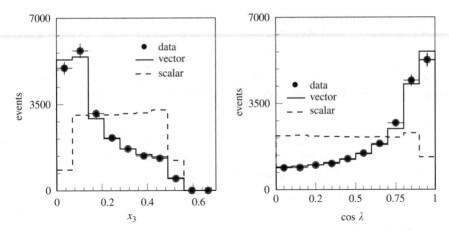

FIG. 9.2. Energy distribution of the lowest energy jet and the Ellis–Karliner angle in e^+e^- annihilation at the Z resonance. Figure from Schmelling(1995a).

$$S_v^S = \frac{x_g^2}{(1 - x_q)(1 - x_{\bar{q}})} \quad \text{and} \quad S_a^S = S_v^S - 2(1 + x_g) . \tag{9.6}$$

In the absence of quark/gluon identification the three jets are energy ordered as $x_1 > x_2 > x_3$. The leading order cross section for a given configuration is obtained by summing the flavour-identified matrix elements, eqns. 9.5 and 9.6, over all possible assignments of $\{x_1, x_2, x_3\}$ to $\{x_q, x_{\bar{q}}, x_g\}$. The sum over all six permutations yields

$$\frac{1}{\sigma_0} \frac{\mathrm{d}^2\sigma^V}{\mathrm{d}x_1 \mathrm{d}x_2} = \frac{\alpha_s C_F}{\pi} \frac{x_1^3 + x_2^3 + x_3^3}{(1 - x_1)(1 - x_2)(1 - x_3)} \tag{9.7}$$

and

$$\frac{1}{\sigma_0} \frac{\mathrm{d}^2\sigma^S}{\mathrm{d}x_1 \mathrm{d}x_2} = \frac{\alpha_s C_F}{\pi} \left(\frac{x_1^2(1 - x_1) + x_2^2(1 - x_2) + x_3^2(1 - x_3)}{(1 - x_1)(1 - x_2)(1 - x_3)} - \frac{10 \sum a_q^2}{\sum a_q^2 + v_q^2} \right) \tag{9.8}$$

for a spin-1 and a scalar gluon, respectively. On the Z resonance the constant offset in the scalar gluon case takes on the value $10 \sum a_q^2 / \sum a_q^2 + v_q^2 \approx 7.45$. In both cases, the differential cross section is singular for $x_3 \to 0$, proportional to $1/x_3^2$ for the vector gluon and proportional to $1/x_3$ for the scalar gluon. The cross section for the latter is less singular, because the spin flip induced by the emission of a soft scalar gluon results in a final state with antiparallel spins for the two fermions, which has less overlap with the vector state of the initial Z boson. In actual studies, configurations close to these singularities are avoided by restricting the phase space to a singularity-free subspace, for example by imposing the Durham algorithm with $y_{\text{cut}} = 0.008$ to define three-jet final states.

Two independent variables which are sensitive to the difference in the singularity structure of the leading order matrix elements are, for example, the x_3-distribution or the Ellis–Karliner angle λ_{EK} (Ellis and Karliner, 1979), the angle between the highest energy jet and the two lower energy ones in the rest-frame of the two lower energy jets. It is related to the scaled jet energies via $\cos\lambda_{\mathrm{EK}} = (x_2 - x_3)/x_1$. Figure 9.2 shows measurements by the L3 Collaboration compared to the expectation for a vector and a scalar gluon. The data are well described by the more singular behaviour expected for the vector gluon hypothesis, and are in clear disagreement with the scalar gluon hypothesis.

An alternative choice for a pair of test variables is

$$x_1 \in \left[\frac{2}{3}, 1\right] \quad \text{and} \quad Z = \frac{1}{\sqrt{3}}(x_2 - x_3) \in \left[0, \frac{1}{\sqrt{3}}\right]. \tag{9.9}$$

How these variables correlate with different event topologies is illustrated in Fig. 9.3.

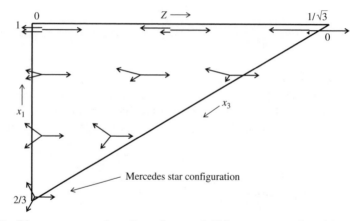

FIG. 9.3. Phase space as function of x_1 and Z for energy-ordered jet configurations, $x_1 > x_2 > x_3$. The arrow length is proportional to the energy. Figure from ALEPH Collab.(1998a).

Like the Ellis–Karliner angle, the Z-variable is also sensitive to the spin of the gluon. Figure 9.4 shows the experimental Z-distribution compared to various theoretical models. Three-jet events were selected using the Durham-metric, projected onto the event plane and the jet energies reconstructed from the jet directions, using the formula

$$E_i = \sqrt{s}\frac{\sin\Psi_{jk}}{\sin\Psi_{12} + \sin\Psi_{23} + \sin\Psi_{31}}, \tag{9.10}$$

with $\{i, j, k\}$ any permutation of $\{1, 2, 3\}$, and Ψ_{jk} the angle between jets j and k. This formula strictly only holds for massless jets. Since the emphasis lies on a

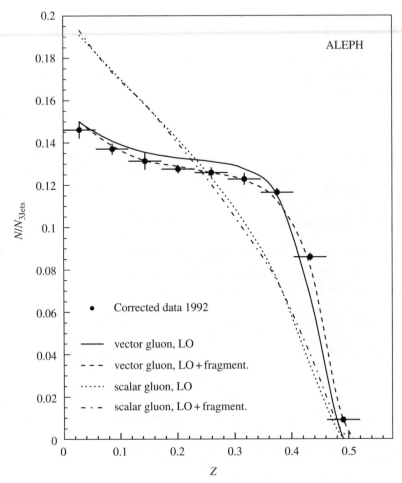

F$_{IG}$. 9.4. Z-Distribution: Plotted are the corrected data with full errors and two
alternative gluon spin models. For both models the leading order (LO) analytical formula and a Monte Carlo simulation including higher order effects
is shown. Figure from ALEPH Collab.(1998a).

comparison between data and theoretical models for massless partons, eqn (9.10)
was taken as the definition of the experimental observables. Both for the vector
and the scalar gluon hypothesis, the results are compared to the predictions from
the leading order matrix elements. Already without taking into account higher
orders and hadronization effects, the qualitative agreement between data and
theory is quite satisfactory for the vector-gluon model. Higher order effects only
induce minor corrections to the theoretical prediction, which further improve the
description of the data by the vector-gluon model. The scalar gluon is clearly
excluded.

The sensitivity of the orientation of the event plane to the spin of the gluon is a particular feature of the different vector and axial-vector couplings of the Z boson to the primary quarks (Körner *et al.*, 1987; Schuler and Körner, 1989). The angular distribution of the normal vector of the event plane with respect to the direction of the incoming beams can be described by

$$\frac{\mathrm{d}\sigma}{\mathrm{d}\cos\theta_n} \sim 1 + A(T)\cos^2\theta_n \;, \tag{9.11}$$

with an asymmetry parameter $A(T)$, which in general is a function of the Thrust T of the event. For vector gluons the prediction is $A(T)_V = -1/3$, for the scalar gluon $A(T)$ increases with decreasing values of T. Experimental results from the DELPHI Collaboration (1992a) are found to be in good agreement with the vector-gluon hypothesis while the scalar-gluon model can be ruled out.

9.2   Flavour independence of strong interactions

Another important test of QCD is the flavour independence of the strong coupling constant. Effectively, this means that all quarks carry the same colour charge. Flavour-universality of the coupling constant is a feature of any non-abelian gauge theory. It is not required in abelian theories such as QED, where different electric charges for different flavours are allowed. The difference comes about because of the self-interaction between the gauge fields in non-abelian theories. To illustrate the point, consider quark–gluon Compton scattering. In QCD the three diagrams shown in Fig. 9.5 contribute, and they all have to add up with the correct weights to yield a gauge invariant scattering amplitude. It follows that the gluon–gluon coupling is linked to the quark–gluon coupling with a fixed ratio determined by the structure of the gauge group, and thus that the strong coupling must be flavour independent. In abelian theories, on the other hand, there is no self-interaction between the gauge fields, and already the sum of the first two diagrams in Fig. 9.5 is gauge invariant. Since there is no link to gauge-field self-interaction, arbitary charges are allowed for different fermions without violating the gauge symmetry.

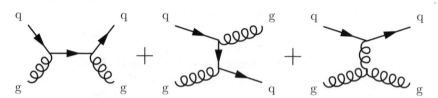

FIG. 9.5. Diagrams contributing to quark–gluon Compton scattering

Indirect evidence for the universality of the strong coupling constant is already obtained from the fact that the measurements of $\alpha_s(M_Z^2)$ from a multitude of reactions and energy scales appear to be consistent with a single common value. More sensitive tests can be performed with dedicated measurements of the strong coupling constant based on data samples where the composition of the active quark flavours can be controlled experimentally. Any observation of a flavour dependence would be a strong indication for physics beyond the Standard Model, either by establishing a dependence between quark flavour and colour charge or by pointing towards new particles that couple to the strong interactions sector.

Table 9.1 *Measurements of ratios of the strong coupling constant for different flavour combinations. The errors are the combined statistical, systematic and theoretical uncertainties of the results.*

Observable	Result	Reference
$\alpha_s(b)/\alpha_s(udsc)$	1.002 ± 0.023	ALEPH Collab.(1995c)
$\alpha_s(uds)/\alpha_s(bc)$	0.971 ± 0.023	ALEPH Collab.(1995c)
$\alpha_s(b)/\alpha_s(udsc)$	1.00 ± 0.05	DELPHI Collab.(1993a)
$\alpha_s(b)/\alpha_s(udsc)$	1.00 ± 0.08	L3 Collab.(1991b)
$\alpha_s(b)/\alpha_s(udsc)$	1.017 ± 0.036	OPAL Collab.(1993b)
$\alpha_s(c)/\alpha_s(udsb)$	0.918 ± 0.115	OPAL Collab.(1993b)
$\alpha_s(s)/\alpha_s(udcb)$	1.158 ± 0.164	OPAL Collab.(1993b)
$\alpha_s(uds)/\alpha_s(cb)$	1.038 ± 0.221	OPAL Collab.(1993b)
$\alpha_s(b)/\alpha_s(udsc)$	0.992 ± 0.016	OPAL Collab.(1995a)
$\alpha_s(uds)/\alpha_s(udscb)$	0.987 ± 0.035	SLD Collab.(1996)
$\alpha_s(c)/\alpha_s(udscb)$	1.012 ± 0.174	SLD Collab.(1996)
$\alpha_s(b)/\alpha_s(udscb)$	1.026 ± 0.065	SLD Collab.(1996)
$\alpha_s(c)/\alpha_s(uds)$	1.036 ± 0.064	SLD Collab.(1999a)
$\alpha_s(b)/\alpha_s(uds)$	1.004 ± 0.041	SLD Collab.(1999a)

Allowing for a flavour dependence of the strong coupling constant, the second order QCD prediction for the differential cross section of a global event shape variable from hadronic Z decays into a quark–antiquark pair of flavour f can be written as

$$\frac{d\sigma(f)}{dx} = A(x, f)\alpha_s(f) + B(x, f)\alpha_s^2(f) \, .$$

The functions $A(x, f)$ and $B(x, f)$ are kinematic functions where the flavour dependence enters through the different quark masses. For b-quarks those purely kinematic effects amount to a correction of typically 5% in the three-jet cross section, which cannot be ignored in any precision test of the flavour independence of α_s. For the other quark flavours mass effects are negligible. The quark mass dependence of $A(x, f)$ has long been known (Ioffe, 1978; Laermann and Zerwas, 1980), the mass effects in the next-to-leading order corrections were determined only recently (Ballestrero *et al.* 1992, 1994; Bernreuther *et al.* 1997, Brandenburg and Uwer 1998, Rodrigo *et al.* 1997).

Table 9.2 *Measurements of the strong coupling constant for all quark flavours. The errors are the total uncertainties.*

Observable	Result	Correlation Coefficients				
$\alpha_s(u)/\alpha_s(incl.)$	0.951 ± 0.209	1.				
$\alpha_s(d)/\alpha_s(incl.)$	0.933 ± 0.195	-0.531	1.			
$\alpha_s(s)/\alpha_s(incl.)$	1.141 ± 0.148	-0.348	-0.386	1.		
$\alpha_s(c)/\alpha_s(incl.)$	0.912 ± 0.091	-0.180	-0.345	-0.051	1.	
$\alpha_s(b)/\alpha_s(incl.)$	1.021 ± 0.026	-0.010	-0.002	-0.036	-0.207	1.

Experimentally, various tagging techniques can be applied to select data samples with different flavour compositions. Requiring for instance a lepton with large transverse momentum relative to the Thrust axis or a displaced secondary vertex in an event yields a b-quark enriched sample. Anti-tagging on lifetime or simply requiring a leading particle in the event with a momentum of more than 70% of the beam momentum enriches light flavours u, d and s. Primary c-quarks can be enriched by selecting jets with a high-momentum charmed meson, like a D*, or by looking for secondary vertices from decays with typical lifetimes of c-mesons. Selecting jets where the most energetic particle is a K_S^0 enhances the fraction of primary s-quarks. A possibility to enhance u-type quarks would be to select events with a hard prompt photon in the final state.

Most of the methods listed above have been applied in one or more measurements to select specific flavour-enriched event samples, where the strong coupling constant then was determined by one of the standard methods based on jet rates or other global event shape variables. Taking the ratio of the flavour-selected α_s and the strong coupling found in the inclusive sample or a complementary data set, most of the otherwise dominant theoretical uncertainties cancel. There remains, however, the uncertainty on the precise flavour composition of the selected events, which is usually estimated by Monte Carlo simulations and is needed to correct a measured ratio of coupling strengths to a ratio of pure flavours. Some results from LEP and SLD are collected in Table 9.1. All ratios are compatible with unity, showing that within the current experimental precision of a few per cent the strong coupling constant indeed is flavour independent.

Ultimately, one would like to express all results concerning the flavour independence of α_s through the ratios $\alpha_s(i)/\alpha_s(incl.)$ with i=u,d,s,c,b. This information can be obtained from any set of five data samples with sufficiently different flavour composition. A first analysis of this kind was performed by the OPAL Collaboration (1993b). In addition to an untagged sample, data sets with different flavour composition were selected using high-p_T leptons, D*, K_S^0 and simple momentum-based leading particle tags. From these the α_s-values for the individual quark flavours were unfolded. All ratios are compatible with unity. The result together with the full correlation matrix (Biebel and Mättig, 1994) is given in Table 9.2.

Exercises for Chapter 9

9–1 Assume negligible fermion masses and parity conservation to show that angular momentum conservation predicts an angular distribution $(1+\cos^2\Theta)$ for e^+e^- annihilation into a pair of spin-1/2 particles, and $(1-\cos^2\Theta)$ if the final state particles have spin 0. $^{(\star)}$

9–2 In e^+e^- annihilation the spin of the gluon can also be inferred from the differential cross section of event shape variables. Use the leading oder matrix elements to calculate the Thrust distribution for a spin-1 and a spin-0 gluon. $^{(\star\star)}$

9–3 In an experiment, the direction of a jet is often more faithfully reconstructed than its directly measured energy. Assuming massless particles and working in the C.o.M. frame, derive expression eqn (9.10) for the jet energies in terms of the inter-jet opening angles, for an e^+e^- annihilation event into three jets.

10

TESTS OF THE GAUGE STRUCTURE OF QCD: COLOUR FACTORS

In Section 2.2 we have already learned that the probabilities for a quark to radiate a gluon, or a gluon to split into a gluon or a quark pair, are related to the strong coupling constant and some factors which are determined by the underlying gauge group. In this chapter, we will present experimental tests which have shown that these factors are indeed consistent with the expectations, namely, that SU(3) is the gauge group for the theory of strong interactions. Before we start to describe the various measurements, let us summarize once more what are the ingredients.

For a general gauge theory based on a simple Lie group the couplings of the fermion fields to the gauge fields and the gauge-field self-interactions in the non-abelian case are determined by the coupling constant and the Casimir operators of the gauge group. Measuring the eigenvalues of these operators, called colour factors, probes the underlying structure of the theory in a gauge invariant way. Considering the case where N_F and N_A are the dimensions of the fundamental and adjoint representations of the gauge group with structure constants f^{abc} and generators T_{ij}^a, the following relations hold:

$$\sum_{a=1}^{N_A} \left(T^a T^{\dagger a} \right)_{ij} = \delta_{ij} C_F \ , \quad \sum_{i,j=1}^{N_F} T_{ij}^a T_{ji}^{\dagger b} = \delta^{ab} T_F \ , \quad \sum_{a,b=1}^{N_A} f^{abc} f^{*abd} = \delta^{cd} C_A \ ,$$

(10.1)

where $a, b, \ldots (i, j, \ldots)$ represent gauge (fermion) field indices and C_F, C_A and T_F are the colour factors. The operators $C_F \delta_{ij}$ and $C_A \delta^{cd}$ are the Casimir operators of the fundamental and adjoint representation of the group, respectively.

As a standard normalization condition $T_F = 1/2$ is chosen. Then for SU(N_c) one finds

$$C_A = N_c \ , \ C_F = \frac{N_c^2 - 1}{2N_c} \ .$$

(10.2)

In the case of QCD, we have three fundamental colours, therefore $N_c = 3$, and from that we derive $C_A = 3$ and $C_F = 4/3$. In Ex. (2-2) and (2-3) it is shown that the probability for a quark to radiate a gluon is proportional to $C_F = 4/3$, independent of its colour state. Similarly, the probability for a gluon to split into two gluons or into two quarks is shown to be proportional to $C_A = 3$ and $T_F = 1/2$, respectively, again not depending on the colour state of the decaying gluon. These relations are illustrated in Fig. 10.1. In the case of a gluon splitting

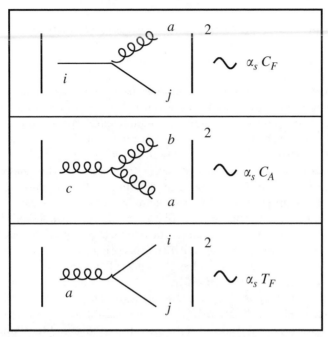

FIG. 10.1. Relations between basic vertices and colour factors. Figure from Dissertori(1998).

into a quark pair, an additional factor n_f has to be taken into account, if the gluon can split into n_f different quark flavours, and if the final state is not differentiated with respect to its flavour content. For example, at LEP we have $n_f = 5$, since the energies are large enough to produce any of the six quarks apart from the top quark, which is too heavy.

The physical quantities to be tested are

$$f_A = \frac{C_A}{C_F} \quad \text{and} \quad f_T = \frac{T_F}{C_F}, \tag{10.3}$$

as well as $\alpha_s C_F$. Note that f_A and f_T are independent of the normalization chosen for the group representation, and that any new choice of normalization affecting C_F can be absorbed into a redefinition of α_s. Remember that the standard definition is $\alpha_s = g^2/(4\pi)$. The ratios f_A and f_T can be interpreted in the following manner: f_A measures the relative strength of the gluon–gluon coupling with respect to the quark–gluon coupling, whereas f_T determines the ratio of the number of possible colour degrees of freedom carried by quarks over the number of gluons. This follows from the relation

$$T_F N_A = C_F N_F, \tag{10.4}$$

derived in Appendix A.1, where the dimension of the adjoint representation of the gauge group, N_A, corresponds to the number of gluons, and the dimension of the fundamental representation, N_F, equals the number of colour degrees of freedom of the quarks.

Now, what is the real interest in a measurement of these colour factors? A very stringent test of QCD would be given by a simultaneous measurement of the strong coupling constant α_s and the colour factor ratios, as, apart from the quark masses, the former is the only free parameter of the theory, and the latter show whether the dynamics is indeed described by an unbroken SU(3) symmetry. Although it is known that quarks come in three 'colours', the relation between these internal degrees of freedom and the dynamics of strong interactions is not fixed a priori. Assuming that all three colours are charges of strong interactions suggests a simple Lie group such as SU(3), SO(3) or an abelian U(1)$_3$. Only the additional input that three quarks or a quark–antiquark pair can exist in a colour neutral state singles out SU(3). Accepting that quarks transform as SU(3) triplets, it is still conceivable that not all internal degrees of freedom contribute to the dynamics of QCD. In other words, only a subgroup of a global SU(3) colour symmetry is also a local gauge symmetry. In this case, subgroups of SU(3) such as SU(2), SO(2) or U(1) become possible candidates for the gauge symmetry. However, one would have to introduce additional mechanisms which force three quarks or $q\bar{q}$-pairs into colour neutral states. Going one step further, one can also imagine strong interactions to be described by a spontaneously broken SU(3) symmetry. The resulting massive gauge bosons would lead to a dynamical structure which would deviate from the SU(3) expectation. Finally, deviations can also be caused by the existence of new physics which couples to the strong interactions sector. An example for the latter is the case of a light gluino, the supersymmetric partner of the gluon, which at $\mathcal{O}(\alpha_s^2)$ effectively contributes three additional fermionic degrees of freedom in e^+e^- annihilation processes.

All these different hypotheses are characterized by a specific set of values for f_A, f_T and n_f. Therefore measuring these numbers should allow to rule out some of the hypotheses. Practically all of the measurements have been performed at LEP, because the statistics is very large and the experimental conditions are very clean. In general, a cross section for e^+e^- annihilation into hadronic final states has the structure

$$\sigma = f\left(\alpha_s C_F, \frac{C_A}{C_F}, n_f \frac{T_F}{C_F}\right). \tag{10.5}$$

In order to proceed, we have to find specific processes and observables with a cross section that has large sensitivity to the colour factors.

10.1 Three-jet variables

We denote as *three-jet variables* those quantities for which the perturbative prediction starts at $\mathcal{O}(\alpha_s)$. Examples are event shape distributions such as Thrust, jet masses, jet broadenings or the differential two-jet rate, which have been discussed already in detail in previous chapters. For a general event shape distribution y, which vanishes in the limit of perfect two-jet topologies, the differential cross section can be written as:

$$\frac{1}{\sigma_{\text{had}}} \frac{\mathrm{d}\sigma}{\mathrm{d}y} = a_{\text{s}}(\mu^2) A(y) + a_{\text{s}}^2(\mu^2) \left(B(y) + b_0 A(y) \ln \frac{\mu^2}{s} \right) + \mathcal{O}(a_{\text{s}}^3) . \quad (10.6)$$

Here σ_{had} is the total hadronic cross section, s is the C.o.M. energy squared, and the following redefinition of the running coupling constant has been adopted:

$$\frac{C_F \alpha_{\text{s}}(\mu^2)}{2\pi} \equiv a_{\text{s}}(\mu^2) = \frac{a_{\text{s}}(s)}{w} \left(1 - \frac{b_1}{b_0} \frac{a_{\text{s}}(s)}{w} \ln w \right) , \quad w = 1 + b_0 a_{\text{s}}(s) \ln \frac{\mu^2}{s} .$$
$$(10.7)$$

Using the above definitions of f_A, f_T and n_f, the first two coefficients b_0 and b_1 of the QCD β-function are

$$b_0 = \frac{11}{6} f_A - \frac{2}{3} n_f f_T \quad \text{and} \quad b_1 = \frac{17}{6} f_A^2 - \left(\frac{5}{3} f_A + 1 \right) n_f f_T . \quad (10.8)$$

Note that when talking about 'colour factors' in the following, we will actually mean the colour factor ratios f_A and f_T.

The coefficient functions $A(y)$ and $B(y)$ are obtained by integrating the fully differential Ellis–Ross–Terrano (ERT) matrix elements (Ellis *et al.*, 1981). Whereas $A(y)$, which results from the integration over the matrix elements for single gluon bremsstrahlung off quarks, is colour factor independent, $B(y)$ can be decomposed as

$$B(y) = B_F(y) + f_A B_A(y) + n_f f_T B_T(y) . \quad (10.9)$$

The function $B_F(y)$ gets contributions from double gluon bremsstrahlung, which occurs at a rate proportional to C_F^2, the function $B_A(y)$ accounts mainly for processes with the triple-gluon coupling and $B_T(y)$ for processes with gluon splitting into quark pairs.

Figure 10.2 shows typical Feynman graphs to be taken into account for the calculation of the cross section of a three-jet variable. The upper left graph is one of the two diagrams (the other one is simply obtained by attaching the gluon to the other quark line) which determine $A(y)$, and it is rather easy to see that the only colour factor dependence at this leading order has to be of the form $\alpha_{\text{s}} C_F A(y)$. The other two diagrams in the upper row are examples for radiative corrections to the leading order term. Very simply speaking, these corrections are taken into account by replacing the bare coupling with a renormalized running coupling, giving $a_{\text{s}}(\mu^2) A(y)$. However, in a full next-to-leading order calculation also graphs of the type shown in the second row are considered.

From the last equations it becomes clear that information on the colour factor ratios enters only in next-to-leading order via the coefficient function $B(y)$,

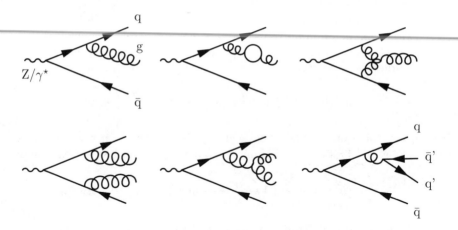

FIG. 10.2. Feynman diagrams of contributions to three-jet variables

which means that the sensitivity to these ratios will not be very large. However, additional dependence at $\mathcal{O}(a_s^3)$ and higher orders enters through the running coupling, mainly via b_0, if the renormalization scale μ^2 is chosen to be different from the hard scale s (see Ex. 10-1). For event shape variables it has been found (ALEPH Collab., 1991a; OPAL Collab., 1992a) that a rather small scale $\mu^2 \ll s$ has to be used in order to achieve a good description of the data. In this case missing higher orders, which appear to be important, are mimicked by terms generated by the expansion of the running coupling.

For several event shape variables it is possible to resum the leading and next-to-leading logarithms $\ln y$ to all orders in a_s, restoring the choice $\mu^2 \approx s$ as the natural one. In those cases a function $g(b_0\, a_s(\mu^2)\ln y)$ is added to the expression in eqn (10.6). Thus b_0 enters again in connection with the leading terms, which introduces a large correlation between the estimates of f_A and f_T. For example, in the case of the differential two-jet rate the leading terms in third order are of the form

$$a_s^3\left[-L^5 - \left(\frac{10}{3}b_0 - \frac{15}{2}\right)L^4 + \mathcal{O}(L^3)\right],\qquad(10.10)$$

with $L = -\ln y_3$. Summarizing, it can be stated that three-jet variables are suited for measuring $a_s(M_Z^2)$ and a function of f_A and f_T, namely b_0.

This is illustrated in Fig. 10.3, which shows the kind of information that can be expected from a colour factor measurement based on three-jet variables compared to analyses based on four-jet variables. While the latter will give a compact confidence region such as the ellipse in the (f_A, f_T)-plane, the former will limit the possible combinations of colour factors only to the band corresponding to a narrow range of values around b_0.

It is worth noting that b_0 also contains information on n_f. As a reminder, it is the number of fermions which give contributions to loop corrections and to

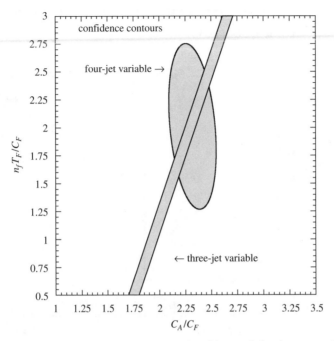

FIG. 10.3. Confidence level contours in the $(f_A, n_f f_T)$ plane as obtained in measurements based on three- and four-jet variables

gluon splitting processes. At LEP $n_f = 5$ is expected, but if additional fermionic degrees of freedom with colour charge exist, then a sizeable effect on b_0 should be observed. As already mentioned, a possible candidate for such additional fermions would be a very light gluino in the mass range $m_{\tilde{g}} \lesssim 1.5$ GeV, which is predicted by particular supersymmetric extensions of the Standard Model (Fayet, 1976; Farrar, 1995). For example, such a gluino would contribute additional fermion loops as depicted in Fig. 10.2.

A first colour factor analysis of three-jet variables has been performed by the ALEPH collaboration (1998a), based on about 110000 three-jet events. A theoretical prediction of the form eqn (10.6) with corrections for hadronization effects has been fitted to the measured two-dimensional distribution of two linearly independent combinations of jet energies. In order to get additional orthogonal information on the colour factors and the strong coupling constant, they also measured the two-jet rate, which in second order perturbation theory is the complement to the three-jet and four-jet rate:

$$\frac{\sigma_2}{\sigma_{\text{had}}} = 1 - \frac{\sigma_3}{\sigma_{\text{had}}} - \frac{\sigma_4}{\sigma_{\text{had}}}. \tag{10.11}$$

Here, the sensitivity to the gauge structure comes from the $\mathcal{O}(\alpha_s^2)$ contributions to the three- and four-jet rates.

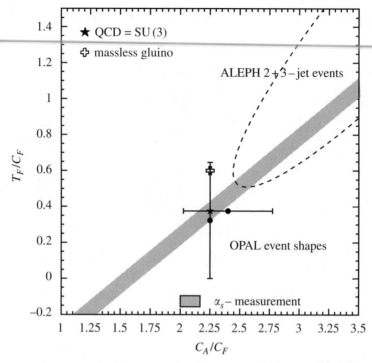

FIG. 10.4. Colour factor measurements based on three-jet variables. The band indicates the expected sensitivity from measurements based on the running of α_s. Figure from Dissertori(1998).

The dominant error source were hadronization uncertainties as well as uncertainties because of the variation of the renormalization scale. The experimental uncertainties were found to be rather small. The final result (with statistical and systematic errors) was

$$\alpha_s(M_Z^2)C_F = 0.210 \pm 0.016_{\text{stat}} \pm 0.048_{\text{syst}}$$
$$C_A/C_F = 4.49 \pm 0.75_{\text{stat}} \pm 1.12_{\text{syst}}$$
$$T_F/C_F = 2.01 \pm 0.49_{\text{stat}} \pm 0.86_{\text{syst}} .$$

This measurement had still rather large errors. However, the confidence level for the SU(3) expectation with $n_f = 5$ was 52%, whereas abelian gauge groups with $C_A = 0$ and $T_F > 0$ were found to be inconsistent with the measurement, having a confidence level below 2×10^{-6}. It is worth noting that the results were highly correlated, for example, the correlation coefficient between the fitted values for C_A/C_F and T_F/C_F is $\rho = 0.96$. The origin of this large correlation can be traced back to the fact that the main sensitivity is to b_0, which is a linear combination of f_A and f_T. The result is shown in Fig. 10.4.

In an analysis by the OPAL collaboration (1995b) the event shape distributions for Thrust, heavy jet mass and the total and wide jet broadenings have been studied. For these variables not only the full next-to-leading order predictions are known, but also the resummation of all leading and next-to-leading logarithms. Very similarly to the techniques of the α_s-measurements from event shape variables, as described in Chapter 8, they employed these variables for a simultaneous fit of the strong coupling constant and one of the colour factors at a time, fixing the others to the QCD expectation. A fit with all colour factors as free parameters did not converge. Again, this can be understood as a consequence of the fact that sensitivity is mainly to b_0. The results, with statistical and systematic errors, were

$$C_A = 3.2 \pm 0.1_{\text{stat}} \pm 0.5_{\text{syst}} \quad \text{with} \quad \alpha_s(M_Z^2) = 0.112 \pm 0.012$$
$$C_F = 1.2 \pm 0.0_{\text{stat}} \pm 0.3_{\text{syst}} \quad \text{with} \quad \alpha_s(M_Z^2) = 0.122 \pm 0.019$$
$$n_f = 4.3 \pm 0.3_{\text{stat}} \pm 3.0_{\text{syst}} \quad \text{with} \quad \alpha_s(M_Z^2) = 0.113 \pm 0.012 \,.$$

The result for n_f had been obtained by fitting $(n_f T_F)$, and then fixing $T_F = 1/2$. Of course the result can also be interpreted as a fit for T_F by fixing $n_f = 5$. The major contributions to the systematic error are hadronization and renormalization scale uncertainties. The results are displayed in Fig. 10.4. The colour factors found in this analysis are nicely consistent with the QCD expectations. However, concerning the measurement of n_f, the errors were still too large to strongly confirm or rule out the light gluino scenario.

10.2 Four-jet variables

One of the obvious motivations for measuring the colour factors is to test the non-abelian nature of the gauge group employed to construct the theory of strong interactions. A consequence of this non-abelian nature is the direct coupling between gluons, in contrast to QED, where photons do not couple directly to each other. So the goal should be to look for evidence of such a gluon–gluon coupling. When looking at the Feynman diagrams of Fig. 10.2 we see that in e^+e^- annihilations such a gluon–gluon interaction will only contribute to final states with at least four partons. Of course, as discussed in the previous chapter, there are less direct effects of the triple-gluon vertex on the running of α_s and on the two- and three-jet cross sections, if the four partons of the final state are clustered into a three-particle final state by some jet-clustering algorithm. But it is clear that when looking at four-jet events, we should be able to resolve more directly the gluon–gluon interaction.

So the steps to proceed are the following: we have to cluster the event to four jets, defined by some resolution parameter y_{cut}^{4j}, construct an observable X out of the four jet-momenta, compute the perturbative cross section for this observable as a function of the colour factor ratios f_A and f_T, correct for hadronization and detector effects, and finally fit this corrected prediction to the data by varying the colour factor ratios.

The perturbative prediction for the differential cross section in this observable X, which we will generally call *four-jet variable*, is given by

$$\frac{1}{\sigma_{\text{had}}} \frac{\mathrm{d}\sigma^{4j}}{\mathrm{d}X} = a_{\mathrm{s}}^2 \left(D_F(X) + \frac{C_A}{C_F} D_A(X) + n_f \frac{T_F}{C_F} D_T(X) \right) + \mathcal{O}(a_{\mathrm{s}}^3) . \quad (10.12)$$

The functions $D_y(X)$, $y = F, A, T$ are obtained by integrating the fully differential ERT matrix elements computed from diagrams such as depicted in the second row of Fig. 10.2. The function $D_F(X)$ gets its contributions from double gluon-bremsstrahlung, $D_A(X)$ from graphs containing the triple-gluon vertex and $D_T(X)$ from graphs with a gluon splitting into quarks. At the time when the measurements of four-jet variables were performed, only the leading order expression eqn (10.12) was known. Only recently the next-to-leading order $\mathcal{O}(a_{\mathrm{s}}^3)$ contributions have been fully calculated (Dixon and Signer 1997; Nagy and Trócsányi 1997, 1999, and references therein), and re-analyses of the LEP1 data set using these improved calculations have just been completed (Bravo 2001; OPAL Collab. 2001).

The lack of knowledge of the higher order corrections causes the total four-jet cross section to depend very strongly on the renormalization scale μ^2, which introduces a large theoretical uncertainty. Otherwise, this observable would be well suited for testing the triple-gluon vertex, since in an abelian theory with $C_A = 0$ we expect fewer four-jet events than in QCD, as follows from eqn (10.12). In order to handle this problem, the LEP experiments employed a different kind of observable. They have measured angular correlations between the four jet momenta, and normalized theory and measurement to the respective total number of four-jet events, thus testing only the shape of the distributions. As will be shown next, the distributions of such angular correlations are sensitive to contributions from different types of Feynman graphs.

For gluon radiation off quarks one finds that the gluon is preferentially polarized in the plane of the splitting process (Olsen *et al.*, 1980). On the other hand, for a gluon splitting into two gluons there is a positive correlation between the plane spanned by the two new gluons and the polarization of the branching one. Finally, in case a gluon splits into two quarks, the plane defined by the momenta of the two quarks is anticorrelated with the polarization of the splitting gluon. So we conclude that for four-jet events induced by a gluon splitting into a $q\bar{q}$-pair, the distribution of the angle between the plane defined by the two primary quarks and the plane defined by the two secondary quarks should be enhanced around $90°$. However, in a non-abelian theory we have contributions also from the triple-gluon interaction, and in this case the favoured angle between the two planes spanned by the primary and secondary partons is rather small. Therefore, the shape of the distribution of this angle is sensitive to the colour factors.

Experimentally, it is very difficult to distinguish between jets induced by the primary and the secondary partons. However, because of the $1/E$-characteristic of radiated gluons we expect the two secondary partons to be less energetic than the two primary quarks. So we arrive at the definition of the angular correlation

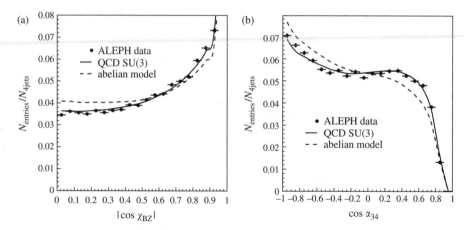

FIG. 10.5. Predicted and measured distributions of the Bengtsson–Zerwas angle
(a) and the angle between the two lowest energetic jets α_{34} (b) from four-jet
events. The data are corrected for hadronization and detector effects, the
errors are statistical only.

variable called *Bengtsson–Zerwas angle* (Bengtsson and Zerwas, 1988; Bengtsson,
1989),

$$\chi_{\mathrm{BZ}} = \sphericalangle[(\boldsymbol{p}_1 \times \boldsymbol{p}_2), (\boldsymbol{p}_3 \times \boldsymbol{p}_4)] = \frac{(\boldsymbol{p}_1 \times \boldsymbol{p}_2) \cdot (\boldsymbol{p}_3 \times \boldsymbol{p}_4)}{|(\boldsymbol{p}_1 \times \boldsymbol{p}_2)| \, |(\boldsymbol{p}_3 \times \boldsymbol{p}_4)|} \, , \tag{10.13}$$

where $\boldsymbol{p}_i$, $i = 1, \ldots, 4$ are the energy-ordered ($E_1 \geq E_2 \geq E_3 \geq E_4$) momenta of
the four partons (jets).

In Fig. 10.5(a) we see the predicted distributions for the Bengtsson–Zerwas
angle for an abelian model and for QCD. The abelian model shows some enhancement around 90° with respect to the non-abelian prediction. The data were
obtained with the ALEPH detector and corrected for hadronization and detector
effects. They clearly favour the QCD prediction over the abelian model.

Next we describe a second test variable, the (generalized) *Nachtmann–Reiter
angle* (Nachtmann and Reiter, 1982), which is based on the following consider-
ations. Again we name the momenta of the primary partons as $\boldsymbol{p}_1$ and $\boldsymbol{p}_2$, and
let $\boldsymbol{p}_3$ and $\boldsymbol{p}_4$ be the momenta of the secondary partons, originating from the
decay of an intermediate gluon. Note that we do not consider double gluon radi-
ation for the moment. Then we choose the particular phase-space configuration
$\boldsymbol{p}_1 + \boldsymbol{p}_2 = \boldsymbol{p}_3 + \boldsymbol{p}_4 = \boldsymbol{0}$ with $E_1 = E_2 \gg E_3 = E_4$. Thus, the two higher-energetic
jets are back-to-back, and so are the two lower-energetic ones. In this configu-
ration the intermediate gluon is at rest and has zero helicity w.r.t. the direction
of the primary partons, and hence helicity components ± 1 in any orthogonal
direction. Strictly speaking this is only exactly true if the two primary partons
are radiated along the direction of the incoming e^+e^- pair. However, it is still
true to leading order in an expansion in E_3/E_1 (Nachtmann and Reiter, 1982).

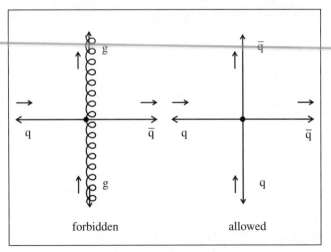

FIG. 10.6. Four-jet configurations for a particular phase-space point as explained in the text. The arrows indicate the direction of the momenta and
spins.

The consequences of such a configuration are illustrated in Fig. 10.6. The
primary (secondary) partons are drawn horizontally (vertically). If the pair of
secondary partons is radiated at 90° as shown in the figure, it has to carry
helicity ±1. On the left-hand side the intermediate gluon decays into two further
gluons. Since these are massless bosons, they carry helicity ±1 each, and their
total helicity has to be 0 or ±2. Therefore, the process as drawn in Fig. 10.6 is
forbidden, whereas it would be allowed if the secondary partons were radiated
along the direction of the primary pair. On the right-hand side, the gluon splits
into a $q\bar{q}$-pair, and the total helicity of this system of two spin-1/2 particles
can be ±1, so this configuration is allowed. On the other hand, the decay of
the secondary quarks parallel to the primary ones would be suppressed. Since in
QCD we have more gluon decays into two gluons rather than into a $q\bar{q}$-pair, we
expect a smaller relative rate of events of the configuration shown in Fig. 10.6
than in a theory without the triple-gluon vertex.

Now, we can define the Nachtmann–Reiter angle θ_{NR} as

$$\theta_{\mathrm{NR}} = \sphericalangle[\boldsymbol{p}_1, \boldsymbol{p}_3] \quad , \quad \theta_{12} \approx \theta_{34} \approx 180° \, , \tag{10.14}$$

which is sensitive to the relative contributions from $q\bar{q}gg$ and $q\bar{q}q\bar{q}$ events to
the discussed four-jet configurations. However, it is clear that the requirement
$\theta_{12} \approx \theta_{34} \approx 180°$ restricts the phase space too much, and experimentally we
would end up with an extremely small number of events. Therefore a *generalized*
Nachtmann–Reiter angle $\theta_{\mathrm{NR}}^{\star}$ has been proposed (Bethke *et al.*, 1991),

$$\theta_{\mathrm{NR}}^{\star} = \sphericalangle[(\boldsymbol{p}_1 - \boldsymbol{p}_2), (\boldsymbol{p}_3 - \boldsymbol{p}_4)] \quad , \quad 0° \leq \theta_{\mathrm{NR}}^{\star} \leq 90° \, , \tag{10.15}$$

which contains θ_{NR} as a special case. So, we have to measure the angle between the difference of momenta of the two higher energetic jets and the two softer ones.

At this point we have to put a warning. In the discussion of the Bengtsson–Zerwas and the Nachtmann–Reiter angle we did not consider the contributions from double gluon bremsstrahlung, which lead to the same final state as events with a g $\rightarrow$ gg splitting. It could happen that other hypothetical gauge groups lead to distributions very similar to the QCD prediction. So we conclude that these two angles are rather sensitive to the relative contribution from q$\bar{\mathrm{q}}$q$\bar{\mathrm{q}}$ events, but in order to obtain clear evidence for the triple-gluon vertex we have to find additional discriminating variables.

As an example for such an alternative variable we present the *Körner–Schierholz–Willrodt angle* Φ_{KSW} (Körner *et al.*, 1981). Again the jets are energy ordered, and we define

$$\Phi_{\mathrm{KSW}} = \angle[(\boldsymbol{p}_1 \times \boldsymbol{p}_4), (\boldsymbol{p}_2 \times \boldsymbol{p}_3)] \quad , \quad 0° \leq \Phi_{\mathrm{KSW}} \leq 180° . \qquad (10.16)$$

In theories without the triple-gluon vertex the planes orthogonal to the vectors $\boldsymbol{p}_1 \times \boldsymbol{p}_4$ and $\boldsymbol{p}_2 \times \boldsymbol{p}_3$ are uncorrelated, and because of phase-space restrictions the angle Φ_{KSW} between these two planes is found preferentially around 90°. However, if there is a triple-gluon vertex, then the pole structure of the propagator for the intermediate gluon leads to a preference for small angles between the two secondary gluons, and a correlation between the planes is induced. Because of the energy ordering, the planes turn out to be antiparallel most of the time. A detailed derivation can be found in the paper by Körner *et al.* (1981). In order to be invariant under exchange of the first and second jet, as well as of the third and fourth jet, a modified definition of the Körner–Schierholz–Willrodt angle has been used by the experiments, namely

$$\Phi_{\mathrm{KSW}} = 1/2 \left\{ \angle[(\boldsymbol{p}_1 \times \boldsymbol{p}_4), (\boldsymbol{p}_2 \times \boldsymbol{p}_3)] + \angle[(\boldsymbol{p}_1 \times \boldsymbol{p}_3), (\boldsymbol{p}_2 \times \boldsymbol{p}_4)] \right\} . \qquad (10.17)$$

Finally, a simplified version of the Körner–Schierholz–Willrodt angle is obtained by looking at the angle between the two lowest energetic jets. So we arrive at the definition of a fourth angular variable (DELPHI Collab., 1993*b*),

$$\alpha_{34} = \angle[\boldsymbol{p}_3, \boldsymbol{p}_4] \quad , \quad 0° \leq \alpha_{34} \leq 180° . \qquad (10.18)$$

In analogy to Φ_{KSW}, the angle α_{34} distinguishes between the relative contributions from double gluon radiation processes and gluon splitting into gluon pairs. Gluon radiation from the two primary quarks occurs more or less independently, and because of the collinear character of bremsstrahlung and the energy ordering of the four jets we expect rather large angles between the secondary partons. Gluon splitting into secondary partons on the other hand will lead to rather small opening angles. The data as shown in Fig. 10.5(b) are again in very good agreement with the QCD prediction and disfavour an abelian model.

For these last two angular distributions we did not consider the contributions from qqqq events, which might lead to a reduced distinction between QCD and other theories. We conclude that an experiment should measure not only one of the angular variables described above, but rather a sensible combination of some of them, in order to obtain good sensitivity to all kinds of possible processes.

Indeed, the various experiments have employed different sets of variables and combination techniques. The DELPHI collaboration (1993b) has performed a binned least-squares fit of eqn (10.12) to the two-dimensional distribution in the variables $\theta^\star_{\text{NR}}$ and α_{34} in order to find estimates of the colour factor ratios. These two variables are sensitive to different types of graphs as we have learned above, and fitting directly the two-dimensional distribution takes into account the correlation between them. A similar technique was applied by the OPAL collaboration (1995c). However, there a three-dimensional distribution was measured by using also the angle χ_{BZ}. Later it has been shown by the DELPHI collaboration (1997b) that the sensitivity of these angular distributions to the colour factors can be further improved by tagging two of the four jets as originating from b or c quarks. Their method gives an efficiency of about 12% to tag both primary jets correctly and a purity of 70%, whereas with energy ordering only in 42% of all events do the two most energetic jets originate from the primary quarks. Finally, the ALEPH collaboration (1997b) has measured all four angular variables and fitted them simultaneously to the theoretical prediction eqn (10.12). This measurement will be described in more detail in the next section.

Generally, in these measurements the prediction at parton level was corrected for hadronization effects by means of bin-by-bin correction factors. Similarly, the data were corrected to correspond to a hadron level distribution without any detector-specific distortions, again by employing bin-by-bin corrections. Both types of corrections were obtained from Monte Carlo simulations. The statistical uncertainties on the measured colour factors are sizeable, despite the large LEP1 data samples. This is due to the rather strong phase-space restriction when asking for four jets with a certain y_{cut} resolution parameter, as well as due to limited sensitivity to the colour factors. The latter is particularly true for T_F/C_F. Applying a heavy-quark tagging algorithm reduces the sample further. The statistics available ranged from about 10000 four-jet events with b-tagging to about 170000 four-jet events in the ALEPH analysis (1997b). The dominant systematic errors arise from uncertainties in the hadronization corrections and from estimates of the unknown higher order contributions.

In the early days of LEP, the ALEPH collaboration (1992c) followed a different approach based on a maximum likelihood fit of selected four-jet events to the theoretical prediction for the five-fold differential four-jet cross section. The idea is the following. If the overall event orientation is not measured, then an event with four massless particles in the final state is characterized by five independent variables (see Ex. 10-2), for example the five scaled invariant masses $y_{ij} = m^2_{ij}/s$ of pairs of partons i and j with $i = 1, 2$; $j = 2, 3, 4$; $i < j$. If we evaluate the matrix elements as a function of these five variables, we exploit all available in-

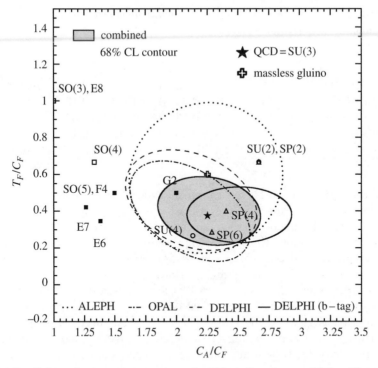

FIG. 10.7. Colour factor measurements based on four-jet variables. The shaded area indicates the average of all measurements. Figure from Dissertori(1998).

formation contained in our event sample, which is important in case of limited statistics. In contrast, working with the angular distributions discussed above we do lose some information, since they cannot represent a complete basis in the five-dimensional space of kinematic four-jet variables. Part of the information is lost by integrating out some regions of the phase space.

In practice, the colour factors were determined from the data by maximizing the so-called *likelihood function*

$$\ln \mathcal{L} = \sum_i \ln \frac{\sigma_i(C_A/C_F, T_F/C_F)}{\sigma_{\text{tot}}(C_A/C_F, T_F/C_F)} \tag{10.19}$$

with respect to C_A/C_F and T_F/C_F. The sum runs over all selected four-jet events. For each event σ_i denotes the four-jet cross section, which again is of the form eqn (10.12), evaluated for the particular momentum configuration of this event. Since no parton-type identification was performed, all permutations of parton-type assignments to the four jets had to be considered. Here σ_{tot} is the total cross section, the ratio $\sigma_i/\sigma_{\text{tot}}$ is the probability density to observe the given set of five kinematic variables y_{ij} in a particular event as function of the colour factors. Only after the fit the results were corrected for detector resolution

Table 10.1 *Table of results of colour factor measurements using four-jet events. $n_f = 5$ is assumed everywhere. The first error is statistical, the second systematic*

Measurement	C_A/C_F	T_F/C_F	Correlation
ALEPH 92	$2.24 \pm 0.32 \pm 0.25$	$0.58 \pm 0.17 \pm 0.21$	$+0.04$
DELPHI 93	$2.12 \pm 0.29 \pm 0.20$	$0.46 \pm 0.13 \pm 0.13$	-0.30
DELPHI 97,b-tag	$2.51 \pm 0.25 \pm 0.13$	$0.38 \pm 0.09 \pm 0.05$	0.0
OPAL 95	$2.11 \pm 0.16 \pm 0.28$	$0.40 \pm 0.11 \pm 0.14$	-0.45
Combined	2.27 ± 0.27	0.40 ± 0.12	-0.16

and hadronization effects. The measurement suffered from small statistics of only about 4000 events. In later analyses a least-squares approach was taken instead of the maximum likelihood method, since for large data sets it is much less time consuming. Furthermore, the correction procedure in a maximum likelihood fit is not as straightforward as in a least-squares approach, where first the distributions are corrected, and only then the fit is performed.

Table 10.1 summarizes the results of the various measurements, which are mentioned above and which were based on four-jet variables only. A graphical representation is given in Fig. 10.7. The average has been obtained using the method proposed by Schmelling (1995b, 2000). Excellent agreement with the expectation from QCD is found, or in other words: the existence of the triple-gluon vertex is confirmed experimentally. Other hypotheses such as an abelian model are definitely ruled out. The more precise measurements also disfavour a light gluino hypothesis.

The results found for the colour factors can be translated into a measurement for the number of gluons N_A, or equivalently, the dimension of the adjoint representation of the gauge group, by employing eqn (10.4),

$$N_A = C_F \, N_F / T_F = 7.5 \pm 2.3 \, .$$

Here, we have assumed the dimension of the fundamental representation N_F to be $N_F = 3$, that is, we assume that quarks carry three colour degrees of freedom, which is a well established assumption as explained in the introductory chapters. The experimental result is consistent with the expected number of eight colour degrees of freedom for the gluons.

10.3   Combination of three- and four-jet variables

In the previous two sections we have learned how the different types of observables are differently sensitive to the colour factors. In Fig. 10.3 the expected shapes of the confidence level contours have been anticipated, and the measurements described later confirmed the expectations. The interesting observation is that the overlap of the contours for three- and four-jet variables is rather small, therefore a combined analysis should help to constrain further the allowed region in the colour factor plane.

The ALEPH experiment (1997b) has performed such a combined measurement. As three-jet variable the distribution of the event shape variable $-\ln y_3$ has been employed, where y_3 is the minimum distance scale y_{ij}, computed according to the Durham prescription, after clustering an event to three jets. This variable is also called differential two-jet rate, and the resummation of leading and next-to-leading logarithms is available for it, as has already been discussed in previous chapters. As four-jet variables all four angular distributions defined in Section 10.2 have been used, with jets ordered in energy and no heavy flavour tagging applied. In total, 2.7 million hadronic events have been analysed, giving a large sample of about 170000 four-jet events. A least-squares fit has been performed simultaneously to all five one-dimensional distributions by taking into account the correlations. The result is

$$a_s(M_Z^2) = 0.0244 \pm 0.0003_{\text{stat}} \pm 0.0009_{\text{syst}}$$
$$C_A/C_F = 2.20 \pm 0.09_{\text{stat}} \pm 0.13_{\text{syst}}$$
$$T_F/C_F = 0.29 \pm 0.05_{\text{stat}} \pm 0.06_{\text{syst}}$$

with a correlation of $\rho(C_A/C_F, T_F/C_F) = 0.47$. The systematic uncertainties include contributions from estimates of unknown higher order terms, from hadronization effects and biases introduced by the detector simulation, and finally from the estimation of mass effects. The graphical representation of the result is given in Fig. 10.8. Note that also for this analysis method new measurements, based on NLO calculations, have been published very recently (Bravo 2001; OPAL Collab. 2001).

Combining the results above with the measurements from four-jet variables, we obtain

$$C_A/C_F = 2.27 \pm 0.18$$
$$T_F/C_F = 0.35 \pm 0.09$$

with a correlation of $\rho(C_A/C_F, T_F/C_F) = 0.25$. This also improves the precision of the measurement of the number of gluons,

$$N_A = C_F N_F/T_F = 8.5 \pm 2.2 \ .$$

The fact that a precise simultaneous measurement of both colour factor ratios, based on jet physics, does not go beyond a relative precision of about 8% for C_A/C_F and about 26% for T_F/C_F is indicative for the difficulties which have to be faced in these types of measurements. They arise from the uncertainties of the theoretical description of hadronic jet production, as well as the limited statistical sensitivity.

10.4 Information from the running of α_s

The QCD β-function, Section 3.4, which governs the running of α_s, is known up to third order in α_s. It is also a function of the colour factors, and the explicit

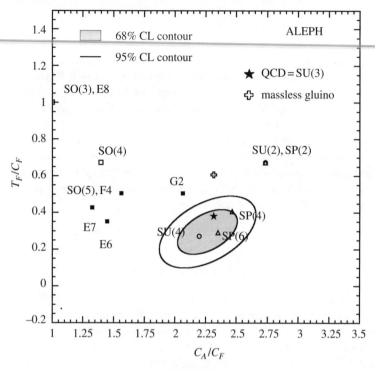

FIG. 10.8. Results of the combined analysis of three- and four-jet variables by ALEPH in terms of the colour factors C_A/C_F and T_F/C_F. Figure from ALEPH Collab.(1997*b*).

dependence of the first two coefficients on these factors is given in eqn (10.8). We have already discussed in detail that three-jet variables are sensitive to the running of α_s and thus help to constrain the allowed region in the colour factor plane. However, remember that this sensitivity is not a leading order effect. Furthermore, we have presented measurements of three-jet variables at one energy scale only. A more explicit study of the running of α_s is achieved by measurements of the coupling constant over a large energy range, and we expect that such an explicit test should help to restrict even further the confidence regions for the colour factors.

The problem is that practically none of the published α_s-determinations could be used as direct input for the study of the colour factor dependence of the running, because they had assumed SU(3) as the underlying gauge group. Therefore, in order to obtain a consistent check of the colour factors, all the measurements would have to be repeated without this implicit assumption. This is practically impossible for most of the cases. Exceptions are the measurements of α_s from R_l and R_τ, c.f. Section 8. These are conceptually very simple and span a large energy range. Furthermore, the theoretical predictions for these two observables

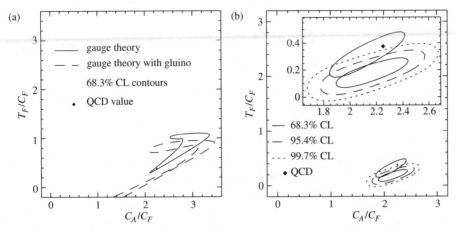

FIG. 10.9. Confidence contours for the colour factor ratios from a study of the running of α_s from R_l and R_τ (left). On the right side also the results from four-jet studies are included. Ellipses with three contours show the results when including gluinos. Figures from Csikor and Fodor(1997).

have been calculated to third order in α_s, and the explicit dependence on the colour factors is known; c.f. Appendix D.

The idea is the following. For a given set of colour factor ratios (C_A/C_F, T_F/C_F) the strong coupling constant a_s (eqn 10.7) is determined by minimizing

$$\chi^2 = \frac{\left(R_\tau^{\mathrm{meas}} - R_\tau^{\mathrm{theo}}\right)^2}{\sigma^2(R_\tau)} + \sum_i \left[\frac{\left(R_l^{\mathrm{meas}} - R_l^{\mathrm{theo}}\right)_i^2}{\sigma_i^2(R_l)}\right] . \tag{10.20}$$

The theoretical predictions are functions of the strong coupling constant and the colour factors, $R_{l,\tau}^{\mathrm{theo}} = R_{l,\tau}^{\mathrm{theo}}(a_\mathrm{s}, C_A/C_F, T_F/C_F)$. The sum goes over measurements R_l^{meas} at various energy scales, where of course a_s has to be evaluated at the corresponding energy scale, using the β-function for the probed parameter set (C_A/C_F, T_F/C_F). The χ^2 minimization is repeated for a large number of pairs (C_A/C_F, T_F/C_F), ending up in a confidence contour in the colour factor plane for which a good α_s determination could be obtained. The final confidence regions are calculated by not only taking into account the experimental errors $\sigma_i(R_l), \sigma(R_\tau)$, but also considering theoretical uncertainties of R_l^{theo} and R_τ^{theo}, which arise from uncertainties in the quark masses, the Higgs mass and the contributions from higher order and non-perturbative terms.

A first analysis (Schmelling and St. Denis, 1994; ALEPH Collab., 1998a) was based on experimental results from the ALEPH experiment only, which at that time were $R_\tau = 3.645 \pm 0.024$ and $R_l = 20.746 \pm 0.073$, measured at the peak of the Z resonance. Within the experimental and theoretical uncertainties QCD was found to be perfectly compatible with the data. Later a more extended analysis was presented by Csikor and Fodor (1997), where they used a large set of measurements over a wide range of energy scales, from 5 GeV up to 91 GeV. In the spirit of the discussion presented in Section 10.3, they also combined their results with the LEP measurements of colour factors from four-jet events. The outcome of their study is displayed in Fig. 10.9. Again, the SU(3) prediction for the colour factor ratios lies well within the 68% confidence regions, whereas a light gluino model is strongly disfavoured.

10.5 Information from jet fragmentation

In previous chapters, we have learned that hadron production can be described by a so-called parton shower, which is a chain of successive bremsstrahlung processes, followed by hadron formation which cannot be described perturbatively. Since bremsstrahlung is directly proportional to the coupling of the radiated gluon to the radiator, which can be a quark or another gluon, we expect the ratio of gluon multiplicities from a gluon and a quark source to be equal to the ratio of the colour factors, $C_A/C_F = 9/4$, in the limit of very large energies where phase-space and non-perturbative effects become negligible. Of course the radiated gluons give rise to the production of hadrons, so the increased radiation from gluons should be reflected in a higher hadron multiplicity in gluon induced jets and also in a stronger scaling violation of the gluon fragmentation function.

The shortcomings of a measurement based on the above idea are the following: At finite energies there might be perturbative as well as non-perturbative effects which cause the ratio of hadron multiplicities to be different from C_A/C_F. Another important aspect is the scale at which quark and gluon jets are compared to each other. As it turns out, simply taking the jet energy is not a good choice. Considerations based on calculations of colour coherence rather suggest to use a transverse-momentum-like scale, such as

$$Q = E_{\text{jet}} \sin(\theta/2) , \qquad (10.21)$$

where E_{jet} is the jet energy and θ the angle to the closest jet. Using such a scale, the hadron multiplicities of quark and gluon jets can be written as

$$\langle N_{\text{q}} \rangle(Q) = N_{\text{q}}^0 + N_{\text{PT}}(Q)$$
$$\langle N_{\text{g}} \rangle(Q) = N_{\text{g}}^0 + N_{\text{PT}}(Q)\, r(Q) . \qquad (10.22)$$

Here $N_{\text{q,g}}^0$ are non-perturbative terms introduced to account for the differences in the transition to hadrons for quarks and gluons. When measuring the ratio of the derivatives of the above expression with respect to the scale, these terms do not contribute.

The perturbative prediction N_{PT} for hadron multiplicities in quark jets has been calculated within the framework of the modified leading logarithmic approximation (see, for example, Ellis $et\ al.$, 1996b),

$$N_{\text{PT}}(Q) = K\,(\alpha_{\text{s}}(Q))^b\,\exp\left(\frac{c}{\sqrt{\alpha_{\text{s}}(Q)}}\right)\left[1 + \mathcal{O}(\sqrt{\alpha_{\text{s}}})\right] . \qquad (10.23)$$

The ratio $r(Q)$ has been calculated by Gaffney and Mueller (1985), and it is directly proportional to C_A/C_F.

The experimental difficulty lies in the need to distinguish between quark and gluon jets. The simplest method relies on the fact that in three-jet events the gluon jet is predominantly the lowest energy jet, which is simply a consequence of the $1/E$ behaviour of the bremsstrahlung spectrum. Therefore, we have to select three-jet events, order the three jets in energy, classify the lowest energetic jet as the gluon jet, and finally correct the measurement for the impurities, which obviously arise from events where the third jet in energy is not induced by a gluon. An improvement of this simple method is obtained by looking for b-hadrons in the jets. If two out of the three jets are identified as originating from a b-quark, the third one has to be a gluon jet by definition. However, this method suffers from a strong reduction in statistics.

The DELPHI experiment has carried out a measurement of C_A/C_F (1999b), based on the ideas outlined above. They indeed observe a clear excess of the hadron multiplicity in gluon jets with respect to quark jets, from which they extract

$$C_A/C_F = 2.246 \pm 0.062_{\text{stat}} \pm 0.080_{\text{syst}} \pm 0.095_{\text{theo}} ,$$

which is of the same precision as the most accurate measurements based on three- and four-jet variables.

10.6 Limits on new physics

In the previous sections we have already stated several times that the various tests of the colour factors all gave results consistent with the expectations from SU(3), and that other hypotheses such as an abelian model without a gluon self-interaction are ruled out. Another extension beyond standard QCD, which has caused much more theoretical interest than possible deviations from SU(3), is the strong interaction sector of supersymmetry. A subclass of supersymmetric models predicts a gluino $\tilde{g}$, the supersymmetric fermionic partner of the gluon, with masses around a few GeV. The existence of such a gluino should manifest itself in the measurements described in this chapter, since it would contribute to loop corrections and thus alter the running of the strong coupling constant, and it could lead to final states of the type $q\bar{q}\tilde{g}\tilde{g}$ via the splitting of an intermediate gluon into a gluino pair. Possible new bound states containing gluinos have not been searched for in the three- and four-jet measurements at LEP.

In Sections 10.1 and 10.2 we had already anticipated that the gluino hypothesis is disfavoured by the measurements of the colour factors. In leading order the contribution of a massless gluino can be parametrized by changing the expectation for T_F/C_F from 0.375 to 0.6, leaving C_A/C_F unchanged. This point in the colour factor plane lies outside the confidence contours of the most precise measurements.

The ALEPH collaboration has used its combined analysis of three- and four-jet variables for a more detailed study of the gluino hypothesis. Instead of assuming five fermionic degrees of freedom and fitting for the colour factors and the strong coupling constant, the argument was turned around by assuming SU(3) to be the correct gauge group, which fixes the values of the colour factors, and fitting for α_s and n_f. The results were $\alpha_s(M_Z^2) = 0.1162 \pm 0.0012_{\text{stat}} \pm 0.0040_{\text{syst}}$ and $n_f = 4.24 \pm 0.29_{\text{stat}} \pm 1.15_{\text{syst}}$. The measurement of $\alpha_s(M_Z^2)$ is in agreement with the world average, and the result for n_f is consistent with the expectation of five. At leading order a massless gluino would lead to an excess above five of three units. However, mass effects can lower this excess. From this measurement, ALEPH computed an upper limit on the excess of $\Delta n_f < 1.9$ at 95% confidence level, from which they deduced a lower limit on the gluino mass of $m_{\tilde{g}} > 6.3$ GeV.

A caveat in the ALEPH result was that the gluino exclusion limit was based on arguments valid in leading order only. This could be overcome by the analysis of Csikor and Fodor (1997), because by then the three-loop calculations for the β-function as well as for R_l and R_T, fully including gluino contributions, had been completed. So they could repeat their analysis of the colour factors by assuming the existence of the gluino, which leads to different confidence contours in the colour factor plane, as shown in Fig. 10.9. Still agreement with SU(3) would be expected, but the results show that the SU(3) point is not covered by the confidence region any more, ruling out the SU(3)+gluino hypothesis. A more quantitative analysis showed that, when combining their analysis with the findings from four-jet studies, they can rule out gluinos with masses below $m_{\tilde{g}} = 5$ GeV at more than 99% confidence level.

Exercises for Chapter 10

10–1 Show that by expanding the expression for the running coupling constant in powers of $\alpha_s(\mu^2)$ and inserting it into eqn (10.6), a dependence on b_0 at third and higher orders appears.

10–2 Show that a final state of four particles is uniquely characterized by a set of five kinematic variables, if the event orientation is integrated over. Find such a set of variables which are Lorentz-invariant.

11

LEADING-LOG QCD

The classical approach to arrive at theoretical predictions for experimental observables is to go order by order in perturbation theory. An alternative is provided by so-called *leading-log* QCD, where some contributions are resummed to all orders. The modified leading-log approximation (MLLA) for example is essentially equivalent to the coherent parton shower picture, where interference effects between subsequent emissions are taken into account through angular ordering, that is, decreasing emission angles, as the parton shower evolves. Theoretical predictions in the framework of MLLA (Azimov *et al.*, 1986a; Fong and Webber, 1991) can be calculated analytically for various observables. Note that the MLLA predictions still refer to the parton level. The connection to the actually observed hadron level is established by the hypothesis of local parton–hadron duality (LPHD) (Azimov *et al.*, 1985b), which assumes that the cross section at hadron level is proportional to the one at parton level.

In the following sections we will first describe some measurements which support the generic picture of a QCD cascade and then turn to specific results that probe details of the MLLA prediction.

11.1 The structure of the parton shower

In previous sections we have described the evolution of an initial quark–antiquark pair into a multihadron final state using the picture of a parton shower, where an initial four-momentum is distributed between a large number of final state particles in a self-similar cascading process. Of course the interesting question is whether experimental evidence for such a self-similar structure can be found.

In fact, corroborating evidence is obtained from the shape of the charged particle multiplicity distribution. Based on rather general assumptions one expects from a cascading model that the multiplicity distribution should be described by a log-normal distribution (Carius and Ingelman, 1990; Szwed and Wrochna, 1990), which is in good agreement with the experimental findings (ALEPH Collab., 1991b; OPAL Collab., 1992b).

Another piece of evidence comes from the study of intermittency, that is, non-Poissonian fluctuations of particle multiplicities in restricted phase-space intervals. Here the assumption of a self-similar cascading mechanism leads to the prediction, that the size of the fluctuations as measured by the so-called *factorial moments* should grow proportionally to an inverse power of the size of the phase-space interval (Bialas and Peschanski, 1986). Experimental results

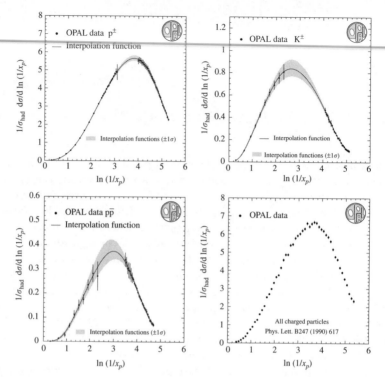

FIG. 11.1. Distributions $1/\sigma \; \mathrm{d}\sigma/\mathrm{d}\xi$, $\xi = \ln(1/x_p)$, of charged particles from hadronic Z decays. Figure from Schmelling(1995a).

(ALEPH Collab., 1992d; DELPHI Collab., 1990; OPAL Collab., 1991b) confirm this behaviour.

Having established the validity of the generic parton shower picture, the next step is to check predictions which are specific to the dynamics of QCD, such as coherence effects in subsequent steps in the parton cascade.

11.2   Momentum spectra

Experimental evidence showing the existence of coherence effects can be obtained from the study of inclusive momentum spectra. Here the consequence of angular ordering is understandable from a simple model where all emission processes happen with the same transverse momentum relative to the mother parton. Having the transverse momentum fixed, decreasing emission angles imply increasing total momenta in subsequent emission processes. This forced increase does not exist in the absence of coherence. In other words, coherence effects lead to a characteristic suppression of low momentum particles (Azimov et al., 1986a).

This effect is observable in the variable $\xi = \ln(1/x_p)$, where $x_p = 2p/\sqrt{s}$ is the scaled momentum of a final state particle. Coherence effects lead to a

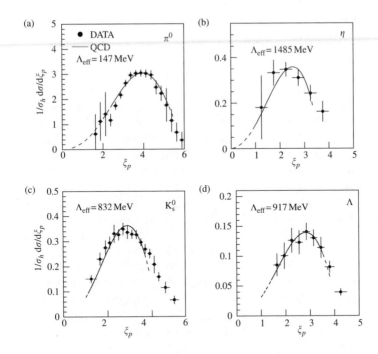

FIG. 11.2. Distributions $1/\sigma_\mathrm{h}\, d\sigma_\mathrm{h}/d\xi$, $\xi = \xi_p$ of neutral particles from hadronic Z decays. Figure from Schmelling(1995a).

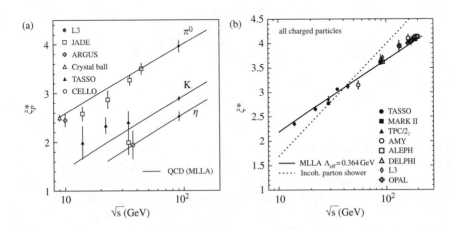

FIG. 11.3. The peak position $\xi^\star$ for different particle types as function of the C.o.M. energy. Figure on the left from Schmelling(1995a).

Table 11.1 *Peak positions ξ^* as function of the particle mass*

Particle	Mass/GeV	ξ^*	Reference
π^0	0.135	4.11 ± 0.18	L3 Collab. (1991c)
π^0	0.135	3.96 ± 0.13	L3 Collab. (1994)
$\pi^\pm$	0.140	3.81 ± 0.02	OPAL Collab. (1994a)
all charged	0.22	3.618 ± 0.028	ALEPH Collab. (1992a)
all charged	0.22	3.67 ± 0.10	DELPHI Collab. (1992b)
all charged	0.22	3.71 ± 0.05	L3 Collab. (1991c)
all charged	0.22	3.603 ± 0.042	OPAL Collab. (1990)
$K^\pm$	0.494	2.63 ± 0.04	OPAL Collab. (1994a)
K^0_S	0.498	2.63 ± 0.04	ALEPH Collab. (1994a)
K^0_S	0.498	2.62 ± 0.11	DELPHI Collab. (1992b)
K^0_S	0.498	2.89 ± 0.05	L3 Collab. (1994)
K^0_S	0.498	2.91 ± 0.04	OPAL Collab. (1991c)
η	0.547	2.60 ± 0.15	L3 Collab. (1992a)
η	0.547	2.52 ± 0.10	L3 Collab. (1994)
$p, \bar{p}$	0.938	3.00 ± 0.09	OPAL Collab. (1994a)
$\Lambda\bar{\Lambda}$	1.116	2.67 ± 0.14	ALEPH Collab. (1994a)
$\Lambda\bar{\Lambda}$	1.116	2.82 ± 0.25	DELPHI Collab. (1992b)
$\Lambda\bar{\Lambda}$	1.116	2.83 ± 0.13	L3 Collab. (1994)
$\Lambda\bar{\Lambda}$	1.116	2.77 ± 0.05	OPAL Collab. (1992c)
Ξ^-	1.321	2.57 ± 0.11	OPAL Collab. (1992c)

suppression at large values of ξ such that $d\sigma/d\xi$ develops a maximum ξ^*. Details of the theoretical prediction can be found in the article by Azimov *et al.* (1986a); see also Ex. (3-34). The important point is that according to LPHD the observed cross section is proportional to an analytically known function. The maximum ξ^* of that function is independent of any constant factors between parton and hadron level and exhibits a dependence on the C.o.M. energy, E_{cm}, which allows a direct test of perturbative QCD in the framework of the MLLA:

$$\xi^*(E_{cm}) = \frac{1}{2} \ln \frac{E_{cm}}{2\Lambda_{eff}} + \frac{101}{324} \sqrt{3 \ln \frac{E_{cm}}{2\Lambda_{eff}}} . \qquad (11.1)$$

The variation of the peak position is predicted with only one free parameter Λ_{eff} which determines the cut-off of the perturbative phase as described by a parton shower. As this cut-off is expected to grow with the mass of the final state particles, the peak position ξ^* should shift to smaller values (larger momenta) for heavier particles. The leading term in eqn (11.1) predicts an energy evolution with the logarithm of the square-root of the C.o.M. energy. This energy dependence is characteristic for a coherent parton shower with angular ordering. An incoherent parton shower would lead to an energy evolution of the position of the maximum with the logarithm of the C.o.M. energy.

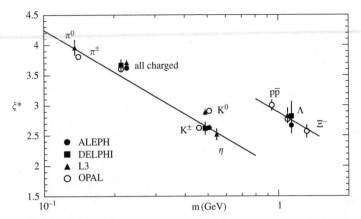

FIG. 11.4. The peak position $\xi^\star$ for different particle types at $E_{\mathrm{cm}} = 91.2$ GeV as a function of the particle mass

Measurements of the ξ-distribution obtained at $E_{\mathrm{cm}} = 91.2$ GeV for charged and neutral particles are shown in Fig. 11.1 and Fig. 11.2. The predicted maximum is clearly visible and the distributions are well described by the MLLA+ LPHD theoretical prediction. Looking at different particle species one also sees how the position of the maximum varies with the particle mass. Using the parameters Λ_{eff} determined at $E_{\mathrm{cm}} = 91.2$ GeV, the energy evolution of the peak position $\xi^\star$ is predicted by QCD without additional free parameters. Figure 11.3 (left) demonstrates how this works for identified neutral particles.

A similar analysis can be performed for the inclusive distributions of all charged particles. Details are described in the review by one of the authors (Schmelling, 1995a). The results are collected in Table 11.2 and displayed in Fig. 11.3 (right). Also here the energy evolution is in good agreement with the expectation based on the MLLA and clearly incompatible with the assumption of an incoherent parton shower which would predict a much steeper slope.

Finally, it is quite instructive to study the peak position as a function of the particle mass. The results from a compilation of such measurements (Schmelling, 1995a) done at LEP is shown in Fig. 11.4 and Table 11.1. The data for mesons and baryons appear to line up on two distinct trajectories when plotting $\xi^\star$ versus the particle mass. The effective mass for the inclusive measurement of all charged particles was taken as the weighted average of the pion, kaon and proton mass, with the weights given by the average multiplicity of the respective particle type as measured by the OPAL Collaboration (1994a).

11.3 Particle multiplicities

Another piece of experimental evidence comes from the measurement of charged particle multiplicities. Although the average multiplicity cannot be calculated perturbatively, its energy evolution is predicted in the framework of MLLA and

Table 11.2 *Peak positions and mean charged particle multiplicities*

$\sqrt{s}/\text{GeV}$	ζ^*	$\langle n_{\text{ch}} \rangle$	Reference
12.0	—	8.40 ± 0.70	JADE Collab. (1983)
12.3	—	8.70 ± 0.60	PLUTO Collab. (1980b)
14.0	2.353 ± 0.043	9.30 ± 0.41	TASSO Collab.(1990, 1989)
17.0	—	9.40 ± 0.70	PLUTO Collab. (1980b)
22.0	—	11.20 ± 1.00	PLUTO Collab. (1980b)
22.0	2.651 ± 0.041	11.30 ± 0.47	TASSO Collab.(1990, 1989)
27.6	—	12.00 ± 0.80	PLUTO Collab. (1980b)
29.0	2.866 ± 0.060	12.80 ± 0.60	TPC/2γ Collab.(1988, 1987)
29.0	—	12.87 ± 0.30	HRS Collab. (1986)
30.0	—	13.10 ± 0.70	JADE Collab. (1983)
30.6	—	12.30 ± 0.80	PLUTO Collab. (1980b)
35.0	3.063 ± 0.024	13.59 ± 0.46	TASSO Collab.(1990, 1989)
35.0	—	13.60 ± 0.70	JADE Collab. (1983)
44.0	3.120 ± 0.054	15.08 ± 0.47	TASSO Collab.(1990, 1989)
52.0	—	15.99 ± 0.23	TOPAZ Collab. (1988)
55.0	—	16.85 ± 0.27	TOPAZ Collab. (1988)
55.0	3.147 ± 0.093	—	AMY Collab. (1990a)
57.0	—	17.19 ± 0.49	AMY Collab. (1990b)
91.2	3.618 ± 0.028	20.85 ± 0.24	ALEPH Collab.(1992a, 1991b)
91.2	3.670 ± 0.100	20.71 ± 0.77	DELPHI Collab.(1992b, 1991)
91.2	3.710 ± 0.050	20.79 ± 0.52	L3 Collab.(1991c, 1992b)
91.2	3.603 ± 0.042	21.40 ± 0.43	OPAL Collab.(1990, 1992b)
130.0	—	23.84 ± 0.73	DELPHI Collab. (1996b)
133.0	3.944 ± 0.059	24.04 ± 0.44	ALEPH Collab. (1998b)
133.0	3.940 ± 0.121	23.40 ± 0.65	OPAL Collab. (1996b)
161.0	4.098 ± 0.093	26.75 ± 0.78	ALEPH Collab. (1998b)
161.0	—	25.46 ± 0.58	DELPHI Collab. (1998a)
161.0	4.000 ± 0.050	24.46 ± 0.40	OPAL Collab. (1997a)
172.0	4.040 ± 0.089	26.45 ± 0.78	ALEPH Collab. (1998b)
172.0	—	26.52 ± 0.76	DELPHI Collab. (1998a)
172.0	4.031 ± 0.053	25.77 ± 1.05	OPAL Collab. (2000a)
183.0	4.110 ± 0.073	26.44 ± 0.60	ALEPH Collab. (1998b)
183.0	—	27.05 ± 0.42	DELPHI Collab. (2000a)
183.0	4.075 ± 0.044	27.04 ± 0.49	L3 Collab. (1998)
183.0	4.087 ± 0.033	26.85 ± 0.59	OPAL Collab. (2000a)
189.0	4.081 ± 0.028	27.42 ± 0.34	ALEPH Collab. (2000a)
189.0	—	27.47 ± 0.45	DELPHI Collab. (2000a)
189.0	4.124 ± 0.038	26.95 ± 0.53	OPAL Collab. (2000a)
196.0	4.145 ± 0.031	27.42 ± 0.49	ALEPH Collab. (2000a)
200.0	4.127 ± 0.034	27.83 ± 0.52	ALEPH Collab. (2000a)
200.0	—	27.58 ± 0.49	DELPHI Collab. (2000a)
206.0	—	27.98 ± 0.23	ALEPH Collab. (2001)

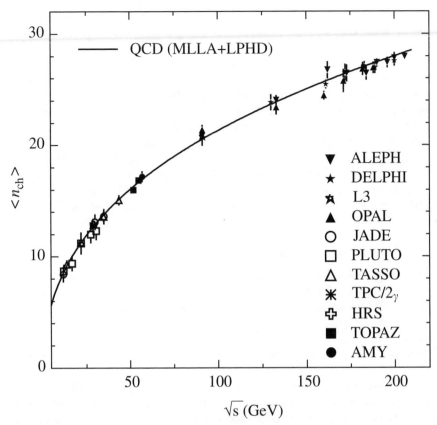

FIG. 11.5. Mean charged particle multiplicity from e^+e^- annihilation into
    hadronic final states for C.o.M. energies between $E_{cm} = 12$ GeV to $E_{cm} = 206$
    GeV

LPHD (Kunszt *et al.*, 1989) as a function of the strong coupling constant. The
energy dependence of the mean multiplicity is thus understood as a consequence
of the running of α_s:

$$\langle n_{ch}(E_{cm}) \rangle = K_{LPHD} \cdot \alpha_s^b(E_{cm}) \cdot \exp\left(\frac{a}{\sqrt{\alpha_s(E_{cm})}}\right) . \qquad (11.2)$$

The numerical values of the coefficients a and b in eqn (11.2) are predicted by
QCD as $a = \sqrt{6\pi}12/23$ and $b = 407/828$. Fixing α_s by external measurements,
the only free parameter in eqn (11.2) is K_{LPHD}.

Experimental data for the mean charged particle multiplicity $\langle n_{ch} \rangle$ from
e^+e^- annihilation into hadrons are collected in Table 11.2. The numbers refer
to the convention where all particles with a mean lifetime below 1ns are forced

to decay while the others are assumed to be absolutely stable. In some cases the
published data were corrected for differences in the analysis (Schmelling, 1995a).

For the entire range of C.o.M. energies between 12 GeV and 206 GeV, that is,
for all available data above the production threshold for b-quarks, one finds excellent agreement between experimental data and the QCD expectation. The energy
dependence of the mean charged particle multiplicity is perfectly described by
using the leading order expression for the running coupling constant $\alpha_s(\mu^2)$, fixing its value at the scale of the Z mass to the global average $\alpha_s(M_Z^2) = 0.118$, and
adjusting only the phenomenological parameter K_{LPHD}. The results are shown
in Fig. 11.5. Note that a priori such a good fit could not be expected, since the
strong coupling constant appearing in eqn (11.2) should be understood as an
effective coupling. The good fit thus indicates that higher order corrections may
be small.

11.4   Isolated hard photons

So far we have established that the parton shower is a self-similar cascading
process with coherence properties in agreement with the predictions of the modified leading logarithmic approximation. Further information about its structure
beyond self-similarity and angular ordering can be obtained from more detailed
studies, such as the study of isolated hard photons or subjet multiplicities.

Isolated hard photons probe the early parton showering stage. Being emitted
with a large momentum transfer, those photons test the short distance properties
of the multi-parton system which forms the shower. For a given formulation of
the parton shower in terms of gluon radiation from quarks and gluons, also the
properties of a photon emitted in the cascade are fixed. As both are massless
vector particles, the only difference between a photon and a gluon is that the
former couples to the electric charge of the quarks and the latter to the colour
charge. There are no new free parameters. As the photon is blind to colour
charges, once emitted, it penetrates without further interaction the complicated
colour fields of the parton shower, thereby producing a kind of X-ray picture of
the cascade.

Experimental results were presented by the LEP Collaborations (OPAL Collab., 1992b; ALEPH Collab., 1993; L3 Collab., 1992c). Figure 11.6 shows how a
measurement of the isolated photon rate as a function of a resolution parame-
ter y_{cut} compares to various parton shower models. Here the parameter y_{cut} is
the distance to the closest jets in the JADE metric. One finds that for large
isolation parameters $y_{\mathrm{cut}} > 0.1$ the production rate is reasonably well described
by the JETSET and HERWIG models and slightly overestimated by ARIADNE.
For small resolution parameters the picture changes. There both ARIADNE and
HERWIG provide a good description of the data while JETSET underestimates
the photon rate. As the various parton shower models only differ in the way in
which next-to-leading logarithmic effects are implemented, these findings demonstrate that the detailed structure of the parton cascade can be studied with
isolated hard photons.

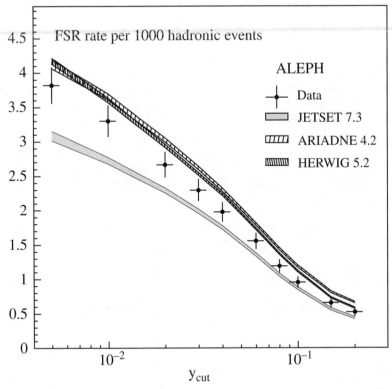

FIG. 11.6. Measured rate of isolated hard photons as function of the separation from the closest jet, compared to Monte Carlo model calculations. Figure from ALEPH Collab.(1993).

As a side remark, it may be worth mentioning that a discrepancy between the data and a particular model does not necessarily imply that the model is conceptually inferior to its competitors. One always has to keep in mind that usually the model parameters are adjusted such that the overall event properties as measured by the final state hadrons are well reproduced by the combined parton showering and hadronization stage. Perturbative and non-perturbative effects are thus entangled.

Measurements with isolated photons, which see through the hadronization phase would allow to tune the description of the perturbative phase independently of the hadronization step and thus offer a way to decouple the parameters describing the two regimes.

11.5 Subjet multiplicities

Another approach to test the structure of the parton shower is by means of subjet multiplicities, defined in the following way: in a first step a hadronic final state is clustered with a jet resolution parameter y_1 into a well defined number of initial jets. Then the same event is looked at with a different resolution parameter $y_0 < y_1$ and the number of clusters studied as a function of y_0. That way these new jets can be uniquely associated with one of the original jets. In the limit $y_0 \to 0$ the subjet multiplicities become equal to the number of particles in the jets. The interest arises because varying y_0 is equivalent to varying the k_T, in the Durham algorithm, at which the event is probed: $k_T^{\min} \approx E_{\mathrm{vis}} \sqrt{y}$. Thus using a large y_0 probes perturbative physics whilst decreasing y_0 allows us to investigate when and to what extent hadronization becomes important. The increase in the number of jets when lowering the jet resolution parameter is predicted in perturbative QCD with a resummation of leading and next-to-leading logarithms (Catani *et al.*, 1992*a*).

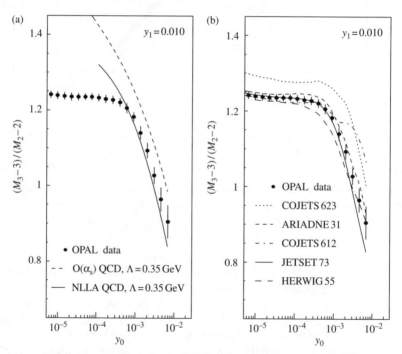

FIG. 11.7. Ratio of subjet multiplicities in two- and three-jet events compared to QCD and model predictions. Figure from OPAL Collab.(1994*b*).

Let M_3 and M_2 be the jet multiplicities at a resolution y_0 when the initial multiplicities at y_1 are 3 and 2 jets, respectively. Figure 11.7 shows experimental results for the ratio of the additional jets $(M_3 - 3)/(M_2 - 2)$ as a function of y_0 (OPAL Collab., 1994*b*). After a sharp rise when going to smaller values of y_0, the ratio starts to level off around $y_0 = 4 \cdot 10^{-4}$. The rise is well described by the perturbative QCD prediction including an all-orders resummation of leading and next-to-leading logarithms. The importance of resumming those logarithms is illustrated by comparing the resummed prediction with a fixed order calculation which fails to describe the data. However, in neither case the levelling off of the experimental data at very small values y_0 is reproduced.

It follows that the dynamics at very low scales is determined by perturbative higher orders beyond the modified leading-log approximation and by non-perturbative effects. The latter are a natural candidate for producing a universal behaviour, independent of the number of jets at the hard scale y_1. It is, therefore, interesting to confront the data with model calculations which take higher order QCD and non-perturbative effects into account. Using models that have been tuned to reproduce the data, one finds that the coherent parton shower models with cluster or string hadronization successfully reproduce the experimental results over the full y_0 range. The models based on an incoherent parton shower and independent fragmentation only reproduce the qualitative trend, but have difficulties in describing the data quantitatively.

11.6   Breit frame analyses

In the previous section the discussion did focus on observables defined in e^+e^- annihilation reactions. Similar analyses can in fact also be performed in electron–proton collisions when choosing an appropriate reference frame. That frame is given by the so-called Breit frame, as discussed in Section 6.2. Viewed from a parton in the proton, which scatters off a virtual photon from the electron, it is the Lorentz frame where the parton's longitudinal momentum before the scattering process is $+Q/2$, and after scattering $-Q/2$. In other words, it is like scattering on an infinitely massive brick-wall. The scattered parton is then essentially equivalent to one of the initial quarks in e^+e^- annihilation into hadrons, that is, the properties of the jets as measured in the hemisphere of the scattered parton can be compared with jet properties in e^+e^- annihilation at a C.o.M. energy $\sqrt{s} = Q$. Analyses of this kind with measurements of momentum spectra and charged particle multiplicities have been performed by the HERA experiments H1 (1997b) and ZEUS (1999c). In general it turned out that results were similar to those obtained in e^+e^- annihilation, showing again the universality of strong interaction phenomena.

12

DIFFERENCES BETWEEN QUARK AND GLUON JETS

Jets initiated by primary quarks and gluons, which differ in both their spins and colour charges, are predicted to have different properties. This has raised significant theoretical and experimental interest ever since the discovery of gluon jets. Following on from early qualitative studies, LEP has heralded a new quantitative era (Gary, 1994; Knowles *et al.*, 1996). In this chapter we begin by reviewing the theoretical expectations and experimental practicalities before discussing the established results.

12.1 Theoretical expectations

An early prediction was that on average the multiplicity of any type of particle in a gluon jet compared to a quark jet should be asymptotically equal to $C_A/C_F = 2.25$ (Brodsky and Gunion, 1976). At NLO this ratio becomes a function of the strong coupling constant α_s and thus energy dependent. If we use $\langle N_q \rangle$ and $\langle N_g \rangle$ to denote these mean multiplicities then at NNLO their ratio is given by (Gaffney and Mueller, 1985)

$$
\frac{\langle N_g \rangle}{\langle N_q \rangle} = \frac{C_A}{C_F} \left\{ 1 - \left(1 + 2\frac{n_f T_F}{C_A} - 4\frac{n_f T_F C_F}{C_A^2} \right) \right.
$$
$$
\left. \times \left[\sqrt{\frac{\alpha_s C_A}{18\pi}} + \left(\frac{25}{8} - \frac{3}{2}\frac{n_f T_F}{C_A} - 2\frac{n_f T_F C_F}{C_A^2} \right) \frac{\alpha_s C_A}{18\pi} \right] \right\} . \tag{12.1}
$$

This prediction has since been refined to account for energy-momentum conservation in the shower (Dremin and Nechitailo, 1994; Eden, 1998) and taken to N³LO (Capella *et al.*, 2000). At the scale of the Z mass the ratio is predicted to be $\langle N_g \rangle / \langle N_q \rangle \approx 1.7$. Also the widths of the multiplicity distributions, σ_g and σ_q, are predicted to differ. At leading order one expects

$$
\sigma_q^2 = \frac{C_A}{C_F}\frac{1}{3}\langle N_q \rangle^2 - \langle N_q \rangle \qquad \sigma_g^2 = \frac{1}{3}\langle N_g \rangle^2 - \langle N_g \rangle . \tag{12.2}
$$

Corrections to this and higher moments are also available (Malaza and Webber, 1986; Dremin and Hwa, 1994). This implies that gluon jets have larger fluctuations in multiplicity by an asymptotic factor $\sigma_g/\sigma_q = \sqrt{C_A/C_F} = 3/2$. The relative factor C_A/C_F is the ratio of the colour charges of a gluon and a quark. It arises because the evolution of a quark-initiated jet is dominated by the emission of the first virtual gluon which then effectively gives a gluon jet.

The higher average multiplicity in gluon jets means that they must have a softer fragmentation function than quark jets. At small x only the primary parton's colour charge is important, since long wavelength gluons do not see the different spins, and so both fragmentation functions have the same shape to NLLA, a Gaussian in $\ln(1/x)$, but the relative heights of the peaks differ by a factor C_A/C_F (Dokshitzer et $al.$, 1991; Fong and Webber, 1991). In order to conserve momentum the greater number of soft particles in a gluon jet then leads to a relative suppression in the number of hard particles.

The angular sizes of quark and gluon jets also differ (Einhorn and Weeks, 1978). This can be seen using the Sterman–Weinberg definition eqn (3.226) of the two-jet fraction, $f_2 = 1 - f_3$, in e^+e^- annihilation. If the two jets are required to contain a fraction $(1 - \epsilon)$ of the total energy inside back-to-back cones of half angle δ, then the prediction for small ϵ is given by

$$\delta_q \sim \exp\left[-\frac{(1 - f_2)\pi}{4C_F\alpha_s(s)\ln(1/\epsilon)}\right] \quad \text{and} \quad \delta_g = \delta_q^{C_F/C_A} . \tag{12.3}$$

The gluon jet result follows by substituting C_A for C_F in eqn (3.226). Thus at the same energy, gluon jets are wider than quark jets. Actually this prediction for gluon jets highlights a problem as it applies to a colour singlet gg state such as might be produced in the decay of some heavy, $J^{PC} = 0^{-+}$ resonance, $X \to$ gg. Here the properties of the gluon can be inferred, perhaps with the aid of an axis such as the Thrust axis, from those of the hadronic system as a whole. That is, there is no need of a jet finding algorithm. This situation is not typical of most experiments. At LEP the decay $Z \to$ gg is forbidden by the Landau–Yang theorem so that gluon jets must be identified as, for example, the 'third' jet in a sample of $e^+e^- \to q\bar{q}g$ events.

12.2 Extracting quark and gluon jet properties

To date LEP has dominated the quantitative study of quark and gluon jet properties. Therefore we focus on the issues which arise in the analysis of e^+e^- annihilation events. It is straightforward to identify and measure the properties of quark jets in either the inclusive or two-jet sub-sample of hadronic events. Furthermore, the energy of the jet is well defined as $\sqrt{s}/2$ and the angle to its colour connected partner fixed at $180°$. This means that the quark jet properties are only measured at one scale, defined by the beam energy, which is larger than that of gluon jets in the same experiment. In order to obtain results at lower scales one can consider $e^+e^- \to q\bar{q}\gamma$ events (DELPHI Collab., 1996c) or extract them as a 'by-product' of the gluon jet analysis.

The analysis of gluon jets is more subtle. The starting point is obtaining a sample of three-jet events from which one can identify and extract the gluon jet's properties. These events are assumed to contain one gluon jet and two quark jets, where the quark and antiquark initiated jets are not distinguished. This is typically done using, for example, the Durham jet clustering algorithm with a fixed cut-off $y_{\text{cut}} = 0.008$ to select events with three jets. This excludes events with four or more jets. Though less common at LEP, cone based jet finders have also been used so as to facilitate comparisons with results from hadron–hadron colliders (OPAL Collab., 1994c). Jets are then classified using energy ordering $E_1 > E_2 > E_3$ and other tags. Allowing for a difference in the distributions of some observable X_j arising from light (q=uds) and heavy (Q=cb) quark jets the measured distribution for one such sample of jets j can be decomposed as

$$X_j = \mathcal{P}(j = \text{g})X_\text{g} + \mathcal{P}(j = \text{q})X_\text{q} + \mathcal{P}(j = \text{Q})X_\text{Q} . \qquad (12.4)$$

Here the probabilities, or purities, satisfy $\mathcal{P}(j = \text{g}) + \mathcal{P}(j = \text{q}) + \mathcal{P}(j = \text{Q}) = 1$ and $\sum_j \mathcal{P}(j = \text{g}) = 1$. Thus given a number of such measurements with differing quark and gluon contents the intrinsic distributions can be unfolded (OPAL Collab., 1993c). Here, a first difficulty lies in both finding samples containing different quark–gluon admixtures and in defining the respective probabilities.

At LEP1 energies it is important to distinguish between light quark jets and b-jets. The properties of the latter are dominated by the decay products of the b-hadron which contributes about half of the jet's multiplicity. Comparisons of b-quark and gluon jets at a scale of 24 GeV show that their multiplicities, angular widths and fragmentation functions are very similar to one another, whilst significant differences are found in comparison to light (uds) quark jets (OPAL Collab., 1996a).

To illustrate the basic method we consider the symmetric, Y-shaped, events defined by $E_1 > E_2 \approx E_3 \sim 22$ GeV, equivalent to two inter-jet angles being in the range $150 \pm 10°$. These were popularized in the early LEP studies. Recalling the characteristic, soft, bremsstrahlung spectrum of gluons, the energy ordering allows us to say that jet 1 is almost certainly not a gluon. By using symmetry we then have $\mathcal{P}(2 = \text{g}) = \mathcal{P}(3 = \text{g}) \approx \frac{1}{2}$. Two enriched sub-samples of jets can then be obtained by tagging the presence of a quark in one of the low energy jets and thereby tagging the other low energy jet as a gluon. The tag may either be for heavy flavours, asking for a displaced vertex, impact parameter or prompt lepton, or light flavours, by the presence of a high momentum track with, for example, $x_E > 0.45$. This allows the properties of light quark, heavy quark and gluon jets to be measured and separated. Of course quark tagging inevitably introduces some biases into the jet samples, but these can be assessed and corrected for using Monte Carlo studies. Another popular set of events are the threefold symmetric events defined by $E_1 \approx E_2 \approx E_3 \approx 30$ GeV, equivalent to inter-jet angles being in the range $120 \pm 10°$: these are the so-called 'Mercedes' events. In the absence of any tagging, symmetry allows us to infer that for each jet $\mathcal{P}(j = \text{q}) + \mathcal{P}(j = \text{Q}) = \frac{2}{3} = 2\mathcal{P}(j = \text{g})$. Given the large final data sets available from LEP1 it has also been possible to study three-jet events in which two jets are tagged as containing heavy flavours leaving the third as an almost pure gluon jet.

In the above discussion it remains for us to fully quantify the probabilities being used. The most common approach is based upon the use of reconstructed

Monte Carlo events. This allows a three-jet event at detector level to also be re-constructed as a three-jet event at the hadron and parton levels. Schemes based upon minimizing angular separations between jets can then be used to match the detector level jets to individual hadron level jets and then to one of the three parton level jets. At the parton level the jets are classified as (anti)quark or gluon according to the primary, q$\bar{\text{q}}$g, partons they contain. Thus, the proba-bility for a particular reconstructed jet to be a quark or gluon can be specified. Unfortunately this is not possible in all cases. Events which are reconstructed as three jets at the detector level may appear as two jets at the parton level. The angular matching scheme may also prove ambiguous in some cases. Or it may not be possible to assign probabilities at the parton level if, for example, the original q$\bar{\text{q}}$-pair go into the same jet. A degree of effort has gone into developing matching schemes which minimize the effects of such ambiguities, and thus the systematic error on the results (ALEPH Collab., 1996b). As an alternative to the elaborate matching schemes an elegant solution is provided by the use of the three-jet matrix element eqn (3.112) (Fodor, 1991; ALEPH Collab., 1997c; OPAL Collab., 2000b). This can be used to define probabilities,

$$\mathcal{P}(3 = \text{g}) \propto \frac{x_1^2 + x_2^2}{(1 - x_2)(1 - x_3)} \quad \text{etc.} \quad \text{and} \quad \sum_{j=1}^{3} \mathcal{P}(j = \text{g}) = 1, \qquad (12.5)$$

which can be applied at the detector, hadron or parton level, thereby avoiding the ambiguities of the matching schemes. Only the usual detector and hadronization corrections need to be applied. If higher order matrix elements are considered then new definitions become possible.

Once the quark and gluon jet properties have been measured a second diffi-culty arises. As noted earlier the theoretical calculations are for idealized, back-to-back q$\bar{\text{q}}$ and gg events. These jets are uniquely characterized by the scale $Q = E_{\text{jet}} = \sqrt{s}/2$. To some extent this is also true when an analysis only se-lects jets from specific phase-space configurations, such as the Y-shaped events. However, in general a jet is sensitive to its event's overall topology and the jet's energy (and type) is insufficient to characterize its properties.

The topology dependence of jet properties can be understood by recalling our discussions of soft gluons, which dominate a jet's multiplicity and small-x fragmentation function; see Section 3.7 and Section 4.2.6. These long wavelength quanta do not see individual partons but colour–anticolour dipoles. Figure 3.28 illustrates the colour flow in an e$^+$e$^- \rightarrow$ q$\bar{\text{q}}$g event. Focussing on the quark we see that as the inter-jet angle to the gluon, θ_{qg}, gets smaller the qg-dipole shrinks and the colour charge of the quark is more effectively shielded by the anticolour charge on the gluon. This suppression of the radiation suggests the use of a transverse-momentum-like energy scale to characterize the quark jet:

$$Q_{\text{qg}} \equiv E_{\text{q}} \sin(\theta_{\text{qg}}/2)|_{\theta_{\text{qg}}=\pi} = E_{\text{q}}$$
$$\approx Q_{\text{T}} \equiv E_{\text{jet}} \sin(\theta_{\text{min}}/2) . \qquad (12.6)$$

The approximation follows because the gluon jet most often is the nearest jet to a quark. This makes it reasonable to use Q_T which is more convenient in practice. In the case of the gluon several topological energy scales have been suggested, such as $\bar{Q}_g = \sqrt{Q_{gq}Q_{g\bar{q}}}$ (ALEPH Collab., 1997c), Q_T (DELPHI Collab. 1999b, 2000b; OPAL Collab. 2000b) and $Q'_T = \sqrt{(q \cdot g)(\bar{q} \cdot g)/(2q \cdot \bar{q})}$ (DELPHI Collab., 2000b). Similar scales have been previously encountered in eqn (4.37) and the definition of the Durham jet clustering algorithm.

The reduction in a jet's phase space implied by the use of a topological energy scale essentially arises from the same physics that leads to angular ordering. That is, the expected topology dependence of quark and gluon jet properties is largely a consequence of colour coherence. Since this is built into many modern, coherent, Monte Carlo generators, see Chapter 4, they should prove capable of describing the data and allowing the extraction of reliable detector corrections, etc.

An alternative approach to the energy scale problem is to look for experimental situations which better match the idealized, theoretical definition of a gluon jet as one half of an $X \to gg$ event. One such option has been to utilize the radiative decays $\Upsilon(1S) \to \gamma gg$, $\sqrt{s_{gg}} = 5\,\mathrm{GeV}$ (CLEO Collab., 1997) and $\Upsilon(3S) \to \gamma(\chi_{b2} \to gg)$, $\sqrt{s_{gg}} = 10.3\,\mathrm{GeV}$ (CLEO Collab., 1992) to study colour singlet pairs of gluon jets. A second option (Dokshitzer $et\ al.$, 1988; Gary, 1994) is to define the gluon to be the 'opposite hemisphere' in those rare $e^+e^- \to Q\bar{Q}g$ events in which two jets in the same hemisphere, as defined by the Thrust axis, both satisfy a heavy flavour tag (OPAL Collab. 1996c, 1998a, 1999a).

12.3 Experimental properties of quark and gluon jets

In the following we will discuss a large spectrum of experimental results which all show that gluon-initiated jets indeed are different from quark jets, and in particular, that the observed differences can be understood in the framework of QCD.

12.3.1 *Topology dependence of jet properties*

The first study to demonstrate the need to use topology dependent scales looked at the average multiplicity of quark and gluon jets as a function of both their energy and the angular separation, $\theta_{\min}$, to the closest of the other two jets (ALEPH Collab., 1997c). For a fixed E_{jet} the jet's purity, as used in the unfolding, is essentially a constant independent of $\theta_{\min}$ which is allowed to span the whole of the available phase space. Figure 12.1 shows the results for pure light quark jets plotted as a function of the jet-energy scale $Q_E = E_{\mathrm{jet}}$ and the topological scale Q_T, eqn (12.6). Every point integrates over an energy range of $\Delta E_{\mathrm{jet}} = 5\,\mathrm{GeV}$ and an angular interval $\Delta\theta_{\min} = 18°$. It is clear that the wide dispersion shown in the results at fixed energy are removed by using the topological scale. A similar result was found for gluon jets using the combined scale $\bar{Q}_g$.

Topological energy scales open up the whole of the three-jet phase space for study. The subsequent increase in the available statistics has helped to fuel a

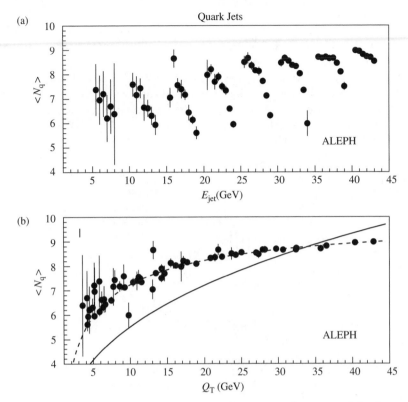

FIG. 12.1. The average multiplicity in quark jets as a function of the jet-energy
scale $Q_E = E_{jet}$ (top) and the topological scale $Q_{min} = Q_T$ (bottom). For
clarity, in the top graph jet energies are shown offset within each ΔE_{jet} bin
such that jets in lower θ_{min} bins are shown with slightly lower energies. Figure
from ALEPH Collab.(1997c).

growth in their popularity. Studies of inclusive and identified particle multiplicities, scaling violations in fragmentation functions etc. have since adopted the
scale $Q_T = E_{jet} \sin(\theta_{min}/2)$ for both quark and gluon jets (DELPHI Collab.
1999b, 2000b; OPAL Collab. 2000b). The consistency of the data sets in these
analyses, both internally and, in the case of quark jets, also with experimental
results obtained at lower C.o.M. energies, as well as the agreement with theoretical expectations has given convincing evidence for the utility of these topological
energy scales.

12.3.2 Multiplicities

A main focus of experimental attention has been the measurement of quark and
gluon jets' (average) charged particle multiplicities and the testing of eqn (12.1)
(Gary, 1994). Whilst early results suggested that gluon jets have a higher mul-

tiplicity than quark jets, though by a significantly lower ratio than C_A/C_F, it
was not until LEP that reliable, quantitative measurements became available.
The first LEP results, based on the study of Y-shaped events and the use of b-
quark tagging, yielded values such as $\langle N_g \rangle / \langle N_q \rangle = 1.27 \pm 0.07$, for jets of energy
24 GeV (OPAL Collab., 1993c), which confirmed earlier qualitative studies. How-
ever, these results used the jet's energy as its scale and more significantly were
sensitive to the choice of jet algorithm used in the analysis (DELPHI Collab.,
1996c; OPAL Collab., 1995d).

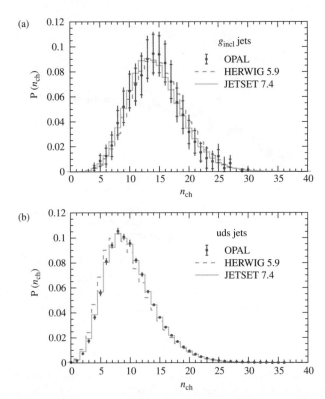

FIG. 12.2. Charged particle multiplicity distributions for inclusively defined
gluon jets of energy 41.8 GeV (top) and light (uds(c)) quark jets of energy
45.6 GeV (bottom), compared to predictions from coherent Monte Carlo event
generators. The error bars are the statistical and systematic errors combined
in quadrature; the smaller bars give the statistical error alone. Figure from
OPAL Collab.(1998a).

As noted above, one way to avoid the side-effects of a jet algorithm is to
use the inclusive definition of a gluon as the whole of the hemisphere opposite
to one containing two tagged heavy quark jets. A corresponding definition for a

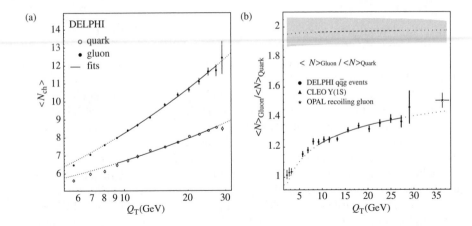

FIG. 12.3. Average charged particle multiplicities (left) and multiplicity ratios (right) for light quark and gluon jets as a function of the topological scale Q_T. The shaded band indicates the ratio of the derivatives. Figure from DELPHI Collab.(1999b).

quark is as one half of a hadronic event. Figure 12.2 shows the measured charged particle multiplicity distributions of light (uds) quark jets of energy 45.6 GeV and gluon jets of energy 41.8 GeV (OPAL Collab., 1998a). Only light quarks are included so as to better correspond to the analytic calculations which employ massless kinematics. The gluons have a larger mean multiplicity, shorter tail and more Gaussian-like distribution than quarks. After using a Monte Carlo model to correct for the slightly higher quark jet energy, the following ratios for the first moments of the distributions are obtained at 41.8 GeV.

$$
\begin{aligned}
&\text{Mean:} && \mu = \langle N \rangle && \mu_g/\mu_q = 1.471 \pm 0.024 \pm 0.043 \\
&\text{Width:} && \sigma = \sqrt{\langle (N - \mu)^2 \rangle} && \sigma_g/\sigma_q = 1.055 \pm 0.046 \pm 0.055 \\
&\text{Skewness:} && \gamma = \langle (N - \mu)^3 \rangle / \sigma^3 && \gamma_g/\gamma_q = 0.47 \pm 0.16 \pm 0.21 \\
&\text{Kurtosis:} && \kappa = \langle (N - \mu)^4 \rangle / \sigma^4 - 3 && \kappa_g/\kappa_q = 0.19 \pm 0.37 \pm 0.33
\end{aligned}
\tag{12.7}
$$

The N^3LO, analytic prediction for the ratio of mean multiplicities is in modest agreement with the data, being about 15% higher. Better agreement is found with the predictions of the coherent Monte Carlo event generators, also for the higher moments. This suggests that it is important to fully treat energy–momentum conservation in the calculations. Further results are available for the normalized factorial and cumulant factorial moments of these distributions together with their ratios (OPAL Collab., 1998a). Again fair agreement is found with coherent Monte Carlo event generators and with analytic predictions at NNLO.

The value of the ratio of mean multiplicities is predicted to vary with scale beyond LO. In (DELPHI Collab., 1996c) such a dependence on the energy of the partons identified within three-jet events was established. To do this for inclusively defined jets, whose scale is $\sqrt{s}/2$, inevitably involves the comparison of results from different experiments. An alternative is to study the variation of the quark and gluon jet properties over three-jet phase space as a function of a topological energy scale (DELPHI Collab., 1999b; OPAL Collab., 2000b). Figure 12.3 (left) shows the mean multiplicities of quark and gluon jets as a function of Q_T. Gluon jets show a stronger growth than quark jets with increasing scale, an effect which is stronger at larger scales. This effect is confirmed by the multiplicity ratio as a function of the scale, Fig. 12.3 (right). The higher order, analytic prediction for the shape of the multiplicity ratio, eqn (12.1), is in qualitative agreement with the data but remains too high, particularly so at lower scales. The shaded band in Fig. 12.3 (right) is the ratio of derivatives of the average multiplicities, which has an almost constant value 1.97 ± 0.10 (DELPHI Collab., 1999b). A similar measurement by the OPAL Collaboration (2000b) gave 2.27 ± 0.20. The ratio of the derivatives is expected to be less sensitive to hadronization corrections than the ratio of the multiplicities and is predicted to have the asymptotic value C_A/C_F and a value around 1.92 at present scales (Capella $et~al.$, 2000).

An indirect approach to quark and gluon jet multiplicities which avoids some of the need to assign particles to jets is provided by a study of the total multiplicity in a three-jet event (DELPHI Collab., 1999b). Based upon MLLA calculations this is predicted to be (Dokshitzer $et~al.$, 1988)

$$N_{3\mathrm{jet}} = 2N_q \left(\sqrt{\frac{q \cdot \bar{q}}{2}} \right) + N_g(Q'_T) \tag{12.8}$$

$$= N_{e^+e^-}(\sqrt{2q \cdot \bar{q}}) + R(Q'_T) \left[\frac{N_{e^+e^-}(2Q'_T)}{2} - \Delta N_0 \right] .$$

In the second formula we have expressed the quark jet multiplicity in terms of the total event multiplicity and the gluon jet multiplicity in terms of the quark jet multiplicity. The factor $R(Q'_T)$ is given by eqn (12.1) and ΔN_0 is an empirical offset designed to account for a larger leading particle multiplicity in quark jets. A good simultaneous fit of eqn (12.8) to the total and three-jet multiplicities, as a function of the inter-jet angles, is possible with $C_A/C_F = 2.246 \pm 0.139$ (DELPHI Collab., 1999b). As a spin-off of this approach one can rearrange eqn (12.8) to give a definition of the gluon multiplicity at the scale Q'_T. Adopting this definition gives similar results to those quoted above.

A useful variant of the standard investigations into the multiplicity distributions in events are provided by the subjet multiplicities (ALEPH Collab., 2000b; DELPHI Collab., 1998b) introduced in Section 11.5. Using a jet clustering algorithm, an event that is found to contain three jets at a resolution parameter y_1 can be re-clustered at a second smaller cut-off, $y_0 < y_1$, and the new number of subjets measured. As explained before, varying y_0 from y_1 to zero allows to probe the transition from the perturbative physics to the hadronization phase.

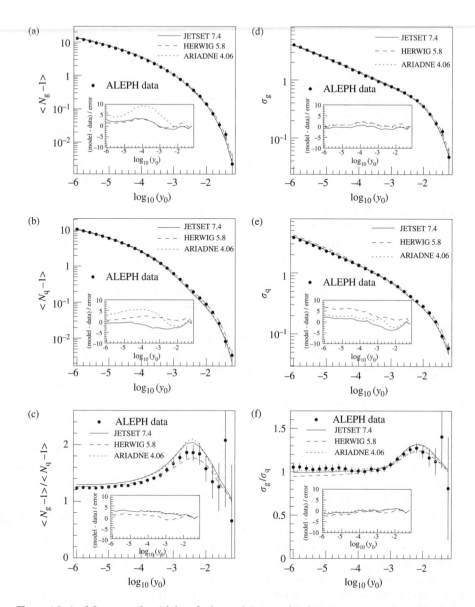

FIG. 12.4. Mean and width of the subjet multiplicities in quark and gluon jets as a function of the subjet resolution parameter y_0. Also shown are the gluon/quark ratios of these quantities. The data are compared to predictions from various coherent Monte Carlo event generators. Figure from ALEPH Collab.(2000b).

Figure 12.4 shows the means and widths of the subjet multiplicity distributions for identified quark and gluon jets together with their ratios (ALEPH Collab., 2000b). The data were obtained from three-jet events, defined using the Durham (k_T) algorithm with a primary resolution parameter $y_1 = 0.1$ ($k_T = 28.8\,\text{GeV}$). The results are for jets integrated over the whole available phase space: consequently the quark jets have a higher average energy (31.6 GeV) than the gluon jets (28 GeV). The average multiplicities minus one are shown in order to focus attention on the additional, radiated partons in the jet. As required by their definition $\langle N_{\text{q,g}}(y_0) - 1 \rangle$ both rise steadily as y_0 decreases, from 0 at $y_0 = y_1$ to close to (one less than) the particle multiplicities at $y_0 \lesssim 10^{-6}$ ($k_T = 0.1\,\text{GeV}$). A similar growth is seen in the widths $\sigma_{\text{q,g}}(y_0)$ as y_0 decreases from y_1. Looking at the ratio $R_\mu(y_0) \equiv \langle N_{\text{g}}(y_0) - 1 \rangle / \langle N_{\text{q}}(y_0) - 1 \rangle$ as a function of y_0 we find that it climbs to a peak value ≈ 1.87 for $y_0 \sim 0.003$ ($k_T = 5\,\text{GeV}$) before falling back to ≈ 1.24 in the individual particle limit. The LLA prediction is $C_A/C_F = 2.25$. Similarly $R_\sigma(y_0) \equiv \sigma_{\text{g}}(y_0)/\sigma_{\text{q}}(y_0)$ climbs to a peak value ≈ 1.28 for $y_0 \sim 0.006$ ($k_T \sim 7\,\text{GeV}$) before plateauing at ≈ 1.05. The LLA prediction is $\sqrt{C_A/C_F} = 1.5$.

Even though the numerical values of the experimental results are sensitive to the details of the event and jet selection (DELPHI Collab., 1998b), the measurements can nevertheless be compared to a number of predictions. At the hadron level coherent Monte Carlo event generators, shown in Fig. 12.4, provide fair descriptions of all the distributions which span up to three orders of magnitude. However, if comparisons are made to the final state parton level predictions then deviations are seen for $y_0 \lesssim 10^{-3}$ ($k_T = 2.9\,\text{GeV}$), indicating that below this scale hadronization corrections start to become important. For the mean subjet multiplicities two analytic calculations are available. The first is based on the leading order (LO) $\mathcal{O}(\alpha_\text{s}^2)$ matrix elements, the second one is improved by the inclusion of summed leading and next-to-leading $\ln(y_1/y_0)$ logarithms (LO+NLLA) (Seymour, 1996). This LO+NLLA calculation contains higher order terms than found in the parton shower implementations but does not rigorously impose energy–momentum conservation. The summed results may be expected to offer a better description of the small y_0 region, where the $\ln(y_1/y_0)$ terms become important, but this must be counterbalanced by the expected growing importance of the missing hadronization corrections. The LO prediction only describes the average subjet multiplicities for the highest y_0 values. In contrast, the LO+NLLA result improves upon the parton level Monte Carlo event generator results, providing a reasonable description of the data over the extended range $10^{-5} < y_0 < 10^{-1}$. The description is better for gluons than the (massive) quark jets. Concerning the ratio R_μ, the general level of agreement is not as good and both the LO and LO+NLLA predictions are only adequate for high y_0 values. For the widths of the subjet multiplicity distributions, $\sigma_{\text{q,g}}(y_0)$, analytic predictions are only available in LLA whilst for their ratio, R_σ, a y_0 independent NLLA calculation is available (Malaza and Webber, 1984), but see also Dokshitzer (1993). However, the agreement between these calculations and the experimental results is generally not as good as for the mean values.

12.3.3   *Jet profiles*

As noted earlier, differences are expected in the transverse sizes of quark and gluon jets. This has been investigated using a number of variables, including the rapidity and transverse momentum distributions of particles within a jet (OPAL Collab., 1999a) and the jet broadening (ALEPH Collab., 2000b; DEL-PHI Collab., 1998b). The latter is defined, in analogy with the event shape variables wide and narrow jet broadening (Catani *et al.*, 1992b), to be

$$B_{\text{jet}} = \frac{\sum_i |\boldsymbol{p}_i \times \boldsymbol{n}_{\text{jet}}|}{\sum_i |\boldsymbol{p}_i|} = \frac{\sum_i |\boldsymbol{p}_{\text{T}i}|}{\sum_i |\boldsymbol{p}_i|} , \tag{12.9}$$

where $\boldsymbol{n}_{\text{jet}}$ is the jet direction. It is essentially an energy weighted average over the angles of the particles in a jet with respect to the jet axis. Figure 12.5 shows the measured jet broadening distributions for quark and gluon jets (ALEPH Collab., 2000b). The data is based on three-jet events selected using the Durham algorithm with $y_{\text{cut}} = 0.1$.

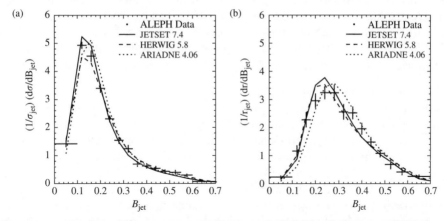

FIG. 12.5. Jet broadening as measured in quark (left) and gluon jets (right), as measured over three-jet phase space, compared to predictions from coherent Monte Carlo event generators. Figure from ALEPH Collab.(2000b).

The jet broadening for quarks clearly is more tightly distributed about lower values than is the case for gluons, indicating that quark jets are indeed narrower than gluon jets. However, we should caution that these quark jets have on average 14% more energy than the gluon jets and that the effect of a boost, though small in this instance, is to collimate a jet. At large B_{jet}, pQCD effects are expected to dominate and indeed the data are described by coherent Monte Carlo event generators at the parton level. At smaller B_{jet}, hadronization affects the partonic jets to giver broader hadron level jets which are in better overall agreement with

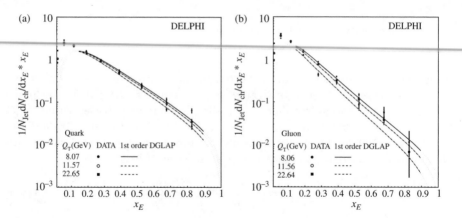

FIG. 12.6. The quark and gluon fragmentation functions at three values of the topological scale Q_{T}. The curves are the result of a simultaneous fit satisfying the time-like DGLAP equations. Figure from DELPHI Collab.(2000b).

the data; these are shown in Fig. 12.5. The scale dependence of the average jet broadening has also been investigated (DELPHI Collab., 1998b). It is found that $\langle B_{\mathrm{jet}}\rangle(Q_{\mathrm{T}})$ is approximately constant, with $\langle B^{\mathrm{g}}_{\mathrm{jet}}\rangle/\langle B^{\mathrm{q}}_{\mathrm{jet}}\rangle \approx 1.5$, reflecting the fact that for particles in a jet p_{L} and p_{T} approximately scale with Q_{T}. In contrast $\langle B_{\mathrm{jet}}\rangle(E_{\mathrm{jet}})$ falls with E_{jet}, indicating that higher energy jets are narrower.

12.3.4 Fragmentation functions

Early LEP measurements using light quark and gluon jet samples extracted from fixed topology events have confirmed the theoretical expectations for the fragmentation functions (ALEPH Collab., 1996b; DELPHI Collab., 1998b; OPAL Collab., 1995c). Later LEP studies have exploited the full three-jet phase space using the topological scales, such as Q_{T}, to study the evolution of the fragmentation functions (DELPHI Collab., 2000b). The correct choice of scale is important as it specifies both the strength of the coupling and the size of the available phase space, which we expect to be reduced by angular ordering effects. We can have faith in this approach since the quark fragmentation function, extracted from three-jet events as a function of Q_{T}, is in good agreement with those obtained from $\mathrm{e}^{+}\mathrm{e}^{-}$ experiments run at lower C.o.M. energies. As Fig. 12.6 shows, the gluon distribution is significantly softer than the quark distribution with a higher peak at small values of $x_E = 2E_{\mathrm{jet}}/\sqrt{s}$ and a faster, approximately exponential, fall off in the region $x_E \to 1$. Furthermore, the fragmentation functions get softer with increasing Q_{T}. That is, above $x_E \sim 0.1$ the fragmentation functions decrease as Q_{T} increases whilst below $x_E \sim 0.1$ they increase. This softening is most pronounced for gluon jets.

The results in the small-x region, which is dominated by wide angle soft gluons, are somewhat sensitive to the choices of jet algorithm and topological scale variable. Thus, as yet, only qualitative studies are justified. The peak regions of the quark and gluon fragmentation functions are both well described by Gaussian distributions in $\ln(1/x)$; c.f. Section 11.2. Furthermore, the fitted peak position $\xi^\star = \ln(1/x_{\mathrm{peak}})$ is compatible with a $0.5\ln(Q_T/Q_0)$ behaviour, that is, the same scale dependence that was seen in the overall fragmentation function for e^+e^- events at different C.o.M. energies.

In the large-x region, $x_E > 0.15$, the fragmentation functions are stable against the details of the jet algorithm. However, it is still important to use Q_T as the scale when testing the scaling violations predicted by the time-like DGLAP equations, eqn (3.285), (Nason and Webber, 1994). The result is an excellent simultaneous fit, over wide x_E and Q_T ranges, of the quark and gluon fragmentation functions as measured in a single experiment. Furthermore, the ratio of the relative rates of change of the gluon and quark fragmentation functions allows the extraction of $C_A/C_F = 2.26 \pm 0.16$ (DELPHI Collab., 2000b); see Ex. (12-2).

12.3.5 *Particle content*

In perturbative QCD the evolution with energy scale of a jet is described by a parton (or dipole) shower. The hadronization, which converts final state partons at a low, fixed virtuality into hadrons, does not directly depend on the initial scale of the jet. Indeed, the standard hadronization models make no distinction between the partons produced at the end of a quark or a gluon initiated shower; see Section 3.8. Consequently, perturbative predictions for the differences between quark and gluon jets are universal and apply equally well to any identified particles just as to the final state partons. Thus, for example, eqn (12.1) is expected to hold for all types of identified particles. Furthermore, the relative rates of particles within a quark or gluon jet are not predicted to vary with the jet's scale. That said, small differences are anticipated due to neglected effects within the pQCD calculations. These include effects coming from unequal phase spaces due to different particle masses and perhaps more significantly the 'leading particle effect' coming from the presence of flavour in a quark initiated jet. In the most extreme example, since b-quarks are only rarely produced in a shower (Seymour, 1995; Abbaneo *et al.*, 2001a), the number of B-hadrons occuring in gluon jets compared to quark jets will be much lower than predicted by eqn (12.1).

The above expectation of universality is generally believed to be robust. There are, however, some unorthodox hadronization schemes which suggest the production of glueballs and other exotics that favour enhanced isoscalar production, for example, of η or η' mesons in gluon jets; for some discussion see (ALEPH Collab., 2000c) or (Knowles and Lafferty, 1997).

Studies of identified particle production in quark and gluon jets have focused on multiplicity ratios, such as $R_{\mathrm{h}}^{\mathrm{g}}(Q) \equiv \langle N_{\mathrm{h}}^{\mathrm{g}}(Q)\rangle/\langle N_{\mathrm{ch}}^{\mathrm{g}}(Q)\rangle$ and $R_{\mathrm{h}}^{\mathrm{q}}(Q)$, $R_{\mathrm{h}}(Q) \equiv \langle N_{\mathrm{h}}^{\mathrm{g}}(Q)\rangle/\langle N_{\mathrm{h}}^{\mathrm{q}}(Q)\rangle$ and $R_{\mathrm{h}}' \equiv R_{\mathrm{h}}^{\mathrm{g}}/R_{\mathrm{h}}^{\mathrm{q}} = R_{\mathrm{h}}/R_{\mathrm{ch}}$, where the index 'h' denotes an identified hadron and 'ch' all final state charged particles. Perturbative QCD predicts that $R_{\mathrm{h}}^{\mathrm{g}}(Q)$ and $R_{\mathrm{h}}^{\mathrm{q}}(Q)$ are independent of the scale Q and also $R_{\mathrm{h}}(Q) \sim R_{\mathrm{ch}}(Q)$ so that the double ratio $R_{\mathrm{h}}'(Q) \approx 1$. The advantage to measuring ratios is that many sources of systematic error cancel. In particular, there is a reduction in the errors associated with the unfolding and the choice of the jet's scale which are both common to the numerator and denominator. A residual sensitivity to the details of the jet algorithm does nevertheless remain; see, for example, (OPAL Collab., 1999b). Measured values of the double ratio are given in Table 12.1.

Table 12.1 *The double multiplicity ratios of identified particles in quark and gluon jets. The first error is statistical, the second is systematic.*

$R'_{\pi^\pm}$ $= 0.997 \pm 0.009 \pm 0.013$	DELPHI Collab.(2000c)
$= 1.016 \pm 0.010 \pm 0.010$	OPAL Collab.(1998b)
R'_{π^0} $= 1.013 \pm 0.017 \pm 0.040$	OPAL Collab.(2000b)
R'_{η} $= 1.088 \pm 0.079 \pm 0.086$	OPAL Collab.(2000b)
$R'_{K^\pm}$ $= 0.86 \ \ \pm 0.31$	ARGUS Collab.(1996)
$= 0.942 \pm 0.019 \pm 0.016$	DELPHI Collab.(2000c)
$= 0.948 \pm 0.017 \pm 0.028$	OPAL Collab.(1998b)
$R'_{K^0_S}$ $= 1.13 \ \ \pm 0.09 \ \ \pm 0.13$	DELPHI Collab.(1997c)
$= 1.10 \ \ \pm 0.02 \ \ \pm 0.02$	OPAL Collab.(1999b)
$= 0.947 \pm 0.014 \pm 0.039$	OPAL Collab.(2000b)
$R'_{p,\bar{p}}$ $= 1.58 \ \ \pm 0.10$	ARGUS Collab.(1996)
$= 1.205 \pm 0.041 \pm 0.025$	DELPHI Collab.(2000c)
$= 1.100 \pm 0.024 \pm 0.027$	OPAL Collab.(1998b)
$R'_{\Lambda,\bar{\Lambda}}$ $= 1.40 \ \ \pm 0.30 \ \ \pm 0.23$	DELPHI Collab.(1997c)
$= 1.41 \ \ \pm 0.04 \ \ \pm 0.04$	OPAL Collab.(1999b)

Despite the variety of jet definitions employed and relatively large errors some conclusions can be drawn. There is no evidence for enhanced production of the isoscalar η, the double ratio is consistent with unity. That is, the increased multiplicity seen in gluon jets compared to quark jets is the same as that for charged particles as a whole. However, there is evidence for an excess production of baryons, both Lambdas and protons, in gluon jets and a more complex pattern of deviations for neutral and charged kaons. At lower C.o.M. energies, $\sqrt{s} \sim$ 10 GeV, resonant $e^+e^- \rightarrow \Upsilon(1S) \rightarrow ggg$ events are seen to contain ≈ 2.5 times the number of baryons than continuum $e^+e^- \rightarrow q\bar{q}$ events; no enhancement was seen for mesons (ARGUS Collab., 1996; CLEO Collab., 1985). It has also been observed that a depletion in the baryon yield and increase in the kaon yield for quark jets can be traced to the presence of heavy flavours in the quark sample (DELPHI Collab., 2000c). Furthermore, studies show that as expected $R'_{\pi^0,\eta,K^0}(Q_T)$ all show no significant dependence on the scale Q_T (OPAL Collab., 2000b). It has also been demonstrated that $R^{q,g}_{K^0,\Lambda}(E_{\rm jet})$ are all independent of the jet energy E_{jet} (OPAL Collab., 1998b).

The momentum spectra in quark and gluon jets as a function of $x_p = p_h/p_{\text{jet}}$ have been measured for identified pions, kaons and (anti)protons (DELPHI Collab., 2000c). These confirm that the gluon jet's momentum spectra are all softer than in an equivalent quark jet. Furthermore, the excess of protons has been traced to the region of moderately large momenta $0.16 < x_p < 0.50$. A Gaussian fit to the $\xi = \ln(1/x_p)$ distribution for identified particles shows that the peak position $\xi^\star$ grows with the jet scale Q_T in a way that is consistent with the logarithmic growth seen for $\xi^\star$ as measured with all charged particles; see Section 11. In quark jets it has been found, at a given Q_T, that $\xi_\pi^\star > \xi_K^\star \approx \xi_p^\star$, whilst in gluon jets $\xi_\pi^\star > \xi_K^\star > \xi_p^\star$. Also comparing the individual peak positions for π, K and p measured in quark and gluon jets of equal Q_T, one finds $\xi_g^\star > \xi_q^\star$, in qualitative agreement with the pQCD prediction (Fong and Webber, 1989),

$$\xi_g^\star - \xi_q^\star \approx \frac{1}{12}\left(1 + \frac{n_f}{C_A^3}\right) \simeq 0.1 . \tag{12.10}$$

A similar shift in $\xi^\star$ has been seen in the charged particle multiplicity distribution (DELPHI Collab., 2000b).

As noted earlier, the pQCD calculations of identified hadron spectra are expected to receive corrections due to particle masses and other non-perturbative effects. One way to account for these corrections is to use Monte Carlo event generators with their exact kinematics and detailed hadronization models. The simulations also allow the specifics of the jet definition to be taken into account. It is therefore common to compare the data on rates and momentum spectra to Monte Carlo predictions. This has provided mixed results, however, it appears that the HERWIG cluster scheme is less successful at describing the data than for example the JETSET model (ALEPH Collab., 2000c; DELPHI Collab., 2000c; OPAL Collab., 1999b).

Exercises for Chapter 12

12–1 A set of onefold symmetric events is defined with two inter-jet angles equal to $150°$ and a set of threefold symmetric events with all inter-jet angles equal to $120°$. At the Z, what are the energies and Q_T scales of the jets in the two event types?

12–2 By considering the time-like DGLAP equations, show that the ratio of the relative sizes of the scaling violations in the fragmentation functions of quarks and gluons approach a constant,

$$\frac{\partial \ln D_g(x,\mu^2)/\partial \ln \mu^2}{\partial \ln D_q(x,\mu^2)/\partial \ln \mu^2} \xrightarrow{x \to 1} \frac{C_A}{C_F} ,$$

in the limit $x \to 1$. $^{(\star)}$

13

FRAGMENTATION

Fragmentation is the process by which the primary partons produced in a hard subprocess convert into the sprays of hadrons seen by experiments. We have learnt that this process is usefully separated into two stages, according to the momentum transfers involved. Perturbative QCD is applicable to the transition from the highly virtual primary partons to a set of low-virtuality final state partons. This is pictured as a cascading process that is dominated by the collinear and soft emissions of gluons and mainly, though not exclusively, light quark–antiquark pairs. By contrast, models are used to describe the non-perturbative transition from these final state partons to hadrons which then may be decayed according to tables or further models. Fortunately the calculable, perturbative stage is responsible for much of an event's structure, even including to some extent the momentum spectrum of the produced hadrons. That said, important details depend on the specific type of hadron produced and at present this can only be modelled. Thus, for example, whilst we may be able to predict something of the momentum spectrum of kaons, we find it more difficult to say how many will be produced in a given situation. A partial exception is the total rate of hadrons containing heavy quarks, c and b, which are not believed to be produced during hadronization.

Whilst many types of particles are produced in a high energy hadronic event, only a few live long enough to be detected directly. In a typical experiment these are $e^{\pm}$, $\mu^{\pm}$, $\pi^{\pm}$, $K^{\pm}$, p, $\bar{p}$ and γ, which are either absolutely stable or effectively so, having lifetimes longer than $\sim 10^{-8}$ s, which permits them to traverse the detector before they decay. Photons can be detected with an electromagnetic calorimeter, neutrons and K_L^0 mesons can in some cases still be recorded with a hadron calorimeter, while neutrinos escape undetected, c.f. Section 5.2. The presence of all other particles can only be inferred from their long-lived decay products.

Extrapolating particle tracks towards the primary production vertex, experiments with high-resolution tracking detectors, ideally silicon strip detectors, may still be capable of seeing the characteristic displaced vertices associated with weakly decaying particles of modest lifetimes, such as $\tau^+\tau^-$ leptons, K_S^0 mesons, hyperons, that is, baryons containing one or more strange quarks, and the lightest c- and b-hadrons. This is still relatively easy in the case of the K_S^0 or the hyperons, since these particles may travel tens of centimetres before decaying. On the other hand, reconstructing decay vertices of c- or b-hadrons, which travel at most a few millimetres, can be quite demanding. Apart from finding a secondary vertex, or more generally, since the primary decay may be into one or more short-lived daughter particles, a sequence of secondary vertices, one also needs to identify the final state particles in order to reconstruct the exact decay chain. Thus, it is also important to have a detector with good particle identification capabilities.

Particles with strong and electromagnetic decays travel no discernible distance from their point of production, and thus have to be reconstructed by other means. The basic technique for finding candidates is based upon identifying sets of putative decay products whose invariant masses are calculated to lie within a given interval about the expected resonance mass. For example, to find $K(982)^{*0}$ candidates via their decay to $K^+\pi^-$ one would consider all the possible pairings of opposite-sign charged tracks in an event. Using the measured three-momenta and assigning kaon and pion masses to the tracks one can calculate the resulting invariant mass. The distribution of these masses M should contain a Breit–Wigner peak,

$$\frac{\mathrm{d}\sigma}{\mathrm{d}M} \sim \frac{\Gamma}{(M - M_{\mathrm{R}})^2 + \Gamma^2/4} \, , \tag{13.1}$$

coming from the correct pairings of genuine decay products on top of a combinatorial background, which may be parameterized by a simple polynomial or another smooth function. The parameters Γ and M_{R} in eqn (13.1) are the width and the peak position of the resonance. For very narrow resonances, such as η or J/Ψ, the finite detector resolution often leads to simple Gaussian peaks.

The removal and control of the combinatorial background poses the greatest experimental challenge to finding short-lived heavy hadrons. Methods for the reduction of background, with preservation of signal, include the use of particle identification. In the example above, a good separation of pion and kaon tracks would allow many wrong combinations to be discarded. In addition kinematic cuts may be applied. For example, decay products from heavy resonances may be emitted with large transverse momenta and in general have a harder momentum spectrum. Seen from the heavy hadron's rest frame, the genuine decay products are also emitted with larger angles than the bulk of the 'background hadrons'. The selection and optimization of such experimental cuts is clearly dependent on the sought-for hadron, but in general the narrower and more distinctive a decay channel the better the chances of finding a significant signal.

Many kinds of studies are possible with identified particles observed in the fragmentation of high energy quarks or gluons. In the following, we will focus on results obtained in e^+e^- annihilation at the scale of the Z and above. Apart from multiplicities and momentum spectra of identified particles, measurements are also available for two-particle correlations. Here, one can study kinematical correlations in rapidity, azimuthal angle and polar angle, but also Bose–Einstein correlations for identical particles. Another interesting subject is quantum number correlations such as for strangeness or baryon number. Together, these results place important constraints on the hadronization models and give us insights into the underlying physics. Tests of perturbative QCD are also possible, especially for heavy quark containing hadrons.

13.1 Identified particles

We now review briefly two of the most important properties of identified particles, namely their multiplicities and their momentum spectra. Most of the results quoted here are based on the high statistics data collected at LEP and SLC from hadronic Z decays.

13.1.1 *Multiplicities*

The breadth and quality of the available data is indicated in Tables 13.1 and 13.2. A convention which is applied commonly in the data is to treat all particles with lifetimes greater than, for example, 10^{-8} s as absolutely stable, while all others are assumed to have decayed. For a determination of the pion multiplicities this would imply that they include the daughter particles from K_S^0 decays. Note that this treatment may differ between experiments so that correction factors must be applied before making any direct comparisons. A related point concerns the origins of the lower lying hadrons. For example, decay chains such as $B^{\star\star} \to B\pi \to D\pi\pi \to K\pi\pi\pi$ make it clear that there are many sources of a particular final state particle other than direct production in the hadronization, and care has to be exercised to correctly account for double counting when interpreting the data.

The data in Tables 13.1 and 13.2 illustrate a generic behaviour of hadronic events, namely the dominance of pion production with approximately equal numbers of charged pions and photons, coming from the decays $\pi^0 \to \gamma\gamma$, in the final state. Kaons and baryons contribute a further $\approx 10\%$ each to the total multiplicity.

The rates of heavy, c and b, quark containing hadrons are essentially fully accounted for by perturbative production, either in the initial Z decay or subsequent gluon splittings. The measured rate for heavy quark production in the parton shower is compatible with the pQCD prediction (Seymour, 1995). As a case in point consider Table 13.3. From the known branching ratios of the Z into heavy quarks, the production rates of heavy quarks from gluon splitting in the perturbative phase and the number of $c/\bar{c}$-quarks produced in the decay chain of bottom quarks one can estimate the expected number of charm quarks, and thus charm hadrons, per hadronic Z decay. The direct measurement is in good agreement with the expectation.

Looking in a little more detail we see that where measurements have proved possible there is evidence for quite substantial production of orbitally excited mesons and baryons. For example, the known tensor (2^{++}) to vector (1^{--}) rates, f_2/ρ^0, f_2'/ϕ and $K_2^\star/K^\star$, are all $\approx 20\%$. Presumably many other higher states are produced, but just as their large widths makes them elusive to find, it also makes the relevance of their fleeting existence questionable. A suppression of strangeness containing states is evident for both light and heavy hadrons.

Table 13.1 *The average multiplicities of mesons produced in the hadronic decay of a Z. The sum of particle and antiparticle rates is implied: for example, π^+ represents also π^-. A dagger (†) indicates internal conflict within the experimental measurements. References are quoted with A,D,L,O,M,S used as shorthand for ALEPH, DELPHI, L3, OPAL, MARKII and SLD. For the definition of x_E see Section 13.1.2.*

Particle	Average		References/Remarks	
charged	21.05	± 0.12	ADLMO(1995d, 1999c, 1991c, 1990, 1995e)	
γ	20.97	± 1.15	O(1998c)	
π^+	16.99	± 0.20	ADOS(1998c, 1998c, 1994a, 1999b)	
π^0	9.82	± 0.23	ADLO(1997d, 1996d, 1996, 2000b)	
η	0.95	± 0.07	LO(1996, 1998c)	
η	0.282	± 0.022	A(1998a)	$0.1 < x_E$
η'	0.17	± 0.02	LO(1997b, 1998c)†	
η'	0.064	± 0.014	A(1998a)	$0.1 < x_E$
ρ^+	2.40	± 0.43	O(1998c)	
ρ^0	1.242	± 0.093	AD(1996c, 1999d)	
ϕ	0.098	± 0.003	ADOS(1996c, 1996e, 1998d, 1999b)†	
ω	1.083	± 0.088	ALO(1996c, 1997b, 1998c)	
$a_0(980)$	0.27	± 0.11	O(1998c)	
$f_0(980)$	0.147	± 0.011	DO(1999d, 1998d)	
$f_2(1270)$	0.169	± 0.021	DO(1999d, 1998d)	
$f_2'(1525)$	0.012	± 0.006	D(1999d)	
K^+	2.25	± 0.05	ADOS(1998c, 1998c, 1994a, 1999b)	
K^0	2.020	± 0.024	ADLOS(1994a, 1995a, 1997c, 2000b, 1999b)	
$K^{\star+}(892)$	0.714	± 0.044	ADO(1998a, 1995a, 1993d)	
$K^{\star0}(892)$	0.739	± 0.022	ADOS(1998a, 1996e, 1997b, 1999b)	
$K_2^{\star0}(1430)$	0.073	± 0.023	D(1999d)†	
$K_2^{\star0}(1430)$	0.19	± 0.07	O(1995f)†	$x_E < 0.3$
D^+	0.187	± 0.014	ADO(1994b, 1993c, 1996d)	
D^0	0.462	± 0.026	ADO(1994b, 1993c, 1996d)	
$D^{\star+}$	0.1833	± 0.0081	ADO(1994b, 1993c, 1998e)	
D_s^+	0.131	± 0.021	O(1996d)	
D_{s1}^+	2.9	± 0.7	O(1997c)	$0.6 < x_E$
$B_{d,u,s}^{\star}$	0.28	± 0.3	D(1995b)	
J/ψ	0.00386	± 0.00024	ADLO(1992b, 1994, 1993, 1996e)	
χ_{c1}	0.0075	± 0.0030	DL(1994, 1993)	
$\psi(3685)$	0.0036	± 0.0006	DO(1994, 1996e)	
$\Upsilon(1,2,3S)$	0.00010	± 0.00006	O(1996f)	

Table 13.2 *The average multiplicities of baryons produced in the hadronic decay of a Z. The sum of particle and antiparticle rates is implied: for example, Σ^+ also represents $\bar\Sigma$. A dagger (†) indicates internal conflict within the experimental measurements. References are quoted with A,D,L,O,S used as shorthand for ALEPH, DELPHI, L3, OPAL and SLD.*

Particle	Average	References/Remarks
p	1.04 ± 0.04	ADOS(1998c, 1998c, 1994a, 1999b)
Δ^{++}	0.088 ± 0.014	DO(1995c, 1995g)†
Λ	0.374 ± 0.003	ADLOS(1994a, 1993a, 1997c, 1997d, 1999b)
$\Lambda(1520)$	0.0225 ± 0.0028	DO(2000d, 1997d)
Σ^+	0.107 ± 0.010	DLO(1995d, 2000, 1997e)
Σ^-	0.082 ± 0.007	DDO(1995d, 2000d, 1997e)
Σ^0	0.078 ± 0.008	ADLO(1998a, 1996f, 2000, 1997e)
$\Sigma^+(1385)$	0.0237 ± 0.0014	ADO(1998a, 1995d, 1997d)
$\Sigma^-(1385)$	0.0237 ± 0.0016	ADO(1998a, 1995d, 1997d)
Ξ^-	0.0264 ± 0.0008	ADO(1998a, 1995d, 1997d)
$\Xi^0(1530)$	0.0058 ± 0.0004	ADO(1998a, 1995d, 1997d)†
Ω^-	0.0016 ± 0.0003	ADO(1998a, 1996f, 1997d)
Λ_c^+	0.078 ± 0.017	O(1996d)

This suggests that only about an eighth of the light quarks produced during hadronization are strange quarks. Multiply-strange states appear to be even further suppressed, as is the production of baryons compared to mesons.

Another area where attention has focused is the dependence of the production rates on a particle spin. All other things being equal the rate might be expected to be proportional to the number of states for a given spin J, $2J + 1$. Whilst there is some evidence for this behaviour, the situation is not clear cut, since it is hard to determine the contributions made by directly produced hadrons and those coming from higher resonance decays.

Specific mention should be made of the contribution from b-hadrons, especially since the branching ratio of the Z into b-quarks is larger than 20%. The average charged particle multiplicity of a b-hadron decay is $\langle n_{\rm ch}(b)\rangle = 4.955 \pm 0.062$ (Abbaneo *et al.*, 1998). Since the total charged multiplicity in $Z \to b\bar{b}$ events is not drastically different from light-quark events, this implies that in a $Z \to b\bar{b}$ event about 50% of the particles come from the two b-hadrons, whilst the fraction of tracks from primary b-quarks is 10% for the natural mix of $Z \to q\bar{q}$ events. Thus it is imperative to have a good understanding of b-hadron decays in order to be able to draw firm conclusions concerning details of the hadronization process from the data. Unfortunately, the present decay tables for b-hadrons are still somewhat sparse, which forces us to model their decays. We expect, however, that with upcoming results from the B-factories which are already running or under construction, this situation will improve significantly.

Table 13.3 *Expected yield of charm particles compared to experimental measurements for hadronic decays of the Z*

---

expected number of charm (anti)quarks per event

Br(Z → c$\bar{c}$)	0.1702 ± 0.0034	(Abbaneo *et al.*, 2001*a*)
+pQCD(g → c$\bar{c}$)	0.0299 ± 0.0039	(Abbaneo *et al.*, 2001*a*)
+ [(Br(Z → b$\bar{b}$)	[(0 .2165 ± 0.0007	(Abbaneo *et al.*, 2001*a*)
+pQCD(g → b$\bar{b}$))	+0.0025 ± 0.0005)	(Abbaneo *et al.*, 2001*a*)
×rate(b → c/$\bar{c}$X)]	×1.226 ± 0.060]	(Abbaneo *et al.*, 2001*b*),
	0.4686 ± 0.0219	
×2	0.9372 ± 0.0438	

measured number of charm particles per event

$\sum$ D$^+$, D^0, D^0_s, ψ, Λ^+_c, . . .	≥0. 862 ± 0.040	Table 13.1, Table 13.2

13.1.2 Momentum spectra

The momentum spectrum, often also referred to as the *fragmentation function*, is usually quoted as the differential rate $\sigma^{-1}\mathrm{d}\sigma/\mathrm{d}x$, where $x = x_E$ or $x = x_p$ is the energy or momentum fraction of a particle relative to that of the beam or jet. The normalization to the total hadronic cross section removes the need for an absolute cross section measurement, thereby eliminating a possible source of systematic error. Note, however, that the momentum spectra are steeply falling functions with x, so that care has to be exercised when interpreting the measurements, usually given in the form of histograms with finite bin width, as differential cross sections at a particular value x (Lafferty and Wyatt, 1995). The total multiplicities per hadronic event, reported above, are obtained by integrating the measured fragmentation functions and extrapolating into any unmeasured regions of x. When this can not be done reliably, limits on the measured x-range are quoted.

A large quantity of experimental data is available for the fragmentation functions of identified particles produced in e$^+$e$^-$ collisions over a range of C.o.M. energies. For a compilation of results we refer the reader to the literature (Lafferty *et al.*, 1995; Böhrer, 1997). Similar data have been produced from a Breit frame analysis, see Ex. (6-3), of the HERA data (H1 Collab., 1997*b*; ZEUS Collab., 1999*c*). The results, even from different kinds of interactions, are quite consistent, showing that the hadron formation in hard scattering processes proceeds in a universal way. The information extracted about the structure of the parton shower is discussed in detail in Section 11.

The fragmentation functions of hadrons containing a heavy quark, c or b, are of special interest. This is because the large quark masses assure the importance of perturbation theory, allowing QCD to be tested, and because of their significance for other interactions. In practice, non-perturbative effects must also be included. To see this, assume that the fragmentation function can be written as the convolution of a perturbative (PT) and a non-perturbative (NP) piece, $D(x) = d_{PT} \otimes d_{NP}$. It follows that the average x is given by $\langle x \rangle = \langle x \rangle_{PT} \times \langle x \rangle_{NP}$; see Ex. (13-1). The NLO calculation of $d_{PT}(x)$ gives a sharply forward-peaked distribution which vanishes at $x = 1$. It predicts $\langle x \rangle_{PT} \approx 0.8$ for $E_{jet} = M_Z/2$. Comparing this with the value measured for b-quarks at LEP, $\langle x \rangle = 0.702 \pm 0.008$ (Abbaneo $et\ al.$, 1998), one not only sees that non-perturbative effects are not negligible, but also that the non-perturbative fragmentation function has to be rather 'hard', that is, peaked towards large values of x. A simple argument (Suzuki, 1977; Bjorken, 1978) shows that due to the heavy quark's inertia this is actually expected. The argument goes as follows. Suppose the heavy quark Q acquires a light quark to form a system of rest mass $m_Q + m_q$. The heavy quark's energy fraction is $m_Q/(m_Q + m_q) \approx 1 - m_q/m_Q$ which is unaffected by the boost into the overall C.o.M. system. As the mass of the light quark, which is inversely proportional to its de Broglie wavelength, determines the size of the hadron, $R_0 \sim 1/m_q$, one thus expects $\langle x \rangle_{NP} \approx 1 - R_0^{-1}/m_Q$, with R_0 the typical size of a hadron. Heavier hadrons are expected to have harder fragmentation functions. In the literature several forms are posited for the non-perturbative fragmentation function. The one most commonly used is that due to Peterson $et\ al.$ (1983),

$$d_{NP}(x) = \frac{N}{x}\left(1 - \frac{1}{x} - \frac{\epsilon_Q}{1-x}\right)^{-2} \quad \text{with} \quad N \approx \frac{4\sqrt{\epsilon_Q}}{\pi}, \qquad (13.2)$$

where the parameter ϵ_Q can be interpreted as $\epsilon_Q = R_0^{-2}/m_Q^2$. For eqn (13.2) one finds $\langle x \rangle_{NP} = 1 - \sqrt{\epsilon_Q} \approx 1 - R_0^{-1}/m_Q$.

A large volume of data is available for c-hadrons, mainly coming from experiments at the $\Upsilon(4S)$ resonance (PDG, 2000). This allows the fragmentation functions of individual hadrons to be measured. The data can be described by eqn (13.2) provided that ϵ_Q is tuned separately for each hadron. In general, the heavier the meson or baryon the harder is their momentum spectrum, with orbitally excited mesons being particularly hard. The fragmentation function for b-hadrons has been measured at LEP1. Here the available statistics only permit a measurement averaged over the naturally ocurring mix of b-hadrons. Again eqn (13.2) can describe the data but one should be aware of differences in the definition of x and of distinctions made between primary and secondary b-hadrons cascading down from excited states. At the Z, the relative rates of primary b-hadrons are roughly given by B : B* : B** $\approx 1 : 3 : 2$. The experimental result for $\epsilon_c/\epsilon_b \approx 10$ is consistent with the mass ratio $(m_b/m_c)^2$.

13.2   Inter-jet soft gluons and colour coherence

In this section we shall investigate predictions for inter-jet coherence effects; that is, the distribution of particles lying between hard jets. Theoretically, the approach is to regard the hard partons as acting as an antenna and predict the distribution of radiated, soft gluons using the techniques of Section 3.7. These studies complement those of intra-jet coherence efects in Section 11.2.

13.2.1   *The string effect*

The classical test of inter-jet coherence is the 'string effect' seen in e^+e^- annihilation to three jets. This predicts a depletion in the flow of particles, or energy, lying in the event plane between the q and $\bar{q}$ relative to that between the q and g or $\bar{q}$ and g. As the historical name suggests, it was first predicted on the basis of the Lund string hadronization model: here the gluon draws the string in the direction of its motion and away from the $q\bar{q}$-axis, causing the depletion (Andersson *et al.*, 1980*b*). Later a pQCD explanation was advanced (Azimov *et al.*, 1985*a*); see eqn (3.343) and Fig. 3.28.

The earliest measurements, carried out at $\sqrt{s} = \mathcal{O}(30\,\text{GeV})$, used energy ordering to identify the gluon jet and made comparison to Monte Carlo event generators based upon string and independent fragmentation models (JADE Collab., 1985; TASSO Collab., 1985; TPC/2γ Collab., 1985). Whilst the string model was favoured, the tests were open to criticism of kinematical biases associated with the difference between narrow quark jets and wide gluon jets. A second, more conclusive, series of tests has since been carried out at LEP1 (ALEPH Collab., 1998*a*; L3 Collab., 1995; OPAL Collab., 1991*d*).

The basic method is to select three-jet events that are well contained within the detector, using, for example, a typical resolution parameter $y_{\text{cut}} = 0.008$ for the Durham jet finder. Events with four or more jets are excluded. If the jets are energy ordered, $E_1 > E_2 > E_3$, then jet 1 is almost certainly a quark or an antiquark jet, and in $\mathcal{O}(70\%)$ of all cases jet 3 originates from a gluon. In the absence of further constraints on the jets' energies it is useful to study the particle or energy flow in the event plane as a function of scaled angular variables, such as $\alpha = (\phi - \phi_1)/(\phi_2 - \phi_1)$, where $\phi_{1,2,3}$ are the angles of the jets in the event plane and ϕ the angle of the particle or energy deposit. Concentrating on the central regions, $0.3 < \alpha < 0.7$, the ratio of the integrated particle or energy flow between jets 1 and 2, in most cases the quark jets, and jets 1 and 3, dominantly one quark and one gluon jet, has been measured to be significantly below unity. In a refinement of this analysis prompt lepton tags may be used to identify heavy flavour jets and increase the gluon purity of the third jet to $\mathcal{O}(90\%)$, thereby making clearer the effect of colour coherence. Asking for a lepton inside a jet enhances the fraction of heavy flavours and thus suppresses the gluon contamination. Such flavour tagging is essential in the case of symmetric events where the jet energies are almost equal so that energy ordering no longer distinguishes between quarks and gluons.

This data has been compared to a variety of Monte Carlo models and analytical predictions with the conclusion that colour coherence is required to describe the data. A significant finding was that using JETSET(PYTHIA) it is necessary to include both angular ordering in the shower and the string hadronization model in order to describe the data. The interplay between these contributions has been evaluated in Khoze and Lönnblad (1990). In practice, the inter-jet particles, chiefly pions, have energies of only a few hundred MeV. What the data suggest then is that pQCD enables us to calculate the strength of the underly-

ing colour field which produces the hadrons. Furthermore, the hadronization is
a soft, not a hard, process which does not disrupt our prediction, that is, local
parton hadron duality works (Khoze and Ochs, 1997).

A second test, proposed in (Azimov *et al.*, 1986b), is the relative depletion
of particles between the q$\bar{\text{q}}$ system in q$\bar{\text{q}}\gamma$ compared to the particle flow in q$\bar{\text{q}}$g
events; see Ex. (13-3). Since the production cross section for q$\bar{\text{q}}\gamma$ events is suppressed by the ratio of electromagnetic and strong coupling, $e_q^2 \alpha_{em}/(C_F \alpha_s)$, this
study requires large integrated luminosities. Given that, and by carefully matching the event selection cuts, it is possible to compare inter-quark regions that are
kinematically alike. Differences arising from the altered primary quark flavour
mix and different contents of the opposite hemispheres have been shown to be
minimal. Again, early studies (JADE Collab., 1988b; MARKII Collab., 1986;
TPC/2γ Collab., 1986) have been followed by studies at LEP1 (DELPHI Collab., 1996c; L3 Collab., 1995; OPAL Collab., 1995h). Once more the conclusion
is that only fully coherent Monte Carlo simulations are capable of describing the
measured ratio of particle flows, $\rho(\text{q}\bar{\text{q}}\gamma)/\rho(\text{q}\bar{\text{q}}\text{g}) \sim 0.65$.

Colour coherence is often summarized by the phrase 'angular ordering of successive parton branchings'. However, it goes further than this and also prescribes
the azimuthal distribution of soft gluons about the emitter, which preferably lie
in the plane between the emitter and its colour partner, see eqn (4.23) and
Fig. 4.2. Experimentally one expects to find an asymmetry in the momentum
flow for a quark or antiquark jet within a three-jet event such that the low momentum particles tend to lie on the side towards the gluon. Evidence for this
effect has actually been seen in the LEP1 data.

13.2.2 *Colour coherence in hadron–hadron collisions*

The complicated nature of hadron–hadron collisions has resulted in fewer tests
of soft gluon coherence. One exception is a study of the distribution of the lowest transverse-energy jet, assumed to be a gluon jet in three-jet events at the
TEVATRON (CDF Collab., 1994; D0 Collab., 1999b). The basic idea is to consider an underlying QCD two-to-two hard scattering and derive the distribution
of radiated soft gluons using the same insertion current technique as developed
in Section 3.7. It is also possible to use the exact three-jet matrix elements
(Mangano and Parke, 1991) which give the same results in the soft gluon limit.
These predictions must be weighted by the appropriate parton density functions,
summed over all contributing subprocesses, integrated over the observed phase
space and finally LPHD invoked: a task naturally done using an event generator. The main difference with the situation encountered in e$^+$e$^-$ annihilation is
that the colour flows connect incoming and outgoing partons, see, for example,
Fig. 4.4, so that coherence must be taken into account in both the initial and
final state radiation (Ellis *et al.*, 1987; Odagiri, 1998).

Events are selected containing three jets which are energy ordered, $E_1 >$
$E_2 > E_3$, and the angular distribution of jet 3 about jet 2 is measured. Studies
show that jet 3 is most likely to be initiated by a gluon, whilst jet 2 is most likely

to be the colour partner of jet 3. Working in the η–ϕ plane, useful variables in addition to the pseudorapidity of jet 3, η_3, are R and β defined by

$$R = \sqrt{(\eta_3 - \eta_2)^2 + (\phi_3 - \phi_2)^2} \,, \quad \beta = \tan^{-1}\left[\frac{\text{sign}(\eta_2)(\phi_3 - \phi_2)}{(\eta_3 - \eta_2)}\right] . \quad (13.3)$$

Here β is the angle of jet 3 about jet 2 with zero defined to be in the direction of the beam nearest to jet 2. For $\beta = 0, \pi$ jet 3 lies within the event plane, spanned by the colliding beams and jet 2, and for $\beta = \pm\pi/2$ it points out of the event plane. Colour coherence predicts enhancements for $\beta = \pi$ and to a lesser extent also for $\beta = 0$. This is seen in the data for both centrally produced jets (CDF Collab., 1994; D0 Collab., 1999b) and more clearly for forward jets (D0 Collab., 1999b). An observed broadening of the η_3 distribution is also predicted. Comparison to Monte Carlo simulations show that the incoherent ISAJET or PYTHIA with angular ordering switched off are incapable of describing the data, whilst the coherent parton shower models HERWIG or PYTHIA reproduce the data.

A related study has investigated the distribution of hadrons in W+jet events (D0 Collab., 1999b). Theoretically this is a simpler situation to analyse since at lowest order only two processes contribute: $q\bar{q}' \rightarrow$ Wg and qg $\rightarrow$ q$'$W ($\bar{q}$g $\rightarrow$ $\bar{q}'$W) (Khoze and Stirling, 1997). The selected events each contained a centrally produced W and at least one jet with $|\eta_{\text{jet}}|, |y_{\text{W}}| < 0.7$, with modest transverse energies, $E_T(\text{jet}), P_T(\text{W}) \gtrsim 10\,\text{GeV}$. In cases with more than one jet the most energetic jet in the azimuthal hemisphere opposite the W was selected. Two annular regions were then defined in the η–ϕ plane such that $0.7 < R < 1.5$ with $R = \sqrt{(\Delta\eta)^2 + (\Delta\phi)^2}$, and pseudorapidity and azimuthal angle measured with respect to those of either the W or the jet. The calorimetric activity was then measured as a function of the angle β, defined as in eqn (13.3), in the annulus. By comparing the results for the colourless W, which does not contribute any soft gluons, and the jet it allows minimization of detector effects and contaminations from the underlying event. In the event plane the data show more activity around the jet than around the W, in accord with expectations from colour coherence and good agreement with the coherent Monte Carlo generators.

In conclusion, several studies clearly establish the need for colour coherence in the description of e^+e^- and hadron–hadron collisions. At LEP the large body of data having a bearing on inter- and intra-jet coherence effects strongly disfavours Monte Carlo models which employ either incoherent showers or independent hadronization. As a consequence the COJETS Monte Carlo generator is used only rarely now, other than as a straw man, and its development has ceased. At the TEVATRON the inadequacy of Monte Carlo programs which use independent fragmentation models is less established and so ISAJET is still employed, thanks in part to the large number of subprocesses that it supports.

13.3   Two-particle correlations

When studying distributions for groups of two or more particles in hadronic
systems, it is interesting to look for deviations of those distributions from the
expectations based on single-particle properties. Very generally, we can measure
a quantity such as

$$C(p_1, p_2) = \frac{P_{\mathrm{j}}(p_1, p_2)}{P_{\mathrm{s}}(p_1) P_{\mathrm{s}}(p_2)} \,, \tag{13.4}$$

where $P_{\mathrm{j}}(p_1, p_2)$ is the joint probability to observe some property p_1 of particle
1 together with a property p_2 of particle 2, normalized to the single probabilities
P_{s}. For uncorrelated particles we would have $C = 1$. We shall call deviations
from this expectation *multi-particle correlations*. Almost all of the experimental
investigations have been performed for two-particle systems. What can be at the
origin of such correlations?

QCD parton shower and hadron dynamics, such as soft gluon coherence,
resonances and final state interactions, can induce correlations. Local quantum number conservation can lead to correlations of particles which are close
in phase space. Finally, quantum mechanical effects such as Bose–Einstein and
Fermi–Dirac correlations should be considered. Correlations have been studied
in particular within the context of baryon production, which in general is less
well understood. Single particle spectra can be reproduced more or less well by
the most important phenomenological models, namely string and cluster hadronization. However, correlation studies allow for a deeper look into the production dynamics of baryons, which are quite differently modelled, as described
in Section 3.8. Baryons, and in particular strange baryons, have the additional
advantage to probe directly the production mechanism. In general, the heavier
the particle the more often it stems directly from the hadronization phase and
not from decays of other hadrons.

From baryon number conservation we expect that a baryon is always accom-
panied by an antibaryon, so, possible pairings are for example $\mathrm{p\bar{p}}$, $\Lambda\bar{\Lambda}$ or $\Lambda\bar{\mathrm{p}}$.
However, flavour conservation in strong interactions suggests that the baryon
pairs may be preferentially particle–antiparticle pairs. If we finally assume that
the conservation of quantum numbers occurs rather locally, and that the concept of local parton–hadron duality holds, then we expect that the baryons are
produced close by in phase space.

A summary of recent measurements with a comprehensive list of references
can be found in the review by Knowles and Lafferty (1997).

13.3.1   *Proton–antiproton correlations*

We start with the analysis of proton–antiproton correlations, which has been
performed at LEP and also at lower energies. The main experimental issues are
particle identification, the subtraction of backgrounds from definitely uncorrelated pairs, and the definition of suitable reference systems and observables.

Particle identification is typically performed by measuring the specific ionization energy loss, which for particular regions of momenta allows the separation

of protons from pions, kaons and electrons. Backgrounds in a selected sample of proton–antiproton pairs can arise from particle misidentification or from the grouping of particles actually belonging to different baryon–antibaryon pairs. It is found that the distribution of so called like-sign pairs, that is, pp or $\bar{p}\bar{p}$, reproduces well the distribution of those backgrounds, and thus is used for background subtraction.

Most of the events in e^+e^- annihilation are two-jet events. In this case, we assume that hadrons are produced out of a colour field of the form of a flux tube stretched between the primary quark and antiquark, which fly apart in opposite directions. Therefore, a relevant reference system should contain an axis which coincides with the direction of the flux tube. An estimate for this axis is given for example by the Thrust or the Sphericity axis, as described in previous chapters. Typical observables which have been measured are rapidity, azimuthal and polar angle with respect to this axis.

Figure 13.1 (left) (ALEPH Collab., 1998a) shows the corrected like-sign subtracted rapidity distribution for antiprotons, given that a proton has been found in a certain rapidity region, as indicated by the shaded horizontal bar. We observe a clear local compensation of baryon number. Generally, it has been determined that for a given rapidity y_p of the proton there is about a 70% probability that the antiproton is found within $|y_p - y_{\bar{p}}| \leq 1$. A marginally weaker short-range $\Lambda\bar{p}$ correlation has been measured, too. In contrast, when pp pairs are found in an event, almost no correlation between the rapidity of the two baryons is observed.

Qualitatively these features are reproduced by the cluster and string hadronization models, as implemented in the HERWIG and JETSET Monte Carlo programs, respectively. However, quantitatively the strength of the correlation is overestimated by HERWIG and by JETSET without the 'popcorn mechanism', c.f. Section 3.8.3. The popcorn model is a convenient explanation for reduced correlations between a generated baryon–antibaryon pair, since in this model additional mesons can be created between the two baryons.

A very interesting test of hadronization models can be performed by measuring the angle $\theta^\star$ between the axis of the baryon–antibaryon pair in its rest frame and the event axis. As mentioned above, in string models the event axis is aligned with the colour field, and since in these models most of the momentum transferred from the string to the particles is longitudinal rather than transverse, the baryons should remember the direction of the colour field. Thus, their momentum difference, or equivalently the axis formed by the pair in its rest frame, is expected to form a small angle with the event axis. The angle $\theta^\star$, measured in the data, peaks at small values as we observe in Fig. 13.1 (right), and this is reproduced by JETSET, which incorporates the Lund string model. On the other hand, the HERWIG model fails to describe the data, because in the cluster hadronization model the clusters decay isotropically.

In summary, we have seen that the expected local compensation of baryon number is observed in the data by studies of proton–antiproton correlations. A measurement of the angle between the baryon pair and the event axis in the pair's

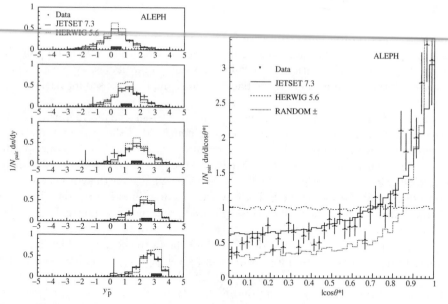

FIG. 13.1. Left: Like-sign subtracted rapidity distribution of antiprotons, given a tagging proton in the region marked by the shaded horizontal bar. Right: Distribution of the angle between the proton and the Sphericity axis in the proton–antiproton rest frame. Superimposed are the predictions of JETSET and HERWIG as well as the distribution for randomly chosen unlike-sign pairs taken from JETSET. Figures from ALEPH Collab.(1998a).

rest frame strongly disfavours cluster hadronization models where no memory of the direction of the event axis is retained.

13.3.2 Strangeness correlations

Now we turn our discussion to strange hadrons. The most frequently studied strange hadrons are K_S^0 mesons and Λ baryons. K_S^0 mesons are detected via their decay $K_S^0 \to \pi^+\pi^-$, and the Λ baryons via $\Lambda \to p\pi^-$ or $\bar{\Lambda} \to \bar{p}\pi^+$.

Strangeness is conserved in strong interactions; therefore, strange hadrons are pair-produced during hadronization. If we look at the probability that a hadron h_1 is accompanied by a hadron h_2, defined as

$$P(h_1, h_2) = 2\frac{\langle n_{h_1, h_2}\rangle}{\langle n_{h_1} + n_{h_2}\rangle} \, , \tag{13.5}$$

then we expect this number on average to be larger for strange–antistrange hadron pairs than for hadron pairs of equal strangeness. Experimentally, the expectation values indicated by the brackets are determined as averages over a large number of events with hadrons of type h_i. The measurements give $P(\Lambda, \bar{\Lambda}) = 49 \pm 6\%$, whilst $P(\Lambda, \Lambda) = 13 \pm 1\%$, $P(\Lambda, K_S^0) = 17 \pm 2\%$ and

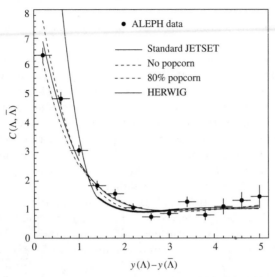

FIG. 13.2. Two-particle correlations for strange baryons as a function of rapidity difference. Figure from ALEPH Collab.(1998a).

$P(\mathrm{K_S^0}, \mathrm{K_S^0}) = 29 \pm 4\%$. So our expectation is confirmed by the data. In order to understand why $P(\Lambda, \bar{\Lambda})$ is not closer to unity, it should be remembered that strange hadrons are produced not only in strong interactions, but also during weak decays of hadrons containing heavy quarks (b, c). The fact that $P(\mathrm{K_S^0}, \mathrm{K_S^0})$ lies somewhere between $P(\Lambda, \bar{\Lambda})$ and $P(\Lambda, \Lambda)$ can be understood since neutral K-mesons decay only with a probability of 50% as $\mathrm{K_S^0}$-states.

The two main sources for the production of strange–antistrange hadron pairs are: leading hadrons associated with the s$\bar{\mathrm{s}}$ pair generated in the hard interactions (for example, Z $\rightarrow$ s$\bar{\mathrm{s}}$), and pairs produced locally out of the event's colour field. In order to distinguish them, we should look for phase space correlations. To first approximation we expect a long-range correlation because of the former production mechanism, and a strong short-range correlation in rapidity with respect to the Thrust axis, as defined above for proton–antiproton correlations, because of the latter. In Fig. 13.2 a measurement of the correlation of $\Lambda\bar{\Lambda}$ pairs as a function of their rapidity difference is shown. The correlation function is defined as

$$C(y_1, y_2) = N_{\mathrm{had}} \frac{n(y_1, y_2)}{n(y_1) n(y_2)} \ , \qquad (13.6)$$

where $y_{1(2)}$ is the rapidity of the first (second) strange hadron, N_{had} the total number of hadronic events, and n refers to the number of events with one or two hadrons found in a specific rapidity range. For uncorrelated pairs we have $C(y_1, y_2) = 1$. The normalization is measured by taking hadrons from different events which are definitely uncorrelated.

The measurement shows that a strong positive short-range correlation is present for $\Lambda\Lambda$ pairs. Similar measurements for $K_S^0 K_S^0$ and ΛK_S^0 pairs show a weaker positive correlation at small and very large rapidity differences, and for $\Lambda\Lambda$ pairs a short-range anticorrelation is observed. The cluster hadronization model overestimates the short-range correlation, whereas the Lund string model with a fraction of 50% of all baryons produced by the popcorn mechanism describes the data well. Increasing this fraction, which reduces the correlation, or switching the popcorn mechanism off leads to a somewhat worse agreement with data, but the rather large errors prevent us from making more quantitative statements. Measurements of the overall production rates of strange hadrons confirm the preference for a 50% popcorn probability.

In summary, we can state that the expected short-range correlations for strange hadrons have been observed in the experiments, and that something like the popcorn mechanism, which in general decreases the correlations, is needed in order to obtain a good description of the data.

13.3.3 *Bose–Einstein correlations*

From quantum mechanics we known that a system of identical particles with integer spin (bosons) has to obey Bose–Einstein statistics, that is to say, the wave function of the system has to be symmetric under the exchange of any two of the particles. Hadronic final states are made out of many bosons (mesons) and fermions (baryons). In particular, pions and kaons are produced very copiously, and usually their momenta are measured rather precisely. So, we might wonder if it is possible to observe effects predicted by statistical quantum mechanics in high energy collisions.

Indeed, when looking at the two-particle differential cross section for like-sign pion production (identical bosons), an enhancement with respect to the unlike-sign cross section is observed when the two pions have similar momenta, as predicted by Bose–Einstein statistics. This was first reported experimentally when studying pairs of charged pions produced in $p\bar{p}$ collisions (Goldhaber *et al.*, 1960). At LEP these effects have been studied in detail, since several million hadronic Z decays are at hand, with about 17 charged pions and a few kaons per event.

In the solutions of Ex. (13-4) and Ex. (13-5) it is shown that the two-pion correlation function can be cast into the form

$$C(1,2) = 1 + \lambda|\tilde{\rho}(\Delta\boldsymbol{k})|^2 . \tag{13.7}$$

The correlation function is usually defined as the ratio of two-particle cross sections for like-sign and unlike-sign particle pairs. The function $\tilde{\rho}$ is proportional to the Fourier transform of the spatial density distribution $\rho(\boldsymbol{x})$ of the pion source, and $0 \leq \lambda \leq 1$ is a measure of the degree of coherence in the source, with $\lambda = 0$ for a completely coherent source, and $\lambda = 1$ for a chaotic source. How this, at first glance, counter-intuitive assignment comes about is shown in the exercises.

In the discussion of the exercises, $C(1,2)$ is derived as a function of Δk only, the difference of the three-momenta of the two bosons. More generally, it should be a function of Δk and the energy difference ΔE, which would measure the time structure of the source. However, in the experimental studies the simplification $C(1,2) = f(Q^2)$ has been adopted, where Q^2 is the Lorentz invariant four-momentum difference of the pions, $Q^2 = (\Delta k)^2 - (\Delta E)^2$. This simplifies the measurements, but makes the interpretation of the results less straightforward. As shown in the discussion of Ex. (13-6), the correlation function can be written as

$$C(1,2) = 1 + \lambda e^{-(\Delta k)^2 \sigma^2} \tag{13.8}$$

if we assume a spherically symmetric source with a Gaussian form,

$$\rho(x) = \frac{1}{(\sqrt{2\pi}\sigma)^3} e^{-\frac{r^2}{2\sigma^2}} , \tag{13.9}$$

where $r^2 = x^2 + y^2 + z^2$. Generalizing this result to

$$C(1,2) = 1 + \lambda e^{-Q^2 \sigma^2} , \tag{13.10}$$

the interpretation of σ as the spatial extension of the source is less obvious. The ansatz $C(1,2) = f(Q^2)$ is valid only for a very restricted class of sources, such as those for which $\rho(x) = \rho(t^2 - x^2)$. For more realistic sources we also expect more complicated space–time distributions. For example, in the case of an expanding source the specific pion pairs will mainly probe the region of the source which, for a certain time period, expanded parallel to the direction of their momenta, and the extracted size parameter cannot give information about the overall size and shape of the source. Finally, if the model for the source distribution is not appropriate, the interpretation of the experimentally measured parameter λ as an indicator for coherence might also be wrong. A very thorough discussion of these problems can be found in the article by Bowler (1985).

So much about the limitations of the physics interpretation of the measurements of Bose–Einstein effects. Since practically all experimental tests are based on the model of eqn (13.10), we are going to stick to it for further discussions. Later, we will see that these measurements have phenomenological implications, so they are nevertheless of importance.

Experimentally, a quantity

$$r_{+-}(Q) = \frac{N_{++}(Q)}{N_{+-}(Q)} \tag{13.11}$$

can be measured, where $N_{++}(Q)$ ($N_{+-}(Q)$) is the number of like-sign (unlike-sign) charged particle pairs as a function of the four-momentum difference. This quantity should give a good approximation of the correlation function $C(1,2)$. Measurements have been performed for $\pi^\pm\pi^\pm$, $K^\pm K^\pm$ and $K_S^0 K_S^0$ pairs, as well as for $\pi^\pm\pi^\pm\pi^\pm$ and $\pi^+\pi^-\pi^\pm$ triplets, by extending the definition of $r(Q)$ accordingly. A review may be found in Knowles and Lafferty (1997) and references therein.

The major experimental problem is to define a reference sample which is completely free of Bose–Einstein correlations. Using unlike-sign charged pion pairs might give a good approximation, but the production of pions during the hadronization phase most probably does not occur independently pion per pion. Therefore, even unlike-sign pairs can be affected by Bose–Einstein correlation effects. A solution to this problem is obtained by constructing pairs of pions which come from different events, thus being, for sure, uncorrelated. Then the measured quantity is

$$r_{\mathrm{mix}}(Q) = \frac{N_{++}(Q)}{N_{\mathrm{mix}}(Q)}, \qquad (13.12)$$

where $N_{\mathrm{mix}}(Q)$ is the number of event-mixed pairs as a function of Q. Care has to be taken in order to measure the momenta of the pions from different events with respect to a well defined common reference frame, such as the one given by the eigenvectors of the Sphericity tensor.

At this stage there might still be some correlations present, which arise from energy–momentum conservation or resonance decays. If a good Monte Carlo simulation of the overall properties of hadronic final states is available, then a double ratio

$$R_{+-}(Q) = \frac{r^{\mathrm{data}}_{+-}(Q)}{r^{\mathrm{MC}}_{+-}(Q)} \qquad \text{or} \qquad R_{\mathrm{mix}}(Q) = \frac{r^{\mathrm{data}}_{\mathrm{mix}}(Q)}{r^{\mathrm{MC}}_{\mathrm{mix}}(Q)} \qquad (13.13)$$

can be measured. Here r^{MC} is the ratio measured with the simulated events. If this simulation correctly accounts for all physics effects apart from Bose–Einstein correlations, then by construction the above double ratios should be close to 1, apart from a possible enhancement at low Q due to Bose–Einstein correlations.

In Fig. 13.3 a measurement of these double ratios by the ALEPH collaboration (1998a) is shown. Here, the correlation functions are first measured for charged particle pairs, and then corrected for the pion purity and for residual Coloumb repulsion/attraction effects. A clear enhancement is observed below $Q = 0.5\,\mathrm{GeV}$, which is interpreted as evidence for Bose–Einstein correlations. In the distribution for $R_{+-}(Q)$ some residual effects from resonances (for example, around $Q = 0.75\,\mathrm{GeV}$) can be seen. Furthermore, $R_{\mathrm{mix}}(Q)$ is not flat at large Q as expected, and the maximal enhancement is not the same as in $R_{+-}(Q)$. These problems are probably due to an inadequate Monte Carlo simulation. It does not simulate final state strong interactions, and not all particle production rates and resonance shapes are well modelled. The imperfections are accounted for by the systematic uncertainties on the final measurement results.

Finally, the measured two-particle correlation is fitted with a function as given by eqn (13.10). The fit results for σ are converted to a dimension of length, using the conversion factor $\hbar c = 0.197\,\mathrm{GeV\,fm}$. The results are typically within

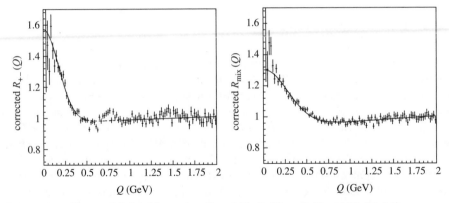

FIG. 13.3. Corrected double ratios $R_{+-}(Q)$ (left) and $R_{\text{mix}}(Q)$ (right), as measured by the ALEPH experiment (ALEPH Collab., 1998a). The curves represent fits of a model based on a Gaussian spherically symmetric source.

the range 0.5 - 0.7 fm, which is of the order of the proton size, $\mathcal{O}(1\,\text{fm})$. This is also the relevant scale for non-perturbative strong interactions. The results for the chaoticity parameter λ vary much more strongly, and for some measurements they are even larger than 1. This indicates shortcomings of the fitted model and/or of the experimental technique to obtain a correlation-free reference sample.

As already outlined above, the interpretation of the fitted parameters is not completely unambiguous, but the observation of the Bose–Einstein correlations has phenomenological implications. For example, in hadronic Z decays these correlations, which in principle are relevant only for identical particles, lead to a general collimation of the jets. All particles are brought closer together in momentum space, which then leads to distortions in the $\pi^+\pi^-$ mass spectra, especially at low momentum where the multiplicity is largest. If this effect is not taken into account, then the simulation of $\pi^+\pi^-$ resonances such as the $\rho(770)^0$ and the $f_0(982)$ will not be able to describe the measured shapes.

Lately, another implication of Bose–Einstein correlations has gained theoretical and experimental interest. In the process $e^+e^- \to W^+W^- \to$ hadrons at LEP2 correlations might not only arise between pions stemming from the same W, but also between pions coming from different Ws. Such an effect could result in a shift of the reconstructed W mass in multihadronic W decays, because the reconstructed jet momenta are distorted; see, for example, Lönnblad and Sjöstrand (1998) and references therein. Since there is no unambiguous experimental proof yet for the existence of correlations between pions from different W bosons, the measured W mass is assigned a systematic uncertainty of the order of 25 MeV. Yet another effect, related to strong interactions in hadronic final states, which could possibly influence the measurement of the W mass, is discussed in the following section.

13.4   Colour reconnection

Throughout most of the book we have discussed the production of hadrons by the separation of colour charges within a colour singlet system, such as $Z \to q\bar{q} \to$ hadrons. However, an interesting situation arises when two colour singlet systems decay close-by in space–time. There the question arises of how the radiation from the two colour singlets is affected by the overlap of the two systems, and at what level, perturbative and/or non-perturbative, the distribution of the final state hadrons is determined. Such situations can arise in weak decays such as $B \to J/\Psi + X$, in $Z \to q\bar{q}$ events provided that there are two perturbative $g \to q\bar{q}$ vertices in the parton shower before string fragmentation or at least three clusters in the case of cluster fragmentation, in W-pair production in e^+e^- annihilation, where both Ws decay hadronically into quark pairs, as well as in $e^+e^- \to ZZ$, $e^+e^- \to ZH$, $pp/p\bar{p} \to W^\pm H$, $H \to W^+W^-$, $t\bar{t} \to bW^+\bar{b}W^-$, etc.

The case which recently has been studied in most detail is the one of fully hadronic W decays at LEP, $e^+e^- \to W^+W^- \to q_1\bar{q}_2 q_3\bar{q}_4$. In these processes a reconfiguration of the colour systems could lead to a change in the spatial and momentum distribution of the final state hadrons, which could ultimately lead to a bias in the measured W mass, if these effects are not reproduced by the Monte Carlo simulations used to calibrate the measurement. This problem has been addressed for the first time by Gustafson *et al.* (1988). A very detailed discussion of perturbative as well as non-perturbative aspects followed, in the article by Sjöstrand and Khoze (1994). Here we only give a brief summary of the findings therein. As a starting point we know that the perturbative parton emission is initiated by two separate colour dipoles, $\widehat{q_1\bar{q}_2}$ and $\widehat{q_3\bar{q}_4}$. Because of their large width, $\Gamma_W = 2.09$ GeV, we expect the Ws to decay at an average distance of $\mathcal{O}(0.05\text{ fm})$ from their production point at LEP energies of about 200 GeV. Therefore, energetic gluons with $E_g \gg \Gamma_W$ have wavelengths much smaller than the separation of the two decay vertices, and they should be emitted almost incoherently from the two dipoles. A spontaneous rearrangement of colour dipoles into $\widehat{q_1\bar{q}_4}$ and $\widehat{q_3\bar{q}_2}$ should be suppressed by $1/N_c^2 = 1/9$, from simple counting of possible colour states. Nevertheless, if it happens, we typically expect an enhancement of particle production between the jets from different Ws.

Strictly speaking, a spontaneous rearrangement of colour dipoles is not what is meant by colour rearrangement in the framework of perturbative QCD. There, interference effects between the two radiating systems have to be calculated. In Sjöstrand and Khoze (1994) it is shown that such interference effects can only occur at $\mathcal{O}(\alpha_s^2)$, that is, when a gluon is radiated from both colour dipoles formed by the original quark pairs. The detailed calculation shows that such effects are suppressed by $\alpha_s^2/(N_c^2-1)$. In addition, because of the finite separation of the two W bosons, only contributions from soft gluons with $E_g \lesssim \Gamma_W$ are to be expected, which means that only a few low-energy particles should be affected, changing the multiplicity by $\Delta N^{\text{recon}}/N^{\text{no-recon}} \lesssim \mathcal{O}(10^{-2})$. The effect on the total cross section for W pair production is estimated to be even smaller than that. So it is concluded that colour reconnection at the perturbative level should be negligible.

The situation is different when going to the hadronization phase, which occurs at distance scales of the order of 1 fm, much larger than the typical separation of the decaying Ws. Therefore, larger effects could be expected. The problem is that for this level only phenomenological models exist, rather than calculations from first principles. For example, within the picture of the Lund string model, we know that at the end of the perturbative phase, which governs the showering of partons, strings are spanned within the original colour dipoles, with possible kinks due to radiated gluons. Now we could imagine that the original string configuration is changed if there is a substantial overlap of the strings (or colour fields) from the two dipoles. Hadron production then occurs from the two rearranged strings, leading to a change in the final state as discussed above.

Sjöstrand and Khoze (1994) have studied the phenomenological consequences of different models, depending on the detailed space–time description of the string. The probability for string rearrangement may be completely fixed by the model, or may be a free parameter to be determined from the data. Hadron distributions such as rapidity or multiplicities are changed more or less strongly by different models. In any case, it turns out that the finite detector resolution and acceptance tend to wash out the effects. In addition, string overlap may be substantial only for a particular subset of all possible WW decay topologies, which results in statistical limitations.

As an example, the four LEP experiments have measured the charged particle multiplicity in fully hadronic W decays and subtracted twice the multiplicity from semi-leptonic W events. For the latter no colour reconnection can occur. Within their uncertainties the differences are found to be consistent with zero, that is, from the data there is no indication for sizeable colour reconnection effects. However, the errors are too large to really discriminate between models without reconnection and models with reconnection, which predict only a small effect for soft particles. In addition, this soft region is the one most affected by detector effects. A more promising approach could be the study of particle flow between jets assigned to the same W and between jets from different Ws. There, observables can be constructed which appear to be more discriminating than the charged particle multiplicity. These studies are ongoing. For an overview of the present situation we refer the interested reader to the paper by Jong (2001).

The most important effect is seen on the measurement of the W mass. From measurements based on Monte Carlo models with or without colour reconnection a systematic uncertainty of $\Delta M_W = 40$ MeV is assigned, which dominates the current uncertainty of the LEP measurement from fully hadronic decays of $\Delta M_W(\text{tot}) = 62$ MeV and $\Delta M_W(\text{stat}) = 31$ MeV (Barberio et al., 2001). A reduction or at least a solid understanding of this uncertainty could be obtained by finding other observables which clearly discriminate between different reconnection scenarios, or by restricting the mass measurements to topologies and/or reconstruction methods which are the least sensitive to possible effects. An indication that reconnection effects might not be large, after all, is given by the fact that the difference of the W masses measured from fully hadronic and semileptonic decays is consistent with zero, $\Delta M_W(q\bar{q}q\bar{q} - q\bar{q}\ell\bar{\nu}_\ell) = 18 \pm 46$ MeV (Barberio et al., 2001).

Exercises for Chapter 13

13–1 Confirm that if $D(x) = d_{PT} \otimes d_{NP}$ then $\langle x \rangle = \langle x \rangle_{PT} \langle x \rangle_{NP}$.

13–2 The popular Peterson et al. fragmentation function, eqn (13.2), can be derived from the independent fragmentation model by applying old fashioned perturbation theory to the process $Q \rightarrow (Q\bar{q}) + q$, where the momentum p of the heavy quark Q is shared between a heavy meson $Q\bar{q}$ and a leftover quark q. With ΔE the energy used for the formation of the meson and the leftover quark this yields

$$d_{NP}(x) \propto (\text{phase space}) \frac{|\langle h_{Q\bar{q}} + q | \mathcal{H} | Q \rangle|^2}{(\Delta E)^2} .$$

Given that the (longitudinal) phase space is proportional to $1/x$ and assuming that the matrix element is constant, derive eqn (13.2) and show that $\epsilon_Q = R^{-2}/m_Q^2$.

13–3 Use eqns (3.342) and (3.343) to calculate for symmetric $\gamma^* \rightarrow q\bar{q}g$ and $\gamma^* \rightarrow q\bar{q}\gamma$ events the relative probabilities for a soft gluon to be produced directly between the q and the $\bar{q}$, directly between the q and the g or γ and perpendicular to the event plane.

13–4 Show that the two-particle correlation function for two identical bosons (for example pions) behaves like $1 + \cos(\Delta\boldsymbol{k} \cdot \Delta\boldsymbol{x})$, where $\Delta\boldsymbol{k}$ ($\Delta\boldsymbol{x}$) is the difference of their momenta (space coordinates). Use the plane wave approximation for the particle wave functions. Then introduce a spatial density distribution $\rho(\boldsymbol{x})$ for the source of pion production, and show that in this case the Bose–Einstein enhancement effect in the correlation function is proportional to the square of the Fourier transform $\hat{\rho}(\Delta\boldsymbol{k})$ of this source distribution. Some hints towards the solution may be found in the article by Bowler (1985). (★)

13–5 Discuss the implication for Bose–Einstein correlations of coherence between pion sources, and show that the correlation function is enhanced for randomly fluctuating sources. (For some hints see, again, Bowler, 1985.)

13–6 Derive the two-particle correlation function for a spherically symmetric Gaussian source (in three space dimensions), using the results of Ex. (13-4).

14

SUMMARY

In the chapters of this book we have tried to give an overview how QCD evolved from the early beginnings in hadron spectroscopy to the precision measurements in hard interactions that became available with high energy collider experiments. Today QCD is a well established part of the Standard Model of elementary particle physics, experimentally tested with a precision at the per cent level for many results.

A major problem and challenge in the study of QCD is the confinement property, that is, colour charges are not observed in asymptotic free particle states. Nevertheless, it has been possible to probe the structure of the QCD Lagrangian with great accuracy, showing that strong interactions are indeed described by a non-abelian gauge theory based on an $SU(3)$ symmetry. The consequences for hadron production processes in high energy interactions as studied at all types of high energy colliders are well understood. Today multi-hadron production is understood to originate from a coherent parton showering process.

Still, many open questions remain, in particular in the low energy regime where the strong coupling becomes so large that perturbation theory breaks down. This field could be treated only very superficially in this book, but it has to be emphasized that also here significant progress has been made in the past years. The field is still rapidly evolving, closely coupled to progress in analytical as well as numerical methods and high performance computing. So, despite the fact that this monograph has become a rather large volume, there are still many subjects that could not be covered.

We nevertheless hope that we have managed to give the reader some insight into how QCD evolved to its current status and how it can be tested in high energy reactions between elementary particles. Historically we are currently in a transition phase. The LEP programme has finished and new information from e^+e^- annihilations has to await the startup of a proposed linear collider or another, similar, facility. For the near future we can expect another set of precision measurements of the nucleon structure function from lepton–nucleon scattering experiments at DESY and CERN. These experiments will perform high-accuracy measurements of the nucleon structure functions, including the spin structure, which will allow a reliable extrapolation to energies in the multi-TeV range. Thus they will provide important input for a quantitative understanding of the physics at the Large Hadron Collider, LHC, at CERN.

QCD is and will remain a very active field of research, and we are looking forward to new measurements and theoretical developments, which will shed further light on the structure of baryonic matter and the interactions between its constituents. The spectrum ranges from studies of bound states at low energy scales, over few-body interactions at high energies, to many-body systems at the extreme conditions of a quark–gluon plasma, as most likely existed in the early universe. Today, the foundations for understanding strong interactions are firmly established, but that is certainly not the end of the story. Much exciting physics still lies ahead.

APPENDIX A

ELEMENTS OF GROUP THEORY

Group theory can be viewed as the mathematical description of symmetries. Since symmetries often play a key rôle in physics, group theory enters into the discussion in many places. Here we will give a short summary of the most important aspects of group theory, with particular emphasis on Lie groups. For further studies the reader is referred to dedicated textbooks (Jones, 1990; Tung, 1985; Cahn, 1984) and references given therein.

A.1 Basics of Lie groups

A group, G, is a set of elements, $\{A, B, C, \ldots\}$, together with a binary operation, the group multiplication $\circ$, which satisfy the following requirements: the product of any two elements is again a member of the group, a property known as closure,

$$A \circ B = C \in G, \tag{A.1}$$

the multiplication is associative,

$$A \circ (B \circ C) = (A \circ B) \circ C, \tag{A.2}$$

there exists a unique group element, the identity E, which leaves all group elements invariant under the group multiplication

$$A \circ E = E \circ A = A, \tag{A.3}$$

and for each group element A there exists a unique inverse A^{-1}, such that

$$A \circ A^{-1} = A^{-1} \circ A = E. \tag{A.4}$$

Note that in general the multiplication is not commutative, that is, $A \circ B \neq B \circ A$. Only for the special case of *abelian* groups one has $A \circ B = B \circ A$.

Two groups are isomorphic to one another if there exists a one-to-one mapping f between them, which preserves the group's product structure: $f(A) \circ f(B) = f(A \circ B)$. An isomorphism which maps the group elements onto square matrices and the group multiplication onto ordinary matrix multiplication is called a representation of the group. From the general properties of the matrix multiplication it is evident that the defining properties of a group can be satisfied. In fact, it can be shown that infinitely many representations exist for every group, which differ in the dimensionality of the matrices. The lowest dimensional non-trivial representation is also referred to as the fundamental or defining representation of a group and denoted by the index F.

Interpreting matrices as transformations of a vector, for example, rotations or reflections, illustrates the connection between group theory and symmetry. It also shows how different representations of a group realize the same symmetry in different-dimensional spaces. A common example are transformations which preserve the orthonormality of a set of N basis vectors. Working on complex numbers, these are $N \times N$ matrices U, which satisfy $U^\dagger U = 1$. These matrices form the fundamental representation of the unitary group of dimension N, U(N). Using the unitarity constraint it is easy to see that $|\det(U)| = 1$; if in addition the matrices have $\det(U) = +1$, then we have the special unitary group, SU(N), a sub-group of U(N). Real $N \times N$ matrices O satisfying $O^T O = 1$ form the orthogonal group of dimension N, O(N), and if in addition $\det(O) = +1$ the special orthogonal group, SO(N).

The special unitary and orthogonal groups are examples of so-called *Lie groups*. These are defined as continuous groups, in the sense that it is possible to parameterize each group element in terms of a set of real parameters $\theta_a, a = \{1, 2, \ldots\}$, in such a way that the product of two group elements is parameterized by an analytic function of their parameters. The number of parameters is specific to a given group. A familiar example is the Euler-angle parameterization of the spatial rotations group SO(3), which has three parameters.

If we arrange that $\theta_a = 0$ parameterizes the identity and introduce transformations $T^a, a = \{1, 2, \ldots\}$, associated with the individual parameters, the so-called *generators* of the group, then the continuity property allows group elements in the neighbourhood of the identity to be written as

$$U(\delta\theta) = 1 + i\,\delta\theta_a T^a + \mathcal{O}(\delta\theta^2)\,. \tag{A.5}$$

Note that here and in the following, even if not explicitly written, summation over identical indices is implied. The factor $i = \sqrt{-1}$ has been introduced for later convenience. By applying a sequence of infinitesimal steps we can write a 'macroscopic' group element as

$$U(\theta) = \lim_{n \to \infty} \left(1 + i\,\frac{\theta_a}{n} T^a\right)^n = e^{i\theta_a T^a}\,. \tag{A.6}$$

Thus, within a connected group, each element can be reached from the identity and written as an exponential of a linear combination of the generators. All information about the structure of the group and its symmetry can be extracted from the generators. This is an enormous simplification for the study of the properties of the group, since one only has to deal with a finite number of generators rather than with a continuum of group elements.

An open question at this point is the actual choice and normalization of the generators. Evidently, given one set of generators the group can equally be constructed from any linear combination of those generators with suitably transformed parameters. While the group as such is invariant with respect to the actual choice for the generators, the study of the group properties is greatly simplified when using a set of generators which most clearly reflects the internal symmetries of the group.

As a heuristic argument, consider the change of an, in general complex valued, vector v under an infinitesimal transformation with any of the generators T^a, given by $dv^a = \delta T^a v$, with δ an infinitesimal global scale parameter. A measure for the total change of v is the scalar product $dv^{a\dagger} dv^a$, which for a good choice of the generators should on average, that is, when integrating over all vectors v, be the same for all generators. Likewise, the scalar product of changes from independent generators, $dv_a^\dagger dv^b$, should vanish on average. Formally these conditions translate to

$$\langle (T^a v)^\dagger (T^b v) \rangle \sim \int d^N v \ \mathrm{Tr}\left\{ vv^\dagger T^{a\dagger} T^b \right\} \sim \mathrm{Tr}\left\{ T^{a\dagger} T^b \right\} = T_R \delta^{ab} . \qquad (A.7)$$

The first proportionality is simply the definition of the average, the second one holds because the integral over the direct product of two vectors is proportional to the unit matrix when integrating over the entire space. By construction $M^{ab} = \mathrm{Tr}\left\{ T^{a\dagger} T^b \right\}$ is a positive-demidefinite hermitian matrix. Hermiticity follows from

$$\left(M^{ba} \right)^\star = \mathrm{Tr}\left\{ T^{b\dagger} T^a \right\}^\star = \mathrm{Tr}\left\{ T^{b\dagger} T^a \right\}^\dagger = \mathrm{Tr}\left\{ T^{a\dagger} T^b \right\} = M^{ab} , \qquad (A.8)$$

where the star denotes complex conjugation, and positive-semidefiniteness from the fact that the quadratic form

$$\sum_{ab} v_a^\dagger M^{ab} v_b = \sum_{ab} \mathrm{Tr}\left\{ (v_a^\dagger T^{a\dagger})(T^b v_b) \right\} = \mathrm{Tr}\left\{ S^\dagger S \right\} \quad \text{with} \quad S = \sum_a T^a v_a$$

is non-negative for every non-zero vector v. Thus, it is always possible to satisfy eqn (A.7) by finding an appropriate transformation of a given set of generators. The proportionality constant T_R, the so-called Dynkin-index, is a normalization parameter that can be chosen freely for one representation of the group. It is then fixed for all other respresentations.

In terms of the exponentiated form, the product of two group elements is given by the Baker–Campbell–Hausdorff formula. As is easily demonstrated by expanding the exponentials to second order and not assuming commutativity of the factors one obtains

$$\exp(i\,\theta) \exp(i\,\phi) = \exp\left\{ i\,(\theta + \phi) - \frac{1}{2!}[\theta, \phi] + \cdots \right\} . \qquad (A.9)$$

Here θ and ϕ are linear combinbations of the generators with real coefficients. The commutator, $[\theta, \phi] = \theta\phi - \phi\theta$, on the right-hand side of eqn (A.9) takes care of the fact that in general the product of two generators does not commute. In order that the group closes, the right-hand side of eqn (A.9) must be expressible in the form of eqn (A.6), which constrains the commutator of any two generators to be a linear combination of the group's generators with real-valued coefficients f^{abc},

$$[T^a, T^b] = \mathrm{i} f^{abc} T^c . \tag{A.10}$$

The coefficients f^{abc} are called *structure constants* and serve to define the Lie algebra of the group's generators. If the f^{abc} vanish, then the generators and hence group elements commute and the group is abelian. Expanding out the commutators

$$[[T^a, T^b], T^c] + [[T^b, T^c], T^a] + [[T^c, T^a], T^b] = 0 , \tag{A.11}$$

one sees that the structure constants satisfy the Jacobi identity

$$f^{abd} f^{dce} + f^{bcd} f^{dae} + f^{cad} f^{dbe} = 0 . \tag{A.12}$$

By construction, the structure constants are manifestly antisymmetric in the first two indices. If the group is semi-simple and compact, then one can also find a basis for the generators where the structure constants are completely antisymmetric in all indices. For an in-depth discussion of these concepts the reader is referred to the literature. Here, we will only show how this property comes about for some groups that frequently appear in physics, such as SU(N) or SO(N).

The generators of these groups are also known as quantum mechanical operators acting on state vectors and thus are associated with physical observables. Therefore, they are represented by hermitian matrices, the eigenvalues of which correspond to observable quantum numbers. For hermitian matrices, which satisfy $T^{a\dagger} = T^a$, eqn (A.7) becomes $\mathrm{Tr}\left\{T^a T^b\right\} = T_R \delta^{ab}$, one obtains from eqn (A.10)

$$\mathrm{Tr}\left\{[T^a, T^b] T^c\right\} = \mathrm{Tr}\left\{\mathrm{i} f^{abd} T^d T^c\right\} = \mathrm{i} f^{abd} T_R \delta^{dc} \tag{A.13}$$

and thus, expanding the commutator in eqn (A.13)

$$f^{abc} = -\frac{i}{T_R} \mathrm{Tr}\left\{T^a T^b T^c - T^b T^a T^c\right\} . \tag{A.14}$$

We see that with the choice of eqn (A.7) for the generators the structure constants form a completely antisymmetric tensor.

It is clear that there is a very close relationship between the generators of a group and their Lie algebra, and the structure of the group itself. In particular, if we can understand the action of the generators on the group's basis states, we can understand the action of the group itself. Therefore, we concentrate on investigating the matrix representations of the Lie algebra. In many applications to high energy physics, wave functions of physical states are represented by vectors and the generators act as operators transforming those states.

The algebra defined by eqn (A.10) does not have a unique realization. Suppose we have an N-dimensional representation of the group generated by a set of $N \times N$

matrices $T^a(R)$ which satisfy eqn (A.10). If the generators are represented by hermitian matrices, then one easily sees that the generators $T^a(\bar{R}) = -T^{a\star}(R)$ also form a representation of the group. With $T^{a\star} = (T^a)^T$ and thus $T^a(\bar{R}) = -(T^a)^T(R)$ one finds

$$[T^a(\bar{R}), T^b(\bar{R})] = [T^b(R), T^a(R)]^T = i f^{bac}(T^c)^T(R) = i f^{abc}T^c(\bar{R}) . \quad (A.15)$$

Note that the minus sign in the definition of the generators $T^a(\bar{R})$ results in states transforming according to the complex conjugate representation having the opposite sign eigenvalues as the states transforming under $T^a(R)$. Thus, if R is used for particle fields, then $\bar{R}$ is the natural representation for the antiparticles.

If the two representations $T^a(R)$ and $T^a(\bar{R})$ are equivalent, that is, if a nonsingular matrix S exists such that $S T^a(R) S^{-1} = T^a(\bar{R})$ for all a, then it is called a real representation. For example, all representations of SU(2) are real; a fact which relies on the eigenvalues of any of the generators occurring in opposite sign pairs $\pm\lambda$. On the other hand, the fundamental representation of SU(3) and its complex conjugate are not equivalent. In flavour SU(3) the quarks transform under the former, the antiquarks under the latter representation.

An important realization of the algebra is the adjoint representation A, defined through the matrices

$$[T(A)^a]_{bc} = i f^{abc} . \quad (A.16)$$

That these matrices satisfy eqn (A.10) follows straightforwardly from the Jacobi identity eqn (A.12). The dimension of the adjoint representation is the same as the number of real parameters required to specify an element of the group. As such, it also becomes the representation under which the gauge fields transform in a gauge theory.

An important quantity to characterize a representation R of a group is the eigenvalue C_R of the quadratic Casimir operator,

$$T^2(R) = \sum_a T^a(R) T^{a\dagger}(R) . \quad (A.17)$$

For the special case of hermitian generators the value of C_R can be determined by simply evaluating the sum in eqn (A.17). As shown below, since that sum commutes with all generators, and thus also with all group elements, it is by Schur's first lemma proportional to the identity matrix. For the commutator $[T^2, T^b]$ of the Casimir operator with an arbitrary generator one finds

$$\sum_a [T^a T^a, T^b] = \sum_a T^a[T^a, T^b] + [T^a, T^b]T^a = \sum_{a,c} i f^{abc}(T^a T^c + T^c T^a). \quad (A.18)$$

Since the f^{abc} are completely antisymmetric in all indices, whilst the term in parenthesis is symmetric under exchange of a and c, the sum vanishes and T^2 has to be proportional to the unit matrix.

In the context of QCD the eigenvalues C_F and C_A associated with the fundamental and the adjoint representation, respectively, are also known as colour factors. An important relation is obtained in connection with eqn (A.7). Comparing the definition of the Dynkin index T_F and the eigenvalue C_F, both for the fundamental representation of the group, one has

$$\sum_{ij} T^a_{ij} T^{b\dagger}_{ji} = T_F \delta^{ab} \quad \text{and} \quad \sum_{a,j} T^a_{ij} T^{a\dagger}_{jk} = C_F \delta_{ik}. \tag{A.19}$$

Summing over all indices, both expressions yield the same result, and one obtains

$$T_F \cdot N_A = C_F \cdot N_F , \tag{A.20}$$

where N_A and N_F are the dimensions of the adjoint and the fundamental representations, respectively.

In practice we often encounter situations where two particles, which transform according to representations R_1 and R_2 of the same group, form a combined system which also exhibits the symmetry of the group. The generators acting on the combined system are then given by the tensor product

$$T^a_{R_1 \otimes R_2} = T^a_{R_1} \otimes \mathbf{1}_{R_2} + \mathbf{1}_{R_1} \otimes T^a_{R_2} . \tag{A.21}$$

In the product-terms the first factor always acts on the particle transforming according to R_1 and the second one on the other particle. It is immediately seen that the quantum numbers of such a system, that is, the eigenvalues of the generators that are simultaneously diagonalizable, are the sums of the quantum numbers of the constituents.

This makes it simple to construct the so-called *root diagram* of a multi-particle system, which displays the elements of a representation in a picture using the eigenvalues of the simultaneously diagonalizable generators, that is, the quantum numbers of the states, as orthogonal axes. An example is given in Fig. A.1 for the combination of an SU(3) triplet and antitriplet, that is, quark and antiquark. The possible quark states are represented by the corners of the downward pointing triangle, the antiquark states are obtained by its mirror image under reflection at the horizontal axis. Shifting the centre of the quark triangle to the corners of the antiquark triangle then yields a figure whose corners define all possible quark–antiquark states.

A deeper result which will not be derived here, but whose essence should be familiar from the Clebsch–Gordan series of angular momentum, is that the tensor products can be reduced into closed subsets of elements which transform into one another under the action of the group, but which do not involve the remaining elements. This decomposition into so-called *irreducible representations* is depicted on the right-hand side of Fig. A.1. The dark dots show the quantum numbers of the octet, the open dot in the centre is a singlet state which is invariant under SU(3) transformations. Note that at the centre of the root diagram there are three states built from the same quark flavours, only two of which

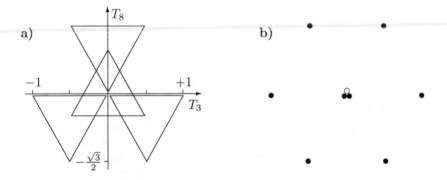

FIG. A.1. The root diagram for the addition of the triplet and antitriplet of SU(3) multiplets and its decomposition into irreducible representations

also have the same quantum numbers. Irreducible respresentations are associated with physical systems. In the above example, the black dots are associated with the meson octet. As a further example, the baryon decuplet can be constructed in a similar way as two of the irreducible representations of the tensor product of three quarks.

A.2   The U(N) and SU(N) groups

Since unitary groups provide the basis for the gauge theories of the Standard Model we consider them in a little more detail. As already mentioned, the defining representation for the unitary group U(N) consists of $N \times N$ complex matrices satisfying

$$U^\dagger U = 1 \quad \rightarrow \quad |\det(U)| = 1 \ . \tag{A.22}$$

A general complex $N \times N$ matrix has $2N^2$ real parameters. Since the unitarity condition imposes N^2 constraints, the number of real-valued parameters for U is only N^2. By applying a common phase factor $\exp\{i\,\theta_0\}$ we can arrange that $\det(U) = +1$. Imposing this extra condition gives the Special Unitary group, SU(N). Thus we can decompose the unitary group into a direct product of two groups, U(N)=U(1)⊗SU(N). In terms of generators we may write

$$U = \exp(i\,\theta_0 + i\,\theta_a T^a) \ , \tag{A.23}$$

where the index a runs over the $N^2 - 1$ generators of SU(N), equal to the number of real parameters for a member of SU(N). The U(1) and SU(N) components can be treated separately. Specializing to an infinitesimal θ, the SU(N) generators are easily seen to satisfy:

$$UU^\dagger = 1 + i\,\delta\theta_a(T^a - T^{a\dagger}) + \cdots = 1 \quad \rightarrow \quad T^a = T^{a\dagger} \tag{A.24}$$

and

$$\det(U) = \exp(\text{Tr}\{\ln U\}) = 1 + i\,\delta\theta_a \text{Tr}\{T^a\} + \cdots = 1 \quad \rightarrow \quad \text{Tr}\{T^a\} = 0 \,. \quad \text{(A.25)}$$

One sees that the generators are traceless hermitian matrices. If the factor i were not included in eqn (A.5), the generators would be anti-hermitian. Note that these constraints are sufficient to give a general description of how to build a set of generators for SU(N). The starting point is the set of Pauli matrices. The non-diagonal ones can be formed by putting a single 1 or an i into one position of the upper triangular matrix. Hermiticity then defines the entire matrix. This procedure already gives $N(N-1)$ traceless hermitian matrices, that is, one only needs to find another $N-1$ such matrices, to have all generators of SU(N). Those missing matrices can be chosen as diagonal matrices proportional to $\text{diag}(1, 1, \cdots, 1, -m, 0, \cdots, 0)$. Here the first m positions on the diagonal, $m = 1, \ldots, N-1$, are filled with unit elements. The next element is then set to $-m$ in order to have a traceless matrix, and the rest of the diagonal is filled with zeros. In the case of SU(3) the diagonal matrices constructed according to this scheme are $\text{diag}(1, -1, 0)$ and $\text{diag}(1, 1, -2)$. In general, for SU(N) we have $N-1$ generators which can be simultaneously diagonalized. In terms of physics this means that a quantum state with an SU(N) symmetry is characterized by $N-1$ quantum numbers.

In practical applications, such as the evaluation of Feynman diagrams, one is often faced with the problem of summing over representation matrices of a symmetry group. In many cases one has to deal with SU(N). Therefore, we list here a collection of identities valid for SU(N) (MacFarlane $et\ al.$, 1968), which are useful for performing such calculations. Note that the summation convention is used throughout.

Let T^a denote the generators for the fundamental respresentation of SU(N). The commutation relations are given by the totally antisymmetric structure constants f_{abc} of the group:

$$[T^a, T^b] = i\,f^{abc}T^c \tag{A.26}$$

A corresponding relation also exists for the anticommutator

$$\{T^a, T^b\} = \frac{1}{N}\delta_{ab}I_N + d^{abc}T^c \,, \tag{A.27}$$

with a totally symmetric tensor d^{abc}. Here, I_N is the N-dimensional unit matrix. For SU(2) one has $f^{abc} = \varepsilon^{abc}$ and $d^{abc} = 0$. Some useful formulae for the T^a are

$$T^a T^b = \frac{1}{2}\left[\frac{1}{N}\delta_{ab}I_N + (d^{abc} + if^{abc})T^c\right] \,, \tag{A.28}$$

$$T^a_{ij}T^a_{kl} = \frac{1}{2}\left(\delta_{il}\delta_{jk} - \frac{1}{N}\delta_{ij}\delta_{kl}\right) \,, \tag{A.29}$$

$$\text{Tr}\{T^a\} = 0 \,, \tag{A.30}$$

$$\text{Tr}\{T^a T^b\} = \frac{1}{2}\delta_{ab} \,, \tag{A.31}$$

$$\mathrm{Tr}\left\{T^a T^b T^c\right\} = \frac{1}{4}(d^{abc} + i f^{abc}) , \tag{A.32}$$

$$\mathrm{Tr}\left\{T^a T^b T^a T^c\right\} = -\frac{1}{4N}\delta_{bc} . \tag{A.33}$$

In addition one has the Jacobi identities

$$f^{abe} f^{ecd} + f^{cbe} f^{aed} + f^{dbe} f^{ace} = 0 , \tag{A.34}$$
$$f^{abe} d^{ecd} + f^{cbe} d^{aed} + f^{dbe} d^{ace} = 0 , \tag{A.35}$$

which can be written in a compact form by introducing the $(N^2 - 1)$-dimensional matrices $(F^a)_{bc} = -i f^{abc}$ and $(D^a)_{bc} = d^{abc}$ as

$$\left[F^a, F^b\right] = i f^{abc} F^c \quad \text{and} \quad \left[F^a, D^b\right] = i f^{abc} D^c . \tag{A.36}$$

Further useful identities involving products and traces are

$$f^{abe} f^{cde} = \frac{2}{N}(\delta_{ac}\delta_{bd} - \delta_{ad}\delta_{bc}) + (d^{ace} d^{bde} - d^{bce} d^{ade}) , \tag{A.37}$$

$$f^{abb} = \mathrm{Tr}\left\{F^a\right\} = 0 , \tag{A.38}$$

$$d^{abb} = \mathrm{Tr}\left\{D^a\right\} = 0 , \tag{A.39}$$

$$f^{acd} f^{bcd} = \mathrm{Tr}\left\{F^a F^b\right\} = N\delta_{ab} , \tag{A.40}$$

$$f^{acd} d^{bcd} = 0 , \tag{A.41}$$

$$d^{acd} d^{bcd} = \mathrm{Tr}\left\{D^a D^b\right\} = \frac{N^2 - 4}{N}\delta_{ab} , \tag{A.42}$$

$$F^a F^a = N I_{N^2 - 1} , \tag{A.43}$$

$$\mathrm{Tr}\left\{F^a F^b F^c\right\} = i\frac{N}{2} f^{abc} , \tag{A.44}$$

$$\mathrm{Tr}\left\{D^a F^b F^c\right\} = \frac{N}{2} d^{abc} , \tag{A.45}$$

$$\mathrm{Tr}\left\{D^a D^b F^c\right\} = i\frac{N^2 - 4}{2N} f^{abc} , \tag{A.46}$$

$$\mathrm{Tr}\left\{D^a D^b D^c\right\} = \frac{N^2 - 12}{2N} d^{abc} . \tag{A.47}$$

A.3 Colour factors

Colour factors are defined as the eigenvalues of the quadratic Casimir operator for a given representation, with the normalization usually fixed through the Dynkin index T_F of the fundamental representation. Below, we give a table of the ratios T_F/C_F and C_A/C_F, with C_F and C_A the colour factors of the fundamental and the adjoint representation, respectively, for all semi-simple Lie groups.

Group	T_F/C_F	C_A/C_F
SU(N)	$N/(N^2-1)$	$2N^2/(N^2-1)$
SO(N)	$2/(N-1)$	$(2N-4)/(N-1)$
Sp(2N)	$2/(2N+1)$	$(4N+4)/(2N+1)$
$U(1)_N$	N	0
G2	$1/2$	2
F4	$3/2$	2
E6	$18/13$	$9/26$
E7	$24/19$	$8/19$
E8	1	1

The groups SU(N) have been discussed in detail before. SO(N) are the rotation groups in N dimensions, and Sp(2N) the so-called symplectic groups. Since the generators of the latter are not represented by hermitian matrices, they are of lesser importance in the context of quantum field theory. $U(1)_N$ are the groups of phase transformations. All these groups come with an index N that defines the dimensionality of the fundamental representation. In addition to those, there are only five so-called *exceptional* groups G2, F4, E6, E7, E8, which are also listed in the above table.

APPENDIX B

BUILDING BLOCKS OF THEORETICAL PREDICTIONS

B.1 The Feynman rules of QCD

Feynman diagrams provide a very useful pictorial device in which each component of a diagram represents a part of the algebraic expressions for the corresponding S-matrix amplitude. When using the Feynman rules, to go from a diagram to an algebraic expression, you are advised to pay careful attention to the directions of the momenta and the order of any indices. These details are important because they impact on the relative signs of the various terms which in turn help to ensure gauge invariance.

External quarks and gluons correspond to basis spinors and polarization vectors as shown below. Ghosts are scalars and therefore have trivial unit basis states.

Internal particles correspond to propagators, which are colour diagonal, as shown. The sign of the infinitesimal, imaginary part $i\epsilon$ is chosen so as to ensure causality. In the case of the gluon the Lorentz tensor $d^{\mu\nu}(p)$ depends on the choice of the gauge fixing term and the gauge parameter ξ. Two common choices are:

$$d^{\mu\nu}(p) = \begin{cases} -\eta^{\mu\nu} + (1-\xi)\dfrac{p^\mu p^\nu}{p^2} & \text{covariant gauge,} \\ -\eta^{\mu\nu} + \dfrac{(p^\mu n^\nu + n^\mu p^\nu)}{n \cdot p} - (n^2 + \xi p^2)\dfrac{p^\mu p^\nu}{(n \cdot p)^2} & \text{physical gauge.} \end{cases}$$

$$(B.1)$$

In the physical gauges ghosts do not appear in the Feynman diagrams, but in the covariant gauges they are required in order to preserve unitarity; see Section 3.3.3.

Particle interactions are represented by vertices. We have the gluon–quark, gluon–ghost, triple-gluon and quartic-gluon vertices corresponding to the following algebraic factors:

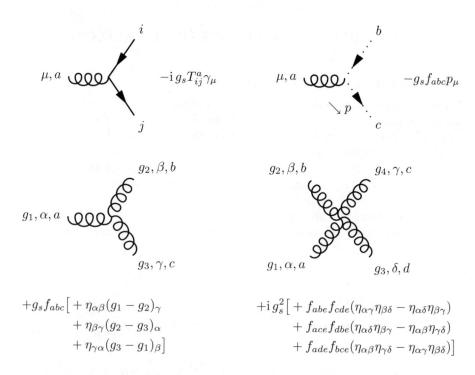

$$+g_s f_{abc}\big[+\eta_{\alpha\beta}(g_1 - g_2)_\gamma$$
$$+ \eta_{\beta\gamma}(g_2 - g_3)_\alpha$$
$$+ \eta_{\gamma\alpha}(g_3 - g_1)_\beta\big]$$

$$+\mathrm{i}\,g_s^2\big[+ f_{abe}f_{cde}(\eta_{\alpha\gamma}\eta_{\beta\delta} - \eta_{\alpha\delta}\eta_{\beta\gamma})$$
$$+ f_{ace}f_{dbe}(\eta_{\alpha\delta}\eta_{\beta\gamma} - \eta_{\alpha\beta}\eta_{\gamma\delta})$$
$$+ f_{ade}f_{bce}(\eta_{\alpha\beta}\eta_{\gamma\delta} - \eta_{\alpha\gamma}\eta_{\beta\delta})\big]$$

In all these graphs the convention is that all momenta are outgoing and so sum to zero. In the gluon–ghost vertex the momentum is that of the outgoing ghost. In the gluon self-couplings, observe that both vertices are symmetric upon interchange of all the labels on any pair of legs. Implicit in each of these vertices is a four-momentum conserving δ-function. Each internal line is accompanied by an integral over its four-momentum. This results in an overall four-momentum conserving δ-function which is absorbed into the phase space definition.

Since quarks and ghosts are fermionic, for every closed loop involving them in a diagram, an additional factor -1 must be included. Furthermore, when a pair of identical fermions is present in the external state of two diagrams, then the 'crossed' diagram acquires a minus sign relative to the 'uncrossed' diagram, again to account for the anticommutativity of fermions. Finally, when n identical particles are present in the final state of a set of diagrams, then a symmetry factor $1/n!$ must be included in the amplitude.

Note that we have suppressed the spinor indices on the quark propagators and vertices. The correct ordering of these terms is given by working backwards along the individual fermion lines. This prescription will also give the correct ordering of any colour matrices.

For completeness we also include some relevant Feynman rules from the standard electroweak theory. Here the propagators for the vector bosons, $V = \gamma$, $W^{\pm}$ or Z, are given by

$$\frac{i}{p^2 - M_V^2 + i\epsilon} \left(-\eta^{\mu\nu} + (1 - \xi)\frac{p^\mu p^\nu}{p^2 - \xi M_V^2 + i\epsilon} \right) , \quad (B.2)$$

whilst their couplings to fermions take the form

$$- i e \kappa \gamma_\mu (v_f + a_f \gamma_5) . \quad (B.3)$$

The individual coefficients appearing in this expression are collected in the following table:

Boson	κ	v_f	a_f
γ	e_f	1	0
Z	$1/(2\sin\theta_w \cos\theta_w)$	$I_3^f - 2e_f \sin^2\theta_w$	$-I_3^f$
$W^{\pm}$	$V_{ff'}/(2\sqrt{2}\sin\theta_w)$	1	-1

Here for the fermion f, e_f is its electric charge measured in units of the positron charge $e > 0$; I_3^f is its third component of weak isospin, $I_3 = +\frac{1}{2}$ for up-type quarks or neutrinos, and $I_3 = -\frac{1}{2}$ for down-type quarks or charged leptons. For the corresponding antiparticles the signs are reversed. For charged current interactions involving quarks, the coefficients $V_{ff'}$ are the respective elements of the Cabbibo–Kobayashi–Maskawa matrix, the dominant elements of which are $V_{ud} \approx 0.975 \approx V_{cs}$, $V_{us} \approx 0.222 \approx -V_{cd}$ and $V_{tb} \approx 1$. For leptons one effectively has $V_{\nu_\ell \ell'} = \delta_{\ell\ell'}$. The parameter θ_w is the weak mixing angle with $\sin^2\theta_w \approx 0.223$.

B.2 Phase space and cross section formulae

Once the amplitude squared for a process has been evaluated, it is necessary to include the flux factor and the (differential) phase space in order to obtain the (differential) cross section. We consider the general process $p_a + p_b \to p_1 + \cdots + p_n$ for which the cross section is given schematically by

$$d\sigma = \frac{1}{\text{flux}} \times |\mathcal{M}|^2 \times d\Phi_n . \quad (B.4)$$

Here it should be understood that the cross section and phase space are typically multi-differential quantities. For head-on collisions the flux factor is given by

$$\begin{aligned} \text{flux} &= 4\sqrt{(p_a \cdot p_b)^2 - (m_a m_b)^2} \\ &= 4|\boldsymbol{p}_a^\star|\sqrt{s} = 4|\boldsymbol{p}_a^{\text{lab}}|m_b \qquad s = (p_a + p_b)^2 \quad (B.5) \\ &\approx 2s . \end{aligned}$$

In the second line the flux is given in terms of the C.o.M. momenta, $\boldsymbol{p}_a^\star = -\boldsymbol{p}_b^\star$, and the laboratory variables $\boldsymbol{p}_a^{\text{lab}}$ and $p_b^{\text{lab}} = (m_b, \boldsymbol{0})$. The third line is appropriate in the limit of negligible particle masses. In the case of a particle decay the flux factor is given by twice the decaying particle's mass, flux $= 2M$.

A differential element of the Lorentz invariant n-body phase space for the outgoing particles is given by

$$d\Phi_n(p_a + p_b : p_1, \ldots, p_n)$$

$$= (2\pi)^4 \delta^{(4)} \left(p_a + p_b - \sum_{i=1}^{n} p_i \right) \prod_{i=1}^{n} \begin{cases} \dfrac{d^4 p_i}{(2\pi)^3} \delta^{(+)}(p_i^2 - m_i^2) \\ \dfrac{d^3 \boldsymbol{p}_i}{(2\pi)^3 2E_i} \end{cases} . \qquad (B.6)$$

In the second version the on mass-shell δ-function has been explicitly integrated out and the positive energy solution $E_i = +\sqrt{\boldsymbol{p}_i^2 + m_i^2}$ selected. Using eqn (B.6) and eqn (B.4) it is easy to verify that the dimensionality of the phase space is given by $3n$, whilst the mass dimension of $|\mathcal{M}|^2$ must be $4 - 2n$.

In many practical situations the incoming particles are unpolarized and the spins of the final state particles are not measured. The same applies for their colours. To take this into account one has to sum the amplitude squared over the spins and colours of the outgoing particles and average over the spins and colours of the incoming particles. Thus, in eqn (B.4) we use

$$|\mathcal{M}|^2 \longrightarrow \overline{\sum} |\mathcal{M}|^2 \equiv \prod_{R=a,b} \frac{1}{2N_R} \times \sum_{\text{spin,colour}} |\mathcal{M}|^2 , \qquad (B.7)$$

where the colour degeneracy is $N_R = N_c$ for a quark or an antiquark and $N_c^2 - 1$ for a gluon and where we allow two spin polarizations for the external fermions and massless external gluons or photons.

APPENDIX C

DIMENSIONAL REGULARIZATION

C.1 Integration in non-integer dimensions

Dimensional regularization is the preferred method in QCD for rendering ultra-violet divergent loop integrals finite. The basic idea is to work in $D = 4 - 2\epsilon$ space–time dimensions. Then, given suitable definitions, we evaluate the loop momentum integrals with any divergences appearing as poles in $1/\epsilon$. This renders the theory finite, for $D < 4$, so that we can carry out the renormalization procedure and afterwards take the limit $\epsilon \to 0$.

In D dimensions the structure of the QCD Lagrangian is unaltered: it contains the same kinetic and interaction terms and, therefore, has the same Feynman rules. There is only one change, the replacement $g_s \to g_s \mu^\epsilon$, where μ is an arbitrary unit mass ('tHooft, 1973). This is needed to ensure that each term in the Lagrangian density has the correct mass dimension; see Ex. (3-17).

Before explaining the method, it is useful to introduce a few standard manipulations which makes the final integrals easier to carry out. We illustrate this approach using the following typical integral which arises in the calculation of the fermion self-energy,

$$I = g_s^2 \mu^{4-D} \int \frac{\mathrm{d}^D k}{(2\pi)^D} \frac{\gamma^\mu (\not{k} + \not{p} + m)\gamma_\mu}{[(k+p)^2 - m^2 + i\epsilon][k^2 + i\epsilon]} . \tag{C.1}$$

For the moment, we have not set $D = 4$ but left it free. We have also introduced an arbitrary mass μ which serves to preserve the canonical dimension of the integral for $D \neq 4$. This integral has a superficial degree of divergence $D - 3$, obtained by counting the number of powers of the loop momentum in the integrand, suggesting a potential linear divergence in $D = 4$ dimensions. At the expense of introducing extra integrals, eqn (C.1) is simplified by combining the two terms in the denominator using the identity

$$\frac{1}{A_1^{n_1} A_2^{n_2} \cdots A_k^{n_k}}$$
$$= \frac{\Gamma(n_1 + n_2 + \cdots + n_k)}{\Gamma(n_1)\Gamma(n_2)\cdots\Gamma(n_k)} \int_0^1 \mathrm{d}\alpha_1 \cdots \mathrm{d}\alpha_k \frac{\alpha_1^{n_1-1}\alpha_2^{n_2-1}\cdots\alpha_k^{n_k-1}\delta(1 - \sum_i \alpha_i)}{(\alpha_1 A_1 + \cdots + \alpha_k A_k)^{n_1+n_2\cdots+n_k}} . \tag{C.2}$$

Here the exponents $\{n_i\}$ need not be integer. The $\{\alpha_i\}$ are known as Feynman parameters. Applying this result to eqn (C.1), and at the same time integrating out the δ-function, gives

$$I = g_s^2 \mu^{4-D} \int \frac{\mathrm{d}^D k}{(2\pi)^D} \int_0^1 \mathrm{d}\alpha \frac{\gamma^\mu (\slashed{k} + \slashed{p} + m)\gamma_\mu}{(\alpha[(k+p)^2 - m^2 + \mathrm{i}\epsilon] + (1-\alpha)[k^2 + \mathrm{i}\epsilon])^2}$$

$$= g_s^2 \mu^{4-D} \int \frac{\mathrm{d}^D k}{(2\pi)^D} \int_0^1 \mathrm{d}\alpha \frac{\gamma^\mu (\slashed{k} + \slashed{p} + m)\gamma_\mu}{((k+\alpha p)^2 - \alpha m^2 + \alpha(1-\alpha)p^2 + \mathrm{i}\epsilon)^2} . \quad \text{(C.3)}$$

In the second line we have 'completed the square', which after shifting the momentum variable, $k^\mu \to k^\mu - \alpha p^\mu$, yields

$$I = g_s^2 \mu^{4-D} \int \frac{\mathrm{d}^D k}{(2\pi)^D} \int_0^1 \mathrm{d}\alpha \frac{\gamma^\mu [\slashed{k} + (1-\alpha)\slashed{p} + m]\gamma_\mu}{(k^2 - A + \mathrm{i}\epsilon)^2}$$

$$= g_s^2 \mu^{4-D} \int_0^1 \mathrm{d}\alpha\, \gamma^\mu [(1-\alpha)\slashed{p} + m]\gamma_\mu \int \frac{\mathrm{d}^D k}{(2\pi)^D} \frac{1}{(k^2 - A + \mathrm{i}\epsilon)^2} . \quad \text{(C.4)}$$

Here, we introduced $A = \alpha m^2 - \alpha(1-\alpha)p^2$. This change of variable and the re-ordering of the integrals is legitimate because we will choose D to make the integral convergent. The k^μ term vanished because the integrand is isotropic and no longer has a preferred direction. The p^μ dependence is now via p^2 in A. This means that the apparent linear divergence of eqn (C.1) is in reality only a logarithmic divergence.

At this point you are reminded that in Minkowski space $k^2 = E^2 - \mathbf{k}^2$, so that the temporal and spatial components are not on an equal footing. To remedy this situation we transform to Euclidean space, $E \mapsto \mathrm{i}k_0$, so that $k^2 \to -k_E^2 = k_0^2 + \mathbf{k}^2$. For the case at hand this gives

$$\int_{-\infty}^{+\infty} \frac{\mathrm{d}E}{2\pi} \int \frac{\mathrm{d}^{D-1}k}{(2\pi)^{D-1}} \frac{1}{(E^2 - \mathbf{k}^2 - A + \mathrm{i}\epsilon)^2}$$

$$= \mathrm{i} \int_{-\infty}^{+\infty} \frac{\mathrm{d}k_0}{2\pi} \int \frac{\mathrm{d}^{D-1}k}{(2\pi)^{D-1}} \frac{1}{(-k_0^2 - \mathbf{k}^2 - A + \mathrm{i}\epsilon)^2}$$

$$= \mathrm{i}(-1)^2 \int \frac{\mathrm{d}^D k_E}{(2\pi)^D} \frac{1}{(k_E^2 + A - \mathrm{i}\epsilon)^2} . \quad \text{(C.5)}$$

A subtlety in this manipulation is the rôle played by the infinitesimal $\mathrm{i}\epsilon$ in the denominator. Essentially, we have used a closed contour in the complex E-plane that goes along the real axis, down the complex axis and closes in the first and third quadrants. Now the integrand has poles at $k^0 = \pm(\sqrt{\mathbf{k}^2 - A} - \mathrm{i}\epsilon)$ which, thanks to the $\mathrm{i}\epsilon$ term ($\epsilon > 0$), lie just outside the contour and thereby ensure the equality of the two integrals in eqn (C.5). Following these manipulations the example integral eqn (C.1) becomes

$$I = \mathrm{i}\, g_s^2 \mu^{4-D} \int_0^1 \mathrm{d}\alpha\, \gamma^\mu [(1-\alpha)\slashed{p} + m]\gamma_\mu \int \frac{\mathrm{d}^D k_E}{(2\pi)^D} \frac{1}{(k_E^2 + A - \mathrm{i}\epsilon)^2} . \quad \text{(C.6)}$$

We now explain the method of dimensional regularization as applied to eqn (C.6). First, we introduce polar coordinates whilst still keeping D free, which yields

$$I = ig_s^2 \mu^{4-D} \int_0^1 d\alpha \, \gamma^\mu \left[(1-\alpha)\slashed{p} + m \right] \gamma_\mu \int \frac{d^{D-1}\Omega}{(2\pi)^D} \int dk_E \frac{k_E^{D-1}}{(k_E^2 + A - i\epsilon)^2} \, .$$
(C.7)

Since by design the required integral eqn (C.6) is isotropic, the angular integrals can be treated separately. The D-dimensional expression for the angular integrals is given in terms of the Euler Γ-function, eqn (C.24),

$$\int d^{D-1}\Omega = \frac{2\pi^{D/2}}{\Gamma(D/2)} \, .$$
(C.8)

This result coincides with the standard expressions, $2\pi, 4\pi, 2\pi^2, \ldots$ for positive integers $n = 2, 3, 4, \ldots$ However, thanks to the use of the Γ-function, the result is analytic in D so that we can use analytic continuation to define the result for non-integer and even complex values of D. The derivation of this result can be found in Ex. (3-18). Finally, there is the k integral which we treat as a regular integral. It is of the standard Euler β-function form:

$$\int dk_E \frac{k_E^{D-1}}{(k_E^2 + A)^n} = \frac{\Gamma(D/2)\Gamma(n - D/2)}{2\Gamma(n)} A^{D/2-n} \, .$$
(C.9)

Combining eqns (C.8) and (C.9), with $n = 2$, allows eqn (C.6) to be written as

$$I = ig_s^2 \mu^{4-D} \frac{\Gamma(2 - D/2)}{(4\pi)^{D/2}\Gamma(2)} \int_0^1 d\alpha \, \gamma^\mu \left[(1-\alpha)\slashed{p} + m \right] \gamma_\mu A(\alpha)^{D/2-2}$$

$$= i \frac{g_s^2}{(4\pi)^2} \Gamma(2 - D/2) \int_0^1 d\alpha \, \gamma^\mu \left[(1-\alpha)\slashed{p} + m \right] \gamma_\mu \left(\frac{A(\alpha)}{4\pi\mu^2} \right)^{D/2-2} \, .$$
(C.10)

For future reference the basic D-dimensional integral is given by (Bollini et al., 1973)

$$\int \frac{d^D k}{(2\pi)^D} \frac{(k^2)^n}{(k^2 - A)^m} = i(-1)^{n-m} \frac{A^{n-m+D/2}}{(4\pi)^{D/2}} \frac{\Gamma(n + D/2)\Gamma(m - n - D/2)}{\Gamma(D/2)\Gamma(m)} \, .$$
(C.11)

The procedure leading to this result was illustrated for the case of a scalar integrand. If the integrand depends on one of the components of k_E, say k_1, then we write the integral as

$$\int d^D k f(k_1, k^2) = \int d^{D-1}k \, dk_1 f(k_1, k^2) = \int d^{D-2}\Omega \, ds \, s^{D-2} dk_1 f(k_1, k_1^2 + s^2) \, .$$
(C.12)

In this way, the integral over k_1 is treated as a normal integral and the D-dimensional treatment is reserved for the remaining 'isotropic' components of k.

Before investigating the $D \to 4$ limit of eqn (C.10) we must first deal with the γ-matrices. This is discussed in Section C.2. Using the D-dimensional algebra of γ-matrices it is easy to show that

$$\gamma^\mu \left[(1 - \alpha)\not{p} + m\right] \gamma_\mu = -(D - 2)(1 - \alpha)\not{p} + Dm . \tag{C.13}$$

Substituting this result in eqn (C.10) and at the same time writing $D = 4 - 2\epsilon$ gives

$$
\begin{aligned}
I &= \mathrm{i}\, \frac{g_s^2}{(4\pi)^2} \Gamma(\epsilon) \int_0^1 \mathrm{d}\alpha \left[-(2 - 2\epsilon)(1 - \alpha)\not{p} + (4 - 2\epsilon)m\right] \left(\frac{A(\alpha)}{4\pi\mu^2}\right)^{-\epsilon} \\
&= \mathrm{i}\, \frac{\alpha_s}{4\pi} \frac{\Gamma(1 + \epsilon)}{\epsilon} \int_0^1 \mathrm{d}\alpha \left[-2(1 - \alpha)\not{p} + 4m + \epsilon(2(1 - \alpha)\not{p} - 2m)\right] \left(\frac{A(\alpha)}{4\pi\mu^2}\right)^{-\epsilon} .
\end{aligned}
\tag{C.14}
$$

In the second line we used eqn (C.25) to make explicit the pole associated with the $D \to 4$, $\epsilon \to 0$ limit. Using eqn (C.26) together with $x^\epsilon = \mathrm{e}^{\epsilon \ln x}$, we can now investigate the $\epsilon \to 0$ limit of eqn (C.14), which becomes

$$
\begin{aligned}
I &= \mathrm{i}\, \frac{\alpha_s}{4\pi} \int_0^1 \mathrm{d}\alpha \Bigg\{ \left[-2(1 - \alpha)\not{p} + 4m\right] \left[\frac{1}{\epsilon} - \gamma_{\mathrm{E}} + \ln(4\pi) - \ln\left(\frac{A(\alpha)}{\mu^2}\right)\right] \\
&\qquad + 2(1 - \alpha)\not{p} - 2m \Bigg\} + \mathcal{O}(\epsilon) \\
&= \mathrm{i}\, \frac{\alpha_s}{4\pi} \Bigg\{ (-\not{p} + 4m) \left(\frac{1}{\epsilon} - \gamma_{\mathrm{E}} + \ln(4\pi)\right) + \not{p} \left[1 + 2\int_0^1 \mathrm{d}\alpha(1 - \alpha) \ln\left(\frac{A(\alpha)}{\mu^2}\right)\right] \\
&\qquad - m \left[2 + 4\int_0^1 \mathrm{d}\alpha \ln\left(\frac{A(\alpha)}{\mu^2}\right)\right] \Bigg\} .
\end{aligned}
\tag{C.15}
$$

In this expression it may be noted that the inclusion of the mass μ takes care of the dimensions in the logarithm. What we find is that the ultraviolet divergence in eqn (C.1) is now isolated as a simple $1/\epsilon$ pole. Experience will confirm that $1/\epsilon$ always occurs in the combination

$$\Delta_\epsilon \equiv \frac{1}{\epsilon} + \ln(4\pi) - \gamma_{\mathrm{E}} . \tag{C.16}$$

There remains a finite part given in terms of tedious but calculable α-integrals.

It must be admitted that our approach to dimensional regularization has been a little cavalier. That our results hold is thanks to the work of others ('tHooft and Veltman, 1972). In essence, what we have done is to first identify those dimensions, $D < 4$ in the example above, for which the desired integral is finite; this means free of both ultraviolet, $k \to \infty$, and, if massless particles appear in the loop, infrared, $k, k \cdot p \to 0$, divergences. The integral is computed and then expressed as an analytic function of D which can be used to continue the integral into the vicinity of $D = 4$.

C.2 D-dimensional γ-matrix algebra

In accord with the discussion of Section 3.3.1, also in the general case of D dimensions we require $\gamma_0^\dagger = +\gamma_0$ and $\gamma_i^\dagger = -\gamma_i$ for $i = 1, 2, 3, \ldots$; the same Clifford algebra $\{\gamma^\mu, \gamma^\nu\} = 2\eta^{\mu\nu}\mathbf{1}$; and the linearity and cyclicity of traces. As these rules essentially coincide with those assumed earlier, we can use the same manipulations on the traces of γ-matrices in D as in 4 dimensions. One word of warning is to remember that $\eta^\mu_{\ \mu} = D$, which leads to extra terms proportional to $D - 4$ appearing in some results; compare eqn (3.83) and Ex. (3-19). Finally, remembering that each trace is proportional to the trace of the D-dimensional unit matrix, we define $\mathrm{Tr}\,\{\mathbf{1}\} = f(D)$ where f is any well behaved function of D subject to the boundary condition $f(4) = 4$. The simplest choice is $f(D) = 4$. Since in practical applications we will always take the limit $\epsilon \to 0$, any difference $f(D) - 4 = \mathcal{O}(\epsilon)$ can only contribute to divergent graphs, $\propto 1/\epsilon^n$, and, as we shall learn, the additional terms are equivalent to a change in the finite part of the counterterms and so unobservable by renormalization group invariance.

Whilst it does not arise in pure unpolarized QCD calculations, for the sake of radiative corrections to chiral weak processes, we mention the treatment of γ_5. The usual properties of γ_5 are: $\gamma_5^\dagger = \gamma_5$, $(\gamma_5)^2 = \mathbf{1}$ and $\{\gamma_5, \gamma_\mu\} = 0$. Unfortunately, if we require results that are analytic functions of D then the anti-commutativity property obliges $\mathrm{Tr}\,\{\gamma_5\gamma_{\mu_1} \cdots \gamma_{\mu_n}\} = 0$ for any n. However, in $D = 4$ dimensions we can realize γ_5 as $\gamma_5 = \mathrm{i}\,\gamma_0\gamma_1\gamma_2\gamma_3$ and obtain the result $\mathrm{Tr}\,\{\gamma_5\gamma_\mu\gamma_\nu\gamma_\sigma\gamma_\tau\} = \mathrm{i}\,\epsilon_{\mu\nu\sigma\tau}$. This conflict highlights the fact that γ_5 is intrinsic to four dimensions. One resolution is to use the $D = 4$ definition of γ_5 and modify the anticommutators ('tHooft and Veltman, 1972) to obey

$$\gamma_5 = \mathrm{i}\,\gamma_0\gamma_1\gamma_2\gamma_3 \quad \Longrightarrow \quad \begin{cases} \gamma_5\gamma_\mu = -\gamma_\mu\gamma_5 & \mu = 0, 1, 2, 3, \\ \gamma_5\gamma_\mu = +\gamma_\mu\gamma_5 & \text{otherwise.} \end{cases} \tag{C.17}$$

The price of this solution is the loss of Lorentz invariance. Thus, when γ_5 is present we must treat separately the sets of components $\mu < 4$ and $\mu \geq 4$.

C.3 D-dimensional phase space

The generalization of the n-body phase space to D dimensions is straightforward,

$$\mathrm{d}\Phi_n = (2\pi)^D\delta^{(D)}\left(p_a + p_b - \sum_{i=1}^n p_i\right)\prod_{i=1}^n \begin{cases} \dfrac{\mathrm{d}^D p_i}{(2\pi)^{(D-1)}}\delta^{(+)}(p_i^2 - m_i^2) \\[2mm] \dfrac{\mathrm{d}^{D-1}\boldsymbol{p}_i}{(2\pi)^{(D-1)}2E_i} \end{cases}. \tag{C.18}$$

Again the $\delta^{(+)}(x) = \Theta(x^0)\delta(x)$ ensures that we only include contributions from positive-energy particles. The case of $n = 1$ is particularly simple,

$$\mathrm{d}\Phi_1 = 2\pi\delta^{(+)}(p_1^2 - m_i^2)\Big|_{p_1 = p_a + p_b}. \tag{C.19}$$

We will illustrate the use of eqn (C.18) for the case $n = 2$, such as might occur in a two-to-two scattering, which gives

$$
\mathrm{d}\Phi_2(Q \to p_1 + p_2) = \int \frac{\mathrm{d}^{D-1}p_1}{(2\pi)^{D-1}2E_1} \int \frac{\mathrm{d}^{D-1}p_2}{(2\pi)^{D-1}2E_2}(2\pi)^D \delta^{(D)}(Q - p_1 - p_2)
$$

$$
= \frac{1}{4(2\pi)^{D-2}} \int \frac{\mathrm{d}^{D-1}p_1}{E_1 E_2} \delta(Q^0 - E_1 - E_2) . \tag{C.20}
$$

Here, we have $E_i = \sqrt{\boldsymbol{p}_i^2 + m_i^2}$ and, after integrating out the spatial momentum components of the second final state particle, $\boldsymbol{p}_2 = \boldsymbol{Q} - \boldsymbol{p}_1$. We shall now specialize to the C.o.M. frame, $Q^\mu = (\sqrt{\hat{s}}, \mathbf{0})$, which is generally the most convenient, and assume that there is an explicit dependence on the longitudinal component, $p_L = p \cos\theta^\star$, of the outgoing momentum in the integrand. The phase-space integral becomes

$$
\int \mathrm{d}\Phi_2 = \frac{1}{4(2\pi)^{D-2}} \int \frac{\mathrm{d}p_L\, \mathrm{d}^{D-2}p_T}{E_1 E_2} \delta(\sqrt{\hat{s}} - E_1 - E_2)
$$

$$
= \frac{1}{4(2\pi)^{D-2}} \int \frac{\mathrm{d}p_L\, \mathrm{d}^{D-3}\Omega\, \mathrm{d}p_T\, p_T^{D-3}}{E_1 E_2} \delta\left(p_T - \sqrt{\hat{p}^2 - p_L^2}\right) \frac{E_1 E_2}{p_T \sqrt{\hat{s}}}
$$

$$
= \frac{1}{4(2\pi)^{D-2}} \frac{2\pi^{(D_2)/2}}{\Gamma((D-2)/2)} \int_{-\hat{p}}^{+\hat{p}} \frac{\mathrm{d}p_L}{\sqrt{\hat{s}}} (\hat{p}^2 - p_L^2)^{(D-4)/2}
$$

$$
= \frac{1}{8\pi} \frac{\hat{p}}{\sqrt{\hat{s}}} \left(\frac{4\pi}{\hat{p}^2}\right)^\epsilon \frac{1}{\Gamma(1-\epsilon)} \int_{-1}^{+1} \mathrm{d}\cos\theta\, \sin^{-2\epsilon}\theta
$$

$$
= \frac{1}{4\pi} \frac{\hat{p}}{\sqrt{\hat{s}}} \left(\frac{\pi}{\hat{p}^2}\right)^\epsilon \frac{1}{\Gamma(1-\epsilon)} \int_0^1 \mathrm{d}v\, [v(1-v)]^{-\epsilon} . \tag{C.21}
$$

In the final line we have changed variables to $v = (1 + \cos\theta)/2$, which proves useful for some applications. If the integrand is a scalar, that is, does not depend explicitly on p_L, then we can reduce eqn (C.21) to

$$
\int \mathrm{d}\Phi_2 = \frac{1}{4\pi} \frac{\hat{p}}{\sqrt{\hat{s}}} \left(\frac{\pi}{\hat{p}^2}\right)^\epsilon \frac{\Gamma(1-\epsilon)}{\Gamma(2-2\epsilon)} . \tag{C.22}
$$

We will also need eqn (C.18) for the case $n = 3$, which can be written as

$$
\int \mathrm{d}\Phi_3 = \frac{Q^2}{2(4\pi)^3} \left(\frac{4\pi}{Q^2}\right)^{2\epsilon} \frac{1}{\Gamma(2-2\epsilon)} \int_0^1 \mathrm{d}x_1 \int_{1-x_1}^1 \mathrm{d}x_2 \frac{1}{[(1-x_3)(1-x_2)(1-x_3)]^\epsilon} . \tag{C.23}
$$

    At this point we also mention the number of spin-polarization states which should be used when averaging the matrix element squared in D dimensions. Given the standard choice $\mathrm{Tr}\{\mathbf{1}\} = 4$, the (anti)quarks as usual should be taken to have two spin-polarization states. On the other hand, it is conventional to give massless gluons $D - 2 = 2(1 - \epsilon)$ spin-polarization states.

C.4   Useful mathematical formulae

Here we collect some useful mathematical results. Further discussion can be found in the standard mathematical physics texts, such as the book by Arfken and Weber (1995).

We make frequent use of the Euler Γ-function, which can be defined by the convergent integral

$$\Gamma(z) = \int_0^\infty dt \ t^{z-1} e^{-t} \quad \mathcal{R}e\{z\} > 0 \ . \tag{C.24}$$

Integration by parts will confirm the important identity

$$\Gamma(1+z) = z\Gamma(z) \ . \tag{C.25}$$

Thus, for positive, integer values of z, $\Gamma(z) = (z-1)!$, which explains the alternative name 'factorial function'. Equation (C.25) can also be used to shift the argument and define the Γ-function when $\mathcal{R}e\{z\} < 0$. This also shows that there are simple poles at $z = 0, -1, -2, \ldots$ We also need the expansion

$$\Gamma(1+\epsilon) = 1 - \gamma_E \epsilon + \left(\frac{\pi^2}{12} + \frac{1}{2}\gamma_E^2\right)\epsilon^2 + \mathcal{O}(\epsilon^3) \ , \tag{C.26}$$

where $\gamma_E = 0.577\,215\,664\,901 \cdots$ is the Euler–Mascheroni constant. We will also often use the related Euler β-function integral,

$$\int_0^1 dx \ x^m (1-x)^n = \frac{\Gamma(1+m)\Gamma(1+n)}{\Gamma(2+m+n)} \quad \mathcal{R}e\{m,n\} > -1 \ . \tag{C.27}$$

APPENDIX   D

R_γ, R_l AND R_T FOR ARBITRARY COLOUR FACTORS

This chapter contains a compilation of the ingredients that go into the theoretical prediction for R_l and R_T. All expressions are given for arbitrary colour factors, which allows to evaluate not only the QCD-SU(3) predictions, but also the predictions for alternative theories with an unbroken gauge symmetry based on a simple Lie group. This is needed, for example, by any analysis which aims at a measurement of the colour factors from R_l and R_T. Keeping the colour factors, it is convenient to redefine the coupling constant such that the amplitude for gluon emission from a quark is independent of the gauge group of the theory. Absorbing a factor 2π as well yields the redefined coupling

$$a_{\mathrm{s}} = \frac{\alpha_{\mathrm{s}} C_F}{2\pi} \,. \tag{D.1}$$

The predictions of the theory for n_f quark degrees of freedom then can be expressed as function of the free parameter a_{s} and the variables

$$f_A = \frac{C_A}{C_F} \,, \quad f_T = \frac{T_F}{C_F} \quad \text{and} \quad f_n = n_f \frac{T_F}{C_F} \,. \tag{D.2}$$

All expressions apply for the $\overline{\mathrm{MS}}$ renormalization scheme and cover at least the dominant contributions. In some cases, the higher order expressions are known but are not quoted here, since the main objective of this section is to provide simple expressions that allow a fast evaluation of the respective effects.

D.1   The running coupling constant and masses

The variation of the strong coupling constant a_{s} and renormalized masses $\overline{m}$ with the renormalization scale of the theory is described by a coupled system of differential equations,

$$\frac{\mathrm{d}a_{\mathrm{s}}}{\mathrm{d}\ln\mu^2} = -b_0 a_{\mathrm{s}}^2 - b_1 a_{\mathrm{s}}^3 - b_2 a_{\mathrm{s}}^4 \cdots \tag{D.3}$$

$$\frac{\mathrm{d}\ln\overline{m}}{\mathrm{d}\ln\mu} = -g_0 a_{\mathrm{s}} - g_1 a_{\mathrm{s}}^2 \cdots \,. \tag{D.4}$$

The parameters b_i and g_i depend on the specific theory. The leading coefficients (Jones, 1974; Caswell, 1974; Tarasov et al., 1980; Tarrach, 1981; Nachtmann and Wetzel, 1981) are, c.f. Section 3.4.5,

$$b_0 = \frac{11}{6} f_A - \frac{2}{3} f_n \tag{D.5}$$

$$b_1 = \frac{17}{6} f_A^2 - \frac{5}{3} f_A f_n - f_n \qquad (D.6)$$

$$b_2 = \frac{2857}{432} f_A^3 - \frac{1415}{216} f_A^2 f_n + \frac{79}{108} f_A f_n^2 - \frac{205}{72} f_A f_n + \frac{11}{18} f_n^2 + \frac{1}{4} f_n \quad (D.7)$$

and

$$g_0 = 3 \qquad (D.8)$$

$$g_1 = \frac{3}{4} + \frac{97}{12} f_A - \frac{5}{3} f_n . \qquad (D.9)$$

Equation (D.3) determines how the strong coupling constant evolves for a fixed number of active flavours, whereas in practical applications one often has to relate a value of α_s from a scale μ_4 with $n_f = 4$ active quark flavours to the measurement at a scale μ_5 with $n_f = 5$ flavours. The treatment of flavour thresholds is described in Bernreuther and Wetzel (1982), Bernreuther (1983), Marciano (1984), and Rodrigo and Santamaria (1993), c.f. Section 3.4.5. With $a_s = a_s(n_f, \mu)$, the coupling constant $a_s(\pm) = a_s(n_f \pm 1, \mu)$ for a different number of flavours, but at the same energy scale μ, can be expressed as a power series in the original coupling. To $\mathcal{O}(a_s^3)$ the expansion is given by

$$a_s(\pm) = a_s \mp a_s^2 \frac{4}{3} f_T \bar{L} +$$

$$a_s^3 \left[\left(\frac{4}{3} f_T \bar{L} \right)^2 \mp \left(\frac{10}{3} f_A f_T + 2 f_T \right) \bar{L} \mp \left(\frac{8}{9} f_A f_T - \frac{17}{12} f_T \right) \right] , \quad (D.10)$$

where $\bar{L} = \ln(\overline{m}(\overline{m})/\mu)$ is the logarithm of the ratio between the fixed point of the $\overline{MS}$ running mass of the extra quark flavour $\overline{m}(\overline{m})$ and the matching scale μ. Note that the matching condition eqn (D.10) implies that two measurements at the same energy scale with different numbers of active flavours, in general, will see a different coupling strength. Only for a point μ close to $\overline{m}(\overline{m})$ is the coupling continuous, as one would naively expect. The numerical value of the point of continuity depends on the order of the perturbative expansion. Up to NLO, it coincides with $\overline{m}(\overline{m})$.

In the context of arbitrary colour factors it would be preferable to express eqn (D.10) as a function of the pole masses M of the quarks rather than the $\overline{MS}$ running masses $\overline{m}$, since the latter already absorb part of the radiative corrections of the specific theory. To leading order the pole mass M is related to the running mass according to

$$\overline{m}(M) = \frac{M}{1 + 2 a_s(M)} . \qquad (D.11)$$

From the leading order term, eqn (D.4), one obtains

$$\overline{m}(\mu) = \overline{m}(M) \left(\frac{\mu}{M} \right)^{-a_s g_0} , \qquad (D.12)$$

and the fixed-point condition $\overline{m}(\mu) = \mu$ immediately yields

$$\overline{m}(\overline{m}) = M(1 + 2a_{\mathrm{s}})^{1/(1 + a_{\mathrm{s}}g_0)} \ . \tag{D.13}$$

To leading order in the strong coupling one thus has

$$\bar{L} = \ln \frac{\overline{m}(\overline{m})}{\mu} = \ln \frac{M}{\mu} + 2a_{\mathrm{s}} + \mathcal{O}(a_{\mathrm{s}}^2) \ , \tag{D.14}$$

which is sufficient to rewrite the third order matching condition, eqn (D.10), as a function of $L = \ln M/\mu$.

$$a_{\mathrm{s}}(\pm) = a_{\mathrm{s}} \mp a_{\mathrm{s}}^2 \frac{4}{3} f_T L +$$

$$a_{\mathrm{s}}^3 \left[\left(\frac{4}{3} f_T L \right)^2 \mp \left(\frac{10}{3} f_A f_T + 2 f_T \right) L \mp \left(\frac{8}{9} f_A f_T + \frac{5}{4} f_T \right) \right] \ . \tag{D.15}$$

With these ingredients, the evolution of the strong coupling constant from the $\mathcal{O}(1 \text{ GeV})$ scale upwards can be realized by using $n_f = 3$ up to $\mu = 2M_{\mathrm{c}}$, then $n_f = 4$ up to $\mu = 2M_{\mathrm{b}}$ and $n_f = 5$ until the top mass threshold $\mu = 2M_{\mathrm{t}}$. At each flavour threshold the matching condition eqn (D.15) has to be applied. The theoretical error of the procedure may be estimated by varying the matching scale between M and $2M$.

D.2   Theoretical predictions for R_γ

The QCD corrections both for R_T and R_l are related to the QCD correction δ^0 of R_γ,

$$R_\gamma = \frac{\sigma(e^+e^- \to \gamma \to \text{hadrons})}{\sigma_{\text{Born}}(e^+e^- \to \gamma \to \mu^+\mu^-)} = 3 \sum_q e_q^2 (1 + \delta_0) \tag{D.16}$$

which is known to order a_{s}^3 (Gorishny *et al.*, 1991; Surguladze and Samuel, 1991),

$$\delta_0 = K_1 a_{\mathrm{s}} + K_2 a_{\mathrm{s}}^2 + \left(K_3 + R_3 + T_3 \frac{\left(\sum_q e_q \right)^2}{3 \sum_q e_q^2} \right) a_{\mathrm{s}}^3 \ . \tag{D.17}$$

For the strong coupling constant taken at the C.o.M. energy of the hadronic system the coefficients are

$$K_1 = \frac{3}{2} \ , \tag{D.18}$$

$$K_2 = -\frac{3}{8} + f_A \left(\frac{123}{8} - 11\zeta_3 \right) - f_n \left(\frac{11}{2} - 4\zeta_3 \right) \ , \tag{D.19}$$

$$K_3 = -\frac{69}{16} - f_A \left(\frac{127}{8} + \frac{143}{2}\zeta_3 - 110\zeta_5 \right) + f_A^2 \left(\frac{90445}{432} - \frac{2737}{18}\zeta_3 - \frac{55}{3}\zeta_5 \right)$$

$$-f_n \left(\frac{29}{8} - 38\zeta_3 + 40\zeta_5 \right) - f_A f_n \left(\frac{3880}{27} - \frac{896}{9}\zeta_3 - \frac{20}{3}\zeta_5 \right)$$

$$+f_n^2 \left(\frac{604}{27} - \frac{152}{9}\zeta_3 \right) , \tag{D.20}$$

$$R_3 = -\frac{\pi^2}{8} \left(\frac{11}{3}f_A - \frac{4}{3}f_n \right)^2 , \tag{D.21}$$

$$T_3 = \frac{d^{abc}d^{abc}}{C_F^3} \left(\frac{11}{24} - \zeta_3 \right) . \tag{D.22}$$

The numerical values of the Riemann ζ functions are $\zeta_3 \approx 1.2020569$ and $\zeta_5 \approx 1.0369278$. The coefficients d^{abc} are the symmetric structure constants of the gauge group. For SU(N) type theories one has $d^{abc}d^{abc}/C_F^3 = 16f_A - 6f_A^2$.

D.3 The theoretical prediction for R_l

The theoretical prediction for R_l is obtained from that for R_γ by taking into account quark mass effects and the fact that, in the coupling of the primary quarks to the Z, vector and axial-vector currents contribute differently (Hebbeker, 1991). The prediction can be written as

$$R_l = \frac{\Gamma(Z \to \text{hadrons})}{\Gamma(Z \to \text{leptons})} = R_l^{\text{EW}}(1 + \delta_0 + \delta_v + \delta_m + \delta_t) . \tag{D.23}$$

Here, R_l^{EW} is the purely electroweak prediction without QCD corrections, δ_0 is the QCD correction for the case of massless quarks which is common to the vector and the axial current, while δ_v is an additional term which only contributes to the vector current. The two remaining terms are mass corrections, δ_m, a contribution to the leading order coefficients which mainly comes from the axial couplings of the quarks to the Z, and δ_t, a second order correction in the axial current due to the large mass splitting between top and bottom quark masses.

Using the effective parameterization of both the top and the Higgs mass dependence from the TOPAZ0 program (Montagna *et al.* 1993*a*, 1993*b*) given by Hebbeker *et al.* (1994), one obtains

$$R_l^{\text{EW}} = 19.999 \left(1 - 2.2 \cdot 10^{-4} \ln \left(\frac{M_H}{M_Z} \right)^2 \right) \left(1 - 4.7 \cdot 10^{-4} \left(\frac{M_t}{M_Z} \right)^2 \right) . \tag{D.24}$$

For $M_t = 150$ GeV and $M_H = 300$ GeV this expression reproduces the value $R_l^{\text{EW}} = 19.963$ given by Passarino (1993) based on TOPAZ0 for the same mass parameters. Note that the measured value of the top mass is $M_t \approx 175$ GeV and that the mass of the Higgs particle is expected to be $M_H < 200$ GeV.

The dominant part of the QCD correction is the same as for R_γ, with the exception of the contribution proportional to T_3 where only the vector current contributes,

$$\delta_0 = K_1 a_s + K_2 a_s^2 + (K_3 + R_3)a_s^3 . \tag{D.25}$$

Introducing v_q and a_q, the vector and axial couplings of quarks q to the Z,

$$v_q = I_3^q - 2e_q \sin^2 \theta_w \quad \text{and} \quad a_q = I_3^q , \tag{D.26}$$

the contribution from T_3 can be written as

$$\delta_v = (T_3 a_s^3) \cdot \left(\frac{(\sum_q v_q)^2}{3 \sum_q v_q^2} \right) \frac{\sum_q v_q^2}{\sum_q v_q^2 + a_q^2} = (T_3 a_s^3) \cdot \frac{(\sum_q v_q)^2}{3 \sum_q v_q^2 + a_q^2} . \tag{D.27}$$

The second fraction is the relative contribution of the vector current to the total cross section, which here is expressed simply as a function of the electroweak couplings. Due to threshold effects which go proportional to $(3 - \beta^2)\beta/2$ for the vector current and β^2 for the axial coupling (Djouadi et al., 1990), there is also a slight dependence on the quark masses. On the Z resonance these effects are small, and within the precision of these calculations, they can be ignored.

The leading order mass correction δ_m, expressed as a function of the pole mass of the quarks, is given by

$$\delta_m = \frac{a_s}{\sum_q v_q^2 + a_q^2} \sum_q 18 \frac{M_q^2}{M_Z^2} \left(v_q^2 - a_q^2 \ln \frac{M_q^2}{M_Z^2} \right) . \tag{D.28}$$

An improved mass correction can be obtained by absorbing large logarithms into running masses $\overline{m}_q$ (Chetyrkin and Kühn, 1990; Chetyrkin et al., 1992). For a determination of the strong coupling from R_l, however, the difference to the leading order term is negligible.

Details about the top mass correction δ_t can be found in the literature (Kniehl and Kühn 1989, 1990). The leading order term comes from an incomplete cancellation between two triangle diagrams Z $\rightarrow$ gg, where via the axial current the Z couples to a b-quark or a t-quark loop. The colour structure of this contribution is of the type $(T_{ji}^a T_{ij}^b)(T_{lk}^a T_{kl}^b) = T_F^2 N_A$, or, because of the identity $N_A = N_F C_F / T_F$, equal to the product $T_F C_F N_F$. Setting the number of quark degrees of freedom to $N_F = 3$, the colour factor dependent correction becomes

$$\delta_t = -a_s^2 f_T \frac{a_b^2}{\sum_q v_q^2 + a_q^2} \left[\frac{37}{12} - 12 \ln \frac{M_Z}{M_t} - \frac{14}{27} \left(\frac{M_Z}{M_t} \right)^2 \right] . \tag{D.29}$$

Note that only the fraction of the cross section with b-quark production contributes to δ_t. The next-to-leading order correction to δ_t, proportional to a_s^3, is known and amounts to 15% of this leading order correction (Chetyrkin and Tarasov, 1994).

D.4   The theoretical prediction for R_T

The theoretical prediction for R_T is also related to R_γ. Detailed discussions can be found in the literature (Braaten et al. 1992; Le Diberder and Pich 1992a,

1992*b*; Pich 1992). Here, we will present only a short summary. Similar to the cases of R_γ and R_l the starting point is

$$R_\tau = \frac{\Gamma(\tau^- \to \nu_\tau \text{hadrons})}{\Gamma(\tau^- \to \nu_\tau \bar{\nu}_e e^-)} = R_\tau^{\text{EW}}(1 + \delta_{\text{EW}} + \delta_0 + \delta_{\text{np}}) . \tag{D.30}$$

Here $R_\tau^{\text{EW}} = 3.0582$ denotes the purely electroweak expectation, which is modified by a residual correction $\delta_{\text{EW}} = 0.001$. The dominant correction is δ_0, which can be calculated in perturbative QCD. The additional term δ_{np} covers the non-perturbative corrections.

The main difference from the case of R_l is the fact that the hadronic system produced in τ decays is not at a fixed mass, but rather exhibits a mass spectrum ranging from M_π to M_τ. As a consequence, the QCD correction to the hadronic width is obtained by integrating the correction to R_γ over the mass spectrum. Expressing the running coupling constant through its value at the scale M_τ and turning the integral over the mass spectrum into a contour integral, one obtains (Le Diberder and Pich 1992*a*, 1992*b*; Pich 1992)

$$\delta_0 = K_1 A_1 + K_2 A_2 + K_3 A_3 + \cdots , \tag{D.31}$$

with

$$A_n = \frac{1}{2\pi i} \oint_{|s|=M_\tau^2} \frac{ds}{s} \left(1 - 2\frac{s}{M_\tau^2} + 2\frac{s^3}{M_\tau^6} - \frac{s^4}{M_\tau^8}\right) a_s^n(-s)$$

$$= \frac{1}{\pi} \mathcal{R}e \int_0^\pi d\phi \left(1 + 2e^{i\phi} - 2e^{3i\phi} + e^{4i\phi}\right) a^n(M_\tau^2 e^{i\phi}) \tag{D.32}$$

where $a_s(-s)$ and $a_s(M_\tau^2)$ are related via eqn (D.3). From the experimental data, there are indications (Braaten *et al.*, 1992; ALEPH Collab., 1998*d*; OPAL Collab., 1999*c*) that the non-perturbative corrections are slightly negative and below 1%, $\delta_{\text{np}} = -0.005 \pm 0.005$.

APPENDIX E

SCALING VIOLATIONS IN FRAGMENTATION FUNCTIONS

Here, the explicit expressions of the various functions are given, which are needed for a NLO analysis of scaling violations in fragmentation functions. As before, we maintain generality and give the expressions for arbitrary colour factors.

E.1 Definitions

The leading order splitting kernels, which also appear in the NLO expressions, are

$$P_{qq}(x) = \frac{1 + x^2}{1 - x} , \tag{E.1}$$

$$P_{qg}(x) = \frac{1 + (1 - x)^2}{x} , \tag{E.2}$$

$$P_{gg}(x) = -x^2 + x - 2 + \frac{1}{x(1 - x)} , \tag{E.3}$$

$$P_{gq}(x) = x^2 + (1 - x)^2 . \tag{E.4}$$

Also, the following quantities are used frequently,

$$S_1(x) = -\text{Li}_2(1 - x) = \int_0^{1-x} \frac{dz}{z} \ln(1 - z) , \tag{E.5}$$

$$S_2(x) = -\text{Li}_2\left(\frac{1}{1 + x}\right) + \frac{1}{2} \ln^2 x - \ln^2(1 + x) + \frac{\pi^2}{6} . \tag{E.6}$$

All functions given below apply for any theory with spin-1/2 fermions interacting via massless spin-1 gauge bosons. The choice of a specific theory is made by substituting the appropriate colour factors C_F, C_A and T_F. For QCD (SU(3)) one has the values $C_F = 4/3$, $C_A = 3$ and $T_F = 1/2$. Below, the ratios

$$a_s = \frac{\alpha_s C_F}{2\pi} , \quad f_A = \frac{C_A}{C_F} \quad \text{and} \quad f_n = n_f \frac{T_F}{C_F} \tag{E.7}$$

will be used. The NLO kernels $P(x)$ at a scale $\sqrt{s}$ are a power series to second order in the strong coupling. In the following sections they are always given for the renormalization scale $\mu_R = \sqrt{s}$, that is,

$$P(x) = a_s A(x) + a_s^2 B(x) \quad \text{with} \quad a_s = a_s(s) . \tag{E.8}$$

The translation to arbitrary renormalization scales is done in the usual way by

$$P(x) = a_{\rm s}(\mu_r^2)A(x) + a_{\rm s}^2(\mu_r^2)\left(B(x) + A(x)b_0 \ln \frac{\mu_r^2}{s}\right)$$ (E.9)

with

$$b_0 = \frac{11}{6}f_A - \frac{2}{3}f_n .$$ (E.10)

Some of the splitting kernels quoted below are singular at $x = 1$ and thus require regularization when substituted into the evolution equations. Here, the so-called (+)-scheme has to be used, indicated by a subscript (+) at the respective function and defined through

$$\int_x^1 {\rm d}z[f(z)]_{(+)}g(z) = \int_x^1 {\rm d}z\,[f(z)(g(z) - g(1))] - g(1)\int_0^x {\rm d}zf(z) .$$ (E.11)

E.2   The flavour non-singlet case

Collecting the formulae for the evolution of the non-singlet parts of the fragmentation functions from the literature (Curci *et al.*, 1980) one finds the following building blocks for the NLO part of the splitting kernels,

$$P_F(x) = 2P_{\rm qq}(x)\left[\ln x \ln(1 - x) - \ln^2 x\right] - 5 + 5x$$
$$+ \left(\frac{3}{1 - x} - 5 - 3x\right)\ln x + \left(\frac{1}{2} + \frac{x}{2}\right)\ln^2 x$$ (E.12)

$$P_G(x) = P_{\rm qq}(x)\left(\frac{67 - 3\pi^2}{18} + \frac{11}{6}\ln x + \frac{1}{2}\ln^2 x\right)$$
$$+ \frac{20}{3} - \frac{20x}{3} + (1 + x)\ln x$$ (E.13)

$$P_n(x) = -P_{\rm qq}(x)\left(\frac{10}{9} + \frac{2}{3}\ln x\right) - \frac{4}{3} + \frac{4x}{3}$$ (E.14)

$$P_A(x) = S_2(x)P_{\rm qq}(-x) + 2 - x + (1 + x)\ln x .$$ (E.15)

Combining these functions with the appropriate colour factors yields

$$P_{\rm qq}^{\rm NS}(x) = P_F(x) + f_A P_G(x) + f_n P_n(x) ,$$ (E.16)

$$P_{\rm q\bar q}^{\rm NS}(x) = (2 - f_A)P_A(x)$$ (E.17)

and

$$P^+(x) = a_{\rm s}P_{\rm qq}(x) + a_{\rm s}^2\left(P_{\rm qq}^{\rm NS}(x) + P_{\rm q\bar q}^{\rm NS}(x)\right) ,$$ (E.18)

from which the NLO splitting function governing the evolution of non-singlet fragmentation functions is finally obtained as

$$P^{\rm NS}(x) = \left[P^+(x)\right]_{(+)} + a_{\rm s}^2 2\delta(1 - x)\int_0^1 {\rm d}z P_{\rm q\bar q}^{\rm NS}(z) .$$ (E.19)

E.3   The flavour-singlet case

The splitting kernels for the flavour-singlet parts of the fragmentation functions (Furmanski and Petronzio, 1980) are slightly more involved. The NLO contributions for the fermion–fermion splitting are:

$$
R_{qq}(x) = 2S_2(x)P_{qq}(-x) - 1 + x - \left(\frac{3}{2} - \frac{x}{2}\right)\ln x + \left(\frac{1}{2} + \frac{x}{2}\right)\ln^2 x
$$

$$
+ P_{qq}(x)\left(\frac{3}{2}\ln x - 2\ln^2 x + 2\ln x \ln(1-x)\right) \tag{E.20}
$$

$$
S_{qq}(x) = \frac{14}{3} - \frac{14}{3}x + P_{qq}(x)\left(\frac{67 - 3\pi^2}{18} + \frac{11}{6}\ln x + \frac{1}{2}\ln^2 x\right)
$$

$$
- S_2(x)P_{qq}(-x) \tag{E.21}
$$

$$
T_{qq}(x) = -\frac{40}{9x} - \frac{52}{3} + \frac{28x}{3} + \frac{112x^2}{9} - \left(10 + 18x + \frac{16x^2}{3}\right)\ln x
$$

$$
+ (2 + 2x)\ln^2 x - P_{qq}(x)\left(\frac{10}{9} + \frac{2}{3}\ln x\right) \tag{E.22}
$$

For the fermion–gluon splitting one has:

$$
R_{qg}(x) = -\frac{1}{2} + \frac{9x}{2} - \left(8 - \frac{x}{2}\right)\ln x + 2x\ln(1-x) + \left(1 - \frac{x}{2}\right)\ln^2 x
$$

$$
- P_{qg}(x)\left(\frac{4\pi^2}{3} - \ln^2(1-x) - 4\ln x \ln(1-x) + 8S_1(x)\right) \tag{E.23}
$$

$$
S_{qg}(x) = S_2(x)P_{qg}(-x) + \frac{62}{9} - \frac{35x}{18} + \frac{44x^2}{9} + \left(2 + 12x + \frac{8x^2}{3}\right)\ln x
$$

$$
- 2x\ln(1-x) - (4+x)\ln^2 x + P_{qg}(x)\left(\frac{17 + 21\pi^2}{18}\right.
$$

$$
\left. - 3\ln x - \frac{1}{2}\ln^2 x - \ln^2 x(1-x) + 8S_1(x)\right) \tag{E.24}
$$

The gluon–fermion splitting is described by:

$$
R_{gq}(x) = -2 + 3x - (7 - 8x)\ln x - 4\ln(1-x) + (1 - 2x)\ln^2 x
$$

$$
- 2P_{gq}(x)\left(5 - \pi^2 - \ln x + \ln(1-x) + \ln^2 x(1-x) - 8S_1(x)\right) \tag{E.25}
$$

$$
S_{gq}(x) = 2S_2(x)P_{gq}(-x) - \frac{40 + 152x - 166x^2}{9x} - \frac{4 + 76x}{3}\ln x + 4\ln(1-x)
$$

$$
+ (2 + 8x)\ln^2 x + P_{gq}(x)\left(\frac{178 - 21\pi^2}{9} - 16S_1(x) - \frac{4}{3}\ln x\right.
$$

$$
\left. + \frac{10}{3}\ln(1-x) - \ln^2 x + 8\ln x \ln(1-x) + 2\ln^2(1-x)\right) \tag{E.26}
$$

$$T_{gq}(x) = -\frac{8}{3} - P_{gq}(x)\left(\frac{16}{9} + \frac{8}{3}\ln x + \frac{8}{3}\ln(1-x)\right) \tag{E.27}$$

Finally, for the gluon–gluon splitting one finds:

$$R_{gg}(x) = \frac{92}{9x} - 4 + 12x - \frac{164x^2}{9}$$
$$+ \left[\frac{16}{3x} + 10 + 14x + \frac{16x^2}{3}\right]\ln x + (2+2x)\ln^2 x \tag{E.28}$$

$$S_{gg}(x) = 2S_2(x)P_{gg}(-x) - \frac{134}{18x} + \frac{27}{2} - \frac{27x}{2} + \frac{134x^2}{18}$$
$$+ \left(\frac{44}{3x} + \frac{11}{3} - \frac{25x}{3}\right)\ln x - (4+4x)\ln^2 x$$
$$+ P_{gg}(x)\left(\frac{67-3\pi^2}{9} + \frac{22}{3}\ln x - 3\ln^2 x + 4\ln x \ln(1-x)\right) \tag{E.29}$$

$$T_{gg}(x) = -\frac{29}{9x} + 2 - 2x + \frac{29x^2}{9}$$
$$- \left(\frac{4}{3} + \frac{4x}{3}\right)\ln x - P_{gg}(x)\left(\frac{20}{9} + \frac{8}{3}\ln x\right) \tag{E.30}$$

From these the complete NLO unregularized splitting kernels for the evolution of the flavour-singlet fragmentation function are constructed as follows:

$$\hat{P}_{QQ}^{S}(x) = a_s P_{qq}(x) + a_s^2\left(R_{qq}(x) + f_A S_{qq}(x) + f_n T_{qq}(x)\right) \tag{E.31}$$

$$P_{QG}^{S}(x) = a_s P_{qg}(x) + a_s^2\left(R_{qg}(x) + f_A S_{qg}(x)\right) \tag{E.32}$$

$$P_{GQ}^{S}(x) = a_s 2 f_n P_{gq}(x) + a_s^2\left(f_n R_{gq}(x) + f_n f_A S_{gq}(x) + f_n f_n T_{gq}(x)\right) \tag{E.33}$$

$$\hat{P}_{GG}^{S}(x) = a_s 2 f_A P_{gg}(x) + a_s^2\left(f_n R_{gg}(x) + f_A f_A S_{gg}(x) + f_n f_A T_{gg}(x)\right) \tag{E.34}$$

Here, the diagonal parts of the splitting function are singular at $x = 1$. The regularized splitting kernels for the evolution equations are given by

$$P_{GG}^{S}(x) = \frac{1}{x}\left[x\hat{P}_{GG}^{S}(x)\right]_{(+)} - \delta(1-x)\int_0^1 dz\, z P_{GQ}^{S}(z) \tag{E.35}$$

$$\text{and} \quad P_{QQ}^{S}(x) = \frac{1}{x}\left[x\hat{P}_{QQ}^{S}(x)\right]_{(+)} - \delta(1-x)\int_0^1 dz\, z P_{QG}^{S}(z). \tag{E.36}$$

E.4   Fragmentation functions and hadron spectra

The relation between fragmentation functions, given at a factorization scale μ_F, and the final hadron spectra at a C.o.M. energy $\sqrt{s}$, is given by a convolution of the fragmentation functions with coefficient functions for the longitudinal and

transverse part of the inclusive cross sections. In NLO those functions (Altarelli
~~et al., 1979b) take into account the effect of the first hard gluon emission on the~~
inclusive particle spectrum. They are given by:

$$P_G^L(z, s, \mu_F) = a_s(\mu_F^2) 2 \frac{1-z}{z} \tag{E.37}$$

$$P_Q^L(z, s, \mu_F) = a_s(\mu_F^2) \tag{E.38}$$

$$P_G^T(z, s, \mu_F) = a_s(\mu_F^2) \left\{ \frac{1 + (1-z)^2}{z} [\ln(1-z) + 2 \ln z] \right.$$

$$\left. -2 \frac{1-z}{z} - \frac{1 + (1-z)^2}{z} \ln \frac{\mu_F^2}{s} \right\} \tag{E.39}$$

$$P_Q^T(z, s, \mu_F) = \delta(1-z) \left[1 + a_s(\mu_F^2) \left(\frac{2\pi^2}{3} - \frac{9}{2} \right) \right]$$

$$+ a_s(\mu_F^2) \left[\frac{3}{2}(1-z) + 2 \frac{1+z^2}{1-z} \ln z - \frac{3}{2} \left(\frac{1}{1-z} \right)_{(+)} \right.$$

$$\left. + (1+z^2) \left(\frac{\ln(1-z)}{1-z} \right)_{(+)} - \ln \frac{\mu_F^2}{s} \left(\frac{1+z^2}{1-z} \right)_{(+)} \right] \tag{E.40}$$

Note that to leading order all terms proportional to a_s can be neglected and the
hadron spectra are directly given by the fragmentation functions. Also, only the
transverse part contributes for quarks.

E.5   Electroweak weight functions

The electroweak weight functions $w_i(s)$ specify the relative fractions of the produced quark flavours as a function of the C.o.M. energy,

$$w_i(s) = \frac{r_i(s)}{2r_u(s) + 3r_d(s)} \quad \text{with} \quad i = \{u, d, s, c, b\} . \tag{E.41}$$

The relative cross sections $r_i(s)$ are given by the electroweak theory:

$$r_i(s) = e_i^2 + \frac{s}{(M_Z^2 - s)^2 + M_Z^2 \Gamma_Z^2}$$

$$\times \left[2e_i v_e v_i \frac{M_Z^2 - s}{4 s_w^2 c_w^2} + (v_e^2 + a_e^2)(v_i^2 + a_i^2) \frac{s}{16 s_w^4 c_w^4} \right] \tag{E.42}$$

Here $s_w = \sin \theta_w$ and $c_w = \cos \theta_w$ are the sine and cosine of the weak mixing
angle θ_w, $a_i = I_3^i$ and $v_i = I_3^i - 2e_i s_w^2$ the axial and the vector coupling to the
Z, e_i the fermion charge and I_3^i the third component of the weak isospin. The
subscript e denotes the electron.

SOLUTIONS

-1 The differential cross section for a $2 \rightarrow 2$ process of two incoming particles with four-momenta p_1 and p_2, which are scattered into outgoing particles p_3 and p_4, is given by the expression

$$\mathrm{d}\sigma = |\mathcal{M}|^2 \frac{1}{4\sqrt{(p_1 \cdot p_2)^2 - m_1^2 m_2^2}} (2\pi)^4 \delta^4(p_1 + p_2 - p_3 - p_4) \frac{\mathrm{d}^3 \boldsymbol{p}_3}{(2\pi)^3 2E_3} \frac{\mathrm{d}^3 \boldsymbol{p}_4}{(2\pi)^3 2E_4} \, .$$

(F.1)

Here, the first term is the scattering matrix element, the second one with the square-root is the flux of incoming particles, then comes a four-dimensional δ-function for energy–momentum conservation and finally the phase-space factors for the outgoing particles. The total cross section is then obtained by integrating over all possible kinematic configurations. The various powers of 2π come from the normalization of the wavefunctions as given in momentum space. Note that the phase-space factors for the final state particles are essentially just the volume elements of the four-momenta subject to the constraint that the particles be on mass-shell, with $E > 0$,

$$\frac{\mathrm{d}^3 \boldsymbol{p}}{2E} = \mathrm{d}^4 p \, \delta \left(p^2 - m^2 \right) \, .$$

The matrix element can be calculated in perturbation theory using the Feynman rules of Section B.1 with the electric charge of the respective fermions instead of the gauge coupling g_s. The amplitude $\mathcal{M}$ for t-channel scattering of an electron with another spin-1/2 particle carrying a charge $e \cdot e_f$ as shown in Fig. F.1, modulo irrelevant phase factors, thus, is given by

$$\mathcal{M} = e_f e^2 \left[\bar{u}(p_3) \gamma^\mu u(p_1) \right] \frac{1}{Q^2} \left[\bar{u}(p_4) \gamma_\mu u(p_2) \right] \, .$$

Note that the spinors are normalized such that the polarization sums are

$$\sum_{\text{spins}} u\bar{u} = m + \not{p} \quad \text{and} \quad \sum_{\text{spins}} v\bar{v} = m - \not{p} \, .$$

In the case of unpolarized particles the proper matrix element $|\mathcal{M}|^2$ is obtained by summing over final state and averaging over initial state polarizations, that is,

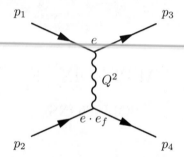

FIG. F.1. Born level diagram for electromagnetic scattering of an electron and a different spin-1/2 fermion with charge $e \cdot e_f$

$$
|\mathcal{M}|^2 = \frac{1}{4} \sum_{\text{spins}} \mathcal{M} \mathcal{M}^\star
$$

$$
= \frac{e_f^2 e^4}{4Q^4} \sum_{\text{spins}} \left[\bar{u}(p_3) \gamma^\mu u(p_1) \bar{u}(p_1) \gamma^\nu u(p_3) \right] \left[\bar{u}(p_4) \gamma_\mu u(p_2) \bar{u}(p_2) \gamma_\nu u(p_4) \right] .
$$

Applying the usual trace rules, the approximation $m_1 = 0$ which is perfectly adequate for the case of electron-nucleon scattering, and setting $m_2 = M$, one obtains

$$
|\mathcal{M}|^2 = \frac{e_f^2 e^4}{4Q^2} \text{Tr} \left\{ \not{p}_3 \gamma^\mu \not{p}_1 \gamma^\nu \right\} \text{Tr} \left\{ \not{p}_4 \gamma_\mu \not{p}_2 \gamma_\nu + M^2 \gamma_\mu \gamma_\nu \right\} .
$$

Using the identities

$$
\text{Tr} \left\{ \not{p}_3 \gamma^\mu \not{p}_1 \gamma^\nu \right\} = 4 \left(p_3^\mu p_1^\nu + p_3^\nu p_1^\mu - \eta^{\mu\nu} (p_1 \cdot p_3) \right) \quad \text{and} \quad \text{Tr} \left\{ \gamma^\mu \gamma^{nu} \right\} = 4 \eta^{\mu\nu} ,
$$

the result simplifies to

$$
|\mathcal{M}|^2 = e_f^2 e^4 \frac{8}{Q^4} \left[(p_1 \cdot p_2)(p_3 \cdot p_4) + (p_1 \cdot p_4)(p_2 \cdot p_3) - M^2 (p_1 \cdot p_3) \right] .
$$

Further calculations now are conveniently done in the laboratory system. With the explicit form of the four-momenta, $p_1 = (E, \boldsymbol{p}_1)$, $p_2 = (M, \boldsymbol{0})$, $p_3 = (E', \boldsymbol{p}_3)$, $p_4 = (E_4, \boldsymbol{p}_4)$ and the scattering angle θ between $\boldsymbol{p}_1$ and $\boldsymbol{p}_3$ the various scalar products appearing above become

$$
p_1 \cdot p_2 = EM ,
$$
$$
p_2 \cdot p_3 = E'M ,
$$
$$
p_1 \cdot p_3 = EE'(1 - \cos\theta) ,
$$
$$
p_3 \cdot p_4 = E'M + EE'(1 - \cos\theta) ,
$$

$$p_1 \cdot p_4 = E'M - EE'(1 - \cos\theta) \,,$$

which leads to

$$|\mathcal{M}|^2 = e_f^2 e^4 \frac{8}{Q^4} EE'M^2 \left\{ (1 + \cos\theta) + \frac{E - E'}{M}(1 - \cos\theta) \right\} \,.$$

For elastic scattering processes one can substitute

$$\frac{E - E'}{M} = \frac{\nu}{M} = \frac{Q^2}{2M^2}$$

and with the usual trigonometric relations

$$1 + \cos\theta = 2\cos^2\frac{\theta}{2} \quad \text{and} \quad 1 - \cos\theta = 2\sin^2\frac{\theta}{2}$$

one finally obtains for the matrix element

$$|\mathcal{M}|^2 = e_f^2 e^4 \frac{16}{Q^4} EE'M^2 \left\{ \cos^2\frac{\theta}{2} + \frac{Q^2}{2M^2}\sin^2\frac{\theta}{2} \right\} \,.$$

The next step in the calculation of the cross section is the determination of the phase-space factors in eqn (F.1). Evaluation of the flux factor and integration over p_4, using the constraint of energy conservation, yields

$$d\sigma = |\mathcal{M}|^2 \frac{1}{16EME'E_4} \frac{1}{(2\pi)^2} d^3p_3 \delta(E + M - E' - E_4) \,,$$

where E_4 now is a function of the remaining variables,

$$E_4 = \sqrt{M^2 + (p_1 - p_3)^2} = \sqrt{M^2 + E^2 + E'^2 + Q^2 - 2EE'} \,.$$

Here, the relation between scattering angle θ and momentum transfer Q^2 has been used, $Q^2 = 2EE'(1 - \cos\theta)$. For the integration over p_3 the available phase space can be parameterized in a variety of ways. The most common one is the use of spherical coordinates, which, using that the scattered particle is assumed to be massless, leads to $d^3p_3 = E'^2 dE' d\cos\theta d\phi$. Instead of integrating over $\cos\theta$ one can also do it over dQ^2. With the Jacobian $2EE'$ this leads to

$$d^3p_3 = \frac{E'}{2E} d\phi dE' dQ^2.$$

Now, carrying out the integration over dE' in order to get rid of the δ-function, and assuming azimuthal symmetry, which gives another factor of 2π from the integral over $d\phi$, yields

$$\frac{d\sigma}{dQ^2} = \frac{1}{64\pi M^2 E^2} |\mathcal{M}|^2 \,,$$

and finally substituting the matrix element calculated earlier, together with the definition of the fine structure constant $\alpha_{\text{em}} = e^2/(4\pi)$, then gives the differential cross section for elastic scattering between two spin-1/2 fermions:

$$\frac{d\sigma}{dQ^2} = \frac{4\pi\alpha_{\text{em}}^2 e_f^2}{Q^4} \frac{E'}{E} \left\{ \cos^2\frac{\theta}{2} + \frac{Q^2}{2M^2}\sin^2\frac{\theta}{2} \right\}.$$

2-2 Quarks can come in a superposition of three basic colour states $|R\rangle$, $|G\rangle$ and $|B\rangle$ which are represented by orthogonal unit vectors, that is,

$$|q\rangle = r|R\rangle + g|G\rangle + b|B\rangle = r\begin{pmatrix} 1 \\ 0 \\ 0 \end{pmatrix} + g\begin{pmatrix} 0 \\ 1 \\ 0 \end{pmatrix} + b\begin{pmatrix} 0 \\ 0 \\ 1 \end{pmatrix}.$$

Gluons describe transitions between colour states. In the basis introduced by the Gell-Mann matrices the eight gluon states are given by

$$\begin{aligned}
\lambda_1 &= |G\rangle\langle R| + |R\rangle\langle G| \\
\lambda_2 &= i|G\rangle\langle R| - i|R\rangle\langle G| \\
\lambda_3 &= |R\rangle\langle R| - |G\rangle\langle G| \\
\lambda_4 &= |B\rangle\langle R| + |R\rangle\langle B| \\
\lambda_5 &= i|B\rangle\langle R| - i|R\rangle\langle B| \\
\lambda_6 &= |B\rangle\langle G| + |G\rangle\langle B| \\
\lambda_7 &= i|B\rangle\langle G| - i|G\rangle\langle B| \\
\lambda_8 &= \frac{1}{\sqrt{3}}|R\rangle\langle R| + \frac{1}{\sqrt{3}}|G\rangle\langle G| - \frac{2}{\sqrt{3}}|B\rangle\langle B|.
\end{aligned}$$

Intuitively these expressions can be read as superpositions of colour-anticolour states, with the ket-vector representing the colour part and the bra-vector the anticolour component. From the above expressions the probability that a red quark emits a gluon is obtained by incoherently adding the contributions of the transitions from red to red, red to green and red to blue. Interference terms do not appear since the final states are distinguishable through the colour state of the emitted gluon. Remembering that the QCD couplings are given by $T^a = \lambda_a/2$ one then finds:

$$\begin{aligned}
\mathcal{A}(R \to \text{qg}) &= A(R \to R) + A(R \to G) + A(R \to B) \\
&= |\langle R|\tfrac{1}{2}\lambda_3|R\rangle|^2 + |\langle R|\tfrac{1}{2}\lambda_8|R\rangle|^2 + |\langle G|\tfrac{1}{2}\lambda_1|R\rangle|^2 + \\
&\quad |\langle G|\tfrac{1}{2}\lambda_2|R\rangle|^2 + |\langle B|\tfrac{1}{2}\lambda_4|R\rangle|^2 + |\langle B|\tfrac{1}{2}\lambda_5|R\rangle|^2
\end{aligned}$$

$$= \frac{1}{4}\left(1 + \frac{1}{3} + 1 + 1 + 1 + 1\right) = \frac{4}{3}$$

$$\mathcal{A}(G \to qg) = A(G \to R) + A(G \to G) + A(G \to B) = \frac{4}{3}$$

$$\mathcal{A}(G \to qg) = A(B \to R) + A(B \to G) + A(B \to B) = \frac{4}{3}$$

-3 The gluon splitting amplitudes are given directly by the structure constants of SU(3) eqn (2.38). Because the f_{abc} are totally antisymmetric in their indices, only splittings into mutually different colour states are allowed. Since the final states are distinguishable the individual amplitudes have to be summed incoherently. One finds the following contributions:

$$\mathcal{A}(1 \to gg) = 2(f_{123}^2 + f_{147}^2 + f_{156}^2) \qquad\qquad = 3\,,$$
$$\mathcal{A}(2 \to gg) = 2(f_{213}^2 + f_{246}^2 + f_{257}^2) \qquad\qquad = 3\,,$$
$$\mathcal{A}(3 \to gg) = 2(f_{312}^2 + f_{345}^2 + f_{376}^2) \qquad\qquad = 3\,,$$
$$\mathcal{A}(4 \to gg) = 2(f_{458}^2 + f_{417}^2 + f_{426}^2 + f_{435}^2) = 3\,,$$
$$\mathcal{A}(5 \to gg) = 2(f_{548}^2 + f_{516}^2 + f_{534}^2 + f_{527}^2) = 3\,,$$
$$\mathcal{A}(6 \to gg) = 2(f_{678}^2 + f_{615}^2 + f_{624}^2 + f_{637}^2) = 3\,,$$
$$\mathcal{A}(7 \to gg) = 2(f_{768}^2 + f_{714}^2 + f_{736}^2 + f_{725}^2) = 3\,,$$
$$\mathcal{A}(8 \to gg) = 2(f_{845}^2 + f_{867}^2) \qquad\qquad\qquad = 3\,.$$

The factor of two in the above equations takes into account that for each amplitude the one with the second and the third index exchanged also contributes. One finds a perfect symmetry for all gluon splitting processes into secondary gluons.

The splitting of a gluon into secondary quark pairs is again governed by the quark–gluon couplings as defined through the Gell-Mann matrices. Each matrix element is directly proportional to the amplitude of the gluon described by that particular matrix splitting into a defined quark–antiquark state. Squaring the amplitudes in order to get the relative probabilities shows perfect symmetry of physics with respect to colour.

$$\mathcal{A}(1 \to q\bar{q}) = \left|\frac{\lambda^1_{12}}{2}\right|^2 + \left|\frac{\lambda^1_{21}}{2}\right|^2 \qquad = \frac{1}{2}$$

...

$$\mathcal{A}(7 \to q\bar{q}) = \left|\frac{\lambda^7_{23}}{2}\right|^2 + \left|\frac{\lambda^7_{32}}{2}\right|^2 \qquad = \frac{1}{2}$$

$$\mathcal{A}(8 \to q\bar{q}) = \left|\frac{\lambda^8_{11}}{2}\right|^2 + \left|\frac{\lambda^8_{22}}{2}\right|^2 + \left|\frac{\lambda^8_{33}}{2}\right|^2 = \frac{1}{2}$$

3–1 Proceeding directly:

$$
\begin{aligned}
F_{\mu\nu} &= \partial_\mu A_\nu + i g_s A_\mu A_\nu - (\mu \leftrightarrow \nu) \\
&\longrightarrow \partial_\mu \left[U A_\nu U^{-1} + i g_s^{-1} (\partial_\nu U) U^{-1} \right] \\
&\quad + i g_s \left[U A_\mu U^{-1} + i g_s^{-1} (\partial_\mu U) U^{-1} \right] \left[U A_\nu U^{-1} + i g_s^{-1} (\partial_\nu U) U^{-1} \right] \\
&\quad - (\mu \leftrightarrow \nu) \\
&= U \left[(\partial_\mu A_\nu) + i g_s A_\mu A_\nu \right] U^{-1} - U \left[A_\mu (\partial_\nu U^{-1}) + A_\nu (\partial_\mu U^{-1}) \right] \\
&\quad - i g_s^{-1} \left[(\partial_\mu \partial_\nu U) U^{-1} - (\partial_\mu U)(\partial_\nu U^{-1}) - (\partial_\nu U)(\partial U^{-1}) \right] \\
&\quad - (\mu \leftrightarrow \nu) \\
&= U F_{\mu\nu} U^{-1}
\end{aligned}
$$

In the third line we made use of the identity supplied and regrouped the terms so as to make manifest the symmetry under $\mu \leftrightarrow \nu$ of the second and third group of terms. These latter terms cancel thereby explicitly confirming eqn (3.8).

3–2 Let the four-momentum of the virtual particle in the laboratory frame be Q^μ. In the particle's rest frame it then has the form $Q^{*\mu} = (\sqrt{Q^2}, \mathbf{0})$. According to the energy–time uncertainty relation this particle lives for a time $t^* = 1/\sqrt{Q^2}$ and thus travels the four-distance $x^{*\mu} = (1, \mathbf{0})/\sqrt{Q^2}$. The rest frame and laboratory frame are related by a boost in the direction $\mathbf{Q}$ with $\beta = |\mathbf{Q}|/Q^0$ and $\gamma = Q^0/\sqrt{Q^2}$, so that

$$x^{*\mu} \longrightarrow x^\mu = (\gamma, \beta\gamma\hat{\mathbf{Q}})\frac{1}{\sqrt{Q^2}} = (Q^0, \mathbf{Q})\frac{1}{Q^2} \quad \text{or} \quad x^\mu = \frac{Q^\mu}{Q^2} .$$

Here, we assumed Q^μ is time-like, the same result holds if Q^μ is space-like. This result can be applied iteratively to a chain of virtual particles by summing their displacements to build up a space–time picture of a scattering event.

3–3 The hadronic tensor $H^{\mu\nu}$ is a rank-2 Lorentz tensor. Ignoring spin degrees of freedom, the free indices can only come from the four-momenta p^μ and q^μ and the constant tensors $\eta^{\mu\nu}$ and $\epsilon^{\mu\nu\sigma\tau}$. In addition, any rank-2 tensor constructed from these ingredients can be multiplied by a scalar function of Lorentz invariant quantities. Using the notation from Section 3.2, the most general expression is

$$H^{\mu\nu} = -F_1\eta^{\mu\nu} + F_2\frac{p^\mu p^\nu}{p \cdot q} + \mathrm{i}\,F_3\epsilon^{\mu\nu}{}_{\sigma\tau}\frac{p^\sigma q^\tau}{p \cdot q} + F_4'\frac{p^\mu q^\nu}{p \cdot q} + F_5'\frac{q^\mu p^\nu}{p \cdot q} + F_6\frac{q^\mu q^\nu}{p \cdot q}\,.$$

The constraints supply two vector equations which must hold for arbitrary p^μ and q^μ, which is only possible if certain relationships hold amongst the $\{F_i\}$, which allow F_4', F_5' and F_6 to be eliminated in favour of F_1 and F_2:

$$q_\mu \cdot H^{\mu\nu} = -q^\nu F_1 + F_2 p^\nu + 0 + F_4' q^\nu + (q^2/p \cdot q)F_5' p^\nu + (q^2/p \cdot q)F_6 q^\nu$$

$$= 0 \implies \begin{cases} 0 = -F_1 + F_4' + (q^2/p \cdot q)F_6 & q^\nu \\ 0 = +F_2 + (q^2/p \cdot q)F_5' & p^\mu \end{cases}$$

$$H^{\mu\nu} \cdot q_\nu = -F_1 q^\mu + F_2 p^\mu + 0 + (q^2/p \cdot q)F_4' p^\mu + F_5' q^\mu + (q^2/p \cdot q)F_6 q^\mu$$

$$= 0 \implies \begin{cases} 0 = -F_1 + F_5' + (q^2/p \cdot q)F_6 & q^\mu \\ 0 = +F_2 + (q^2/p \cdot q)F_4' & p^\mu \end{cases}$$

The contractions with the $\epsilon_{\mu\nu\sigma\tau}$ vanish because of its antisymmetry.

3-4 For the azimuthally integrated, invariant phase space element of the final state lepton one has

$$\int_\phi \frac{\mathrm{d}^3\boldsymbol{\ell}'}{2\pi E_{\ell'}} = \int_\phi \frac{\boldsymbol{\ell}'^2}{E_{\ell'}}\mathrm{d}|\boldsymbol{\ell}'|\mathrm{d}\cos\theta\,\frac{\mathrm{d}\phi}{2\pi} = |\boldsymbol{\ell}'|\,\mathrm{d}E_{\ell'}\,\mathrm{d}\cos\theta \approx E_{\ell'}\mathrm{d}E_{\ell'}\,\mathrm{d}\cos\theta\,.$$

Now, it is most convenient to work in the target hadron's rest frame, with $s - M_\mathrm{h}^2 = 2M_\mathrm{h}E_\ell = Q^2/(xy)$ and the different kinematical variables given by:

$$Q^2 = 2E_\ell E_{\ell'}(1 - \cos\theta)\,, \quad x = \frac{E_\ell E_{\ell'}(1 - \cos\theta)}{M_\mathrm{h}(E_\ell - E_{\ell'})}\,, \quad \nu = E_\ell - E_{\ell'}\,, \quad y = \frac{E_\ell - E_{\ell'}}{E_\ell}\,.$$

From this the Jacobians for the change of variables are evaluated as

$$\left|\frac{\partial(Q^2,\nu)}{\partial(E_{\ell'},\cos\theta)}\right| = 2E_\ell E_{\ell'} \qquad = \frac{(s - M_\mathrm{h}^2)}{M_\mathrm{h}}E_{\ell'} \qquad = \frac{Q^2}{xyM_\mathrm{h}}E_{\ell'}$$

$$\left|\frac{\partial(Q^2,x)}{\partial(E_{\ell'},\cos\theta)}\right| = 2\frac{E_\ell^2 E_{\ell'}^2(1-\cos\theta)}{M_\mathrm{h}(E_\ell - E_{\ell'})^2} = 2x^2\frac{(s-M_\mathrm{h}^2)}{Q^2}E_{\ell'} = \frac{2x}{y}E_{\ell'}$$

$$\left|\frac{\partial(x,y)}{\partial(E_{\ell'},\cos\theta)}\right| = \frac{E_\ell}{M_\mathrm{h}(E_\ell - E_{\ell'})} = \frac{1}{M\nu}E_{\ell'} = \frac{2x}{Q^2}E_{\ell'}\,,$$

and the transformed cross sections are obtained using the rule

$$\mathrm{d}E_{\ell'}\,\mathrm{d}\cos\theta = \left|\frac{\partial(a,b)}{\partial(E_{\ell'},\cos\theta)}\right|^{-1}\mathrm{d}a\,\mathrm{d}b\,,$$

where (a, b) denotes any of the alternative pairs of variables.

3–5 Referring to eqn (3.36) and noting that $\epsilon_\lambda \cdot q = 0$ we have

$$H_\lambda = -\epsilon_\lambda \cdot \epsilon_\lambda^\star F_1 + (p \cdot \epsilon_\lambda)(p \cdot \epsilon_\lambda^\star)\frac{F_2}{p \cdot q} + \epsilon_{\mu\nu\sigma\tau}\epsilon_\lambda^\mu \epsilon_\lambda^{\star\nu} p^\sigma q^\tau \frac{F_3}{p \cdot q} .$$

Noting that $\epsilon_\pm \cdot \epsilon_\pm^\star = -1$ and $\epsilon_\pm \cdot p = 0$ whilst $\epsilon_0 \cdot \epsilon_0^\star = +1$ and $\epsilon_0 \cdot p = M_h \sqrt{\nu^2 + Q^2}/Q$ we derive

$$H_\pm = F_1 \mp F_3\sqrt{1 + \frac{Q^2}{\nu^2}} \approx F_1 \mp F_3 > 0$$

$$H_0 = \frac{F_2}{2x}\left(1 + \frac{Q^2}{\nu^2}\right) - F_1 \approx \frac{F_2}{2x} - F_1 > 0$$

The approximations follow since $Q^2/\nu^2 = (2xM_h)^2/Q^2 \ll 1$. Since the scalar polarization $\epsilon_s^\mu \propto q^\mu$ and $q^\mu H_{\mu\nu} = 0$ then $H_s = 0$. Again, in the limit $Q^2/\nu^2 \ll 1$ we have $\epsilon_s \approx \epsilon_0$ so that we can anticipate that $H_0 \approx H_s = 0$. This inability to absorb a scalar boson implies the Callan–Gross relation $F_2 = 2xF_1$.

3–6 Recalling that the weak current couples u to $(\cos\theta_c d + \sin\theta_c s)$, and ignoring any contributions from heavier quarks, the structure functions are given by

$$x^{-1}F_2^{\bar{\nu}h} = \cos^2\theta_c \bar{d} + u + \sin^2\theta_c \bar{s} \quad \text{and} \quad x^{-1}F_2^{\nu h} = \cos^2\theta_c d + \bar{u} + \sin^2\theta_c s .$$

Then using the constraints for the third component of the weak isospin and the electric charge of the hadron h,

$$T_3(h) = \int_0^1 dx \left\{-\frac{1}{2}[d(x) - \bar{d}(x)] + \frac{1}{2}[u(x) - \bar{u}(x)]\right\}$$

$$Q(h) = \int_0^1 dx \left\{-\frac{1}{3}[d(x) - \bar{d}(x)] + \frac{2}{3}[u(x) - \bar{u}(x)] - \frac{1}{3}[s(x) - \bar{s}(x)]\right\} ,$$

one immediately obtains

$$\frac{1}{2}\int_0^1 \frac{dx}{x}\left[F_2^{\bar{\nu}h}(x) - F_2^{\nu h}(x)\right] = 2T_3(h) + \sin^2\theta_c[3Q(h) - 4T_3(h)] .$$

This is the original Adler sum rule (Adler, 1963). If charm production is allowed kinematically, then the term proportional to $\sin^2\theta_c$ is absent.

3–7 Crossing symmetry gives three relationships: g → gg gives $V_{gg}(z) = V_{gg}(1 - z)$, q → qg gives $V_{gq}(z) = V_{qq}(1 - z)$ and g → q$\bar{q}$ gives $V_{qg}(z) = V_{qg}(1 - z)$. Drell–Levy–Yan crossing also gives three relationships: $V_{gg}(z) = -zV_{gg}(1/z)$, $V_{qq}(z) = -zV_{qq}(1/z)$ and $V_{qg}(z) = +zV_{gq}(z)$. Away from $z = 1$ the plus-prescriptions and δ-functions can be ignored. It is then a trivial matter to verify the eight leading order relationships.

3–8 Writing the convolutions schematically as products ($\otimes$) of splitting kernel and p.d.f., then substituting the expressions for q_i^{NS}, q_{ij}^{NS}, Σ and g into eqn (3.282) yields

$$\mu^2 \frac{\partial}{\partial\mu^2} q_i^{\mathrm{NS}} = \frac{\alpha_{\mathrm{s}}}{2\pi} \left[P_{\mathrm{qq}}^{\mathrm{NS}} - P_{\bar{\mathrm{q}}\mathrm{q}}^{\mathrm{NS}} \right] \otimes q_i^{\mathrm{NS}} , \tag{F.2}$$

$$\mu^2 \frac{\partial}{\partial\mu^2} q_{ij}^{\mathrm{NS}} = \frac{\alpha_{\mathrm{s}}}{2\pi} \left[P_{\mathrm{qq}}^{\mathrm{NS}} + P_{\bar{\mathrm{q}}\mathrm{q}}^{\mathrm{NS}} \right] \otimes q_{ij}^{\mathrm{NS}} , \tag{F.3}$$

$$\mu^2 \frac{\partial}{\partial\mu^2} \begin{pmatrix} \Sigma \\ g \end{pmatrix} = \frac{\alpha_{\mathrm{s}}}{2\pi} \begin{pmatrix} [P_{\mathrm{qq}}^{\mathrm{NS}} + P_{\bar{\mathrm{q}}\mathrm{q}}^{\mathrm{NS}} + n_f(P_{\mathrm{qq}}^{\mathrm{S}} + P_{\bar{\mathrm{q}}\mathrm{q}}^{\mathrm{S}})] & 2n_f P_{\mathrm{qg}} \\ P_{\mathrm{gq}} & P_{\mathrm{gg}} \end{pmatrix} \otimes \begin{pmatrix} \Sigma \\ g \end{pmatrix} . \tag{F.4}$$

The equations for the flavour-singlet and the gluon p.d.f.s do not decouple.

3–9 With the definition

$$F_2^\gamma(x,\mu^2) = x \sum_{\mathrm{q}} e_{\mathrm{q}}^2 \left[q(x,\mu^2) + \bar{q}(x,\mu^2) \right] ,$$

the evolution equation is directly obtained from the sum of the DGLAP equations for quarks and antiquarks, weighted by the square of their electric charges:

$$\mu^2 \frac{\partial F_2^\gamma}{\partial\mu^2}(x,\mu^2) = \frac{\alpha_{\mathrm{s}}(\mu^2)}{2\pi} \int_x^1 \mathrm{d}z \left[P_{\mathrm{qq}}^{(0)}(z) F_2^\gamma\left(\frac{x}{z},\mu^2\right) + P_{\mathrm{qg}}^{(0)}(z) G\left(\frac{x}{z},\mu^2\right) 2\sum_{\mathrm{q}} e_{\mathrm{q}}^2 \right].$$

In the limit $x \to 0$, the term proportional to F_2^γ on the right hand side can be neglected. Writing out explicitly $P_{\mathrm{qg}}^{(0)}(z)$, eqn (3.50), the equation becomes

$$\mu^2 \frac{\partial F_2^\gamma}{\partial\mu^2}(x,\mu^2) = 2T_F \sum_{\mathrm{q}} e_{\mathrm{q}}^2 \frac{\alpha_{\mathrm{s}}(\mu^2)}{2\pi} \int_0^1 \mathrm{d}z \left[z^2 + (1-z)^2 \right] G\left(\frac{x}{z}\right) .$$

With the ansatz $G(x) \sim x^n(1-x)^m$ this has been used to extract the gluon density from the evolution of F_2^γ (Ellis et al., 1994).

3–10 In the limit of small values for $|t|$ and M_{h}^2 the required results follow immediately from

$$2q \cdot (p - p') = (q + p - p')^2 - q^2 - (p - p')^2 = M_X^2 + Q^2 - t \approx M_X^2 + Q^2$$
$$\text{and} \quad 2q \cdot p = (q + p)^2 - q^2 - p^2 = W^2 + Q^2 - M_{\mathrm{h}}^2 \approx W^2 + Q^2 .$$

The smallest value of $-t$ occurs when the incoming hadron carries on undeflected. The actual calculations are most conveniently performed in the C.o.M. system, where one finds

$$-t_{\min} = -(p - p')_{\min}^2 = -2M_{\mathrm{h}}^2 + 2\left(E^\star E'^\star - \sqrt{(E^{\star 2} - M_{\mathrm{h}}^2)(E'^{\star 2} - M_{\mathrm{h}}^2)} \right)$$

$$\approx -2M_{\rm h}^2 + M_{\rm h}^2 \frac{(E^{\star 2} + E'^{\star 2})}{E^\star E'^\star} = M_{\rm h}^2 \frac{(E^\star - E'^\star)^2}{E^\star E'^\star} \ .$$

Substituting $E^\star = (W^2 + M_{\rm h}^2 + Q^2)/2W$ and $E'^\star = (W^2 + M_{\rm h}^2 - M_X^2)/2W$ then gives the result

$$t_{\rm min} \approx M_{\rm h}^2 \frac{(Q^2 + M_X^2)^2}{(W^2 + M_{\rm h}^2 + Q^2)(W^2 + M_{\rm h}^2 - M_X^2)} \approx M_{\rm h}^2 \frac{(Q^2 + M_X^2)^2}{W^4} \ .$$

3–11 Using $\rho(s,t) = \mathcal{R}e[\mathcal{M}(s,t)]/\mathcal{I}m[\mathcal{M}(s,t)] \approx \rho(s,0)$, eqn (3.66), one has

$$\frac{d\sigma_{\rm el}}{dt}(s, t = 0) = \frac{1}{16\pi\lambda}\Big|(i + \rho)\,\mathcal{I}m\big[\mathcal{M}(s,t)\big]\Big|^2_{t=0} = \frac{1}{16\pi}(1 + \rho^2)\sigma_{\rm tot}^2 \ .$$

In practice, a cross section is related to the number of events seen using the beam luminosity, $\sigma = N/\mathcal{L}$. The non-linear relationship is useful because it allows us to avoid the need to know the luminosity. If $N_{\rm el}$ and $N_{\rm inel}$ are the numbers of elastic and inelastic events then

$$\sigma_{\rm tot} = \frac{16\pi}{(1 + \rho^2)} \frac{(dN_{\rm el}/dt)_{t=0}}{(N_{\rm el} + N_{\rm inel})} \ .$$

Here we assume that the detection efficiencies for the two event types are the same.

3–12 With the substitutions $p_z = p\cos\theta$ and $E = \sqrt{p^2 + m^2} \approx p$, one immediately obtains

$$y = \frac{1}{2}\ln\left(\frac{E + p_z}{E - p_z}\right) \approx \ln\sqrt{\frac{1 + \cos\theta}{1 - \cos\theta}} = \ln\left(\frac{\cos\theta/2}{\sin\theta/2}\right) = -\ln\left(\tan\frac{\theta}{2}\right) = \eta \ .$$

3–13 Using the identity

$$(E + p_z)(E - p_z) = E^2 - p_z^2 = m^2 + \boldsymbol{p}_T^2 = m_T^2$$

one can express the rapidity, eqn (3.55), of a particle as

$$y = \frac{1}{2}\ln\frac{E + p_z}{E - p_z} = \frac{1}{2}\ln\frac{(E + p_z)^2}{(E + p_z)(E - p_z)} = \ln\frac{E + p_z}{m_T} = \ln\frac{m_T}{E - p_z} \ ,$$

and thus

$$E \pm p_z = m_T e^{\pm y} \ .$$

From energy–momentum conservation one has

$$E_1 + E_2 = x_a p_a + x_b p_b \quad \text{and} \quad p_{z_1} + p_{z_2} = x_a p_a - x_b p_b \ ,$$

which are easily solved for x_a and x_b.

14 There are many subprocesses which contribute to di-jet production, using the abbreviation $f_1 f_2$ for $f_{h_1}(x_1) f_{h_2}(x_2)$, the full expression is

$$
d\sigma = \int_0^1 dx_1 dx_2 \Big\{ g_1 g_2 \Big[d\hat{\sigma}(gg \to gg) + \sum_q d\hat{\sigma}(gg \to q\bar{q}) \Big]
$$

$$
+ \sum_{f=q,\bar{q}} [g_1 f_2 + f_1 g_2] d\hat{\sigma}(gq \to gq) + \sum_{q,\bar{q}' \neq \bar{q}} [q_1 \bar{q}'_2 + \bar{q}_1 q'_2] d\hat{\sigma}(q\bar{q}' \to q\bar{q}')
$$

$$
+ \sum_q [q_1 \bar{q}_2 + \bar{q}_1 q_2] \Big[d\hat{\sigma}(q\bar{q} \to q\bar{q}) + \sum_{q' \neq q} d\hat{\sigma}(q\bar{q} \to q'\bar{q}') + d\hat{\sigma}(q\bar{q} \to gg) \Big]
$$

$$
+ \sum_{q,q' \neq q} [q_1 q'_2 + \bar{q}_1 \bar{q}'_2] d\hat{\sigma}(q\bar{q}' \to q\bar{q}') + \sum_q [q_1 q_2 + \bar{q}_1 \bar{q}_2] d\hat{\sigma}(qq \to qq) \Big\}.
$$

The gg → gg matrix element is very strongly forward–backward peaked where the $1/\hat{t}^2$ and $1/\hat{u}^2$ poles dominate. This gives the classical $1/\sin^4(\theta^\star/2)$ Rutherford scattering formula, characteristic of spin-1 particle exchange. By contrast, non-diagonal scatterings, $q\bar{q} \to gg$, $gg \to q\bar{q}$ and $q\bar{q} \to q'\bar{q}'$ have lesser $1/\hat{t}$ or $1/\hat{s}^2$ poles and make little contribution. Figure F.2 shows the ratio of the matrix element squared to that for gg → gg. Since we do not distinguish between outgoing partons the matrix elements were symmetrized under $\hat{t} \leftrightarrow \hat{u}$. These ratios are reasonably constant as a function of $\cos\theta^\star$. In the limit $\cos\theta^\star \to 1$, that is $\hat{t} \to 0, \hat{u} \to -\hat{s}$, it is easy to verify, by identifying the coefficients of the leading $1/\hat{t}$ poles, that the ratios become $4/9$, $(4/9)^2$ or 0.

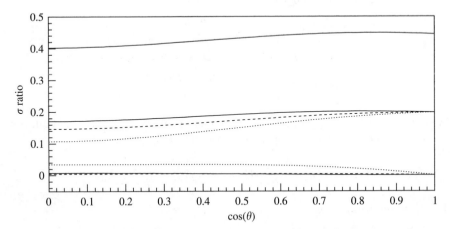

FIG. F.2. The ratio of partonic scattering matrix elements squared to that for gg → gg as a function of the cosine of the partons' C.o.M. frame scattering angle; from top to bottom: gq → gq, q\bar{q} → q\bar{q}, qq' → qq' = q\bar{q}' → q\bar{q}', qq → qq, q\bar{q} → gg, gg → q\bar{q} and q\bar{q} → q'\bar{q}'.

This approximation allows a significant simplification of the di-jet cross section,

$$d\sigma = \int_0^1 dx_1 dx_2 \left\{ g_1 g_2 [1+0] + \sum_{f=q,\bar{q}} [g_1 f_2 + f_1 g_2] \frac{4}{9} \right.$$

$$+ \sum_{q,\bar{q}' \neq \bar{q}} [q_1 \bar{q}_2' + \bar{q}_1 q_2'] \left(\frac{4}{9}\right)^2 + \sum_q [q_1 \bar{q}_2 + \bar{q}_1 q_2] \left[\left(\frac{4}{9}\right)^2 + 0 + 0\right]$$

$$\left. + \sum_{q,q' \neq q} [q_1 q_2' + \bar{q}_1 \bar{q}_2'] \left(\frac{4}{9}\right)^2 + \sum_q [q_1 q_2 + \bar{q}_1 \bar{q}_2] \left(\frac{4}{9}\right)^2 \right\} \hat{\sigma}(gg \to gg)$$

$$= \int_0^1 dx_1 dx_2 \left[g_1 + \frac{4}{9} \sum_{f=q,\bar{q}} f_1 \right] \left[g_2 + \frac{4}{9} \sum_{f=q,\bar{q}} f_2 \right] d\hat{\sigma}(gg \to gg)$$

$$= \int_0^1 dx_1 dx_2 f_1^{\text{eff}}(x_1) f_2^{\text{eff}}(x_2) d\hat{\sigma}(gg \to gg)$$

The factor $4/9$ is the colour factor ratio C_F/C_A.

3–15 Given k^μ and n^μ, the most general form for a polarization sum is

$$T_{\mu\nu} = A\eta_{\mu\nu} + Bk_\mu k_\nu + Cn_\mu n_\nu + Dk_\mu n_\nu + En_\mu k_\nu ,$$

with Lorentz scalar coefficient functions $A, \dots, E$. The conditions on the contractions with k^μ and n^μ yield:

$$k^\mu T_{\mu\nu} = Ak_\nu + 0 + C(n \cdot k)n_\nu + 0 + E(n \cdot k)k_\nu$$

$$= 0 \implies \begin{cases} 0 = A + E(n \cdot k) & k_\nu \\ 0 = C(n \cdot k) & n_\nu \end{cases}$$

$$n^\mu T_{\mu\nu} = An_\nu + B(n \cdot k)k_\nu + Cn^2 n_\nu + D(n \cdot k)n_\nu + En^2 k_\nu$$

$$= 0 \implies \begin{cases} 0 = A + Cn^2 + D(n \cdot k) & k_\nu \\ 0 = B(n \cdot k) + En^2 & n_\nu \end{cases}$$

We have used $k^2 = 0$, appropriate to an on mass-shell gluon. Equivalent equations can be obtained by contracting with k^ν and n^ν. This gives:

$$T_{\mu\nu} = A \left[\eta_{\mu\nu} - \frac{1}{k \cdot n}(k_\mu n_\nu + n_\mu k_\nu) + \frac{n^2}{(k \cdot n)^2} k_\mu k_\nu \right]$$

The last condition gives

$$-2 = \eta^{\mu\nu} T_{\mu\nu} = 4A \implies A = -1 ,$$

which confirms eqn (3.121).

3–16 Averaging over the orientation of the event plane is equivalent to using $L^{\mu\nu} = e^2(-\eta^{\mu\nu}Q^2 + Q^\mu Q^\nu)/3$, eqn (3.111), for the lepton tensor, which, because of gauge invariance, can be simplified to $-e^2\eta^{\mu\nu}Q^2/3$. The hadronic matrix element is given by

$$\mathcal{M}_\mu = ee_q\hat{g}_s T^a \bar{u}(q)\left[1\frac{(\not{q}+\not{g})}{(q+g)^2}\gamma_\mu - \gamma_\mu\frac{(\not{\bar{q}}+\not{g})}{(\bar{q}+g)^2}1\right]v(\bar{q})$$

$$= ee_q\hat{g}_s T^a \bar{u}(q)\left[\frac{\not{g}\gamma_\mu}{2g\cdot q} - \frac{\gamma_\mu\not{g}}{2g\cdot\bar{q}}\right]v(\bar{q}),$$

where we used the Dirac equation to eliminate the $\not{q}$ and $\not{\bar{q}}$ terms. It satisfies $Q^\mu\mathcal{M}_\mu = 0$. Summing over the external particle spins, the first term squared gives

$$-(ee_q\hat{g}_s)^2\frac{\mathrm{Tr}\left\{T^aT^a\right\}}{(2g\cdot q)^2}\mathrm{Tr}\left\{\not{q}\not{g}\gamma_\mu\not{q}\gamma^\mu\not{g}\right\} = (ee_q\hat{g}_s)^2\frac{C_F N_c}{(2g\cdot q)^2}2\mathrm{Tr}\left\{\not{q}\not{g}\not{\bar{q}}\not{g}\right\}$$

$$= 4(ee_q\hat{g}_s)^2\frac{C_F N_c}{(2g\cdot q)^2}2(g\cdot q)2(g\cdot\bar{q}) = 4(ee_q\hat{g}_s)^2 C_F N_c\frac{g\cdot\bar{q}}{g\cdot q}.$$

The third term squared is obtained from this result by exchanging q^μ and $\bar{q}^\mu$. The interference term is given by

$$2(ee_q\hat{g}_s)^2\frac{\mathrm{Tr}\left\{T^aT^a\right\}}{2(g\cdot q)2(g\cdot\bar{q})}\mathrm{Tr}\left\{\not{q}\not{g}\gamma_\mu\not{\bar{q}}\not{g}\gamma^\mu\right\} = 2(ee_q\hat{g}_s)^2\frac{C_F N_c}{4(g\cdot q)(g\cdot\bar{q})}4(g\cdot\bar{q})\mathrm{Tr}\left\{\not{q}\not{g}\right\}$$

$$= 8(ee_q\hat{g}_s)^2 C_F N_c.$$

Combining these results yields for the spin and orientation averaged matrix element,

$$\sum|\mathcal{M}|^2 = \frac{e^2Q^2}{3}\frac{1}{Q^4}4(ee_q\hat{g}_s)^2 C_F N_c\left[\frac{g\cdot\bar{q}}{g\cdot q} + \frac{g\cdot q}{g\cdot\bar{q}} + 2\right]$$

$$= \frac{4}{3}\frac{(e^2e_q\hat{g}_s)^2}{Q^2}C_F N_c\frac{(g\cdot Q)^2}{(g\cdot q)(g\cdot\bar{q})} = \frac{4}{3}\frac{(e^2e_q\hat{g}_s)^2}{Q^2}C_F N_c\frac{x_g^2}{(1-x_{\bar{q}})(1-x_q)}.$$

Note the factor $e^2Q^2/3$ from the lepton tensor. Including the flux factor, $1/(2Q^2)$, and the three-body phase space element, the differential cross section is given by

$$d^2\sigma = \frac{1}{2Q^2}\frac{4}{3}\frac{(e^2e_q\hat{g}_s)^2}{Q^2}C_F N_c\frac{x_g^2}{(1-x_{\bar{q}})(1-x_q)}\frac{Q^2}{2(4\pi)^3}dx_q dx_{\bar{q}}$$

$$= \frac{4\pi\alpha_{em}^2}{3Q^2}e_q^2 N_c\frac{\hat{\alpha}_s}{4\pi}C_F\frac{x_g^2}{(1-x_{\bar{q}})(1-x_q)}dx_q dx_{\bar{q}}$$

$$\implies \frac{d^2\sigma}{dx_q dx_{\bar{q}}} = \sigma_0\frac{\hat{\alpha}_s}{4\pi}C_F\frac{x_g^2}{(1-x_q)(1-x_{\bar{q}})},$$

where $\hat{\alpha}_s = \hat{g}_s^2/(4\pi)$ is the strong coupling for a scalar gluon. As an aside, if the vector boson coupling to the primary $q\bar{q}$-pair has an axial-vector coupling, as is relevant for Z exchange, then only the sign of the interference term changes.

3–17 In natural units the action is dimensionless, which requires $[\int d^D x \mathcal{L}(x)] = 1$. Here we use square brackets to indicate the mass dimensions. It follows that the mass dimension, equivalent to inverse length dimension, of the Lagrangian density must be D. Looking at the kinetic terms we can determine the mass dimensions of the fields, and keeping in mind that $[\partial] = 1$, we find:

$$
\begin{aligned}
{[\bar\psi \partial \psi]} &= D = 2\,[\psi] + 1 &\implies& \quad [\psi] = (D-1)/2\,, \\
{[\partial_\mu A^{a\nu} \partial^\mu A^a_\nu]} &= D = 2\,[A] + 2 &\implies& \quad [A] = (D-2)/2\,. \\
{[\partial^\mu \eta^{a\dagger} \partial_\mu \eta^a]} &= D = 2\,[\eta] + 2 &\implies& \quad [\eta] = (D-2)/2\,.
\end{aligned}
$$

Next consider any interaction term, for example:

$$
[g_s \bar\psi A \psi] = D = [g_s] + 2[\psi] + [A] = [g_s] + (D-1) + (D/2 - 1)
$$
$$
\implies \quad [g_s] = 2 - D/2 = (4-D)/2 = \epsilon\,.
$$

All interaction terms give the same result. Thus, a consistent, dimensionless action with a dimensionless gauge coupling g_s can be achieved using the replacement $g_s \to g_s \mu^\epsilon$, where the arbitrary parameter μ has mass dimension one.

3–18 With the D-dimensional volume element in Cartesian and polar coordinates

$$
d^D x = d^{D-1}\Omega\, r^{D-1} dr
$$

the integral is given by

$$
\prod_{i=1}^{D} \int_{-\infty}^{+\infty} dx_i\, e^{-x_i^2/2} = \int d^{D-1}\Omega \int_0^\infty dr\, r^{D-1} e^{-r^2/2}
$$

$$
\prod_{i=1}^{D} \sqrt{2\pi} = \int d^{D-1}\Omega \int_0^\infty dz\, (2z)^{D/2-1} e^{-z}
$$

$$
(2\pi)^{D/2} = 2^{D/2-1}\Gamma(D/2) \int d^{D-1}\Omega
$$

$$
\implies \quad \int d^{D-1}\Omega = \frac{2\pi^{D/2}}{\Gamma(D/2)}\,.
$$

On the left-hand side we have re-expressed the exponential of a sum as a product of exponentials and used the standard result for the Gaussian integral. On the right-hand side we have transformed to spherical polar coordinates and, after the change of variables, $z = r^2/2$, recognized the integral definition of the Γ-function. The final result naturally generalizes to non-integer dimensions D.

3–19 The derivation follows the usual procedure based upon application of the Clifford algebra, $\gamma^\mu\gamma^\nu = 2\eta^{\mu\nu}\mathbf{1} - \gamma^\nu\gamma^\mu$, but now remembering that $\eta_\mu{}^\mu = D$ so that $\gamma_\mu\gamma^\mu = D$. One finds:

$$\slashed{a}\slashed{a} = a_\mu a_\nu \gamma^\mu\gamma^\nu = a_\mu a_\nu \frac{1}{2}(\gamma^\mu\gamma^\nu + \gamma^\nu\gamma^\mu) = a_\mu a_\nu \eta^{\mu\nu}\mathbf{1}$$

$$= +a^2\mathbf{1}\,,$$

$$\gamma_\mu\slashed{a}\gamma^\mu = \gamma_\mu(2a^\mu\mathbf{1} - \gamma^\mu\slashed{a}) = 2\slashed{a} - D\slashed{a}$$

$$= -(D-2)\slashed{a}\,,$$

$$\gamma_\mu\slashed{a}\slashed{b}\gamma^\mu = \gamma_\mu\slashed{a}(2b^\mu\mathbf{1} - \gamma^\mu\slashed{b}) = 2\slashed{b}\slashed{a} - \gamma_\mu\slashed{a}\gamma^\mu\slashed{b} = 2(2a\cdot b\mathbf{1} - \slashed{a}\slashed{b}) + (D-2)\slashed{a}\slashed{b}$$

$$= +4a\cdot b\mathbf{1} + (D-4)\slashed{a}\slashed{b}\,,$$

$$\gamma_\mu\slashed{a}\slashed{b}\slashed{c}\gamma^\mu = \gamma_\mu\slashed{a}\slashed{b}(2c^\mu\mathbf{1} - \gamma^\mu\slashed{c}) = 2\slashed{c}\slashed{a}\slashed{b} - \gamma_\mu\slashed{a}\slashed{b}\gamma^\mu\slashed{c}$$

$$= 2\slashed{c}(2a\cdot b\mathbf{1} - \slashed{b}\slashed{a}) + (4a\cdot b\mathbf{1} + (D-4)\slashed{a}\slashed{b})\slashed{c}$$

$$= -2\slashed{c}\slashed{b}\slashed{a} - (D-4)\slashed{a}\slashed{b}\slashed{c}\,.$$

Observe that these reduce to eqn (3.83) when $D = 4$.

3–20 Starting from the defining equation $m_0 = Z_m m$ and taking into account that the scale dependence of Z_m comes from the scale dependence of the gauge coupling g_s, using the chain rule one finds

$$0 = \mu\frac{dm_0}{d\mu} = Z_m\mu\frac{\partial m}{\partial\mu} + m\mu\frac{\partial Z_m}{\partial g_s}\frac{\partial g_s}{\partial\mu} = m\left[\gamma_m Z_m + \beta_g\frac{\partial Z_m}{\partial g_s}\right]\,.$$

Now use the expansions $Z_m = 1 + \sum_{n\geq 1} b_n/\epsilon^n$, $\beta_g = g_s^2 a_1' - g_s\epsilon$ and $\gamma_m = A$ and follow the procedure used in eqn (3.167). Note, no powers of ϵ are needed in γ_m. Substituting the expansions into the defining differential equation, sorting according to powers of ϵ and requiring that the resulting coefficients vanish yields the result for γ_m and a recursion for the higher order coefficients,

$$\gamma_m(g_s) = g_s b_1' \quad\text{and}\quad b_n' = b_1' b_{n-1} + g_s a_1' b_{n-1}',\quad n\geq 2\,.$$

Referring to eqn (3.155), and taking into account that $\alpha_s = g_s^2/4\pi$, the leading b-coefficients are given by

$$b_1 = -\left(\frac{g_s}{4\pi}\right)^2 3C_F + \left(\frac{g_s}{4\pi}\right)^4\left[\frac{5}{3}T_F n_f - \frac{3}{4}C_F - \frac{97}{12}C_A\right]C_F\,,$$

$$b_2 = -\left(\frac{g_s}{4\pi}\right)^4\left[2T_F n_f - \frac{9}{2}C_F - \frac{11}{2}C_A\right]C_F\,,$$

from which eqn (3.170) for γ_m can be immediately confirmed. For $n = 2$ the recurrence relation can only be tested to $\mathcal{O}(g_s^3)$, yielding

$$-4\left[2T_F n_f - \frac{9}{2}C_F - \frac{11}{2}C_A\right]C_F = -6C_F\left[-3C_F + 2\left(\frac{2}{3}T_F n_f - \frac{11}{6}C_A\right)\right],$$

where a common factor $(g_s/4\pi)^3$ has been removed. You are invited to repeat the exercise for γ_A where one must also allow for ξ dependence.

3–21 Differentiating eqn (3.174) with respect to Q^2 and using eqn (3.171) gives:

$$\beta_0 \frac{Q^2}{\Lambda_{QCD}^2}\frac{d\Lambda_{QCD}^2}{dQ^2} = \beta_0 - Q^2\frac{d\alpha_s}{dQ^2}\left[-\frac{1}{\alpha_s^2} + \frac{\beta_1}{\beta_0}\left(\frac{1}{\alpha_s} - \frac{\beta_1}{\beta_0 + \beta_1\alpha_s}\right)\right]$$

$$= \beta_0 + \alpha_s^2(\beta_0 + \beta_1\alpha_s)\left[\frac{-\beta_0}{\alpha_s^2(\beta_0 + \beta_1\alpha_s)}\right]$$

$$= 0.$$

You are invited to confirm a similar result for $\overline{m}_0$ in eqn (3.180).

3–22 The leading order β-function is given by eqn (3.171) with a modified β_0. Using eqn (3.179) we have

$$\text{gluons} + \text{gluinos:} \quad 12\pi\beta_0 = +11\cdot 3 - 2\cdot 3$$

$$\text{quarks} + \text{squarks:} \quad 12\pi\beta_0 = \left(-4\cdot\frac{1}{2} - 2\times 1\cdot\frac{1}{2}\right)n_f$$

$$\text{Combined:} \quad 12\pi\beta_0 = 27 - 3n_f,$$

where n_f is the number of complete families that contribute.

3–23 At $\mathcal{O}(\alpha_s)$ the corrections to the leading order gq$\bar{q}$ vertex $-i\,g_s T_{ij}^a\gamma^\mu$ are given by self-energy corrections to the gluon, given in Fig. 3.17, eqn (3.134) and eqn (3.137), the self-energy corrections to the outgoing quarks, given in Fig. 3.20 and eqn (3.132), and the vertex corrections, given in Fig. 3.18 and eqn (3.138). Remembering again the conventional factor $\frac{1}{2}$ due to the wavefunction renormalization, and collecting the respective terms, the sum of the divergent parts is proportional to

$$\frac{1}{2}\Bigg[\left(-\frac{4}{3}T_F n_f - \frac{(13-3\xi)}{6}\right)(g^2\eta^{\mu\sigma} - g^\mu g^\sigma)\frac{1}{g^2}\left(-\eta_{\sigma\nu} + (1-\xi)\frac{g_\sigma g_\nu}{g^2}\right)\gamma^\nu$$

$$-C_F[\xi\slashed{q} - (3+\xi)m]\frac{(\slashed{q}+m)}{q^2 - m^2}\gamma^\mu - \gamma^\mu\frac{(-\slashed{q}+m)}{\bar{q}^2 - m^2}C_F[-\xi\slashed{q} - (3+\xi)m]\Bigg]$$

$$+ \left[\xi\left(C_F - \frac{C_A}{2}\right) + \frac{3(1+\xi)}{4}C_A\right]\gamma^\mu$$

$$= \frac{1}{2}\Bigg[\left(-\frac{4}{3}T_F n_f - \frac{(13-3\xi)}{6}\right)\left(\eta_\nu^\mu - \frac{g^\mu g_\nu}{g^2}\right)\gamma^\nu - C_F\left(\xi - 3m\frac{(\slashed{q}+m)}{q^2-m^2}\right)\gamma^\mu$$

$$-\gamma^\mu C_F \left(\xi + 3m \frac{(\slashed{q} - m)}{\bar{q}^2 - m^2} \right) \right] + \left[\xi C_F + \frac{(3 + \xi)}{4} C_A \right] \gamma^\mu$$

$$= \left(\frac{11}{6} C_A - \frac{2}{3} T_F n_f \right) \gamma^\mu ,$$

where we have extracted a common factor $-i\, g_s T_{ij}^a (\alpha_s/4\pi)\Delta_\epsilon$. In the second line we used the Dirac equation to eliminate the $(\slashed{q} + m)$, $(\slashed{q} - m)$ and $\slashed{q} = (\slashed{q} + m) + (\slashed{q} - m)$ terms. The resultant coefficient of the divergence is the leading term of the QCD β-function, eqn (3.169). The divergence is cancelled by adding the contributions from the counterterms, eqn (3.149).

3–24 Rather than calculate directly it is simpler to obtain the tensor describing $\gamma^\star q \to q'$ from that describing $\gamma^\star \to q\bar{q}$ using crossing, eqn (3.95). That is, substitute $\bar{q}^\mu \to -q'^\mu$ in eqn (3.84) and add an overall minus sign, since there is no longer a closed quark loop in the squared matrix element, to obtain

$$\sum \mathcal{M}(\gamma_\mu^\star q \to q') \mathcal{M}^\star(\gamma_\nu^\star q \to q') = e^2 e_q N_c \left[q_\mu q'_\nu + q'_\mu q_\nu - (q \cdot q')\eta_{\mu\nu} \right] \mathrm{Tr}\,\{\mathbf{1}\}\,.$$

This has the couplings and colour factors restored. Next, we average over the incoming quark's spins and colours, $2N_c$, include the one-body phase space factor $d\Phi_1 = 2\pi\delta(q'^2)$ using eqn (3.234) and divide by the conventional factor $4\pi e^2$. Finally, we eliminate q'_μ in favour of $q_\mu + q_\mu^\gamma$ and q_μ in favour of p_μ/y to obtain

$$\hat{H}_{\mu\nu}^{(\gamma q)} \equiv \frac{1}{4\pi e^2} \int d\Phi_1 \overline{\sum} \mathcal{M}(\gamma_\mu^\star q \to q')\mathcal{M}^\star(\gamma_\nu^\star q \to q')$$

$$= y e_q^2 \left\{ \frac{1}{2} \left(-\eta_{\mu\nu} + \frac{q_\mu^\gamma q_\nu^\gamma}{q^{\gamma 2}} \right) + \frac{y}{p \cdot q^\gamma} \left(p_\mu - \frac{p \cdot q^\gamma}{q^{\gamma 2}} q_\mu^\gamma \right) \left(p_\nu - \frac{p \cdot q^\gamma}{q^{\gamma 2}} q_\nu^\gamma \right) \right\} \delta(y - x)\,.$$

Here we used $2q \cdot q^\gamma = -q^\gamma \cdot q^\gamma$ to obtain the expected tensor structure. Convoluting this expression with the p.d.f.s, eqn (3.228), the resultant expression can be compared with eqn (3.36) to give

$$2x F_1^{(\gamma h)}(x) = F_2^{(\gamma h)}(x) = x \sum_{f=q,\bar{q}} e_f^2 f_h(x) \quad \text{and} \quad F_3^{(\gamma h)}(x) = 0\,.$$

3–25 The identity

$$\left(\frac{1 + z^2}{1 - z} \right)_+ - (1 + z^2) \left(\frac{1}{1 - z} \right)_+ = \frac{3}{2}\delta(1 - z)$$

can easily be proven by using the definition of the plus-prescription eqn (3.248) on the left-hand side. One obtains

$$\frac{1 + z^2}{1 - z} - \delta(1 - z) \int_0^1 dy \frac{1 + y^2}{1 - y} - \frac{1 + z^2}{1 - z} + (1 + z^2)\delta(1 - z) \int_0^1 dy \frac{1}{1 - y}$$

$$= \delta(1-z) \int_0^1 dy \left(\frac{2}{1-y} - \frac{1+y^2}{1-y} \right) = \delta(1-z) \int_0^1 dy \, (1+y) = \delta(1-z)\frac{3}{2} \,.$$

Here, we have used $(1+z^2)\delta(1-z) = 2\delta(1-z)$.

3–26 Using the hint one finds

$$\int_0^1 dx \frac{f(x)}{(1-x)^{1-\epsilon}} = \int_0^1 dx \frac{f(x) - f(1)}{(1-x)^{1-\epsilon}} + f(1) \left. \frac{(1-x)^\epsilon}{-\epsilon} \right|_0^1$$

$$= \int_0^1 dx \frac{f(x) - f(1)}{(1-x)} \left[1 + \epsilon \ln(1-x) + \cdots \right] - \frac{f(1)}{-\epsilon}$$

$$= \int_0^1 dx f(x) \left[\left(\frac{1}{1-x} \right)_+ - \frac{\delta(1-x)}{-\epsilon} \right] + \mathcal{O}(\epsilon) \,.$$

In the second line, the coefficient of ϵ is finite, as can be confirmed by Taylor expanding $f(x)$ about $x = 1$. In the third line we recognise the distribution function $1/(1-x)_+$ and place the $f(1)/\epsilon$ term back under the integral using a δ-function. Since $f(x)$ is arbitrary the identity is established and we can change $-\epsilon \mapsto \epsilon$. In a similar way, again starting with negative ϵ, one finds

$$\int_0^1 dx f(x) \frac{x^{-\epsilon}}{(1-x)^{1-\epsilon}}$$

$$= \int_0^1 dx \left[1 - \epsilon \ln x + \cdots \right] \frac{f(x)}{(1-x)^{1-\epsilon}}$$

$$= \int_0^1 dx \frac{f(x) - f(1)}{1-x} \left[1 + \epsilon \ln(1-x) + \cdots \right] + f(1) \int_0^1 \frac{dx}{(1-x)^{1-\epsilon}}$$

$$- \epsilon \int_0^1 dx f(x) \frac{\ln x}{1-x} \left[1 + \epsilon \ln(1-x) + \cdots \right]$$

$$= \int_0^1 dx f(x) \left[\left(\frac{1}{1-x} \right)_+ + \epsilon \left(\frac{\ln(1-x)}{1-x} \right)_+ - \frac{\delta(1-x)}{-\epsilon} - \epsilon \frac{\ln x}{1-x} \right] + \mathcal{O}(\epsilon^2) \,.$$

Using the arbitrariness of $f(x)$ and substituting $-\epsilon \mapsto \epsilon$ then gives eqn (3.263).

3–27 To obtain the matrix element squared appropriate to $-\eta_{\mu\nu}\hat{H}^{\mu\nu}(\gamma^*g \to q\bar{q})$ apply the crossing $g^\mu \to -g^\mu$ to eqn (3.207), which gives

$$\sum |\mathcal{M}(\gamma^*g \to q\bar{q})|^2 = \text{Tr}\,\{1\}e^2 e_q^2 (g_s\mu^\epsilon)^2 \text{Tr}\,\{T^a T^a\} 2(1-\epsilon)$$

$$\times \left[(1-\epsilon) \left(\frac{g \cdot \bar{q}}{g \cdot q} + \frac{g \cdot q}{g \cdot \bar{q}} \right) - \frac{Q^2(q \cdot \bar{q})}{(g \cdot q)(g \cdot \bar{q})} - 2\epsilon \right] \,.$$

Then, introduce the variables

$$\hat{s} = (2\hat{p}_{\text{out}})^2 = 2q \cdot \bar{q} = \frac{Q^2}{z}(1 - z) , \quad 2g \cdot \bar{q} = \frac{Q^2}{z}v \quad \text{and} \quad 2g \cdot q = \frac{Q^2}{z}(1 - v) ,$$

where $v = (1 + \cos\theta^\star)/2$ and $z = x/y$. The variable x has its usual meaning, eqn (3.39), and y is the momentum fraction of the struck parton with $g^\mu = yp^\mu$. Then, use $\text{Tr}\{1\} = 4$, include an average over the $D - 2 = 2(1 - \epsilon)$ transverse gluon polarizations, and the $N_c^2 - 1$ colours, write $\text{Tr}\{T^aT^a\} = T_F(N_c^2 - 1)$, divide by the conventional factor $4\pi e^2$ and integrate over the two-body phase space, eqn (C.21), to obtain

$$\hat{H}_\Sigma^{(\gamma g)} \equiv \frac{1}{4\pi e^2} \int d\Phi_2 \overline{\sum} |\mathcal{M}(\gamma^\star g \to q\bar{q})|^2$$

$$= e_q^2 \alpha_s T_F 4 \frac{1}{4\pi} \frac{\hat{p}_{\text{out}}}{\sqrt{\hat{s}}} \left(\frac{\pi\mu^2}{\hat{p}_{\text{out}}^2}\right)^\epsilon \frac{1}{\Gamma(1 - \epsilon)}$$

$$\times \int_0^1 dv \, v^{-\epsilon}(1 - v)^{-\epsilon} \left\{(1 - \epsilon)\left[\frac{v}{1 - v} + \frac{1 - v}{v}\right] - 2\frac{z(1 - z)}{v(1 - v)} - 2\epsilon\right\}$$

$$= e_q^2 \frac{\alpha_s}{2\pi} T_F \left(\frac{4\pi\mu^2}{\hat{s}}\right)^\epsilon \frac{1}{\Gamma(1 - \epsilon)}$$

$$\times \frac{\Gamma^2(1 - \epsilon)}{\Gamma(1 - 2\epsilon)} \left\{-\frac{2(1 - \epsilon)^2}{\epsilon(1 - 2\epsilon)} + \frac{4z(1 - z)}{\epsilon} - \frac{2\epsilon}{(1 - 2\epsilon)}\right\}$$

$$= 2e_q^2 \frac{\alpha_s}{2\pi} T_F \left(\frac{4\pi\mu^2}{Q^2} \frac{z}{1 - z}\right)^\epsilon \frac{\Gamma(1 - \epsilon)}{\Gamma(1 - 2\epsilon)} \left\{-\frac{1}{\epsilon}\left[z^2 + (1 - z)^2\right] + \mathcal{O}(\epsilon)\right\}$$

$$= 2e_q^2 \frac{\alpha_s}{2\pi} P_{qg}^{(0)}(z) \left\{-\frac{1}{\epsilon}\left(\frac{4\pi\mu^2}{Q^2}\right)^\epsilon \frac{\Gamma(1 - \epsilon)}{\Gamma(1 - 2\epsilon)} + \ln\left(\frac{1 - z}{z}\right) + \mathcal{O}(\epsilon)\right\} .$$

The integral was evaluated using eqn (C.27) and the result manipulated using eqn (C.25). The matrix element appropriate to the calculation of $g_\mu g_\nu \hat{H}^{\mu\nu}$ is given by

$$g_\mu \mathcal{M}(\gamma^{\star\mu}g \to q\bar{q}) = -\mathrm{i}\, ee_q g_s \mu^\epsilon T^a \bar{u}(q) \left\{\slashed{g}\frac{(\slashed{g} - \slashed{\bar{q}})}{(g - \bar{q})^2}\gamma_\sigma + \gamma_\sigma \frac{(\slashed{q} - \slashed{g})}{(q - g)^2}\slashed{g}\right\} v(\bar{q})\epsilon^\sigma(g)$$

$$= -\mathrm{i}\, ee_q g_s \mu^\epsilon T^a \bar{u}(q) \left\{\frac{\slashed{g}}{g \cdot \bar{q}}\bar{q}_\sigma - q_\sigma \frac{\slashed{g}}{g \cdot q}\right\} v(\bar{q})\epsilon^\sigma(g) .$$

As a preliminary to squaring the amplitude, in the second line we have used $\slashed{g}\slashed{g} = g^2 = 0$ and swapped the order of $\slashed{g}\gamma_\sigma$ and $\gamma_\sigma\slashed{g}$ so as to be able to use $\slashed{g}v(\bar{q}) = 0$ and $\bar{u}(q)\slashed{g} = 0$, respectively. Upon squaring and summing over spins, using $-\eta^{\sigma\sigma'}$ for the gluon, the diagonal terms vanish since $q^2 = 0 = \bar{q}^2$, leaving only the cross-term

$$\sum |g_\mu\mathcal{M}(\gamma^{\star\mu}g \to q\bar{q})|^2 = 2e^2 e_q^2 (g_s\mu^\epsilon)^2 \text{Tr}\{T^aT^a\} \frac{q \cdot \bar{q}}{g \cdot qg \cdot \bar{q}} \text{Tr}\{\slashed{q}\slashed{g}\slashed{\bar{q}}\slashed{g}\}$$

$$= 2e^2 e_q^2 (g_s\mu^\epsilon)^2 \text{Tr}\{T^aT^a\}(2q \cdot \bar{q})\text{Tr}\{1\} .$$

This contribution to the longitudinal component of the hadron tensor is free of singularities. Next, we include the average over the gluon's spin and colour, divide by the conventional factor $4\pi e^2$ and integrate over the phase space using eqn (C.22), since the integrand is a constant. The final result is

$$
\hat{H}_L^{(\gamma g)} \equiv \frac{1}{4\pi e^2} \int d\Phi_2 \sum |g_\mu \mathcal{M}(\gamma^{\star\mu} g \to q\bar{q})|^2
$$

$$
= e_q^2 \alpha_s T_F \hat{s} \frac{4}{(1-\epsilon)} \times \frac{1}{8\pi} \left(\frac{4\pi\mu^2}{\hat{s}}\right)^\epsilon \frac{\Gamma(1-\epsilon)}{\Gamma(2-2\epsilon)}
$$

$$
= e_q^2 \frac{\alpha_s}{2\pi} T_F Q^2 \frac{(1-z)}{z} \left(\frac{4\pi\mu^2}{Q^2} \frac{z}{1-z}\right)^\epsilon \frac{\Gamma(1-\epsilon)}{(1-\epsilon)\Gamma(2-2\epsilon)}
$$

$$
= e_q^2 \frac{\alpha_s}{2\pi} T_F Q^2 \frac{(1-z)}{z} + \mathcal{O}(\epsilon) .
$$

Note that in any summation over quark flavours it should be remembered that these expressions for $\hat{H}_\Sigma^{(\gamma g)}$ and $\hat{H}_L^{(\gamma g)}$ contain the contributions from one flavour of quark and antiquark.

3–28 Using eqn (3.275), the NLO contribution of a single quark to F_3 is given by

$$
\int_0^1 dx\, F_3^{(Wq)}(x, Q^2)
$$

$$
\propto \frac{\alpha_s}{2\pi} \int_0^1 dx \int_x^1 \frac{dz}{z} q\left(\frac{x}{z}, \mu_F^2\right) \left[\left(P_{qq}^{(0)}(z) \ln\left(\frac{Q^2}{\mu_F^2}\right) + C_3^{(Wq)}(z)\right)\right]
$$

$$
= \frac{\alpha_s}{2\pi} \int_0^1 dz \left[\left(P_{qq}^{(0)}(z) \ln\left(\frac{Q^2}{\mu_F^2}\right) + C_3^{(Wq)}(z)\right)\right] \times \int_0^1 dy\, q(y)
$$

$$
= -\frac{\alpha_s}{2\pi} \int_0^1 dz\, C_F (1 + z) \times \int_0^1 dy\, q(y)
$$

$$
= -\frac{3C_F}{4} \frac{\alpha_s}{\pi} \times \int_0^1 dy\, q(y) .
$$

To decouple the two integrals, we first swapped their order, $\int_0^1 dx \int_x^1 dz/z \to \int_0^1 dz \int_0^x dx/z$, and then changed variables from x to $y = x/z$. In line two we used the fact that $P_{qq}^{(0)}$ is a pure plus-function so that its integral vanishes. Likewise in the $\overline{MS}$ scheme most of the integral of $C_3^{(Wq)}$, eqn (3.276), vanishes, leaving an expression which equals $C_3^{(Wq)}$ in the DIS scheme, eqn (3.279). The NLO correction is proportional to the LO expression. Summing this result over the appropriate combination of p.d.f.s then gives the one-loop correction to the GLS sum rule (Altarelli *et al.*, 1978; Bardeen *et al.*, 1978). The final result including the leading order term is

$$
I_{GLS} = \int_0^1 dx\, F_3^{(Wq)}(x, Q^2) = 3\left\{1 - \frac{3C_F}{4}\frac{\alpha_s}{\pi} + \mathcal{O}(\alpha_s^2)\right\} .
$$

This result is independent of the factorization scheme.

3-29 Direct application of the leading order DGLAP equations, eqn (3.49), gives

$$
\mu^2 \frac{\partial}{\partial \mu^2} \int_0^1 dx\, x \left\{ g(x, \mu^2) + \sum_{f=q,\bar{q}} f(x, \mu^2) \right\} = \frac{\alpha_s}{2\pi} \int_0^1 dx\, x \int_x^1 \frac{dz}{z}
$$

$$
\times \left\{ \left[P_{gg}^{(0)}(z) + 2n_f P_{qg}^{(0)}(z) \right] g\left(\frac{x}{z}, \mu^2\right) + \left[P_{gq}^{(0)}(z) + P_{qq}^{(0)}(z) \right] \sum_{f=q,\bar{q}} f\left(\frac{x}{z}, \mu^2\right) \right\}
$$

$$
= \frac{\alpha_s}{2\pi} \left\{ \int_0^1 dz\, z \left[P_{gg}^{(0)}(z) + 2n_f P_{qg}^{(0)}(z) \right] \times \int_0^1 dy\, y g(y, \mu^2) \right.
$$

$$
\left. + \int_0^1 dz\, z \left[P_{gq}^{(0)}(z) + P_{qq}^{(0)}(z) \right] \times \int_0^1 dy \sum_{f=q,\bar{q}} y f(y, \mu^2) \right\}.
$$

In the last equation we have employed the properties of the Mellin transform to decouple the convolution, c.f. Ex. (3-28). Since the result should hold for arbitrary gluon and summed (anti)quark p.d.f.s, the individual coefficients must vanish. For the gluon, direct calculation gives

$$
0 = \int_0^1 dz\, z \left[P_{gg}^{(0)}(z) + 2n_f P_{qg}^{(0)}(z) \right]
$$

$$
= \int_0^1 dz \left\{ 2C_A \left[\frac{z^2 - 1}{(1-z)} + (1-z) + z^2(1-z) \right] + C_{gg} z \delta(1-z) \right.
$$

$$
\left. + 2n_f T_F z [z^2 + (1-z)^2] \right\}
$$

$$
= -\frac{11}{6} C_A + C_{gg} + \frac{2}{3} T_F n_f
$$

$$
\implies \quad C_{gg} = \frac{11 C_A - 4 T_F n_f}{6}.
$$

The coefficient of the (anti)quark p.d.f.s is identically zero with the previously determined value of $C_{qq} = 3/2$.

3-30 The only subtlety in calculating the anomalous dimension is to remember to treat correctly the $1/(1-z)_+$ term. This gives

$$
\gamma_{qq}^{(0)}(n) \equiv C_F \int_0^1 dz\, z^{n-1} \left[\frac{1+z^2}{(1-z)_+} + \frac{3}{2} \delta(1-z) \right]
$$

$$
= C_F \left\{ \frac{3}{2} + \int_0^1 dz \frac{(z^{n-1} - 1) + (z^{n+1} - 1)}{1-z} \right\}
$$

$$
= C_F \left\{ \frac{3}{2} - \int_0^1 dz \left[z^n + z^{n-1} + 2(z^{n-2} + \cdots + 1) \right] \right\}
$$

$$= C_F \left\{ \frac{3}{2} + \frac{1}{n(n+1)} - 2 \sum_{m=1}^{n} \frac{1}{m} \right\} .$$

For $n = 1$ the result is $\gamma_{qq}^{(0)}(1) = 0$. In eqn (F.2), at one-loop level, we have $P_{qq}^{(0)} = P_{qq}^{NS(0)}$ and $P_{\bar{q}q}^{(0)} = 0$. In carrying out the Mellin transform it is useful to introduce a δ-function which leads to

$$\int_0^1 dx\, x^{n-1} \mu^2 \frac{\partial V}{\partial \mu^2}(x, \mu^2) = \frac{\alpha_s}{2\pi} \int_0^1 dx\, x^{n-1} \int_0^1 dz \int_0^1 dy\, \delta(x - yz) P_{qq}^{(0)}(z) V(y)$$

$$\mu^2 \frac{\partial}{\partial \mu^2} \int_0^1 dx\, x^{n-1} V(x, \mu^2) = \frac{\alpha_s}{2\pi} \int_0^1 dz\, z^{n-1} P_{qq}^{(0)}(z) \int_0^1 dy\, y^{n-1} V(y)$$

$$\implies \quad \mu^2 \frac{\partial \tilde{V}}{\partial \mu^2}(n, \mu^2) = \frac{\alpha_s}{2\pi} \gamma_{qq}^{(0)}(n) \tilde{V}(n, \mu^2) .$$

Here, we have $V = q_i - \bar{q}_i$. Alternatively, one can use the usual form of eqn (F.2) and proceed by reversing the order of the x and z integrals. The solution to this differential equation is given by

$$\int_{\tilde{V}(n,\mu_0^2)}^{\tilde{V}(n,\mu^2)} \frac{dV'}{V'} = \gamma_{qq}^{(0)}(n) \int_{\mu_0^2}^{\mu^2} \frac{dt}{t} \frac{\alpha_s(t)}{2\pi} = \frac{\gamma_{qq}^{(0)}(n)}{2\pi \beta_0} \int_{\mu_0^2}^{\mu^2} \frac{dt}{t \ln(t/\Lambda^2)}$$

$$\implies \quad \tilde{V}(n, \mu^2) = \tilde{V}(n, \mu_0^2) \exp \left\{ \frac{\gamma_{qq}^{(0)}(n)}{2\pi \beta_0} \ln \left(\frac{\ln(\mu^2/\Lambda^2)}{\ln(\mu_0^2/\Lambda^2)} \right) \right\}$$

$$\text{or} \quad = \tilde{V}(n, \mu_0^2) \left(\frac{\alpha_s(\mu_0^2)}{\alpha_s(\mu^2)} \right)^{\frac{\gamma_{qq}^{(0)}(n)}{2\pi \beta_0}} .$$

Since $\gamma_{qq}^{(0)}(1) = 0$, the quantity $\tilde{V}(1, \mu^2) = \int_0^1 dx\, [q_i(x, \mu^2) - \bar{q}_i(x, \mu^2)]$ is a constant. This, for example, implies that if the net strangeness in a proton is zero at one scale, then it is zero at any scale.

3–31 At one-loop order, using eqns (3.291)–(3.294) to obtain the anomalous dimensions, the $n = 2$ equations are given by

$$\mu^2 \frac{\partial}{\partial \mu^2} \begin{pmatrix} \tilde{\Sigma}(2, \mu^2) \\ \tilde{g}(2, \mu^2) \end{pmatrix} = \frac{\alpha_s}{2\pi} \begin{pmatrix} -\frac{4}{3} C_F & , +\frac{2}{3} n_f T_F \\ +\frac{4}{3} C_F & , -\frac{2}{3} n_f T_F \end{pmatrix} \begin{pmatrix} \tilde{\Sigma}(2, \mu^2) \\ \tilde{g}(2, \mu^2) \end{pmatrix} .$$

These equations are diagonalized by the linear combinations

$$\tilde{\Sigma}(2, \mu^2) + \tilde{g}(2, \mu^2) \quad \text{eigenvalue} \quad 0$$
$$\tilde{\Sigma}(2, \mu^2) - \frac{n_f T_F}{2 C_F} \tilde{g}(2, \mu^2) \quad \text{eigenvalue} \quad -(\tfrac{4}{3} C_F + \tfrac{2}{3} n_f T_F) < 0 .$$

The diagonalized equations can then be integrated as in Ex. (3-30) to yield

$$\tilde{\Sigma}(2, \mu^2) + \tilde{g}(2, \mu^2) = \tilde{\Sigma}(2, \mu_0^2) + \tilde{g}(2, \mu_0^2) \quad \text{and}$$

$$\tilde{\Sigma}(2,\mu^2) - \frac{n_f T_F}{2C_F}\tilde{g}(2,\mu^2) = \left[\tilde{\Sigma}(2,\mu_0^2) - \frac{n_f T_F}{2C_F}\tilde{g}(2,\mu_0^2)\right]\left(\frac{\alpha_s(\mu_0^2)}{\alpha_s(\mu^2)}\right)^{\frac{-(4C_F+2n_f T_F)}{6\pi\beta_0}}.$$

The first equation implies momentum conservation, eqn (3.362). The solution to the second equation gives the ratio of momentum carried by (anti)quarks and gluons at asymptotic scales.

$$\frac{\int_0^1 dx\, x\Sigma(x,\mu^2)}{\int_0^1 dx\, xg(x,\mu^2)} \xrightarrow{\mu^2\to\infty} \frac{n_f T_F}{2C_F} = \frac{3n_f}{16}\bigg|_{n_f=5} \approx 1\,.$$

Above the b-quark threshold gluons carry approximately half of the parent hadron's momentum.

3–32 In DLLA, gluons drive the evolution of the p.d.f.s and splitting functions can be approximated by their small-z limits. Thus, using the one-loop expression for $\alpha_s(Q^2)$ in eqn (3.49) the gluon pdf's evolution is described by

$$Q^2\frac{\partial g(x,Q^2)}{\partial Q^2} = \frac{1}{2\pi\beta_0\ln(Q^2/\Lambda^2)}\int_x^1\frac{dy}{y}\frac{2C_A}{(x/y)}g(y,Q^2)$$

$$-x\frac{\partial}{\partial x}\left(Q^2\ln(Q^2/\Lambda^2)\frac{\partial[xg(x,Q^2)]}{\partial Q^2}\right) = \frac{C_A}{\pi\beta_0}\,xg(x,Q^2)$$

$$\frac{\partial^2[xg(x,Q^2)]}{\partial\ln(1/x)\partial\ln(\ln(Q^2/\Lambda^2))} = \frac{C_A}{\pi\beta_0}\,xg(x,Q^2)\,.$$

Defining $L_x = \ln(1/x)$, $L_\alpha = \ln(\ln(Q^2/\Lambda^2))$ and $c = C_A/\pi\beta_0$, this equation and the solution, eqn (3.301), become

$$\frac{\partial^2(e^{2\sqrt{cL_xL_\alpha}})}{\partial L_x\,\partial L_\alpha} = \frac{\partial}{\partial L_x}\left(\sqrt{\frac{cL_x}{L_\alpha}}e^{2\sqrt{cL_xL_\alpha}}\right)$$

$$= c\left(1+\frac{1}{2\sqrt{cL_xL_\alpha}}\right)e^{2\sqrt{cL_xL_\alpha}}$$

$$\approx ce^{2\sqrt{cL_xL_\alpha}}\,.$$

3–33 Near $n = 1$, eqn (3.291) gives

$$\gamma_{gg}^{(0)}(n) = \frac{2C_A}{n-1} + \kappa + \mathcal{O}(n-1)\quad\text{with}\quad \kappa = \frac{C_A}{6} - n_f\frac{2T_F}{3}\,.$$

Introducing the notation $L_x = \ln(1/x)$ and $L_\alpha = \ln(\alpha_s(\mu_0^2)/\alpha_s(\mu^2))$ and using eqn (3.288) in eqn (3.286) gives

$$xg(x,\mu^2) =$$

$$= \frac{1}{2\pi i} \int_{c-i\infty}^{c+i\infty} \frac{dn}{x^{n-1}} \tilde{g}(n,\mu^2) \left(\frac{\alpha_s(\mu_0^2)}{\alpha_s(\mu^2)} \right)^{\frac{1}{2\pi\beta_0} \left[\frac{2C_A}{n-1} + \kappa + \cdots \right]}$$

$$= \frac{1}{2\pi i} \int_{c-i\infty}^{c+i\infty} dn\, \tilde{g}(n,\mu^2) \exp\left\{ (n-1)L_x + \frac{1}{2\pi\beta_0} \left[\frac{2C_A}{n-1} + \kappa + \cdots \right] L_\alpha \right\}$$

$$\approx \frac{1}{2\pi i} \int_{c-i\infty}^{c+i\infty} dn\, \tilde{g}(n_0,\mu^2) \exp\left\{ \sqrt{\frac{4C_A}{\pi\beta_0} L_\alpha L_x} + \sqrt{\frac{\pi\beta_0 L_x^3}{C_A L_\alpha}} (n-n_0)^2 \right.$$

$$\left. + \cdots + \frac{\kappa}{2\pi\beta_0} L_\alpha + \cdots \right\}$$

$$\approx \tilde{g}(n_0,\mu_0^2) \left(\frac{C_A L_\alpha}{16\pi^3 \beta_0 L_x^3} \right)^{1/4} \exp\left\{ \sqrt{\frac{4C_A}{\pi\beta_0} L_\alpha L_x} + \frac{\kappa}{2\pi\beta_0} L_\alpha \right\} .$$

In line three we expand about the saddle point at $n_0 = \sqrt{C_A L_\alpha/(\pi\beta_0 L_x)}$. Inspection shows that the contour of steepest descent has constant real part, $n - n_0 = iy$, and evaluating the resulting Gaussian integral gives the solution. This solution assumes that there are no poles in $\tilde{g}(n,\mu_0^2)$ to the right of n_0. For example, if $xg(x,\mu_0^2) \sim x^{-\epsilon}$, then $\tilde{g}(n,\mu_0^2) \sim (n-1-\epsilon)^{-1}$ and this behaviour would dominate $g(x,\mu^2)$. The above solution applies also to the full gluon and singlet quark evolution but with a modified value of $\kappa = C_A/6 + n_f 2T_F(C_A - 2C_F)/(3C_A)$.

3–34 The multiplicity is given by the $n = 1$ moment, so that

$$\langle N \rangle = \tilde{D}(1,Q^2) \equiv \int_0^1 dx\, D(x,Q^2) \sim \exp\left(\frac{1}{\beta_0} \sqrt{\frac{2C_A}{\pi\alpha_s(Q^2)}} \right) .$$

Higher order corrections to this formula are available (Mueller, 1983). In anticipation of the result for the shape write

$$\xi^\star = \frac{1}{4\beta_0\alpha_s(Q^2)}, \quad \sigma^2 = \frac{1}{24\beta_0} \sqrt{\frac{2\pi}{C_A\alpha_s^3(Q^2)}} \quad \text{and} \quad C = \frac{1}{\beta_0} \sqrt{\frac{2C_A}{\pi\alpha_s(Q^2)}} .$$

As in Ex. (3-33) we find

$$xD(x,Q^2)$$

$$\sim \frac{1}{2\pi i} \int_{c-i\infty}^{c+i\infty} \frac{dn}{x^{n-1}} \exp\left(C - \xi^\star(n-1) + \frac{\sigma^2}{2}(n-1)^2 + \cdots \right)$$

$$\sim \frac{1}{2\pi i} \int_{c-i\infty}^{c+i\infty} dn \exp\left(C - \frac{(L_x - \xi^\star)^2}{2\sigma^2} + \frac{\sigma^2}{2} \left[(n-1) + \frac{(L_x - \xi^\star)}{\sigma^2} \right]^2 \right)$$

$$\sim \exp\left(-\frac{(L_x - \xi^\star)^2}{2\sigma^2} \right) .$$

In line two we have moved the $x^{-(n-1)}$ term into the exponential and completed the square. The quadratic $(n-1)$ and C terms just contribute to the fragmentation function's overall normalization. Higher order corrections distort the shape of this Gaussian in $\ln(1/x)$ which becomes downward skewed and platykurtic (Fong and Webber, 1991).

3–35 The phase space element is Lorentz invariant so we can work in any convenient frame. Choosing $p^\mu = E(1,0,0,1)$ and $n^\mu = (1,0,0,-1)$ with $k_\perp^\mu = (0, k_{\perp 1}, k_{\perp 2}, 0)$ eqn (3.315) becomes

$$g^\mu = \left((1-z)E - \frac{k_\perp^2}{4(1-z)E}, -k_{\perp 1}, -k_{\perp 2}, (1-z)E + \frac{k_\perp^2}{4(1-z)E} \right),$$

which has the on mass-shell constraint, $g^2 = 0$, built into the energy. In this form we have

$$\frac{\partial g_\parallel}{\partial z} = -E + \frac{k_\perp^2}{4(1-z)^2 E} = -\frac{E_g}{(1-z)}.$$

Assuming azimuthal symmetry the gluon phase space element thus becomes

$$\frac{\mathrm{d}^3 g}{(2\pi)^3 2E_g} = \frac{1}{8\pi^2} \frac{\mathrm{d}g_\parallel}{E_g} \frac{\mathrm{d}\phi}{2\pi} \frac{\mathrm{d}g_\perp^2}{2} = \frac{1}{16\pi^2} \frac{\mathrm{d}z}{1-z} \mathrm{d}k_\perp^2.$$

3–36 The manipulations for $g \to q\bar{q}$ are a direct analogue of those for $q \to qg$ given in Section 3.6.7:

$$\mathcal{M}_j^{(n+1)} = g_s T_{jk}^a \bar{v}(\bar{q}) \gamma_\mu \frac{(\not{g} - \not{\bar{q}})}{(g-\bar{q})^2} \mathcal{M}_k^{(n)}$$

$$\implies \sum |\mathcal{M}^{(n+1)}|^2$$

$$= g_s^2 T_{jk}^a T_{k'j}^a \mathrm{Tr} \left\{ \cdots \frac{(\not{g} - \not{\bar{q}})}{(g-\bar{q})^2} \left(-\gamma_\mu \not{q} \gamma^\mu + \frac{[\not{n}\not{q}\not{g} + \not{g}\not{q}\not{n}]}{n \cdot g} \right) \frac{(\not{g} - \not{\bar{q}})}{(g-\bar{q})^2} \cdots \right\}$$

$$= g_s^2 T_F \frac{(N_c^2 - 1)}{N_c} \delta_{kk'} \frac{2}{n \cdot g} \frac{1}{2g \cdot \bar{q}}$$

$$\times \mathrm{Tr} \left\{ \cdots [n \cdot (g - \bar{q})(\not{g} - \not{\bar{q}}) + (g \cdot \bar{q})\not{n} + (n \cdot \bar{q})\not{q}] \cdots \right\}$$

$$= 2g_s^2 T_F \frac{(N_c^2 - 1)}{N_c} \delta_{kk'} \frac{(1-z)}{k_\perp^2}$$

$$\times \mathrm{Tr} \left\{ \cdots \left([z^2 + (1-z)^2] \not{g} + (2z-1)\not{k}_\perp + \frac{k_\perp^2}{n \cdot g}\not{n} \right) \cdots \right\}$$

$$= (N_c^2 - 1) 2g_s^2 T_F [z^2 + (1-z)^2] \frac{(1-z)}{k_\perp^2} \frac{\delta_{kk'}}{N_c} \mathrm{Tr} \left\{ \cdots \not{g} \cdots \right\}$$

$$= (N_c^2 - 1)2g_s^2 \hat{P}_{qg}(z)\frac{(1-z)}{k_\perp^2}\frac{1}{N_c}\sum |M^{(n)}|^2 \, .$$

In the final expression the colour factor has been arranged such that the factors of $(N_c^2 - 1)$ and $1/N_c$ compensate for the averaging over colour polarizations when the incoming gluon is replaced by an incoming quark.

3–37 In the hadron's C.o.M. frame the colliding quark and antiquark momenta are given by $p_1^\mu = x_1\sqrt{s}(1,0,0,1)/2$ and $p_2^\mu = x_2\sqrt{s}(1,0,0,-1)/2$, so that $Q^\mu = p_1^\mu + p_2^\mu = \sqrt{s}(x_1+x_2,0,0,x_1-x_2)/2$. Therefore (assuming $Q^2 = \hat{s}$) $Q^2 = x_1 x_2 s$, or $\tau = x_1 x_2$, and $y = \frac{1}{2}\ln[(Q^0 + Q_z)/(Q^0 - Q_z)] = \frac{1}{2}\ln(x_1/x_2)$, which gives $x_1 = \sqrt{\tau}e^{+y}$ and $x_2 = \sqrt{\tau}e^{-y}(=\tau/x_1)$. In eqn (3.330) change variables from $(x=)x_1$ to y, $dx/x = dy$, to obtain the result

$$\frac{d^2\sigma}{dQ^2 dy} = \sigma_{DY}^{(0)}\,\tau\sum_q \left[q_{h_1}(\sqrt{\tau}e^{+y})\bar{q}_{h_2}(\sqrt{\tau}e^{-y}) + \bar{q}_{h_1}(\sqrt{\tau}e^{+y})q_{h_2}(\sqrt{\tau}e^{-y})\right]e_q^2 \, .$$

3–38 The exercise is tedious but not difficult. We focus on the $q_{h_1}\bar{q}_{h_2}$ term in eqn (3.334) and assume an arbitrary factorization scheme. Working to $\mathcal{O}(\alpha_s)$ we find

$$\int dx_1 dx_2\, e_q^2\left(\delta(1-\hat{\tau}) + \frac{\alpha_s}{2\pi}\left[2P_{qq}^{(0)}(\hat{\tau})\ln\left(\frac{Q^2}{\mu_F^2}\right) + H_{q\bar{q}}^F(\hat{\tau})\right]\right)$$

$$\times\left(q_{h_1}(x_1) - \frac{\alpha_s}{2\pi}\int\frac{dz_1}{z_1}\sum_{f=q,g}f_{h_1}\left(\frac{x_1}{z_1}\right)\left(P_{qf}^{(0)}(z_1)\left[\Delta_\epsilon - \ln\left(\frac{\mu_F^2}{\mu^2}\right)\right] - R_f^F(z_1)\right)\right)$$

$$\times\left(\bar{q}_{h_2}(x_2) - \frac{\alpha_s}{2\pi}\int\frac{dz_2}{z_2}\sum_{f=q,g}f_{h_2}\left(\frac{x_2}{z_2}\right)\left(P_{qf}^{(0)}(z_2)\left[\Delta_\epsilon - \ln\left(\frac{\mu_F^2}{\mu^2}\right)\right] - R_f^F(z_2)\right)\right)$$

$$=\int dx_1 dx_2\, e_q^2\left\{q_{h_1}(x_1)\bar{q}_{h_2}(x_2)\left(\delta(1-\hat{\tau}) + \frac{\alpha_s}{2\pi}\left[2P_{qq}^{(0)}(\hat{\tau})\ln\left(\frac{Q^2}{\mu_F^2}\right) + H_{q\bar{q}}^F(\hat{\tau})\right]\right)\right.$$

$$-\frac{\alpha_s}{2\pi}\int\frac{dz}{z}\delta(1-\hat{\tau})\left(q_{h_1}(x_1)\sum_{f=q,g}f_{h_2}\left(\frac{x_2}{z}\right)\left(P_{qf}^{(0)}(z)\left[\Delta_\epsilon - \ln\left(\frac{\mu_F^2}{\mu^2}\right)\right] - R_f^F(z)\right)\right.$$

$$\left.\left. + \bar{q}_{h_2}(x_2)\sum_{f=q,g}f_{h_1}\left(\frac{x_1}{z}\right)\left(P_{qf}^{(0)}(z)\left[\Delta_\epsilon - \ln\left(\frac{\mu_F^2}{\mu^2}\right)\right] - R_f^F(z)\right)\right)\right\} \, .$$

The triple integral in the second line is simplified using the following manipulation,

$$\int dx_1 dx_2 \frac{dz}{z}\delta\left(1 - \frac{\tau}{x_1 x_2}\right)q_{h_1}(x_1)f_{h_2}\left(\frac{x_2}{z}\right)P(z)$$

$$=\int dx_1 \frac{dz}{z}\frac{\tau}{x_1}q_{h_1}(x_1)f_{h_2}\left(\frac{\tau}{x_1 z}\right)P(z)$$

$$= \int dx_1 dx_2 \, q_{h_1}(x_1) f_{h_2}(x_2) \mathcal{P}\left(\frac{\tau}{x_1 x_2} = \hat{\tau}\right) .$$

First, we eliminated x_2 using $\delta(1 - \tau/x_1 x_2) = (\tau/x_1)\delta(x_2 - \tau/x_1)$, then we eliminated z in favour of $x_2 = \tau/(x_1 z)$. This then gives the first term in eqn (3.334).

$$\int dx_1 dx_2 \, e_q^2 \left\{ q_{h_1}(x_1)\bar{q}_{h_2}(x_2)\left(\delta(1 - \hat{\tau}) + \frac{\alpha_s}{2\pi}\left[2P_{qq}^{(0)}(\hat{\tau})\ln\left(\frac{Q^2}{\mu_F^2}\right) + H_{q\bar{q}}^F(\hat{\tau})\right]\right)\right.$$

$$- \frac{\alpha_s}{2\pi}\left(q_{h_1}(x_1)\sum_{f=\bar{q},g} f_{h_2}(x_2)\left(P_{qf}^{(0)}(\hat{\tau})\left[\Delta_\epsilon - \ln\left(\frac{\mu_F^2}{\mu^2}\right)\right] - R_f^F(\hat{\tau})\right)\right.$$

$$\left.\left. + \bar{q}_{h_2}(x_2)\sum_{f=q,g} f_{h_1}(x_1)\left(P_{qf}^{(0)}(\hat{\tau})\left[\Delta_\epsilon - \ln\left(\frac{\mu_F^2}{\mu^2}\right)\right] - R_f^F(\hat{\tau})\right)\right)\right\}$$

Picking out the $q_{h_1}\bar{q}_{h_2}$ piece and collecting terms gives

$$\int dx_1 dx_2 \, q_{h_1}(x_1)\bar{q}_{h_2}(x_2) \, e_q^2 \left\{\delta(1 - \hat{\tau})\right.$$

$$\left. + \frac{\alpha_s}{2\pi}\left(-2P_{qq}^{(0)}(\hat{\tau})\left[\Delta_\epsilon - \ln\left(\frac{Q^2}{\mu}\right)\right] + H_{q\bar{q}}^F(\hat{\tau}) + R_q^F(\hat{\tau}) + R_{\bar{q}}^F(\hat{\tau})\right)\right\},$$

which coincides with the original expression provided that

$$H_{q\bar{q}}^F(z) = H_{q\bar{q}}(z) - R_q^F(z) - R_{\bar{q}}^F(z) .$$

In the DIS scheme the finite terms are $R_q^{DIS} = C_2^{(\gamma q)} = R_{\bar{q}}^{DIS}$ and $R_g^{DIS} = C_2^{(\gamma g)}$. Thus, the form of eqn (3.334) is the same with $f_h^{\overline{MS}}(x, \mu_F^2) \to f_h^{DIS}(x, \mu_F^2)$ and the new coefficient functions

$$H_{q\bar{q}}^{DIS}(z) = H_{q\bar{q}}(z) - 2C_2^{(\gamma q)}(z) ,$$

$$H_{qg}^{DIS}(z) = H_{qg}(z) - C_2^{(\gamma g)}(z) .$$

-39 Starting from,

$$\mathcal{M}_i^{(N+1)}(\bar{q}, k) = g_s T_{ji}^a \epsilon^*(k)_{a\mu} \mathcal{M}_j^{(N)}(\bar{q} + k)\frac{[-(\slashed{\bar{q}} + \slashed{k}) + m]}{(\bar{q} + k)^2 - m^2}\gamma^\mu \bar{v}(\bar{q})$$

follow exactly the same steps as for the quark case. The equivalent expression for radiation off a hard(er) gluon is given by

$$\mathcal{M}^{(N+1)}(g, k) = i g_s f_{abc} \mathcal{M}(g + k)_{\lambda'}^c \frac{\sum_T \epsilon_T^*(g + k)^{\lambda'} \epsilon_T(g + k)^{\lambda}}{2g \cdot k}$$

$$\times \left[(k-g)_\lambda \eta_{\mu\nu} + (2g+k)_\mu \eta_{\nu\lambda} - (g+2k)_\nu \eta_{\lambda\mu} \right] \epsilon_{T'}^\star(g)^{b\nu} \epsilon^\star(k)^{a\mu} \, ,$$

which reduces to the form of eqn (3.338) upon using $\epsilon(p) \cdot p = 0$, appropriate to physical polarizations, the orthonormality condition $\epsilon_T^\star(g+k) \cdot \epsilon_{T'}(g) = -\delta_{T,T'} + \mathcal{O}(\omega)$ and neglecting $\mathcal{O}(\omega)$ terms in the numerator.

3–40   The results follow by applying the definitions in eqn (3.339) to an arbitrary state; for example,

$$\hat{T}_q^2 | \cdots ; q, i; \cdots \rangle = \hat{T}_q^a \left(T_{ij}^a | \cdots ; q, j; \cdots \rangle \right) = T_{ij}^a T_{jk}^a | \cdots ; q, k; \cdots \rangle$$
$$= C_F \delta_{ik} | \cdots ; q, k; \cdots \rangle = C_F | \cdots ; q, i; \cdots \rangle \, .$$

Using $|q, i; \bar{q}, j\rangle \propto \delta_{ij}$, $|q, i; \bar{q}, j; g, b\rangle \propto T_{ij}^b$ and $|g_1, b; g_2, c; g_3, d\rangle \propto f_{bcd}$, application of the definitions gives

$$(\hat{T}_q + \hat{T}_{\bar{q}}) | q, i; \bar{q}, j\rangle \propto T_{ij}^a - T_{ij}^a$$
$$(\hat{T}_q + \hat{T}_{\bar{q}} + \hat{T}_g) | q, i; \bar{q}, j; g, b\rangle \propto T_{ik}^a T_{kj}^b - T_{kj}^a T_{ik}^b - \mathrm{i} f_{abc} T_{ij}^c$$
$$(\hat{T}_{g_1} + \hat{T}_{g_2} + \hat{T}_{g_3}) | g_1, b; g_2, c; g_3, d\rangle \propto f_{abe} f_{ecd} + f_{ace} f_{bed} + f_{ade} f_{bce} \, ,$$

all of which are zero, either manifestly or due to the closure property eqn (A.10) or the Jacobi identity eqn (A.12).

3–41   To obtain the leading behaviour of a matrix element (squared) in the soft-gluon limit, retain those terms with the most factors of g^μ in the denominator and least in the numerator. In this instance, eqn (3.107), substitute $q \cdot Q \approx \bar{q} \cdot Q \approx q \cdot \bar{q} \approx Q^2/2$ and $Q^\mu = q^\mu + \bar{q}^\mu$ into the tensor and compare the result with eqn (3.85):

$$H_{\mu\nu}(\mathrm{q\bar{q}g}) = \frac{4(ee_q)^2 g_s^2 C_F N_c}{(q \cdot g)(\bar{q} \cdot g)} \times 2(q \cdot \bar{q})[q_\mu \bar{q}_\nu + \bar{q}_\mu q_\nu - (Q^2/2)\eta_{\mu\nu}]$$
$$= g_s^2 C_F \frac{q \cdot \bar{q}}{(q \cdot g)(\bar{q} \cdot g)} \times H_{\mu\nu}(\mathrm{q\bar{q}}) \, .$$

4–1   The result follows from four-momentum conservation, in particular p_- conservation with $(E - p_z) = (p_\perp^2 + m^2)/(E + p_z)$,

$$\left(P_+, -\boldsymbol{q}_\perp, \frac{q_\perp^2 - t_b}{P_+} \right) = \left(zP_+, \boldsymbol{0}, -\frac{t_a}{zP_+} \right) + \left((1-z)P_+, +\boldsymbol{q}_\perp, \frac{q_\perp^2 + t_{a'}}{(1-z)P_+} \right) \, .$$

In a chain of branchings, since $x = z_1 z_2 \cdots z_n$, we require $z \geq x$ and thus $z_{\min} = x$. To obtain an upper bound we work in the strongly ordered limit $t_b \ll t_a$, so that

$$0 \leq \frac{t_a}{z} - \frac{t_{a'}}{1-z} \implies z \leq 1 - \frac{t_{a'}}{t_a + t_{a'}} .$$

The upper limit is largest when $t_{a'}$ is smallest, so that we have

$$[z_{\min}, z_{\max}] = [x, 1 - \epsilon^s_{a'}(t_a)] \quad \text{with} \quad \epsilon^s_{a'}(t_a) = \frac{t_0^{a'}}{(t_a + t_0^{a'})} < \epsilon_{a'}(t_a) .$$

Naïvely, the requirement that $q_\perp^2 \geq 0$ gives $t_b < t_a/z$ and $t_{a'} < t_a(1-z)/z$. However, the first limit does not guarantee strong ordering, $t_b < t_a$, which therefore has to be imposed dynamically.

4–2 The t-derivative of $\ln \Pi'_a(x, t, t_s)$ can be read off from its definition. Proceeding to $\Pi_a(x, t, t_s)$, we obtain

$$t\frac{\partial \Pi_a}{\partial t} = -\Pi_a \left\{ \frac{1}{f_{a/h}(x,t)} t\frac{\partial}{\partial t} f_{a/h}(x,t) + \sum_{a \to cc'} \int_{\epsilon^s_c(t)}^{1-\epsilon^s_{c'}(t)} dz \frac{\alpha_s(z,t)}{2\pi} \bar{P}_{ca}(z) \right\}$$

$$= -\frac{\Pi_a}{f_{a/h}(x,t)} \int_x^{1-\epsilon^s_{a'}(t)} \frac{dz}{z} \frac{\alpha_s(z,t)}{2\pi} \bar{P}_{ab}(z) f_{b/h}(x/z,t) ,$$

where we have used eqn (4.4) to eliminate $t\partial f_{a/h}/\partial t$. Thus both derivatives are given by

$$t\frac{\partial \ln \Pi_a}{\partial t} = t\frac{\partial \ln \Pi'_a}{\partial t} = - \int_x^{1-\epsilon^s_{a'}(t)} \frac{dz}{z} \frac{\alpha_s(z,t)}{2\pi} \bar{P}_{ab}(z) \frac{f_{b/h}(x/z,t)}{f_{a/h}(x,t)} .$$

Furthermore, $\Pi_a(x, t_s, t_s) = \Pi'_a(x, t_s, t_s) = 1$ so that $\Pi_a(x, t, t_s) = \Pi'_a(x, t, t_s)$ for all t by integration of the above equation.

Concerning the interpretation, the form $\Pi'_a(x, t, t_s)$ satisfies the following equation for the probability of parton a evolving from the scale t to $t_s < t$ without any resolvable radiation,

$$\mathcal{P}(t + \delta t \to t_s \mid \text{no res. rad.}) = \mathcal{P}(t \to t_s \mid \text{no res. rad.})$$

$$\times \left[1 - \frac{1}{f_{a/h}(x,t)} \int_x^{1-\epsilon_{a'}(t)} \frac{dz}{z} \frac{\alpha_s(t,z)}{2\pi} \bar{P}_{ab}(z) \frac{\delta t}{t} f_{b/h}\left(\frac{x}{z}, t\right) + \mathcal{O}(\delta t^2) \right] ;$$

c.f. eqn (4.14). The term in square brackets gives the fraction of partons of type a at scale $t + \delta t$ which did not come from a branching $b \to aa'$ in the interval t to $t + \delta t$. That is, it is the fraction of partons of type a which evolved from $t + \delta t$ to t without any resolved radiation.

4–3 Referring to Fig. F.3 and using the uncertainty principle, Ex. (3-2), the distance travelled by the intermediate parton prior to emission is given by $d_\parallel =$

$(p_i + k)/(p_i + k)^2 \approx p_i/(2p_i \cdot k) = 1/(2\omega(1 - \cos\theta_{ik}))$. Likewise, the separation between partons i and j at the time of emission is given by $d_\perp = d_\parallel \tan\theta_{ij} = \tan\theta_{ij}/(2\omega(1 - \cos\theta_{ik}))$. This should be compared to the transverse wavelength of the emitted gluon, $\lambda_\perp = 1/k_\perp = 1/(\omega\sin\theta_{ik})$, where we assume that the gluon is sufficiently soft that parton i is essentially undeflected. In order to resolve the charges on i and j we require $\lambda_\perp < d_\perp$. Substituting the expressions derived above we find

$$\lambda_\perp < d_\perp \quad \Rightarrow \quad \frac{1}{\sin\theta_{ik}} < \frac{\tan\theta_{ij}}{2(1 - \cos\theta_{ik})} \quad \Rightarrow \quad 2\tan\frac{\theta_{ik}}{2} < \tan\theta_{ij} \quad \Rightarrow \quad \theta_{ik} < \theta_{jk} \ .$$

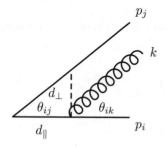

FIG. F.3. The radiation of a soft gluon from two colour-connected partons

4–4 Using the angles defined in Fig. 4.2 we parameterize the directions of the hard partons and soft gluon using $\boldsymbol{n}_i = (0,0,1)$, $\boldsymbol{n}_j = (\sin\theta_{ij}, 0, \cos\theta_{ij})$ and $\boldsymbol{n}_k = (\sin\theta_{ik}\cos\phi, \sin\theta_{ik}\sin\phi, \cos\theta_{ik})$, which implies

$$\cos\theta_{jk} = \boldsymbol{n}_j \cdot \boldsymbol{n}_k = \sin\theta_{ij}\sin\theta_{ik}\cos\phi + \cos\theta_{ij}\cos\theta_{ik} \ .$$

The corresponding four-momenta are given by $p_i = E_i(1, \beta_i\boldsymbol{n}_i)$, $p_j = E_j(1, \beta_j\boldsymbol{n}_j)$ and $k = \omega(1, \boldsymbol{n}_k)$ so that eqn (4.22) becomes

$$\begin{aligned}
W_{ij}^{(i)}(k) &= \frac{1}{2\zeta_{ik}}\left[\frac{(1 - \beta_i\cos\theta_{ik}) - (1 - \beta_i^2)}{1 - \beta_i\cos\theta_{ik}} + \frac{(1 - \beta_i\beta_j\cos\theta_{ij}) - (1 - \beta_i\cos\theta_{ik})}{1 - \beta_j\cos\theta_{jk}}\right] \\
&= \frac{\beta_i}{2\zeta_{ik}}\left[\frac{\beta_i - \cos\theta_{ik}}{1 - \beta_i\cos\theta_{ik}} + \frac{\cos\theta_{ik} - \beta_j\cos\theta_{ij}}{1 - \beta_j\cos\theta_{jk}}\right] \ ,
\end{aligned}$$

which is eqn (4.23). The azimuthal average of the first term is trivial, as it does not depend on ϕ. The second term depends on ϕ via $\cos\theta_{jk}$. To treat this second term convert the integral from 0 to 2π into a contour integral around the unit circle using $z = e^{i\phi}$, giving $d\phi = -i z^{-1}dz$ and $\cos\phi = (z + z^{-1})/2$, and thus

$$\int_0^{2\pi} \frac{d\phi}{2\pi}\frac{1}{(A - B\cos\theta)} = \oint \frac{dz}{2\pi i}\frac{2}{(2Az - B - Bz^2)} = \frac{i}{\pi B}\oint \frac{dz}{(z - z_-)(z - z_+)}$$

$$= \frac{2}{B(z_+ - z_-)} = \frac{1}{\sqrt{A^2 - B^2}} \ ,$$

where we have introduced $A = (1 - \beta_j \cos\theta_{ij} \cos\theta_{ik}) > 0$ and $B = \beta_j \sin\theta_{ij} \sin\theta_{ik}$. The poles occur at $z_\pm = [A \pm \sqrt{A^2 - B^2}]/B$ of which only z_- is within the unit circle ($z_+ = 1/z_-$), it has residue $2\pi i / (z_- - z_+)$. Finally,

$$
\begin{aligned}
A^2 - B^2 &= (1 - \beta_j \cos\theta_{ij} \cos\theta_{ik})^2 - \beta_j^2 \sin^2\theta_{ij} \sin^2\theta_{ik} \\
&= 1 - 2\beta_j \cos\theta_{ij} \cos\theta_{ik} + \beta_j^2 \left[\cos^2\theta_{ij}(1 - \sin^2\theta_{ik}) - \sin^2\theta_{ij}\sin^2\theta_{ik}\right] \\
&= 1 - \cos^2\theta_{ik} + (\cos\theta_{ik} - \beta_j \cos\theta_{ij})^2 - \beta_j^2 \sin^2\theta_{ik} \\
&= (1 - \beta_j^2)\sin^2\theta_{ik} + |\cos\theta_{ik} - \beta_j \cos\theta_{ij}|^2
\end{aligned}
$$

which leads to the quoted result, eqn (4.25), for the azimuthal average.

4–5 We proceed in exactly the same way as for the general case:

$$
\begin{aligned}
\frac{1}{2} J \cdot J^\dagger &= -\hat{T}_i \cdot \hat{T}_j W_{ij} - \hat{T}_i \cdot \hat{T}_l W_{il} - \hat{T}_j \cdot \hat{T}_l W_{jl} \\
&= \hat{T}_i \cdot (\hat{T}_i + \hat{T}_l) W_{ij}^{(i)} + (\hat{T}_j + \hat{T}_l) \cdot \hat{T}_j W_{ij}^{(j)} - \hat{T}_i \cdot \hat{T}_l W_{il} - \hat{T}_j \cdot \hat{T}_l W_{jl} \\
&= \hat{T}_i^2 W_{ij}^{(i)} - \hat{T}_i \cdot \hat{T}_l \left[(W_{il}^{(i)} - W_{ij}^{(i)}) + W_{il}^{(l)}\right] \\
&\quad + \hat{T}_j^2 W_{ij}^{(j)} - \hat{T}_j \cdot \hat{T}_l \left[(W_{jl}^{(j)} - W_{ij}^{(j)}) + W_{jl}^{(l)}\right] \\
&= C_i W_{ij}^{(i)} - \hat{T}_i \cdot \hat{T}_l \left[\tilde{W}_{il}^{(i)} + W_{il}^{(l)}\right] + C_j W_{ij}^{(j)} - \hat{T}_j \cdot \hat{T}_l \left[\tilde{W}_{jl}^{(j)} + W_{jl}^{(l)}\right]
\end{aligned}
$$

Next, for the terms in square brackets, we replace p_i and p_j by $p_s = p_i + p_j$, and obtain

$$
\begin{aligned}
\frac{1}{2} J \cdot J^\dagger &= C_i W_{ij}^{(i)} + C_j W_{ij}^{(j)} - (\hat{T}_i + \hat{T}_j) \cdot \hat{T}_l \left[\tilde{W}_{sl}^{(s)} + W_{sl}^{(l)}\right] \\
&= C_i W_{ij}^{(i)} + C_j W_{ij}^{(j)} + C_s \tilde{W}_{sl}^{(s)} + C_l W_{sl}^{(l)} ,
\end{aligned}
$$

where in the last line we have introduced $\hat{T}_s = \hat{T}_i + \hat{T}_j = -\hat{T}_l$.

4–6 Within the braces we recognize the coefficient of $\ln(1 - z)$ as $\alpha_s \beta_0$, eqn (3.173). Now using the lowest order expression for the running coupling we find

$$
\alpha_s\left[(1 - z)Q^2\right] = \alpha_s(Q^2) - \beta_0 \ln(1 - z)\alpha_s^2(Q^2) + \mathcal{O}(\alpha_s^3) ,
$$

so that making the argument $(1 - z)Q^2$ in the LO expression for α_s gives the $\ln(1 - z)$ contribution of the NLO term. Likewise, using

$$
\Lambda_{\mathrm{MC}} = \exp\left(\frac{C_A(67 - 3\pi^2) - 20T_F n_f}{3(11C_A - 4T_F n_f)}\right) \Lambda_{\overline{\mathrm{MS}}}
$$

gives the constant part of the NLO term. A similar result holds for the g $\to$ gg splitting function. Together they motivate the use of transverse momentum

4–7   First we change variables from (x_1, x_3) to $\left(z = x_3, k_\perp^2 = (1-x_1)(1-x_3)s_{\rm dip}\right)$ in eqn (4.36) to obtain

$$d\sigma = C_{mn}\frac{\alpha_{\rm s}}{2\pi}\frac{x_1^m + z^n}{1-z}dz\frac{dk_\perp^2}{k_\perp^2} \xrightarrow{\;x_1 \to 1\;} \frac{\alpha_{\rm s}}{2\pi}C_{mn}\frac{1+z^n}{1-z}dz\frac{dk_\perp^2}{k_\perp^2}\;,$$

with $x_1 = 1 - k_\perp^2/(1-z)s_{\rm dip}$. For emission off a quark $n = 2$ and the $\bar{P}_{\rm qq}^{(0)}(z)$ splitting function is immediately apparent. For emission off a gluon $n = 3$ but we must be careful to include also the emission off the neighbouring, colour connected dipole with $z \to 1 - z$, which then yields

$$\frac{1+z^3}{1-z} + \frac{1+(1-z)^3}{z} = 2\left[\frac{1-z}{z} + \frac{z}{1-z} + z(1-z)\right] = \bar{P}_{\rm gg}^{(0)}(z)\;.$$

4–8   The invariant mass squared of the first daughter dipole is given by $s_{12} \equiv (p_1 + p_2)^2 = (Q - p_3)^2 = Q^2(1 - 2p_3 \cdot Q/Q^2) = s_{\rm dip}(1 - x_3)$, where Q^μ is the parent dipole's four-momentum and $x_3 = 2p_3 \cdot Q/Q^2$. We then use eqn (4.37) to show that $e^{-2y_1}k_{1\perp}^2/s_{\rm dip} = (1-x_3)^2$ and hence $s_{12} = e^{-y_1}k_{1\perp}\sqrt{s_{\rm dip}}$. Likewise we find $s_{23} = e^{+y_1}k_{1\perp}\sqrt{s_{\rm dip}}$.

Now, in a dipole branching, inspection of eqn (4.37) shows that the largest $k_\perp^{\rm max} = \sqrt{s_{\rm dip}}/2$ occurs for $x_2 = 1$ and $x_1 = \frac{1}{2} = x_3$: dropping the factor $\frac{1}{2}$ gives the triangular phase space approximation in Fig. 4.5. That is, the apex of the triangle given by the invariant mass squared of the dipole. The apexes of the sub-triangles in this $y - \ln(k_\perp/\Lambda)$ plot are given by the intersections of

$$\left.\begin{array}{l}\ln(k_\perp/\Lambda) + y = \ln(\sqrt{s_{\rm dip}}/\Lambda)\\[4pt]\ln(k_\perp/\Lambda) - y = \ln(k_{1\perp}/\Lambda) - y_1\end{array}\right\} \implies \left\{\begin{array}{l}k_\perp^2 = e^{-y_1}k_{1\perp}\sqrt{s_{\rm dip}}\\[4pt]y = \left[y_1 + \ln(\sqrt{s_{\rm dip}}/k_{1\perp})\right]/2\end{array}\right.$$

for the 1–2 dipole and

$$\left.\begin{array}{l}\ln(k_\perp/\Lambda) - y = \ln(\sqrt{s_{\rm dip}}/\Lambda)\\[4pt]\ln(k_\perp/\Lambda) + y = \ln(k_{1\perp}/\Lambda) + y_1\end{array}\right\} \implies \left\{\begin{array}{l}k_\perp^2 = e^{+y_1}k_{1\perp}\sqrt{s_{\rm dip}}\\[4pt]y = \left[y_1 - \ln(\sqrt{s_{\rm dip}}/k_{1\perp})\right]/2\end{array}\right.$$

for the 2–3 dipole.

4–9   We will assume that it is parton 1 which branched so that $Q^2 = s_{12}$ and $t = (E_1 + E_2)^2(1 - \cos\theta_{12}) = (E_1 + E_2)^2 s_{12}/(2E_1 E_2)$. Referring to Ex. (4-8), $Q^2 =$

$e^{-y}k_\perp\sqrt{s_{\text{dip}}}$, so that $\ln(Q^2/\Lambda^2) = \ln(\sqrt{s_{\text{dip}}}/\Lambda) + \ln(k_\perp/\Lambda) - y$. Thus ordering in Q^2 is equivalent to ordering in $\ln(k_\perp/\Lambda) - y$, which is diagonally downwards from the left-hand edge in Fig. 4.5.

Furthermore, referring to eqn (4.37) we have $E_1 = x_1\sqrt{s_{\text{dip}}}/2 = [\sqrt{s_{\text{dip}}} - e^{+y}k_\perp]/2$ and $E_3 = x_3\sqrt{s_{\text{dip}}}/2 = [\sqrt{s_{\text{dip}}} - e^{-y}k_\perp]/2$, so that $E_2 = \sqrt{s_{\text{dip}}} - (E_1 + E_3) = (e^{+y} + e^{-y})k_\perp/2$. This gives

$$t = \frac{1}{2^2}(\sqrt{s_{\text{dip}}} + e^{-y}k_\perp)^2 \frac{2e^{-y}k_\perp\sqrt{s_{\text{dip}}}}{(\sqrt{s_{\text{dip}}} - e^{+y}k_\perp)(e^{+y} + e^{-y})k_\perp}$$

$$= \frac{(\sqrt{s_{\text{dip}}} + e^{-y}k_\perp)^2}{2(1 + e^{2y})(1 - e^{+y}k_\perp/\sqrt{s_{\text{dip}}})} \xrightarrow{k_\perp \ll \sqrt{s_{\text{dip}}}} \frac{s_{\text{dip}}}{2(1 + e^{2y})} \,,$$

so that for $k_\perp \ll \sqrt{s_{\text{dip}}}$, lines of constant t correspond to lines of constant y. At larger $k_\perp$ the numerator requires y to increase for constant t and the contours curve to the right. That is, t ordering moves 'horizontally' outwards in Fig. 4.5.

5–1 The square of the C.o.M. energy is given by

$$s = (p_b + p_t)^2 = (E_b + E_t)^2 - (\boldsymbol{p}_b + \boldsymbol{p}_t)^2$$

with p_b and p_t the four-vectors of the colliding beam and target particles, respectively. The collider-mode is defined by $\boldsymbol{p}_b = -\boldsymbol{p}_t$, $E_b = E_t$ with $\sqrt{s} = 14$ TeV. In the fixed target case one has $\boldsymbol{p}_t = 0$ and $E_t = m$, with $m = 0.938$ GeV/c^2 the mass of the proton, giving

$$s = (E_b + m)^2 - \boldsymbol{p}_b^2 = (E_b^2 - \boldsymbol{p}_b^2) + 2E_b m + m^2 = 2(E_b m + m^2)\,.$$

It follows that a beam energy

$$E_b = \frac{s}{2m} - m \approx 104478 \text{ TeV}$$

is required to match the collider. The relativistic γ-factor is given by

$$\gamma = \frac{E_{\text{tot}}}{m_{\text{tot}}} = \frac{E_b + E_t}{\sqrt{s}} = \frac{\sqrt{s}}{2m} \approx 7463\,.$$

As a consequence of the large Lorentz boost, the secondary particles from such an interaction would be emitted, typically, within a cone of 0.1 mrad in the forward direction.

5–2 The transfer matrix $M_d(L)$ of a drift space is determined by the fact that the direction is not changed while the transverse displacement x is modified by $x'L$,

$$M_d(L) = \begin{pmatrix} 1 & L \\ 0 & 1 \end{pmatrix}\,.$$

The matrix $M_q(f)$ for a focusing quadrupole with focal length f is derived from the condition that a parallel beam is deflected to the focal point independent of

its displacement. In the thin-lens approximation there is no displacement at the
quadrupole, and one finds

$$M_q(f) = \begin{pmatrix} 1 & 0 \\ -1/f & 1 \end{pmatrix} .$$

A defocusing quadrupole is described by the same matrix with a negative value
for the focal length. The combined system of a focusing and a defocusing quadrupole separated by a drift space is then obtained by the product of the respective
matrices, giving

$$M = M_q(-f_2)M_d(L)M_q(f_1) = \begin{pmatrix} 1 - L/f_1 & d \\ 1/f_2 - 1/f_1 - L/f_1 f_2 & 1 + L/f_2 \end{pmatrix} .$$

The entire system has a net focusing effect if the lower left field of the matrix is
negative, that is, for

$$f_1 - f_2 < L.$$

The transfer matrix for a sequence which starts with a defocusing quadrupole
and ends with a focusing gives the focusing condition $f_2 - f_1 < L$. Thus, setting
$f_1 = f_2$ yields an accelerator structure which focuses the beam in both transverse
dimensions.

5–3 In a homogeneous magnetic field the momentum $p = |\boldsymbol{p}|$ and the radius of
curvature r are related by the condition that the Lorentz and centripetal force
are equal,

$$e|\boldsymbol{v} \times \boldsymbol{B}| = evB = \frac{mv^2}{r} , \qquad \text{that is,} \qquad p = eBr .$$

Here $\boldsymbol{v}$ is the velocity of the particle. The momentum, or more precisely the
ratio p/e thus can be measured from the radius of curvature of the trajectory in
the magnetic field. The minimum information needed to determine r are three
points on a segment of the circle. For simplicity let's assume that the points are
evenly spaced, subtending an angle θ, and that the secant from the first to the
third measurement has the length L. It follows

$$\sin\frac{\theta}{2} = \frac{L}{2r} \qquad \text{and} \qquad s = r\left(1 - \cos\frac{\theta}{2}\right) ,$$

where s is the sagitta of the trajectory. With $r \gg L$ one finds

$$s = r\left(1 - \sqrt{1 - \frac{L^2}{4r^2}}\right) \approx \frac{L^2}{8r} ,$$

that is, an expression where the radius of curvature is given as a function of the
observables L and s. The length L is fixed through the positioning of the detector

elements, s is the result of a position measurement which can be assumed to have a Gaussian error. The uncertainty of $1/p$ is thus proportional to the standard deviations of s, and standard error propagation yields

$$\mathrm{d}\left(\frac{1}{p}\right) \propto \mathrm{d}\left(\frac{1}{r}\right) = \text{const.} \qquad \rightarrow \qquad \frac{\mathrm{d}p}{p} \sim p \ .$$

–4 A particle with charge q and momentum $\boldsymbol{p}$ entering a magnetic field with flux density $\boldsymbol{B}$ feels a Lorentz force which deflects it perpendicularly to its velocity and the direction of the field. Denoting the unit vector in the direction of the particle by $\boldsymbol{n}$, the direction of $\boldsymbol{B}$ by $\boldsymbol{b}$ and $p = |\boldsymbol{p}|$ one has

$$q \cdot (\boldsymbol{v} \times \boldsymbol{B}) = \frac{\mathrm{d}\boldsymbol{p}}{\mathrm{d}t} = v\frac{\mathrm{d}\boldsymbol{p}}{\mathrm{d}s} \quad \text{or} \quad qB\,(\boldsymbol{n} \times \boldsymbol{b}) = p\frac{\mathrm{d}\boldsymbol{n}}{\mathrm{d}s} \ .$$

Here $\boldsymbol{v}$ is the velocity of the particle and s the longitudinal coordinate along its trajectory. In case the magnetic field is orthogonal to $\boldsymbol{v}$ the cross product in the above expression is again a unit vector and the deflection angle ϕ is obtained by

$$\mathrm{d}\phi = \frac{|\mathrm{d}\boldsymbol{n}|}{|\boldsymbol{n}|} = \frac{qB}{p}\mathrm{d}s \quad \text{giving} \quad \phi = \frac{q}{p}\int Bds \ .$$

With $q = ze$, where e is the proton charge, p in GeV and $\int Bds$ in Tm=Vs/m, one arrives at the convenient expression

$$\phi = 0.3\ z\ \frac{(\int Bds)/(\text{Tm})}{p/\text{GeV}} \ .$$

Thus, for a field integral $\int Bds = 1.5$ Tm the deflection angle of a 100 GeV pion is only 4.5 mrad. In order to determine the charge of the pion with a significance of three standard deviations one has to measure this angle with a precision of 1.5 mrad. Given two points and a lever arm of $L = 1$ m, the angle of $\phi = 4.5$ mrad leads to a transverse displacement of $D = L \cdot \phi = 4.5$ mm between the two chambers, which has to be measured with an error of $\sigma_D = 1.5$ mm. Thus the required position resolution in each of the two chambers is $\sigma = 1.5\ \text{mm}/\sqrt{2} \approx 1$ mm.

Ignoring the small logarithmic correction, the presence of multiple scattering leads to a Gaussian smearing of the deflection angle,

$$\sigma(\phi)_{\text{MS}} = \frac{13.6\ \text{MeV}}{pc}\sqrt{X} \ ,$$

where X is the thickness of the scatterer in units of radiation lengths. For $p = 5$ GeV one has

$$\sigma(\phi)_{\text{MS}} = 0.00272\sqrt{X}$$

which becomes equal to the contribution $\sigma(\phi)_D = 0.0015$ from the chamber resolution for $X \approx 0.3$. Given this amount of material in front of the second chamber, the momentum resolution is limited by multiple scattering for $p <$ ~~5 GeV and by the chamber resolution for larger momenta.~~

The above example illustrates how the combined effect of multiple scattering and chamber resolution on a momentum measurement can be parameterized. Adding both contributions to the uncertainty in the deflection angle in quadrature yields

$$\sigma^2\left(\frac{z}{p}\right) = \frac{1}{(0.3 \int Bds)^2}\sigma^2(\phi) = \frac{1}{(0.3 \int Bds)^2}\left(\sigma_D^2 + \left(\frac{0.0136}{p}\right)^2 X\right) ,$$

with p in units of GeV, $\int Bds$ in Tm and X in units of radiation lengths. The general structure is of the form

$$\sigma^2\left(\frac{1}{p}\right) \sim a^2 + \frac{b^2}{p^2} ,$$

where the constant term depends on the spatial resolution of the tracking system and the p-dependent term on the amount of material built into it.

5–5 Inside a calorimeter the energy of a primary particle is distributed over N secondaries. The readout system then yields a signal which is proportional to the number of those secondary particles. Since the number of particles created in a shower is subject to Poissonian fluctuations one has

$$dE \sim dN \sim \sqrt{N} \sim \sqrt{E} \quad \text{or} \quad \frac{dE}{E} \sim \frac{1}{\sqrt{E}} .$$

5–6 The relation between particle mass m, momentum p and velocity v is given by

$$\beta = \frac{v}{c} = \frac{pc}{E} = \frac{p}{\sqrt{p^2 + m^2c^2}} .$$

The difference in the arrival times of a 1 GeV/c pion and a kaon travelling over a distance of $L = 2$ m is thus

$$\Delta t = \frac{L}{c}\left(\frac{1}{\beta_K} - \frac{1}{\beta_\pi}\right) = \frac{L}{c}\left(\sqrt{1 + \frac{m_K^2c^2}{p^2}} - \sqrt{1 + \frac{m_\pi^2c^2}{p^2}}\right) \approx 700 \text{ ps} .$$

A separation at the level of three standard deviations thus requires a time resolution around $\sigma_t = 230$ ps.

5–7 The transfer matrix formalism introduced earlier to describe the beam transport system of a particle accelerator can equally be applied to conventional optics. Here, we are dealing with Cherenkov light emitted from a relativistic particle, which falls onto a focusing lens with focal length f and is viewed at a distance f behind the lens. At the z-position of the lens it can be described by state vectors $(x, x')^T$ and $(y, y')^T$ which specify the transverse position (x, y) and the directions $x' = dx/dz$ and $y' = dy/dz$ of the photons. The resulting image can be determined by multiplying the state vectors with the transfer matrices of the lens and that of a drift space of length f. For the x-direction one gets

$$\begin{pmatrix} 1 & f \\ 0 & 1 \end{pmatrix} \begin{pmatrix} 1 & 0 \\ -1/f & 1 \end{pmatrix} \begin{pmatrix} x \\ x' \end{pmatrix} = \begin{pmatrix} fx' \\ x' - x/f \end{pmatrix} ;$$

an analogous result is obtained for y. The image of the photons is then given by the coordinates $(x_I, y_I) = (fx', fy')$. For a particle travelling along the optical axis of the imaging system, the photons are emitted on the cone $(x', y') = (\sin\theta \cos\phi, \sin\theta \sin\phi)$, and the image is a ring around a centre which corresponds to the direction of the particle. The radius of the ring is proportional to the product $f \sin\theta$, that is, given the radius and the index of refraction of the radiator medium one can extract the velocity of the particle.

6–1 We have to show that $T^{(m+1)} \to T^{(m)}$, with

$$T^{(m+1)} = \max_{n'} \frac{\sum\limits_{i=1}^{m+1} |\boldsymbol{p}_i \cdot \boldsymbol{n}'|}{\sum\limits_{i=1}^{m+1} |\boldsymbol{p}_i|} , \quad T^{(m)} = \max_{n} \frac{\sum\limits_{i=1}^{m} |\boldsymbol{p}_i \cdot \boldsymbol{n}|}{\sum\limits_{i=1}^{m} |\boldsymbol{p}_i|}$$

if one of the $(m+1)$ particles' energy vanishes, or if two of the $(m+1)$ particles are collinear.

Let us consider the first case. For massless particles we have to study the situation $E_k \to 0 \Rightarrow |\boldsymbol{p}_k| \to 0$. For simplicity we choose $k = m + 1$. Now it is easy to see that the numerator as well as the denominator show the expected behaviour, namely

$$\sum_{i=1}^{m+1} |\boldsymbol{p}_i| = \sum_{i=1}^{m} |\boldsymbol{p}_i| + |\boldsymbol{p}_k| \to \sum_{i=1}^{m} |\boldsymbol{p}_i|$$

$$\sum_{i=1}^{m+1} |\boldsymbol{p}_i \cdot \boldsymbol{n}'| = \sum_{i=1}^{m} |\boldsymbol{p}_i \cdot \boldsymbol{n}'| + |\boldsymbol{p}_k \cdot \boldsymbol{n}'| \to \sum_{i=1}^{m} |\boldsymbol{p}_i \cdot \boldsymbol{n}'|$$

for $|\boldsymbol{p}_k| \to 0$. Since now both denominator and numerator are the same as for $T^{(m)}$, it is clear that the maximum under variation of $\boldsymbol{n}'$ has to be found for the same direction $n_{\max}$ as in the case of m particles. This optimal direction is generally referred to as the Thrust axis n_T. So in this limit we have $T^{(m+1)} = T^{(m)}$.

Next we have to test the collinear safety, so we have to study the situation $\boldsymbol{p}_j \parallel \boldsymbol{p}_k$. Let us call $\boldsymbol{p} = \boldsymbol{p}_j + \boldsymbol{p}_k$. Because of the properties of parallel vectors,

$$|\boldsymbol{p}_j \cdot \boldsymbol{n}'| + |\boldsymbol{p}_k \cdot \boldsymbol{n}'| = |\boldsymbol{p} \cdot \boldsymbol{n}'| \quad , \quad |\boldsymbol{p}_j| + |\boldsymbol{p}_k| = |\boldsymbol{p}| \quad ,$$

it follows again that the denominator as well as the numerator reduce to the m-particle case, therefore $T^{(m+1)} \to T^{(m)}$.

6–2   The definition of rapidity is

$$y = \frac{1}{2} \ln \frac{E + p_z}{E - p_z} \, .$$

Now let us apply a Lorentz transformation to the particles' momentum, in particular a boost along the z-axis with velocity β. If the four-momentum in the original frame is $p = (E, p_x, p_y, p_z)$, then the transformed momentum $p' = (E', p'_x, p'_y, p'_z)$ is

$$E' = \gamma(E - \beta p_z) \, , \quad p'_x = p_x \, , \quad p'_y = p_y \, , \quad p'_z = \gamma(p_z - \beta E) \, ,$$

with $\gamma^2 = (1 - \beta^2)^{-1}$. From the above equations it follows that

$$E' + p'_z = \gamma(1 - \beta)(E + p_z) \, , \quad E' - p'_z = \gamma(1 + \beta)(E - p_z) \, .$$

Thus the rapidity y' in the boosted frame is

$$y' = \frac{1}{2} \ln \frac{E' + p'_z}{E' - p'_z} = \frac{1}{2} \ln \frac{E + p_z}{E - p_z} + \frac{1}{2} \ln \frac{1 - \beta}{1 + \beta} = y + f(\beta) \, ,$$

that is, the boosted rapidity is the same as the original one up to a function of β, only. The difference in rapidities of two particles is invariant under Lorentz boosts, since $f(\beta)$ cancels out in the difference. Identifying the jet directions with the directions of primary scattered partons, it follows that one actually does not need to know the boost of the two-parton system along the z-axis in order to infer the C.o.M. scattering angle.

6–3   We start our discussion by defining the relevant momenta in the laboratory frame, where the z-axis coincides with the proton's direction of flight. Thus the proton has four-momentum $p = (E_p, 0, 0, p_z)$, and the exchanged virtual photon has the four-momentum $q^{\mathrm{Lab}} = (q_0^{\mathrm{Lab}}, \boldsymbol{q}^{\mathrm{Lab}})$ with $(q^{\mathrm{Lab}})^2 = -Q^2 < 0$. As a first step we apply a boost to these momenta in order to transform them into the rest frame of the proton. Such a transformation exists, since the proton has a finite mass $p^2 = M^2 > 0$. The necessary boost velocity β_1 is found by the requirement $p'_z = 0$, where p'_z is the z-component of the proton's momentum in its rest frame,

$$0 = p'_z = \gamma_1(p_z - \beta_1 E_{\mathrm{p}}) \ \Rightarrow \ \beta_1 = p_z/E_{\mathrm{p}} \in [-1,1]\,,$$

with $\gamma_1^2 = (1-\beta_1^2)^{-1}$. Applying this transformation to both particles, we find new four-momenta $p'_z = (M,0,0,0)$ and $q = (q_0, q_x, q_y, q_z) = (q_0, \boldsymbol{q})$. Next a spatial rotation is applied such that the photon's momentum takes on the form $q' = (q'_0, 0, 0, q'_z)$, with $(q')^2 = (q^{\mathrm{Lab}})^2 < 0$. The zero component remains unchanged under spatial rotations, hence $q'_0 = q_0$, and $|q'_z| = |\boldsymbol{q}|$. The four-momentum of the proton obviously is invariant. Finally, again a Lorentz boost is applied, such that the boosted photon momentum becomes $q'' = (0,0,0,q''_z)$. This is the definition of the Breit frame of reference. Such a boost exists,

$$0 = q''_0 = \gamma_2(q'_0 - \beta_2 q'_z) \ \Rightarrow \ \beta_2 = q'_0/q'_z \in [-1,1]\,.$$

The last statement is true because of

$$(q')^2 = (q'_0)^2 - (q'_z)^2 < 0 \ \Rightarrow \ |q'_0| < |q'_z|\,.$$

Let us briefly discuss the sign of the photon momentum and the last boost. If we start out with the proton moving in the positive and the electron in the negative z-direction, then after the boost into the proton's rest frame the photon will have a negative z-component of its momentum, $q'_z < 0$. Hence, we see that $\beta_2 < 0$. The new proton momentum is $p''_z = (\gamma_2 M, 0, 0, -\gamma_2 \beta_2 M)$. This can be rewritten using

$$1 - \beta_2^2 = 1 - \frac{(q'_0)^2}{(q'_z)^2} = \frac{Q^2}{(q'_z)^2}\,,$$

and, hence,

$$\gamma_2 = \frac{|q'_z|}{Q} \quad \text{and} \quad \gamma_2 \beta_2 = \frac{\nu}{Q}\frac{|q'_z|}{q'_z} = -\frac{\nu}{Q}\,.$$

In the last equation we have introduced the definition $\nu = q'_0 = E_{\mathrm{e}}^{\mathrm{in}} - E_{\mathrm{e}}^{\mathrm{out}}$, which is the energy loss of the scattered electron in the proton's rest frame. So, finally, we find the proton four-momentum as

$$p'' = \left(\gamma_2 M, 0, 0, \frac{\nu}{Q}M\right) = \left(\gamma_2 M, 0, 0, \frac{1}{2}\frac{Q}{x}\right)\,,$$

where the definition of the variable $x = Q^2/(2M\nu)$ from Section 3.2.2 has been substituted.

For the following considerations we restrict ourselves to the parton model. In this model the proton is built out of partons, each parton i carrying a fraction x_i of the proton's longitudinal momentum. The actual scattering process between the electron and the proton is described as incoherent scattering between the virtual photon and one of the partons. As was shown in Section 3.2.2, to leading order one has $x_i = x$.

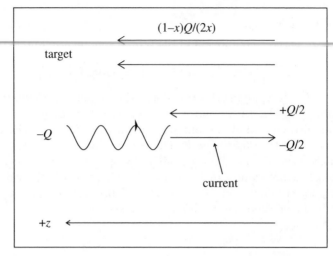

FIG. F.4. Configuration of longitudinal momenta in the Breit frame of reference

Now we see that in the Breit frame of reference, to first approximation, the scattering process occurs between a photon with momentum

$$q'' = (0, 0, 0, q''_z) = (0, 0, 0, -Q)$$

moving in the negative z-direction, and a (massless) parton with momentum

$$p''_{\text{par,in}} = \left(\frac{1}{2}Q, 0, 0, \frac{1}{2}Q \right)$$

moving in the positive z-direction. The zero-component is obtained from $(p''_{\text{par,in}})^2 = 0$. The scattered (outgoing) parton has to have momentum

$$p''_{\text{par,out}} = p''_{\text{par,in}} + q'' = \left(\frac{1}{2}Q, 0, 0, -\frac{1}{2}Q \right) .$$

This configuration of momenta is represented graphically in Fig. F.4. The hadrons produced from the outgoing parton can have a maximum longitudinal momentum with respect to the proton of $-\frac{1}{2}Q$, and all particles carrying negative longitudinal momentum have to stem from this parton and lie in the so-called *current hemisphere*, as depicted in Fig. F.5. The other partons in the proton, which do not participate in the scattering, obviously have to carry longitudinal momentum $(1 - x)Q/(2x)$. Therefore, all the hadrons produced out of this remaining system will have momenta within the so-called *target hemisphere*, that is, positive longitudinal momenta with respect to the proton's momentum, with the limit set by $p_{\text{L}}^{\text{max}} = (1 - x)Q/(2x)$, as shown in Fig. F.5.

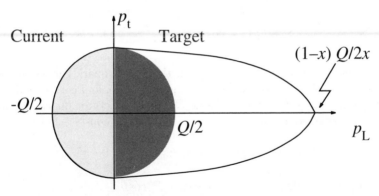

FIG. F.5. The current and the target hemisphere for the Breit frame of reference. The variables are the momentum transfer Q, the Bjorken scaling variable x and the longitudinal (transverse) momentum p_t (p_L). Figure from ZEUS Collab.(1999c).

-4 For massless particles energy–momentum conservation gives the following constraints:

$$E_p + E_e = E'_e + \sum_a E_a , \quad E_p - E_e = E'_e \cos\theta_e + \sum_a E_a \cos\theta_a , \qquad (F.5)$$

where the sum runs over all hadrons a in the final state. From there we derive

$$2E_e = \sum_a E_a (1 - \cos\theta_a) + E'_e (1 - \cos\theta_e) . \qquad (F.6)$$

Now, we define $\Sigma = \sum_a E_a (1 - \cos\theta_a)$, and insert $\Sigma = 2E_e - E'_e (1 - \cos\theta_e)$ in the expression for y obtained from the electron method, eqn (6.11),

$$y_e = 1 - \frac{E'_e}{2E_e} (1 - \cos\theta_e) = \frac{\Sigma}{2E_e} = y_h . \qquad (F.7)$$

The transverse momentum in the initial state is zero, therefore, the final state has to satisfy

$$E'_e \sin\theta_e + P_{\perp,h} = 0 , \quad P_{\perp,h} = \sum_a E_a \sin\theta_a . \qquad (F.8)$$

Starting again from the expression in the electron method, eqn (6.12), it is now easy to express Q^2 in terms of the hadron observables:

$$Q^2 = \frac{E'^2_e \sin^2\theta_e}{1 - y_e} = \frac{P^2_{\perp,h}}{1 - y_h} = Q^2_h . \qquad (F.9)$$

The expressions for the sigma method are simply obtained by replacing the factor $2E_e$ in eqn (F.7) by eqn (F.6), and replacing y_h by y_Σ in the expression for Q^2,

$$y_\Sigma = \frac{\Sigma}{\Sigma + E_e'\left(1 - \cos\theta_e\right)} \quad , \quad Q_\Sigma^2 = \frac{E_e'^2 \sin^2\theta_e}{1 - y_\Sigma} . \tag{F.10}$$

Finally, when going to the parton model, the struck quark carries momentum and energy xE_p (E_q) in the incoming (outgoing) state, and correspondingly the non-interacting spectator quarks $(1-x)E_p$. Since spectator quarks do not pick up any transverse momentum, the energy–momentum constraints turn out to be

$$xE_p + (1-x)E_p + E_e = E_q + E_e' + (1-x)E_p \tag{F.11}$$

$$xE_p + (1-x)E_p - E_e = E_q \cos\theta_q + E_e' \cos\theta_e + (1-x)E_p \tag{F.12}$$

$$\text{and} \quad E_e' \sin\theta_e = -E_q \sin\theta_q . \tag{F.13}$$

Combining eqn (F.11) and eqn (F.12), and taking the definition of Σ from eqn (F.6) we arrive at

$$\Sigma = E_q \left(1 - \cos\theta_q\right) . \tag{F.14}$$

From eqn (F.13) we obtain

$$E_q^2 = \frac{(E_e' \sin\theta_e)^2}{\sin^2\theta_q} = \frac{P_{\perp,h}^2}{\sin^2\theta_q} . \tag{F.15}$$

By combining eqns (F.14) and (F.15) we arrive at a quadratic equation for $\eta = \cos\theta_q$,

$$(1+a)\eta^2 - 2\eta + 1 - a = 0 , \quad a = \frac{\Sigma^2}{P_{\perp,h}^2} . \tag{F.16}$$

This equation has two solutions, $\eta_1 = 1$ and

$$\eta_2 = \cos\theta_q = \frac{P_{\perp,h}^2 - \Sigma^2}{P_{\perp,h}^2 + \Sigma^2} \tag{F.17}$$

which is also defined as the hadronic angle $\cos\theta_h$.

7–1 Let us use the notation of Fig. 2.1. The incoming lepton, a positron in Fig. F.6, has four-momentum l, the incoming and struck quark, a d- ($\bar{u}$-) quark in Fig. F.6 left (right), has the four-momentum p_q, and the outgoing lepton, an antielectron neutrino in Fig. F.6, carries the four-momentum l'. We have learned that the deep inelastic lepton–proton scattering can be described as an incoherent sum over elastic lepton–quark scatterings. Since we will neglect all masses, we have to consider an elastic scattering of two massless particles. Furthermore, let us consider the scattering in the C.o.M. frame where $l = -p_q$, and thus $E = E_q$

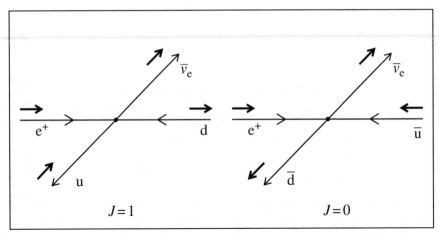

FIG. F.6. Configuration of momenta and spins in a positron–quark scattering process

and $E = E'$. The C.o.M. scattering angle $\theta_e^\star$ of the lepton is then given by the following ratio of Lorentz scalars:

$$\frac{p_q \cdot l'}{p_q \cdot l} = \frac{E_q E' - p_q \cdot l'}{E_q E - p_q \cdot l} = \frac{EE' + l \cdot l'}{EE + l \cdot l}$$

$$= \frac{E^2(1 + \cos\theta_e^\star)}{2E^2} = \frac{1}{2}(1 + \cos\theta_e^\star) = \cos^2\frac{\theta_e^\star}{2}.$$

The momentum transfer is defined as $q = l - l'$, the inelasticity is $y = (p \cdot q)/(p \cdot l)$, where p is the proton momentum. The incoming quark carries a fraction x of the proton's momentum, that is, $p_q = xp$. Using these relations we find

$$\cos^2\frac{\theta_e^\star}{2} = \frac{p_q \cdot l'}{p_q \cdot l} = 1 - \frac{p_q \cdot q}{p_q \cdot l} = 1 - \frac{xp \cdot q}{xp \cdot l} = 1 - y.$$

Let us now consider the specific case of a charged current interaction as depicted in Fig. F.6, which occurs via the exchange of a W^+. Remember that only (right)left-handed (anti)particles participate in a weak interaction. This is indicated by the small arrows showing the spin vectors which are antiparallel to the momentum for particles and parallel for antiparticles. The left figure shows a positron scattering off a down quark, leading to an outgoing antielectron neutrino and a u-quark. The total spin of the incoming system is $J = +1$. Angular momentum conservation requires the same total spin for the outgoing system. From spin algebra we know that the amplitude (projection) of two spin-1 states at an angle θ is given by $(1 + \cos\theta)/2 = \cos^2(\theta/2)$. Therefore the cross section for e^+d scattering is proportional to $\cos^4(\theta_e^\star/2) = (1 - y)^2$. On the other hand, if the positron scatters off a $\bar{u}$-quark, Fig. F.6 (right), then the total spin of the system is $J = 0$, and the outgoing particle is emitted with an isotropic angular distribution.

7-2 For simplicity, let us neglect the proton mass and also the contribution from the longitudinal structure function, that is, $R_L = 0$. Furthermore, we will use the notation $\kappa = (M_W^2/(Q^2 + M_W^2))^2$. Then the neutrino–proton deep inelastic scattering cross section according to eqn (7.14) is given by

$$\frac{\mathrm{d}^2\sigma^{\nu p}}{\mathrm{d}x\,\mathrm{d}Q^2} = \frac{G_F^2}{4\pi x}\,\kappa\,\left[Y_+\,F_2^{\nu p}(x,Q^2) + Y_-\,xF_3^{\nu p}(x,Q^2)\right]\,,$$

where we have used the definition $Y_\pm = 1 \pm (1 - y)^2$. From simple crossing symmetry we expect that the charged current cross section for e^+p scattering is of the same form, that is,

$$\frac{\mathrm{d}^2\sigma_{CC}^{e^+p}}{\mathrm{d}x\,\mathrm{d}Q^2} = \frac{G_F^2}{4\pi x}\,\kappa\,\left[Y_+\,F_2^{e^+p}(x,Q^2) + Y_-\,xF_3^{e^+p}(x,Q^2)\right]\,.$$

We assume that the lepton beam is fully polarized, that is, all positrons are right-handed. Now we shall find out how these structure functions are related to each other. First, we insert the leading order expressions for $F_{2,3}^{\nu p}$ in terms of the parton distribution functions, given by eqn (7.16):

$$\frac{\mathrm{d}^2\sigma^{\nu p}}{\mathrm{d}x\,\mathrm{d}Q^2} = \frac{G_F^2}{\pi}\,\kappa\,\left[(d(x) + s(x)) + (1-y)^2\,(\bar{u}(x) + \bar{c}(x))\right]\,.$$

Then, we compare this to the expression for the e^+p cross section in terms of the parton distribution functions, given in eqns (7.6) and (7.7):

$$\frac{\mathrm{d}^2\sigma_{CC}^{e^+p}}{\mathrm{d}x\,\mathrm{d}Q^2} = \frac{G_F^2}{\pi}\,\kappa\,\left[(\bar{u}(x) + \bar{c}(x)) + (1-y)^2\,(d(x) + s(x))\right]\,.$$

From this, the relations $F_2^{e^+p} = F_2^{\nu p}$ and $xF_3^{e^+p} = -xF_3^{\nu p}$ can be deduced. The inverted $(1 - y)^2$ behaviour with respect to quarks and antiquarks can be understood along exactly the same lines as discussed in Ex. (7-1). The e^+p scattering is discussed there. In neutrino–proton scattering the incoming neutrino is left-handed. Therefore, when scattering off a right-handed $\bar{u}$-quark, the total spin is $J = 1$, leading to a $\cos^4(\theta/2) = (1 - y)^2$ angular dependence for the outgoing particles. However, when scattering off a left-handed down-type quark, the net spin is $J = 0$, which results in a flat angular dependence.

7-3 To start, we write down the F_2 structure functions for neutral current electron–proton and electron–neutron scattering:

$$F_2^{e^-p}(x) = x\left[\frac{4}{9}\,(u + \bar{u} + c + \bar{c}) + \frac{1}{9}\,(d + \bar{d} + s + \bar{s})\right]\,,$$

$$F_2^{e^- n}(x) = x \left[\frac{4}{9} \left(d + \bar{d} + c + \bar{c} \right) + \frac{1}{9} \left(u + \bar{u} + s + \bar{s} \right) \right].$$

We omit the argument of the parton distribution functions. The second equality is obtained assuming isospin invariance, that is, $u = u^{\mathrm{P}} = d^{\mathrm{n}}$ and $d = d^{\mathrm{P}} = u^{\mathrm{n}}$. The nucleon structure function is defined as

$$
\begin{aligned}
F_2^{e^- N}(x) &= \frac{1}{2} \left(F_2^{e^- P}(x) + F_2^{e^- n}(x) \right) \\
&= x \left[\frac{5}{18} \left(u + \bar{u} + d + \bar{d} \right) + \frac{1}{9} \left(s + \bar{s} \right) + \frac{4}{9} \left(c + \bar{c} \right) \right] \\
&= x \left[\frac{5}{18} \left(q + \bar{q} \right) - \frac{1}{6} \left(s + \bar{s} \right) + \frac{1}{6} \left(c + \bar{c} \right) \right],
\end{aligned}
$$

where we have introduced the short-hand notation $(q + \bar{q}) = (u + \bar{u} + d + \bar{d} + s + \bar{s} + c + \bar{c})$. The structure function for neutrino–nucleon scattering has been given in eqn (7.18). Assuming $s = \bar{s}$ and using the above short-hand notation, it can be cast in the simple form $F_2^{\nu N}(x) = x\,(q + \bar{q})$. Now, it is easy to see that the ratio is given by

$$
\begin{aligned}
\frac{F_2^{e^- N}}{F_2^{\nu N}} &= \frac{5}{18} \frac{(q + \bar{q}) - \frac{3}{5}(s + \bar{s}) + \frac{3}{5}(c + \bar{c})}{q + \bar{q}} \\
&= \frac{5}{18} \left[1 - \frac{3}{5} \frac{s + \bar{s} - (c + \bar{c})}{q + \bar{q}} \right] \approx \frac{5}{18} \left[1 - \frac{3}{5} \frac{s + \bar{s}}{q + \bar{q}} \right].
\end{aligned}
$$

The last approximation is valid because of the small charm contribution to the sea. This result can also be rewritten as an expression for $x(s + \bar{s})$, by simply inserting $F_2^{\nu N} = x(q + \bar{q})$ and inverting the last equation:

$$x(s + \bar{s}) = \frac{5}{3} F_2^{\nu N} - 6 F_2^{e^- N}.$$

7–4 In Ex. (7-2) we have seen that the neutrino–nucleon DIS cross section is given by

$$\frac{d^2 \sigma^{\nu(\bar{\nu})}}{dx\,dQ^2} = \frac{G_F^2}{4\pi x} \kappa \left[Y_+ \, F_2^{\nu(\bar{\nu})}(x, Q^2) \pm Y_- \, x F_3^{\nu(\bar{\nu})}(x, Q^2) \right].$$

We rewrite it as a differential cross section in y, using $Q^2 = sxy$ and $dQ^2 = sx\,dy$. In the proton rest frame we have $s = 2ME_\nu$. When integrating over x, we find

$$\frac{1}{E_\nu} \frac{d\sigma^{\nu(\bar{\nu})}}{dy} = \frac{G_F^2}{2\pi} M\kappa \left[Y_+ \, \tilde{F}_2^{\nu(\bar{\nu})}(Q^2) \pm Y_- \int_0^1 dx\, x F_3^{\nu(\bar{\nu})}(x, Q^2) \right],$$

with $\tilde{F}_2^{\nu(\bar{\nu})}(Q^2) = \int_0^1 dx\, F_2^{\nu(\bar{\nu})}(x, Q^2)$. Now let us study the limit $y \to 0$. We have $\lim_{y \to 0} Y_+ = 2$, $\lim_{y \to 0} Y_- = 0$, $\lim_{y \to 0} Q^2 = 0$ and $\lim_{y \to 0} \kappa = 1$. Therefore,

$$\lim_{y \to 0} \frac{1}{E_\nu} \frac{d\sigma^{\nu(\bar\nu)}}{dy} = \frac{G_F^2}{\pi} M \tilde{F}_2^{\nu(\bar\nu)}(Q^2 \to 0) = C \,,$$

where the constant C is independent of energy and the same for neutrinos and antineutrinos. The neutrino flux Φ is related to the number N of observed events by $dN^{\nu(\bar\nu)}/dy = \Phi^{\nu(\bar\nu)} d\sigma^{\nu(\bar\nu)}/dy$. Measuring the ratio $E_{\bar\nu}dN^\nu/(E_\nu\, dN^{\bar\nu})$ in the limit $y \to 0$ the constant C cancels, thus giving the ratio of neutrino and antineutrino fluxes. Experimentally the limit is studied by extrapolating the E_{had} dependence of the counting rate to $E_{\mathrm{had}} = 0$, because of $y = E_{\mathrm{had}}/E_\nu$.

7–5 The constraints for the proton of total electric charge 1, baryon number 1, and zero net strangeness and charm can be formulated as integrals over the full x range of the parton distribution functions, weighted by the relevant charge:

$$\int_0^1 dx \left[\frac{2}{3} \left(u - \bar u + c - \bar c \right) - \frac{1}{3} \left(d - \bar d + s - \bar s \right) \right] = 1 \,,$$

$$\int_0^1 dx \left[\frac{1}{3} \left(u - \bar u + c - \bar c \right) + \frac{1}{3} \left(d - \bar d + s - \bar s \right) \right] = 1 \,,$$

$$\int_0^1 dx \, (s - \bar s) = \int_0^1 dx \, (c - \bar c) = 0 \,.$$

Now simply insert the constraints of the last line into the first two equations and solve the remaining equation system. This results in

$$\int_0^1 dx \, (u - \bar u) = \int_0^1 dx \, u_{\mathrm v} = 2 \quad, \quad \int_0^1 dx \, (d - \bar d) = \int_0^1 dx \, d_{\mathrm v} = 1 \,.$$

8–1 With n an arbitrary unit vector, one finds

$$n^T W n = \frac{\sum_{\boldsymbol p}(n^T \boldsymbol p)(\boldsymbol p^T n)}{\sum_{\boldsymbol p} \boldsymbol p^2} = \frac{\sum_{\boldsymbol p} p_{\mathrm L}^2}{\sum_{\boldsymbol p} \boldsymbol p^2} = 1 - \frac{\sum_{\boldsymbol p} p_{\mathrm T}^2}{\sum_{\boldsymbol p} \boldsymbol p^2} \,,$$

where $p_{\mathrm L}$ and $p_{\mathrm T}$ are the longitudinal and transverse components of $\boldsymbol p$ with respect to n. Minimization of the sum of the squares of the transverse momenta, which is the defining property of Sphericity, then corresponds to finding the direction n which maximizes $n^T W n$. With the constraint that n be a unit vector one obtains

$$\nabla_n \left(n^T W n + \lambda(1 - n^T n) \right) = 0 \,,$$

where λ is a Lagrange multiplier taking care of the normalization constraint. The solution is given by

$$W n = \lambda n \,,$$

that is, n is an eigenvector of W with eigenvalue λ. Since W is a symmetric 3×3 matrix, there are three eigenvectors n_i, $i = 1, 2, 3$, which form an orthogonal basis. It follows

$$\sum_{i=1}^{3} p_{\mathrm{L}}^2(n_i) = p^2$$

and thus

$$\sum_{i=1}^{3} n_i^T W n_i = \frac{\sum_p p^2}{\sum_p p^2} = \lambda_1 + \lambda_2 + \lambda_3 = 1 \ .$$

Sorting the eigenvalues according to $\lambda_1 \geq \lambda_2 \geq \lambda_3$, the Sphericity S can be written as

$$S = \frac{3}{2}(1 - \lambda_1) = \frac{3}{2}(\lambda_2 + \lambda_3) \ .$$

The eigenvector n_1 to λ_1 defines the *Sphericity axis*. For an ideal two-jet event one has $\lambda_1 = 1$ and $S = 0$. For a perfectly spherical event all eigenvalues have the same value $\lambda_i = 1/3$ and the Sphericity attains the value $S = 1$.

8–2 For a collinear safe event-shape variable the value must remain unchanged if any of the contributing momentum vectors is split into two collinear ones. Since Sphericity is quadratic in the momenta, this requirement is not met. To illustrate this, consider the case that one momentum vector which is parallel to the Sphericity axis is split like

$$p \rightarrow \frac{p}{2} + \frac{p}{2}.$$

Since p is parallel to the event axis, the direction which minimizes the sum of the transverse momenta does not change under this operation. The value of S, however, is affected. Since

$$p^2 \rightarrow 2 \left(\frac{p}{2}\right)^2 = \frac{p^2}{2} \ ,$$

the denominator of the defining expression eqn (8.22) for S becomes smaller, that is, the Sphericity is increased.

8–3 The C-parameter is collinear safe if the linear momentum tensor Θ with components

$$\Theta_{ij} = \frac{1}{\sum_p |p|} \sum_p \frac{p_i p_j}{|p|}$$

also has this property. Splitting any of the momenta p into a sum of collinear ones,

$$p = \sum_k \alpha_k p \quad \text{with} \quad \sum_k \alpha_k = 1 \ , \qquad \alpha_k > 0 \ \forall k \ ,$$

one sees immediately that the normalization in the denominator has the desired property. It remains to be checked what happens to the individual terms of the sum in the numerator. One finds

$$\frac{p_i p_j}{|\boldsymbol{p}|} = \sum_k \frac{\alpha_k^2 p_i p_j}{\alpha_k |\boldsymbol{p}|} = \frac{p_i p_j}{|\boldsymbol{p}|} \sum_k \alpha_k \ ,$$

and since the sum of the coefficients α_k is normalized by construction, it follows that Θ is collinear safe.

8–4 For an ideal two-jet event the momentum flow is only along one direction in space. Since in the C.o.M. system the total momentum vanishes and since the C-parameter is collinear safe, it can be assumed without loss of generality that one has only a two-particle event, one particle with momentum $\boldsymbol{p}$ and the other one with momentum $-\boldsymbol{p}$. The linear momentum tensor then becomes

$$\Theta = \frac{1}{2|\boldsymbol{p}|} \frac{2\boldsymbol{p}\boldsymbol{p}^T}{|\boldsymbol{p}|} = \boldsymbol{n}\boldsymbol{n}^T \ ,$$

that is, the direct product of the unit-vector $\boldsymbol{n}$ along the direction of $\boldsymbol{p}$ with itself. One verifies immediately that the first eigenvector of Θ is $\boldsymbol{n}$ with eigenvalue $\lambda_1 = 1$. The remaining two eigenvectors then have to be orthogonal to $\boldsymbol{n}$, that is, they have eigenvalues $\lambda_2 = \lambda_3 = 0$. Since the C-parameter is the sum of all products of two different eigenvalues it will be zero.

For an isotropic event, the momentum flow into every element of solid angle is the same. With a total momentum

$$\sum_{\boldsymbol{p}} |\boldsymbol{p}| = P$$

one thus has

$$\frac{\mathrm{d}P}{\mathrm{d}\Omega} = \frac{P}{4\pi}.$$

With θ the polar angle and ϕ the azimuth, one obtains the Cartesian components of the momentum flow as

$$\frac{\mathrm{d}P_1}{\mathrm{d}\Omega} = \frac{P}{4\pi} \sin\theta \cos\phi \ , \quad \frac{\mathrm{d}P_2}{\mathrm{d}\Omega} = \frac{P}{4\pi} \sin\theta \sin\phi \quad \text{and} \quad \frac{\mathrm{d}P_3}{\mathrm{d}\Omega} = \frac{P}{4\pi} \cos\theta \ ,$$

and the linear momentum tensor becomes

$$\Theta_{ij} = \frac{1}{P} \int_0^{2\pi} \mathrm{d}\phi \int_{-1}^1 \mathrm{d}\cos\theta \ \frac{\mathrm{d}P_i/\mathrm{d}\Omega \ \mathrm{d}P_i/\mathrm{d}\Omega}{\mathrm{d}P/\mathrm{d}\Omega} \ .$$

Substituting the above expressions the total momentum P cancels and all that remains are integrals over the solid angle. The off-diagonal elements of Θ_{ij} vanish, the diagonal elements, that is, the eigenvalues of the momentum tensor, are $\Theta_{ii} = 1/3$, $i = 1, 2, 3$. It follows that $C = 1/3$.

9–1 Assume that in the C.o.M. system the electron and positron move along the
z-axis. If mass effects can be neglected, they are most conveniently described
through helicity eigenstates, that is, as states with the spin either parallel or
antiparallel to the momentum vector. Assume that the annihilation proceeds
parity conserving via a virtual photon. Since the coupling of the fermions to the
photon is helicity conserving, the intermediate state which then decays into a
pair of secondary particles can be described through the angular momentum state
$|i\rangle = |J, m\rangle = |1, \pm 1\rangle$. In the C.o.M. system the final state from the decay consists
of a pair of back-to-back particles, which are emitted along a new direction z'.
The angle between z and z' is denoted by Θ. If the final state particles are
spin-1/2 fermions, the angular momentum state is $|f_f\rangle = |J, m'\rangle = |1, \pm 1\rangle$, for
spin-0 particles it has to be $|f_b\rangle = |J, m'\rangle = |1, 0\rangle$. Note that the final state has a
rotated quantization axis compared to the initial state. The angular distribution
is then obtained by the quantum mechanical overlap of the initial and the final
state,

$$\frac{d\sigma}{d\cos\Theta} \sim \sum_{i,f} |\langle i|f\rangle|^2 \, ,$$

where the sum runs over all possible initial and final states. From the spin alge-
bra the scalar products $\langle i|f\rangle$ are given by the so-called d-functions, $d^J_{m\,m'}$. One
obtains

$$\frac{d\sigma}{d\cos\Theta} \sim |d^1_{1\,1}|^2 + |d^1_{-1\,1}|^2 + |d^1_{1\,-1}|^2 + |d^1_{-1\,-1}|^2 = 1 + \cos^2\Theta$$

for spin-1/2 and

$$\frac{d\sigma}{d\cos\Theta} \sim |d^1_{1\,0}|^2 + |d^1_{-1\,0}|^2 = \sin^2\Theta = 1 - \cos^2\Theta$$

for spin-0 particles.

9–2 The differential cross sections for three-jet production are given by eqn (9.5) for
the case of a vector gluon and eqn (9.6) for the scalar gluon. Lines of constant
cross section σ in the x_q–$x_{\bar{q}}$ plane for both cases are shown in Fig. F.7. The
lines are equidistant in $\ln\sigma$. Only the triangle above $x_{\bar{q}} = 1 - x_q$ is kinematically
allowed. Note that energy–momentum conservation gives the constraint $x_q + x_{\bar{q}} +
x_g = 2$. For energy ordered configurations $x_1 > x_2 > x_3$, there are six regions
with different mappings from $\{x_1, x_2, x_3\}$ to $\{x_q, x_{\bar{q}}, x_g\}$. They, too, are pictured
in Fig. F.7.

	1	2	3	4	5	6
x_1	x_g	$x_{\bar{q}}$	$x_{\bar{q}}$	x_g	x_q	x_q
x_2	$x_{\bar{q}}$	x_g	x_q	x_q	x_g	$x_{\bar{q}}$
x_3	x_q	x_q	x_g	$x_{\bar{q}}$	$x_{\bar{q}}$	x_g

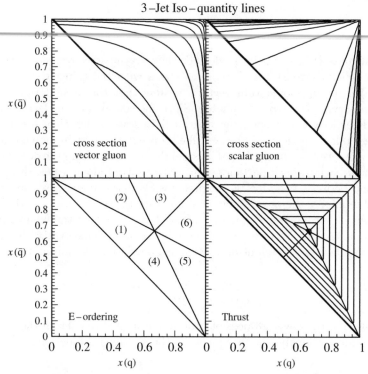

FIG. F.7. Three-jet kinematics as a function of the scaled quark and antiquark energies. The upper row shows lines of constant cross section for the vector (left) and the scalar gluon (right). The lower left plot shows the regions corresponding to a fixed mapping between energies and parton types when performing energy ordering of the jets, and the lower right finally displays lines of constant Thrust.

The border lines between different regions are given by $x_{\bar{q}} = x_q$, $x_{\bar{q}} = 1 - x_q$, $x_{\bar{q}} = 1 - x_q/2$ and $x_{\bar{q}} = 2(1 - x_q)$. For a given three-jet configuration the variable Thrust T is given by

$$T = \max_{\boldsymbol{n}} \frac{\sum_{\boldsymbol{p}} |\boldsymbol{p} \cdot \boldsymbol{n}|}{\sum_{\boldsymbol{p}} |\boldsymbol{p}|} = x_1 .$$

The Thrust-distribution is obtained by integration over the differential cross section

$$\frac{\mathrm{d}\sigma}{\mathrm{d}T} = \int \mathrm{d}x_q \mathrm{d}x_{\bar{q}} \, \frac{\mathrm{d}^2\sigma}{\mathrm{d}x_q \mathrm{d}x_{\bar{q}}} \, \delta(T - T(x_q, x_{\bar{q}})) ,$$

where the δ-function assures that only those regions of the phase space contribute where the Thrust has a fixed value T. Since the function $T(x_q, x_{\bar{q}})$ changes for different parts of the phase space, the integral has to be done separately for those regions. Because of the symmetry of $T(x_q, x_{\bar{q}})$ as well as that of the differential cross section about the line $x_q = x_{\bar{q}}$ one only has to consider region (1), with $T = x_g$, and regions (2) and (3) with $T = x_{\bar{q}}$.

For doing the integration it is most convenient to start with the $x_{\bar{q}}$. Using the energy constraint $x_q + x_{\bar{q}} + x_g = 2$ one finds

$$\frac{d\sigma}{dT} = 2 \int_{2(1-T)}^{1-T/2} dx_q \left. \frac{d^2\sigma}{dx_q dx_{\bar{q}}} \right|_{x_{\bar{q}}=2-T-x_q} + 2 \int_{2(1-T)}^{T} dx_q \left. \frac{d^2\sigma}{dx_q dx_{\bar{q}}} \right|_{x_{\bar{q}}=T} .$$

Substituting the expressions eqn (9.5) and eqn (9.6) for the vector and scalar gluon, respectively, the final results including coupling constant and cross section normalization are

$$\frac{1}{\sigma_0} \frac{d\sigma^V}{dT} = \frac{\alpha_s C_F}{2\pi} \frac{1}{1-T} \left\{ (12 - 24T + 9T^2) + \frac{4 - 6T + 6T^2}{T} \ln \frac{2T-1}{1-T} \right\}$$

and

$$\frac{1}{\sigma_0} \frac{d\sigma^S}{dT} = \frac{\alpha_s C_F}{2\pi} \left\{ 2 \ln \frac{2T-1}{1-T} + \frac{1}{1-T} - 9(1-T) - 10 \frac{\sum a_q^2}{\sum a_q^2 + v_q^2} (3T-2) \right\} .$$

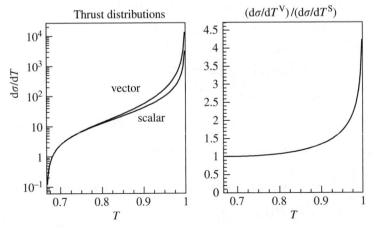

FIG. F.8. Thrust distributions for vector and scalar gluons. The left plot shows the differential cross sections for both cases, the right frame is the ratio of the two. The cross sections are normalized such that they are the same for small values of T.

The differential cross sections are displayed in Fig. F.8, normalized to the prediction of the vector gluon for $T \to 2/3$, where the ratio of the two predictions approaches the limiting value $\sigma^S/\sigma^V = 1/2 - 5 \sum a_q^2 / 12 \sum (a_q^2 + v_q^2)$. The prediction for the vector gluon is more strongly peaked in the two-jet region $T \to 1$ than the one for the scalar gluon.

9-3 This is Lamy's theorem. In the C.o.M. frame one has $P = p_1 + p_2 + p_3 = 0$, which implies that all momenta are within a plane. Assuming massless particles, the cross products of p_1 and p_2 with P give

$$p_1 \times P = 0 \implies E_1 E_2 \sin \Psi_{12} + E_1 E_3 \sin \Psi_{31} = 0$$
$$p_2 \times P = 0 \implies E_2 E_1 \sin \Psi_{12} + E_2 E_3 \sin \Psi_{23} = 0 ,$$

where Ψ_{ij} is the opening angle between p_i and p_j. Combining these two equations yields

$$(E_1 + E_2) \sin \Psi_{12} + E_3 (\sin \Psi_{13} + \sin \Psi_{23}) = 0 ,$$

and with the energy-conservation constraint $E_1 + E_2 = \sqrt{s} - E_3$ we find

$$E_3 = \sqrt{s} \frac{\sin \Psi_{12}}{\sin \Psi_{12} + \sin \Psi_{23} + \sin \Psi_{31}} .$$

The corresponding expressions for E_1 and E_2 are obtained by cyclic permutation of the indices.

10-1 First we write down the expansion of $a_s(\mu^2)$ in terms of $a_s(s)$. Introducing $L = \ln \mu^2/s$ and thus $\omega = 1 + a_s(s) b_0 L$, we find up to $\mathcal{O}(a_s^3(s))$

$$a_s(\mu^2) = \frac{a_s(s)}{\omega} \left(1 - \frac{b_1}{b_0} \frac{a_s(s)}{\omega} \ln \omega \right)$$
$$= a_s(s) \left[1 - a_s b_0 L + a_s^2 (b_0^2 L^2 - b_1 L) \right] + \mathcal{O}(a_s^4(s)) .$$

Similarly, we obtain for $a_s^2(\mu^2)$

$$a_s^2(\mu^2) = a_s^2(s) \left[1 - 2 b_0 a_s(s) L \right] + \mathcal{O}(a_s^4(s)) .$$

Now, inserting these expressions into eqn (10.6),

$$\frac{1}{\sigma_{\text{had}}} \frac{d\sigma}{dy} = a_s(s) A(y) - a_s^2(s)(b_0 L) A(y) + a_s^3(s)(b_0^2 L^2 - b_1 L) A(y)$$
$$+ a_s^2(s) B(y) + a_s^2(s)(b_0 L) A(y)$$
$$- a_s^3(s) 2(b_0 L) B(y) - a_s^3(s) 2(b_0^2 L^2) A(y) + \mathcal{O}(a_s^4(s)) ,$$

we see that the dependence on $(b_0 L)$ cancels out in $\mathcal{O}(a_s^2(s))$. However, there remains a dependence in the next higher order, namely

$$\frac{1}{\sigma_{\text{had}}} \frac{d\sigma}{dy} = a_s(s) A(y) + a_s^2(s) B(y)$$
$$- a_s^3(s) b_0 L \left[b_0 L A(y) + 2 B(y) + \frac{b_1}{b_0} A(y) \right] + \mathcal{O}(a_s^4(s)) .$$

10–2 With four particles in the final state we have 16 variables to start with, namely four times the four components of the four-momenta. If the particle masses are known, then there are four mass-shell constraints, $E_i^2 = \boldsymbol{p}_i^2 + m_i^2$, $i = 1, \ldots, 4$, which reduce the number of independent variables to 12. Next we have four equations from energy–momentum conservation, which gives a further reduction to eight. The overall event orientation can be described by three angles, such as the Euler angles, and if we integrate over them, we end up with five independent kinematic variables.

In order to get a list of such variables, we write down energy–momentum conservation in terms of the four-momenta,

$$p_1^\mu + p_2^\mu + p_3^\mu + p_4^\mu = p_{\text{tot}}^\mu ,$$

with for example $p_{\text{tot}} = (E_{\text{cm}}, \boldsymbol{0})$ at LEP. Taking the square of both sides yields

$$2p_1 \cdot p_2 + 2p_1 \cdot p_3 + 2p_1 \cdot p_4 + 2p_2 \cdot p_3 + 2p_2 \cdot p_4 + 2p_3 \cdot p_4 + \sum_{i=1}^{4} m_i^2 = p_{\text{tot}}^2 .$$

If we now introduce $s = p_{\text{tot}}^2$ and the Lorentz-invariant quantities

$$y_{ij} = (p_i + p_j)^2/s = (2p_i \cdot p_j + m_i^2 + m_j^2)/s = m_{ij}^2/s ,$$

we can rewrite the previous result as

$$y_{12} + y_{13} + y_{14} + y_{23} + y_{24} + y_{34} = 1 + 2\sum_{i=1}^{4} m_i^2/s .$$

We see that we have six variables, where one can be expressed by the other five, which thus can be taken as the required set of independent variables. One possible choice is $y_{12}, y_{13}, y_{14}, y_{23}, y_{24}$.

12–1 Applying eqn (9.10) to the onefold symmetric events, the hardest jet has $E_1 = M_Z \cdot \sin 60°/(\sin 60° + 2\sin 150°) \approx 42.3 \,\text{GeV}$ and $Q_{T1} = E_1 \cdot \sin(150°/2) \approx 40.9 \,\text{GeV}$. The two lower energy jets have $E = E_2 = E_3 = M_Z \cdot \sin 150°/(\sin 60° + 2\sin 150°) = (M_Z - 2E_1)/2 \approx 24.5 \,\text{GeV}$ and $Q_{T2} = Q_{T3} = E \cdot \sin(60°/2) \approx 12.2 \,\text{GeV}$. In the threefold symmetric events, $E = E_1 = E_2 = E_3 = M_Z/3 \approx 30.4 \,\text{GeV}$ and $Q_{T1} = Q_{T2} = Q_{T3} = E \sin(120°/2) \approx 26.3 \,\text{GeV}$.

12–2 The quark and gluon fragmentation functions evolve according to eqn (3.285). At high x, only the $z \to 1$ parts of the splitting functions contribute and of these only the singular terms in $P_{gg}(z)$ and $P_{qq}(z)$ are important. Referring to eqs.(3.268) and (3.50) we have

$$P_{qq}(z) \xrightarrow{z\to 1} C_F \left[\frac{2}{(1-z)_+} + \frac{3}{2}\delta(1-z) \right]$$

$$P_{gg}(z) \xrightarrow{z \to 1} 2C_A \frac{1}{(1-z)_+} + \left. \frac{11C_A - 4n_f T_F}{6} \delta(1-z) \right|_{C_A=3, n_f=3}$$

$$= C_A \left[\frac{2}{(1-z)_+} + \frac{3}{2}\delta(1-z) \right] .$$

The case $C_A = 3$ is appropriate to colour SU(3), whilst $n_f = 3$ is justified since heavy flavours essentially do not contribute in the evolution of the parton shower. Thus, in the limit $x \to 1$, we have

$$\mu^2 \frac{\partial D_f}{\partial \mu^2}(x, \mu^2) = C_f \frac{\alpha_s}{2\pi} \int_x^1 \frac{dz}{z} \left[\frac{2}{(1-z)_+} + \frac{3}{2}\delta(1-z) \right] D_f\left(\frac{x}{z}, \mu^2\right) ,$$

where $C_f = C_F$ for $f = $ q and $C_f = C_A$ for $f = $ g. If we approximate $D_f(x/z, \mu^2)$ in the integrand by $D_f(x, \mu^2)$ we obtain

$$\mu^2 \frac{\partial D_f}{\partial \mu^2}(x, \mu^2) \propto C_f D_f(x, \mu^2) \implies \frac{\partial \ln D_f(x \to 1, \mu^2)}{\partial \ln \mu^2} \propto C_f ,$$

where the proportionality constant is independent of D_f. This result allows the colour charge ratio, C_A/C_F, to be extracted by comparing scaling violations seen for leading particles in quark and gluon jets; for more details see (Nason and Webber, 1994; DELPHI Collab., 1998b).

13–1 The result follows immediately if the convolution is written as

$$D(x) = d_{PT}(x) \otimes d_{NP}(x) = \int_0^1 dy \int_0^1 dz\, \delta(x - yz) d_{PT}(y) d_{NP}(z) .$$

Integrating over x and getting rid of the δ-function, the remaining integrals factorize, thereby proving the result:

$$\langle x \rangle = \int_0^1 dx\, x D(x) = \int_0^1 dy \int_0^1 dz \int_0^1 dx\, \delta(x - yz) d_{PT}(y) d_{NP}(z)$$

$$= \int_0^1 dy \int_0^1 dz (yz) d_{PT}(y) d_{NP}(z) = \langle x \rangle_{PT} \langle x \rangle_{NP} .$$

13–2 Taking $x = x_p$ ($x = x_E$ gives the same result) and using relativistic kinematics the change in energy is given by:

$$\Delta E = \sqrt{\mathbf{p}^2 + m_Q^2} - \sqrt{x_p^2 \mathbf{p}^2 + M_{Q\bar{q}}^2} - \sqrt{(1-x_p)^2 \mathbf{p}^2 + m_q^2}$$

$$\approx |\mathbf{p}| + \frac{m_Q^2}{2|\mathbf{p}|} - x_p|\mathbf{p}| - \frac{M_{Q\bar{q}}^2}{2x_p|\mathbf{p}|} - (1-x_p)|\mathbf{p}| - \frac{m_q^2}{2(1-x_p)|\mathbf{p}|}$$

$$\approx \frac{m_Q^2}{2|\boldsymbol{p}|}\left(1 - \frac{(M_{Q\bar{q}}/m_Q)^2}{x_p} - \frac{(m_q/m_Q)^2}{1 - x_p}\right).$$

By approximating the mass of the hadron with the mass of the heavy quark, $M_{Q\bar{q}} \approx m_Q$, and equating the mass of the light quark with the inverse of the typical hadron size, $m_q = R_0^{-1}$, the required form for the fragmentation function follows.

3–3 We work in the x–y plane and use the explicit four-vectors $q^\mu = (1, 1/2, +\sqrt{3}/2, 0)$, $\bar{q}^\mu = (1, 1/2, -\sqrt{3}/2, 0)$ and $v^\mu = (1, -1, 0, 0)$ for the quark, antiquark and vector boson, respectively. Without loss of generality, we have assumed unit energy for all three particles. In the three cases, the soft gluon momenta are then given by $k^\mu = \omega(1, 1, 0, 0)$, $k^\mu = \omega(1, -1/2, \sqrt{3}/2, 0)$ and $k^\mu = \omega(1, 0, 0, 1)$. Substituting these into eqn (3.342) and eqn (3.343) gives

$$q\bar{q}\gamma : \qquad \frac{3}{\omega^2}\{4C_F : C_F : C_F\} \qquad\qquad\qquad = \frac{3}{\omega^2}\left\{\frac{16}{3} : \frac{4}{3} : \frac{4}{3}\right\}$$

$$q\bar{q}g : \qquad \frac{3}{\omega^2}\{4C_F - C_A : 2C_A + C_F : C_A/2 + C_F\} = \frac{3}{\omega^2}\left\{\frac{7}{3} : \frac{22}{3} : \frac{17}{6}\right\}.$$

3–4 In order to simplify the following discussion we describe pions using the plane wave approximation. Hence the wave function or amplitude $A(1)$ for a pion (spin zero) with momentum $\boldsymbol{k}_1$ at some space point $\boldsymbol{x}_a$ at time $t = 0$ is given by $A(1) = \exp(\mathrm{i}\,\boldsymbol{k}_1 \cdot \boldsymbol{x}_a)$, up to a normalization and some arbitrary phase Φ_a. For a system of two distinguishable pions (for example, pions of opposite charge) the joint amplitude is

$$A^{+-}(1, 2) = \mathrm{e}^{\mathrm{i}\,\boldsymbol{k}_1 \cdot \boldsymbol{x}_a}\, \mathrm{e}^{\mathrm{i}\,\boldsymbol{k}_2 \cdot \boldsymbol{x}_b}\, \mathrm{e}^{\mathrm{i}\,\Phi_a}\, \mathrm{e}^{\mathrm{i}\,\Phi_b} = A(1)A(2),$$

where the wave function of the second pion is characterized by a momentum $\boldsymbol{k}_2$ and space coordinates $\boldsymbol{x}_b$. For indistinguishable (like-sign) pions the joint amplitude has to be symmetric in $\boldsymbol{x}_a$ and $\boldsymbol{x}_b$ because of Bose–Einstein statistics; therefore,

$$A^{++,--}(1, 2) = \frac{1}{\sqrt{2}}\left[\mathrm{e}^{\mathrm{i}\,\boldsymbol{k}_1 \cdot \boldsymbol{x}_a}\mathrm{e}^{\mathrm{i}\,\boldsymbol{k}_2 \cdot \boldsymbol{x}_b} + \mathrm{e}^{\mathrm{i}\,\boldsymbol{k}_2 \cdot \boldsymbol{x}_a}\mathrm{e}^{\mathrm{i}\,\boldsymbol{k}_1 \cdot \boldsymbol{x}_b}\right]\mathrm{e}^{\mathrm{i}\,\Phi_a}\mathrm{e}^{\mathrm{i}\,\Phi_b}.$$

Let us now introduce a correlation function $C(1, 2)$, defined according to

$$C(1, 2) = \frac{|A^{++,--}(1, 2)|^2}{|A^{+-}(1, 2)|^2} = \frac{|A^{++,--}(1, 2)|^2}{|A(1)A(2)|^2}.$$

Inserting the expressions for the various amplitudes, the correlation function is found to be

$$C(1, 2) = 1 + \frac{1}{2}\left[\mathrm{e}^{\mathrm{i}\,(\boldsymbol{k}_1 - \boldsymbol{k}_2)\cdot(\boldsymbol{x}_a - \boldsymbol{x}_b)} + \mathrm{e}^{-\mathrm{i}\,(\boldsymbol{k}_1 - \boldsymbol{k}_2)\cdot(\boldsymbol{x}_a - \boldsymbol{x}_b)}\right]$$

$$= 1 + \cos(\Delta \boldsymbol{k} \cdot \Delta \boldsymbol{x}) \, ,$$

where we have defined $\Delta \boldsymbol{k} = \boldsymbol{k}_1 - \boldsymbol{k}_2$ and $\Delta \boldsymbol{x} = \boldsymbol{x}_a - \boldsymbol{x}_b$. Already, at this stage, it is clear that for an arbitrary $\Delta \boldsymbol{x} \neq 0$ the correlation gets enhanced when $\Delta \boldsymbol{k} \to 0$.

These conclusions are not drastically altered if we introduce wave packets instead of plane waves, as long as the spatial smearing is very narrow, that is, the wave function in momentum space becomes

$$\phi(\boldsymbol{k}_1) = \int \mathrm{d}^3 x \, \tilde{\phi}(\boldsymbol{x}) \, \mathrm{e}^{\mathrm{i} \boldsymbol{k}_1 \cdot \boldsymbol{x}} \, ,$$

and the spatial wave function $\tilde{\phi}(\boldsymbol{x})$ is centred around $\boldsymbol{x}_a$ with very small spread around that point. Then it is easy to see that a newly defined correlation function

$$C(1, 2) = \frac{|\phi^{++,--}(\boldsymbol{k}_1, \boldsymbol{k}_2)|^2}{|\phi^{+-}(\boldsymbol{k}_1, \boldsymbol{k}_2)|^2}$$

has again the behaviour

$$C(1, 2) \to 1 + \cos(\Delta \boldsymbol{k} \cdot \Delta \boldsymbol{x})$$

for narrow spatial wave packets centred around $\boldsymbol{x}_a$ and $\boldsymbol{x}_b$. Here $\phi^{+-}, \phi^{++,--}$ are the wave functions in momentum space for unlike-sign and like-sign pions, obtained by smearing the amplitudes A^{+-} and $A^{++,--}$ with the product of spatial wave functions.

For the remaining discussion we return to the plane wave approximation. Now we make the additional assumption that the probability or cross section for production of a pion at a point $\boldsymbol{x}_a$ is proportional to the amplitude squared for the simple free plane wave, that is, $P(1) \propto |A(1)|^2$, and similarly $P(1, 2) \propto |A(1, 2)|^2$, which is the probability for the production of two pions at points $\boldsymbol{x}_a$ and $\boldsymbol{x}_b$. If the source for pion production has some spatial density distribution $\rho(\boldsymbol{x})$, then we obtain the total cross section σ for pion production by integrating over the full source extension, and the correlation function is rewritten as

$$C(1, 2) = \frac{|\sigma^{++,--}(\boldsymbol{k}_1, \boldsymbol{k}_2)|^2}{|\sigma^{+-}(\boldsymbol{k}_1, \boldsymbol{k}_2)|^2} \, ,$$

where

$$\sigma^{ss}(\boldsymbol{k}_1, \boldsymbol{k}_2) \propto \iint \mathrm{d}^3 x_a \mathrm{d}^3 x_b \, \rho(\boldsymbol{x}_a) \rho(\boldsymbol{x}_b) \, |A^{ss}(1, 2)|^2 \quad .$$

The index ss stands for like- ($ss = ++, --$) or unlike-sign pions ($ss = +-$). Filling in the expressions for the amplitudes we get

$$C(1, 2) = 1 + \frac{1}{N^2} \left(\int \mathrm{d}^3 x_a \, \rho(\boldsymbol{x}_a) \, \mathrm{e}^{\mathrm{i} \Delta \boldsymbol{k} \cdot \boldsymbol{x}_a} \right) \left(\int \mathrm{d}^3 x_b \, \rho(\boldsymbol{x}_b) \, \mathrm{e}^{-\mathrm{i} \Delta \boldsymbol{k} \cdot \boldsymbol{x}_b} \right)$$

$$= 1 + \frac{1}{N^2} \hat{\rho}(\Delta \boldsymbol{k}) \hat{\rho}^\star(\Delta \boldsymbol{k}) \Rightarrow$$

$$C(1,2) = 1 + |\tilde{\rho}(\Delta \boldsymbol{k})|^2 , \qquad (\text{F.18})$$

where

$$N = \int \mathrm{d}^3 x \, \rho(\boldsymbol{x}) , \quad \hat{\rho}(\Delta \boldsymbol{k}) = \int \mathrm{d}^3 x \, \rho(\boldsymbol{x}) \, \mathrm{e}^{\mathrm{i} \Delta \boldsymbol{k} \cdot \boldsymbol{x}}$$

and $\tilde{\rho} = \hat{\rho}/N$, $\rho(\boldsymbol{x}) = \rho^\star(\boldsymbol{x})$. Thus, we have found that the enhancement factor in the two-pion correlation function is proportional to the Fourier transform of the spatial source distribution with respect to the momentum difference. Since the normalization is with respect to the single-pion cross section, the correlation function approaches its maximum value of 2 when $\Delta \boldsymbol{k} \to 0$.

13–5 In the previous example we have ignored any time dependence, and furthermore we have assumed that the various sources emit pions incoherently. A thorough treatment of the inclusion of coherence effects and time dependence can be found in (Bowler, 1985). Here we give a simplified discussion, starting by considering only two point sources. If these emit coherently, then the single-pion amplitude is changed to

$$A(1) = f_a \, \mathrm{e}^{\mathrm{i} \boldsymbol{k}_1 \cdot \boldsymbol{x}_a} + f_b \, \mathrm{e}^{\mathrm{i} \boldsymbol{k}_1 \cdot \boldsymbol{x}_b} \quad ,$$

where f_a, f_b describe the two sources with possibly different intensities and arbitrary time dependence. The joint amplitude is given by

$$A(1,2) = \left(f_a \, \mathrm{e}^{\mathrm{i} \boldsymbol{k}_1 \cdot \boldsymbol{x}_a} + f_b \, \mathrm{e}^{\mathrm{i} \boldsymbol{k}_1 \cdot \boldsymbol{x}_b} \right) \left(f_a \, \mathrm{e}^{\mathrm{i} \boldsymbol{k}_2 \cdot \boldsymbol{x}_a} + f_b \, \mathrm{e}^{\mathrm{i} \boldsymbol{k}_2 \cdot \boldsymbol{x}_b} \right) \quad .$$

The important observation is that this amplitude is already Bose-symmetric, and therefore the two-pion production probability is

$$P(1,2) \propto \left(f_a^2 + f_b^2 + \left[f_a^\star f_b \, \mathrm{e}^{-\mathrm{i} \boldsymbol{k}_1 \cdot (\boldsymbol{x}_a - \boldsymbol{x}_b)} + \text{c.c.} \right] \right) \cdot$$
$$\left(f_a^2 + f_b^2 + \left[f_a^\star f_b \, \mathrm{e}^{-\mathrm{i} \boldsymbol{k}_2 \cdot (\boldsymbol{x}_a - \boldsymbol{x}_b)} + \text{c.c.} \right] \right) , \qquad (\text{F.19})$$

which is just the product of the two single probabilities. Here c.c. are the complex conjugate terms. Thus the correlation function $C(1,2) = P(1,2)/(P(1)P(2))$ will not have an oscillating term with $\Delta \boldsymbol{k} \cdot \Delta \boldsymbol{x}$, and no 'Bose–Einstein' enhancement effect will be seen. For completely coherent sources this might be a rather counter-intuitive result, but it simply reflects the fact that for coherent emission there is no additional symmetrization of the wave function, which could produce an extra enhancement.

Next we consider the case that the sources fluctuate randomly, that is, the time dependence of the coherent emitters is of chaotic nature. Now the relevant quantity to look at is the time-averaged probability, $\langle P(1,2) \rangle$. It is sufficient to observe that all terms in eqn (F.19) which are proportional to $f_{a,b}$ or $f_{a,b}^\star$ will average out to zero, and only those terms with quadratic or quartic dependence on the emission amplitude will remain. Hence, the result must be

$$\langle P(1,2) \rangle \propto \left(f_a^2 + f_b^2 \right)^2 + 2 f_a^2 f_b^2 \cos(\Delta \boldsymbol{k} \cdot \Delta \boldsymbol{x}) \,.$$

Since $\langle P(1) \rangle = \langle P(2) \rangle \propto f_a^2 + f_b^2$, the correlation function becomes

$$C(1,2) = \frac{\langle P(1,2) \rangle}{\langle P(1) \rangle \langle P(2) \rangle} = 1 + \frac{2 f_a^2 f_b^2}{\left(f_a^2 + f_b^2 \right)^2} \cos(\Delta \boldsymbol{k} \cdot \Delta \boldsymbol{x}) \,.$$

In the special case of two sources of equal strength, $f_a = f_b$, we finally find

$$C(1,2) = 1 + \frac{1}{2} \cos(\Delta \boldsymbol{k} \cdot \Delta \boldsymbol{x}) \rightarrow \frac{3}{2} \quad \text{for} \quad \Delta \boldsymbol{k} \rightarrow 0 \,.$$

It can be shown that in the case of n random sources of equal strength the above expression generalizes to

$$C(1,2) \rightarrow 1 + \left(1 - \frac{1}{n} \right) \quad \text{for} \quad \Delta \boldsymbol{k} \rightarrow 0 \,.$$

Thus, in the case of a large number of coherent sources the correlation function approaches a maximum of 2 for $\Delta \boldsymbol{k} \rightarrow 0$, just as we have found previously for the simplest case of incoherent emission from two sources and neglecting any time dependence.

It is rather obvious that any realistic description of a physical source should take account of both coherence effects and time dependence such as random fluctuations. The net effect should be that the effective strength of the Bose–Einstein enhancement lies somewhere between 0 and 1. Comparisons to measurements should therefore be based on a parameterization such as

$$C(1,2) \propto 1 + \lambda |\tilde{\rho}(\Delta \boldsymbol{k})|^2 \,, \tag{F.20}$$

where $\tilde{\rho}$ has been defined in the previous exercise, and $0 \leq \lambda \leq 1$.

13–6 The density distribution for a Gaussian source in three space dimensions, with different widths $\sigma_i, i = 1, 2, 3$, for every dimension is given by

$$\rho(\boldsymbol{x}) = \prod_{i=1}^{3} \rho_i(x_i) = \prod_{i=1}^{3} \frac{1}{\sqrt{2\pi}\sigma_i} e^{-\frac{x_i^2}{2\sigma_i^2}} \,.$$

The individual densities satisfy $\int \mathrm{d}^3x\, \rho(\boldsymbol{x}) = 1$. In order to get the two-pion correlation function as defined in eqn (F.18), we have to compute the Fourier transform of this density:

$$\tilde{\rho}(\Delta \boldsymbol{k}) = \int \mathrm{d}^3x\, \rho(\boldsymbol{x})\, e^{i\,\Delta \boldsymbol{k} \cdot \boldsymbol{x}} = \prod_{i=1}^{3} \left(\int \mathrm{d}x_i\, \rho_i(x_i)\, e^{i\,\Delta k_i x_i} \right) = \prod_{i=1}^{3} \tilde{\rho}_i(\Delta k_i) \,.$$

The Fourier transform of a Gaussian distribution can be found in every standard collection of integrals,

$$\tilde{\rho}_i(\Delta k_i) = e^{-\frac{1}{2}k_i^2\sigma_i^2} \rightarrow \tilde{\rho}(\Delta \mathbf{k}) = \exp\left(-\frac{1}{2}\sum_{i=1}^{3}(\Delta k_i)^2\sigma_i^2\right).$$

When squaring this, we finally find a parameterization for the correlation function, as proposed in eqn (F.20),

$$C(1,2) = 1 + \lambda \exp\left(-\sum_{i=1}^{3}(\Delta k_i)^2\sigma_i^2\right).$$

The special case of a spherically symmetric source is obtained for $\sigma_1 = \sigma_2 = \sigma_3 = \sigma$,

$$C(1,2) = 1 + \lambda e^{-(\Delta \mathbf{k})^2\sigma^2}, \quad (\Delta \mathbf{k})^2 = \sum_{i=1}^{3}(\Delta k_i)^2.$$

REFERENCES

Abbaneo, D. *et al.* (1998). The LEP/SLD Heavy Flavour Working Group, LEPHF/98-01.

Abbaneo, D. *et al.* (2000). The LEP Collab.s ALEPH, DELPHI, L3 and OPAL, the LEP Elektroweak Working Group and SLD Heavy Flavour and Electroweak Groups, CERN-EP/2000-016.

Abbaneo, D. *et al.* (2001*a*). The LEP Collab.s ALEPH, DELPHI, L3 and OPAL, the LEP Elektroweak Working Group and SLD Heavy Flavour and Electroweak Groups, CERN-EP/2001-021.

Abbaneo, D. *et al.* (2001*b*). The Collab.s ALEPH, CDF, DELPHI, L3, OPAL and SLD, CERN-EP-2001-050.

Abbas, A. (2000). *arXiv:* hep-ph/0009242.

Abramovskii, V. A. *et al.* (1972). In *XVI Int. Conf. on High Energy Physics, Chicago–Batavia* (ed. J. D. Jackson, A. Roberts, and R. Donaldson), Volume 1, pp. 389. NAL (Batavia).

Adler, S. L. (1963). *Phys. Rev.*, **143**, 1144.

AFS Collab., Akesson, T. *et al.* (1983). *Phys. Lett.*, **B123**, 133.

Aitchison, I. J. R. and Hey, A. J. G. (1989). *Gauge theories in particle physics: a practical introduction.* Hilger, Bristol.

ALEPH Collab., Buskulic, D. *et al.* (1992*a*). *Z. Phys.*, **C55**, 209.

ALEPH Collab., Buskulic, D. *et al.* (1992*b*). *Phys. Lett.*, **B295**, 396.

ALEPH Collab., Buskulic, D. *et al.* (1993). *Z. Phys.*, **C57**, 17.

ALEPH Collab., Buskulic, D. *et al.* (1994*a*). *Z. Phys.*, **C64**, 361.

ALEPH Collab., Buskulic, D. *et al.* (1994*b*). *Z. Phys.*, **C62**, 1.

ALEPH Collab., Buskulic, D. *et al.* (1995*a*). *Nucl. Instr. and Meth.*, **A360**, 481.

ALEPH Collab., Buskulic, D. *et al.* (1995*b*). *Phys. Lett.*, **B357**, 487. Erratum : (1995). *ibid.*, **B364**, 247.

ALEPH Collab., Buskulic, D. *et al.* (1995*c*). *Phys. Lett.*, **B355**, 381.

ALEPH Collab., Buskulic, D. *et al.* (1995*d*). *Z. Phys.*, **C69**, 15.

ALEPH Collab., Buskulic, D. *et al.* (1996*a*). *Z. Phys.*, **C71**, 357.

ALEPH Collab., Buskulic, D. *et al.* (1996*b*). *Phys. Lett.*, **B384**, 353.

ALEPH Collab., Buskulic, D. *et al.* (1996*c*). *Z. Phys.*, **C69**, 379.

ALEPH Collab., Buskulic, D. *et al.* (1997*a*). *Z. Phys.*, **C73**, 409.

ALEPH Collab., Barate, R. *et al.* (1997*b*). *Z. Phys.*, **C76**, 1.

ALEPH Collab., Barate, R. *et al.* (1997*c*). *Z. Phys.*, **C76**, 191.

ALEPH Collab., Barate, R. *et al.* (1997*d*). *Z. Phys.*, **C74**, 451.

ALEPH Collab., Barate, R. *et al.* (1998*a*). *Phys. Rept.*, **294**, 1.

ALEPH Collab., Barate, R. *et al.* (1998*b*). ALEPH 98-025. conf. note CONF 98-014.

ALEPH Collab., Barate, R. *et al.* (1998c). *Eur. Phys. J.*, **C5**, 205.

ALEPH Collab., Barate, R. *et al.* (1998d). *Eur. Phys. J.*, **C4**, 409.

ALEPH Collab., Barate, R. *et al.* (1999). ALEPH 99-023. conf. note CONF 99-018.

ALEPH Collab., Barate, R. *et al.* (2000a). ALEPH 2000-044. conf. note CONF 2000-027.

ALEPH Collab., Barate, R. *et al.* (2000b). *Eur. Phys. J.*, **C17**, 1.

ALEPH Collab., Barate, R. *et al.* (2000c). *Eur. Phys. J.*, **C16**, 613.

ALEPH Collab., Barate, R. *et al.* (2001). ALEPH 2001-007. conf. note CONF 2001-004.

ALEPH Collab., Decamp, D. *et al.* (1991a). *Phys. Lett.*, **B257**, 479.

ALEPH Collab., Decamp, D. *et al.* (1991b). *Phys. Lett.*, **B273**, 181.

ALEPH Collab., Decamp, D. *et al.* (1992c). *Phys. Lett.*, **B284**, 151.

ALEPH Collab., Decamp, D. *et al.* (1992d). *Z. Phys.*, **C53**, 21.

Altarelli, G. (1982). *Phys. Rept.*, **81**, 1.

Altarelli, G. *et al.* (1978). *Nucl. Phys.*, **B143**, 521.

  Erratum : (1978). *ibid.*, **B146**, 544.

Altarelli, G. *et al.* (1979a). *Nucl. Phys.*, **B157**, 461.

Altarelli, G. *et al.* (1979b). *Nucl. Phys.*, **B160**, 301.

Altarelli, G. *et al.* (1989). In *Z physics at LEP1* (ed. G. Altarelli, R. Kleiss, and C. Verzegnassi), Number 89-08, Geneva. CERN.

Altarelli, G. *et al.* (1996). In *Physics at LEP2* (ed. G. Altarelli, T. Söstrand, and F. Zwirner), Number 96-01, Geneva. CERN.

Altarelli, G. *et al.* (2000). *Nucl. Phys.*, **B575**, 313.

Altarelli, G. and Parisi, G. (1977). *Nucl. Phys.*, **B126**, 298.

Amati, D. *et al.* (1980). *Nucl. Phys.*, **B173**, 429.

Amati, D. and Veneziano, G. (1979). *Phys. Lett.*, **B83**, 87.

AMY Collab., Li, Y. K. *et al.* (1990a). *Phys. Rev.*, **D41**, 2675.

AMY Collab., Zheng, H. W. *et al.* (1990b). *Phys. Rev.*, **D42**, 737.

Andersson, B. *et al.* (1979a). *Z. Phys.*, **C1**, 105.

Andersson, B. *et al.* (1979b). *Phys. Lett.*, **B85**, 417.

Andersson, B. *et al.* (1980a). *Z. Phys.*, **C6**, 235.

Andersson, B. *et al.* (1980b). *Phys. Lett.*, **B94**, 211.

Andersson, B. *et al.* (1982). *Nucl. Phys.*, **B197**, 45.

Andersson, B. *et al.* (1983a). *Phys. Rept.*, **97**, 31.

Andersson, B. *et al.* (1983b). *Z. Phys.*, **C20**, 317.

Andersson, B. *et al.* (1985). *Phys. Scripta*, **32**, 574.

Andersson, B. *et al.* (1989a). *Nucl. Phys.*, **B328**, 76.

Andersson, B. *et al.* (1989b). *Z. Phys.*, **C43**, 625.

Andersson, B. *et al.* (1990). *Nucl. Phys.*, **B339**, 393.

Andersson, B. *et al.* (1996a). *Nucl. Phys.*, **B463**, 217.

Andersson, B. *et al.* (1996b). *Z. Phys.*, **C71**, 613.

Andersson, B. and Gustafson, G. (1980). *Z. Phys.*, **C3**, 223.

Andersson, B. and Hofmann, W. (1986). *Phys. Lett.*, **B169**, 364.

Anselmino, M. *et al.* (1993). *Rev. Mod. Phys.*, **65**, 1199.

Appelquist, T. and Carazzone, J. (1975). *Phys. Rev.*, **D11**, 2856.

Arfken, G. B. and Weber, H. J. (1995). *Mathematical methods for physicists*. Academic Press, New York.

ARGUS Collab., Albrecht, H. *et al.* (1996). *Phys. Rept.*, **276**, 223.

Artru, X. (1983). *Phys. Rept.*, **97**, 147.

Artru, X. and Bowler, M. G. (1988). *Z. Phys.*, **C37**, 293.

Artru, X. and Mennessier, G. (1974). *Nucl. Phys.*, **B70**, 93.

Askew, A. J. *et al.* (1993). *Phys. Rev.*, **D47**, 3775.

Aurenche, P. *et al.* (1994). *Comput. Phys. Commun.*, **83**, 107.

Aurenche, P. *et al.* (1996). $\gamma\gamma$ physics. In *Physics at LEP2* (ed. G. Altarelli, T. Sjöstrand, and F. Zwirner), Number 96-01, vol. 1, Geneva. CERN.

Aurenche, P. *et al.* (1999). *Eur. Phys. J.*, **C9**, 107.

Aurenche, P. *et al.* (2000). *Eur. Phys. J.*, **C13**, 347.

Azimov, Ya. I. *et al.* (1985*a*). *Phys. Lett.*, **B165**, 147.

Azimov, Ya. I. *et al.* (1985*b*). *Z. Phys.*, **C27**, 65.

Azimov, Ya. I. *et al.* (1986*a*). *Z. Phys.*, **C31**, 213.

Azimov, Ya. I. *et al.* (1986*b*). *Sov. J. Nucl. Phys.*, **43**, 95.

Balitsky, Y. Y. and Lipatov, L. N. (1978). *Sov. J. Nucl. Phys.*, **28**, 822.

Ball, P. *et al.* (1995). *Nucl. Phys.*, **B452**, 563.

Ball, R. D. and Forte, S. (1994). *Phys. Lett.*, **B336**, 77.

Ball, R. D. and Forte, S. (1995). *Phys. Lett.*, **B351**, 313.

Ball, R. D. and Forte, S. (1999). *Phys. Lett.*, **B465**, 271.

Ball, R. D. and Landshoff, P. V. (2000). *J. Phys.*, **G26**, 672.

Ballestrero, A. *et al.* (1992). *Phys. Lett.*, **B294**, 425.

Ballestrero, A. *et al.* (1994). *Nucl. Phys.*, **B415**, 265.

Barberio, E. *et al.* (2001). The LEP Collab.s ALEPH, DELPHI, L3, OPAL and the LEP W Working Group, LEPEWWG/MASS/2000-01, ALEPH 2001-023 PHYSIC 2001-004, DELPHI 2001-007 PHYS 887, L3 Note 2642, OPAL TN-681, http://www.cern.ch/LEPEWWG/lepww/mw/Winter01/.

Bardeen, W. A. *et al.* (1976). *Phys. Rev.*, **D13**, 2364.

Bardeen, W. A. *et al.* (1978). *Phys. Rev.*, **D18**, 3998.

Barnett, R. M. (1976). *Phys. Rev. Lett.*, **36**, 1163.

Bartels, J. (1993). *Phys. Lett.*, **B298**, 204.

Bartels, J. *et al.* (1996). *Phys. Lett.*, **B373**, 215.

Bartels, J. and Lotter, H. (1993). *Phys. Lett.*, **B309**, 400.

Basham, C. L. *et al.* (1978*a*). *Phys. Rev.*, **D17**, 2298.

Basham, C. L. *et al.* (1978*b*). *Phys. Rev. Lett.*, **41**, 1585.

Bassetto, A. *et al.* (1980). *Nucl. Phys.*, **B163**, 477.

Bassetto, A. *et al.* (1983). *Phys. Rept.*, **100**, 201.

Bassler, U. and Bernardi, G. (1995). *Nucl. Instr. and Meth.*, **A361**, 197.

BCDMS Collab., Benvenuti, A. C. *et al.* (1987). *Phys. Lett.*, **B195**, 91.

BEBC Collab., Allasia, D. *et al.* (1984). *Phys. Lett.*, **B135**, 231.

BEBC Collab., Allasia, D. *et al.* (1985). *Z. Phys.*, **C28**, 321.

BEBC Collab., Jones, G. T. *et al.* (1994). *Z. Phys.*, **C62**, 601.

Becchi, C. *et al.* (1974). *Phys. Lett.*, **B52**, 344.

Bengtsson, M. (1989). *Z. Phys.*, **C42**. 75.

Bengtsson, M. *et al.* (1986). *Phys. Lett.*, **B179**, 164.

Bengtsson, M. and Zerwas, P. M. (1988). *Phys. Lett.*, **B208**. 306.

Berera, A. and Soper, D. E. (1994). *Phys. Rev.*, **D50**, 4328.

Berger, E. L. and Fox, G. C. (1973). *Phys. Lett.*, **B47**, 162.

Bernreuther, W. (1983). *Annals Phys.*, **151**, 127.

Bernreuther, W. *et al.* (1997). *Phys. Rev. Lett.*, **79**, 189.

Bernreuther, W. and Wetzel, W. (1982). *Nucl. Phys.*, **B197**, 228.

Bertolini, S. and Marchesini, G. (1982). *Phys. Lett.*, **B117**, 449.

Bethke, S. (2000). *J. Phys.*, **G26**, R27.

Bethke, S. *et al.* (1991). *Z. Phys.*, **C49**. 59.

Bialas, A. and Peschanski, R. (1986). *Nucl. Phys.*, **B273**, 703.

Biebel, O. and Mättig, P. (1994). private communication.

Bjorken, J. D. (1966). *Phys. Rev.*, **148**, 1467.

Bjorken, J. D. (1969). *Phys. Rev.*, **179**, 1547.

Bjorken, J. D. (1970). *Phys. Rev.*, **D1**, 1376.

Bjorken, J. D. (1973). In *Proceedings of the SLAC Summer Institute on Particle Physics*, Volume 1, pp. 1. SLAC.

Bjorken, J. D. (1978). *Phys. Rev.*, **D17**, 171.

Bjorken, J. D. (1992). *Phys. Rev.*, **D45**, 4077.

Bjorken, J. D. and Paschos, E. A. (1969). *Phys. Rev.*, **185**, 1975.

Blazey, G. C. and Flaugher, B. L. (1999). *Ann. Rev. Nucl. Part. Sci.*, **49**, 633.

Block, M. M. and Halzen, F. (2001). *Phys. Rev.*, **D63**, 114004.

Blum, W. and Rolandi, L. (1994). *Particle detection with drift chambers.* Springer-Verlag, Heidelberg, Germany.

Böhrer, A. (1997). *Phys. Rept.*, **291**, 107.

Bollini, C. G. *et al.* (1973). *Nuovo Cim.*, **16A**, 423.

Botje, M. (2000). *Eur. Phys. J.*, **C14**, 285.

Bowler, M. G. (1981). *Z. Phys.*, **C11**, 169.

Bowler, M. G. (1984). *Z. Phys.*, **C22**, 155.

Bowler, M. G. (1985). *Z. Phys.*, **C29**, 617.

Braaten, E. *et al.* (1992). *Nucl. Phys.*, **B373**, 581.

Brandenburg, A. and Uwer, P. (1998). *Nucl. Phys.*, **B515**, 279.

Braun, V. M. and Kolesnichenko, A. V. (1987). *Nucl. Phys.*, **B283**, 723.

Bravo, S. (2001). *Measurements of the strong coupling constant and the QCD colour factors using four-jet observables from hadronic Z decays in ALEPH.* Ph. D. thesis, Universitat Autònoma de Barcelona, Spain.

Breitenlohner, P. and Maison, D. (1977). *Commun. Math. Phys.*, **52**, 11,39,55.

Brodkorb, T. *et al.* (1989). *Z. Phys.*, **C44**, 415.

Brodsky, S. J. *et al.* (1983). *Phys. Rev.*, **D28**, 228.

Brodsky, S. J. and Farrar, G. R. (1975). *Phys. Rev.*, **D11**, 1309.

Brodsky, S. J. and Gunion, J. F. (1976). *Phys. Rev. Lett.*, **37**, 402.

Browder, T. E. (1996). *Results on QCD from the CLEO-II Experiment*. In *Proceedings of the XXVIII International Conference on High Energy Physics*, Warsaw. *arXiv:* hep-ex/9701002.

Buras, A. J. *et al.* (1977). *Nucl. Phys.*, **B131**, 308.

Burrows, P. N. *et al.* (1996). *Phys. Lett.*, **B382**, 157.

Cahn, R. N. (1984). *Semi-simple Lie algebras and their representations*. Benjamin/Cummings Publishing, Menlo Park, California.

Cahn, R. N. and Goldhaber, G. (1989). *The experimental foundations of particle physics*. Cambridge University Press, Cambridge.

Callan, C. G. (1972). *Phys. Rev.*, **D5**, 3202.

Capella, A. *et al.* (2000). *Phys. Rev.*, **D61**, 074009.

Capitani, S. *et al.* (2002). *Nucl. Phys. Proc. Suppl.*, **106**, 299.

Carius, S. and Ingelman, G. (1990). *Phys. Lett.*, **B252**, 647.

Casher, A. *et al.* (1979). *Phys. Rev.*, **D20**, 179.

Caswell, W. E. (1974). *Phys. Rev. Lett.*, **33**, 244.

Caswell, W. E. and Wilczek, F. (1974). *Phys. Lett.*, **B49**, 291.

Catani, S. (1991). Invited talk given at the *17th Workshop of INFN Eloisatron Project, QCD at 200 TeV*, Erice, Italy, Jun 11-17, 1991; CERN-TH 6281/91.

Catani, S. *et al.* (1986). *Nucl. Phys.*, **B264**, 588.

Catani, S. *et al.* (1990*a*). *Phys. Lett.*, **B242**, 97.

Catani, S. *et al.* (1990*b*). *Phys. Lett.*, **B234**, 339.

Catani, S. *et al.* (1991*a*). *Nucl. Phys.*, **B366**, 135.

Catani, S. *et al.* (1991*b*). *Nucl. Phys.*, **B361**, 645.

Catani, S. *et al.* (1991*c*). *Nucl. Phys.*, **B349**, 635.

Catani, S. *et al.* (1992*a*). *Nucl. Phys.*, **B383**, 419.

Catani, S. *et al.* (1992*b*). *Phys. Lett.*, **B295**, 269.

Catani, S. *et al.* (1993). *Nucl. Phys.*, **B407**, 3.

Catani, S. *et al.* (2000). In *Workshop on standard model physics (and more) at the LHC* (ed. G. Altarelli and M. L. Mangano), Number 2000-004, Geneva. CERN. *arXiv:* hep-ph/0005025.

Catani, S. and Seymour, M. H. (1997). *Nucl. Phys.*, **B485**, 291. Erratum : (1997). *ibid.*, **B510**, 503.

CCFR Collab., Bazarko, A. O. *et al.* (1995). *Z. Phys.*, **C65**, 189.

CCFR Collab., Kim, J. H. *et al.* (1998). *Phys. Rev. Lett.*, **81**, 3595.

CCFR Collab., Seligman, W. G. *et al.* (1997). *Phys. Rev. Lett.*, **79**, 1213.

CDF Collab., Abe, F. *et al.* (1989). *Phys. Rev. Lett.*, **62**, 613.

CDF Collab., Abe, F. *et al.* (1992). *Phys. Rev. Lett.*, **68**, 1104.

CDF Collab., Abe, F. *et al.* (1994). *Phys. Rev.*, **D50**, 5562.

CDF Collab., Abe, F. *et al.* (1996). *Phys. Rev. Lett.*, **77**, 438.

CDF Collab., Abe, F. *et al.* (1998). *Phys. Rev. Lett.*, **81**, 5754.

CDF Collab., Affolder, T. *et al.* (2000). *Phys. Rev. Lett.*, **84**, 845.

CDHS Collab., Abramowicz, H. *et al.* (1982). *Z. Phys.*, **C15**, 19.

CELLO Collab., Behrend, H. J. *et al.* (1982). *Phys. Lett.*, **B110**, 329.

Cheng, Hai-Yang (1988). *Phys. Rept.*, **158**, 1.

Chetyrkin, K. G. *et al.* (1992). *Phys. Lett.*, **B282**, 221.

Chetyrkin, K. G. *et al.* (1996*a*). *Phys. Rept.*, **277**, 189.

Chetyrkin, K. G. *et al.* (1996*b*). *Nucl. Phys.*, **B482**, 213.

Chetyrkin, K. G. and Kühn, J. H. (1990). *Phys. Lett.*, **B248**, 359.

Chetyrkin, K. G. and Tarasov, O. V. (1994). *Phys. Lett.*, **B327**, 114.

Chodos, A. and Thorn, C. B. (1974). *Nucl. Phys.*, **B72**, 509.

Chudakov, A. E. (1955). *Isv. Akad. Nauk. S.S.S.R. Ser. Fiz.*, **19**, 650.

Chun, S. B. and Buchanan, C. D. (1987). *Phys. Rev. Lett.*, **59**, 1997.

Chun, S. B. and Buchanan, C. D. (1993). *Phys. Lett.*, **B308**, 153.

Chyla, J. (2001). *arXiv:* hep-ph/0102100.

Ciafaloni, M. (1988). *Nucl. Phys.*, **B296**, 49.

Ciafaloni, M. *et al.* (1999*a*). *Phys. Rev.*, **D60**, 114036.

Ciafaloni, M. *et al.* (1999*b*). *JHEP*, **9910**, 017.

CLEO Collab., Alam, M. S. *et al.* (1992). *Phys. Rev.*, **D46**, 4822.

CLEO Collab., Alam, M. S. *et al.* (1997). *Phys. Rev.*, **D56**, 17.

CLEO Collab., Ammar, R. *et al.* (1998). *Phys. Rev.*, **D57**, 1350.

CLEO Collab., Behrends, S. *et al.* (1985). *Phys. Rev.*, **D31**, 2161.

Clerbaux, B. (1998). *Electroproduction élastique de mésons ρ á HERA*. Ph. D. thesis, Univ. Libre de Bruxelles, Bruxelles, Belgium, DESY-THESIS-1999-001.

Close, F. E. (1979). *An introduction to quarks and partons*. Academic Press, London.

Close, F. E. and Roberts, R. G. (1993). *Phys. Lett.*, **B316**, 165.

Coleman, S. R. and Gross, D. J. (1973). *Phys. Rev. Lett.*, **31**, 851.

Coleman, S. R. and Weinberg, E. (1973). *Phys. Rev.*, **D7**, 1888.

Collins, J. C. (1986). *Renormalization*. Cambridge University Press, Cambridge.

Collins, J. C. (1998). *Phys. Rev.*, **D57**, 3051.

 Erratum : (2000). *ibid.*, **D61**, 019902.

Collins, J. C. *et al.* (1985). *Nucl. Phys.*, **B250**, 199.

Collins, J. C. *et al.* (1993). *Phys. Lett.*, **B307**, 161.

Collins, J. C. and Landshoff, P. V. (1992). *Phys. Lett.*, **B276**, 196.

Collins, J. C. and Soper, D. E. (1981). *Nucl. Phys.*, **B193**, 381.

 Erratum : (1983). *ibid.*, **B213**, 545.

Collins, J. C. and Soper, D. E. (1982). *Nucl. Phys.*, **B197**, 446.

Collins, J. C. and Soper, D. E. (1987). *Ann. Rev. Nucl. Part. Sci.*, **37**, 383.

Collins, P. D. B. (1977). *An introduction to Regge theory and high energy physics*. Cambridge University Press, Cambridge, UK.

Combridge, B. L. *et al.* (1977). *Phys. Lett.*, **B70**, 234.

Combridge, B. L. and Maxwell, C. J. (1984). *Nucl. Phys.*, **B239**, 429.

Conrad, J. M. *et al.* (1998). *Rev. of Mod. Phys.*, **70**, 1341.

Corcella, G. *et al.* (2001). *JHEP*, **0101**, 010. *arXiv:* hep-ph/0107071, hep-ph/9601212.

Cowan, G. D. (1998). *Statistical data analysis*. Oxford University Press, Oxford.

Csikor, F. and Fodor, Z. (1997). *Phys. Rev. Lett.*, **78**, 4335.

CTEQ Collab., Lai, H. L. *et al.* (1997). *Phys. Rev.*, **D55**, 1280.

CTEQ Collab., Lai, H. L. *et al.* (2000). *Eur. Phys. J.*, **C12**, 375.

Curci, G. *et al.* (1980). *Nucl. Phys.*, **B175**, 27.

D0 Collab., Abbott, B. *et al.* (1999*a*). *Phys. Rev. Lett.*, **82**, 2451.

D0 Collab., Abbott, B. *et al.* (1999*b*). *Phys. Lett.*, **B464**, 145.

D0 Collab., Abbott, B. *et al.* (2000*a*). *Phys. Rev.*, **D61**, 032004.

D0 Collab., Abbott, B. *et al.* (2000*b*). *Phys. Rev. Lett.*, **84**, 2792.

D0 Collab., Abbott, B. *et al.* (2001). *Phys. Lett.*, **B513**, 292.

Davies, C. T. H. *et al.* (1996). *Nucl. Phys. Proc. Suppl.*, **47**, 409.

DELPHI Collab., Abreu, P. *et al.* (1990). *Phys. Lett.*, **B247**, 137.

DELPHI Collab., Abreu, P. *et al.* (1991). *Z. Phys.*, **C50**, 185.

DELPHI Collab., Abreu, P. *et al.* (1992*a*). *Phys. Lett.*, **B274**, 498.

DELPHI Collab., Abreu, P. *et al.* (1992*b*). *Phys. Lett.*, **B275**, 231.

DELPHI Collab., Abreu, P. *et al.* (1993*a*). *Phys. Lett.*, **B307**, 221.

DELPHI Collab., Abreu, P. *et al.* (1993*b*). *Z. Phys.*, **C59**. 357;
  DELPHI Collab., Seitz, A., Paper submitted to the *XXVII International Conference on High Energy Physics*, Glasgow, ICHEP94 Ref. gls0180, 1994.

DELPHI Collab., Abreu, P. *et al.* (1993*c*). *Z. Phys.*, **C59**, 533.
  Erratum : (1995). *ibid.*, **C65**, 709.

DELPHI Collab., Abreu, P. *et al.* (1994). *Phys. Lett.*, **B341**, 109.

DELPHI Collab., Abreu, P. *et al.* (1995*a*). *Z. Phys.*, **C65**, 587.

DELPHI Collab., Abreu, P. *et al.* (1995*b*). *Z. Phys.*, **C68**, 353.

DELPHI Collab., Abreu, P. *et al.* (1995*c*). *Phys. Lett.*, **B361**, 207.

DELPHI Collab., Abreu, P. *et al.* (1995*d*). *Z. Phys.*, **C67**, 543.

DELPHI Collab., Abreu, P. *et al.* (1996*a*). *Z. Phys.*, **C73**, 11.

DELPHI Collab., Abreu, P. *et al.* (1996*b*). *Phys. Lett.*, **B372**, 172.

DELPHI Collab., Abreu, P. *et al.* (1996*c*). *Z. Phys.*, **C70**, 179.

DELPHI Collab., Adam, W. *et al.* (1996*d*). *Z. Phys.*, **C69**, 561.

DELPHI Collab., Abreu, P. *et al.* (1996*e*). *Z. Phys.*, **C73**, 61.

DELPHI Collab., Adam, W. *et al.* (1996*f*). *Z. Phys.*, **C70**, 371.

DELPHI Collab., Abreu, P. *et al.* (1997*a*). *Phys. Lett.*, **B398**, 194.

DELPHI Collab., Abreu, P. *et al.* (1997*b*). *Phys. Lett.*, **B414**. 401.

DELPHI Collab., Abreu, P. *et al.* (1997*c*). *Phys. Lett.*, **B401**, 118.

DELPHI Collab., Abreu, P. *et al.* (1998*a*). *Phys. Lett.*, **B416**, 233.

DELPHI Collab., Abreu, P. *et al.* (1998*b*). *Eur. Phys. J.*, **C4**, 1.

DELPHI Collab., Abreu, P. *et al.* (1998*c*). *Eur. Phys. J.*, **C5**, 585.

DELPHI Collab., Abreu, P. *et al.* (1999*a*). *Phys. Lett.*, **B456**, 322.

DELPHI Collab., Abreu, P. *et al.* (1999*b*). *Phys. Lett.*, **B449**, 383.

DELPHI Collab., Abreu, P. *et al.* (1999*c*). *Eur. Phys. J.*, **C6**, 19.

DELPHI Collab., Abreu, P. *et al.* (1999*d*). *Phys. Lett.*, **B449**, 364.

DELPHI Collab., Abreu, P. *et al.* (2000*a*). *Eur. Phys. J.*, **C18**, 203.

DELPHI Collab., Abreu, P. *et al.* (2000*b*). *Eur. Phys. J.*, **C13**, 573.

DELPHI Collab., Abreu, P. *et al.* (2000*c*). *Eur. Phys. J.*, **C17**, 207.

DELPHI Collab., Abreu, P. *et al.* (2000*d*). *Phys. Lett.*, **B475**, 429.

Dissertori, G. (1998). *Nucl. Phys. B (Proc. Suppl.)*, **65**, 43.

Dixon, L. J. and Signer, A. (1997). *Phys. Rev. Lett.*, **78**, 811.

Djouadi, A. *et al.* (1990). *Z. Phys.*, **C46**, 411.

Dokshitzer, Yu. L. (1977). *Sov. Phys. JETP*, **46**, 641.

Dokshitzer, Yu. L. (1990). Workshop on Jets at LEP and HERA, Durham.

Dokshitzer, Yu. L. (1993). *Phys. Lett.*, **B305**, 295.

Dokshitzer, Yu. L. (1996). In *1995 European school of high energy physics* (ed. N. Ellis and M. Neubert), Number 96-04, Geneva, pp. 59. CERN.

Dokshitzer, Yu. L. (1999). *arXiv:* hep-ph/9911299.

Dokshitzer, Yu. L. *et al.* (1980). *Phys. Rept.*, **58**, 269.

Dokshitzer, Yu. L. *et al.* (1988). *Sov. J. Nucl. Phys.*, **47**, 881.

Dokshitzer, Yu. L. *et al.* (1989). In *Perturbative QCD* (ed. A. H. Mueller), pp. 241. World Scientific, Singapure.

Dokshitzer, Yu. L. *et al.* (1991). *Basics of perturbative QCD*. Editions Frontières, Gif-Sur-Yvette, France.

Dokshitzer, Yu. L. and Webber, B. R. (1995). *Phys. Lett.*, **B352**, 451.

Donnachie, A. and Landshoff, P. V. (1987). *Phys. Lett.*, **B191**, 309. Erratum : (1987). *ibid.*, **B198**, 590.

Donnachie, A. and Landshoff, P. V. (1992). *Phys. Lett.*, **B296**, 227.

Donnachie, A. and Landshoff, P. V. (1994). *Z. Phys.*, **C61**, 139.

Donoghue, J. F. *et al.* (1979). *Phys. Rev.*, **D20**, 2759.

Doyle, A. T. *et al.* (1999). In *Monte Carlo generators for HERA physics* (ed. A. Doyle, G. Grindhammer, G. Ingelman, and H. Jung), Number DESY-PROC-1999-02, Hamburg. DESY.

Drell, S. D. and Yan, T. M. (1971). *Annals Phys.*, **66**, 578.

Dremin, I. M. and Hwa, R. C. (1994). *Phys. Rev.*, **D49**, 5805.

Dremin, I. M. and Nechitailo, V. A. (1994). *Mod. Phys. Lett.*, **A9**, 1471.

E665 Collab., Adams, M. R. *et al.* (1996). *Phys. Rev.*, **D54**, 3006.

E706 Collab., Apanasevich, L. *et al.* (1998). *Phys. Rev. Lett.*, **81**, 2642.

E866/NuSea Collab., Hawker, E. A. *et al.* (1998a). *Phys. Rev. Lett.*, **80**, 3715.

E866/NuSea Collab., Peng, J. C. *et al.* (1998b). *Phys. Rev.*, **D58**, 092004.

Eden, P. (1998). *JHEP*, **9809**, 015.

Einhorn, M. B. and Weeks, B. G. (1978). *Nucl. Phys.*, **B146**, 445.

El-Khadra, A. X. (1996). XXXIst Rencontres de Moriond, 16-23 March. *arXiv:* hep-ex/9608220.

Ellis, J. R. *et al.* (1996a). *Phys. Rev.*, **D54**, 6986.

Ellis, J. R. and Jaffe, R. L. (1974). *Phys. Rev.*, **D9**, 1444. Erratum : (1974). *ibid.*, **D10**, 1669.

Ellis, J. R. and Karliner, I. (1979). *Nucl. Phys.*, **B148**, 141.

Ellis, R. K. *et al.* (1981). *Nucl. Phys.*, **B178**, 421.

Ellis, R. K. *et al.* (1987). *Nucl. Phys.*, **B286**, 643. Erratum : (1987). *ibid.*, **B294**, 1180.

Ellis, R. K. *et al.* (1994). *Nucl. Phys.*, **B420**, 517. Erratum : (1995). *ibid.*, **B433**, 498.

Ellis, R. K. *et al.* (1995). *Phys. Lett.*, **B348**, 582.

Ellis, R. K. *et al.* (1996*b*). *QCD and collider physics.* Cambridge University Press, Cambridge, UK.

Ellis, R. K. *et al.* (1997). *Nucl. Phys.*, **B503**, 309.

Ellis, R. K. and Veseli, S. (1998). *Nucl. Phys.*, **B511**, 649.

Ernström, P. *et al.* (1997). *Z. Phys.*, **C76**, 515.

Ernström, P. and Lönnblad, L. (1997). *Z. Phys.*, **C75**, 51.

Faddeev, L. D. and Popov, V. N. (1967). *Phys. Lett.*, **B25**, 29.

Fadin, V. S. and Lipatov, L. N. (1998). *Phys. Lett.*, **B429**, 127.

Farhi, E. (1977). *Phys. Rev. Lett.*, **39**, 1587.

Farrar, G. R. (1995). *Phys. Rev.*, **D51**, 3904.

Fayet, P. (1976). *Phys. Lett.*, **B64**, 159.

Feynman, R. P. (1972). *Photon-hadron interactions.* Addison-Wesley, Menlo Park, California.

Field, R. D. and Feynman, R. P. (1978). *Nucl. Phys.*, **B136**, 1.

Field, R. D. and Wolfram, S. (1983). *Nucl. Phys.*, **B213**, 65.

Flynn, J. M. (1996). ICHEP conference, Warsaw, 25-31 July. *arXiv:* hep-lat/9611016.

Fodor, Z. (1991). *Phys. Lett.*, **B263**, 305.

Fong, C. P. and Webber, B. R. (1989). *Phys. Lett.*, **B229**, 289.

Fong, C. P. and Webber, B. R. (1991). *Nucl. Phys.*, **B355**, 54.

Fox, G. C. and Wolfram, S. (1978). *Phys. Rev. Lett.*, **41**, 1581.

Fox, G. C. and Wolfram, S. (1980). *Nucl. Phys.*, **B168**, 285.

Frautschi, S. C. (1971). *Phys. Rev.*, **D3**, 2821.

Frenkel, J. and Taylor, J. C. (1976). *Nucl. Phys.*, **B116**, 185.

Fritzsch, H. *et al.* (1973). *Phys. Lett.*, **B47**, 365.

Frixione, S. *et al.* (1996). *Nucl. Phys.*, **B467**, 399.

Frixione, S. *et al.* (1999). *Nucl. Phys.*, **B542**, 311.

Froissart, M. (1961). *Phys. Rev.*, **123**, 1053.

Furmanski, W. and Petronzio, R. (1980). *Phys. Lett.*, **B97**, 437.

Furmanski, W. and Petronzio, R. (1982). *Z. Phys.*, **C11**, 293.

Furry, W. H. (1937). *Phys. Rev.*, **51**, 125.

Gaffney, J. B. and Mueller, A. H. (1985). *Nucl. Phys.*, **B250**, 109.

Gaisser, T. K. *et al.* (1986). *Phys. Lett.*, **B166**, 219.

Garcia Canal, C.A. and Sassot, R. (2000). In *AIP Conf. Proc. 531*, pp. 199.

Gary, W. J. (1994). *Phys. Rev.*, **D49**, 4503.

Gell-Mann, M. (1994). *The quark and the jaguar.* W. H. Freeman and Company, New York.

Georgi, H. and Politzer, H. D. (1976). *Phys. Rev.*, **D14**, 1829.

Giele, W. T. *et al.* (1993). *Nucl. Phys.*, **B403**, 633.

Giele, W. T. *et al.* (1996). *Phys. Rev.*, **D53**, 120.

Girone, M. and Neubert, M. (1996). *Phys. Rev. Lett.*, **76**, 3061.

Glazov, A. A. (1998). *Measurement of the proton structure functions $F_2(x, Q^2)$ and $F_L(x, Q^2)$ with the H1 detector at HERA.* Ph. D. thesis, Humboldt-Universität, Berlin, Germany.

Glück, M. *et al.* (1995). *Z. Phys.*, **C67**, 433.

Glück, M. *et al.* (1998). *Eur. Phys. J.*, **C5**, 461.

Goldhaber, G. *et al.* (1960). *Phys. Rev.*, **120**, 300.

Gorishny, S. G. *et al.* (1991). *Phys. Lett.*, **B259**, 144.

Gottfried, K. (1967). *Phys. Rev. Lett.*, **18**, 1174.

Gottschalk, T. D. (1984). *Nucl. Phys.*, **B239**, 325, 349.

Gottschalk, T. D. (1986). *Nucl. Phys.*, **B277**, 700.

Gottschalk, T. D. and Morris, D. A. (1987). *Nucl. Phys.*, **B288**, 729.

Gribov, L. V. *et al.* (1983). *Phys. Rept.*, **100**, 1.

Gribov, V. N. (1961). *J.E.T.P. Lett.*, **41**, 667.

Gribov, V. N. (1978). *Nucl. Phys.*, **B139**, 1.

Gribov, V. N. and Lipatov, L. N. (1972). *Sov. J. Nucl. Phys.*, **15**, 438 [781].

Gross, D. J. (1976). Applications of the renormalization group to high-energy physics. In *Les Houches 1975, Proceedings, Methods in field theory* (ed. R. Balian and J. Zinn-Justin), Amsterdam, pp. 141.

Gross, D. J. and Llewellyn Smith, C. H. (1969). *Nucl. Phys.*, **B14**, 337.

Gross, D. J. and Wilczek, F. (1973). *Phys. Rev.*, **D8**, 3633.

Grupen, C. (1996). *Particle detectors.* Cambridge University Press, Cambridge, UK.

Gustafson, G. (1986). *Phys. Lett.*, **B175**, 453.

Gustafson, G. *et al.* (1988). *Phys. Lett.*, **B209**, 90.

Gustafson, G. and Nilsson, A. (1991). *Nucl. Phys.*, **B355**, 106.

Gustafson, G. and Pettersson, U. (1988). *Nucl. Phys.*, **B306**, 746.

Gyulassy, M. *et al.* (1979). *Phys. Rev.*, **C20**, 2267.

Gyulassy, M. and Wang, X-N. (1991). *Phys. Rev.*, **D44**, 3501.

H1 Collab., Ahmed, T. *et al.* (1994). *Nucl. Phys.*, **B429**, 477.

H1 Collab., Aid, S. *et al.* (1995*a*). *Nucl. Phys.*, **B449**, 3.

H1 Collab., Ahmed, T. *et al.* (1995*b*). *Phys. Lett.*, **B346**, 415.

H1 Collab., Aid, S. *et al.* (1996). *Nucl. Phys.*, **B470**, 3.

H1 Collab., Adloff, C. *et al.* (1997*a*). *Nucl. Phys.*, **B497**, 3.

H1 Collab., Adloff, C. *et al.* (1997*b*). *Nucl. Phys.*, **B504**, 3.

H1 Collab., Adloff, C. *et al.* (1998). *Eur. Phys. J.*, **C5**, 625.

H1 Collab., Adloff, C. *et al.* (1999*a*). *Nucl. Phys.*, **B545**, 21.

H1 Collab., Adloff, C. *et al.* (1999*b*). *Eur. Phys. J.*, **C6**, 575.

H1 Collab., Adloff, C. *et al.* (2000). *Eur. Phys. J.*, **C13**, 609.

H1 Collab., Adloff, C. *et al.* (2001). *Eur. Phys. J.*, **C19**, 289.

Hagedorn, R. (1965). *Nuovo Cim. (Suppl.)*, **3**, 147.

Haidt, D. (1995). *Precision Tests of the Standard Electroweak Model.* Ed. P. Langacker, World Scientific, Singapore.

Hamberg, R. and van Neerven, W. L. (1992). *Nucl. Phys.*, **B379**, 143.

Hamer, C. J. and Peierls, R. F. (1973). *Phys. Rev.*, **D8**, 1358.

Hebbeker, T. (1991). Aachen PITHA 91/8. revised version.

Hebbeker, T. *et al.* (1994). *Phys. Lett.*, **B331**, 165.

Hebecker, A. (2000). *Phys. Rept.*, **331**, 1.

Hinchliffe, I. and Manohar, A. V. (2000). *Ann. Rev. Nucl. Part. Sci.*, **50**, 643.

HRS Collab., Derrick, M. *et al.* (1986). *Phys. Rev.*, **D34**, 3304.

Huston, J. *et al.* (1998). *Phys. Rev.*, **D58**, 114034.

Iancu, E. *et al.* (2000). *arXiv:* hep-ph/0011241.

Ingelman, G. and Schlein, P. E. (1985). *Phys. Lett.*, **B152**, 256.

Ioffe, B. L. (1978). *Phys. Lett.*, **B78**, 277.

Jacquet, F. and Blondel, A. (1979). In *An ep facility for Europe* (ed. U. Amaldi), Hamburg.

JADE Collab., Bartel, W. *et al.* (1983). *Z. Phys.*, **C20**, 187.

JADE Collab., Bartel, W. *et al.* (1985). *Phys. Lett.*, **B157**, 340.

JADE Collab., Bartel, W. *et al.* (1986). *Z. Phys.*, **C33**, 23.

JADE Collab., Bethke, S. *et al.* (1988*a*). *Phys. Lett.*, **B213**, 235.

JADE Collab., Biebel, O. *et al.* (1999). *Phys. Lett.*, **B459**, 326.

JADE Collab., Ould-Saada, F. *et al.* (1988*b*). *Z. Phys.*, **C39**, 1.

JADE Collab., Movilla Fernandez, P. A. *et al.* (1998). *Eur. Phys. J.*, **C1**, 461.

Jaffe, R. L. and Llewellyn-Smith, C. H. (1973). *Phys. Rev.*, **D7**, 2506.

James, F. (1980). *Rept. Prog. Phys.*, **43**, 1145.

Jones, D. R. T. (1974). *Nucl. Phys.*, **B75**, 531.

Jones, H. F. (1990). *Groups, representations and physics*. Hilger, Bristol.

Jong, P. de (2001). *arXiv:* hep-ex/0103018.

Kataev, A. L. *et al.* (2000). *Nucl. Phys.*, **B573**, 405.

Keller, S. and Laenen, E. (1999). *Phys. Rev.*, **D59**, 114004.

Kharraziha, H. and Lönnblad, L. (1998). *JHEP*, **9803**, 006.

Khoze, V. A. *et al.* (2000). *Eur. Phys. J.*, **C18**, 167.

Khoze, V. A. and Lönnblad, L. (1990). *Phys. Lett.*, **B241**, 123.

Khoze, V. A. and Ochs, W. (1997). *Int. J. Mod. Phys.*, **A12**, 2949.

Khoze, V. A. and Stirling, W. J. (1997). *Z. Phys.*, **C76**, 59.

Kilgore, W. B. and Giele, W. T. (1997). *Phys. Rev.*, **D55**, 7183.

Kinoshita, T. (1962). *J. Math. Phys.*, **3**, 650.

Kleinknecht, K. (1999). *Detectors for Particle Radiation*. Cambridge University Press, Cambridge, UK.

Kleiss, R. (1986). *Phys. Lett.*, **B180**, 400.

Kleiss, R. *et al.* (1989). Monte Carlos for electroweak physics. In *Z physics at LEP1* (ed. G. Altarelli, R. Kleiss, and C. Verzegnassi), Number 89-08, vol. 3, Geneva. CERN.

Kniehl, B. A. and Kühn, J. H. (1989). *Phys. Lett.*, **B224**, 229.

Kniehl, B. A. and Kühn, J. H. (1990). *Nucl. Phys.*, **B329**, 547.

Knowles, I. G. (1990). *Comput. Phys. Commun.*, **58**, 271.

Knowles, I. G. *et al.* (1996). In *Physics at LEP2: QCD event generators* (ed. G. Altarelli, T. Söstrand, and F. Zwirner), Number 96-01, Geneva. CERN. *arXiv:* hep-ph/9601212.

Knowles, I. G. and Lafferty, G. D. (1997). *J. Phys.*, **G23**, 731.

Koba, Z. *et al.* (1972). *Nucl. Phys.*, **B40**, 317.

Kobel, M. (1992). XXVIIth Rencontres de Moriond, 22-28 March.

Konishi, K. *et al.* (1979). *Nucl. Phys.*, **B157**, 45.

Körner, J. G. *et al.* (1981). *Nucl. Phys.*, **B185**. 365.

Körner, J. G. *et al.* (1987). *Phys. Lett.*, **B188**, 272.

Körner, J. G. *et al.* (1989). *Int. J. Mod. Phys.*, **A4**, 1781.

Kovchegov, Yu. V. (1999). *Phys. Rev.*, **D60**, 034008.

Kubar-André, J. and Paige, F. E. (1979). *Phys. Rev.*, **D19**, 221.

Kulesza, A. and Stirling, W. J. (1999). *Nucl. Phys.*, **B555**, 279.

Kumano, S. and Londergan, J. T. (1991). *Phys. Rev.*, **D44**, 717.

Kunszt, Z. *et al.* (1989). QCD. In *Z physics at LEP1* (ed. G. Altarelli, R. Kleiss, and C. Verzegnassi), Number 89-08, Geneva. CERN.

Kuraev, E. A. *et al.* (1977). *Sov. Phys. JETP*, **45**, 199.

Kuti, J. and Weisskopf, V. F. (1971). *Phys. Rev.*, **D4**, 3418.

L3 Collab., Adeva, B. *et al.* (1991*a*). *Phys. Lett.*, **B263**, 551.

L3 Collab., Adeva, B. *et al.* (1991*b*). *Phys. Lett.*, **B271**, 461.

L3 Collab., Adeva, B. *et al.* (1991*c*). *Phys. Lett.*, **B259**, 199.

L3 Collab., Adriani, O. *et al.* (1992*a*). *Phys. Lett.*, **B286**, 403.

L3 Collab., Adeva, B. *et al.* (1992*b*). *Z. Phys.*, **C55**, 39.

L3 Collab., Adriani, O. *et al.* (1992*c*). *Phys. Lett.*, **B292**, 472.

L3 Collab., Adriani, O. *et al.* (1993). *Phys. Lett.*, **B317**, 467.

L3 Collab., Acciarri, M. *et al.* (1994). *Phys. Lett.*, **B328**, 223.

L3 Collab., Acciarri, M. *et al.* (1995). *Phys. Lett.*, **B345**, 74.

L3 Collab., Acciarri, M. *et al.* (1996). *Phys. Lett.*, **B371**, 126.

L3 Collab., Acciarri, M. *et al.* (1997*a*). *Phys. Lett.*, **B404**, 390.

L3 Collab., Acciarri, M. *et al.* (1997*b*). *Phys. Lett.*, **B393**, 465.

L3 Collab., Acciarri, M. *et al.* (1997*c*). *Phys. Lett.*, **B407**, 389.

 Erratum : (1998). *ibid.*, **B427**, 409.

L3 Collab., Acciarri, M. *et al.* (1998). *Phys. Lett.*, **B444**, 569.

L3 Collab., Acciarri, M. *et al.* (2000). *Phys. Lett.*, **B479**, 79.

Ladinsky, G. A. and Yuan, C. P. (1994). *Phys. Rev.*, **D50**, 4239.

Laenen, E. and Levin, E. (1994). *Ann. Rev. Nucl. Part. Sci.*, **44**, 199.

Laermann, E. *et al.* (1980). *Z. Phys.*, **C3**, 289.

Laermann, E. *et al.* (1991). *Z. Phys.*, **C52**, 352.

Laermann, E. and Zerwas, P. M. (1980). *Phys. Lett.*, **B89**, 225.

 Erratum : (1980). *ibid.*, **B91**, 487.

Lafferty, G. D. *et al.* (1995). *J. Phys.*, **G21**, A1.

Lafferty, G. D. and Wyatt, T. R. (1995). *Nucl. Instr. and Meth.*, **A355**, 541.

Landshoff, P. V. (1974). *Phys. Rev.*, **D10**, 1024.

Landshoff, P. V. and Polkinghorne, J. C. (1971). *Nucl. Phys.*, **B28**, 240.

Larin, S. A. *et al.* (1997). *Phys. Lett.*, **B400**, 379.

Larin, S. A. and Vermaseren, J. A. M. (1991). *Phys. Lett.*, **B259**, 345.

Le Diberder, F. and Pich, A. (1992*a*). *Phys. Lett.*, **B289**, 165.

Le Diberder, F. and Pich, A. (1992*b*). *Phys. Lett.*, **B286**, 147.

Lee, T. D. and Nauenberg, M. (1964). *Phys. Rev.*, **133**, B1549.

Leibbrandt, G. (1987). *Rev. Mod. Phys.*, **59**, 1067.

Llewellyn-Smith, C. H. (1972). *Phys. Rept.*, **3**, 261.

~~Long, K. *et al.* (2001). *arXiv:* hep-ph/0109092.~~

Lönnblad, L. (1992). *Comput. Phys. Commun.*, **71**, 15.

Lönnblad, L. (1996*a*). *Z. Phys.*, **C65**, 285.

Lönnblad, L. (1996*b*). *Nucl. Phys.*, **B458**, 215.

Lönnblad, L. and Kniehl, B. A. (1992). In *Workshop on photon radiation from quarks* (ed. S. Cartwright), Geneva, pp. 109. CERN.

Lönnblad, L. and Sjöstrand, T. (1998). *Eur. Phys. J.*, **C2**, 165.

Low, F. E. (1975). *Phys. Rev.*, **D12**, 163.

MacFarlane, A. J. *et al.* (1968). *Commun. Math. Phys.*, **11**, 77.

Malaza, E. D. and Webber, B. R. (1984). *Phys. Lett.*, **B149**, 501.

Malaza, E. D. and Webber, B. R. (1986). *Nucl. Phys.*, **B267**, 702.

Mangano, M. L. and Parke, S. J. (1991). *Phys. Rept.*, **200**, 301.

Marchesini, G. *et al.* (1981). *Nucl. Phys.*, **B181**, 335.

Marchesini, G. and Webber, B. R. (1988). *Phys. Rev.*, **D38**, 3419.

Marchesini, G. and Webber, B. R. (1990). *Nucl. Phys.*, **B330**, 261.

Marciano, W. J. (1984). *Phys. Rev.*, **D29**, 580.

MARKII Collab., Abrams, G. S. *et al.* (1990). *Phys. Rev. Lett.*, **64**, 1334.

MARKII Collab., Sheldon, P. D. *et al.* (1986). *Phys. Rev. Lett.*, **57**, 1398.

Martin, A. (1963). *Phys. Rev.*, **129**, 1432.

Martin, A. D. *et al.* (1998). *Eur. Phys. J.*, **C4**, 463.

McLerran, L. D. and Venugopalan, R. (1999). *Phys. Rev.*, **D59**, 094002.

Meyer, T. (1982). *Z. Phys.*, **C12**, 77.

Mishra, S. R. and Sciulli, F. (1989). *Ann. Rev. Nucl. Part. Sci.*, **39**, 259.

MIT-SLAC Collab., Bloom, E. D. *et al.* (1970). SLAC-PUB-796.

MIT-SLAC Collab., Miller, G. *et al.* (1972). *Phys. Rev.*, **D5**, 528.

Montagna, G. *et al.* (1993*a*). *Nucl. Phys.*, **B401**, 3.

Montagna, G. *et al.* (1993*b*). *Comput. Phys. Commun.*, **76**, 328.

Morris, D. A. (1987). *Nucl. Phys.*, **B288**, 717.

Mueller, A. H. (1981). *Phys. Rept.*, **73**, 237.

Mueller, A. H. (1983). *Nucl. Phys.*, **B213**, 85. (1984). *ibid.*, **B241**, 141.

Mueller, A. H. (1991). *J. Phys.*, **G17**, 1443.

Mueller, A. H. (1993). *Phys. Lett.*, **B308**, 355.

Mueller, A. H. (1994). *Nucl. Phys.*, **B415**, 373.

Mueller, A. H. and Qiu, J. (1986). *Nucl. Phys.*, **B268**, 427.

Müller, A.-S. (2000). *Precision Measurements of the LEP Beam Energy for the Determination of the W Boson Mass.* Ph. D. thesis, Universität Mainz, Germany, published by Shaker Verlag, Aachen.

Nachtmann, O. and Reiter, A. (1982). *Z. Phys.*, **C16**. 45.

Nachtmann, O. and Wetzel, W. (1981). *Nucl. Phys.*, **B187**, 333.

Nagy, Z. and Trócsányi, Z. (1997). *Phys. Rev. Lett.*, **79**, 3604.

Nagy, Z. and Trócsányi, Z. (1999). *Phys. Rev.*, **D59**, 014020. Erratum : (2000). *ibid.*, **D62**, 099902.

Nason, P. and Webber, B. R. (1994). *Nucl. Phys.*, **B421**, 473.

Erratum : (1996). *ibid.*, **B480**, 755.

Nikolaev, N. N. and Zakharov, V. I. (1975). *Phys. Lett.*, **B55**, 397.

NMC Collab., Amaudruz, P. *et al.* (1991). *Phys. Rev. Lett.*, **66**, 2712.

NMC Collab., Arneodo, M. *et al.* (1995). *Phys. Lett.*, **B364**, 107.

Nussinov, S. (1975). *Phys. Rev. Lett.*, **34**, 1286.

Odagiri, K. (1998). *JHEP*, **9810**, 006.

Odorico, R. (1984). *Comput. Phys. Commun.*, **32**, 173.

Erratum: (1985). *ibid.*, **34**, 437.

Odorico, R. (1990). *Comput. Phys. Commun.*, **59**, 527.

Olsen, H. A. *et al.* (1980). *Phys. Lett.*, **B89**, 221.

OPAL Collab., Akrawy, M. Z. *et al.* (1990). *Phys. Lett.*, **B247**, 617.

OPAL Collab., Alexander, G. *et al.* (1991*a*). *Z. Phys.*, **C52**, 543.

OPAL Collab., Akrawy, M. Z. *et al.* (1991*b*). *Phys. Lett.*, **B262**, 351.

OPAL Collab., Alexander, G. *et al.* (1991*c*). *Phys. Lett.*, **B264**, 467.

OPAL Collab., Akrawy, M. Z. *et al.* (1991*d*). *Phys. Lett.*, **B261**, 334.

OPAL Collab., Acton, P. D. *et al.* (1992*a*). *Z. Phys.*, **C55**, 1.

OPAL Collab., Acton, P. D. *et al.* (1992*b*). *Z. Phys.*, **C53**, 539.

OPAL Collab., Acton, P. D. *et al.* (1992*c*). *Phys. Lett.*, **B291**, 503.

OPAL Collab., Acton, P. D. *et al.* (1993*a*). *Z. Phys.*, **C59**, 1.

OPAL Collab., Akers, R. *et al.* (1993*b*). *Z. Phys.*, **C60**, 397.

OPAL Collab., Acton, P. D. *et al.* (1993*c*). *Z. Phys.*, **C58**, 387.

OPAL Collab., Acton, P. D. *et al.* (1993*d*). *Phys. Lett.*, **B305**, 407.

OPAL Collab., Akers, R. *et al.* (1994*a*). *Z. Phys.*, **C63**, 181.

OPAL Collab., Akers, R. *et al.* (1994*b*). *Z. Phys.*, **C63**, 363.

OPAL Collab., Akers, R. *et al.* (1994*c*). *Z. Phys.*, **C63**, 197.

OPAL Collab., Akers, R. *et al.* (1995*a*). *Z. Phys.*, **C65**, 31.

OPAL Collab., Akers, R. *et al.* (1995*b*). *Z. Phys.*, **C68**, 519.

OPAL Collab., Akers, R. *et al.* (1995*c*). *Z. Phys.*, **C65**, 367.

OPAL Collab., Akers, R. *et al.* (1995*d*). *Z. Phys.*, **C68**, 179.

OPAL Collab., Akers, R. *et al.* (1995*e*). *Z. Phys.*, **C68**, 203.

OPAL Collab., Akers, R. *et al.* (1995*f*). *Z. Phys.*, **C67**, 389.

OPAL Collab., Alexander, G. *et al.* (1995*g*). *Phys. Lett.*, **B358**, 162.

OPAL Collab., Akers, R. *et al.* (1995*h*). *Z. Phys.*, **C68**, 531.

OPAL Collab., Alexander, G. *et al.* (1996*a*). *Z. Phys.*, **C69**, 543.

OPAL Collab., Alexander, G. *et al.* (1996*b*). *Z. Phys.*, **C72**, 191.

OPAL Collab., Alexander, G. *et al.* (1996*c*). *Phys. Lett.*, **B388**, 659.

OPAL Collab., Alexander, G. *et al.* (1996*d*). *Z. Phys.*, **C72**, 1.

OPAL Collab., Alexander, G. *et al.* (1996*e*). *Z. Phys.*, **C70**, 197.

OPAL Collab., Alexander, G. *et al.* (1996*f*). *Phys. Lett.*, **B370**, 185.

OPAL Collab., Ackerstaff, K. *et al.* (1997*a*). *Z. Phys.*, **C75**, 193.

OPAL Collab., Ackerstaff, K. *et al.* (1997*b*). *Phys. Lett.*, **B412**, 210.

OPAL Collab., Ackerstaff, K. *et al.* (1997*c*). *Z. Phys.*, **C76**, 425.

OPAL Collab., Alexander, G. *et al.* (1997*d*). *Z. Phys.*, **C73**, 569.

OPAL Collab., Alexander, G. *et al.* (1997*e*). *Z. Phys.*, **C73**, 587.

OPAL Collab., Ackerstaff, K. *et al.* (1998*a*). *Eur. Phys. J.*, **C1**, 479.

OPAL Collab., Ackerstaff, K. *et al.* (1998*b*). Opal PN299, paper submitted to the *XXIX International Conference on High Energy Physics*, Vancouver.

OPAL Collab., Ackerstaff, K. *et al.* (1998*c*). *Eur. Phys. J.*, **C5**, 411.

OPAL Collab., Ackerstaff, K. *et al.* (1998*d*). *Eur. Phys. J.*, **C4**, 19.

OPAL Collab., Ackerstaff, K. *et al.* (1998*e*). *Eur. Phys. J.*, **C1**, 439.

OPAL Collab., Abbiendi, G. *et al.* (1999*a*). *Eur. Phys. J.*, **C11**, 217.

OPAL Collab., Ackerstaff, K. *et al.* (1999*b*). *Eur. Phys. J.*, **C8**, 241.

OPAL Collab., Ackerstaff, K. *et al.* (1999*c*). *Eur. Phys. J.*, **C7**, 571.

OPAL Collab., Abbiendi, G. *et al.* (2000*a*). *Eur. Phys. J.*, **C16**, 185.

OPAL Collab., Abbiendi, G. *et al.* (2000*b*). *Eur. Phys. J.*, **C17**, 373.

OPAL Collab., Abbiendi, G. *et al.* (2001). *Eur. Phys. J.*, **C20**, 601.

Owens, J. F. (1978). *Phys. Lett.*, **B76**, 85.

Paige, F. E. and Protopopescu, S. (1986). In *Super Collider Physics, Oregon workshop on super high energy physics* (ed. D. Soper), Eugene, USA, pp. 41. World Scientific, Singapure.

Palmer, R. B. (1990). *Ann. Rev. Nucl. Part. Sci.*, **40**, 529.

Parisi, G. (1978). *Phys. Lett.*, **B74**, 65.

Parisi, G. and Petronzio, R. (1979). *Nucl. Phys.*, **B154**, 427.

Passarino, G. (1993). *Phys. Lett.*, **B313**, 213.

PDG (2000). *Eur. Phys. J.*, **C15**, 1.

Peskin, M. E. and Schroeder, D. V. (1995). *An introduction to quantum field theory.* Addison-Wesley, Menlo Park, California.

Peterson, C. *et al.* (1983). *Phys. Rev.*, **D27**, 105.

Phaf, L. and Weinzierl, S. (2001). *JHEP*, **0104**, 006.

Pi, H. (1992). *Comput. Phys. Commun.*, **71**, 173.

Pich, A. (1992). CERN-TH.6738.

Pich, A. (2000). *Int. J. Mod. Phys.*, **A15S1**, 157.

Plothow-Besch, H. (1993). *Comput. Phys. Commun.*, **75**, 396.

PLUTO Collab., Berger, C. *et al.* (1980*a*). *Phys. Lett.*, **B97**, 459.

PLUTO Collab., Berger, C. *et al.* (1980*b*). *Phys. Lett.*, **B95**, 313.

Poggio, E. C. and Quinn, H. R. (1976). *Phys. Rev.*, **D14**, 578.

Politzer, H. D. (1974). *Phys. Rept.*, **14**, 129.

Ramond, P. (1990). *Field theory: a modern primer (Second edn.).* Addison-Wesley, Redwood City, CA.

Richardson, P. (2001). *JHEP*, **0111**, 029.

Rodrigo, G. *et al.* (1997). *Phys. Rev. Lett.*, **79**, 193.

Rodrigo, G. and Santamaria, A. (1993). *Phys. Lett.*, **B313**, 441.

Rujula, A. de *et al.* (1974). *Phys. Rev.*, **D10**, 1649.

Sakurai, J. J. and Schildknecht, D. (1972). *Phys. Lett.*, **B40**, 121.

Salam, G. P. (1998). *JHEP*, **9807**, 019.

Santiago, J. and Yndurain, F. J. (1999). *Nucl. Phys.*, **B563**, 45.

Schmelling, M. (1994). *Nucl. Instr. and Meth.*, **A340**, 400.

Schmelling, M. (1995*a*). *Phys. Scripta*, **51**, 683.

Schmelling, M. (1995*b*). *Phys. Scripta*, **51**, 676.

Schmelling, M. (1996). Status of the strong coupling constant. In *Proceedings of the XXVIII International Conference on High Energy Physics*, Warsaw. *arXiv:* hep-ex/9701002.

Schmelling, M. (2000). *arXiv:* hep-ex/0006004.

Schmelling, M. and St. Denis, R. (1994). *Phys. Lett.*, **B329**, 393.

Schuler, G. A. and Körner, J. G. (1989). *Nucl. Phys.*, **B325**, 557. Erratum : (1997). *ibid.*, **B507**, 547.

Schwitters, R. *et al.* (1975). *Phys. Rev. Lett.*, **35**, 1320.

Seymour, M. H. (1995). *Nucl. Phys.*, **B436**, 163.

Seymour, M. H. (1996). *Phys. Lett.*, **B378**, 279.

Seymour, M. H. (1998). *Nucl. Phys.*, **B513**, 269.

Shifman, M. A. *et al.* (1979). *Nucl. Phys.*, **B147**, 385, 448, 519.

Sjöstrand, T. (1984). *Nucl. Phys.*, **B248**, 469.

Sjöstrand, T. (1985). *Phys. Lett.*, **B157**, 321.

Sjöstrand, T. (1994). *Comput. Phys. Commun.*, **82**, 74.

Sjöstrand, T. *et al.* (2001). *Comput. Phys. Commun.*, **135**, 238.

Sjöstrand, T. and Khoze, V. A. (1994). *Z. Phys.*, **C62**, 281.

Sjöstrand, T. and van Zijl, M. (1987). *Phys. Rev.*, **D36**, 2019.

Slavnov, A. A. (1972). *Theor. Math. Phys.*, **10**, 99.

SLD Collab., Abe, K. *et al.* (1995). *Phys. Rev.*, **D51**, 962.

SLD Collab., Abe, K. *et al.* (1996). *Phys. Rev.*, **D53**, 2271.

SLD Collab., Abe, K. *et al.* (1999*a*). *Phys. Rev.*, **D59**, 012002.

SLD Collab., Abe, K. *et al.* (1999*b*). *Phys. Rev.*, **D59**, 052001.

SMC Collab., Adams, D. *et al.* (1994). *Phys. Lett.*, **B329**, 399.

Speer, E. R. (1974). *J. Math. Phys.*, **15**, 1.

Spitz, A. *et al.* (1999). *Phys. Rev.*, **D60**, 074502.

Stein, E. *et al.* (1995). *Phys. Lett.*, **B343**, 369.

Sterman, G. (1976). *Phys. Rev.*, **D14**, 2123.

Sterman, G. and Weinberg, S. (1977). *Phys. Rev. Lett.*, **39**, 1436.

Stevenson, P. M. (1981). *Phys. Rev.*, **D23**, 2916.

Sudakov, V. V. (1956). *Sov. Phys. JETP*, **3**, 65.

Surguladze, L. R. and Samuel, M. A. (1991). *Phys. Rev. Lett.*, **66**, 560. Erratum : (1991). *ibid.*, **66**, 2416.

Suzuki, M. (1977). *Phys. Lett.*, **B71**, 139.

Symanzik, K. (1971). *Commun. Math. Phys.*, **23**, 49.

Symanzik, K. (1973). *Commun. Math. Phys.*, **34**, 7.

Szwed, R. and Wrochna, G. (1990). *Z. Phys.*, **C47**, 449.

Tarasov, O. V. *et al.* (1980). *Phys. Lett.*, **B93**, 429.

Tarrach, R. (1981). *Nucl. Phys.*, **B183**, 384.

TASSO Collab., Althoff, M. *et al.* (1985). *Z. Phys.*, **C29**, 29.

TASSO Collab., Brandelik, R. *et al.* (1980). *Phys. Lett.*, **B97**, 453.

TASSO Collab., Braunschweig, W. *et al.* (1989). *Z. Phys.*, **C45**, 193.

TASSO Collab., Braunschweig, W. *et al.* (1990). *Z. Phys.*, **C47**, 187.

Taylor, J. C. (1971). *Nucl. Phys.*, **B33**, 436.

Taylor, R. E. (1975). In *Procs of the 1975 Int. Symp. on lepton photon interactions* (ed. T. W. Kirk). Stanford Univ., California.

'tHooft, G. (1973). *Nucl. Phys.*, **B61**, 455.

'tHooft, G. (1974). *Nucl. Phys.*, **B72**, 461.

'tHooft, G. and Veltman, M. J. G. (1972). *Nucl. Phys.*, **B44**, 189.

TOPAZ Collab., Ohnishi, Y. *et al.* (1993). *Phys. Lett.*, **B313**, 475.

TOPAZ Collab., Yamauchi, M. *et al.* (1988). In *Proceedings of the XXIVth International Conference on High Energy Physics*, Munich, 4-10 August.

Tournefier, E. (1998). *arXiv:* hep-ex/9810042.

TPC/2γ Collab., Aihara, H. *et al.* (1985). *Z. Phys.*, **C28**, 31.

TPC/2γ Collab., Aihara, H. *et al.* (1986). *Phys. Rev. Lett.*, **57**, 945.

TPC/2γ Collab., Aihara, H. *et al.* (1987). *Phys. Lett.*, **B184**, 299.

TPC/2γ Collab., Aihara, H. *et al.* (1988). LBL-23737. Update by G. Cowan, private communication.

Treiman, S. B. *et al.* (1972). *Lectures on current algebra and its applications.* Princeton University Press, Princeton.

Trentadue, L. and Veneziano, G. (1994). *Phys. Lett.*, **B323**, 201.

Tung, Wu-Ki (1985). *Group theory in physics.* World Scientific Publishing, Singapore.

UA1 Collab., Arnison, G. *et al.* (1986). *Phys. Lett.*, **B172**, 461.

UA1 Collab., Albajar, C. *et al.* (1988). *Nucl. Phys.*, **B309**, 405.

UA1 Collab., Albajar, C. *et al.* (1996). *Phys. Lett.*, **B369**, 46.

UA2 Collab., Appel, J. A. *et al.* (1985*a*). *Phys. Lett.*, **B165**, 441.

UA2 Collab., Appel, J. A. *et al.* (1985*b*). *Phys. Lett.*, **B160**, 349.

UA2 Collab., Ansari, R. *et al.* (1987). *Phys. Lett.*, **B194**, 158.

UA2 Collab., Alitti, J. *et al.* (1991). *Phys. Lett.*, **B257**, 232.

UA2 Collab., Bagnaia, P. *et al.* (1985*c*). *Phys. Lett.*, **B154**, 338.

UA5 Collab., Alner, G. J. *et al.* (1987). *Nucl. Phys.*, **B291**, 445.

UA6 Collab., Werlen, M. *et al.* (1999). *Phys. Lett.*, **B452**, 201.

UA8 Collab., Bonino, R. *et al.* (1988). *Phys. Lett.*, **B211**, 239.

Uematsu, T. (1978). *Phys. Lett.*, **B79**, 97.

Virchaux, M. and Milsztajn, A. (1992). *Phys. Lett.*, **B274**, 221.

Wang, X-N. (1993). *Phys. Rev.*, **D47**, 2754.

Webber, B. R. (1984). *Nucl. Phys.*, **B238**, 492.

Webber, B. R. (1986). *Ann. Rev. Nucl. Part. Sci.*, **36**, 253.

Webber, B. R. (1995). *arXiv:* hep-ph/9510283.

Weinberg, S. (1960). *Phys. Rev.*, **118**, 838.

Weinberg, S. (1973*a*). *Phys. Rev. Lett.*, **31**, 494.

Weinberg, S. (1973*b*). *Phys. Rev.*, **D8**, 3497.

Wiedemann, H. (1993). *Particle accelerator physics*, Volume I. Springer-Verlag, Heidelberg, Germany.

Wiedemann, H. (1995). *Particle accelerator physics*, Volume II. Springer-Verlag, Heidelberg, Germany.

Willenbrock, S. (1989). In *TASI, Boulder CO.*, pp. 323. Boulder ASI.

Wilson, K. G. (1974). *Phys. Rev.*, **D10**, 2445.

Windmolders, R. (1999). *Nucl. Phys. Proc. Suppl.*, **79**, 51.

Witten, E. (1977). *Nucl. Phys.*, **B120**, 189.

Yang, Chen-Ning and Mills, R. L. (1954). *Phys. Rev.*, **96**, 191.

ZEUS Collab., Breitweg, J. *et al.* (1999*a*). *Eur. Phys. J.*, **C11**, 427.

ZEUS Collab., Breitweg, J. *et al.* (1999*b*). *Eur. Phys. J.*, **C7**, 609.

ZEUS Collab., Breitweg, J. *et al.* (1999*c*). *Eur. Phys. J.*, **C11**, 251.

ZEUS Collab., Breitweg, J. *et al.* (2000*a*). *Eur. Phys. J.*, **C12**, 411.

ZEUS Collab., Breitweg, J. *et al.* (2000*b*). *Phys. Lett.*, **B487**, 53.

ZEUS Collab., Derrick, M. *et al.* (1993). *Phys. Lett.*, **B315**, 481.

ZEUS Collab., Derrick, M. *et al.* (1995). *Phys. Lett.*, **B363**, 201.

Zijlstra, E. B. and van Neerven, W. L. (1992). *Nucl. Phys.*, **B383**, 525.

INDEX